NANOFABRICATION USING FOCUSED ION AND ELECTRON BEAMS

OXFORD SERIES ON NANOMANUFACTURING

Nanofabrication Using Focused Ion and Electron Beams:
Principles and Applications
Edited by Ivo Utke, Stanislav Moshkalev, and Phillip Russell

Nanofabrication Using Focused Ion and Electron Beams

Principles and Applications

Edited by
Ivo Utke, Stanislav Moshkalev, AND *Phillip Russell*

OXFORD
UNIVERSITY PRESS

OXFORD
UNIVERSITY PRESS

Oxford University Press

Oxford University Press, Inc., publishes works that further
Oxford University's objective of excellence
in research, scholarship, and education.

Oxford New York

Auckland   Cape Town   Dar es Salaam   Hong Kong   Karachi
Kuala Lumpur   Madrid   Melbourne   Mexico City   Nairobi
New Delhi   Shanghai   Taipei   Toronto

With offices in

Argentina   Austria   Brazil   Chile   Czech   Republic   France   Greece
Guatemala   Hungary   Italy   Japan   Poland   Portugal   Singapore
South Korea   Switzerland   Thailand   Turkey   Ukraine   Vietnam

Published by Oxford University Press, Inc.
198 Madison Avenue, New York, New York 10016

www.oup.com

Library of Congress Cataloging-in-Publication Data

Nanofabrication using focused ion and electron beams : principles and applications /
edited by Ivo Utke, Stanislav Moshkalev, Phillip Russell.
p. cm. – (Nanomanufacturing series; v. 1)
ISBN 978-0-19-973421-4
1. Nanostructured materials. 2. Nanotechnology. 3. Electron beams–Industrial applications.
4. Ion bombardment–Industrial applications. I. Utke, Ivo. II. Moshkalev, Stanislav.
III. Russell, Phillip, 1955- IV. Title. V. Series.
TA418.9.N35N2525 2011
620.1'15—dc23   2011028174

1 3 5 7 9 8 6 4 2

Printed in China on acid-free paper

CONTENTS

PART II
APPLICATIONS

PART III
PROSPECTIVES

FOREWORD

Electron microscopy was one of the great scientific advances of the twentieth century. From their primitive beginnings in the 1930s transmission, and scanning, electron microscopes (EM) developed into instruments capable of supplying topographic, structural, crystallographic, and chemical information at resolutions down to the atomic level. In fact, without the electron microscope, research in fields ranging from semiconductor technology, to nanoscience, and to structural biology would be difficult or impossible.

But everyone who has ever used an electron microscope knows that electron beams cause harm as well as doing good. All EMs, from the most primitive early examples to contemporary state-of-the-art machines, contaminate the samples they image. The interaction of electrons with the ever-present hydrocarbons within the microscope leads to reactions in which carbon is deposited from the hydrocarbon while the volatile organic residues are pumped away. Because deposition occurs at every point touched by the beam, samples rapidly become covered with dots, lines, and rectangles of carbon, which reduce image contrast, degrade resolution, and lower the sensitivity of microanalysis.

This excellent new book challenges our perception of contamination to demonstrate how this unwanted interaction can now be employed to provide controlled deposition onto, or erosion into, the specimen. One key step in this transformation of a problem into an opportunity is to employ specially formulated "precursor" gases, which when they interact with electrons either deposit a material of interest—such as tungsten—onto the surface, or which release a chemical—such as fluorine—which can etch the surface. The electrons mainly responsible for these reactions are the secondary electrons generated at the sample surface, so control of the deposition or etching reaction can therefore be obtained by adjusting the rate at which secondary electrons are generated as compared to the rate at which the precursor gas molecules arrive. As the contributors to this book demonstrate, when used with appropriate physical insight this technology is capable of rapidly fabricating on demand an infinite variety of nano-scale structures of accurately defined size and shape using nothing more than their standard microscope.

It is likely that although electrons were the natural choice for the twentieth century, the twenty-first century will be the era of ion beams. Scanning ion microscopy is a more recent scientific advancement; it was introduced around 1985 with its liquid Ga-ion metal source.

It became very soon a popular nanopatterning instrument due to its local milling capability in almost any material. Ion beams do not necessitate the introduction of a precursor molecule and allowed to make cross-section inside complex layered samples, lamellae for transmission electron microscopy, or to trim magnetic heads. Even more recently, the introduction of combined electron and ion scanning microscopes permitted the observation of the milling process *in situ*. But soon the merits of gas-assistance were also exploited in scanning ion beam microscopes since significant enhancement in milling could be demonstrated and deposition of three-dimensional structures was realized. Gas-assisted focused ion beam processing seemed almost to outperform its electron beam counterpart with respect to yield and metal content. However, it turned out that both are complementary. Electron beams are "clean" compared to the current Ga-ion beam, in the sense that they do not alter the sample by implanting source ions or by amorphizing surfaces. Electrons trigger solely a chemical dissociation reaction, while ion beams have simultaneously a sputter effect due to their large mass. This leads to higher yields, differing resolution, and differing deposit compositions, which are favorable in other fields of nanotechnology prototyping where damage can be tolerated.

All these aspects as well as numerous novel developments in FIB processing are also addressed in this book. Attention is given to a large number of prototyping activities in industry and academia, achieved with either focused ion beams or focused electron beams, or with both together.

Summarizing, the transformation of a classic problem into an innovative new technology has wide importance for nanoscience and technology, and anyone planning to fabricate nano-structures should obtain and study this invaluable road map to the future of nano-fabrication without delay.

David Joy, March 2010
Distinguished Professor at The University of Tennessee, Knoxville, TN
Distinguished Scientist at Oak Ridge National Laboratory, USA
Editor-in-chief, SCANNING

PREFACE

It is not once nor twice
but times without number
that the same ideas
make their appearance in the world.

Aristotle

During the last decade, bottom-up and top-down nanotechnology activities multiplied enormously in fundamental and applied research. Top-down prototyping employing focused charged particle beams (electrons, ions, and, more recently, clusters) is one of the most rapidly growing areas due to the successful introduction of new powerful tools, such as combined scanning focused ion beam and electron beam microscopes, and due to the acceptance of gas-injection systems that now routinely permit a sub-10-nm resolution for nanostructuring and material analysis. They are essential parts of many research laboratories in industry and academia.

In the metaphorical sense of the Greek philosopher Aristotle, we face the reinvention of pencil and sculpting tools to change bulk material at the nanometer scale. And indeed, charged particle beam deposition and etching have become very attractive since these are minimally invasive methods with an unprecedented combination of resolution *and* flexibility in shape and material. Alterations are performed only where the beam is focused on the sample—an integrated circuit device that needs a test contact in the second metallic layer, a prototype sensor which needs fine trimming of dimensions, a photomask needing local repair, or simply a cross section for observation—the neighboring device structures remain unaffected. Furthermore, the process is maskless and does not include wet chemistry, which is of advantage for flexible and rapid prototyping of sensible structures.

The number of articles published in this area annually now exceeds thousands. This book probably is the first attempt to present fundamentals and numerous application examples of nanofabrication using focused ion and electron beams in one volume, bringing together many well-known experts working worldwide in this field. The advantages, limitations, potentials, and prospects are highlighted from a fundamental viewpoint as well as from the many engineering applications.

In the introductory part of this book, we attempt to overview the historical evolution of gas-assisted FEB and FIB systems, from the first phenomenological findings, the development of visions and their materialization into today's three-dimensional industrial and academic nanofabrication and prototyping.

The "Fundamentals and Models" part of the book includes a number of contributions that comprehensively describe the fundamentals involved in focused charged particle beam nanostructuring: the choice of molecules, electron and ion beam induced molecule dissociation chemistries; surface science aspects, design and configuration of equipment; Monte Carlo and continuum simulation of deposition, etching, and milling; and basic solutions for nanoprototyping. Emerging technologies as processing by cluster beams are also discussed.

In the "Applications" part of the book, we tried to collect a comprehensive number of contributions demonstrating the wealth of opportunities that were realized by FEB/FIB milling, etching, and deposition in nanolithography, nanoprototyping, and nanocharacterization of different nanostructured materials and devices. The main focus in this part is on the practical details of using focused ion and electron beams with gas assistance (deposition and etching) and without gas assistance (milling/cutting) for fabrication of devices for the fields of nanoelectronics, nanophotonics, nanomagnetics, functionalized scanning probe tips, nanosensors, and other types of NEMS (nanoelectromechanical systems). Special attention is given to strategies designed to overcome limitations of the techniques (e.g., due to damaging produced by energetic ions interacting with matter), in particular involving multi-step processes and multi-layer materials.

In the final part of the book, we attempt a summary of emerging trends in the field of focused ion and electron beam nanofabrication and to look beyond the present achievements: new sources of focused beams, the purity of deposited material, and the ultimate minimum dimensions.

It is our hope that the discussion of general fundamentals and specific processes within one volume will be a great benefit for all interdisciplinary inclined researchers, engineers, teachers, and students trying to "walk across the borders of fields": surface science, molecule-electron-ion interactions, lithography, applied physics, and engineering; basically to all who are looking for synergies in the exciting field of nanosciences. We endeavoured to make the book understandable to novice readers as well as interesting for advanced scientists in the fields.

We wish to sincerely thank all the authors who contributed to this book with their excellent chapters. We appreciate that they found time in their tight work schedules to share their progress and insights into the field and that they made the book a comprehensive source of up-to-date information in the fundamental and practical aspects of nanoprototyping.

Ivo Utke, Stanislav Moshkalev, Phillip Russell

CONTRIBUTORS

Michael Allan Department of Chemistry, University of Fribourg, Chemin du Musée 9, CH-1700 Fribourg, Switzerland

Nicole Auth Carl Zeiss SMS GmbH – Carl Zeiss Group, 64380 Rossdorf, Germany

Laurent Bernau EMPA, Swiss Federal Laboratories for Materials Science and Technology, Feuerwerkerstrasse 39, 3602 Thun, Switzerland

Igor L. Bolotin Department of Chemistry, University of Illinois at Chicago, 60607-7061, USA

Tristan Bret Carl Zeiss SMS GmbH – Carl Zeiss Group, 64380 Rossdorf, Germany

Victor Callegari Empa, Swiss Federal Institute for Materials Science and Technology, Laboratory for Electronics/Metrology/Reliability, CH-8600 Duebendorf, Switzerland; and ETHZ, Swiss Federal Institute of Technology, Institute of Electronics, CH-8092 Zurich, Switzerland

Marco Cantoni EPFL Ecole Polytéchnique Fédéral, Lausanne, Switzerland

Guangyu Chai Apollo Technology Inc, Lake Mary, FL 32746 USA

Lee Chow University of Central Florida, Department of Physics, Orlando, FL 32816 USA

Rosa Córdoba Departamento de Física de la Materia Condensada, Universidad de Zaragoza, Facultad de Ciencias, Zaragoza, 50009, Spain; and Laboratorio de Microscopías Avanzadas (LMA), Instituto de Nanociencia de Aragón (INA), Universidad de Zaragoza, Zaragoza, 50018, Spain

Martin J. Cryan Photonics Research Group, Electrical and Electronic Engineering, University of Bristol, UK

Jose Maria De Teresa Instituto de Ciencia de Materiales de Aragón, Universidad de Zaragoza-CSIC, Facultad de Ciencias, Zaragoza, 50009, Spain; and Departamento de Física de la Materia Condensada, Universidad de Zaragoza, Facultad de Ciencias, Zaragoza, 50009, Spain; and Laboratorio de Microscopías Avanzadas (LMA), Instituto de Nanociencia de Aragón (INA), Universidad de Zaragoza, Zaragoza, 50018, Spain

Klaus Edinger Carl Zeiss SMS GmbH – Carl Zeiss Group, 64380 Rossdorf, Germany

Amalio Fernández-Pacheco Instituto de Ciencia de Materiales de Aragón, Universidad de Zaragoza-CSIC, Facultad de Ciencias, Zaragoza, 50009, Spain; and Departamento de Física de la Materia Condensada, Universidad de Zaragoza, Facultad de Ciencias, Zaragoza, 50009, Spain; and Laboratorio de Microscopías Avanzadas (LMA), Instituto de Nanociencia de Aragón (INA), Universidad de Zaragoza, Zaragoza, 50018, Spain

Richard G. Forbes University of Surrey, Advanced Technology Institute, Guildford, Surrey GU2 7XH, UK

Stefano Frabboni S3 Center, Nanoscience Institute – CNR, Via Campi 213/a, 41125 Modena, Italy, and Dipartimento di Fisica, Università di Modena e Reggio Emilia, Via Campi 213/a, 41125 Modena, Italy

Vinzenz Friedli Empa, Swiss Federal Laboratories for Materials Science and Technology, Laboratory for Mechanics of Materials and Nanostructures, Feuerwerkerstr. 39, 3602 Thun, Switzerland

Mihai Gabureac EMPA, Swiss Federal Laboratories for Materials Science and Technology, Feuerwerkerstrasse 39, 3602 Thun, Switzerland

Vishwas J. Gadgil MESA+ Institute for Nanotechnology, University of Twente, PO Box 217, 7500 AE Enschede, The Netherlands

Gian Carlo Gazzadi S3 Center, Nanoscience Institute – CNR, Via Campi 213/a, 41125 Modena, Italy

Rogério Gelamo Universidade Federal de Triangulo Mineiro, Uberaba, MG, Brazil

Armin Gölzhäuser Physik Supramolekularer Systeme und Oberflächen, Universität Bielefeld, Universitätsstr. 25, 33615 Bielefeld, Germany

Luke Hanley Department of Chemistry, University of Illinois at Chicago, 60607-7061, USA

Peter Heard Interface Analysis Centre, University of Bristol, UK

Francisco Hernández-Ramírez Catalonia Institute for Energy Research (IREC), Barcelona, Spain; and Department of Electronics, University of Barcelona (UB), Barcelona, Spain

Daniel Ho Photonics Research Group, Electrical and Electronic Engineering, University of Bristol, UK

Gerhard Hobler Vienna University of Technology, A-1040 Vienna, Austria

Patrik Hoffmann Empa, Swiss Federal Laboratories for Materials Science and Technology, Laboratory for Advanced Materials Processing, Feuerwerkerstr. 39, 3602 Thun, Switzerland

Thorsten Hofmann Carl Zeiss SMS GmbH – Carl Zeiss Group, 64380 Rossdorf, Germany

Lorenz Holzer Empa, Swiss Federal Laboratories for Materials Science and Technology, Dübendorf, Switzerland

Michael Huth Physikalisches Institut, Goethe-Universität, Max-von-Laue-Str. 1, 60438 Frankfurt am Main, Germany

Manuel R. Ibarra Instituto de Ciencia de Materiales de Aragón, Universidad de Zaragoza-CSIC, Facultad de Ciencias, Zaragoza, 50009, Spain; and Departamento de Física de la Materia Condensada, Universidad de Zaragoza, Facultad de Ciencias, Zaragoza, 50009, Spain; and Laboratorio de Microscopías Avanzadas (LMA), Instituto de Nanociencia de Aragón (INA), Universidad de Zaragoza, Zaragoza, 50018, Spain

Pavel Ivanov Photonics Research Group, Electrical and Electronic Engineering, University of Bristol, UK

Heinz Jaeckel ETHZ, Swiss Federal Institute of Technology, Institute of Electronics, CH-8092 Zurich, Switzerland

Martin Jenke EMPA, Swiss Federal Laboratories for Materials Science and Technology, Feuerwerkerstrasse 39, 3602 Thun, Switzerland

Román Jiménez-Díaz MIND/IN2UB Department of Electronics, University of Barcelona (UB), Barcelona, Spain

Heung-Bae Kim KITECH (Korea Institute of Industrial Technology), 330–825 Cheonan, Republic of Korea

Hans Koops HaWilKo GmbH, Particle Sources Systems, Ernst Ludwig Strasse 14–16, D-64372 Ober- Ramstadt, Germany

Charlene Lobo University of Technology, Sydney, Australia

Oleg Lupan University of Central Florida, Department of Physics, Orlando, FL 32816 USA Technical University of Moldova, Department of Microelectronics and Semiconductor Devices, Chisinau, MD-2004, Moldova

Stefan Matejcik Department of Experimental Physics, Comenius Unversity, Mlynska dolina F2, 84248 Bratislava, Slovakia

John Melngailis Department of Electrical and Computer Engineering and Institute for Research in Electronics and Applied Physics, University of Maryland, College Park, MD 20742, USA

Johann Michler EMPA, Swiss Federal Laboratories for Materials Science and Technology, Feuerwerkerstrasse 39, 3602 Thun, Switzerland

Kazutaka Mitsuishi Surface Physics and Structure Unit, National Institute for Materials Science, 3–13 Sakura, Tsukuba 305–0005 Japan

John H. Moore Department of Chemistry and Biochemistry, University of Maryland, College Park, Maryland, USA. [Dr. Moore, sadly, passed away during the production of this book.]

Stanislav A. Moshkalev Center for Semiconductor Components, State University of Campinas, C.P. 6101, 13083-870, Campinas, SP, Brazil

Hans Mulders FEI Company, Eindhoven, The Netherlands

Mariana Pojar Dept. of Electronic Systems, Polytechnic School, University of Sao Paulo Brazil

Juan Daniel Prades MIND/IN2UB Department of Electronics, University of Barcelona (UB), Barcelona, Spain

John Rarity Photonics Research Group, Electrical and Electronic Engineering, University of Bristol, UK

Albert Romano-Rodríguez MIND/IN2UB Department of Electronics, University of Barcelona (UB), Barcelona, Spain

Judy Rorison Photonics Research Group, Electrical and Electronic Engineering, University of Bristol, UK

Francisco Rouxinol Center for Semiconductor Components, UNICAMP, C.P. 6101, 13083-870, Campinas, SP, Brazil

Phillip Russell Appalachian State University, 525 Rivers Street, Boone, NC, USA 28608

Antonio Carlos Seabra Dept. of Electronic Systems, Polytechnic School, University of Sao Paulo, Brazil

Urs Sennhauser Empa, Swiss Federal Laboratories for Materials Science and Technology, Laboratory for Electronics/Metrology/Reliability, CH-8600 Duebendorf, Switzerland

Petra Swiderek Institute for Applied and Physical Chemistry, University of Bremen, Leobener Straße/NW2, 28334 Bremen, Germany

Milos Toth University of Technology, Sydney, Australia

Simone Camargo Trippe Dept. of Electronic Systems, Polytechnic School, University of Sao Paulo, Brazil

Ivo Utke Empa, Swiss Federal Laboratories for Materials Science and Technology, Laboratory for Mechanics of Materials and Nanostructures, Feuerwerkerstr. 39, 3602 Thun, Switzerland

Alfredo Vaz Center for Semiconductor Components, UNICAMP, C.P. 6061, 13083-870, Campinas, SP, Brazil

Carla Veríssimo Center for Semiconductor Components, UNICAMP, C.P. 6061, 13083-870, Campinas, SP, Brazil

Heinz Wanzenboeck Vienna University of Technology, Institute of Solid State Electronics, Floragasse, 1040 Wien, Austria

Oliver Wilhelmi FEI Company, Eindhoven, The Netherlands

Adam M. Zachary Department of Chemistry, University of Illinois at Chicago, 60607-7061, USA

NANOFABRICATION USING FOCUSED ION AND ELECTRON BEAMS

THE HISTORICAL DEVELOPMENT OF ELECTRON BEAM INDUCED DEPOSITION AND ETCHING

From Carbonaceous to Functional Materials

Ivo Utke and Hans W. P. Koops

1 UNWANTED SIDE EFFECTS: THE "DISCOVERY" OF GAS ASSISTED FOCUSED ELECTRON BEAM INDUCED PROCESSING

Gas assisted focused electron beam induced processing (FEBIP) comprises two nanostructuring techniques that are employed in electron microscopes: focused electron beam induced deposition (FEBID) and focused electron beam induced etching (FEBIE). Both rely on reversibly adsorbed gas molecules being present on the substrate to be nanostructured, and on the fact that electrons dissociate those adsorbates. Other names are sometimes used in the literature: electron induced deposition (EBID), electron beam assisted chemical vapor deposition (CVD), or electron beam induced etching (EBIE). Curiously, the discovery of FEBIP originated from unwanted side effects during observation with the electron beam, and then, from the wish of scientists to employ the unwanted contamination and etching effects for a useful result.

1.1 HYDROCARBON CONTAMINATION: THE ORIGIN OF FEBID

Deposits due to hydrocarbon contamination are as old as the "free" electron. Back at the very beginning of the twentieth century—at the time when cathode ray beams were studied under

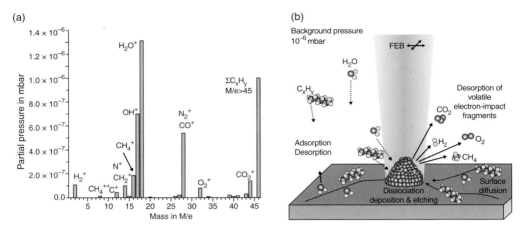

FIGURE I.1: a) Residual gas analysis of background pressure in a high vacuum electron microscope. Of note, the mass measurements required the gas molecules to be ionized; light ionized hydrocarbon fragments suggest the existence of heavier hydrocarbon molecules with M/e > 45, which were cracked during ionization. The sum of the heavier hydrocarbons, $\Sigma C_x H_y$, amounts to a "partial" pressure of around 10^{-6} mbar.

b) Hydrocarbon (contamination) deposition with an electron beam (FEB) which can, nowadays, routinely be focused to sub-10-nm size. The adsorbate supply comes from the background pressure of 10^{-6} mbar of heavy hydrocarbons. An additional supply comes via surface diffusion of adsorbates. Irradiated hydrocarbon adsorbates are dissociated and polymerised. Non-volatile products become the deposit and volatile products are pumped away. Note that electron-impact dissociated water adsorbates can react with the deposited carbon to form volatile products (etching). The emitted secondary electrons are omitted for sake of clarity.

vacuum (Crookes tubes, first vacuum triodes, etc.)—the first written reports on carbon- films due to electron impact appeared [1]. Since the inventions of the first transmission electron microscope (TEM) constructed by Ruska and Knoll in 1931 [2] and the first *scanning* electron microscope (SEM) constructed by von Ardenne in 1938 [3; 4], the literature has continuously reported on electron beam induced hydrocarbon deposition problems on surfaces [5]. At that time those deposits were unwanted nuisance and severely hindered sample observation. Today they witness that focused electron beam induced deposition on surfaces was the first focused particle-beam deposition technique before focused ion beams (FIB), scanning tunneling microscopes (STM), or lasers came into use. But from where did the the hydrocarbon adsorbates originate? Badly cleaned samples and the hydrocarbon molecules present in the gas phase of the microscope chamber [6] were both regarded as possible sources. Figure I.1a shows a typical analysis of the background gas inside the high vacuum chambers of an electron microscope.

The most abundant molecules are water followed by hydrocarbons. Apart from venting during sample transfer, water continuously permeates through polymer-based O-rings and hydrocarbons diffuse from pump oils or from vacuum grease. As a result, the surfaces in the vacuum system and the samples themselves become covered with hydrocarbons and water molecules. Upon electron irradiation, hydrocarbon adsorbates are converted into thin carbonaceous polymer films. In our own microscope chambers pumped with a turbo molecular pump backed by a rotary pump, we measure typical background pressures of 5×10^{-6} mbar and a partial pressure of water of up to 2.2×10^{-6} mbar. This translates to about 0.8 ML/s

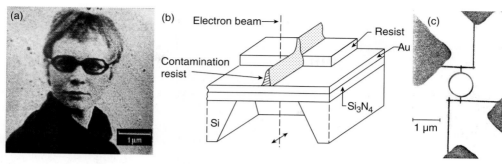

FIGURE I.2: **Examples of contamination deposits.** a) High density image written on a carbon membrane using TEM (Siemens, Elmiskop 101) from Müller in 1971. This figure was published in [17], Copyright Elsevier (1971). The pixel size was 20 nm and the total writing time was 10 min. b) Principle of FEB mask fabrication using backpressure pump oil molecules. The final metal structure was obtained by a successive broad-beam ion etching step. Reprinted with permission from [13] Copyright (1976), American Institute of Physics. c) Bright-field TEM image of fabricated gold ring and conducting wires for an electron wave interference experiment. (Aharonov-Bohm effect). Reprinted with permission from [18]. Copyright (1986), American Institute of Physics.

(monolayers per second) of water molecules impinging on the sample surface [7]. A similar impingement rate is found for the hydrocarbons.

Hydrocarbon adsorbates were found to offer a unique and novel approach for very *high resolution* lithography. The principle is shown in Figure I.1b. The pattern resolution is closely confined to the size of the impinging electron beam and the secondary electrons emitted from the impact site. Due to its large depth of focus, the electron beam could also be used on non-planar samples with high resolution.

By controlling the electron beam position and dwell time (or speed) [8; 9], the first patterns were materialized:

a) Image recording, see fig. I.2a: Grey tone images with a remarkable density and resolution were fabricated: 1'275'000 dpi (dots per inch) with 20 nm dot resolution in 1971 [8]. In terms of storage density, this recording method achieved 250 Gbit/cm² with a "bit depth" of a few grey tones. The recording sample was a simple TEM grid with a carbon membrane. For text patterns, see [10; 11; 8]. Of note, a similar high-density recording using an intense focused electron beam (100 keV), but relying on electron-stimulated desorption of halogene surface atoms instead of adsorbate dissociation, was published in 1990 by Humphreys et al.; they wrote an entry of the *Encyclopaedia Britannica* on an AlF_3 crystal [12].

b) High-resolution etch-mask lithography, see Figure I.2b: Highly ion-sputter-resistant etch masks from hydrocarbon adsorbates enabled the fabrication of metal structures with very high resolution for quantum theory experiments. The structure shown in Figure I.2c served for the investigation of the coherency of electron motion in solid gold. By evaluating the electron conductance through the ring in the presence of a magnetic field, which introduced a controlled phase shift to the electrons in the left or right ring part, modulated electron current flow through the ring could be detected. Also in 1976, Broers [13] fabricated 8 nm gaps in metal electrodes as weak links in a Josephson superconducting quantum interference device (SQUID). The use of FEB deposited material as an etch mask had a recent revival using $W(CO)_6$ and $Me_3PtCpMe$

molecules instead of hydrocarbons, resulting in higher sputter resistance [14; 15]. Thin FEB-deposited films or lines from $W(CO)_6$ or $Me_3PtCpMe$ are also used as protection against damage from Ga-ion beam milling when fabricating cross-sections or cross-section lamellae. The latter enable structural and chemical TEM analysis and are frequently used to investigate failure in semiconductor circuits.

c) Tip deposits, see Figure I.10c: With the invention of the scanning tunneling microscopy and the atomic force microscopy in 1981 and 1986, respectively, three-dimensional pillar deposits became popular in focused electron beam deposition. Such pillars were straightforwardly obtained with a stationary electron beam; the pillar is deposited co-axially to the electron beam within its impinging area. Due to the large depth of focus of a SEM, deposition (or etching) on non-planar substrates was never a problem, and FEB-deposited tips on pre-fabricated cantilever-type AFM sensors improved their functionality. Besides hard and high-aspect ratio AFM tips, today's portfolio comprises magnetic force microscopy tips, scanning near field optical microscopy tips, field emitters; see Figure I.10c and I.10d. For more applications we refer to [16].

Today, pump oil contamination can be avoided by using oil-free turbomolecular pumps backed by oil-free membrane pumps. Also decontaminators which create a plasma inside the SEM chamber are used to remove hydrocarbons from the substrate and chamber surface. Unwanted co-deposition of carbon during FEBID can thus be minimized [19]. Recently, successful SEM chamber cleaning by exposure to ultra-violet light and ozone was demonstrated [132].

1.2 RESIDUAL WATER: THE ORIGIN OF FEBIE

In the 1960s, hydrocarbon contamination deposits were still severely hindering observation in TEM and SEM experiments. Experimental approaches at that time included cooling chambers that cooled the space around the sample [20] and introduction of additional "cleaning" gases, like Ar, O_2, N_2, and air [21]. Both approaches are still in use today. The contamination could be reduced with decreasing temperature or increasing additional gas pressure. An interesting phenomenon was observed: reducing the temperature below (or increasing the cleaning gas pressure above) a critical value caused severe damage on organic, carbon-based samples to such an extent that they were fully destroyed in the irradiated area. It was Heide in 1963 [22] who proposed a chemical etch reaction of the carbon material with fragments of electron-dissociated water adsorbates as a plausible explanation. Furthermore, he showed that deposition of carbonaceous material (due to hydrocarbon adsorbates) and carbon etching (due to water adsorbates) take place simultaneously during electron irradiation in high-vacuum chambers. Which of the two processes dominates depends on the two adsorbate species: their partial pressures, their sticking probabilities, their surface residence times, and their electron dissociation efficiencies. For these details, see Chapter 7 by Lobo and Toth in this volume. Heide observed, that at room temperature, deposition dominated on samples, see Figure I3.a, in chambers with a residual gas composition as shown in Figure I.1a. When the surrounding temperature was lowered, the sum of the hydrocarbon partial pressure rapidly decreased, while the water partial pressure was measured to stay approximately constant inside the chamber until $T \approx -100°C$ (see Figure I.3b). Hence etching started to dominate over deposition in this temperature window (see Figure I.3a). For temperatures below $-100°C$, the

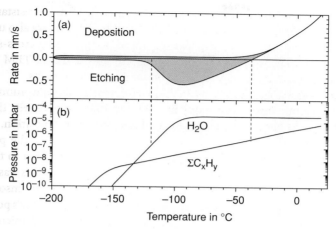

FIGURE I.3: a) Gas-assisted focused electron-beam etching (negative rate) and deposition (positive rate) in an electron microscope. The carbon membrane sample is kept at room temperature inside a sub-chamber. By varying the sub-chamber wall temperature, the pressure is reduced (cooling trap). The irradiated area was 1.6–1.9 mm in diameter and the current density was 0.3–0.4 Acm^{-2} at 60 keV. The shaded region comprises the experimental scatter, depending on beam heating and actual residual gas composition.

b) Partial pressures of water and hydrocarbons versus sub-chamber temperature. The differing decay of partial pressures is responsible for the temperature window of etching in Figure I.3a. Modified from Heide 1963 [22].

water vapor pressure also dropped considerably, and deposition and etching approximately compensate.

The etching by residual water was lithographically exploited in 1963 by Löffler [23] for producing micro-grids by etching carbon foils with water and oxygen in a reducing image projection system, and in 1970 by Holl [24], who etched holes into a cooled Formvar membrane using a TEM, see Figure I.4a. Examples of recent carbon etching experiments are shown in Figure I.4b and c: Here, water molecules were introduced via a gas injection system and directed to the substrate. The resulting increased water adsorbate density on the carbon membrane and diamond samples enabled etching at room temperature.

Water vapor has also been used as additional gas to decrease the carbon content obtained in deposition with organometallic molecules. This works best for noble metals which hardly oxidize, such as gold [26; 27].

2 FEB DEPOSITION AND ETCHING OF FUNCTIONAL MATERIAL

After scientists had first observed and investigated deposition from hydrocarbon contamination, it was their wish to deposit functional materials, like metals or dielectrics. Also,

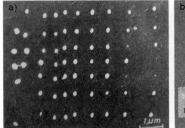 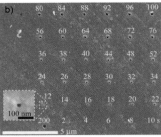 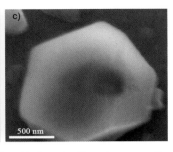

FIGURE I.4: a) Etching with the residual water chamber pressure rendered holes of a diameter as small as 50 nm into a 20 nm thick Formvar membrane, which was cooled to $-100°$C. This figure was published in [24], Copyright Elsevier (1970).

b) SEM image of an array of 6×6 nanoholes etched with injected H_2O (2×10^{20} molecules/cm^{2}s) into a 10 nm carbon membrane using a 25 keV and 985 pA electron beam. The numbers indicate the exposure time in seconds. Insert: 18 nm diameter hole. Reprinted with permission from [25]. Copyright (2008), American Institute of Physics. c) SEM image of a diamond crystal with a pit etched by a 25 keV electron beam and injected water vapor.

the wish to locally remove material from metals or oxides also became stronger. To do so, a gas injection system was needed to supply the precursor molecules. What today looks like a logical consequence in order to move forward was, at the time of the 1970s and 1980s, a "heretical" modification of an electron microscope: molecules containing metal atoms and organic or halogen ligands were willingly introduced and "contaminated" the chamber vacuum. It was a courageous step undertaken by the pioneers of FEBID and their coworkers Chang [28], Koops [29], Matsui [30], and Kunz [31], who introduced organometallic and inorganic precursors into "their" electron microscopes and did the first investigations on the influence of secondary and primary electrons, on material deposition, and on substrate temperature. Their experiments opened the exciting field of FEBIP as we know it today: as a maskless, minimally-invasive, high-resolution, three-dimensional nanoprototyping technology for numerous functional materials and nanodevices and as an interdisciplinary research field in which the fundamentals of surface science, chemistry, and physics meet each other at nanometer scale.

2.1 EARLY GAS INJECTION SYSTEMS

A schematic of an external gas injection system integrated to an SEM is shown in Figure I.5. Gaseous or liquid precursors with large vapor pressure (above about 0.1 mbar) could be injected in a controlled manner by operating closing and dosing valves at the outside of the microscope. The gas injection system could be heated moderately along all its components to generate a temperature-defined precursor-molecule flow. This ensured delivery of a sufficient number of molecules to the substrate. The substrate can be placed inside a sub-chamber to minimize exposure of the microscope chamber to the gas. This sub-chamber has a small access window for the electron beam. Secondary and backscattered electrons will be also detected through this window. The small window assured that very little precursor molecules would leak into the microscope chamber. In addition, differential pumping for safe lowest-vacuum operation of the optics column and electron gun chamber is realized by an additional aperture (tube) with

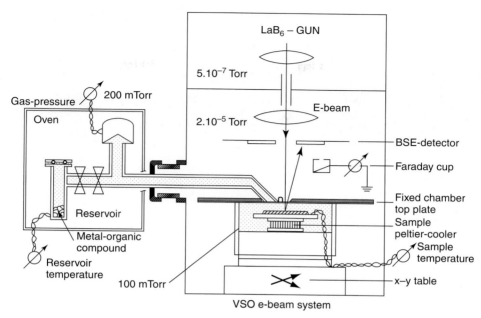

FIGURE I.5: Schematics of an SEM modified with an external gas injection system to perform FEBIP. Reprinted with permission from [32]. Copyright (1988), American Institute of Physics.

low conductance. This sub-chamber concept was already successfully used by Heide in 1962 [21] for Ar, O_2, N_2, CO gas supply into a TEM.

Sub-chambers that contained an internal precursor molecule reservoir were also used (see Figure I.6b). A "valve screw" sealed a tube connecting the reservoir to the environmental sub-chamber. The valve was manually operated with a screw driver tip at the end of the airlock sample loading stick [33].

Alternatively, internal reservoirs with simple hypodermic metal injection needles were integrated into the microscope chamber. The reservoir was closed with a screw and an O-ring seal (see Figure I.6a). In 1986, this arrangement was used as a simple internal gas injection system, which was fixed on the SEM sample holder and air-locked into the chamber without having a closing mechanism. As an example, when using the precursor $W(CO)_6$ which has a saturated vapor pressure of 3×10^{-3} mbar at room temperature, the background pressure inside the microscope chamber during deposition was 2×10^{-5} mbar due to the flow resistance of the needle.

The beauty of such a system lies in its simplicity, facilitating at least three important issues:

a) Filling and emptying such a system in a glove box is extremely rapid, handy, and facilitates a rapid screening of molecules.
b) The short tube length allows low vapor pressure molecules (down to <0.001 mbar) to be tested without heating and does not provide a large tubing surface for possible (catalytic) reactions.
c) The average flow of evaporated molecules could be estimated via the weight loss, i.e., by weighing the needle and reservoir before and after the experiment. This method is much more direct than trying to infer the molecule flux via the chamber background

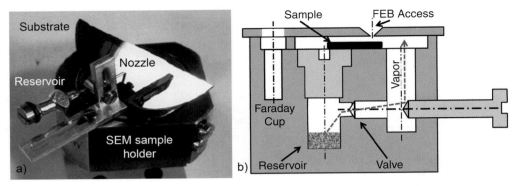

FIGURE I.6: a) Simple internal gas injection system from 1986 fixed on the SEM sample, without closing mechanism. Note that the sample could not be moved with respect to the nozzle once inserted into the SEM. b) SEM integrated environmental sub-chamber from 1995. The sample was fixed inside the sub-chamber and could be moved with respect to the electron beam. The screw, actuated from outside, opened and closed the reservoir. The arrow indicates the molecule flow.

pressure and the pumping speed, and gives easy access to prove if possible non-volatile residues remain in the reservoir due to decomposition of the molecules. In today's commercial gas injection systems, such an easy reservoir access is sacrified for easy automatic open/close operation and long-term molecule storage. Fundamental studies that need the exact knowledge of the molecule flux become difficult using such systems.

The disadvantage of the absence of a closing mechanism is obvious. Any observation scan of the sample or the just-fabricated deposits was accompanied by a thin-film deposition, which might degrade other sensitive structures on the chip. To avoid thin film deposits in the area of interest, a given area on the substrate near some pre-deposited reference markers was used to make adjustments of the beam focus and astigmatism. With the help of the reference markers, the beam was moved to the region of deposition using dedicated electron-beam lithography software, avoiding any prior observation.

More recent developments of GIS and their related molecule flow characteristics will be discussed in Chapter 3 by Friedli et al. in this volume.

2.2 FEB-DEPOSITED MATERIALS AND ETCHED MATERIALS

It is fair to say that the FEBIP pioneers did not start from scratch when choosing molecules for electron-impact deposition. Already in the 1960s and 1970s, vapors of volatile organometallic and inorganic tin derivatives ($Sn(CH_3)_4$, $Sn(C_4H_9)_4$, $SnCl_4$) were used [34; 35; 36] as well as several metal carbonyl molecules, all in *broad* electron-beam impact experiments [37; 38]. Among the first precursor molecules used for FEB metal deposition were $Cr(C_6H_6)_2$ [39], WCl_6, and WF_6 [39]. Often metal carbonyls were exploited for metal deposition [29]. Another group of metal molecules frequently used in FEBID are acetylacetonates (acac), trifluoroacetylacetonates (tfa), and hexafluoroacetylacetonates (hfa). Metal cyclopentadienyls (Cp) were stable, yet volatile, compounds, which were also used frequently. Further precursor molecules known from (plasma enhanced) chemical vapor deposition were used, such as hydrides, silanes, and siloxanes, the latter ones especially for insulator deposition. More recently, inorganic

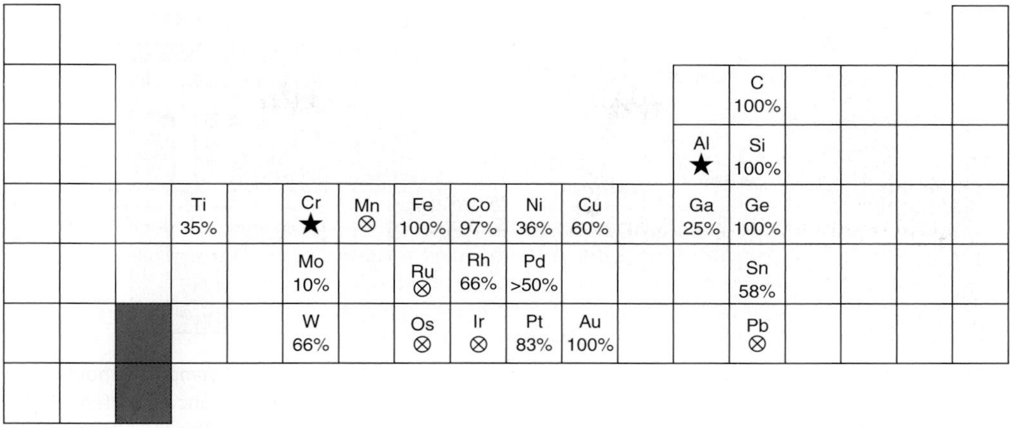

										C 100%				
								Al ★	Si 100%					
Ti 35%		Cr ★	Mn ⊗	Fe 100%	Co 97%	Ni 36%	Cu 60%	Ga 25%	Ge 100%					
Mo 10%			Ru ⊗	Rh 66%	Pd >50%			Sn 58%						
W 66%		Os ⊗	Ir ⊗	Pt 83%	Au 100%			Pb ⊗						

FIGURE I.7: Periodic table of elements obtained by FEBID. The best purity (at.%) obtained so far is indicated; ⊗ means no quantitative information on the composition of the deposit is available; "★" means the material is claimed as being "pure" when deposited, though the claim is not quantitatively substantiated. The remaining atomic percents adding up to 100 are composed of matrix material. Depending on the precursor elements, the matrix is composed of mainly carbon or phosphorous, and can have substantial traces of hydrogen, oxygen, and fluorine. Reprinted with permission from [40]. Copyright (2009), Institute of Physics.

compounds based on fluorophosphines, have been investigated for high metal-content deposits, as well as organic compounds, such as phenanthrene for diamond deposition.

Figure I.7 lists elements that have been studied for FEB deposition, together with the best obtained purities. In addition, *compounds* have been deposited, such as GaN, Fe_3O_4, NiO_2, TiO_2, and SiO_2.

The list of materials that can be (selectively) etched by electron-beam induced molecule dissociation comprises: all carbon allotropes (diamond, amorphous, nanotubes), polymers (PMMA and other resists), GaAs, Si, SiO_2, MoSi (as in the absorbing film of a lithographic mask), SiC, TaN, and Cr. The list of "etchant" molecules—as far as they were published and disclosed—comprises chlorine, bromine, iodine, water, oxygen, hydrogen, XeF_2, and mixtures of the aforementioned molecules.

Research for novel FEBIP molecules is now a very active field and mainly driven by the wish to prototype or repair material at nanoscale for industrial applications. Recently, FEB etching of Al_2O_3 with NOCl was demonstrated [44]. The requirements that have to be respected in the quest for new "FEBIP" molecules are discussed in Chapter 2 by Bret and Hoffmann in this book. Extensive precursor molecule lists for FEB deposition and etching of elements or compounds can be found in recent reviews from Botman et al. [40], Utke et al. [16], van Dorp et al. [41], and Randolph et al. [42].

2.3 INTERNAL STRUCTURE OF FEB-DEPOSITS

The FEB-deposit composition depends on the dissociation channels that an energetic electron can trigger in a surface-adsorbed molecule. To fully release the metal atom, all ligands, which make the metal atom volatile, must be stripped. Due to the continuous energy spectrum from electronvolts (secondary electrons) to kilo-electronvolts (primary electrons) (see Figure I.12),

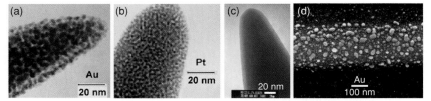

FIGURE I.8: TEM bright-field images of nanocomposite metal/carbon structures of
20-keV-FEB deposits. a) Pillar deposit with 700 pA beam current and the molecule
$(CH_3)_2$-Au(tfa) results in gold crystals (face centred cubic = fcc type) with diameters
of 3–4 nm embedded in a carbonaceous matrix. b) FEB pillar deposit with 200 pA beam
current using $(CH_3)_3$-Pt-Cp; fcc-platinum with crystal size 1.8–2.1 nm embedded in a
carbonaceous matrix. c) Amorphous FEB pillar deposit with 500 pA beam current
obtained from the molecule $Mo(CO)_6$ consisting of Mo, MoO, MoC. d) SEM top view
image of *pure* gold line deposit (25 keV, 300 pA) with percolated grain structure (Au
crystal size up to 60 nm, no carbonaceous matrix) on bulk Si substrate. FEB-deposited
from the AuClPF$_3$ molecule containing no carbon. This was the first pure metal
deposition in an SEM operated at normal high vacuum background pressure of about
10^{-6} mbar. Reprinted with permission from [46]. Copyright (2000), American Institute
of Physics.

bond dissociation is not selective. It can occur for any bond in the molecule, including metal-
ligand bonds as well as intra-ligand bonds. Consequently, volatile ligands and light elements
(e.g. H, O, F—depending on the ligand composition), as well as many non-volatile intermediate
dissociation products can be formed: adsorbate fragments and ligand fragments contribute to
the final deposit composition. We refer to Chapter 4 by Moore and colleagues in this volume
for an in-depth discussion about the numerous dissociation pathways and species that can be
produced in electron-triggered dissociation reactions.

The phenomenological picture of deposition from organometallic molecules further
assumes that the dissociated heavy metal atoms aggregate to nanocrystals in competition to
reticulation (network formation) of the organic ligand fragments. Oxidation of some metals
by the oxygen present in the background gas was also observed. The carbon reticulation
reactions can result in chemically and mechanically very stable amorphous hydrogenated
carbon networks [43]. The carbon network surrounds the metal nanocrystals and will limit their
aggregate size. For this reason, FEB deposits are mostly nanocrystalline with crystals < 4 nm in
a matrix, as shown in Figures I.8a and I.8b. However, if the metal reacts with the ligand elements
(containing carbon, oxygen), an amorphous mixture of pure metal, metal-oxide, and metal-
carbide could be obtained, e.g. for the molecules $Cu(hfa)_2$ and $Mo(CO)_6$ (see Figure I.8c). When
FEBID started, pure metal was only obtained for the WF_6 molecule, which was dissociated in an
ultra-high vacuum SEM [45]. The three "archetypes" of internal structures of metal-containing
deposits obtained by FEBID are shown in Figure I.8: metal-matrix nano-composite, amorphous,
and pure metal.

A number of various effects determining the metal-to-matrix ratio have been identified and
are increasingly well understood:

a) Novel FEBID molecules: carbon-free molecules, such as the class of metal tri-
    fluorophosphines, e.g. $Pt(PF_3)_4$ and $AuClPF_3$, give large metal contents or even
    pure percolated metal crystals, which are not anymore embedded in a matrix (see

Figure I.8d); It seems that the tendency to autocatalysis (molecule dissociation catalyzed by the central metal atom of the molecule) plays a vital role [133, 134];

b) Introduction of additional "reactive" gases via multiple gas-injection systems: gases such as H_2O and O_2 dissociate under irradiation with electrons to reactive radicals which form volatile compounds with carbon or non-volatile stoichiometric oxides with metals. Consequently, they decrease the carbon versus metal content as well as they and improve the purity of an oxide deposition, see also the Chapter F-1 by Melngailis et al. in this volume;

c) Multiple adsorbate systems: when two or more adsorbates are introduced, the metal-to-matrix composition can be tailored by the scan parameters, especially dwell and refresh times [135];

d) Electron beam heating: heat energy adds thermal dissociation channels to the electronically excited dissociation and can lead to purer metal deposits, however, at the expense of resolution;

e) High-vacuum residual gas molecules: many of the gas molecules shown in Figure I.1a impinge at a rate of about 1 monolayer per second on the substrate and thus on the deposit material. If adsorbed and electron-dissociated they can be co-deposited (fragments from hydrocarbons) and thus passivate the metal for autocatalysis or the dissociation products oxidize the metal (radicals from water, oxygen). Strictly speaking, high-vacuum leads to multi-adsorbate conditions, see effect c) above.

The effects identified above illustrate that gas assisted focused electron beam induced processes are cross-disciplinary in nature, and, indeed, intensive joint studies are undertaken today by scientists coming from the fundamental and applied fields of surface science, physics, lithography, and chemistry.

2.4 SHAPES OF FEB-DEPOSITS

E-beam lithography systems are commonly used for control of beam position and exposure time. Such a control is needed when opting for three-dimensional prototyping of functional nanodevices. Unlike resist-based planar e-beam lithography, where the beam exposes a thin resist film, gas-assisted FEBID can deposit three-dimensional high-aspect-ratio structures due to the continuously supplied adsorbates. For example, arch deposits were obtained by depositing both branches "simultaneously" as shown in Figure I.9a. When each branch of the arch is deposited separately, the branches will generally not connect at the top due to various reasons, e.g., beam drift over time due to lack of instrument stability or shadowing of molecule flux owing to the already-grown branch. There is also a specific proximity effect related to FEB deposition of high-aspect-ratio structures; the first deposited structure will bend due to additional side wall deposition from forward-scattered electrons of the second deposited structure [47]. The beam control thus needs to toggle the beam between two branches (see Figure I.9b), or even six branches (as shown in Figure I.9c).

Having such a beam control for parallel exposure allows the deposition of 3D nanodevices with a specific functionality (see Figure I.10). Examples include microwave antennae [49] and photonic crystals. The photonic crystal in Figure I.10b was fabricated with 11 nm accuracy using the above described environmental sub-chamber with $(CH_3)_3$-Pt-Cp [50]. The pillar deposits had only 2 nm edge roughness, a resolution of 150 nm cylinder diameter, and a growth-rate of 20–100 nm/s. The full photonic crystal was deposited in 40 minutes exposure time.

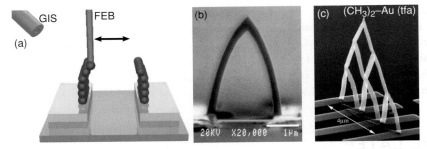

FIGURE I.9: a) Scheme of e-beam toggling between arch branches for simultaneous branch deposition. The width between branches is reduced to form an arch. b) Arch grown in 2 minutes under program control. Reprinted with permission from [48]. Copyright (1995), American Institute of Physics. c) Multiple arches deposited on pre-fabricated gold electrodes. The branches were simultaneously deposited "bottom-up" by FEBID with the molecule $(CH_3)_2$-Au(tfa).

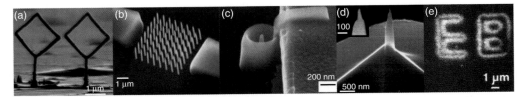

FIGURE I.10: 3D FEB deposits obtained with dedicated beam control. a) Wire electrodes or antennas for 4 μm infrared light, deposited with a stationary beam (for the base pillar) and a toggling beam (for the square) to write both branches "simultaneously" with increasing and decreasing jump width. Modified from [51]. b) Photonic crystal deposited in the gap of a PMMA waveguide in 40 min. The sequence of pillar deposition is dependent on the precursor flux supply. Reprinted from [52], Copyright (2001), with permission from Elsevier. c) Completed field-emission device fully fabricated by FEBID. Reprinted with permission from [53]. Copyright (2004), American Institute of Physics. d) FEB-deposited dual tip on top of a Si-cantilever pyramid: the base tip is amorphous carbon on top of which a gold-containing supertip with 7 nm curvature radius was deposited. Reprinted with permission from [50]. Copyright (1996), American Institute of Physics. e) 500 nm line width FEB etching of (110) silicon with XeF_2. Reprinted with permission from [54]. Copyright (1989), American Institute of Physics.

The flexibility of FEBID for prototyping was demonstrated by Murakami et al. [53] when depositing a complete functional field-emission device (shown in Figure I.10c). First, an insulating SiO_2 aperture was deposited, followed by FEB deposition of a platinum gate. A platinum tip emitter was then deposited in the aperture and the gate was connected to an electrical lead with a FEB-deposited platinum wire. Figure I.10d shows a FEB-deposited dual tip. On top of a Si-cantilever pyramid, an amorphous carbon tip is deposited as a base, on top of which a gold containing tip with 7 nm curvature radius was finally deposited. Since then, scanning-probe-sensor tip-functionalization by FEBID has been demonstrated with respect to magnetic, optical, wear-resistant, or thermo-sensitive probing, depending on the deposited material. Figure I.10e shows an early result of local FEB-etching of Si(110) using XeF_2 with a line-width of 500 nm. In contrast to FIB-milling, etching with FEB is material selective since it is a chemically controlled process and induces negligible or no mechanical damage to the substrate. For this reason, FEB etching is today employed for optical mask repair (see Section 5 in this chapter). Summarizing, with respect to 3D shapes, FEBID offers a large flexibility already recognized early on. The focus of today's research is in depositing 3D shapes with significantly improved material properties and in etching various metals and compounds used in the semiconductor industry.

3 THE LATERAL RESOLUTION IN FEBIP: NANOMETERS AND BEYOND?

Scanning electron microscopes that are able to scan a finely focused beam over a surface by use of magnetic or electrostatic lenses are the workhorses for gas-assisted focused electron-beam induced etching and deposition. The widespread use of SEMs became possible in 1958, when researchers from Cambridge (UK) built the first commercial prototype; see the review by Oatley [55].

3.1 THE SIZE AND INTENSITY OF THE PRIMARY ELECTRON BEAM

The typical primary electron beams used for gas-assisted FEBIP in a SEM have an energy of 1 to 30 kV and a beam current of 1 pA to 20 nA, which can be focused to about 2–100 nm spot size depending on the filament (see Table I.1). Today, spatial frequency analysis of an SEM image from a pre-defined resolution sample can provide an operator independent measurement of the full-width at half-maximum of the electron beam; see [56].

Table I.1 and Figure I.11 compare various electron emitters, see also chapter 1 by Forbes in this volume. Early SEMs used thermal tungsten hairpin cathodes as electron sources. Then tipped tungsten thermal cathodes were used, having 5 times higher brightness than a tungsten hairpin. Lanthanum hexaboride single-crystal heated cathodes provide a fine spot and 10 times higher brightness than tungsten hairpins. Thermal field-emission cathodes (Schottky-Cathodes using Zr-O-W) and cold field-emission cathodes are the most standard electron sources in SEMs of today. Mostly, standard short-focal-distance magnetic lenses are used as the objective lens defining the beam spot size. The cutting-edge photomask repair systems based on gas-assisted FEBIP use the aberration-compensated GEMINI® objective lens at 1 kV electron energy, which allows for an optimum beam aperture of $7 \cdot 10^{-3}$ rad. Comparison of optimum aperture and resolution with conventional magnetic lenses shows that a 50 times higher spot current is obtained at 1 kV with the GEMINI® lens [57].

To get an idea of gas-assisted FEBIP, it is illustrative to calculate the number of electrons arriving on the surface during a given exposure time and to estimate how many electrons will on average hit a molecule when being irradiated. Table I.1 gives a comparison of this data for the various electron sources.

Assuming an electron-impact dissociation cross-section of 10^{-3} nm^2, such focused electron fluxes will translate into molecule dissociation rates of $10^{4...7}$s^{-1} (dissociated molecules per second). In comparison, the acceptable injected gas-molecule fluxes vary approximately between $10^{0...3}$s^{-1} (molecule monolayers per second). According to the ratio of adsorbate dissociation and (intact) adsorbate replenishment, gas-assisted FEB deposition and etching

Table I.1 Source brightness and approximate maximum electron beam current at 2 nm beam size for various electron sources. The corresponding values for the electron flux, number of electrons, and the electron dose were deduced for a 1 microsecond exposure pulse.

	Field emission gun (FEG)	Thermofield Zr-O-W-Cathode	LaB$_6$ cathode	Tungsten tipped cathode	Tungsten Filament
Brightness @ 10 kV [A/(cm^{2}sr)]	2×10^8	2×10^7	10^7	10^6	10^5
Reduced brightness [A/(cm^{2}sr V)]	2×10^4	2×10^3	10^3	10^2	10
Emission current [μA]	8	10	10	5	100
Max. beam current [nA] @ beam size: 2 nm	2	0.2	0.1	0.05	0.005
Electron flux [electrons/nm^{2}s]	3×10^9	3×10^8	1.6×10^8	8×10^7	8×10^6
Number of electrons in a μs pulse (beam size: 2 nm) [electrons/μs][a]	12500	1250	625	312	31
Number of electrons per molecule and μs pulse [electrons/(molecule μs)]	278	28	14	7	0.7
Dose [mC/cm^2] for μs pulse	63.7	6.4	3.2	1.6	0.2

[a] The exposure time per pixel is typically ≤ 1 μs in a raster scan for depositing or etching rectangles. For continuous tip deposits using a stationary beam, the exposure time is typically > 1s, i.e., the number of electrons and the dose will be multiplied by a factor of 10^6.

[b] The molecule diameter was assumed to be 0.3 nm, i.e., for a 2 nm beam size one calculates 45 molecules inside the focus spot.

will proceed in different regimes: electron-limited or adsorbate-limited. The consequences for the shape and rates are detailed in Chapter 6 by Utke on continuum models in this volume.

Interesting electron sources not mentioned in table I.1 are scanning tunneling tips (STM), since they provide primary electrons in the low-energy range of 0–30eV and finely focused tunneling currents of about 50 pA. Basically, the atomically sharp tips constitute a cold field-electron source and allow depositing nanostructures. Using the gas supply system of Figure I.8, tungsten-containing deposits of 10 nm size were obtained in 1988 by McCord et al. [59]. Bruckl et al. used in 1999 the STM for deposition of materials in air with $(CH_3)_2$-Au(tfa) and $(CH_3)_3$-Pt-Cp [60]. Lyubinetsky et al. [61] used (hfa)-Cu-VTMS for STM deposition and identified electrostatically driven and current-driven deposition regimes. Marchi et al. [62] used the inorganic molecule $Rh_2Cl_2(PF_3)_4$. STM has the advantage of tuning the primary electron energy to energies needed for specific dissociation pathways. Disadvantages are that the tunneling current is very sensitive to any "contamination" of the surface and tip (process stability) and that it needs conductive substrates. In UHV and low-temperature conditions, STM is known to be able to manipulate at the molecular scale (flipping, rotation). However, FEBIP closes the gap to a practical industrial scale approach at the expense of atomic resolution.

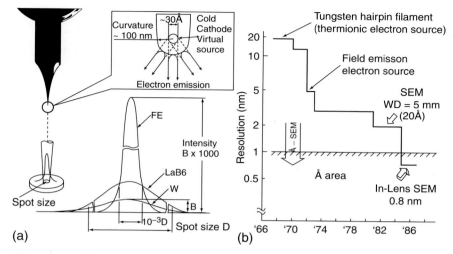

FIGURE I.11: a) Comparison of electron sources W, LaB$_6$, cold field emitter (FE), and b) resolution of standard and "in-lens" SEMs [58]. An "in-lens" is a strong single-field condenser objective lens with very small aperture aberrations; the sample is immersed in the centre plane of the magnetic field. Reprinted with permission from [58].

3.2 NON-LOCALIZED EFFECTS

Elastic and inelastic electron interaction with solid matter was actively investigated in 1930 by Bethe [63] and in 1954 by Lenz [64]. Elastic collisions lead to beam broadening via changes in electron trajectory directions, whereas inelastic collisions manifest as energy losses via ionization (secondary electron generation), plasmon excitation, molecule dissociation, phonon scattering (heat), and photon emission (X-rays) along the primary electron trajectories. A portion of the secondary electrons (SEs), as well as some of the primary electrons (PE), can exit the sample surface as back-scattered electrons (BSE), resulting in an emitted energy spectrum shown in Figure I.12a around the incident beam. The energy of the primary electrons remaining inside the specimen will be converted into heat. Figures I.12b and I.12c sketch the relevant ranges of SEs and BSEs, namely R_{SE} and R_{BSE}. Two SE ranges can be distinguished; the SE1 are generated by incident PEs, whereas the SE2 are generated by the BSEs. Parameterized equations for these ranges in a given element were established in the 1970s by Kanaya and colleagues [65; 66; 67]. Accordingly, the BSE range is proportional to $E_{PE}^{5/3}$, the primary electron energy, and inversely proportional to the electron density of the element, given by its density, molar mass, and atomic number ($\rho Z/M$). For instance, for bulk silicon the electron range extends from approximately 30 nm at 1 keV to 10 µm for 30 keV. For lower energies, other dependencies were more correct, see Reimer's SEM book [68].

A good review on SE emission was given by Seiler in 1983 [69] from whom Figure I.12 was inspired. The generated SEs have generally an initial energy smaller than a few 100 eV. At such energies, the mean free path of an electron is quite short (below 1 nm), and it will suffer from energy losses by numerous collisions. For this reason, only SEs produced near the surface will be able to escape the surface (and will have lost most of their initial energy). The most probable escape depth λ of SEs is independent of the primary electron energy (as long as $E_{PE} >$ first ionization energy), proportional to the first ionization energy, and, approximately inversely proportional to the electron density of the element [67]; for instance, for carbon $\lambda = 11$ nm

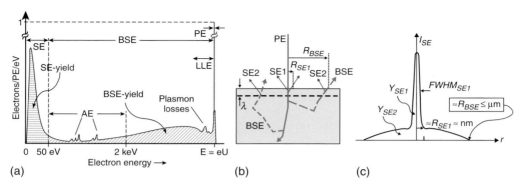

FIGURE I.12: a) Energy spectrum of emitted electrons (SE = secondary electrons, BSE = backscattered electrons, AE = Auger electrons, LLE = low loss electrons) and incident primary electrons (PE). Note that the energy axis is not linear. For 25 keV electrons impinging normally onto Si the following yields (areas below curve) were measured: $Y_{SE} = 8\%$, $Y_{BSE} = 30\%$, $Y_{AE} = 0.1\%$. Typical SE spectra for metals peak at 1 eV < E < 5 eV and have full-widths at half-maximum of 3 eV < FWHM < 15 eV. Insulators have smaller widths and peak energies.

b) Schematized view of secondary electron (SE) generation: SEs generated within the escape depth λ can exit the surface; all other SEs generated along the entire trajectories will remain inside the specimen. SEs can be distinguished according to PE or BSE generated SEs.

c) Schematics of corresponding lateral intensity distribution (SE per area) of emitted secondary electrons and approximate ranges generated by a zero diameter electron beam. The yields (Y) are given by the volume below the curve.

and for gold $\lambda = 1.5$ nm. Seiler distinguishes mean SE escape depths as : $\lambda \approx 0.5$–1.5 nm for metals and $\lambda \approx 10$–20 nm for insulators. Maximum escape depths are in the order of 5λ. Seiler also proposed that the exit range of secondary electrons produced by primary electrons $R_{SE1} \approx \lambda$ to 5λ. It was shown that yields of the different types of SEs, Y_{SE1} and Y_{SE2} are often comparable [68]. Nevertheless, the SE1 distribution peaks sharply while the SE2 distribution gives a large and relatively small intensity background due to the large exit range area of BSEs (see Figure I.12c).

From today's FEBIP perspective, the range concept allows us to attribute halo deposits around low-aspect-ratio deposits to BSE and SE2 [70]. Proximity effects known from resist-based lithography can also be explained using this concept. However, the range concept as presented in Figure I.12b and c is not precise enough when adsorbate dissociation comes into play. Then the spatial distribution of energies of the exiting secondaries needs to be taken into account. We refer the interested reader to an excellent comprehensive book about FEBIP-specific MC simulations from Silvis-Cividjian and Hagen [71] showing that the related FWHMs of the distribution of dissociated adsorbates can be less than 1 nm.

Simple MC software allowing the simulation of primary electron trajectories, yields, and ionization-density distributions in homogeneous bulk structures or with a given geometry and material distribution comprise CASINO (freeware) [72; 73] and MOCASIM (commercial) [72; 73]. A good source for experimental stopping powers and SE and BSE yields was compiled in Napchan's/Joy's database [74], which also now features MC simulations, the details of which were derived in the seminal work of Joy in 1995 [75].

Apart from the above range concept, inelastic electron scattering is by nature delocalized at the angstrom-scale due to Heisenberg's uncertainty relation. As a result, the generation of secondary electrons [76], surface plasmons [77], and dissociated molecules [71] is also delocalized. Delocalization of SEs was estimated to add < 0.5 nm to their distribution (see Figure I.12c), and dissociation of molecules due to delocalized surface plasmons was estimated to be around 1 nm [71]. These estimations compare to the world record of 0.7 nm lateral dot resolution obtained by FEBID which discussed in section 3.4 further below.

3.3 MATERIALS SUPPLY BY MOLECULE POLARIZATION AND STRONG ELECTRIC FIELDS

Electrostatic fields generated by the electron beam can lead to fractal deposit shapes. Fourie suggested that the intense electric fields generated around the impingement point of the FEB on the substrate induced molecule polarization, and thus, enhanced the surface diffusion toward the contaminating spot [78]. An electrostatic origin of enhanced molecule transport was proven by Banhart when irradiating an insulating boron nitride sample with a large beam and observing fractal filament growth from hydrocarbons [79]. The very ends of the fine filaments created the highest electrostatic field gradients, and thus increasingly polarized and attracted molecules for further growth. These results were reproduced on polymer, metal, and ceramic substrates and for organometallic molecules [80; 81; 82]. Obviously, any electrically insulating freestanding pillar deposit can take the role of an electrostatic field generator and cause the above-mentioned enhanced material transport phenomena. Other polarization mechanisms invoke the high electromagnetic field or inelastic collisions with the passing primary electron beam itself [83].

3.4 THE BEST LATERAL RESOLUTION FOR DEPOSITS AND ETCH HOLES

To achieve the best resolution for FEBID, the experimental rules below are suggested:

a) Use the best focused electron beam; i.e., use the best electron optics for probe forming.
b) Use the shortest electron wavelength (high acceleration voltage) for lowest diffraction aberration.
c) Avoid "proximity" exposure by BSE and SE; either use carbon or silicon nitride membranes, which transmit primary electrons without large inelastic energy loss, or use very small exposure times on bulk samples.
d) Use a molecule that upon electron irradiation dissociates into one small non-volatile fragment. Good candidates are monomers, for instance, of metal carbonyls which can auto-catalytically dissociate, releasing volatile CO and the non-volatile core metal atom.
e) Select a short beam exposure pulse (small dose), which deposits less material and avoids dissociation of volatile ligand fragments not yet desorbed from the sample.

Most of the resolution experiments summarized in Table I.2 followed the above "rules." The world record of 0.7 nm for a dot FWHM was obtained for $W(CO)_6$ on a membrane with 200 keV and small exposure time [84]. From Table I.2 distinct differences in resolution can be observed for various deposit types, like dots, lines, tips, and freestanding rods. This is basically due to the fact that tips and rods constitute self-supporting 3D structures which need a given volume. Within this volume SEs will be generated and exit through its surface—thus leading to lateral growth by dissociating surface adsorbates on the deposit. In contrast, a "dot" deposition can, in principle, be interrupted after the first surface adsorbed molecule is dissociated.

Sub-beam resolution for dots and holes on membranes can be achieved with feedback-loop-controlled exposure: For water- and XeF_2-assisted FEB-etching of holes in carbon membranes, the sample current was used for exposure interruption (see Figure I.4b). For $W(CO)_6$-assisted FEB-deposition, the annular dark-field signal generated by deposited mass was shown to have the potential for atomic resolution [85]. The 40–80 nm large diameters of FIB-prefabricated holes in a SiN_x-membrane could be refined down to about 4 nm using a TEOS-assisted FEB-deposition and SEM image scans for control [86]. In principle, for lines or slits on membranes, similar sub-beam diameter resolution could be expected. The limiting factor is the detection sensitivity for the deposited material, as well as the bandwidth of the feedback control—small process rates (low molecule supply) are helpful for these specific deposit (etch) patterns. Sub-beam resolution can also, in principle, be achieved when competing deposition and etching processes take place within the irradiated region; see Chapter 6 by Utke and Chapter 7 by Lobo and Toth in this volume. Lateral resolutions of etched trenches and holes (frequently called "vias" in chip fabrication) in industrially relevant materials are given in Section 5 of this chapter.

4 FROM DAMAGE TO ELECTRON CHEMISTRY CONTROLLED LITHOGRAPHY

In 1960 Christy [102] proposed that two mechanisms be considered during gas-assisted electron deposition: electron-impact dissociation of surface adsorbed molecules leading to material deposition and electron beam damage in the already deposited material. As we will see, both mechanisms subdivide into various other mechanisms or so-called reaction channels. There is a distinction between primary damage and secondary damage (see also Figure I.13). Primary damage is caused by electronic molecule excitation (energy transfer), leading to molecule dissociation (inside bulk material and of surface adsorbates), for instance, via dissociation into ions, dissociation into neutrals, and dissociative electron attachment. Secondary damage comprises all "follow up" reactions of the dissociation fragments produced from the primary damage. A detailed discussion of these reactions can be found in Chapter 4 by Moore et al. in this volume.

A classification of radiation damage in bulk material depending on the type of electron scattering was given by Egerton [103].

4.1 "DAMAGE" IN CONDENSED MOLECULAR FILMS BY ELECTRONS

For electron microscopists, irradiation damage describes all (wanted or unwanted) changes of bulk material, condensed thin films, or resist films upon irradiation with electrons. On a rough time scale proposed by Chapiro in 1962 [104], radiation damage proceeds as follows. Within the first 10^{-18} to 10^{-15}s, energy transfer from the primary and secondary electrons creates highly excited molecules. Within the next 10^{-14} to 10^{-11}s, the highly excited molecules relax via various pathways (reactions), some of them resulting in primary damage (see

Table I.2 Lateral resolutions of several FEB deposit types: dots, lines, freestanding rods, and tips. Dot resolution was not uniquely defined in the literature. Dot diameters obtained visually from TEM pictures are probably larger than the contrast intensity analysis, yielding a well-defined full width at half maximum value.

Resolution (nm)	Year	Molecule	Energy (keV)	Substrate	Ref.	Type
Dots						
20	1971	Hydrocarbons	≈ 100	thin C-film	[17]	
4	2005	D_2GaN_3	200	thin ion-milled Si(100)	[87]	FEB
3.5–7	2003	$W(CO)_6$	200	thin *and* thick Si(110)	[88]	
3.5	2004	Hydrocarbons	20	bulk Si(100)	[89]	
3	2009	$(CH_3)CpPt(CH_3)_3$	30	thin C-film	[90]	
1.5	2004	$W(CO)_6$	200	thin ion-milled Si	[91]	
0.7	2005	$W(CO)_6$	200	SiN (30 nm)	[84]	
Lines						
8	1976	Hydrocarbons	45	10 nm Au/Pd film on 10 nm C-film	[13]	FEB
8	2001	$Ni(C_5H_5)_2$	100	C-film (10 nm)	[92]	
4–9	2003	Hydrocarbons	200	SiN membrane	[93]	
2	2005	$W(CO)_6$	200	SiN membrane	[84]	
Free Rods						
65	1972	Hydrocarbons	25(?)	–	[94]	
15	1988	$W(CO)_6$	120	–	[45]	FEB
10–14	1993	Hydrocarbons	100	–	[95]	
5	2003	Phenanthrene	15	–	[96] [a]	
20	2004	$W(CO)_6$	20	–	[97]	
8	2004	$W(CO)_6$	200	–	[97]	
Tips (apex diameter)						FEB
20	1992	Hydrocarbons	5–35	Pt/Ir-tip	[98]	
20	1993	Hydrocarbons	5–20	Si	[99]	
20	1994	$(CH_3)_2$-Au(tfa)	20	–	[100]	
1 [b]	1995	Hydrocarbons	40	Si AFM tips	[101]	
12	2003	$Co_2(CO)_8$	30	Si bulk	[7]	

[a] The same article notes that the best resolution with phenanthrene-assisted Ga-FIB was about 50 nm.
[b] Post-deposition sharpening with O_2 plasma.

Figure I.13). The outcome of these reactions, e.g. fragments, radicals, vacancies/interstitials, volatile lightweight molecule elements, is temperature *independent*. Recombination of the dissociation/displacement products can take place and will be temperature dependent if not kinetically hindered. The products can also diffuse within the material. Secondary damage occurs within 10^{-10} to 1 s and depends on pressure, temperature, and molecule network. It comprises radical-induced reactions: polymerization, network formation (reticulation),

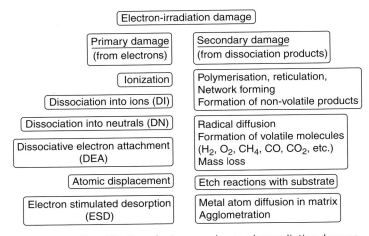

FIGURE I.13: Classification of primary and secondary radiation damage and related reactions. Note that formation of non-volatile products can also proceed via primary reactions.

evaporation of volatile reaction products, and agglomeration by diffusion. The secondary damage can be by far more noticeable, and is, for instance, exploited in lithography by chemically amplified resists. A recent novel exploitation of secondary damage, namely the polymerization in biphenyl-self-assembled monolayers to yield graphene-like membranes is discussed in Chapter 26 by Gölzhäuser in this book.

All the damage mechanisms in Figure I.13 proceed initially at a rate that is proportional to the electron current density j *and* the electron efficiency of this mechanism in a specific material, described by the energy dependent electron-impact cross-section σ. As an example, the displacement cross-section for carbon is about $\sigma_{dp} = 2 \cdot 10^{-22}$ cm^2 (at 100 keV), and with a current density of $j = 100$ A/cm^2 a displacement rate of about $\sigma_{dp} \cdot (j/e_0) = 0.1$s^{-1} is obtained, e_0 being the electron charge. Other damage mechanisms may have smaller or larger cross-sections and proceed at lower or higher rates during irradiation, respectively. Engineering tables were established for a given material where the damage was classified according to the minimum irradiation dose [105; 106]; see Table I.3 for a compilation. The dose is defined as $D = j \cdot t_d$, where t_d is the exposure time needed to cause detectable damage. Such tables were, and are, very useful for engineering tasks, like sterilization of materials, e-beam surface treatment, or e-beam lithography. It is worth mentioning that such minimum dose values do *not* constitute thresholds in a fundamental sense since damage healing always proceeds at much slower rate –if at all- when not thermally activated. It is rather the damage indicator, like diffraction fading, mass loss, or solubility, which imposes a minimum dose in order to be *measurable*. It can also happen that specific effective damage channels get exhausted and that less effective damage channels can be observed in turn, as is observed for PMMA: at low dose of about 100 μC/cm^2 bond scission prevails and renders a positive resist, while at high dose > 1 mC/cm^2 reticulation takes over and renders a negative resist. However, each damage mechanism can have *energy* thresholds below which it is not active. For instance, carbon atom displacements in the network require a minimum energy transfer of around 16–17 eV to the carbon atom by electron collision. Given the mass ratio of a carbon atom and an electron, a ballistic collision transferring such energy would require high primary electron energy of larger than around 90 keV.

With respect to gas-assisted FEB deposition and etching, the generic damage tables provide only limited insight. Instead, detailed studies of FEBIP relevant molecules and adsorbates are

Table I.3 Approximate minimum dose and related (detectable) radiation damage. Note that dose values for the separate damage mechanisms can vary considerably with material. Most observation electron microscopes can supply 100 pA electron current into a spot size of ≤ 10 nm which gives a current density of $j \geq 100\,\text{A/cm}^2$. A one second exposure amounts thus to a dose of 100 C/cm^2. Electrons most efficiently causing the classified damage ("trigger electrons"), SE = secondary electrons ($< 100\,\text{eV}$), PE = primary electrons (1–200 keV), and the most probable damage location are given.

Electron dose [C/cm^2]	Measurable Damage	"Trigger electrons" and Location
Condensed films[a]		
10–100 μ	Molecular chain scission	by SE close to beam impact
≈ 0.5 m	Cross-linking, reticulation	by SE close to beam impact
≈ 5 m	Loss of light-weight atoms, radical formation	by PE/SE close to beam impact
≈ 1	Displacement of heavy atoms	PE at beam impact (E > 200 keV)
≈ 10	Crystal formation[e]	PE energy loss at beam impact
$\approx > 100$	Evaporation, Melting [e]	PE energy loss in excitation volume
Resist Lithography		
1 μ – 100 μ	Chemically amplified resist: bond scission and secondary reactions	PE/SE in electron excitation volume, 2^{ary} reactions delocalized due to matrix diffusion
10 μ – 100 μ	e-beam resist PMMA: bond scission (positive resist)[b]	PE/SE in electron excitation volume
> 1m	polymerisation (negative resist)[c] scission.	
5 m	Collodion[d]: for 60% mass loss	PE/SE in electron excitation volume
Gas assisted FEB deposition		
1–100 μ	Dissociation of adsorbates	PE/SE at surface
100 μ – 1000	Accumulative pulses or continuous exposure: damage as for condensed films	PE/SE in electron excitation volume

[a] Dose: Amino acids < higher carbon compounds < aromatic compounds < phthalocyanines [107].

[b] The irradiated area dissolves rapidly in a "developer solution" due to smaller polymer chain lengths, the non-irradiated resist film remains = positive resist.

[c] The irradiated area remains in a "developer solution" due to longer polymer chains created, the non-irradiated resist film dissolves = negative resist.

[d] A viscous solution of nitrocellulose in ether and alcohol.

[e] Related to beam heating: For crystal formation, melting, and evaporation very high beam currents of up to milliamperes need to be applied. The implanted *power $\sim$ voltage $\times$ current* primarily determines heating, *not* the dose. Due to the low power of some microwatts, beam heating in SEMs / TEMs occurs only for specific sample geometries having reduced heat dissipation due to reduced dimensionality (encapsulated nanocrystals, cylinders, or membranes) *and* low heat conductance materials, see [16].

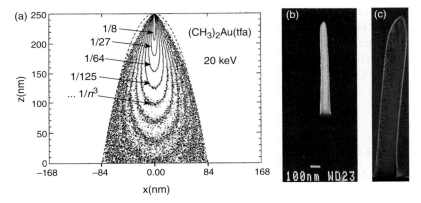

FIGURE I.14: a) MC simulations of spatial energy distribution obtained from 20 keV primary electrons impinging on FEB deposited Au/C material obtained from $(CH_3)_2Au(tfa)$. The scaling of the energy density inside the tip apex is indicated, the initial value being 1 $eV/nm^3/PE$ (PE = primary electron). Modified from Weber [109].

b) SEM image of a pillar deposit (20 keV, 50 pA, from $(CH_3)_2$-Au(tfa), 100 nm diameter), showing compact core material and less-compact material at the surroundings. Reprinted with permission from Weber [109].

c) Secondary electron image from FIB observation. Cross-section of a pillar deposit (25 keV, 1 nA, Co_2CO_8), cylinder diameter 550 nm. The core material has 56 at.% Co content; the outer crust shows pure cobalt grains.

needed to understand how electron-driven dissociation reactions can be controlled by changing irradiation parameters, like beam energy, pulse rates, or electron density.

However, from Table I.3 it becomes clear that FEBID can proceed in a wide dose window: surface-adsorbed molecules will dissociate at doses smaller than 100 $\mu C/cm^2$. There is an accumulation of dose for the already deposited material. Basically, this means that many damage mechanisms listed in Table I.3 could show "visible" outcome.

In addition, Monte Carlo simulations visualize how inhomogeneous the electron dose becomes within the material due to primary electron collisions and related beam broadening. Figure I.14a shows the relative spatial distribution of energy that was consumed in primary damage ionization and SE-generation processes (Bethe continuous slowing-down approximation [63]) for a 20 keV electron beam impinging on a Au/C pillar material (obtained from $(CH_3)_2$-Au(tfa)). Obviously, the dose (and related energy) decays strongly with the radial distance. Looking at Figure I.14b, one could attribute the low-density material surrounding the pillar deposit to a low-dose, cross-linking (polymerization) mechanism of organic radicals. Within the central zone, presumably further high-dose secondary reactions lead to a higher gold content. In contrast, Figure I.14c shows a pillar cross-section where carbon-rich core material is contained in a crust of metal (cobalt) grains [108]. Without going into details, these examples, as well as the example of the amorphous deposit in Figure I.8c, should point out that electron induced "damage" in FEBID is still a collection of puzzling phenomena, which often are difficult to explain with simple physical schemes known from "historical" approaches.

More specifically, high irradiation doses and high energy densities do not guarantee that organic ligands selectively separated from their core metal atoms, since highly electronically excited polyatomic adsorbates have numerous dissociation channels, depending on their molecule structure and their surrounding adsorbate network. Basically, with the energy spectrum shown in Figure I.12a, it is difficult to achieve selective dissociation of the metal-ligand bond. Such selectivity can be achieved only at low (monochromatic) electron energies.

However, such systematic studies for FEBIP relevant molecules are still at the very beginning [110; 111; 112; 136]. They would open a way to turn irradiation "damage" into a controlled dissociation of the adsorbate giving eventually a functional material the shape of which can be controlled at nanometer scale.

4.2 WHICH ELECTRONS DISSOCIATE ADSORBATES?

Generally, the electronic excitation of an adsorbate depends on the energy of the interacting electron, as well as on the electronic orbital states of the specific molecule.

Figure I.15a was published in 1994 by Weber [109] to illustrate which part of the secondary electron energy spectrum can contribute to dissociation processes. Since dissociation into neutrals and ions has energy thresholds around 10 eV, most of the SEs *cannot* contribute to dissociation. Although the maximum of most cross-sections for dissociation into neutrals and ions is around 70–100 eV, primary electrons can also contribute to a large extent to dissociation, provided the decay with energy E is weak like the $\ln(E)/E$ dependence of the ionization cross-sections. Unfortunately, complete sets of cross-sections are not yet available for the majority of FEBIP relevant molecules and fundamental cross-section investigations from the surface science community mostly concentrate on a very limited low energy window of 0–20 eV where dissociative electron attachment processes dominate.

One of the few measurements of deposition rates as a function of primary electron energy within the energy window of 10 eV–1 keV was performed for the aromatic benzene molecule C_6H_6 in 1967 by Kunze et al. [114]. The comparison with the ionization cross-section shows that deposition starts below the ionization threshold energy and suggests that another dissociation mechanism, like dissociative electron attachment, may already be effective at lower electron energy. Recently, in 2009, a similar experiment was performed for the more FEBID-relevant molecule $(CH_3)_3$-Pt-$CpCH_3$ [115]. A focused electron beam with low landing energies, down to 10 eV, was used. The deposition rate was a maximum at 140 eV and over ten times more efficient than at 20 keV. No significant dependence of composition with landing energy was found in the deposits performed at energies between 40 and 1000 eV. This study suggests that dissociation is primarily driven by sub-20-eV secondary electrons. In contrast, another recent FEB experiment with $Cu(hfa)_2$ showed that the primary electron energy (20 and 5 keV) has a distinct influence on the density and composition of the carbonaceous matrix in pillar deposits [116]. The dominance of SEs or PEs with respect to the dissociation rate and deposit

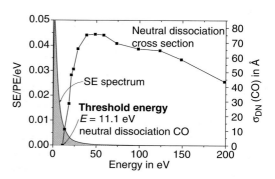

FIGURE I.15: Dissociation of CO molecules by the emitted secondary electron spectrum, adapted from Weber [109]. The electron impact cross-section for dissociation into neutrals was taken from [113]. Only about 30% of the SEs have energies above the threshold and are able to dissociate the molecule into neutrals.

composition will probably vary with the sensitivity of the molecule and the fragments. Here, fundamental understanding is still at the beginning, and interesting insights are to be expected in the future.

A further argument for SE-dominated deposition relates to high-resolution deposits; since the high energy primary beam (>100 keV) can be focused smaller than a nanometer, one should easily obtain small *high-aspect-ratio* structures comparable to the beam size. This is, however, not the case, and one can state that the lateral growth (in diameter) is governed by the SEs. However, the vertical deposition rate, i.e., the molecule dissociation inside the beam spot, can, in principle, still be governed by the primary electrons impinging with a "yield" of 100%.

5 FEBID AND FEBIE: FROM RESEARCH TO MARKET

5.1 RESEARCH: FLEXIBLE PROTOTYPING WITH FEBID MATERIALS

From the beginning, organometallic molecule-assisted FEB deposition had to compete with the purity of deposits of thermal CVD metal films (or plasma assisted CVD of dielectrics). In fact, the FEB-deposits obtained from organometallic adsorbates (under high-vacuum conditions) were nano-composites or amorphous (see Figure I.8), and had comparatively low amounts of metals, i.e., high contents of carbon of > 70 at.%. To date, only a few pure metal FEB-deposits have been achieved; see here the reviews of Botman et al. [40] and Utke et al. [16], and the Chapter F-1 by Melngailis et al. in this volume. However, it was soon recognized that a nano-composite material with the appropriate carbon content can have superior properties to pure metals. Examples comprise Hall sensitivity (see Chapter 16 by Gabureac et al. in this volume), and wear- and chemical-resistant, high-resolution, magnetic force SPM tips [117]; the carbon matrix gave chemical and mechanical protection, whereas the metal nanocrystals gave functionality.

With FEBID, one is able to tailor the physical properties of nanostructured material over several orders of magnitude by changing the deposited metal/carbon ratio. Apart from using inorganic molecules which gave pure Au deposits (see Figure I.8d), other approaches to tailor the metal/carbon ratio have been investigated: beam dwell time variation—increase from 40 to 75 at.% [118]; additional reactive gases—increase from 20 at.% to 50–100 at.% [26; 27]; post-deposition hydrogen/oxygen radical plasma treatment—increase by a factor of 2 [119; 120]; sample temperature—increase from 40 to 80 wt.% [50]; and UHV conditions—increase to 100 at.% [121]. Up to now, only a few molecules have been used for these studies: carbonyls of W, Fe, and Co, and acetylacetonates of Au, Cu, and Pt.

An intense field of research pursued since 1996 covers the conductivity of FEB deposited lines from organometallic molecules [50]. Tailoring the metal-to-carbon ratio of the nanocrystalline compound structures, the resistivity of deposited lines can be varied between insulating to metallic resistivities, $(10^6–10^{-6})\,\Omega$cm. Recent investigations on the electrical conduction mechanisms of FEB-deposited material are discussed in Chapter 22 by Gazzadi and Frabboni, in Chapter 23 by de Teresa et al., and in Chapter 24 by Huth in this volume.

Another intense research field comprises optical materials obtained from FEBID. Here, amorphous dielectric FEB deposits were investigated [122; 123]. The focus was on obtaining UV-transparent materials, i.e., co-deposition of carbon from the molecule or from the background had to be avoided. Additional O_2 injection was used mainly for pure SiO_2

deposition, see Chapter 21 by Wanzenboeck in this volume. Other recent work discusses the use of nano-composite Au-C material for stabilization of laser polarization [137].

Mechanical properties of FEB pillar deposits have been investigated: The elastic modulus could be tuned from diamond-like (0.5 TPa) using an organic molecule [124] down to more amorphous-carbon-like values (approx. 10 GPa) using Cu(hfa)$_2$ [116]. Depending on the metal/carbon content, the high elastic moduli of metals together with an order of magnitude higher yield stress than in metals can be obtained [7]. This is attributed to the stabilizing carbon matrix preventing dislocations from propagating through the structure.

Apart from the flexibility in materials, there is the flexibility in shape, which was already illustrated in figure I.10. Such 3D flexibility gives researchers a large freedom to design fundamental experiments or novel prototype nanodevices.

5.2 THE MARKET: COMBINED FEB/FIB SYSTEMS

Today the trend is toward combined focused electron and focused ion machines with a gas-injection capability that allows for gas-assisted FEB and FIB deposition and etching, as well as gas-less FIB-milling. Complete FEBIP (and FIBIP) machines for laboratory research are now available from several vendors, like the *Lyra* system line from *TESCAN* (using an ion gun from *Orsay Physics*), the *Nova, Helios, Magellan* line from *FEI*, the *Auriga, NVision, Neon, and CrossBeam* line from *Carl Zeiss*, and the *JIB* line from *JEOL*. *Raith* offers separate systems for FEB and FIB: the *e−* and *ionLiNE* systems. These systems are commonly used in nanostructure assembly, nanoscale prototyping, machining, characterization, and analysis of structures below 100 nm. With their unique capability of sub-micrometer resolution in "machining" and flexibility in shape (no mask needed!), they are the "tool" of choice for laboratory basic research of FEBIP and for prototyping in fields of nanophotonics, nanooptics, nanomagnetics, nanobiology, and nanoelectronics.

5.3 THE MARKET: FEBIP

The main reason that FEBIP is not yet embraced as a nanotechnology in industry is still the high cost of the tool (0.5 to 5 Mio €) and the low throughput for single- (or dual-) beam-based machines which render most technical applications non-profitable to the industry. Also, the purity of deposited metals is still an issue. Nevertheless, two market applications are served predominantly by FEBIP companies.

5.3.1 Photomask repair tool

Zeiss SMT (former NaWoTech) is the world market leader of photomask repair tools. Etch and deposition processes are available for most types of thin-film materials used in mask fabrication: Cr for binary masks, "MoSi" (5 to 20% Mo-rich SiN$_x$O$_y$) as used for attenuated phase shift masks, quartz, TaN, SiC, and Si [125], as well as for masks for extreme-ultraviolet lithography (EUV) [138]. The flexibility, ease of use, and yield of the FEB deposition and etch techniques allow for the processing of many different types of materials or "blanks," as required by the downstream customers (mask-shops, chip and memory foundries). The Zeiss MeRiT® mask repair tool, based on FEBIP technology, represents the current state of the art for 65 nm and 45 nm lithography processes, with most leading maskshops worldwide being equipped.

Further important manufacturing issues for photomask and integrated circuit repair are described in Chapter 9 by Hofmann et al. in this book.

5.3.2 AFM tips

The first use of FEBID contamination tips in AFM was shown in 1990 by Akama et al. [126]. The tips were produced straightforwardly from hydrocarbon contamination using continuous spot exposure. Similar results were presented by [98] as well as specially fabricated sideways pointing tips that could be used to measure the sidewall roughness of the reactive ion etching (RIE) process [127]. Tall tips with a rim at the top were used for 3-dimensional AFM metrology to measure the dimension of deep-trench etching in the chip fabrication at IBM by [128]. Finally, these technical applications triggered the founding of *Nanotools* GmbH in Munich in Germany, a company which supplies the sharpest AFM (atomic force microscope) tips to the semiconductor industry for in-line vias inspection as well as specifically shaped tips to research institutes.

5.3.3 SEM gas-injection systems

Apart from the integrated gas-injection systems of combined FEB/FIB system vendors in section 5.2, there are also stand-alone gas supply systems available, for instance from *Orsay Physics*, *Omniprobe*, *Kleindieck*, and *Alemnis*. They feature one to five gas channels, heating systems, and computer control with pneumatically or electrically activated valves. They are reviewed in more detail in Chapter 3 by Friedli et al. in this book. They can be retrofitted to almost any available SEM in a laboratory.

5.4 MASSIVELY PARALLEL CHARGED PARTICLE BEAM SYSTEMS

With respect to the issue of improving throughput, encouraging activities are in progress. Although reported FEBID rates of 100 nm/s for the initial deposition stage of tips out-perform process rates of molecular beam epitaxy (MBE) and, in part, chemical vapor deposition (CVD), the *volume throughput* is low. This is due to the serial nature of the process and the localization of deposition or etching to the beam size or the beam-scanned area.

To overcome the lack of speed of single-beam systems, a reducing image projection system was used back in 1988 by Rüb [129]; 200 nm wide structures were obtained in an image field of 80 μm diameter containing 16,000 square beams in a system. The theoretical resolution was 25 nm.

A very modern approach is the Projection Maskless Lithography (PML2) approach of *IMS Nanofabrication* Vienna addressing the 16 nm node and below. PML2 uses numerous individually addressable electron (or ion) beams working in parallel, thereby pushing the throughput into the 5–10 wafers per hour regime using resist films. A proof-of-concept ion projection multi-beam nanopatterning tool with 43,000 programmable beams and 12.5 nm half pitch resolution capability has been demonstrated [130]. A multi-electron beam tool with 264,000 programmable electron beams is in development.

A similar multi-beam concept is being pursued by *MAPPER*, based in Delft. It uses a fiber-optics system through which potentially more than 100,000 parallel electron beams can be switched through a MEMS aperture plate. Platforms with 110 beams are installed at CEA-Leti in France and in Taiwan [139].

Either of these multi-beam lithography systems could be used for parallel gas-assisted FEBIP provided there is an appropriate gas-injection system. Industrial applications at wafer scale level might include the production of customized chips and nano-optical components. Also, both systems have the potential to match the speed of several wafers per hour when using low sensitivity resists and deposition processes, and are thus valuable tools for circuit development, see Chapter F-1 by Melngailis et al. in this volume.

A third approach proposes the construction of many parallel electron-beam deposition systems by using the construction of miniaturized SEMs with the FEBID technology on top of pre-fabricated VLSI substrates [131]. This technique enables the use of large currents in each beam for deposition and etching, but is still to be proven.

6 CONCLUSIONS

Within about forty years gas-assisted focused electron-beam deposition and etching matured from research to market, from contamination deposition and etching to deposition and etching of functional materials, from millimeter to sub-nanometer structures. Gas-assisted FEBIP has further enormous potential as a minimally invasive, high-resolution, maskless, and single-step nanoprototyping platform when the following issues can be solved:

a) Improved purity for the semiconductor industry;
b) More demonstrations of the superior properties of tailored nano-composite FEBID nanostructures versus standard materials and processes obtained from lithography and CVD;
c) The low-throughput (e.g. by using multi-beam machines and adapted gas injection systems).

The best way to tackle the above issues seems a joint engineering and fundamental approach, gathering interdisciplinary researchers with expertise in surface science, electron induced processes, chemistry, physics, lithography, and microscopy.

Such a common enterprise will truly show the merits of gas assisted FEBIP as a new technology platform and its scientific and industrial impact in nanotechnology.

ACKNOWLEDGMENTS

The authors cordially thank all colleagues with whom they have worked together over the years to develop this novel nanotechnology platform and who gave valuable comments and material to this chapter.

REFERENCES

[1] Gehrcke, E. and R. Seeliger. Über Oberflächenladungen auf Leitern im Vakuum. *Verhandlungen der Deutschen Physikalischen Gesellschaft* 15 (1913): 438–450.

[2]   Ruska, E. The Development of the Electron Microscope and of Electron-Microscopy. *Reviews of Modern Physics* 59.3 (1987): 627–638.

[3]   von Ardenne, M. The scanning electron microscope: Theoretical fundamentals. *Zeitschrift für Physik* 109 (1938): 553–572.

[4]   von Ardenne, M. The scanning electron microscope: Practical construction. *Zeitschrift für technische Physik* 19 (1938): 407–416.

[5]   Stewart, R. L. "Insulating films formed under electron and ion bombardment." *Phys. Rev.* 45 (1934): 488–490.

[6]   Ennos, A. E. The sources of electron-induced contamination in kinetic vacuum systems. *British Journal of Applied Physics* 5 (1954): 27–31.

[7]   Utke, I. (unpublished results).

[8]   Muller, K. H. Speed-controlled electron-microrecorder. 1. *Optik* 33.3 (1971): 296–311.

[9]   Broers, A. N. High-resolution systems for microfabrication. *Physics Today* 32.11 (1979): 38–45.

[10] Möllenstedt, G., and R. Speidel. Elektronenoptischer Mikroschreiber unter Elektronen-mikroskopischer Arbeitskontrolle. *Phys. Blätter* 16 (1960): 192.

[11] Mollenstedt, G., R. Schief, et al. Electron optical micro recorder controlled by television. *Optik* 27.7 (1968): 488–491.

[12] Humphreys, C. J., T. J. Bullough, et al. Electron-beam nano-etching in oxides, fluorides, metals and semiconductors. *Scanning Microscopy* (1990): 185–192.

[13] Broers, A. N., W. W. Molzen, et al. Electron-beam fabrication of 80-angström metal structures. *Applied Physics Letters* 29.9 (1976): 596–598.

[14] Guan, Y. F., J. D. Fowlkes, et al. Nanoscale lithography via electron beam induced deposition. *Nanotechnology* 19.50 (2008): 505302.

[15] Heerkens, C. T. H., M. J. Kamerbeek, et al. Electron beam induced deposited etch masks. *Microelectronic Engineering* 86. 4–6 (2009): 961–964.

[16] Utke, I., P. Hoffmann, et al. Review: Gas-assisted focused electron beam and ion beam processing and fabrication. *Journal of Vacuum Science & Technology* B 26.4 (2008): 1197–1276.

[17] Muller, K. H. Speed-controlled electron-microrecorder. 2. *Optik* 33.4 (1971): 331–343.

[18] Umbach, C. P., S. Washburn, et al. Observation of H/E Aharonov-Bohm interference effects in submicron diameter, normal metal rings. *Journal of Vacuum Science & Technology* B 4.1 (1986): 383–385.

[19] XEI Scientific, see http://www.evactron.com/.

[20] Schott, O. and S. Leisegang. "Objektkühlung im Elektronenmikroskop" *Proc. Eur. Conf. Electron Microsc. Stockholm* (1956): 27–30.

[21] Heide, H. G. Electron microscopic observation of specimens under controlled gas pressure. *J. Cell. Biology* 13 (1962): 147–152.

[22] Heide, H. G. Die Objektverschmutzung im Elektronenmikroskop und das Problem der Strahlenschädigung durch Kohlenstoffabbau. *Zeitschrift für angewandte Physik* 15.2 (1963): 116–128.

[23] Löffler, K. H. PhD thesis: Erzeugung freitragender Mikroobjekte durch elektronenstrahlaktivierten Kohlefolienabbau. Technische Universität Berlin, Fakultät für Allgemeine Ingenieurwissenschaften (1964).

[24] Holl, P. An electron optical micro recorder of high stability and its application. *Optik* 30.3 (1970): 116–137.

[25] Miyazoe, H., I. Utke, et al. Controlled focused electron beam-induced etching for the fabrication of sub-beam-size nanoholes. *Applied Physics Letters* 92.4 (2008): 043124.

[26] Folch, A., J. Servat, et al. High-vacuum versus "environmental" electron beam deposition. *Journal of Vacuum Science & Technology* B 14.4 (1996): 2609–2614.

[27] Molhave, K., D. N. Madsen, et al. Solid gold nanostructures fabricated by electron beam deposition. *Nano Letters* 3.11 (2003): 1499–1503.

[28] Chang, T. H. P. PhD thesis (1967), "Combined microminiature processing and microscopy using a scanning electron probe system." University of Cambridge.

[29] Scheuer, V., H. Koops, et al. Electron beam decomposition of carbonyls on silicon. *Microelectronic Engineering* 5.1–4 (1986): 423–430.

[30] Matsui, S., and K. Mori New selective deposition technology by electron-beam induced surface-reaction. *Japanese Journal of Applied Physics Part 2-Letters* 23.9 (1984): L706–L708.

[31] Kunz, R. R., T. E. Allen, et al. Selective area deposition of metals using low-energy electron-beams. *Journal of Vacuum Science & Technology* B 5.5 (1987): 1427–1431.

[32] Koops, H. W. P., R. Weiel, et al. High-resolution electron-beam induced deposition. *Journal of Vacuum Science & Technology* B 6.1 (1988): 477–481.

[33] Koops, H. W. P., M. Weber, et al. Three-dimensional additive electron-beam lithography. *Proceedings of the SPIE – The International Society for Optical Engineering* 2780 (1996): 388–395.

[34] Baker, A. G. and W.C. Morris Deposition of metallic films by electron impact decomposition of organometallic vapors. *Rev. Sci. Instrum.* 32 (1961): 458–458.

[35] Christy, R. W. Conducting thin films formed by electron bombardment of substrate. *J. Appl. Phys.* 33 (1962): 1884–1888.

[36] Mann, H. T. Method for the deposition of thin films by electron bombardment. US Patent: 3132046 (1964).

[37] Fritz, O. G. Conducting film formed by electron bombardment of tungsten hexacarbonyl vapor in vacuum. *J. Appl. Phys.* 35 (1964): 2272–2272.

[38] Vishnyakov, B. A., and K. A. Osipov. *Production of thin films of chemical compounds by electron beam bombardment.* Tel Aviv: Freund Publishing House, 1972.

[39] Matsui, S., and K. Mori. New selective deposition technology by electron-beam induced surface-reaction. *Journal of Vacuum Science & Technology* B 4.1 (1986): 299–304.

[40] Botman, A., H. Mulders, et al. Creating pure nanostructures from electron-beam-induced deposition using purification techniques: a technology perspective. *Nanotechnology* 20 (2009): 372001.

[41] van Dorp, W. F. and C. W. Hagen. A critical literature review of focused electron beam induced deposition. *J. Appl. Phys.* 104.8 (2008): 081301.

[42] Randolph, S. J., J. D. Fowlkes, et al. Focused, nanoscale electron-beam-induced deposition and etching. *Critical Reviews in Solid State and Materials Sciences* 31.3 (2006): 55–89.

[43] Utke, I., V. Friedli, et al. Resolution in focused electron- and ion-beam induced processing. *Journal of Vacuum Science & Technology* B 25.6 (2007): 2219–2223.

[44] Bret, T., B. Afra, et al. Gas assisted focused electron beam induced etching of alumina. *Journal of Vacuum Science & Technology* B 27.6 (2009): 2727–2731.

[45] Matsui, S. and T. Ichihashi. In situ observation on electron-beam-induced chemical vapor deposition by transmission electron microscopy. *Applied Physics Letters* 53.10 (1988): 842–844.

[46] Utke, I., P. Hoffmann, et al. Focused electron beam induced deposition of gold. *Journal of Vacuum Science & Technology* B 18.6 (2000): 3168–3171.

[47] Molhave, K., D. N. Madsen, et al. Constructing, connecting and soldering nanostructures by environmental electron beam deposition. *Nanotechnology* 15.8 (2004): 1047–1053.

[48] Koops, H. W. P., A. Kaya, et al. Fabrication and characterization of platinum nanocrystalline material grown by electron-beam induced deposition. *Journal of Vacuum Science & Technology* B 13.6 (1995): 2400–2403.

[49] Kretz, J., M. Rudolph, et al. 3-dimensional structurization by additive lithography, analysis of deposits using TEM and EDX, and application to field-emitter tips. *Microelectronic Engineering* 23.1–4 (1994): 477–481.

[50] Koops, H. W. P., C. Schossler, et al. Conductive dots, wires, and supertips for field electron emitters produced by electron-beam induced deposition on samples having increased temperature. *Journal of Vacuum Science & Technology* B 14.6 (1996): 4105–4109.

[51] Koops, H. W. P. Rapid prototyping and structure generation using three-dimensional nanolithography with electron-beam-induced chemical reactions. *Proceedings of the SPIE – The International Society for Optical Engineering* 5116 (2003): 393–401.

[52] Koops, H. W. P., O. E. Hoinkis, et al. Two-dimensional photonic crystals produced by additive nanolithography with electron beam-induced deposition act as filters in the infrared. *Microelectronic Engineering* 57–58 (2001): 995–1001.

[53] Murakami, K. and M. Takai. Characteristics of nano electron source fabricated using beam assisted process. *Journal of Vacuum Science & Technology* B 22.3 (2004): 1266–1268.

[54] Matsui, S., T. Ichihashi, et al. Electron beam induced selective etching and deposition technology. *Journal of Vacuum Science & Technology* B 7.5 (1989): 1182–1190.

[55] Oatley, C. W. The early history of the scanning electron microscope. *J. Appl. Phys.* 53.2 (1982): R1–R13.

[56] Babin, S., M. Gaevski, et al. Technique to automatically measure electron-beam diameter and astigmatism: BEAMETR. *Journal of Vacuum Science & Technology* B 24.6 (2006): 2956–2959.

[57] Boegli, V., H. W. P. Koops, et al. Electron-beam induced processes and their applicability to mask repair. *Proceedings of the SPIE - The International Society for Optical Engineering* 4889 (Part 1&2) (2002): 283–292.

[58] Hitachi company information.

[59] McCord, M. A., D. P. Kern, et al. Direct deposition of 10-nm metallic features with the scanning tunneling microscope. *Journal of Vacuum Science & Technology* B 6.6 (1988): 1877–1880.

[60] Bruckl, H., J. Kretz, et al. Low energy electron beam decomposition of metalorganic precursors with a scanning tunneling microscope at ambient atmosphere. *Journal of Vacuum Science & Technology* B 17.4 (1999): 1350–1353.

[61] Lyubinetsky, I., S. Mezhenny, et al. Scanning tunneling microscope assisted nanostructure formation: Two excitation mechanisms for precursor molecules. *Journal of Applied Physics* 86.9 (1999): 4949–4953.

[62] Marchi, F., D. Tonneau, et al. Direct patterning of noble metal nanostructures with a scanning tunneling microscope. *Journal of Vacuum Science & Technology* B 18.3 (2000): 1171–1176.

[63] Bethe, H. Zur Theorie des Durchgangs schneller Korpuskularstrahlen durch Materie. *Annalen der Physik.* 397.3 (1930): 325–400.

[64] Lenz, F. Zur Streuung mittelschneller Elektronen in kleinste Winkel. *Zeitschrift für Naturforschung* 9a.3 (1954): 185–204.

[65] Kanaya, K., and H. Kawakatsu. Secondary-electron emission due to primary and backscattered electrons. *Journal of Physics D-Applied Physics* 5.9 (1972): 1727–1742.

[66] Kanaya, K., and S. Okayama. Penetration and energy-loss theory of electrons in solid targets. *Journal of Physics D-Applied Physics* 5.1 (1972): 43–58.

[67] Ono, S., and K. Kanaya. The energy dependence of secondary emission based on the range-energy retardation power formula. *J. Phys. D: Appl. Phys.* 12 (1979): 619–632.

[68] Reimer, L. *Scanning electron microscopy: Physics of image formation and microanalysis.* Berlin: Springer, 1998.

[69] Seiler, H. Secondary-electron emission in the scanning electron-microscope. *Journal of Applied Physics* 54.11 (1983): R1–R18.

[70] Bret, T., I. Utke, et al. Electron range effects in focused electron beam induced deposition of 3D nanostructures. *Microelectronic Engineering* 83.4–9 (2006): 1482–1486.

[71] Silvis-Cividjian, N., and C. W. Hagen. *Electron-beam-induced nanometer-scale deposition.* Advances in Imaging and Electron Physics 143 (2006): 1–235.

[72] Reimer, L., and D. Stelter. Fortran-77 Monte-Carlo program for minicomputers using Mott cross-sections. *Scanning* 8.6 (1986): 265–277.

[73] Drouin, D., A. R. Couture, et al. CASINO V2.42: A fast and easy-to-use modeling tool for scanning electron microscopy and microanalysis users. *Scanning* 29.3 (2007): 92–101.

[74] Napchan, E., and D. C. Joy. www.mc-set.com.

[75] Joy, D. C. *Monte Carlo modeling for electron microscopy and microanalysis.* New York: Oxford University Press, 1995.

[76] Joy, D. C. The theory and practice of high-resolution scanning electron microscopy. *Ultramicroscopy* 37.1–4 (1991): 216–233.

[77] Isaacson, M., J. P. Langmore, et al. Determination of the non-localization of the inelastic scattering of electrons by electron microscopy. *Optik (Jena)* 41.1 (1974): 92–96.

[78] Fourie, J. T. Electric effects in contamination and electron-beam etching. *Scanning Electron Microscopy* (1981): 127–134.

[79] Banhart, F. Laplacian growth of amorphous-carbon filaments in a non-diffusion-limited experiment. *Physical Review* E 52.5 (1995): 5156–5160.

[80] Wang, H. Z., Q. Wu, et al. Fractal carbon trees on polymer and metal substrates in TEM observation. *Journal of Materials Science Letters* 19.24 (2000): 2225–2226.

[81] Song, M., K. Mitsuishi, et al. Fabrication of self-standing nanowires, nanodendrites, and nanofractal-like trees on insulator substrates with an electron-beam-induced deposition. *Applied Physics A - Materials Science & Processing* 80.7 (2005): 1431–1436.

[82] Furuya, K. Nanofabrication by advanced electron microscopy using intense and focused beam. *Science and Technology of Advanced Materials* 9.1 (2008).

[83] Aristov, V. V., N. A. Kislov, et al. Direct electron-beam-induced formation of nanometer-scale carbon structures in STEM. I. Nature of 'long-range' growth outside the substrate. *Microscopy, Microanalysis, Microstructures* 3.4 (1992): 313–322.

[84] van Dorp, W. F., B. van Someren, et al. Approaching the resolution limit of nanometer-scale electron beam-induced deposition. *Nano Letters* 5.7 (2005): 1303–1307.

[85] van Dorp, W. F., C. W. Hagen, et al. In situ monitoring and control of material growth for high resolution electron beam induced deposition. *Journal of Vacuum Science & Technology* B 25.6 (2007): 2210–2214.

[86] Danelon, C., C. Santschi, et al. Fabrication and functionalization of nanochannels by electron-beam-induced silicon oxide deposition. *Langmuir* 22.25 (2006): 10711–10715.

[87] Crozier, P. A., J. Tolle, et al. Synthesis of uniform GaN quantum dot arrays via electron nanolithography of D_2GaN_3. *Applied Physics Letters* 84.18 (2004): 3441–3443.

[88] Mitsuishi, K., M. Shimojo, et al. Electron-beam-induced deposition using a subnanometer-sized probe of high-energy electrons. *Applied Physics Letters* 83.10 (2003): 2064–2066.

[89] Guise, O., J. Ahner, et al. Formation and thermal stability of sub-10-nm carbon templates on Si(100). *Applied Physics Letters* 85.12 (2004): 2352–2354.

[90] van Kouwen, L., A. Botman, et al. Focused electron-beam-induced deposition of 3 nm dots in a scanning electron microscope. *Nano Letters* 9.5 (2009): 2149–2152.

[91] Tanaka, M., M. Shimojo, et al. The size dependence of the nano-dots formed by electron-beam-induced deposition on the partial pressure of the precursor. *Applied Physics A -Materials Science & Processing* 78.4 (2004): 543–546.

[92] Jiang, H., C. N. Borca, et al. Fabrication of 2- and 3-dimensional nanostructures. *Int. J. Mod. Phys.* B 15.24–25 (2001): 3207–3213.

[93] Silvis-Cividjian, N., C. W. Hagen, et al. Direct fabrication of nanowires in an electron microscope. *Applied Physics Letters* 82.20 (2003): 3514–3516.

[94] Martin, J. P., and R. Speidel. Self-sustaining microgratings manufactured in the "Stereoscan MK II" scanning electron microscope. Herstellung freitragender Mikrogitter im REM. *Optik* 36.1 (1972): 13–18.

[95] Kislov, N. A. Direct STEM fabrication and characterization of self-supporting carbon structures for nanoelectronics. *Scanning* 15.4 (1993): 212–218.

[96] Fujita, J., M. Ishida, et al. Carbon nanopillar laterally grown with electron beam-induced chemical vapor deposition. *Journal of Vacuum Science & Technology* B 21.6 (2003): 2990–2993.

[97] Shimojo, M., K. Mitsuishi, et al. Application of transmission electron microscopes to nanometre-sized fabrication by means of electron beam-induced deposition. *Journal of Microscopy-Oxford* 214 (2004): 76–79.

[98] Hübner, B., H. W. P. Koops, et al. Tips for scanning tunneling microscopy produced by electron-beam induced deposition. *Ultramicroscopy* 42–44 (1992): 1519–1525.

[99] Schiffmann, K. I. Investigation of fabrication parameters for the electron-beam-induced deposition of contamination tips used in atomic force microscopy. *Nanotechnology* 4.3 (1993): 163–169.

[100] Weber, M., M. Rudolph, et al. Electron-beaming induced deposition for fabrication of vacuum field emitter devices. *Journal of Vacuum Science & Technology* B 13.2 (1995): 461–464.

[101] Wendel, M., H. Lorenz, et al. Sharpened electron beam deposited tips for high resolution atomic force microscope lithography and imaging. *Applied Physics Letters* 67.25 (1995): 3732–3734.

[102] Christy, R. W. Formation of thin polymer films by electron bombardment. *Journal of Applied Physics* 31.9 (1960): 1680–1683.

[103] Egerton, R. F., P. Li, et al. Radiation damage in the TEM and SEM. *MICRON* 35.6 (2004): 399–409.

[104] Chapiro, A. *Radiation chemistry of polymeric systems*. New York: John Wiley & Sons, 1962.

[105] Heger, A., H. Dorschner, et al. *Technologie der Strahlenchemie von Polymeren*. Berlin: Akademie-Verlag, 1990.

[106] Reimer, L. Methods of detection of radiation-damage in electron-microscopy. *Ultramicroscopy* 14.3 (1984): 291–303.

[107] Reimer, L. *Transmission electron microscopy: Physics of image formation and microanalysis.* Berlin: Springer, 1997.

[108] Utke, I., J. Michler, et al. Cross section investigations of compositions and sub-structures of tips obtained by focused electron beam induced deposition. *Advanced Engineering Materials* 7.5 (2005): 323–331.

[109] Weber, M. PhD thesis, Technische Universität Darmstadt, Germany (1994).

[110] Hedhili, M. N., J. H. Bredehoft, et al. Electron-induced reactions of $MeCpPtMe_3$ investigated by HREELS. *Journal of Physical Chemistry* C 113.30 (2009): 13282–13286.

[111] van Dorp, W. F., J. D. Wnuk, et al. Electron induced dissociation of trimethyl(methyl-cyclopentadienyl)platinum(IV): Total cross section as a function of incident electron energy. *Journal of Applied Physics* 106.7 (2009).

[112] Wnuk, J. D., J. M. Gorham, et al. Electron induced surface reactions of the organometallic precursor trimethyl(methylcyclopentadienyl)platinum(IV). *Journal of Physical Chemistry* C 113.6 (2009): 2487–2496.

[113] Cosby, P. C. Electron-impact dissociation of carbon-monoxide. *Journal of Chemical Physics* 98.10 (1993): 7804–7818.

[114] Kunze, D., O. Peters, et al. Polymerisation adsorbierter Kohlenwasserstoffe bei Elektronenbeschuss. *Zeitschrift für angewandte Physik* 22.2 (1967): 69–75.

[115] Botman, A., D. A. M. de Winter, et al. Electron-beam-induced deposition of platinum at low landing energies. *Journal of Vacuum Science & Technology* B 26.6 (2008): 2460–2463.

[116] Friedli, V., I. Utke, et al. Dose and energy dependence of mechanical properties of focused electron-beam-induced pillar deposits from $Cu(C_5HF_6O_2)_2$. *Nanotechnology* 20.38 (2009).

[117] Utke, I., F. Cicoira, et al. Focused electron beam induced deposition of high resolution magnetic scanning probe tips. *Materials Research Society Symposium – Proceedings* 706 (2002): 307–312.

[118] Hoyle, P. C., M. Ogasawara, et al. Electrical-resistance of electron-beam-induced deposits from tungsten hexacarbonyl. *Applied Physics Letters* 62.23 (1993): 3043–3045.

[119] Botman, A., J. J. L. Mulders, et al. Purification of platinum and gold structures after electron-beam-induced deposition. *Nanotechnology* 17.15 (2006): 3779–3785.

[120] Miyazoe, H., I. Utke, et al. Improving the metallic content of focused electron beam-induced deposits by a scanning electron microscopeintegrated hydrogen-argon microplasma generator. *Journal of Vacuum Science & Technology* B 28 (4) (2010): 744–750.

[121] Lukasczyk, T., M. Schirmer, et al. Electron-beam-induced deposition in ultrahigh vacuum: Lithographic fabrication of clean iron nanostructures. *Small* 4.6 (2008): 841–846.

[122] Hoffmann, P., I. Utke, et al. Comparison of fabrication methods of sub-100 nm nano-optical structures and devices. *Proceedings of SPIE – The International Society for Optical Engineering* 5925 (2005): 1–15.

[123] Perentes, A., P. Hoffmann, et al. Focused electron beam induced deposition of DUV transparent SiO2. *Proceedings of SPIE – The International Society for Optical Engineering* 6533 (2007): Q5331–Q5331.

[124] Janchen, G., P. Hoffmann, et al. Mechanical properties of high-aspect-ratio atomic-force microscope tips. *Applied Physics Letters* 80.24 (2002): 4623–4625.

[125] Edinger, K., H. Becht, et al. Electron-beam-based photomask repair. *Journal of Vacuum Science & Technology* B 22.6 (2004): 2902–2906.

[126] Akama, Y., E. Nishimura, et al. New scanning tunneling microscopy tip for measuring surface-topography. *Journal of Vacuum Science & Technology* A 8.1 (1990): 429–433.

[127] Griesinger, U. A., C. Kaden, et al. Investigations of artificial nanostructures and lithography techniques with a scanning probe microscope. *Journal of Vacuum Science & Technology* B 11.6 (1993): 2441–2445.

[128] Lee, K. L., D. W. Abraham, et al. Submicron Si trench profiling with an electron-beam fabricated atomic force microscope tip. *Journal of Vacuum Science & Technology* B 9.6 (1991): 3562–3568.

[129] Rueb, M., H. W. P. Koops, et al. Electron beam induced deposition in a reducing image projector. *Microelectronic Engineering* 9 (1989): 251–254.

[130] Platzgummer, E. Maskless Lithography and Nanopatterning with Electron and Ion Multi-Beam Projection, *Proceedings of SPIE The International Society for Optical Engineering* 7637 (2010) 763703–1.

[131] Koops, H. W. P. Method and devices for producing corpuscular radiation systems. (Verfahren und Vorrichtung zur Herstellung von Korpuskularstrahlsystemen.) Patent. EP000001590825B1 (2004).

[132] Wanzenboeck, H. D., P. Roediger, et al. Novel method for cleaning a vacuum chamber from hydrocarbon contamination, *Journal of Vacuum Science & Technology* A 28.6 (2010): 1413–1420.

[133] Utke, I. and A. Gölzhäuser. Small, minimally-invasive, direct: electrons induce local nano-sized reactions of adsorbed functional molecules. *Angewandte Chemie International Edition* 49 (2010): 9328–9330.

[134] Walz, M.-M., M. Schirmer, et al. Electrons as "Invisible Ink": Fabrication of nanostructures by local electron beam induced activation of SiOx. *Angewandte Chemie International Edition* 49 (2010): 4669–4673.

[135] Bernau, L., M. Gabureac, et al. Tunable Nanosynthesis of Composite Materials by Electron-Impact Reaction. *Angewandte Chemie International Edition* 49 (2010): 8880–8884.

[136] Rosenberg, J. D., S. G., Gorham, et al. Electron beam deposition for nanofabrication: Insights from surface science. *Surface Science* 605.3–4 (2011): 257–266.

[137] Utke, I., M. Jenke, et al. Polarisation stabilisation of vertical cavity surface emitting lasers by minimally invasive focused electron beam triggered chemistry. *Nanoscale* 3 (2011): 2718–2722.

[138] Waiblinger, M., K. Kornikov, et al. E-beam induced EUV photomask repair – a perfect match. *Proceedings of SPIE – The International Society for Optical Engineering* 7545 (2010): 7545P–1.

[139] Wieland, M.J., G. de Boer, et al. MAPPER: High throughput maskless lithography. *Proceedings of SPIE – The International Society for Optical Engineering* 7637 (2010): 76370F–1.

HISTORICAL EVOLUTION OF FIB INSTRUMENTATION AND TECHNOLOGY

From Circuit Editing to Nanoprototyping

Phillip Russell

1 INTRODUCTION

Focused ion beam (FIB) technology is now a very popular technique in semiconductor manufacturing and many research laboratories, where it is increasingly employed for analysis of a wide variety of materials and fast prototyping of micro and nanodevices. Most of current applications in nanotechnologies are the result of developments in recent years. However, the history of FIB instrumentation and technology began a long time ago. This is an example of a complex innovation, where both incremental (expected) improvements and breakthroughs in many areas have finally contributed to creation of a new class of instruments and technologies.

It should be emphasized that any attempt to fully describe the history of a topic as dynamic as that of focused ion beams is limited in many ways. This description is from the perspective of a FIB user and developer active in the field since 1984.

2 DEVELOPMENT OF HIGH-BRIGHTNESS ION SOURCES AND FIB

In the majority of modern FIBs, gallium liquid metal ion sources are used, but various other types of ion sources were tested in the first attempts to process materials locally.

The first application of ion beam sources in integrated circuit fabrication dates to 1974, when Seliger and Fleming at Hughes Research Laboratories used modified ion implanter with gaseous ion source [1] for maskless implantation doping and resist exposure. The beam diameter of 3.5 μm at 60 keV energy was obtained. He beam was used for PMMA exposition, and direct maskless implanation of B in Si was also carried out. The source had low current densities (1–200 μA/cm^2), but the potential of the technique was clearly demonstrated.

Simultaneously, Orloff and Swanson at Oregon Graduate Center worked on development of a gas field-ionization source for microprobe applications [2; 3]. They used liquid nitrogen cooled fine metal field-emission tips to ionize hydrogen or argon. High brightness of the source, better than 10^8 A cm^{-2} sr^{-1}, ion current of 0.2 nA and resolution of ~0.1 μm were achieved. However, this kind of cryogenic ion source was cumbersome and did not find industrial applications.

Another type of ion source based on liquid metal (LMIS, or liquid metal ion source), with similarly high ion brightness and easy to operate, was much more successful. The history of focused ion beam technology, which led to the development of today's modern Ga LMIS based FIB systems, dates back to the early 1600s. Gilbert in 1628 [4] was perhaps the first to observe and record a fluid deflection in an electric field (the electric field was produced via a piece of charged amber). He observed that when a suitably electrically charged piece of amber was brought near a droplet of water, it would form a cone shape and small droplets would be ejected from the tip of the cone: this is the first recorded observation of electrospraying, and of the electric field induced emission from a conical liquid. In 1914, John Zeleny (University of Minnesota) published a description of electrical discharge from liquid points in his paper "The Electrical Discharge from Liquid Points, and a Hydrostatic Method of Measuring the Electric Intensity at Their Surfaces" [5].

In 1961 [6], Krohn was the first to study ion formation in charged liquid drops, seeking applications in rocket propulsion. In this and later studies, he found that liquid Ga and liquid Sn tended to produce ions rather than droplets. This result limited the applicability of liquid metal ion sources (LMIS) for propulsion, but ultimately led to the successful development of the modern LMIS FIB system. He also reported a scanning focused ion beam system using a capillary source of Ga. This was the first demonstration of what evolved into modern focused ion beam, or FIB, systems; however, the use of a capillary source severely limited the brightness of the source.

Later, in 1974, Krohn [7] reported experiments with capillary source of ions and droplets in Ga, Sn, and Pb-Bi, and in 1975 Krohn and Ringo reported development of improved ion source using liquid gallium with high brightness of 0.9×10^5 A cm^{-2} sr^{-1} at 21 kV at 10 μA total current and effective source diameter of 0.2 μm, with a focal spot on target of 15 μm [8]. Cs and Hg were also tested, but showed inferior results.

During the same period, Clampitt and collaborators were also working on development of high brightness LMIS sources exploring different metals like Li, Cs, Sn, Ga, and Hg [9].

In 1979, Seliger et al. reported LMIS Ga$^+$ 57 kV source with resolution of 100 nm at probe, current density and brightness were 1.5 A/cm^2 and 3.3×10^6 A cm^{-2}sr^{-1}. The source was used for milling of 0.1 μm wide lines in a 40 nm thick Au film deposited over Si substrate [10]. This was the first scanning ion microscope successfully employed for direct patterning.

Sir Geoffrey Ingram Taylor was the first to explain the conical shape of a liquid in an electric field in 1964 [11] in his paper "Disintegration of Water Drops in an Electric Field." His final research paper was published in 1969 [12], when he was 83. In it he resumed his interest in electrical activity in thunderstorms, as jets of conducting liquid motivated by electrical fields. The cone from which such jets are observed is called the Taylor cone (or Taylor-Gilbert cone).

This is precisely the Taylor cone of liquid Ga used in the vast majority of modern FIB systems. The same Taylor cone concept is used in electrospinning of polymer materials.

In a real liquid cone such as that utilized in modern Ga source FIB systems, ion emission begins to occur when the electric field on the liquid reaches a value in the range of 10^{10} volts per meter. Fields of this magnitude require a radius of a few nm for reasonable voltages of around 10 kV when the extraction electrode is a few mm from the liquid apex.

Various researches demonstrated that low melting temperature metals such as Ga, In, or Sn can be easily liquified in a LMIS. However, for many applications in semiconductor technology, other elements are more interesting, for example Si, Be, B, or As. These elements can be incorporated into alloys that can also have reduced melting points and thus can be used in the same kind of ion sources. For their separation, mass separators using crossed electric and magnetic fields (ExB) were developed by Seliger and his collaborators in 1981 [13].

A number of other alloy metal sources have also been successfully developed and utilized for implantation of localized dopants in compound semiconductor devices. Systems developed by JEOL Ltd. utillized a mass filter to allow for selective dopant implantation with a 100 kV FIB system (Model JIBL 100). L. Harriott (AT&T Bell Labs) was a major user of this system for selective doping in compound semiconductor device fabrication [14].

A progress in theory of ion beam optics allowed the design of an ion beam column with ultra-high resolution of 10 nm using LMIS sources for use in lithography and implantation in 1985 by Orloff and Sudraud [15, 16].

Earlier, the pointed metal field emitter used in modern field ion microscopes was created by Erwin Müller [17–19]; for more details, see Chapter 1 by Forbes in this volume.

The list of scientists who made important contributions to the development of LMIS and FIB in this initial phase of research should also include the names of Mahoney [20], Mair [21], Anazawa [22], Levi-Setti [23], Gamo, Matsui, Namba [24], and many others.

3 APPLICATIONS OF FIBS

In 1987 J. Melngailis published a detailed review of the applications of focused ion beams [25]. Already at that point of time a number of applications was growing very fast, and in the review the following areas were considered as potentially most promising:

- implantation (mostly for doping in semiconductors, using alloy sources to produce B, As, and Si);
- milling (for mask repair and circuit microsurgery, i.e., for correction of integrated circuits by milling);
- surface chemistry (surface reactions induced by ions impact, both etching using gases like Cl_2 and deposition of metals like Al, W or Au, had already been reported);
- lithography or bulk chemistry (organic and inorganic resists modification);
- material microanalysis (secondary ion mass spectroscopy or SIMS); and
- scanning ion microscopy.

An interesting point is that his estimate of FIB systems in use worldwide at that period of time was about 35, with about 25 of them in Japan, compared to many hundreds in use currently.

Note that all the applications were limited to sub-micron resolution. The description of the great progress achieved in all these areas, making possible now processing at nanoscale, is actually the subject of the present book.

4 MARKET–ORIENTED DEVELOPMENT OF FIB MACHINERY

Even in the first experiments [25], the high potential of FIB for applications in semiconductor industry became very clear, and this spurred the development of the machinery for research and industrial applications. As a result, many important contributions to the development of LMIS based FIB devices were done by a number of private companies, usually closely linked to university labs. Later, in the 1990s, some of these companies experienced impressive growth bringing to market a new class of equipment for industry and research.

The list of companies involved in research and development in this area includes: FEI, JEOL, Orsay Physics, Hitachi, Seico, Carl Zeiss, Raith, Tescan, Micrion, Advantest, Schlumberger, AMAT, Micro Beam, and others. Modern vendors of FIB machines for material processing are listed in Introduction 1 by Koops and Utke in this book.

We will mention here just some of the developments that occurred in this sector, documented by patents.

In 1972, Cohne and Tarasevich of Kollsman Instrument Company were awarded a patent for their "Apparatus Using a Beam of Positive Ions for Controlled Erosion of Surfaces" [26]. Their filing date was June 16, 1969. Although their work was not based on the use of a LMIS, it was worded so as to refer to any positively charged atomic species. The system described includes an electron beam directed at the processed surface for positive charge neutralization and an interferometer to monitor the progress of erosion.

Rusch and Sievers of the Minnesota Mining and Manufacturing Company (3M) filed for a patent for a "Charged Particle Beam Apparatus" in 1975, and patent was issued in 1976 [27]. Included in their patent was the use of focused ions in a scanning focused ion beam configuration. The company 3M was in the business of producing surface analysis equipment (in addition to the film and paper products for which they are best known). The 3M system utillized electrostatic lenses and included an ExB mass filter. The ion source was an early version of a liquid metal ion source (LMIS).

The FIB capability to locally remove material by milling, in order to fabricate various 3D structures with micron and submicron resolution, was recognized and explored in a number of patents for applications in electronics, optoeletronics, and photonics. Puretz et al. from Oregon Graduate Center proposed the use of ion milling to form optical surfaces, e.g., to fabricate semiconductor laser based devices in a body of material [28]. Horton and Tasker from Galileo Electro-Optics Corp. proposed the use of ion beams in fabrication of electron multipliers or microchannel plates [29]. Ito et al. from Hitachi Ltd. [30] proposed a method of processing of rotating workpieces to produce micro-sized objects of complex shape.

One more patent [31] was issued to Ward et al. in 1986, Hughes Aircraft Company, for the invention of a focused ion beam microfabrication column which included an ExB filter and provided focused ion beams of up to 150 keV. The high beam energy, together with the ExB mass filter, was designed to allow the system to be used as a localized doping system, especially for compound semiconductor device fabrication.

During second half of 1980s, intense studies of FIB induced etching and deposition with sub-micron resolution started [32, 33]. The combined use of milling and deposition (the "mill and fill" method) for electronic chips microsurgery (correction of defects) or mask repair was proposed by Tao and Melngailis from MIT, Cambridge [34].

The first generation FEI ion optical columns were glass rod based, similar to the glass rod based electron optical systems used in oscilloscopes by companies such as Techtronics, which also had development and manufacturing facilities in the same area near Portland, Oregon. FEI was a small company known initially for supplying LaB_6 and Field emission tips for electron microscopy suppliers and users. The company (at that time only a few employees) was located in a small space within the Oregon Graduate Center in Hillsborough, Oregon, where both Jon Orlof and Lyn Swanson were employed as faculty members. Their initial systems were purchased and utillized mainly by the semiconductor industry (Intel was located nearby), as well as a few academics. In the early 1980s, FEI and JEOL teamed together with Intel to add an FEI FIB to a JEOL SEM equiped with a voltage contrast spectrometer for measurement of voltage waveforms on internal metal lines. The FIB was used to remove the passivation layer from the device locally at the area of interest [35]. Phil Russell was the JEOL technical person at that time.

Orsay Physics (France) was founded in 1989 by Pierre Sudraud and other researchers and engineers from Paris-Orsay University. Their first commercial FIB system was adapted to a JEOL SEM in 1990, and was referred to as a CrossBeam, as the FIB and electron beams shared the same vacuum chamber, specimen chamber, and stage with a common focal point on a sample at the coincident point of the electron and the ion beam. They have also developed UHV clean room compatible systems. Orsay has built and provided FIB columns and electronics suitable for adaption to a wide range of systems, such as existing SEMs, in-situ fabrication systems such as MBE, CVD, etc. Orsay Physics holds several patents regarding FIB and electron beam systems. Their FIB columns are employed, for example, on the Zeiss Cross Beam platform. Orsay Physics continues to develop hardware for FIB enhancements and are very commonly involved in user development systems.

Micrion, established in Peabody, Massachusetts, developed 50 kV FIB systems primarily for lithography mask repair applications within the semiconductor industry. Nicholas Economou held various positions at Micrion, including president, chief executive officer, and board director. Micrion designed, manufactured, and marketed focused ion beam (FIB) workstations, including FIB instruments used for semiconductor mask repair. The FIB maker had approximately 200 employees. FEI and Micrion merged their operations in 1999, creating the dominant supplier of FIB systems worldwide. The merged companies retained the corporate name FEI. FEI later merged with Philips Electron Optics, retaining the name FEI. The majority shareholder of FEI was Philips Business Electronics, a wholly owned subsidiary of Royal Phillips Electronics.

As of 2010, FEI, headquartered in Hillsboro, Oregon, has manufacturing facilities in the United States (Hillsboro, Oregon, and Peabody, Massachusetts [former facility of Micrion]), Eindhoven Netherlands (Philips Electronic Optics Facilities), and Brno (Czech Republic), along with sales and service representation worldwide.

Among recent new developments, we should also mention that JEOL recently introduced high voltage (100 keV) FIB systems with mass selectable ternary alloy sources (e.g., Au/Si/Be) with integrated mass filter for ion selection. This provided the ability for FIB to be integrated into a III–V semiconductor research or development line as a localized ion implantation system.

In conclusion, the history of ion beam sources design and technologies is about a half century long (since the first experiments in the 1960s with ion cluster propulsion), while experiments with focused ion beams started in the 1970s. The high potential of FIB applications for processing of integrated circuits and mask repair with sub-micron resolution was important for the rapid development of the machinery. Currently, together with the improvement of resolution achieved with modern LMIS sources, beam column design, and the joining of the ion and electron beams in dual beam machines, the area of FIBs applications has expanded dramatically, becoming one of the major areas of nanotechnology.

REFERENCES

[1] Seliger, R. L., and W. P. Fleming. Focused ion beams in microfabrication. *J. Appl. Phys.* 45 (1974): 1416–1422.

[2] Orloff, J., and L. W. Swanson. Study of a field-ionization source for microprobe applications. *J. Vacuum Sci. Technol.* 12 (1975): 1209.

[3] Orloff, J., and L. W. Swanson. Fine-focus ion beams with field ionization. *J. Vacuum Sci. Technol.* 15 (1978): 845–848.

[4] Gilbert, W. *De Magnete, Magneticisque Corporibus, de Magno Magnete Tellure* (On the Magnet and Magnetic Bodies, and on That Great Magnet the Earth). London: Peter Short, 1628.

[5] Zeleny, John. The electrical discharge from liquid points, and a hydrostatic method of measuring the electric intensity at their surfaces. *Phys. Rev.* 3 (1914): 69–91.

[6] Krohn, V. E. Liquid metal droplets for heavy particle propulsion. *Progress in Astronautics and Rocketry* 5. New York: Academic, 1961, 73–80.

[7] Krohn, V. E. Electrohydrodynamic capilliary source of ions and charged droplets. *J. Appl. Phys.* 45 (1974): 1144–1146.

[8] Krohn, V. E., and G. R. Ringo. Ion source of high brightness using liquid metal. *Appl. Phys. Lett.* 27 (1975): 479–481.

[9] Clampitt, R., K. L. Aitken, and D. K. Jefferies. Abstract: Intense field-emission ion souce of liquid metals. *J. Vacuum Sci. Technol.* 12 (1975): 1208.

[10] Seliger, R. L., J. W. Ward, V. Wang, and R. L. Kubena. A high-intensity scanning ion probe with submicrometer spot size. *Appl. Phys. Lett.* 34 (1979): 310.

[11] Taylor, G. I. Disintegration of water drops in an electric field. *Proc. R. Soc. Lond.* A 280.1382 (July 1964): 383–397.

[12] Taylor, G. I. Electrically driven jets. *Proc. Roy. Soc. Lond.* A 313 (1969): 453.

[13] Wang, V., J. W. Ward, and R. L. Seliger. A mass-separating focused-ion-beam system for maskless ion implantation. *J. Vacuum Sci. Technol.* 19 (1981): 1158–1163.

[14] Harriott, L. R., H. Temkin, R. A. Hamm, J. Weiner, and M. B. Panish. A focused ion beam vacuum lithography process compatible with gas source molecular beam epitaxy. *J. Vac. Sci. Technol.* B 7 (1989): 1467.

[15] Orloff, J. and P. Sudraud. Design of a 100 kV, high resolution focused ion beam column with a liquid metal ion source. *Microelectron. Eng.* 3 (1985): 161–165.

[16] Orloff, J. Comparison of optical design approaches for use with liquid metal ion sources. *J. Vacuum Sci. Technol.* B 5 (1987): 175–177.

[17] Müller, E. W. Resolution of the atomic structure of a metal surface by the field ion microscope. *J. Appl. Phys.* 27 (1956): 474.

[18] Müller, E. W., and K. Bahadur. Filed ionization of gases at a metal surface and the resolution of the field ion microscope. *Phys. Rev.* 102 (1956): 624.

[19] Müller, E. W., J. A. Panitz, and B. McLane. The atom-probe field ion microscope. *The Review of Scientific Instruments,* 39 (1968): 83.

[20] Mahoney, J. F., A. T. Yahiku, H. L. Daley, R. D. Moore, and J. Perel. Electrohydrodynamic ion source. *J. Appl. Phys.* 40 (1969): 5101.

[21] Mair, G. L. R. Emission from liquid metal ion sources. *Nucl. Istrum. Meth.* 172 (1980): 567.

[22] Anazawa, N., R. Aihara, M. Okunuki, and R. Shimizu. Development of a gallium ion source scanning ion microscope and its applications. *Scanning Electron Microscopy* IV, AMF O'Hare, Chicago (1982): 1443–1451.

[23] Levi-Setti, R. Proton scanning microscopy: feasibility and promise. *Scanning Electron Microscopy* (1974): 125.

[24] Gamo, K., T. Matsui, and S. Namba. Characteristics of Be-Si-Au ternary alloy liquid metal ion sources Japan. *J. Appl. Phys.* 22 (1983): 153.

[25] Melngailis, J. Focused ion beam technology and applications. *J. Vacuum Sci. Technol.* B 5 (1987): 469–495.

[26] Cohne, M. F., and M. Tarasevich. Apparatus using a beam of positive ions for controlled erosion of surfaces. US patent 3,699,334 (1972).

[27] Rusch, T. W., and J. Sievers. Charged particle beam apparatus. US Patent 3937958 (1976).

[28] Puretz, J., J. H. Orloff, R. K. de Freez, and R. Elliott. Focused ion beam micromachining of optical surfaces in material. US patent 4,698,129 (1987).

[29] Horton, J. R., and W. G. Tasker. Microchannel electron multipliers and method of manufacture. EP patent 0 413 481, A2 (1991).

[30] Ito, F., A. Shimase, S. Haraichi, and J. Azuma. Ion beam processing method and application. US patent 5,223,109 (1993).

[31] Ward, J. W., et al., US Patent 4,563,587 (January 7, 1986).

[32] Ochiai, Y., K. Shihoyama, T. Shiokawa, K. Toyoda, A. Masuyama, K. Garno, and S. Namba. Characteristics of maskless ion beam assisted etching of silicon using focused ion beams. *J. Vac. Sci. Technol.* B 4 (1986): 333.

[33] Melngailis, J. Focused ion beam induced deposition: A review. *Proc. SPIE* 1465 (1991): 36.

[34] Tao, T., and J. Melngailis. Ion beam induced deposition of materials, US patent 5,104,684 (1992).

[35] Puretz, J., J. Orloff, and L. Swanson. Application of focused ion beams to electron beam testing of integrated circuits. *Proceedings of SPIE – The International Society for Optical Engineering* 471 (1984): 38–46.

PART I

FUNDAMENTALS AND MODELS

CHAPTER 1

THE THEORY OF BRIGHT FIELD ELECTRON AND FIELD ION EMISSION SOURCES

Richard G. Forbes

1 INTRODUCTION

This chapter discusses the bright, field-emission-based, charged-particle sources used in modern nanofabrication and nanoscale materials examination. It covers cold field and Schottky electron emitters, the liquid-metal ion source (LMIS), and the gas field ionization source (GFIS). It also provides theoretical background about field electron and ion emission and charged-particle optics. The aim is an overview of these devices and underlying scientific concepts.

1.1 TERMINOLOGY

With all these sources, a high-magnitude electric field (classically negative for electron emission, positive for ion emission) exists at the tip of a pointed, solid or liquid, physical object (the "emitter"). The terms *field emitter* and *field emission* are applied to either polarity or both, depending to context. Remarks applying only to electron emission use the term *field electron emission* (abbreviated as FE), and the FE-based projection microscope is called a *field electron microscope* (FEM). The literature uses various names: thus "FE" can be read as "field electron," "field emission," "field electron emission," or "electron field emission," according to preference or context. "FI" can treated similarly, with "I" denoting "ion," "ionization," or "ion emission."

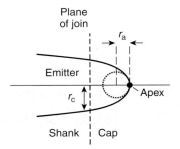

FIGURE 1.1: To illustrate the distinction between *apex radius* r_a and *cross-sectional radius* r_c, when the emitter tip is not nearly hemispherical.

The emitter end-region, where emission occurs, is called its *tip*. The term *apex* refers to a point on axis, at the tip. The apex radius of curvature is r_a; the cross-sectional radius of the tip, when different, is r_c (see Figure 1.1). For a fitted sphere, the term *tip radius* is used, but is denoted by r_a.

For electron emitters, the electric field and the current and related quantities are discussed as positive, even though negative in classical electromagnetism (this is a common convention). For both electron and ion emission, F is the electric field *magnitude*.

1.2 TECHNICAL CONVENTIONS

All equations and quantities use the International System of Quantities (ISQ), established in the 1970s and recently given this name [1]. All equations are dimensionally consistent, with no hidden conversion factors. However, we employ the mixed system of SI units and dimensionally equivalent atomic-level units (in particular, the eV) used in field emission for easy formula evaluation. In Appendix A, Table 1.8 shows relevant fundamental constants, in SI units and in these customary units. Table 1.9 lists basic universal constants needed in emission theory. Universal constants defined in the text are given to seven significant figures. In a logarithm, e.g., $\ln\{x\}$, the notation $\{x\}$ means "express x in specified units (usually SI units) and take its numerical value."

1.3 CHAPTER STRUCTURE

Section 2 provides general background. The science underlying field emitters can be unfamiliar, so Section 2 provides introductions to relevant topics. Sections 3 and 4 consider requirements for charged-particle (CP) machines with field emission sources, and related CP-optical theory. Sections 5 and 6 cover electron sources. Section 7 examines the LMIS, Section 8 the GFIS, and Section 9 other ion sources. Section 10 provides commentary. Appendix A presents numerical data, and Appendix B a list of acronyms.

2 GENERAL BACKGROUND

2.1 ORIGINS OF PRACTICAL FIELD EMISSION

In 1744 Winkler [2] used pointed wires to cause electrical discharges now recognizable as resulting from FE into poor vacuum. Lilienfeld [3] did most to establish early experimental

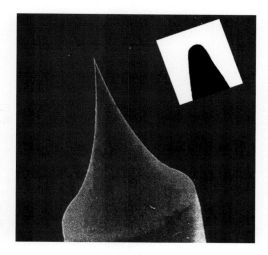

FIGURE 1.2: A tungsten Müller emitter of moderate apex radius (inset: electron micrograph of emitter tip).

phenomenology. Millikan and Lauritsen [4] discovered the empirical law $\ln\{i\} \sim 1/V$ relating emission current i to applied voltage V, and in 1928 Fowler and Nordheim [5; 6] explained it. Kleint [7; 8] describes other early work. However, the main originator of the modern, smooth, clean, pointed metal field emitter (see Figure 1.2) was Erwin W. Müller; we call this device a *Müller emitter.*

Müller invented *field electron microscopy* (FEM) in the 1930s [9], *field ion microscopy* (FIM) in the 1950s [10], and *atom-probe field ion microscopy* (APFIM) in the 1960s [11]. With Kanwar Bahadur, Müller observed individual atoms [12] in 1955, four years before Feynmann's famous lecture [13] "There's Plenty of Room at the Bottom." Many machines of modern nanotechnology use Müller emitters as probes or CP sources, and Müller deserves recognition as the "grandfather of nanoscience." Melmed has written a biographical memoir [14].

2.2 ELECTRON MOTIVE ENERGY

A model is needed for the "motive" energy that determines electron motion. Simple theories disregard atomic structure, and use a Sommerfeld-type model [15]. The emitter surface can be modeled as a sharp step; alternatively, outside the surface, the electron potential energy (PE) can be represented by Schottky's classical planar-surface image PE [16].

When no external field is present, electrons approaching the emitter surface from the inside see a barrier of *zero-field height H*. In simple emission models, the external field F is assumed constant outside the surface. If the planar-surface image PE is used, then the barrier to electron emission is the *Schottky-Nordheim* (SN) *barrier* [16; 17] (see Figure 1.3), where:

$$M^{\mathrm{SN}}(x) = H - eFx - e^2/16\pi\varepsilon_0 x \tag{1}$$

Here, x is distance measured from the Sommerfeld-model step. $M(x)$ can be called the *electron motive energy*. The external field has lowered the zero-field barrier by energy $\Delta M_S = cF^{1/2}$, where c is the *Schottky constant* (see Table 1.9). This barrier lowering is called the *Schottky effect* [16; 18]. For a Fermi-level electron moving *"forwards"* (normal to the emitting surface) in a bulk metal, H is given by the *local thermodynamic work-function*, ϕ, of the emitting surface.

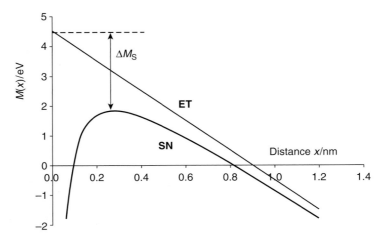

FIGURE 1.3: The electron motive energies $M(x)$ for exact triangular (ET) and Schottky-Nordheim (SN) barriers, for barriers with zero-field height 4.5 eV and barrier field 5 V/nm. The barrier-height reduction ΔM_S by the Schottky effect is approximately 2.7 eV.

2.3 FIELD ELECTRON EMISSION

Field electron emission (FE) involves escape by *wave-mechanical tunneling* through an exact or rounded triangular energy barrier. In field electron microscopy (FEM), electrons tunnel out of the emitter tip and travel in vacuum to a particle detector, where they form a projection image of the emission current density (ECD) variations across the tip. Depending on tip radius, a voltage usually between 2 to 20 kV must be applied between the emitter and the detector (or an "extractor"), to create a surface field F_s strong enough to cause FE. For a Müller emitter with $\phi \sim 4.5$ eV, F_s needs to be of order 5 V/nm.

Figure 1.4a is the FEM image of a tungsten emitter with a (110) crystallographic plane normal to the wire axis (use of Miller indices in field emission is discussed in Appendix D in Ref. 19). This is the usual orientation for tungsten wire, but wires with other orientations can be made. Bright image regions correspond to tip regions where F_s is high or ϕ low (see Equation 40). With a smoothed endform, F_s is relatively uniform across the tip and bright areas correspond to low ϕ.

References 20 to 25 provide general information about FE.

2.4 FIELD IONIZATION

Field ionization (FI) is a wave-mechanical tunneling process in which a high electric field (typically 45 V/nm for helium) causes an electron to tunnel out of an atom or molecule. The classical field ion microscope (FIM) (see Figure 1.5) uses a Müller emitter. The emitter is cooled, typically to near 80 K, and biased positive relative to a particle detector (or an extractor electrode), usually by 10 to 30 kV. An *operating gas* (e.g., helium) is present in the chamber at low pressure. Individual gas atoms[a] are attracted to the emitter by polarization forces, and (after partial thermal accommodation) are field ionized preferentially in high electric fields directly above the atomic nuclei of atoms that protrude from the emitter surface. The electron

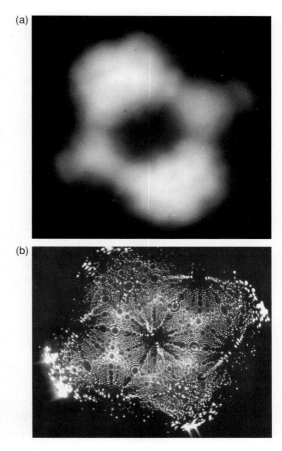

FIGURE 1.4: Micrographs of (110)-oriented tungsten emitters: (a) field electron micrograph; (b) field ion micrograph.

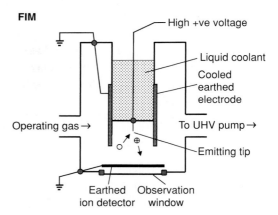

FIGURE 1.5: Schematic diagram showing the functionalities needed in a field ion microscope.

tunnels to an emitter state with energy close above the emitter Fermi level. This process is called *surface field ionization*, to distinguish it from the *free-space field ionization* (FSFI) that can occur in space well away from the emitter, in sufficiently high fields. After surface FI, ions travel across the chamber to the detector, and form an image of the emitter's surface structure, as in Figure 1.4b.

In most cases, an image dot relates to a single protruding surface atom. The dot patterns correlate to the surface crystallography of the quasi-spherical emitter. Roughly 10 percent of surface atoms are seen. Adjacent atoms are resolved in some locations, in particular the tungsten (111) facets.

Helium has an ionization energy greater than all other atoms and molecules. Thus, the field needed for helium surface FI (around 45 V/nm) will ionize approaching contaminant atoms or molecules by FSFI, well away from the surface. In a helium FIM (and with a helium GFIS), contaminants cannot directly hit the tip surface (but they can hit the emitter shank away from the tip).

References 19 and 26–28 provide general information about FI and FIM. GFIS theory relates closely to FIM theory.

2.5 BASIC EMITTER ELECTROSTATICS

In a diode arrangement, electrodes around the emitter are electrically connected, with a voltage of magnitude V between them and the emitter. The field F_a at the emitter apex is related to V by:[b]

$$F_a = \beta_a V, \tag{2}$$

where β is the *voltage-to-surface-field conversion factor* and β_a its apex value. β_a can vary widely, depending on system geometry, but is typically around 10^6 m^{-1} for an old-style FIM or FEM [29].

In FIM/FEM geometry, the phenomenological formula

$$F_a = V/k_M r_a \tag{3}$$

is often used, where r_a is apex radius, and k_M a dimensionless *shape-factor*. A sphere of radius r_a has surface field V/r_a. Thus, Equation 3 relates F_a to the surface field F_{sphere} for a sphere fitted to the emitter tip, by $F_a = F_{sphere}/k_M$. For a conventional Müller emitter well away from its surroundings, k_M is often assumed to be between 5 and 7. The value 5 is often used as a simple approximation.

2.6 PREPARATION OF MÜLLER EMITTERS

The fabrication of reliable field-assisted emission sources involves complex procedures, and operating sources may need occasional or regular "restoration." Development of relevant processes is a very specialized commercial activity.

All sources discussed here utilize Müller emitters. Various preparation techniques exist, and fabrication is always a multi-stage process. Originally, the initial step was to electropolish a thin wire (see Appendix A in Ref. 19), but some applications now use grinding or focused ion beam (FIB) milling. Electropolishing leaves a rough endform that needs to be smoothed and cleaned. The original (1930s) technique used controlled emitter heating, producing a rounded tip of radius r_a, typically 1 μm. Such emitters are too blunt for many purposes. To achieve an apex field of order 50 V/nm by applying voltages less than (say) 30 kV, the emitter must have

r_a somewhat less than 100 nm. With ion emitters, the initially prepared endform is usually smoothed by field evaporation (see Section 2.8).

Emitter shape can also be changed by surface-atom migration or by atom removal by *field-assisted etching*[c] with an active gas present. *Thermal-field (TF) sharpening* (also known as *TF build-up*) can be induced by emitter heating in a high field, and can be *chemically assisted* by an active gas. The emitter may develop facets, or an already faceted shape may change due to growth or retreat of facets [30; 31]. Unplanned field-assisted etching in poor vacuum conditions, and also unplanned TF sharpening, can degrade or destroy an operating emitter.

TF sharpening procedures may involve several stages, and are sensitive to process details (treated as proprietary in commercial contexts). Patents can be informative; References 32–35 are particularly relevant. The theory underlying TF shaping (and LMIS behavior) is *electrical thermodynamics*. A brief introduction follows.

2.7 ELECTRICAL THERMODYNAMICS

All field emission systems have capacitance $\mathfrak{C}$ between the emitter and its surroundings. When a thermodynamic system contains internal capacitance, and external work is done by an electrical battery or equivalent, then an electrical form of thermodynamics is needed.

With a voltage generator connected (and when appropriate thermodynamic parameters, including temperature, are constant), system behavior is determined by the change $\Delta\Psi^{el}$ in a thermodynamic potential Ψ^{el} called the *electrical Gibbs function*. Forbes and Ljepojevic [36; 37] introduced this variant of the Gibbs-function concept [38] in the 1990s, with:

$$\Delta\Psi^{el} = \Delta\mathfrak{F} - V\Delta q, \tag{4}$$

Here $\Delta\mathfrak{F}$ is the change in the system's total Helmholtz free energy $\mathfrak{F}$, V is the (constant) applied voltage, and Δq is the charge passed through the generator. $\Delta\mathfrak{F}$ must include appropriate electrical terms, including a term $\frac{1}{2}V^2\Delta\mathfrak{C}$ representing change in the system's internal capacitative energy. When applied to field emission, taking only the most basic terms, Equation 4 becomes

$$\Delta\Psi^{el} \approx \Delta\mathfrak{F}_{bulk} + \gamma_0\Delta A - \tfrac{1}{2}V^2\Delta\mathfrak{C}, \tag{5}$$

where $\Delta\mathfrak{F}_{bulk}$ is a free-energy change associated with the emitter *volume*, γ_0 is the free energy per unit area of the emitter *surface* (in zero field), and ΔA is the change in surface area.

Equation 5 predicts shaping effects from the rule that system change should make the Gibbs function more negative. In this context, the $\Delta\mathfrak{F}_{bulk}$ term is disregarded, presumably on the grounds that surface-atom migration involves no significant change in internal elastic-strain energy.

Predictions then depend on the applied voltage V. If V is sufficiently small, then the surface-energy term dominates, and Ψ^{el} is made more negative by reducing emitter area; thus the emitter tends to "ball up" and become blunted. On the other hand, if V is sufficiently large, then the capacitance term dominates, and Ψ^{el} is made more negative by increasing capacitance between the emitter and its surroundings; thus, the emitter changes its shape by atomic migration, to "reach out" toward its surroundings (by sharpening and/or by the growth of one or more protrusions).

It is often assumed that the local stress-based criterion $\gamma_0 c_u = \frac{1}{2}\varepsilon_0 F_{ch}^2$, where c_u is the total local surface curvature, can determine a *change-over field* F_{ch}, above which TF sharpening occurs (or blunting ceases). In the electrohydrodynamics (EHD) of charged conducting liquids, a criterion of this form was proposed by Maxwell [39] (see his p. 190), and by Raleigh [40; 41] (also see Refs. 42 and 43). This *Raleigh (stability) criterion* gives the charge level and surface field above which a conducting liquid sphere becomes unstable with respect to deformation. Raleigh proved his criterion valid for a sphere (and should have specified that it be surrounded by a concentric electrode). However, Taylor [44] showed the Raleigh criterion is not valid for an uncharged liquid sphere in a uniform field, and hence is not generally valid as a stability criterion.

Field emitters need a change-over criterion rather than a stability criterion. However, Taylor's result suggests that the Maxwell/Raleigh criterion would not be exactly valid at the Müller emitter apex (though it seems helpful as a rough guide). The detailed science of TF shaping is not fully formulated.

A complication is that γ_0 varies as between different crystallographic faces. Thus, solid emitters become faceted when heated. For the field-free case, this was predicted/explained by Herring [45]. Gas adsorption (or the diffusion of bulk impurities to the surface) may change γ_0 differently on different crystallographic faces, thereby affecting TF shaping. The theory of such effects is not well understood in detail.

2.8 FIELD EVAPORATION

Field evaporation (FEV) is the field-induced removal of a surface metal atom, as an ion. FEV occurs with both solid and liquid metals. Details are material specific. For example, tungsten FEV occurs at a surface field of around 55 to 60 V/nm, gallium FEV at around 15 V/nm. It is difficult to define or measure a FEV field exactly. With a rough solid emitter, FEV takes place selectively at sharp protrusions, where the local surface field is particularly high. Hence, end-form smoothing takes place. FEV is also the LMIS emission mechanism.

With solid metal-element emitters in vacuum at low temperature, the main FEV product is an atomic ion. At higher temperatures, and with liquid emitters, cluster ions may be produced. Most liquid metal emitters produce cluster ions, though some elements (including gallium) produce relatively few (see below).

Even for solid metal elements at low temperature, FEV can be complex, with ions created in a mixture of charge-states. For some elements, FEV has two stages: *escape* into a singly or multiply charged state, followed by one or more *post-field-ionization* (PFI) steps into the next higher charge state. Chemical Group 13 elements tend to escape singly charged, and have low PFI probability; this makes them useful as LMIS ionizants.

The field for atomic FEV can be estimated from *Müller's formula*. This gives the *escape field* (the field for fast escape) into an *n*-fold-charged state as:

$$F_n^M = b_n (\mathfrak{K}_n^0)^{1/2}, \quad \text{with} \tag{6a}$$

$$b_n \equiv 4\pi\varepsilon_0/n^3 e^3 \cong (0.6944616/n^3)\,\text{V nm}^{-1}\,\text{eV}^{-2}, \tag{6b}$$

$$\mathfrak{K}_n^0 \equiv \lambda^0 + \mathfrak{H}_n - n\phi. \tag{6c}$$

The *thermodynamic term* $\mathfrak{K}_n^0$ depends on the zero-field bonding energy λ^0 for the atom, the sum $\mathfrak{H}_n$ of its first n (free-space) ionization energies, and the local work-function ϕ of the *surface left after the atom's removal*. For elements, References 19 and 46 tabulate values of F_n^M. From *Brandon's criterion* [47], the lowest value of F_n^M, for the different n-values, predicts the *zero-Q evaporation field* F^E (the field for which the activation energy Q for FEV is zero).

Equation 6 is known in older literature as the "image-hump formula." The name *Müller's formula* now seems preferable [48]. There is no recent general review of FEV, but References 19, 27, and 48 contain further information.

2.9 FIELD EMITTED VACUUM SPACE-CHARGE

Field emitters (of both polarities) can generate very high current density in the space outside them. This constitutes a *field emitted vacuum space-charge* (FEVSC) that can affect the emission and/or beam properties. FEVSC can reduce the surface field, thus altering the i-V relationship expected in its absence. Mair's terminology [49] is useful: the *Poisson surface field* F_P is the surface field that exists with space-charge present; the *Laplace surface field* F_L is the surface field that would exist, for the same geometry and applied voltage, without space-charge. The terminology can be applied to other parameters. The *field reduction factor* $\theta \equiv F_P/F_L$ assesses the extent of field reduction.

FEVSC can cause beam expansion (but this can be corrected) [50; 51], and can alter spherical aberration coefficients (but this effect seems relatively small) [52; 53]. Effects can also result from *stochastic coulomb particle-particle interactions* (SCPPI) between emitted particles: these can increase both the beam energy-spread (the *Boersch effect* [54]) and the probe radius (see Section 3.5). The latter effect, noted by Loeffler [55; 56] and called *trajectory displacement*, occurs because repulsive interactions between particles alter their trajectories. SCPPI effects are discussed in Section 3.8.

For real geometries, FEVSC effects are best analyzed using numerical methods and Monte-Carlo techniques. This is a very specialized activity. Electron-optical design programs often incorporate FEVSC effects [53; 57].

The one-dimensional (1D) FEVSC case can be treated analytically, using an "average space-charge" model. Theory applies to positive and negative particles, and is useful because it gives insight and provides upper bounds for some effects. 1D-FEVSC theory was initially developed by Stern et al. [6] in 1929, and Barbour et al. [58] in 1953 (see Ref. 59 for other references.)

In a new look at 1D-FEVSC theory, Forbes [59] has introduced a dimensionless parameter ζ that describes the *strength* of FEVSC effects:

$$\zeta = \kappa_{sc} J_P F_P^{-2} V^{1/2}, \tag{7}$$

$$\text{with} \quad \kappa_{sc} \equiv \varepsilon_0^{-1}(m/2\,ze)^{1/2}, \tag{8}$$

where J_P is the emission current density, m the mass of the emitted entity, and ze the magnitude of the mean charge on emitted entities. Illustrative values for κ_{sc} are 1.904×10^5 A^{-1} V$^{3/2}$ for an electron, 6.789×10^7 A^{-1} V$^{3/2}$ for a Ga$^+$ ion.

The field reduction factor $\theta[\equiv F_P/F_L]$ depends on the FEVSC strength via [59]:

$$\theta(\zeta) = [1/(6\zeta^2)][1 \pm \sqrt{S_q}], \quad \text{with} \quad S_q \equiv 1 + 4\zeta^2(4\zeta - 3). \tag{9}$$

The physically valid result uses the negative sign in Equation 9 if $\zeta < \frac{1}{2}$, the positive sign if $\zeta > \frac{1}{2}$. ζ and θ can serve as "engineering indicators" as to whether FEVSC effects may be significant. For example, a LMIS tip of radius $r_a = 1.5$ nm (see Section 7.7) has emission area $A = \pi r_a^2 = 7.1$ nm^2. Taking $i = 2$ μA yields $J_P = i/A = 2.8 \times 10^{11}$ A/m^2. In Equation 8, taking $F_P = 15$ V/nm, $V = 3000$ V, yields $\zeta = 4.7$. Equation 9 then gives $\theta = 0.29$, suggesting strongly that FEVSC affects the LMIS surface field.

This calculation is not quantitatively reliable, because it applies planar FEVSC theory to sharp LMIS emitters, but the conclusion is qualitatively correct. FEVSC does affect LMIS behavior, even at the low emission currents used in FIB machines (see Section 3.8).

FEVSC literature is extensive, but no recent review of FEVSC basic science exists. Reference 59 notes earlier basic work, Reference 50 discusses classical space-charge effects in electron guns.

3 BASIC OPTICAL CONCEPTS

3.1 INTRODUCTION

This section introduces basic concepts of charged-particle (CP) optics. Electron and ion optics are similar because—in the absence of space-charge effects—CP trajectories do not depend on charge-to-mass ratio. Thus, unified treatment of field emission optics is possible.

The electron optics textbook [50] by Hawkes and Kasper (H&K) comprehensively covers most relevant topics. Chapters 45 to 50 cover the optics, emission physics, and design of electron sources and guns (both thermal and field emission). Much of this also applies to field ion emission sources. Useful treatments of field ion emission optics, with further references, are the Orloff, Utlaut, and Swanson textbook [60], and various articles in the second edition [30] of the *Handbook of Charged Particle Optics*, edited by Orloff.

Gas field ionization sources have two special characteristics. First, the ion emission sites can be observed by FIM: this provides firm information about physical source size not normally available for CP sources. Second, except at very low temperatures, the Gaussian image of the physical source is smaller than the cross-over (see Section 8.6); this may affect machine-focusing procedures. Thus, it is necessary to extend conventional treatments of CP optics, as in Reference 61. The "probed" object is called here "the specimen."

3.2 OPTICAL MACHINES

Following the handbook theme, this chapter concentrates on scanning-beam machines. For our purposes,[d] a CP optical machine consists of: (1) a *CP source-module*, which generates an *extracted CP beam*; and (2) *post-source optics*, which use this beam to generate a focused *CP probe* at the specimen, and provide facilities such as beam acceleration, mass and/or energy filtering, blanking, scanning, and aberration correction.

The probe characteristics, especially its radius, depend on the optical properties of both the source-module and the post-source optics. Modern computing power allows analysis of complete systems. But the task is usually split, and this chapter examines the source-module and the post-source optics separately. Primary interest is in source characteristics, but these need to be set in the context of machine behavior.

3.3 MACHINE OPTICAL REQUIREMENTS

A CP optical machine must generate at the specimen a probe with small radius and sufficiently high current density. This puts requirements on both the source-module and the post-source optics.

The source should have high angular intensity on axis and small effective size, and produce a CP beam with small energy spread and no extended tail on the energy distribution. At the specimen, the post-source optics need to produce a de-magnified image (of the source) that is neither blurred unnecessarily by aberrations or other effects, nor weakened unnecessarily by apertures.

The term *source-module* refers to the emitter and immediate surroundings. Many field emission source-modules are variants of the arrangement shown in Figure 1.6. The *emitter* (E) is the physical CP emitter. The *extractor* (X) (sometimes called a *wehnelt*) applies a voltage between itself and the emitter: this generates the field that extracts CPs from the emitter. CPs escape from the extractor via a large hole in its *cap* (C). When a *suppressor* (S) is used, the emitter tip physically protrudes from a small hole at its center, and there will often be a small voltage difference between S and E. The suppressor's usual function is to inhibit various unwanted effects related to the emitter shank and supporting structure. (For example, with an LMIS, the suppressor greatly reduces heating due to high-energy secondary electrons.) In some designs, the suppressor can be used to alter the emitter surface field and provide fine emission control, by varying the voltage between S and E.

There should be no significant asymmetry in the emission intensity distribution from the surface areas that supply the beam entering the machine column. The emitter should be adequately stable in time itself, and the source as a whole needs emission characteristics that are stable on all relevant timescales. Particularly for high-resolution machines, this requires that all current and voltage supplies be highly stable, and that there be no mechanical vibrations that move the emitter relative to the specimen. Obviously, it helps if both emitter fabrication and in-situ restoration (if needed) are reliable well-proven processes.

It is usually impossible to achieve all aims simultaneously; thus, design relies on trade-offs and optimization, which can be particularly complex when SCPPI effects occur [52].

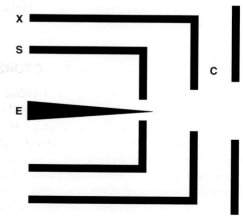

FIGURE 1.6: Schematic diagram showing the three electrodes that may be present in a field-emission-type CP source module: **E**–emitter; **X**–extractor (or "wehnelt"); **S**–suppressor. The "cap" of the extractor is marked **C**.

3.4 THE DIODE APPROXIMATION

For analyzing CP-beam machines, H&K use a *diode approximation* (see their Section 45.2). They consider a source comprising only the emitter and extractor, and assume that beyond the extractor there is a region at the same potential V (relative to the emitter) as the extractor. For analyzing the post-source optics, they assume that space on the emitter side of the extractor is at the same potential V as the extractor, and that the source module can be replaced by a "nearly-point" virtual optical object of appropriate effective radius (ρ_v) and optical characteristics; ρ_v is called here the *source optical radius*. We call the dividing plane (in practice the plane of the extractor cap) *plane M*.

A suppressor electrode does not significantly affect optical principles, although it will modify the source-module electrostatics, and hence may modify optical parameter values. For simplicity, this chapter does not discuss the influence of the suppressor.

Optical machines include an *aperture* that limits the beam. Depending on design philosophy, this can have one of several locations. For understanding its role, one can think of the real aperture as an *effective aperture* of appropriate radius, in plane M.

3.5 OPTICAL RADIUS

Discussions often use the optical *diameters* of the source and probe, and often these diameters are called the optical *size*. Here, optical *radius* is the more natural choice and is always used. Care is needed in comparing results here with treatments using optical diameter (or "size" to mean diameter).

The *optical radius* of the probe (or other optical entity) can be specified in various ways, for example as the radius containing 50% (or 90%) of the incident current, or as the root-mean-square radius. The probe (etc.) *profile* is also significant. The current density distribution is often not strictly Gaussian with respect to radius, particularly for the LMIS (a Holtsmark distribution is perhaps more appropriate {51; 62}. The LMIS probe profile can have a large, relatively flat "tail." This can cause an area of "light erosion," or (in gas-assisted processes) an area of deposition or chemical attack, that extends well beyond the stated probe radius.

Obviously, the effects of CP impact onto a specimen depend both on beam characteristics and specimen properties, in particular surface sputtering characteristics and the modes in which CPs are slowed and/or scattered. The radius of the affected area will always be greater than the probe radius. Further information can be found in the vast literature on the interactions of CPs with solids, including other chapters in this handbook.

3.6 MACHINE FACTORS AFFECTING PROBE RADIUS

The optical factors affecting probe radius ρ_p are of interest. Ideal post-source optics would produce an ideal probe spot of radius $M_{ps}\rho_v$, where ρ_v is source optical radius and M_{ps} the post-source transverse linear magnification. This spot would be a de-magnified image of the virtual source, focused exactly onto the specimen surface. In practice, this ideal spot is enlarged by other effects.

Aperture diffraction, acting by itself on an ideal point source, would generate at the specimen an *Airy disc* of radius $0.610\,\lambda_i/\alpha_i$, where α_i is the beam *convergence half-angle* onto the specimen, and λ_i the incident-beam wavelength [63]. This Airy radius (related to the first zero in the radial intensity distribution) is normally used to estimate an aberration-disc diameter [64]. The aberration-disc *radius* is

$$\rho_{d,i} = 0.305 \, \lambda_i / \alpha_i, \tag{10}$$

(At $\rho_{d,i}$ the intensity distribution function has fallen to approximately half its peak value.) With a coherent extended source, analysis is complicated [60; 63] and beyond the scope of this chapter.

Spherical aberration occurs because CPs moving at different angles to the optical axis are focused to different axial points. By itself, spherical aberration would broaden the point image of a point source into a disc of radius $\rho_{sph,i}$ given, in first-order theory, by:

$$\rho_{sph,i} = \tfrac{1}{4} C_{s,ps,i} \alpha_i^3, \tag{11}$$

where $C_{s,ps,i}$ is the first-order spherical aberration coefficient for the post-source optics, referred to the image side.

Chromatic aberration occurs because CPs moving with different kinetic energies (and hence different velocities parallel to the optical axis) are focused to different axial points. By itself, chromatic aberration would broaden the point image of a point source into a disc of radius:

$$\rho_{chr,i} = \tfrac{1}{2} C_{c,ps,i} \alpha_i \Delta E_{Gauss} / E, \tag{12}$$

where $C_{c,ps,i}$ is the chromatic aberration coefficient for the post-source optics, referred to the image side, and E is the mean total energy of the incident beam. ΔE_{Gauss} is the *full width at half maximum* (FWHM) of the CP *total-energy distribution* (TED), which aberration theory assumes Gaussian.

These aberration coefficients are "referred to the image side" because Equations 11 and 12 contain the convergence angle α_i. Alternative formulae use object-side parameters; the coefficients in these formulae have different but related values.

In the literature, an "effective probe radius" ρ_{eff} is often estimated from the quadrature formula:

$$\rho_{eff} \sim \sqrt{(M_{ps}^2 \rho_v^2 + \rho_{d,i}^2 + \rho_{sph,i}^2 + \rho_{chr,i}^2)}. \tag{13}$$

In machines with field emission sources (but not with the GFIS), SCPPI effects will normally enlarge the probe radius further. To include these, one can add a term (ρ_{PPI}^2) to Equation 13, giving:

$$\rho_p \sim \sqrt{(M_{ps}^2 \rho_v^2 + \rho_{d,i}^2 + \rho_{sph,i}^2 + \rho_{chr,i}^2 + \rho_{PPI}^2)}. \tag{14}$$

In fact, neither formula has any good theoretical basis for real systems [65], and using a single term for SCPPI interactions is gross oversimplification [53]. These formulae are not accurate— but they can signal how important the various effects are.

There is a large, highly mathematical, literature that explores precise methods of estimating probe size, and ways of minimizing it, both in the absence (e.g., Ref. 66) and the presence (e.g., Refs. 51, 67) of SCPPI effects. There is also rapidly growing interest in software-based corrections for column aberrations [68], though at present mainly for transmission and scanning transmission electron microscopes [69].

Obviously, complete discussion needs a theory of source optical radius. Section 4 presents this.

3.7 CURRENT-RELATED OPTICAL CHARACTERISTICS

Several parameters assess an emitter's ability to deliver current. Emission physics determines the *emission current density* J_p. The *mean angular intensity* i/Ω_m assesses the current i passing through a measuring-aperture (of radius r_m and area A_m) placed normal to the beam and well away from the emitter, in a region where electrostatic potential is almost constant—see Figure 1.7. Here, α_m is the half-angle, and Ω_m the solid angle, subtended by this aperture at the position V (a distance L_v away) of the virtual object generated by the source-module. If α_m is small, then $\Omega_\mathrm{m} \approx \pi\,\alpha_\mathrm{m}^2 \approx \pi\,r_\mathrm{m}^2/L_\mathrm{v}^2$.

In the limit of small r_m, the corresponding *differential angular intensity*[e] $K[\equiv \mathrm{d}i/\mathrm{d}\Omega]$ is

$$K \equiv \mathrm{d}i/\mathrm{d}\Omega = \lim_{r_\mathrm{m}\to 0}[i/\Omega_\mathrm{m}] = L_\mathrm{v}^2 \lim_{r_\mathrm{m}\to 0}[i/\pi\,r_\mathrm{m}^2] = L_\mathrm{v}^2 J_\mathrm{m}, \tag{15}$$

where $J_\mathrm{m}[\equiv \lim_{r_\mathrm{m}\to 0}(i/\pi r_\mathrm{m}^2)]$ is the current density in the plane of the measuring-aperture, at its center. An ideal source would have J_m and K constant over a reasonable range of solid angle around the optical axis. This may not always be the case; if so, then the value K_oa on the optical axis, or some average value $\overline{K}$ over directions close to the optical axis, can be quoted. For simplicity, the symbol K_v is used to denote an appropriate value, defined in plane M. K_v is called the *source angular intensity*.

A reduced source angular intensity $K_{\mathrm{R,v}}$ can be defined by:

$$K_{\mathrm{R,v}} = K_\mathrm{v}/V_\mathrm{o}, \tag{16}$$

where V_o is the voltage difference between extractor and emitter. For a given emitter, operated at given apex field and temperature but in different source-module geometries, $K_{\mathrm{R,v}}$ is expected (in the absence of SCPPI effects) to be independent of the related applied voltage. For a LMIS, this was confirmed experimentally by Ishitani et al. [70].

Other parameters commonly used are brightness B and reduced brightness B_R. A generalized "differential-type" brightness can in principle be defined, but cannot be directly measured [50]. A convenient working approach defines an "effective source area" $A_\mathrm{v} = \pi\rho_\mathrm{v}^2$, where ρ_v is the source optical radius (including SCPPI effects, where relevant), and a *source brightness* by:

$$B_\mathrm{v} = K_\mathrm{v}/A_\mathrm{v} = K_\mathrm{v}/\pi\rho_\mathrm{v}^2. \tag{17}$$

The corresponding reduced source brightness (also called normalised source brightness) is:

$$B_{\mathrm{R,v}} = B_\mathrm{v}/V = K_{\mathrm{R,v}}/\pi\rho_\mathrm{v}^2. \tag{18}$$

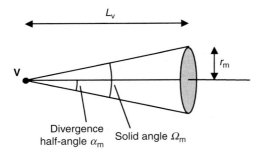

Divergence half-angle α_m Solid angle Ω_m

FIGURE 1.7: To illustrate the definition of differential angular intensity K in the context of a field-emission type CP gun. The position of the virtual optical object generated by the source-module is labelled **V**, and L_v is the distance from **V** to the measurement aperture (shown shaded).

If ρ_v is the radius containing 50% of the current, then $B_{R,v}$ is the "practical brightness" used by Bronsgeest et al. [71]. These authors argue that their "practical brightness" is the only form of brightness that can successfully quantify probe current, and suggest that their definition should replace all other brightness definitions.

The parameter "reduced brightness" was originally introduced because it was thought [65] to be conserved as the beam passed down the column. In reality, with field emission sources, high reduced brightness at the emitter can be degraded by SCPPI effects, and in some cases by a small aperture (see Section 4.6), as the beam passes through the system.

The parameter $B_{R,v}$ is often used as a "catch-all figure of merit" for comparing CP sources. In the author's view, using a single optical parameter for comparisons is not helpful for field-emission-based machines. A better approach uses the reduced source angular intensity $K_{R,v}$, and values of the extraction voltage and whatever source parameters may limit the probe size (usually the source optical radius and/or the energy-spread of the extracted beam). Other figures of merit have been suggested by Swanson [72], Prewett and Mair [73] (see p. 156), and Swanson and Schwind [31] (see p. 22).

3.8 STOCHASTIC COULOMB PARTICLE-PARTICLE INTERACTIONS

In beam regions where current density (and hence CP concentration) are sufficiently high, stochastic coulomb particle-particle interactions (SCPPI) occur. When brighter sources and more intense beams were introduced, it was realized that SCPPI effects could occur all the way along optical columns. Jansen's 1990 monograph [51] is a comprehensive account of SCPPI effects in columns. Reference 67 is a shorter account, with guidelines for assessing when SCCPI effects may be important, and contains many relevant formulae (also see References 53 and 74).

As noted in Section 2.9, SCPPI effects can influence probe radius in several ways. In the source-module, they increase the beam energy-spread and the source optical radius. In post-source optics they can cause further increase in probe radius, even when there are no cross-overs; both energy broadening and trajectory displacement may occur, but the latter seems more significant [52].

As noted earlier, LMIS emission current is FEVSC controlled. For a normal LMIS, it was quickly found that the measured energy spread [75] and source optical radius [76] were anomalously large. Knauer's theoretical work [77; 78] on SCPPI was able to explain the experimental power law linking energy-spread to current, and was influential in getting SCPPI accepted as the basic cause of large LMIS energy spreads. It is now clear that SCPPI can also explain the large LMIS optical radius (see Section 7.7).

SCPPI also affect field-assisted electron emitters. This was first suspected [79; 80] in the 1970s, when Schottky emitter energy distributions were measured. Useful recent discussions [31; 71; 74; 81] of SCPPI effects in Schottky emission record other relevant work.

For the LMIS [52; 60] and for the Schottky emitter [82], a recommended approach for reducing post-source SCPPI effects is to "strip out" the probe beam from the extracted beam as soon as possible, by putting the physical aperture as close to the extractor cap as other constraints allow.

Due to LMIS physics, the only method of alleviating LMIS source-module SCPPI effects is to run the source at as low an emission current as jet stability allows (see Section 7.2); FIB machines usually do this. With Schottky emitters, SCPPI-induced energy broadening is minimized by operating at the lowest value of angular intensity K_v and the largest value of emitter radius r_a that are consistent with beam-current and source-brightness requirements for a particular application [31; 81].

4 THE OPTICS OF FIELD EMISSION SOURCES

4.1 INTRODUCTION

Photon and CP sources have different optics, because (in ray-type theory) light travels in straight lines, but CPs in an electric field do not. The discussion of CP optics here follows H&K and Reference 61, but is modified to cover both electron and ion emission.

The optical model to be used combines: (1) a hypothetical *spherical charged-particle emitter* (SCPE) of radius r_a, fitted to the emitter tip; and (2) a weak lens. For each emission site on its surface, the SCPE forms two different virtual images (usually called the "Gaussian image" and the "cross-over"). The weak lens takes these as virtual objects and generates a second set of virtual images, thereby transforming SCPE optical behavior to the intrinsic behavior of a Müller emitter. These images act as possible (intrinsic) virtual objects for the post-source optics. But their properties are modified by any aberrations occurring within the source-module (significantly so if SCPPI effects occur): in effect, the source-module forms a third set of virtual images that serve as the actual virtual objects for the post-source optics. This behaviour (but not SCPPI effects) is illustrated in Figure 1.9 below. The "principal trajectory" relates to a CP that leaves normal to SCPE surface, with no lateral kinetic energy.

In the approach here, basic issues are discussed using the SCPE; the lens properties are then used to adjust formulae and parameter values to the intrinsic values for an "ideal" Müller emitter. For an emitter with a substantially different shape, further corrections would be needed. In practice, correction factors would be determined by numerical simulation or experimental measurement, or parameters such as the emitter's angular magnification would be determined directly.

Parameters relating to the SCPE are shown primed; those for the weak lens are double-primed; and those for the complete Müller emitter are unprimed. For Müller emitters operating without SCPPI effects, numerical simulations [83–86] justify this approach. It should also work for other emitter types (e.g., carbon nanotubes), but with different properties/parameters for the weak lens.

The "intrinsic" Müller emitter parameters will/may be modified further by source aberrations, including SCPPI effects, to give the effective parameters for the virtual optical source. Again, parameter values may in practice be determined experimentally, or via numerical simulation using a program that includes SCCPI effects.

4.2 THE SPHERICAL CHARGED-PARTICLE EMITTER

The spherical charged-particle emitter (SCPE) has no analogy in photon optics but is key to understanding Müller emitters. SCPE behavior is illustrated in Figure 1.8. Initially assume that the SCPE S_0 of radius r_a is surrounded by a spherical detector D of radius L_D, with $L_D \gg r_a$.

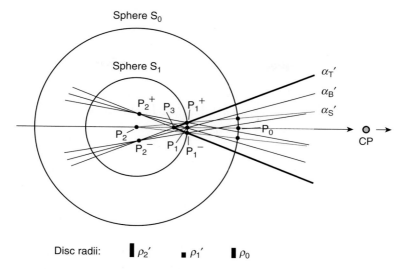

FIGURE 1.8: Schematic diagram showing how a SCPE forms two virtual images. The real physical object is a disc of radius ρ_0 centered at P_0 on sphere S_0. The disc radius subtends a FOHA α_S' at the sphere center P_2. Due to transverse velocity effects (with blurring cone FOHA α_B'), there is a virtual image (the *Gaussian image*) centered at P_1 on S_1, of FORD ρ_1' and FOHA α_T', apparently projected from point P_3. A second virtual image exists near P_2, corresponding to the disc of minimum confusion (the *cross-over*) of FORD ρ_2'. This diagram shows the case where $\alpha_B' > \alpha_S'$, and $\rho_1' < \rho_2'$. To allow detail to be shown, angles are exaggerated.

The emission site

Consider an emission site centered at point P_0 on sphere S_0. For simplicity, treat the site as a small circular disc, tangential to the surface of S_0, with some characteristic intensity *fall-off radius* (FORD) ρ_0, with $\rho_0 \ll r_a$. The term "fall-off radius" makes discussion independent of radius criterion; however, one can think of the disc as emitting roughly 50% of the site current.

In Figure 1.8, a disc diameter lies between the dots on either side of P_0. The emission disc *radius* subtends, at the center P_2 of sphere S_0, an *emission-site fall-off half-angle* (FOHA) α_S', with

$$\alpha_S' = \rho_0/r_a. \tag{19}$$

The unblurred angular distribution

Consider CPs emitted on radial trajectories from points in the emission site. Emission appears to come from a virtual point projection center at P_2, with an appropriate angular intensity distribution. This is the *unblurred angular distribution*; on D it would form a projection image spot of FORD $\alpha_S' L_D$.

The effects of transverse kinetic energy

In reality, the CPs emitted from any particular point on S_0 have a distribution of transverse kinetic energy (KE). Motion in a central force field conserves angular momentum. After travel by a radial distance r_a, 75% of the initial transverse KE κ has been converted to radial KE, and after $10r_a$ around 99% (see Ref. 20). Distant from the emitter a CP effectively moves in a straight line. (The trajectory approximates to a hyperbola, with the line as its asymptote.)

Optical consequences have been analyzed by (among others) Ruska [87], Gomer [88], Wiesner and Everhart [83; 84], and H&K. The CPs emitted (in various directions) from point P_0 appear to diverge from point P_1, on sphere S_1 of radius $r_a/2$. CPs with κ less than some characteristic value κ_c have trajectories lying within a limiting cone of FOHA α_B', as shown in Figure 1.8. α_B' is called here the *blurring FOHA*, and is given by:

$$\alpha_B' = 2(\kappa_c/eF_0 r_a)^{1/2},\tag{20}$$

where F_0 is the field at the surface of S_0. On D, for a SCPE, the emission from a point on S_0 would form a *blurring disc* of FORD $\alpha_B' L_D$.

The embedding approximation

Equation 20 is derived using spherical geometry, in which $F_0 = V/r_a$. In reality, the SCPE models a Müller emitter tip, with apex field given (from Equation 3) by $F_a = V/k_M r_a$. Conversion of initial transverse KE into radial KE mostly takes place relatively close to the emitter. Thus, H&K consider it adequate to identify F_0 with F_a and write:

$$\alpha_B' = 2(\kappa_c/eF_a r_a)^{1/2} = 2(k_M\kappa_c/eV)^{1/2},\tag{21}$$

where V is the voltage difference between extractor and emitter.

The blurred angular distribution

If the CP emission is taken as incoherent (i.e., if each point in an emission site functions as an individual point source), then a CP cone is emitted from each point. Emission from the whole site near P_0 now lies within a cone, of FOHA α_T', with its apex P_3 slightly inside S_1 (see Figure 1.8). This is the "blurred angular distribution." On D, for a SCPE, the site would generate an "image spot" with FORD $\alpha_T' L_D$.

For large distances from the SCPE, and for values of α_B' larger than or comparable with the site FOHA α_S', one can disregard the small differences in projection-center positions and estimate α_T' from the quadrature approximation:

$$\alpha_T' \sim [(\alpha_S')^2 + (\alpha_B')^2]^{1/2}.\tag{22}$$

This formula is inaccurate but illustrative: convolution or numerical simulation would give a better result. If $\alpha_B' \ll \alpha_S'$, the linear approximation $\alpha_T' \sim \alpha_S' + \alpha_B'$ is probably better than formula 22. Given α_T', the *blurring magnification* m_T' (for a SCPE) is defined by:

$$m_T' \equiv \alpha_T'/\alpha_S'.\tag{23}$$

Virtual images generated by the SCPE

As noted, the SCPE generates a virtual image on or near the surface of sphere S_1. This is the emission site's *Gaussian image*. Its FORD is $\rho_1' \approx \rho_0/2$. In respect of this image the SCPE has exerted an *angular magnification* $m' = 1$, and a *transverse linear magnification* M' given by:

$$M' = \rho_1'/\rho_0 \approx \tfrac{1}{2}. \tag{24}$$

In Figure 1.8, the blurring-cone generators are inserted for P_1 and for the points P_1^+ and P_1^- that are the Gaussian images of the emission site's fall-off boundaries. All resulting rays are then projected backwards beyond the sphere center. Two discs of minimum confusion result, one the Gaussian image just described, the other a disc of FORD $\rho_2' = \alpha_B' r_a/2$, near the center of the sphere (P_2). This second disc is a second virtual image, usually called the *cross-over*. From this relationship, using Equation 21:

$$\rho_2' = r_a(k_M\kappa_c/eV_o)^{1/2}. \tag{25}$$

The ratio ρ_2'/ρ_1' $[=\alpha_B'/\alpha_S']$ is called here the *objects size ratio* and is denoted by m_B'.

4.3 THE OPERATION OF THE WEAK LENS

The weak lens represents the optical effects of the emitter shank. As noted, the lens forms virtual images, taking as virtual objects the SCPE-generated Gaussian image and cross-over. Figure 1.9 shows this for the Gaussian image (which is easier to illustrate). The results about weak-lens angular magnification apply to both cases.

For the Gaussian virtual object at/near P_1, the lens has three main effects: (1) it forms a virtual image near E, significantly further away from the emitter apex than P_1; (2) the "angle of departure" (θ) is *compressed* by the lens, from θ_P to θ_E, because the CP trajectories are bent

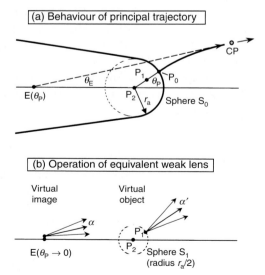

FIGURE 1.9: Schematic diagram showing the (intrinsic) optical behavior of an ideal Müller emitter. The emitter tip is modeled as a spherical charged particle emitter (SCPE). Figure 9a shows a real source at point P_0 on the tip, and the corresponding principal trajectory (trajectory for CPs emitted from P_0 normal to the emitter surface). Figure 9b shows the behavior of the equivalent weak lens. The SCPE forms a virtual Gaussian image of P_0 at point P_1. The weak lens takes this as a virtual object and forms a virtual Gaussian image near point E. The weak lens compresses angles and reduces the lateral size of extended objects (thus, the image of P_1 is closer to the optical axis than P_1 is). In reality, the distance P_2 to E is much greater than this diagram suggests. Similar principles apply to the "cross-over-type" virtual image formed by the SCPE near point P_2.

(a) Behaviour of principal trajectory

(b) Operation of equivalent weak lens

inward by the field distribution associated with the Müller emitter; and (3) the radius ρ_2 of the virtual image near E is less than the radius ρ_2' of the virtual object near P_1.

Angular magnification

For the weak CP lens, the angular magnification m'' is defined by [50]:

$$\tan \theta_E = m'' \theta_P. \tag{26}$$

(This differs from the Helmholtz definition [63] used in photon optics.) The SCPE angular magnification is $m' = 1$, thus (for the Gaussian image) the angular magnification m for the complete Müller emitter is $m = m'm'' = m''$. Thus, for the complete Müller emitter:

$$\tan \theta_E = m \theta_P. \tag{27}$$

At large angles, Equation 27 differs significantly from the Helmholtz angular magnification (m_H) defined by $\tan \theta_E = m_H \cdot \tan \theta_P$. The issue here has been: for a Müller emitter, which parameter (m or m_H) is more nearly constant as a function of angle, particularly at large angles? The best evidence for the superiority of Equation 27 comes from FIM. A central experimental fact of the crystallographic analysis of field ion micrographs is that (on average, disregarding local distortions) the angular magnification m appearing in Equation 27 is very nearly constant for values of θ_P up to about 50° (see Refs. 19, 89, 90). This near-constancy of m is also found in numerical trajectory analyses, both in field electron [85; 86] and in field ion [91] emission.

The ideal Müller emitter

H&K assume that, for a high quality beam, the Abbe sine condition [63] $\sin \theta_E = m \cdot \sin \theta_P$ also needs to hold. With Equation 27, this yields the condition for optimum optical behavior:

$$m \approx \theta_E/\theta_P \approx 1/\sqrt{3} \approx 0.577. \tag{28}$$

A Müller emitter with this m-value can be described as *ideal*.

The *compression factor* $\beta_c [\equiv 1/m]$ is often used instead of m. It is commonly written [19] that β_c is "around 1.5," which implies $m \sim 0.67$. Older work [89] found a typical value equivalent to $m \approx 0.55$. Given variability in emitter preparation, these figures seem consistent (for emitters not specially processed). By contrast, the ZrO/W Schottky emitter end-form has a built-up shape with a flat facet facing forwards [31]; in this case, numerical simulation (with a suppressor present) gave [31] lower m-values, typically around 0.19.

Linear magnification

The weak lens makes virtual images, of radii ρ_1 and ρ_2 respectively, of the SCPE-generated Gaussian image and cross-over. For small objects, transverse linear magnification M'' is defined for the weak lens by:

$$M'' \equiv \rho_1/\rho_1' = \rho_2/\rho_2'. \tag{29}$$

It follows that the objects size ratio for the complete Müller emitter, $m_B [\equiv \rho_2/\rho_1]$, is equal to the objects size ratio $m_B' [\equiv \rho_2'/\rho_1']$ for the SCPE; thus, m_B can replace m_B' in formulae.

As a CP moves away from the emitter, it moves from a region where it effectively sees the potential distribution of a sphere to one where it sees the potential distribution of the complete Müller emitter. H&K note that in CP optics this is equivalent to a change in refractive index, where the ratio of the refractive indexes is given by:

$$\text{ratio} = (V/F_a r_a)^{1/2} = k_M^{1/2}, \tag{30}$$

where k_M is the shape-factor used earlier. So, for the weak lens here, the Smith-Helmholtz formula of photon optics (e.g., Ref. 63, Section 4.5) becomes, certainly for small angles:

$$m''M''k_M^{1/2} = 1. \tag{31}$$

For Müller emitters in conventional FEM/FIM-type arrangements, k_M is considered to lie approximately in the range $5 < k_M < 7$. To derive illustrative values, consider an ideal Müller emitter that has $k_M = 6$. In this case:

$$M'' = 1/(m''k_M^{1/2}) \sim \sqrt{3}/\sqrt{6} = 1/\sqrt{2} \approx 0.707. \tag{32}$$

This complete emitter would (for the Gaussian image) have overall linear magnification $M = M'M'' \approx \sqrt{2}/4 \cong 0.35$.

Blurring magnification

To a good approximation, the ratio α_T'/α_S' is unaffected by the lens. Thus, the blurring magnification m_T for the complete Müller emitter equals m_T' as given by Equation 23, and m_T can replace m_T' in formulae.

4.4 OPTICAL OUTCOMES FOR THE COMPLETE MÜLLER EMITTER

Virtual source radii

From Equations 24 and 32, noting $M = M'M''$, the Gaussian image formed by the complete Müller emitter has FORD:

$$\rho_1 = M\rho_0 = m^{-1}k_M^{-1/2}(\rho_0/2). \tag{33}$$

From Equations 23 and 31, noting $m = m''$, the cross-over formed by the Müller emitter has FORD:

$$\rho_2 = M''\rho_2' = m^{-1}r_a(\kappa_c/eV)^{1/2} = m^{-1}k_M^{-1/2}r_a(\kappa_c/eF_a r_a)^{1/2}. \tag{34}$$

As noted, these images act as possible intrinsic virtual sources for the post-source optics, with the intrinsic radii shown (before considering source aberrations).

A more careful treatment [71] suggests that, if κ_c is interpreted as the mean transverse KE and ρ_2 as the radius containing 50% of the current, then a numerical factor $\sqrt{(\ln 2)}$ [$\cong 0.833$] needs to be introduced onto the right-hand side of Equation 34.

Blurring parameters

Like the principal trajectory in Figure 1.9b, the generators of the blurring cone will be "compressed" toward the optical axis. Hence, for the complete Muller emitter, the blurring FOHA (the FOHA of a cone that contains about 50% of the CPs from an emission *point*) becomes $\alpha_B = m''\alpha_B' = m\alpha_B'$, where α_B' is given by Equation 21. The radius ρ_B of the corresponding spot in plane M is of interest. If plane M is at a distance L_E from the virtual source (near point E in Figure 1.9b), then this *blurring radius* is:

$$\rho_B = \alpha_B L_E = m\alpha_B' L_E = 2m(\kappa_c/eF_a r_a)^{1/2} L_E. \qquad (35)$$

Linear spot magnification

By similar arguments, the *spot radius* ρ_{spot} of the "image spot" projected onto plane M by the emission site near P_0 (which has radius ρ_0) is, using the formula $\alpha_T' L_E$ for the SCPE-related spot radius, and Equations 19 and 33:

$$\rho_{spot} \approx m\alpha_T' L_E = mm_T\alpha_S' L_E = m_T m\rho_0 L_E/r_a \qquad (36)$$

This formula also applies to emission sites/spots in FIM and FEM images, when aberrations are disregarded (hence the name "spot radius"). Thus, one obtains a *linear spot magnification* μ_{spot} as:

$$\mu_{spot} \equiv \rho_{spot}/\rho_0 = m_T m(L_E/r_a). \qquad (37)$$

Note that these "linear" effects in plane M relate to the "angular" magnification properties of the Müller emitter: this is because plane M is not a focal plane of the optical system.

In CP optics (in the absence of SCPPI effects), a spot of radius ρ_{spot} in plane M is the "natural aperture" for an emission site beamlet to enter the post-source optics. If, as in the HeSIM, one wishes to extract the beam from just the central part of an emission site, the effective defining aperture in plane M must have radius significantly less than ρ_{spot}.

4.6 USE OF FIM IMAGES TO ILLUSTRATE OPTICAL EFFECTS

FIM images can provide useful intuitive understanding of optical system effects. Figure 1.10 shows helium-ion FIM images of part of a tungsten emitter, taken [92] at emitter temperatures near 5 K and near 80 K. (Due to the physics of the FIM imaging process, the gas temperatures may be somewhat higher [26].) These FIM images are maps of the current density distribution J_m that would be observed in plane M. Some self-evident points can be made.

(1) A field ion emitter is an assembly of individual atomic-level emission sites, each of which emits a helium ion beamlet. (2) When the ions reach the plane of the extractor cap, these beamlets (at most) overlap only slightly with their nearest neighbors. (3) The temperature-dependent blurring effects, that enter the definition of α_B' in Equation 21 and m_T in Equation 36, are very obvious when Figures 1.10a and 10b are compared.

The blurring effect is large, but the observations are consistent with the prediction (in Table 1.4, later) that, for He+ ions at 78 K, the objects size ratio $m_B \approx 2$. The table suggests that, typically, the blurring radius ρ_B is only slightly smaller that the spot radii ρ_{spot} seen in

(a) Near 80 K   (b) Near 5 K

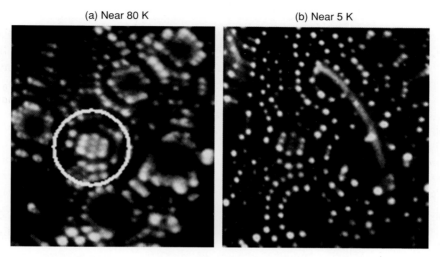

FIGURE 1.10: Field ion micrographs of part of a tungsten emitter, taken at emitter temperatures (a) near 80 K and (b) near 5 K. The W(111) facet is circled. Comparisons show that spot radius ρ_{spot} increases with temperature (and also that intensity-redistribution effects occur at very low temperatures).

the near-80 K image. In the absence of all aberrations, a point emission site at P_0 would be focused by the machine optics at an image point I_0 at the specimen surface. The ions from P_0 pass though plane M with about half of them spread out over the blurring disc of radius ρ_B, the other half outside the disc. Clearly, if the radius of a beam-defining aperture is significantly less than ρ_B (as it is for a HeSIM), then many ions emitted from point P_0 will not reach I_0. Thus, the reduced brightness of the beam reaching I_0 will be less than the reduced brightness of the beam leaving P_0. (For an ideal system, there can be no contribution to the beam intensity at I_0 from points adjacent to P_0.)

This shows unequivocally that, with ion beams, reduced brightness can sometimes be *decreased* by a small aperture. This result is in contrast to the separate effect [71], related to current density variations across an emission site, whereby mean beam brightness is assumed to be *increased* by an aperture (because weak distribution tails are cut off).

5 ELECTRON EMISSION THEORY

5.1 ELECTRON EMISSION REGIMES

Electron emission is usually classified into regimes. Various approaches exist, and nomenclature is confusingly inconsistent. Physics-based criteria are used here to define three regimes, as temperature increases for a given barrier field. These regimes apply when (as is usually the case) the top of the tunneling barrier is significantly above the emitter Fermi level. Names and definitions preferred here are as follows.

1. *Cold field electron emission* (CFE) is a regime in which more than 50% of the emitted electrons escape by tunneling from states below the emitter Fermi level.

2. *Intermediate-temperature electron emission* (ITE) is a regime lying between the CFE and TE regimes, in which the characteristic mode of escape is by tunneling from states above the emitter Fermi level, but below the top of the barrier.
3. *Thermal electron emission* (TE) is a regime in which more than 50% of emitted electrons escape by emission over the barrier, as a result of thermal activation.

This chapter does not use the term "thermal-field" as a regime name, partly because of ambiguous past use, partly because Section 2.7 uses it in a more general sense in relation to emitter shaping.

For a flat emitting surface, the main independent physical variables are the work-function ϕ, the surface field F (best called the *barrier field*), and the emitter's thermodynamic temperature T. The purpose of identifying CFE and TE regimes is to define regions of $\{\phi, F, T\}$ parameter space where specified equations are (approximately) valid. The physical phenomena are, of course, continuous across regime boundaries.

In the CFE regime, the case of theoretically assumed zero temperature (the *zero-temperature limit*), was analyzed by Fowler and Nordheim (FN) in 1928. Emission near room temperature (or below) is often called "cold field electron emission." However, the present author uses CFE to mean the whole regime as defined above (because Equation 40 covers the whole regime), and calls the higher-temperature part of the CFE regime the *above-room-temperature (ART) limit.*

The ART limit has in the past been called the "thermal-field regime." However, the devices called "thermal-field emitters" probably operate, not in the ART limit, but near the boundary of the ITE and TE regimes. (For electron optics, there is little point in operating in the ART limit, because there is little gain in current, but the energy-spread is significantly larger, and the temperature is not really high enough to solve contamination problems—(see Sections 6.2 and 6.4.)

In the TE regime, two limiting situations are recognized here: in the *Richardson limit*, at zero applied field and high temperature, there is no lowering of the tunneling barrier by the Schottky effect, and all electrons escape over the barrier. In the *Schottky limit*, there is significant barrier lowering by the Schottky effect, and many electrons escape by tunneling through the top of the barrier. Emission near this limit is often called *Schottky emission*, and an emitter designed to operate near this limit a *Schottky emitter*. Section 5.3 discusses an alternative (older) TE sub-regime nomenclature system, based on Equation 43.

The term "field-assisted emission sources," used later, covers electron sources that operate in the CFE or ITE regimes, or in the TE regime Schottky limit.

In the CFE and TE regimes, well-recognized (approximate) equations exist for the *emission current density* (ECD) from the surface of a bulk metal. In the ITE regime it is difficult to formulate a theory-based equation (though Jensen has had significant success [24; 93]), and numerical methods are often used. ECD equations fitted to numerical results exist in the vacuum-arc literature [94], but are not used in electron source design. Obviously, numerical methods can be used for the whole temperature range.

5.2 ECD EQUATIONS FOR COLD FIELD ELECTRON EMISSION

In CFE theory, an important role is played by a parameter d, called the *decay width*, which measures how fast the tunneling probability for a particular barrier falls off as the height of the barrier increases. For a SN barrier of zero-field height ϕ (see Section 2.2), d can be approximated as:

$$d \approx 2F/3b\phi^{1/2}. \tag{38}$$

Typically, d has values between 0.1 and 0.3 eV. For illustration, when $\phi = 4.5$ eV, $F = 5$ V/nm, then $d \approx 0.23$ eV. From d, a dimensionless parameter p can be defined by:

$$p = k_B T/d \approx (3bk_B/2)(\phi^{1/2}T/F). \tag{39}$$

The factor $(3\, bk_B/2)$ has the value 8.829618×10^{-4} eV$^{-1/2}$K^{-1} V nm^{-1}.

In the design of CFE-based sources, it is customary to use a Fowler-Nordheim-type equation developed [95] by Murphy and Good (MG) in 1956. This uses the SN barrier model, and describes emission from a metal that is "not particularly sharp" (apex radius greater that 10 nm, say). MG's equation gives the ECD in terms of ϕ, F and T. Due to recent reformulation of CFE theory [25], the following format is preferred here:

$$J^{\text{FE-MG}} = \lambda_T a\phi^{-1}F^2 \exp[-v(f)b\phi^{3/2}/F], \tag{40}$$

where λ_T is a temperature-dependent correction factor discussed below, f is a particular value allocated to a mathematical variable $l,'$ and $v(f)$ is a particular value of a mathematical function $v(l')$ called the *Principal Schottky-Nordheim Barrier Function* [96]. f is a physical parameter called the *scaled barrier field* for a SN barrier of height ϕ, and is given by:

$$f = (e^3/4\pi\,\varepsilon_0)\phi^{-2}F = c^2\phi^{-2}F, \tag{41}$$

where c is the Schottky constant. For reasons discussed elsewhere [25], this format is considered significantly better than the historical one.

In MG's theory, the temperature correction factor λ_T is given (for p less than about 0.7) by

$$\lambda_T = \pi\, p/\sin(\pi\, p). \tag{42}$$

The 50% criterion for the CFE upper boundary corresponds approximately to $p = 0.5$, $\lambda_T = \pi/2$. The values $\phi = 4.5$ eV, $F = 5$ V/nm, give a boundary around 1350 K. Thus, the formal CFE regime extends to quite high temperatures, with the limit depending on the values of ϕ and F. For T near room temperature, λ_T is always of order unity (typically $\lambda_T < 1.05$); thus, it is common to omit λ_T from CFE equations applied to room temperature emission.

Equation 40 is one of a family of approximate equations (Fowler-Nordheim-type equations) used to predict the CFE current density from bulk metals. More-general FN-type equations exist [97] and contain correction factors that do not appear in Equation 40. One factor depends on the emitter material and the crystallography of the emitting surface. In most cases it cannot easily be (and never has been) calculated accurately. For given values of ϕ and F, the present best guess [98; 99] is that (for normal situations, without resonance effects) plausible predictions of CFE current density probably lie between $0.01\ J^{\text{FE-MG}}$ and $10\ J^{\text{FE-MG}}$. These error-factor considerations are not often raised in discussions of theoretical electron-optical brightness.

Another important point is that FN-type equations describe CFE from bulk metals, and do not apply unmodified to other materials, particularly if quantum-confinement effects occur in the emitter. Non-metallic materials, in particular carbon nanotubes (CNTs), are currently attracting interest as possible bright electron sources [100; 101]. Caution should be exercised in applying FN-type equations to such materials.

5.3 ECD EQUATIONS FOR SCHOTTKY EMISSION

In the Schottky limit (SL), it is customary to use a second equation developed [96] by MG in 1956, which gives the ECD J^{SL} as:

$$J^{SL} = \{\pi\, q/\sin(\pi\, q)\}A_{R0}T^2 \exp[cF^{1/2}/k_B T]\exp[-\phi/k_B T], \tag{43}$$

where A_{R0} is the universal theoretical Richardson constant (see Table 1.9). The term $\exp[cF^{1/2}/k_B T]$ relates to Schottky barrier lowering (see Section 2.2). The term $\{\pi\, q/\sin(\pi\, q)\}$ is a "tunneling correction." The parameter q is given in terms of δ^{SN} as below. (δ^{SN} is a mathematical decay parameter related to the theory of tunneling through a shallow SN barrier.)

$$q \equiv \delta^{SN}/k_B T, \tag{44}$$

$$\delta^{SN} = (h_P/2\pi^2 m^{1/2})(4\pi\,\varepsilon_0 e)^{1/4}F^{3/4}. \tag{45}$$

These generate the formula:

$$q = C^{SL\text{-}SN}(F^{3/4}/T), \tag{46}$$

where $C^{SL\text{-}SN}$ has the value 930.8191 V$^{-3/4}$ nm$^{3/4}$ K ($\equiv 1.655256 \times 10^{-4}$ V$^{-3/4}$ m$^{3/4}$ K).

Equation 43 was originally thought valid for $q <\sim 0.7$, and was used [102] in the following mathematically based TE classification system. In the *Richardson (or "thermionic") regime*, both the Schottky term and the tunneling correction can be disregarded. In the *Schottky regime,* the Schottky term must be included, but the tunneling term can be disregarded. In the *extended Schottky regime*, the tunneling correction must also taken into account. The upper boundary of the extended Schottky regime corresponds to breakdown of Equation 43. The boundaries of the extended regime were originally taken [102] to be $0.3 < q < 0.7$. Recent work [31; 103] suggests that, in fact, Equation 43 begins to lose mathematical validity near $q = 0.4$, yielding results 30% too high, as compared with numerical simulations.

5.4 ENERGY CHARACTERISTICS

LMIS-based machines are limited by the source optical radius, but other modern CP-optical machines tend to be limited mainly by chromatic aberration. With some electron beam machines, such problems can be overcome by monochromators and/or aberration corrections, but it helps to have a source with a low energy-spread. This section discusses aspects of electron energy distribution theory.

Distribution types

With field emitters, one must distinguish between the distributions (at emission) of kinetic-energy (KE) components normal to and parallel to the emitting surface, and between these and the distribution of total energy. These are called, respectively, the normal (NED), parallel (PED), and total (TED) energy distributions.

CP motion takes place in an approximately central field. Thus, KE parallel to the surface at emission is converted into KE normal to the surface, as distance from the surface increases. At large distances (in the absence of SCPPI effects), the energy distribution in the direction of motion corresponds to the TED immediately after emission. Chromatic aberration is thus determined by energy-spread of the TED. By contrast, the Section 4.2 blurring effects depend on the PED immediately after emission, in particular on the PED mean value κ_{av} (i.e., *the mean transverse KE at emission*).

Measures of energy-spread

Discussion above uses the qualitative term "energy-spread." This can be specified in various ways, in particular by the standard deviation ΔE_{sd}, by the full width at half maximum (FWHM) ΔE_{FWHM}, or by the energy-width ΔE_{FW50} that contains 50% of the current. An issue is which measure to use.

Chromatic-aberration theory at a certain level assumes the TED is Gaussian. In reality, at emission, before any broadening occurs, the TEDs for most field-assisted emission sources are asymmetric [104]. It has been usual to set ΔE_{Gauss} in Equation 12 equal to the FWHM ΔE_{FWHM} of the measured or predicted field emission TED, assuming that no significant error is involved.

When quantifying SCPPI effects in Schottky emission, it is now thought best [71; 105; 106] to use criteria (for both the FORD and the energy-spread) based on "50% of the current." This makes it possible, when combining contributions to probe radius, to use formulae with a respectable theoretical foundation [51]. Unfortunately, the complex details [71; 105; 106] are too long to present here. References 31 and 81 use this approach. A practical difficulty [106] is determining the "50% width" for a measured distribution.

Energy-parameter estimates

For some specific distributions, analytical relationships exist between the different measures of energy spread [107]. In other cases, results need to be obtained numerically. For electron emission, unaberrated values of ΔE_{FWHM} have been derived numerically by several researchers, most recently by Schwind, Magera and Swanson [81] and by Jensen [24] (see his p. 115). Both groups show that, as T is increased for given values of ϕ and F, ΔE_{FWHM} is expected to have a narrow peak in the middle of the ITE regime (see Figure 1.11), and to then go through a minimum before rising again toward a limiting high-temperature value. Emitter operation needs to avoid the region of $\{\phi, F, T\}$ parameter space where the peak occurs. For a Schottky emitter, the best arrangement is to operate near the minimum on the peak's high-temperature side.

References 31 and 81 show that the peak position moves to lower temperature as work-function decreases (see Figure 1.11). Thus, for fixed operating temperature, a low-work-function emitter has lower FWHM. The low work-function ($\sim$2.9 eV) of the ZrO/W Schottky emitter allows it to take advantage of this effect.

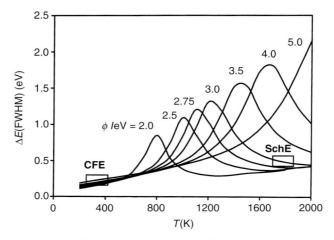

FIGURE 1.11: Plots showing the predicted temperature dependence of ΔE_{FWHM}, for $J=2\times10^7$ A/m^2 and the indicated work-function value, calculated numerically [81] using a Schottky-Nordheim barrier. The small boxes marked "CFE" and "SchE" approximately indicate the ranges where cold field electron emitters and Schottky electron emitters operate. (Energy-spread data provided by G.A. Schwind, FEI Co.)

6 FIELD-ASSISTED ELECTRON SOURCES

6.1 INTRODUCTION

Source comparisons

Electron microscope source types have changed over time. The main options have been: the pointed-tungsten-hairpin thermal electron emitter; the lanthanum (or cerium) hexaboride emitter; and three types of Müller emitter. These are: the room-temperature tungsten emitter (often called a "cold field emitter"); the heated tungsten emitter, sometimes called a "thermal-field emitter," but which is probably better described as a "built-up tungsten Schottky emitter"; and the zirconiated tungsten Schottky emitter (usually called a "ZrO/W Schottky emitter"—see Section 6.4).

Due to data variability, it is surprisingly difficult to make reliable quantitative comparisons of source performance. Table 1.1 shows illustrative electron-optical data, from a textbook [108] revised in 2003. Clearly, the field-assisted emission sources have higher brightness than the older types, and smaller effective source radii; this makes the field-assisted sources attractive for high-resolution microscopy and analysis.

Historical background

Unsuccessful attempts to develop field emitters as technological electron sources were made in the late 1910s and early 1920s, by Lilienfeld [3], who wanted to generate X-rays for medical applications.

Table 1.1 Comparative optical information about various forms of electron source (adapted from Ref. 108).

Source type	Brightness (at 20 kV) A m^{-2} sr^{-1}	Optical radius	Energy spread eV
Tungsten hairpin	10^9	15 to 50 μm	1 to 3
Lanthanum hexaboride	10^{10}	2.5 to 25 μm	1 to 2
Cold tungsten FE	10^{12}	< 2.5 nm	0.3
Hot built-up tungsten FE	10^{12}	< 2.5 nm	1
ZrO/W Schottky emitter	10^{12}	8 to 15 nm	0.3 to 1

Later, due to Müller's early work and the development of electron microscopy [109] by Ruska and others from the 1940s onward, it was realized that the Müller emitter had potential as a bright electron source. Much early development was carried out in the 1950s, in W. P. Dyke's research group at Linfield College, and by the associated "Field Emission Corporation." This work was taken further, mainly in the 1970s and 1980s, by L.W. Swanson's group at the Oregon Graduate Center and the associated FEI Company, and later included LMIS research and development. Source development was also taken up by many commercial companies and numerous university groups.

The first people to build a (scanning transmission) electron microscope with a FE source were Crewe and his Chicago University group [110], in the late 1960s. They initially used a room-temperature tungsten FE source, and imaged individual heavy metal atoms [111].

Müller [112] in the 1950s, and Shrednik [113] in the 1960s, did basic research on zirconium-coated tungsten emitters. Others contributed (see Ref. 31). But the leaders in developing the present ZrO/W Schottky emitter were Swanson, Schwind and their colleagues [33; 79; 114; 115].

6.2 THE COLD TUNGSTEN FE SOURCE

Presumably, tungsten (W) was originally chosen for emitters because it has the highest melting point of any metal (exceeded only by graphite and diamond). This favors it to survive prolonged and repeated high-temperature heating. An FE source must have adequate angular intensity on-axis. As Figure 1.4a shows, the common W(110)-oriented emitter does not. Thus, emitters must be prepared with a different orientation, such as (100), (111) or (310).

A room-temperature FE source has good optical properties, as shown by Crewe [110; 111]. But very good vacuum conditions (less than 10^{-10} Torr^f) are required. Differential pumping may be needed to sustain good vacuum near the source; this complicates design and increases cost.

Even at very low pressure, vacuum-system gas molecules build up in space close to the emitter and on the emitter surface—probably released from the vacuum-system walls as a result of electron impact. Further, the emitter is bombarded by ions derived from electron-beam-induced ionization of these gas molecules. This causes sputtering. Also, it appears that heating bursts associated with this bombardment eventually cause enough local TF sharpening for local protrusions to develop (see Section 2.6). If this happens, then high local emission currents can lead to run-away heating (both resistive and due to the Nottingham effect [23; 116–118]), and to emitter damage or destruction.

To prevent this, the emitter has to be "flash-heated" to very high temperature at regular intervals (typically daily, depending on vacuum conditions), to desorb contaminants and restore and smooth the tungsten surface. Software-controlled heating procedures now exist, and residual-gas effects are not regarded as an issue of concern for modern instruments [119].

More detailed information about cold tungsten FE sources, as currently used, can be found in a recent review by Swanson and Schwind [120].

6.3 THE SEARCH FOR ALTERNATIVES

In practice, these vacuum problems helped stimulate (from the late 1960s onward) a search for alternatives to the cold field emitter. Running the emitter at elevated temperatures reduces the contamination problem. But this also increases the energy-spread, which is a problem for machines limited by chromatic aberration.

The initial response [114] was to look for sources that (a) would "run hot," and (b) could be given better optical properties by TF shaping. Two successful methods were initially found [32; 79; 114]: (1) TF buildup of a carbon-contamination-free (100)-oriented tungsten emitter, at temperatures of order 1600 to 1900 K (this source emits only from the forward-facing (100) facet). (2) Heating of a pre-prepared (100)-oriented tungsten emitter onto which a mixture of zirconium and oxygen has been deposited before heating, and subsequent TF processing. This source emits mainly from a forward-facing (100) facet on which a co-adsorbed mixture with the composition ZrO has formed. The facet has $\phi < 3$ eV (for tungsten, $\phi \sim 4.5$ eV). Both sources have come into use. However, the ZrO/W Schottky emitter has become more popular, probably because the lower work-function leads to lower energy-spread (see Section 2.5.4). Figure 1.12a shows one of the several modern commercial forms.

Swanson's patent [32] for the built-up W emitter and Swanson and Schwind's discussion [31] of how to prepare a ZrO/W emitter and keep it working illustrate how complex the emitter preparation and maintenance can be. They provide insight into the research and development problems involved in making a heated cathode operate reliably. Electron sources based on pointed wires have been under serious technological development for more than 50 years. The complexities of this technology both encourage the search for alternatives, and show how difficult and expensive it is likely to be to develop alternatives into commercially reliable devices. Published comments about alternatives sometimes seem unduly naïve.

6.4 THE ZrO/W SCHOTTKY SOURCE

Reference 31 provides an excellent, detailed state-of-the-art report on the ZrO/W Schottky source, and lists useful references. This section summarizes source properties.

Figure 1.12a shows a common commercial emitter, and Table 1.2 details its properties [31]. In working sources, the emitter tip protrudes from a suppressor, which inhibits Richardson-type TE from the shank, and also protects the shank from bombardment by ionized contaminants. A continuous supply of Zr (and possibly also O) atoms to the tip is provided by surface diffusion from a ZrO_2 reservoir on the shank. The source lifetime is normally determined by the time taken for the reservoir to empty, due to thermal evaporation of Zr into vacuum.

In operation, the source is stable against "intrinsic" TF blunting effects if the apex field is kept sufficiently high [31] (> 0.8 V/nm, for operating temperatures not exceeding 1800 K).

(a)

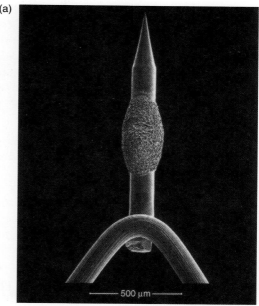

500 μm

(b)

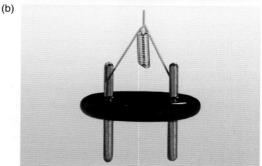

FIGURE 1.12: Common commercial forms of (a) the ZrO/W Schottky emitter and (b) the gallium LMIS. (FEI Co., with permission.)

Contaminant effects are less at these temperatures, but the source is sensitive to oxygen and water adsorption (which lead to work-function increase). The expendable ZrO layer seems to give some protection against ion bombardment and related local TF sharpening The source operates successfully in vacua less stringent than cold FE sources require, provided the H_2O and O_2 partial pressures are kept below about 10^{-9} Torr. When operated correctly, the long-term drift can be as low as 0.2%/hour. Short-term noise (presumably due to emitter surface atom migration) is typically less than 1%.

6.5 OTHER ELECTRON SOURCES

To compete effectively with existing electron sources, any alternative technology would need to be robust against poor vacuum conditions, generate adequate angular intensity on axis, have low energy-spread, be less susceptible to SCPPI effects, and not be unduly difficult to fabricate.

Four alternative point-geometry electron emitters have been investigated in academic research. Some illustrative references are: carbide emitters [121; 122], metallic Müller emitters

Table 1.2 Main operating properties of ZrO/W Schottky cathode
(adapted from Ref. 31).

Operating temperature	1710–1800 K	Emitter tip radius	0.3–1.0 μm
Work-function	2.95 eV	Virtual optical radius	20–40 nm
Vacuum needed (Torr)	$< 10^{-8}$ total $< 10^{-9}$ H$_2$O & O$_2$	Energy spread at cathode at 1800 K	0.4 eV
Cathode life	> 18 000 h	Reduced brightness at 5 kV extraction voltage	5×10^7 to 3×10^8 A/(m^2 sr V)

ending in a single atom [123–126], carbon nanotubes [100; 101], and electron emission from a liquid metal cone [127]. These may have niches (particularly the first three), but at present—for electron beam machines—they seem unlikely to pose a serious challenge to existing technology.

7 LIQUID METAL ION SOURCES

7.1 INTRODUCTION

Physical description

The liquid metal ion source (LMIS) (see Ref. 46) has several technical variants. The source commonly used in FIB machines is the so-called *low-drag blunt-needle LMIS*, called here the *normal LMIS*. As shown in Figure 1.13, ion formation occurs at the tip of a liquid metal cusp/jet drawn out by electrical forces from a roughly conical body of liquid metal, itself formed on top of a Müller emitter with a tip radius of a few μm. With the common gallium (Ga) LMIS, the Müller emitter (often called a "needle" in LMIS literature) is formed from tungsten.

The conical liquid body acts as a local reservoir, and is supplied by thin-film flow along the needle shank, from a main reservoir, as shown in Figure 1.12b. At the emission currents (around 2 μA) normally used in FIB machines, the jet tip has a radius of around 1 to 2 nm. The resulting high current densities (of order 10^{11} A/m^2) immediately above the jet tip cause space-charge effects, which play an important role in LMIS physics.

In principle, many different pure metals or alloys can be used as LMIS ionizants. Gallium is widely used because it has low melting point and vapor pressure, is mostly unreactive, and produces a nearly pure Ga$^+$ beam, and because the heavy Ga$^+$ ion is effective for ion milling.

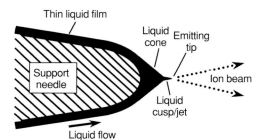

FIGURE 1.13: Schematic diagram of a blunt-needle LMIS in ion generation mode.

Table 1.3 Emission regimes for low-drag blunt-needle Ga LMIS

Current range	Name of regime	status	Energy-spread vs current relation
$< 0.45\,\mu A$	Extinct	–	–
0.45 to $\sim 2\,\mu A$	Intermediate regime	$\sim$ steady	anomalous
~ 2 to $\sim 25\,\mu A$	Most-steady regime	$\sim$ steady	$\Delta E_{FWHM} \sim i^{0.7}$
$>\sim 25\,\mu A$	Upper unsteady regime	unsteady	$\Delta E_{FWHM} \sim i^{0.3}$

A LMIS is a specialized type of electrohydrodynamic (EHD) emitter. All liquid metal EHD (LMEHD) emitters generate two products: charged droplets (by mechanisms similar to those of EHD sprayers of non-metallic liquids [128]), and ions (by a form of FEV). Reference 46 summarizes current understanding (far from complete) of LMEHD emitter physics.

Historical background

LMEHD behavior is a complex interplay between (a) EHD effects on liquid ionizant supply, and (b) ion emission and FEVSC effects. Thus, there are several themes in LMIS history and theory. The EHD phenomena involved are liquid-cone formation, growth of its tip into a liquid jet, and (undesirably) liquid-droplet emission.

Cone formation was reported by Gilbert [129] in 1600, but scientific understanding had to wait until Taylor's mathematical analysis [44] in 1964. (Taylor's results are not completely valid, because he neglected an important atomic-level term [130].) Astonishingly, electrically induced jet formation and droplet emission were first reported (for water and mercury) in 1731, by Gray [131], and were described by Priestly in his 1768 electricity textbook [132]. Thus, the liquid shape in Figure 1.13 is called here a *Gilbert-Gray* (GG) *cone-jet*.

The LMIS had its technical origin in attempts in the 1960s to develop LMEHD thrusters for spacecraft propulsion [133]. Unexpectedly, these generated many ions. In 1974 Krohn [134] suggested using them as bright sources for ion microprobes. Early devices supplied ionizant via a capillary tube (as usual with EHD sources). At UKAEA Culham in the 1970s, R. Clampitt and colleagues were investigating the current-voltage characteristics of an alternative source supplied by liquid flow along a sharp needle. The blunt-needle LMIS was discovered accidentally when S. Ventakesh (an undergraduate on a professional placement year) blew the end off her sharp needle by raising the voltage too high. Afterward , emission currents were higher by a factor of order 100.

This was found [135] to result from GG cone-jet formation on the needle tip, and it was assumed [135] that grooves or roughness on the needle surface assisted liquid flow. Later, grooves or roughness were deliberately created [136; 137]; this reduces viscosity problems by allowing *low-drag* liquid flow.

Emission modes and regimes

Usually, an LMEHD emitter switches repeatedly between ion and droplet emission modes, with details dependent on physical design, ionizant, and liquid flow rate. For any particular design and ionizant, there are several behavioural regimes, related to emission current and/or other source characteristics. Table 1.3 shows the known regimes for a gallium normal LMIS.

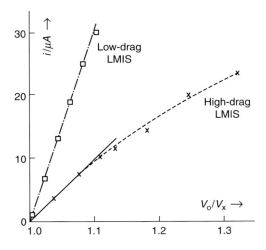

FIGURE 1.14: Current-voltage characteristics for a "rough" (low-drag) (Ω) and a "smooth" (high-drag) ($\times$) gallium LMIS, each with tip radius 3 μm. Marked points are experimental values taken from Reference 144. The diagram shows the effect, on $i - V$ characteristics, of viscous drag between the needle and the liquid ionizant. It also demonstrates that the experimental results can be fitted by theory, as described in the text, using physically reasonable parameter values.

Applications

The main applications of LMEHD emitters are as an ion source, as a space-vehicle engine based on *field emission electric propulsion* (FEEP) [138], and as a LMEHD sprayer (for surface coating [139] or similar purposes). Various geometrical designs and technical variants have been explored (see Ref. 46). As noted above, the source type best suited to FIB machines is the low-drag blunt-needle LMIS ("normal LMIS").

Section 7 describes the properties of the normal LMIS, shows how these relate to underlying physics, and shows why the gallium normal LMIS, operated at around 2 μA, is usually the best LMEHD option for FIB machine use.

There are several books [60; 73; 140; 141] on the LMIS and FIB machine applications, and LMIS theory is covered in several review articles (e.g., Refs. 142 and 143). Reference 46 covers, in more detail than is relevant here, most of the material useful for this chapter. This section aims to provide basic information and explain principles, often leaving the detail to Ref. 46 or other sources.

7.2 CURRENT-VOLTAGE CHARACTERISTICS

The LMIS is driven by a small, negative hydrostatic pressure at the jet tip, which makes liquid ionizant flow toward the tip, where FEV occurs. This negative pressure exists because, at the tip, the electrostatic "Maxwell" stress pulling outward on the liquid surface is slightly greater than the surface-tension stress pulling inward.

In ion emission mode, a normal LMIS has experimental current-voltage characteristics as follow. (1) It turns on at an *onset voltage* V_{on}, and turns off at an *extinction voltage* V_x usually slightly lower than V_{on}. (2) There is an associated "minimum steady emission current" i_x, called here the *extinction current*. (3) Above V_x, the emission current i is nearly linear as a function of voltage V, and is usually well described by Mair's equation (see below). Figure 1.14 shows experimental data [144] for a normal (low-drag) Ga LMIS and—for comparison—for a high-drag Ga LMIS (though without details near the extinction voltage).

Measured emitter (i.e., anode) currents sometimes include components other than the ion emission current, in particular a component due to secondary electrons. In FIB machines a suppressor electrode, if present, will minimize this secondary-electron component. If no precautions have been taken, then true emission currents may be less than measured currents by a factor as much as 2.

Extinction voltage

The simplest way to estimate the extinction voltage V_x is to estimate the *zero-current static collapse voltage* V_c, as follows. If i is taken as zero at the point of collapse, then no FEVSC effects operate. The LMIS cone can then be treated as a body of liquid in which the internal pressure is effectively zero, and on which two main forces act. The total electrostatic (Maxwell) force is along the cone axis and acts to pull the body of liquid away from the needle. The assumed absence of space-charge means that this force is given by the Laplace force $f_{L,x}$ on the body. The *slender body approximation* introduced by Van Dyke [145; 146] in association with Taylor's work [147; 148] gives this as:

$$f_{L,x} = (4\pi\, \varepsilon_0/k_{VD})V_x^2, \tag{47}$$

where k_{VD} is a dimensionless constant used by Van Dyke. k_{VD} is dependent on apparatus geometry, and lies in the range:

$$\ln(2L_{sb}/r_c) < k_{VD} < \ln(4L_{sb}/r_c), \tag{48}$$

where L_{sb} is the effective length of the slender body, and r_c is its radius.

This electrostatic pull is opposed by surface tension forces between the liquid body and the needle. At extinction the liquid-body shape approximates well to a Taylor cone, thus the magnitude $f_{st,x}$ of the resulting force component along the cone axis is:

$$f_{st,x} = 2\pi\, r_b \gamma_0 \cos \phi_T, \tag{49}$$

where r_b is the cone base-radius, γ_0 is the liquid's surface free energy per unit area, and ϕ_T is the half-angle of a Taylor cone ($\phi_T \cong 49.3°$). Hence the liquid body will collapse if the applied voltage is less than the static collapse voltage:

$$V_c = (k_V r_b \gamma \cos \phi_T/2\varepsilon_0). \tag{50}$$

Values of V_c derived using Equation 50 are compatible with experiments on LMIS extinction voltage [149].

This total-force argument is analogous to Taylor's [148] explanations of the collapse of water-cones and soap films. However, analyses [150–153] of the origin of LMIS extinction current suggest there may be an alternative, "steady-steady dynamic," explanation of LMIS extinction. This leads to the prediction of a *steady-state-dynamic collapse voltage* V_d slightly different from V_c. Further, due to hydrodynamic fluctuations, one might expect observed extinction voltages V_x to be slightly higher than either V_c or V_d. However, all these voltages are expected to be close; thus, V_c should (and does [137; 149]) provide an adequate estimate of V_x.

Numerically successful predictions of the onset voltage V_{on} have been made [136; 149] using a local-stress-based criterion equivalent to that discussed in Section 2.7, and a needle model that gives apex field as a function of applied voltage. As noted in Section 2.7, this approach is not formally correct. It presumably works because the criterion is "near enough" to give only small error in the estimation of onset field and voltage.

Mair's equation

Mair's equation [154; 155] describes the i-V characteristics of a low-drag LMIS; it applies to the capillary LMIS and the normal LMIS. Its derivation assumes that, for voltages V above V_x, one must consider a further force on the liquid emitter, due to FEVSC. The force component needed to give momentum to the emitted ions is very small, so in equilibrium the force due to FEVSC almost exactly cancels the increase in Laplace force associated with the voltage increase. A detailed discussion [46] yields:

$$i = (C_M r_b V_x^{-1/2}/2)[(V/V_x)^{3/2} - (V/V_x)^{1/2}]. \tag{51}$$

Here, C_M is a material-specific constant, with:

$$C_M = 3\pi \, (2ze/m_i)^{1/2} \gamma \cos \phi_T, \tag{52}$$

where m_i is the mass, and ze the average charge, for an emitted ion. Table 1.10 lists C_M-values for elements of interest. For large V, i becomes proportional to $V^{3/2}$; this is the signature of space-charge-limited current.

It is a known result that current-voltage equations applying to space-charge-limited emission regimes do not involve the detailed theory of the emission mechanism (in this case FEV). This is why Mair's equation can be derived without involving the fine details of FEV theory.

On writing $V = V_x + \delta V$ in Equation 51, binomial expansion gives Mair's equation in its usual form:

$$i = C_M r_b V_x^{-1/2}[(V/V_x) - 1]. \tag{53}$$

Near extinction, di/dV is predicted to depend only on known constants and the measurable parameters r_b and V_x. Mair [155] tested this by using experimental di/dV values to plot $[(di/dV)/(C_M V_x^{-3/2}]$ against r_b, for gallium, indium and caesium low-drag sources. The result [155] is a convincing straight line, over the r_b-range 1 to 300 μm. Notwithstanding the unfamiliarity of the physics, this theory certainly works. Figure 1.14 is another illustration of the fit obtainable between theory and experiment.

The effect of flow impedance

Experimentally, it is found that flow impedance between the reservoir and the cone base (assumed zero in low-drag theory) reduces the slope of the i-V plot [137], and thus the emission current for given voltage, as shown in Figure 1.14. Detailed theory is given in References 156 and 46. Reference 46 shows that for a low-drag source the "opposing force" (that counteracts the Laplace force) is due to FEVSC above the liquid tip, whereas for a high-drag source the opposing force is the viscous force between the ionizant and the needle shank, and is transmitted through the liquid. Thus, it is no surprise that high-drag sources are hydrodynamically less stable than low-drag sources.

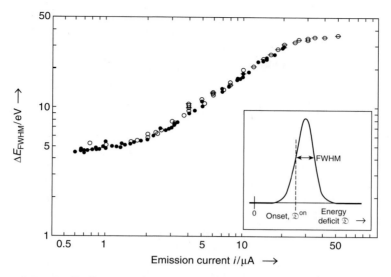

FIGURE 1.15: To illustrate how the energy spread (FWHM) of a gallium LMIS varies with emission current. The inset shows a typical measured ion energy distribution for a LMIS, plotted as a function of energy deficit, and shows "onset" as defined experimentally by a "half-maximum" criterion.

Temperature effects

The main effect of temperature (T) is to change the value of γ_0. Jousten et al. [157] used this to modulate emission current with a chopped laser beam. Mair gave relevant theory [158].

7.3 ION ENERGY DISTRIBUTIONS

The inset to Figure 1.15 shows a notional TED for ions of charge ξe emitted from a normal LMIS. (For gallium $\xi = 1$.) The horizontal axis plots the measured *ion energy deficit* $\mathfrak{D} = \xi eV_s$, where V_s is the voltage difference ($V_{emitter} - V_{collector}$) needed to bring an ion of charge ξe to a halt just outside a collector of work-function ϕ_c. Two distribution features are of interest: the onset energy deficit and the FWHM.

Onset energy deficits

When an ion of charge-state ξe is created at the emitter surface by transferring its electrons directly to the emitter Fermi level, its initial $\mathfrak{D}$-value is the *onset energy deficit* $\mathfrak{D}^{on}$ given by [159]:

$$\mathfrak{D}^{on} = \Lambda^F + \mathfrak{H}_\xi - \xi\phi_c - Q, \tag{54}$$

where Λ^F is the atomic bonding energy with the field present, $\mathfrak{H}_\xi$ is the sum of its first ξ ionization energies, ϕ_c is the collector work-function, and Q is the FEV activation energy (in practice, $Q \approx 0$). For Ga$^+$ this predicts $\mathfrak{D}^{on} \approx 5$ eV.

Figure 1.15 shows the onset as given by a "half-maximum" criterion applied to the high-energy (low-deficit) distribution edge. As an ion moves, SCPPI effects randomly change its energy; thus, the observed peak energy deficit (or something between this and the "half-maximum" deficit) may be a better experimental estimate of $\mathfrak{D}^{on}$. If, as is the case with Ga^+, the peak has a measured deficit close to the prediction for $\mathfrak{D}^{on}$, then this shows that ion formation was a surface process. Measured deficits significantly higher than $\mathfrak{D}^{on}$ indicate either ion formation in space away from the emitter, or energy loss in interactions during motion.

Energy spreads

The FEV natural FWHM is around 1 eV [160; 161]. For a normal LMIS, ΔE_{FWHM} is always much greater (usually at least 5 eV for Ga^+). For some (but not all) ionizants, the variation of ΔE_{FWHM} with i, when plotted logarithmically, has the S-shaped form shown (for Ga) in Figure 1.15. The linear middle region, with $\Delta E_{FWHM} \sim i^\sigma$, where [160; 162] $\sigma \approx 0.7$, is associated with the *most-steady emission regime*.

This regime (which starts for Ga at about 2 μA) has attracted much theoretical interest [61; 77; 78; 163–166]. Accepted thinking is that ΔE_{FWHM} is determined by SCPPI effects close to the emitter, and should have the approximate functional dependences:

$$\Delta E_{FWHM} \sim (mi^2)^{1/3}T^{1/2}. \tag{55}$$

Dependence on a power of the product (mi^2) is expected on dimensional grounds [164], and is found experimentally for Ga, In, Cs, Al, and Li (see Ref. 167 or 46); dependence on the power-value 1/3 is an interpretation[g] of Knauer's work [77; 78]. Clearly, for Ga the result $\sigma \approx 0.7$ is consistent with Equation 55.

As noted, low-drag sources turn off below the extinction current (around 0.45 μA for Ga). Below the most-steady regime, down to extinction, there is an *intermediate regime* where the ΔE_{FWHM} vs i curve flattens, and measured peak energy deficits exhibit anomalous behavior [167]. This regime is not properly understood.

The upper part of the S-curve has a different power-dependence of ΔE_{FWHM} on current. For ionizants with a most-steady regime, there is a critical current [168] i_u (around 25 μA for Ga) at which σ changes from near 0.7 to near 0.3 to 0.4. Above i_u, current noise rapidly increases, probably [168] due to droplet emission. These features mark a change to an *upper unsteady regime*. Mair [168] argues this is linked to growing jet instability. He also argues that some ionizants (notably Au, Sn, and Bi), with σ between 0.35 and 0.39, have no most-steady regime, only an upper unsteady regime.

Distribution shape

The main peak shape was initially assumed Gaussian. Ward et al. [169] suggested representing it by a Holtsmark distribution, because a tail of high-energy-deficit ions was found in experiments [170]. SCPPI effects cannot be excluded, but it is more likely that high-deficit ions are created by free-space field ionization of neutrals [171], and/or by energy-exchange and/or charge-exchange between a fast primary ion and a slow neutral. Analogous charge-exchange events occur when a FIM operates at high background gas pressure [61; 92].

Even when secondary electrons are diverted, a Ga LMIS tip emits light. Spectral analysis shows that it comes from de-excitation of excited Ga neutral atoms [161; 172; 173]. This means that (a) neutral Ga atoms are present in space above the liquid tip, (b) they are excited by the primary ion beam, and (c) primary ions must have corresponding energy losses.

Temperature effects

Raising emitter temperature significantly broadens the energy distribution, and also (for Ga) causes it to develop into a double-hump [74] (or triple-hump [174; 175]) distribution. Reference 46 discusses related physics. For FIB applications, a LMIS should be operated as close above the ionizant melting point as practicable.

7.4 TIME-DEPENDENT PHENOMENA

Measured $i - V$ characteristics are time-averaged characteristics. Particularly at higher currents, LMIS operation can be affected by various EHD instabilities, notably by several modes of droplet emission. In some circumstances, part or all of the liquid jet (and maybe part of the underlying cone) may break off, creating nanodroplets and/or microdroplets. The cone-jet then re-forms and ion emission re-starts. An operating LMIS can also emit liquid globules from the back end of the liquid cone and/or the supporting needle [176]. Some LMIS types, but not the normal LMIS, have a low-emission-current *pulsation regime* [46]. Reference 46 gives further details. Because a LMIS is an EHD device, it is not surprising that these phenomena exist. Their theory is complex, and not fully understood.

Where possible, FIB applications should use an ionizant with a most-steady regime, and work close above the regime's lower boundary. Usually, a Ga normal LMIS, operating at emission currents near 2 μA, is relatively little affected by unwanted EHD phenomena.

7.5 IONIZATION EFFECTS (ELEMENTS)

FEV-produced atomic ions

As noted in Section 2.8, FEV may produce atomic ions in a mixture of charge-states. Reference 46 contains details, references and tabulated results. For many applications, an ionizant that generates only (or predominantly) singly charged ions is best. A simple test calculates:

$$t_1 = (F_2^{\mathrm{M}}/F_1^{\mathrm{M}}) - 1. \tag{56}$$

where F_1^{M} and F_2^{M} are the Müller escape fields for singly and doubly charged ions (see Section 2.8). If $t_1 > 0.5$, the ions produced should be mainly singly charged. As shown in Reference 46 (and partly in Table 1.10 here), this test selects the Group 1 elements Li, Na, K, Cs, Rb (mostly too reactive for ordinary LMIS use) and the Group 13 elements Al, Ga, In, Tl (this list includes all elements recognized as good emitters of singly charged atomic ions). Clearly, this is a Periodic System effect, as the elements selected have a closed sub-shell and one extra electron. This test also picks out Ag, Bi and Se.

Other atomic ion generation mechanisms

Other mechanisms [46] that can generate atomic ions are the free-space field ionization of neutral atoms, collision processes involving a primary-beam ion, and fragmentation of cluster ions or charged droplets. If formed near the optical axis, the resulting ion may enter the column, but with energy much less than primary-beam ions, as noted earlier. Ideally, LMIS operation should be in conditions where these mechanisms have low probability. A reason for requiring low ionizant vapor pressure is to keep down the space concentration of neutrals, and hence inhibit the first two mechanisms.

Cluster ions

Some liquid elements emit cluster ions; often, many different types are emitted together [46]. These cluster ions may result from nanoscale surface processes, or may be generated by a gas-phase mechanism and/or by fragmentation in flight of larger clusters or liquid droplets (possibly induced by ion impact). Cluster ion formation causes problems when the need is for a tightly focused beam of specific ionic composition; thus, ionizants that generate many cluster ions are not suitable for standard FIB applications. Gallium generates very few cluster ions.

7.6 LIQUID ALLOY ION SOURCES

There is continuing interest in generating beams of specific ions (e.g., Dysprosium [177]) for specialized purposes. When an elemental source is impracticable, a liquid alloy ion source can often be made. (The FIB machine then needs an appropriate filter.) Solutions have long existed for the main semiconductor dopants [178–180], but many additional species are now available. Information is easily obtained by typing "Liquid metal ion source LMIS" "element" "chemical symbol" into a web search engine. At the last check, results were found for the following 39 elements: {Ag, Al, As, Au, B, Be, Bi, Ce, Co, Cr, Cs, Cu, Dy, Er, Ga, Ge, Hg, In, K, Li, Mg, Mn, Na, Nb, Nd, Ni, P, Pb, Pd, Pr, Pt, Rb, Sb, Si, Sm, Sn, U, Y, Zn}.

7.7 ION OPTICAL CONSIDERATIONS

To get small probe radius in an FIB machine, the source optical radius ρ_v should ideally be small. This would avoid any need to make the post-source magnification M_{ps} small ($\ll 1$)—thereby increasing aberrations (see Ref. 60, p. 96). In practice, with modern FIB machines, the largest terms on the right-hand side of Equation 14 are the source optical radius, the chromatic aberration term, and the term ρ_{PPI} relating to post-source SCPPI effects.

As noted in Section 3.6, post-source SCPPI effects can be reduced by putting the physical aperture as close to the extractor as practicable. Chromatic aberration is minimized by operating with ΔE_{FWHM} as low as practicable. This is achieved by operating a LMIS at as low an emission current as is practicable.

One can estimate the LMIS tip physical radius by taking the apex field equal to F^E (see Section 2.8), and assuming the pressure inside the liquid tip to be close to zero (see Ref. 46). This yields:

$$r_a \sim r^E = 4\gamma_0/\varepsilon_0(F^E)^2. \qquad (57)$$

where r^E is the apex radius given by Equation 57. For gallium, $\gamma_0 \approx 0.72$ J/m^2 and $F^E \sim 15$ V/nm, which yields $r^E \sim 1.4$ nm. This is comparable with a jet cross-sectional radius (r_c) measurement made with a high resolution transmission electron microscope [181], and is consistent with an intrinsic optical radius (< 1 nm) derived by ray tracing [182] that is based on a numerical LMIS analysis that disregards SCPPI effects.

However, measurements [76] of LMIS optical radius, made (for Ga) using ion optical techniques, yielded the much larger value of $\sim$25 nm. Although "jet wobble" has been suggested as a cause [183], the more plausible (and generally accepted) explanation is that this large value results from trajectory displacement (see Section 3.8). This is consistent with the high FEVSC strength ζ predicted for a LMIS (see Section 2.9), the SCPPI-based explanation of the anomalously large FWHM (see Section 3.8), and Monte-Carlo-based simulations [166; 184; 185] of LMIS optics. Recent calculations, by Radlicka and Lencová [166], predict a source optical radius of $\sim$25 nm, which matches the experimental result. A LMIS turns off or becomes hydrodynamically unstable if one attempts to reduce SCPPI effects by significantly reducing emission current. Thus, it seems impossible to make any large reduction in LMIS source-module SCPPI effects.

Typical ion optical properties for a LMIS are summarized in Table 1.7 in Section 9.

7.8 TECHNOLOGICAL ISSUES

Operating conditions

LMIS technology is well established, with details in the literature. This section notes a few specific points. For reliable operation, source contamination must be avoided. Gallium oxidizes in air, so background gas pressure nearby should be less than about 10^{-8} Torr. Most ionizants do not need fully bakeable systems, but these are essential for reactive ionizants such as Cs.

Source-module design must take account of LMIS metallurgical properties. Sputtering occurs when the beam hits electrodes surrounding the emitter [186]; thus, electrode materials should be chosen to minimize side effects from the resulting source contamination [187]. Contamination can create various difficulties [187], including increase in ionizant melting temperature, increase in hydrodynamic flow resistance, and (especially with alloys) chemical and metallurgical effects that may (for example) change the emitted beam composition beam and/or attack the support needle. All these tend to reduce source lifetime. When contamination has occurred, normal operation can sometimes be restored by heating or by overvolting the source for several seconds.

Secondary electrons

Primary-ion impact onto the aperture or other electrode generates secondary electrons. If allowed to strike the needle shank and supporting structure, these cause heating, and/or excite surface atoms that subsequently emit light and/or X-rays, and/or cause error in emission current measurement. Apart from this, there seems little evidence that secondary electrons affect normal LMIS operation at low emission currents, although this has been disputed [188–190]. Commercial source-modules may include a suppressor electrode to inhibit secondary-electron effects.

8 GAS FIELD IONIZATION SOURCES

8.1 INTRODUCTION

LMIS-based systems are good for nanofabrication but less suitable for scanning ion micro-scopy (SIM). This is partly because heavy ions cause erosion during imaging, but also because emitted droplets can cause unwanted contamination. Incompatibility between milling and imaging was a reason for developing the so-called *dual-beam machine*, which has an ion beam for milling and an electron beam for observation.

This SIM gap has recently been filled by the *helium scanning ion microscope*[h] (HeSIM) [191], based on a new form of helium gas field ionization source[i] (GFIS). A GFIS has the same emission mechanism as a FIM. The idea of a GFIS is not new, but has been re-vitalized by progress in emitter preparation, and by the development of the proprietary *atomic level ion source* (ALIS™) [34; 35; 191; 192]. For SIM, an obvious advantage is that He^+ ions are less destructive than metal ions.

ALIS™ fabrication starts from a rounded (111)-oriented tungsten emitter, and uses TF sharpening to form its tip into a three-sided pyramid that ends in three atoms. The working GFIS beam is drawn, by aperturing, from the central area of one of the three atomic-level emission sites.

Atomically sharp field ion emitters are not new—Fink [123; 193] made an emitter ending in a single atom in 1986. But all such emitters may have a finite working period, terminated when an apex atom leaves its bonding site. The clever features of the ALIS™ are that an in-situ process (apparently chemically assisted) can restore a damaged emitter to its original state, and that this is set up as a "push-button" computer-controlled process. ALIS™ procedure details are proprietary; information here is deduced from patents [34; 35].

An LMIS has physical emitter radius at least 1400 pm. The total emitting area of an ALIS™ is much less—three neighboring atoms on a W(111) facet, giving a physical radius less than 480 pm. This would make space-charge effects (if any) less significant. Also, GFIS physics imposes no minimum emission current. Thus, trajectory displacement effects can be made negligible, and there is no problem of large source optical radius. A carefully designed GFIS may be closer to the perfect point ion source than any competing technology.

For some materials applications, the HeSIM already seems superior to scanning elec-tron microscopy (SEM) [194]. There is potential for improving the HeSIM's resolving power, and the technique is attracting growing interest. Also, GFI sources based on other gases may be possible. For some nanofabrication applications, an FIB machine based on a reliable neon GFIS might provide an interesting alternative to an LMIS-based machine.

Past GFIS development has explored various alternatives (see Ref. 61). GFIS literature discusses mainly technology and characterization. There is some basic work, notably that of Kalbitzer and colleagues (e.g., Ref. 195), but much relevant science is in the older (1960s and 1970s) literature of FIM imaging, and is sometimes overlooked.

This foundational work is summarized in a recent book chapter. Reference 61 covers, in greater detail than relevant here, most of the material useful for the present chapter. Thus, this section aims to provide basic information and explain the main issues involved in GFIS physics and operation, often leaving the detail to Reference 61 or other sources, but providing updates where appropriate.

8.2 GAS FIELD ION EMISSION BASICS

The GFIS and LMIS emission mechanisms differ in two main ways: the GFIS ionizant is in a gas phase close above a solid surface, rather than strongly bound to a liquid surface; and the GFIS ions are formed by field ionization (FI) very slightly above the surface, rather than by FEV at the surface.

With the noble gases normally considered, the ionizant arrives as neutral atoms. Most atoms undergo *surface field ionization* when the gas-atom nucleus is in a thin *ionization layer* just outside a *critical surface* slightly above the needle's tip (about 500 pm above the edge of the metal atom, for the helium-on-tungsten system). The ionization layer thickness is of order 10 pm. In the layer there are strong variations, across the surface, in the generated ion current density. These relate to the existence of disc-like zones of relatively intense ionization centered above the nuclei of metal atoms protruding from the emitter surface. From each such *ionization zone* a narrow ion beam (sometimes called an "ion beamlet" emerges. Figure 1.16 illustrates this process.

Between the critical surface and the emitter surface there is a *forbidden zone* in which no direct FI can occur, because there are no unoccupied emitter states for electrons to tunnel into. For emitter temperatures less than about 100 K, this zone contains a layer of immobile *strongly field-adsorbed* gas atoms. The gas atoms subject to ionization bounce on top of the field-adsorbed layer, which has a role in cooling down incoming gas.

The current density (current per unit area) J_A associated with FI at a point "A" in the critical surface (and at related positions close above A) is given by:

$$J_A = (e/n_1)C_{G,A}P_{e,A}\delta_A. \tag{58}$$

Here: $C_{G,X}$ (called the *gas concentration* at point X) relates to the probability per unit volume of finding the gas atom nucleus very near point X ($C_{G,X}$ is obtained by integrating over all velocities, and is measured in "atoms per unit volume"); $P_{e,X}$ (called the *electron tunneling rate-constant* at point X) is the rate-constant for FI of an atom with its nucleus at point X; and δ_A is a decay-length that indicates how quickly the product $C_G P_e$ falls off with distance outside the critical surface, for point A. The constant n_1 is included for dimensional consistency, and is best read as "1 atom."

The decay length δ_A varies little with position in the critical surface, so the current-density variations across the emitter surface (which lead to FIM images such as Figures 1.10a,b) are mainly determined by the variation, across the emitter surface, in the critical-surface values of

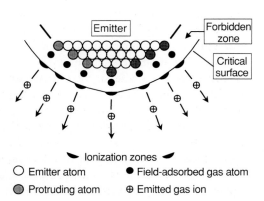

FIGURE 1.16: Schematic diagram illustrating the physical situation at a gas field ion emitter tip. Atoms of the operating gas (not shown) are weakly bound to the emitter by the long-range polarization potential energy well. The atoms "bounce about" on the atomic structure shown, and can be field ionized when a nucleus enters an ionization zone. The firmly field-adsorbed atoms are present only at temperatures below about 100 K.

the product $C_G P_e$. Except at very low gas temperatures, P_e is the dominant factor in this "local contrast" that allows resolution of individual atomic emission sites in FIM images.

The rate-constant P_e can be written formally as:

$$P_e = a_{FI} \exp[-\nu b I^{3/2}/F], \tag{59}$$

where b is the second FN constant, as before, I is the gas-atom ionization energy [$\cong$ 24.587 eV for He], a_{FI} is an "attempt frequency" usually taken as 10^{15} to 10^{16} s^{-1}, F is the local field at the position of the gas-atom nucleus, and ν ("nu") is a correction factor for the tunnelling exponent. ν is around 0.6, but is a function of F and is a sensitive function of position, both across and normal to the surface. These ν-variations determine both FIM resolving ability [26] and the fall-off in P_e with distance outwards from the critical surface. (In normal circumstances, this fall-off determines the energy spread of the emitted ions). A distinctive feature of the ALIS™ is that F and P_e are relatively high above the emission sites, as compared with other sites on the needle tip.

8.3 CURRENT-VOLTAGE CHARACTERISTICS

A GFIS operates in an enclosure filled with ionizant gas at low pressure. If the emitter and the enclosure walls are at the same temperature, then the gas distant from the emitter is in thermodynamic equilibrium and has a well-defined "background" pressure p_{bk} and temperature T_{bk}. The corresponding *background gas flux density* Z_{bk} (count of gas atoms crossing unit area per unit time) is given by:

$$Z_{bk} = n_1 (2\pi \, m_G k_B T_{bk})^{-1/2} p_{bk}. \tag{60}$$

where m_G is the gas atom mass. For convenience, an *equivalent background current density* J_{bk} can be defined by

$$J_{bk} = (e/n_1) Z_{bk} = e(2\pi \, m_G k_B T_{bk})^{-1/2} p_{bk} \equiv a_G p_{bk}, \tag{61}$$

where a_G is defined by Equation 61. For helium at $T_{bk} = 78$ K, a_G has the value 2.39×10^4 A m^{-2} Pa^{-1} ($\equiv 3.19 \times 10^6$ A m^{-2} Torr^{-1}); at $T_{bk} = 300$ K, $a_G = 1.22 \times 10^4$ A m^{-2} Pa^{-1} ($\equiv 1.63 \times 10^6$ A m^{-2} Torr^{-1}).

If the enclosure walls and the emitter are at different temperatures, then effective values of Z_{bk} and J_{bk} can be defined, and will be approximately given by these equations, with T_{bk} set equal to the wall temperature (or to some "effective wall temperature," if different parts of the enclosure are at different temperatures).

Note that, even without the temperature differences just discussed, the interpretation of pressure–gauge readings can sometimes be problematic.[j]

For given emitter temperature T_e (possibly different from T_{bk}), the total emission current i_{tot} from the whole emitter is proportional to J_{bk}, and hence to pressure p_{bk}. For given p_{bk} in the GFIS operating range (less than about 0.01 Pa, 10^{-5} Torr), measured total-current vs voltage (V) characteristics depend on the emitter temperature T_e and have the general form [196; 197] illustrated in Figure 1.17. The apex field F_a is proportional to V; thus, Figure 1.17 is (with suitable change in the horizontal axis) also an i_{tot} vs F_a characteristic.

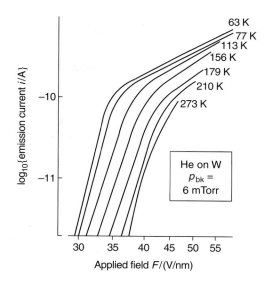

FIGURE 1.17: Current/field characteristics for the helium-on-tungsten system, at various emitter temperatures T_e. (Measured by Southon [196; 197], in conditions where the background gas temperature T_{bk} was greater than T_e).

For the temperatures near 80 K at which the ALIS™ is normally operated, two emission regimes exist, as shown in Figure 1.17. The higher-field regime (with the smaller logarithmic slope) is called the *supply-and-capture* (SAC) *regime*. A GFIS is normally operated near the SAC regime lower boundary, near an F_a-value called here the *best source field* (BSF). A helium ALIS™ uses a (111)-oriented W emitter, and the BSF is thought [61] to be in the range 40 to 45 V/nm. This is effectively the same as the "best image field" for a conventional (110)-oriented W emitter as used in FIM.

With a GFIS, probe current can in principle be changed by adjusting the applied voltage, but this will normally degrade optical performance. Therefore, a GFIS should be operated at a fixed voltage (*the best source voltage* [BSV], which corresponds to the BSF), and probe-current changes should be made by altering background gas pressure p_{bk}.

There is an upper limit to the gas pressure that can gainfully be used, owing to the onset (away from the emitter) of charge-exchange collisions between emitted ions and neutral gas atoms. This would create, in the beam entering the column, "slow" (high-energy-deficit) ions that could not be focused. For a GFIS operating near 80 K, appreciable numbers of charge-exchange events begin to occur at background pressures above around 1 mTorr (around 0.1 Pa). Since the incidence of these events goes as $(p_{bk})^2$ the problem can be avoided by using p_{bk} values well below this. The ALIS™ is usually operated [191] at He pressures below 0.01 Pa; thus, in normal operation there is no problem.

8.4 GAS SUPPLY EFFECTS

The long-range polarization potential energy well

In a field F, a neutral gas atom has a polarization potential energy (PPE) $U = -\frac{1}{2}\alpha_G F^2$, where α_G is the gas-atom polarizability (for helium, $\alpha_G = 0.143$ meV V^{-2} nm^2). At a BSF of 45 V/nm, a helium atom coming directly from space hits the emitter tip with kinetic energy ~0.14 eV. For background gas temperatures between 80 K and 300 K, this is much greater than the equipartition energy $\frac{1}{2}k_B T_{bk}$ (0.0035 eV to 0.013 eV). Thus, after its first bounce on the rough emitter surface, the incoming atom will normally be trapped in the long-range PPE well around the end of the needle, and will then lose energy in further bounces. Similar arguments apply to the part of the emitter shank where the local surface field F_s is high enough.

The PPE well extends into space by several or many tip radii (depending on emitter shape and gas temperature), and has complex local structure close to the emitter surface. The well extends a considerable way down the needle shank. In the well, the neutral-atom concentration is higher than the background value, and increases toward the needle tip.

Typical neutral-gas-atom history

Supply of gas to the GFIS ionization zones is taken to be the same steady-state kinetic process as in FIM [26]. In the SAC regime, the gas near the ionization zones is not in thermodynamic equilibrium with the background gas distribution. In these conditions, the typical neutral-gas-atom history has three main stages—*capture, accommodation,* and *diffusion*—as follows. (1) The gas is captured mainly on the needle shank, and then moves to its tip, heating up as it does so, because it gains kinetic energy from the PPE well. (2) This hot trapped gas then cools, by transferring kinetic energy to the emitter when the atoms bounce, and accumulates into the higher-field regions above the needle tip. (3) As the gas becomes fully accommodated to the emitter at temperature T_e, across-surface diffusion takes place relatively close to the surface, and gas concentrations build toward those characteristic of a thermodynamic equilibrium across the needle tip as a whole. This account of gas behavior is derived from detailed analysis of voltage and temperature dependences in FIM images [26; 92; 198; 199].

The apex-field value F_a controls the mean FI rate-constant, and hence determines the instant in the history at which FI occurs. Thus, the value of F_a determines the distribution of the gas, in space and in energy, at the instant of ionization—and hence the properties of the emitted ion distribution.

Best source field

The GFIS best source field (BSF) is the apex field at which the probe can be focused most sharply, when other controls are set to "best focus." At the time of writing, it is not definitively known what the corresponding gas condition at ionization is. The most plausible hypothesis is that setting the BSF experimentally is a balance between: (a) choosing a field value that allows the gas time to cool to emitter temperature (or near this), and to accumulate into the vicinity of the high-field apex region without suffering prior field ionization at some other position on the emitter surface; and (b) lowering the field to the point where a significant fraction of captured atoms escape un-ionized. Probably, this trade-off would be optimized if FI occurs near the boundary between the "accommodation" and "diffusion" stages. This implies that, at the instant of ionization, the average transverse velocity κ_{av} of a gas atom would be $k_B T_e$ or slightly above. Allowing for differences between the (111)-oriented ALIS™ emitter and the conventional (110)-oriented FIM emitter, Reference 61 concludes that (until better knowledge exists) the helium ALIS™ BSF can be taken as 45 V/nm.

Gas supply theory

In the SAC emission regime, nearly all gas atoms captured by the PPE well are field-ionized at the emitter tip. An empirical parameter, the *effective capture area* A_c, is defined by:

$$i_{SAC} \equiv A_c J_{bk} = A_c a_G p_{bk}, \tag{62}$$

where i_{SAC} is the total emission current. The parameter A_c assesses the ability of an operating field ion emitter to collect gas atoms and supply them to the active emission sites. A_c depends on apex field F_a and emitter temperature T_e, and will also depend on background gas temperature T_{bk} if different from T_e. A_c also depends on emitter shape parameters (such as the shank half-angle) and on the condition of the emitter shank (clean, adsorbate-covered, etc.).

If a Müller emitter tip is nearly hemispherical, effectively with surface area $2\pi r_a^2$, then the *total captured-flux enhancement factor* σ_{Mc}, is defined by:

$$\sigma_{Mc} = A_c/(2\pi r_a^2). \qquad (63)$$

Experiments show that A_c is much greater than the tip area $2\pi r_a^2$ (i.e., $\sigma_{Mc} \gg 1$). For example, the results of Feldman and Gomer [200], for an 80 nm apex-radius emitter, with $F_a = 45$ V/nm, $T_{bk} = T_e = 78$ K, yield $A_c \sim 1$ μm², $\sigma_{Mc} \sim 25$. Due to the complexities and uncertainties of the physical situation, accurate theoretical prediction of A_c is impossible. However, gas supply theory [61] can support the qualitative assertions made earlier, and generate illustrative values.

ALIS™ ion currents

Historically, a GFIS development difficulty was to produce sufficient ion current (in the right place) without having the gas pressure too high. An ALIS™ advantage is that (ideally) it has only three active atomic-level emission sites. Thus, the whole collected gas supply is funneled out as ions through these sites. A fraction of the current from one site is enough to operate a SIM. Ward et al. [191] report that an ALIS™ can generate currents as high as 800 pA, though normal operation apparently uses much smaller probe currents, typically 0.5 to 1 pA for best resolution [201].

There is scope for improving gas supply theory, and for experimental research into how to make emitters with high A_c, but there seems little need at present for this kind of optimization.

8.5 ION ENERGY DISTRIBUTIONS

For a GFIS He⁺ ion energy distribution, several factors influence the FWHM, including apex field, gas temperature, tip radius, and the local environment. For emission near FIM best image field, from a single-atom site in a W(111) facet on a moderately-sharp Müller emitter, Ernst et al. report [202] an FWHM of 414 meV at $T_e = 79$ K, 747 meV at $T_e = 201$ K. For a "sharply sputtered (111)-oriented W emitter," they found 367 meV at $T_e = 79$ K. Values as low as 285 meV have been found [202], but the "sharply sputtered" result seems most appropriate to the ALIS™. This suggests that, when operated at BSF in its "safe" pressure range, with cooling to near 80 K, a helium ALIS™ would have FWHM around 350 meV. Direct measurements [191] gave an FWHM near 1 eV, but the energy analyzer had limited resolution.

8.6 ION OPTICAL CONSIDERATIONS

At present, ALIS™ optical behavior is not finely characterized experimentally, and the theoretical optics of this emitter shape is not known in detail. The limited aims here are to assess whether a HeSIM is limited by the source-module or the post-source optics, and to identify the limiting factors.

The best method of finding accurate optical parameters for an ALIS™ would be by backward ray tracing after numerical simulation of ion trajectories (as done by Wiesner and Everhart [83]), using a realistic emitter shape model. This has not yet been done.

Provisionally, one can use the H&K-type approach described in Section 4 and Reference 61. Thus, one first examines a model that notionally fits a sphere to the ALIS™ emitter, and then considers the effect of its detailed shape. For modeling, Reference 61 chose extraction voltage $V = 20$ kV and apex radius $r_a = 80$ nm; these values are used here. The k_M-value is 5.56, and (for an ideal Müller emitter with angular magnification $m = m'' = 1/\sqrt{3}$) the weak-lens transverse linear magnification $M'' = 0.735$.

Physical emission-site radius

The separation of the nuclei of adjacent W(111)-facet atoms is $s_{111} \approx 448$ pm, and the intensity distribution across an isolated FIM image spot is more like a rounded triangle than a Gaussian [203]. Thus, Reference 61 takes the emission site as having fall-off radius (FORD) $\rho_0 = s_{111}/4 = 112$ pm, the thinking being that roughly 50% of site current is within this radius.

Intrinsic optical radii

As noted in Section 4, a CP source based on a Müller emitter makes *two* virtual images of the physical emission site (the Gaussian image and the cross-over), Table 1.4 records how to derive relevant parameters, including the objects size ratio m_B. As shown earlier, the ratio ρ_2/ρ_1, of the radii of the cross-over (ρ_2) and the Gaussian image (ρ_1), is given by m_B.

For an electron-beam instrument, m_B is usually less than 1, and the operator usually focuses on the cross-over when minimizing probe radius. With a helium GFIS, m_B is much greater than 1 at normal operating temperatures. Because of column aberrations (see below), the optics of GFIS probe-size minimization is not understood at the time of writing, but the numbers in Table 1.4 suggest the operator might attempt to focus on the Gaussian-type virtual object.

Source aberrations

To investigate GFIS geometrical aberrations for the source module, Liu and Orloff [204] have used the *sphere-on-orthogonal cone* (SOC) shape model [205] (known to be a good model for a Müller emitter), and simple extractor geometry. They calculated values of object-side spherical and chromatic aberration coefficients ($C_{s,sm,o}$, $C_{c,sm,o}$). Using their results, Table 1.5 predicts aberration disc radii of 0.7 pm for chromatic aberration (disregarding SCPPI effects), and less than 0.1 pm for spherical aberration. These results are less than those given in Reference 61, for the same emitter shape, because lower estimates are used here for the intrinsic GFIS energy spread ($\Delta E_{FWHM} = 0.35$ eV, rather than 1 eV) and the beam divergence angle ($\alpha_o = 0.2$ mrad, rather than 1 mrad). However, both the old and new disc radii are very small in comparison with the unaberrated optical radii in Table 1.6; thus, these aberrations can be disregarded for Müller emitters well represented by the SOC model (and probably also for a wider range of emitter shapes).

One should next assess the effect of the pyramid formed on the ALIS™ emitter. A simple-minded approach might argue that H&K-type theory can be applied, but with an apex radius

Table 1.4 Illustrative values predicted[a] for the optical parameters of a Müller emitter and its embedded spherical charged-particle emitter (SCPE), for helium gas field ionization. SCPPI effects are not part of these calculations.

Parameter	Symbol and derivation	For embedded SCPE		derivation	For Müller emitter	
		T_g = 78 K	T_g = 300 K		T_g = 78 K	T_g = 300 K
Angles		mrad	mrad		mrad	mrad
Emission site FOHA	$\alpha_S' = \alpha_0 = \rho_0/r_a$	1.40	1.40	$\alpha_S = m''\alpha_S'$	0.81	0.81
Blurring FOHA	$\alpha_B' = 2(k_BT_g/eF_ar_a)^{1/2}$	2.73	5.36	$\alpha_B = m''\alpha_B'$	1.58	3.09
Total FOHA	$\alpha_T' = \alpha_B' \otimes \alpha_S'$	3.07[b]	5.54[b]	$\alpha_T = m''\alpha_T'$	1.77[b]	3.20[b]
Radii		pm	pm		pm	pm
FORD of Gaussian image	$\rho_1' = \rho_0/2$	56	56	$\rho_1 = M''\rho_1'$	41	41
FORD of cross-over	$\rho_2' = \alpha_B'r_a/2$ $= r_a(k_BT_g/eF_ar_a)^{1/2}$	109	214	$\rho_2 = M''\rho_2'$	80	158
Magnifications and related parameters		–	–		–	–
Angular magnification	$m' = \alpha_S'/\alpha_0$	1	1	$m = m''m'$	0.578	0.578
Transverse linear magnification	$M' = \rho_1'/\rho_0$	0.5	0.5	$M = M''M'$	0.367	0.367
Objects size ratio[c]	$m_B' = \rho_2'/\rho_1'$ $= \alpha_B'/\alpha_S'$	1.95	3.83	$m_B = \rho_2/\rho_1$ $= \alpha_B/\alpha_S = m_B'$	1.95	3.83
Blurring magnification	$m_T' = \alpha_T'/\alpha_S'$	2.19	3.96	$m_T = \alpha_T/\alpha_S = m_T'$	2.19	3.96

[a]Estimates assume: SCPE of radius $r_a = 80$ nm; source apex field $F_a = 45$ V/nm; extraction voltage $V_o = 20$ kV; shape-factor $k_M = V_o/r_aF_a = 5.56$; real object disc of FORD $\rho_0 = 112$ pm, that subtends FOHA $\alpha_0 = \rho_0/r_a = 1.40$ mrad at the SCPE center; weak-lens magnifications $m'' = 1/\sqrt{3} = 0.578$, and $M'' = 1/m''k_M^{1/2} = 0.735$. T_g is the mean gas-atom temperature at ionization; the critical transverse kinetic energy at emission (κ_c) is taken as k_BT_g. Results are given to 3 sig. fig. to allow mathematical consistency to be checked; accuracy at this level is not implied.
[b]Estimated via quadrature formula.
[c]Called "blurring ratio" in Ref. 61.

Table 1.5 Contributions to source optical radius ρ_v for a GFIS source-module[a].

Contribution	Disc radius	Contribution	Disc radius	Contribution	Disc radius
Gaussian image	41 pm	Spherical aberration	< 0.1 pm	Chromatic aberration (SCPPI effects)	thought negligible
or Cross-over	80 pm	Chromatic aberration (without SCPPI effects)	0.4 pm	Trajectory displacement	thought negligible

[a]Assumes: emitter as analyzed in Table 2.4, for gas temperature $T_g = 80$ K; $\Delta E_{FWHM} = 0.35$ eV; $V_o = 20$ kV; and object-side data: $\alpha_o = 0.2$ mrad, $C_{s,sm,o} = 2.5$ mm, $C_{c,sm,o} = 200$ μm. α_o is obtained by assuming $\alpha_i = 0.3$ mrad, post-source magnification $M_{ps} = 0.6$, and $V_i/V_o = 3/2$. $C_{s,sm,o}$ and $C_{c,sm,o}$ are obtained by extrapolating the data in Ref. 204 to the case $r_a = 80$ nm, $F_a = 45$ V/nm, using the parameters $n = 0.16$, $\gamma = 3$ to describe emitter shape, and allowing for the different aberration-coefficient definitions used here and in Ref. 204.

Table 1.6 Predicted contributions from post-source effects to HeSIM probe radius[a].

Contribution	Disk radius pm	Contribution	Disk radius pm
Source Gaussian image	25	Chromatic aberration	55
Source cross-over	50	Diffraction	85
Spherical aberration	3	SCPPI	Assumed negligible
		TOTAL[b]	105 to 115

[a]Assumes: data in Table 2.5; post-source magnification $M_{ps} = 0.6$; $\Delta E_{gauss} \approx \Delta E_{FWHM} = 0.35$ eV; probe voltage $V_i = 30$ kV (hence helium atom wavelength $\lambda_i = 0.083$ pm); convergence angle $\alpha_i = 0.3$ mrad; aberration coefficients $C_{s,ps,i} = 300$ mm, $C_{c,ps,i} = 30$ mm (values suggested by J. Orloff—private communication).
[b]Determined by quadrature approximation.

intermediate between the 80 nm used above and an estimate of the physical radius of the top of the ALIS™ pyramid (which would be at most a few nm). This would not affect the radius estimated for the Gaussian virtual source, but would reduce the estimated radius of the cross-over-type virtual source (which behaves as $r_a^{1/2}$). Thus, we can provisionally take the Table 1.4 values as upper limits for the intrinsic source radii (before SCPPI effects are considered).

To assess the likelihood of source-module SCPPI effects, we estimate the FEVSC strength ζ for a GFIS (see Section 2.9). For He$^+$ ions $\kappa_{sc} \cong 1.627 \times 10^7$ A^{-1} V$^{3/2}$. To estimate the ECD J, assume that a current of 10 pA is drawn from a disc of radius 112 pm; this yields $J \sim 2.5 \times 10^8$ A/m^2. Assuming BSF 45 V/nm, and extraction voltage 20 kV, yields $\zeta \sim 0.00029$. This strongly suggests that SCPPI effects will not be significant (though possible energy broadening via single-file interactions [163] has yet to be explored). Thus, we provisionally take the Table 1.4 values of 41 pm and 80 pm as the estimates of source optical radius ρ_v for an operating GFIS. Both values are less than the 150 pm originally assumed by Ward et al. [191].

Fundamental limit on GFIS optical radius

Since the cross-over radius is temperature dependent, it can be reduced by operating at lower emitter and gas temperatures. At sufficiently low temperature the cross-over would be smaller than the Gaussian virtual object, and in operation an HeSIM might then be focussed on the cross-over.

It is of interest to know what puts a lower bound to the GFIS intrinsic optical radius. Reference 61 suggests a limiting factor is the zero-point energy associated with the transverse wave-function of the *helium-atom nucleus*, during field ionization. This puts a lower limit on the transverse KE at emission. (This effect also puts a minimum to the size of FIM image spots [92].) Using a guess (10 nm) at a plausible lower limit on the effective ALIS™ tip radius, a rough estimate [61] puts the lower limit on ρ_v as less than 20 pm. However, possible aberrations associated with the non-spherical shape of the ALIS™ tip need investigation.

Probe radius

To estimate probe radius ρ_p, the ρ_{PPI} term in Equation 14 is assumed to be negligibly small, and Equations 2.10 to 2.12 are used for other aberration disc radii, as described in Table 1.6. Equation 13 then yields the range 105 to 115 pm for ρ_p. The related diameters (210 to 230 pm) are encouragingly close to the value (240 pm) measured [206], using a 25% to 75% edge-rise criterion, for the ORION® helium ion microscope manufactured by Carl Zeiss SMT.

The aberration-disc radius predictions in Table 1.6 are provisional, because actual aberration-coefficient values for the ORION® machine are not publicly available. However, the Table 1.6 values seem illustrative of the performance to be expected from a HeSIM in its present state of development. The machine seems limited mainly by the balance between aperture diffraction and chromatic aberration in the column, with these effects having equal size at an α_i-value (predicted to be) between 0.3 and 0.4 mrad. These conclusions coincide with those illustrated by Hill et al. [192] in their Figure 1.3, although numerical details here are slightly different.

On these figures, there is no evidence that probe-spot radius is being limited either by the source optical radius or by non-optical effects such as mechanical vibration. The main conclusion is that further technical developments might seek to reduce chromatic aberration.

Current-related parameters

With a GFIS, the emission current can be changed by changing the background gas pressure, provided an upper limit is not exceeded (see Section 8.3). As a result, all GFIS current-related parameters can be varied widely. Ion-optical parameters said to be typical of GFIS operating conditions are included in Table 1.7.

9 OTHER FIELD-ASSISTED ION SOURCES

Experimental work on two other field-assisted ion sources has been reported. The *atomic-sized ion source* (ASIS) [207] is a variant of the sharp-needle LMIS. The substrate is a near-atomically sharp tungsten needle, with liquid ionizant supplied by flow along its shank. No liquid cone is involved. Stable currents much less than 2 μA can be emitted. The hope is that current could be made large enough to be useful, with the energy spread significantly less than the LMIS. Hopefully, the FWHM would be closer to the 1.5 eV previously measured [161] for Ga+ emission from a thin liquid film adhering to a tungsten surface. In 2011, no ASIS development work is known to be in progress.

Table 1.7 Typical CP-optical data for source-modules using a ZrO/W Schottky emitter (SchE), a He GFIS and a Ga LMIS, and probe data for related scanning machines.

Property	Unit	ZrO/W SchE + SEM	He GFIS + SIM		Ga LMIS + FIB	
		Value[a]	Value	Origin	Value	Origin
Typical extraction voltage*	kV	5	20	des.[b]	5	meas.[f]
Energy spread* (FWHM)	eV	0.6	0.35	meas.[c]	5	meas.[g]
Source optical radius*	nm	20	< 0.08	calc.	25	meas.[h]
Typical angular intensity	μA sr^{-1}	~ 400	2.5	meas.[d]	10	meas.[i]
Reduced Angular intensity*	A sr^{-1} V^{-1}	$\sim 7 \times 10^8$	10^{-10}	deduc.	2×10^{-9}	deduc.
Source brightness	A m^{-2} sr^{-1}	$\sim 5 \times 10^{11}$	$> 10^{14}$	deduc.	5×10^9	deduc.
Reduced source brightness	A m^{-2} sr^{-1} V^{-1}	$\sim 10^8$	5×10^9	deduc.	10^6	deduc.
Typical emission current	μA	40–400	0.5 to 0.8	meas.[d]	2	meas.[f]
Typical probe current	pA	variable	1	meas.[d]	variable	–
Typical probe voltage	kV	variable	30	des.[b]	variable	–
Best probe radius	nm	0.5	0.12	meas.[e]	3	calc.[j]

Abbreviations: des. = design decision; meas. = measured; calc. = calculated; deduc. = deduced here.
Origins of data are:
[a]L. W. Swanson and G. A. Schwind [31], supplemented by G.A. Schwind, private communication 2010.
[b]Carl-Zeiss SMT Inc., technical specification for ORION® Helium Ion Microscope, 2009.
[c]N. Ernst et al. [202].
[d]B.W. Ward et al. [191].
[e]Carl Zeiss SMT Inc. press release [206].
[f]J. Orloff, M. Utlaut and L. W. Swanson [60].
[g]L.W. Swanson [72].
[h]M. Komuro et al. [76].
[i]L.W. Swanson et al., *J. Vac. Sci. Technol.* 16 (1979): 1864.
[j]P. W. Hawkes and B. Lencová, in *E-nano-newsletter*, Issue 6 (2006), at www.phantomsnet.net/Foundation/newsletters.php.

The *ionic liquid ion source* (ILIS) [208; 209] is broadly similar to a LMIS, but uses as ionizant an organic liquid engineered chemically to have relatively high conductivity. In EHD devices, the cone-jet physics (and effectiveness as a "small ion" generator) depend critically [210] on ionizant conductivity. Because ILIS ionizants are conductive, ILIS physics may be closer to LMIS physics than to the physics of conventional non-metallic electrospray devices.

Presumably, ILIS and LMIS optical characteristics will be broadly similar, but with the ILIS energy-spread somewhat greater (due to the larger mass of emitted ions). This may make small probe radii even more difficult to obtain than with a LMIS.

As regards milling, it seems unlikely that the ILIS will offer serious competition to the LMIS. The ILIS is more likely to find a niche as a focused delivery system for certain molecular ions of chemical interest, and/or as an alternative form of field emission electric propulsion for spacecraft.

An obvious further possibility is an ALIS™-type GFIS that operates with one of the heavier noble gases. It is well known that such gases can be used as FIM operating gases. Probably, better vacuum conditions would be needed, because the lower best-image fields of these gases provide less FSFI protection against contaminants. However, good UHV conditions near the emitter are also needed for electron-beam systems with CFE guns, and technology may be transferable.

The search for high-brightness monochromatic ion sources using noble gases has been well reviewed by Tondare [211].

10 COMMENTARY

Table 1.7 compiles basic optical data for the ZrO/W emitter (as operated in a scanning electron microscope), the GFIS (as operated in a HeSIM), and the normal LMIS (as operated in a FIB). For completeness all the usual parameters are given. Parameters considered most needed when specifying source optical performance are marked with an asterisk.

Although the reduced angular intensity is smaller for the He GFIS than the Ga LMIS, the source optical radius is very much smaller (by a factor of over 30). Thus, the source optical area is much less for the GFIS, and, when operated at typical emission currents, the GFIS emerges as the brighter source.

In reality, these devices are normally used in different contexts, the He GFIS for scanning imaging and the Ga LMIS for milling or local energy deposition. For the He GFIS, the true comparison is with the optical performance of an SEM. The ORION® helium ion microscope seems to generate a probe with radius smaller than those of existing SEMs.

A missing comparison (not yet quantified) is one between the performances of a Ga LMIS and a GFIS based on a heavier inert gas. Qualitatively, one might expect such sources to generate currents similar to those of an He GFIS, but to have slightly greater energy spread and cross-over FORD. The optical performance is thus expected to be intermediate between that of a He GFIS and a Ga LMIS, but closer to the He GFIS.

As already noted, if a heavy-inert-gas GFIS could be developed as a commercially reliable device, then the corresponding FIB machine could well have some interesting applications. This seems one of the more interesting medium-term prospects in this subject area.

APPENDIX A: NUMERICAL DATA

A.1 VALUES FOR CONSTANTS

Table 1.8 gives values of fundamental constants, and Table 1.9 values of other universal constants, used in CP emission theory. This table employs the atomic-level units that are

Table 1.8 The electronvolt (eV) and fundamental constants used in emission physics, given in SI units and in the dimensionally-equivalent atomic-level units (based on the eV) sometimes used in field emission to make equation evaluation easier.

Name	Symbol	SI units		Atomic-level units based on eV	
		Numerical value	Units	Numerical value	Units
Electronvolt	eV	$1.602\ 176\ 5 \times 10^{-19}$	J	1	eV
Elementary (positive) charge	e	$1.602\ 176\ 5 \times 10^{-19}$	C	1	eV V^{-1}
Elementary constant of Amount of substance[a]	n_1	$1.660\ 538\ 7 \times 10^{-24}$	mol	1	entity[a]
Unified atomic mass constant	m_u	$1.660\ 538\ 7 \times 10^{-27}$	kg	$1.036\ 430 \times 10^{-26}$	eV nm^{-2} s^2
Electron mass in free space	m_e	$9.109\ 381\ 9 \times 10^{-31}$	kg	$5.685\ 630 \times 10^{-30}$	eV nm^{-2} s^2
Electric constant	ε_0	$8.854\ 187\ 8 \times 10^{-12}$	F m^{-1}	$5.526\ 350 \times 10^{-2}$	eV V^{-2} nm^{-1}
Electric constant × 4π	$4\pi\varepsilon_0$	$1.112\ 650\ 1 \times 10^{-10}$	F m^{-1}	$0.694\ 461\ 6$	eV V^{-2} nm^{-1}
Planck's constant	h_p	$6.626\ 069\ 8 \times 10^{-34}$	J s	$4.135\ 667 \times 10^{-15}$	eV s
Boltzmann's constant	k_B	$1.380\ 650\ 3 \times 10^{-23}$	J K^{-1}	$8.617\ 342 \times 10^{-5}$	eV K^{-1}

[a]Name used here, not "official."

Table 1.9 Basic universal emission constants. Values are given in the units often used in field emission.

Name	Symbol	Derivation	Expression	Numerical value	Units
Image-force constant	–	–	$1/16\pi\varepsilon_0$	$0.359\ 991\ 1$	eV^{-1} V^2 nm
Sommerfeld supply density[a]	z_S	–	$4\pi e m_e/h_p^3$	$1.618\ 311 \times 10^{14}$	A m^{-2} eV^{-2}
JWKB constant for electron	g_e	–	$4\pi(2m_e)^{1/2}/h_p$	$10.246\ 24$	eV$^{-1/2}$ nm^{-1}
First Fowler-Nordheim constant	a	$z_S(e/g_e)^2$	$e^3/8\pi h_p$	$1.541\ 434 \times 10^{-6}$	A eV V^{-2}
Second Fowler-Nordheim constant	b	$2g_e/3e$	$(8\pi/3)(2m_e)^{1/2}/eh_p$	$6.830\ 890$	eV$^{-3/2}$ V nm^{-1}
Universal theoretical Richardson constant	A_{R0}	$z_S k_B^2$	$4\pi e m_e k_B^2/h_p^3$	$1.201\ 735 \times 10^6$	A m^{-2} K^{-2}
Schottky constant	c	–	$(e^3/4\pi\varepsilon_0)^{1/2}$	$1.199\ 985$	eV V$^{-1/2}$ nm$^{1/2}$

[a]The supply density is the electron current crossing a mathematical plane inside the emitter, per unit area of the plane, per unit area of energy space, when the relevant electron states are fully occupied. In a free-electron model for a bulk conductor, the supply density is isotropic, is the same at all points in energy space, and is given by z_S.

Table 1.10 Physico-chemical data and predicted LMIS basic properties, for the liquid metals of most interest. (See text for remarks on accuracy.)

1	Eq.	OK?	Parameter		Units	Al+	Ga+	Ag+	In+	Sn++	Cs+	Au+	Hg+	Pb+	Bi+
2			proton number		–	13	31	47	49	50	55	79	80	82	83
3			chemical group		–	13	13	11	13	14	1	11	12	14	15
4			relative mass	m_r	–	26.98	69.72	107.87	114.82	118.71	132.90	196.97	200.59	207.19	208.98
5			mass	m	$\times 10^{-25}$ kg	0.448	1.158	1.791	1.907	1.971	2.207	3.271	3.331	3.440	3.470
6			density	ρ	kg/m^3	2385	6100	9330	7030	6980	1840	17360	13550	10678	10050
7			atomic volume	ω	$\times 10^{-2}$ nm^3	1.878	1.898	1.920	2.712	2.824	11.994	1.884	2.458	3.222	3.453
8			surface free energy p.u. area ($F = 0$)	γ_0	mJ/m^2	914	718	966	556	560	70	1169	486	458	378
9			melting temperature	T_{melt}	K	933	303	1234	430	505	302	1336	300[a]	600	544
10			FEV charge-state	z	–	1	1	1	1	2	1	1	1	1	1
11	.6	fair	charge-to-mass ratio	ze/m	MC/kg	3.576	1.384	0.894	0.840	1.626	0.726	0.490	0.481	0.466	0.462
12			evaporation field	F^E	V/nm	19.0	15.2	24.5	13.4	23.1	5.1	52.6	30.5	19.9	18.3
13	.57	fair	apex radius ($\Delta p = 0$)	r^E	nm	1.14	1.40	0.73	1.40	0.47	1.22	0.19	0.24	0.52	0.51
14	.52	good	coefficient in Mair's equation	C_M	μA μm^{-1} kV$^{1/2}$	475	232	251	140	196	16.4	225	92.6	85.9	70.6
16	.56	fair	test parameter	t_1	–	0.8	1.5	0.4	1.4	-0.1	10	0.02	0.2	0.2	0.5
17		fair	max. steady current	i_u	mA	34	25	9	14	6	1.8	1.0	0.7	2.0	1.7

[a] Room-temperature used, because Hg is liquid at room temperature.

customarily used in field emission. They are dimensionally equivalent to the SI units that could be used, but make formula evaluation easier.

A.2 LMIS PROPERTIES

Table 1.10 provides numerical data relevant to LMIS operation. Some comments are needed, particularly about accuracy. (1) Parameters on rows 2 to 10 are well-defined. (2) Results are shown for $z = 1$ (except Sn, for which $z = 2$). (3) Some parameters are shown to 3 significant figures (to facilitate checking), although their accuracy does not justify this; the column headed "OK?" gives an indication of likely accuracy. (4) The zero-Q evaporation field F^E derives from Müller's formula, which may underestimate true values by 20% or more. The derived quantity r^E may be 40% too high, or more. (5) Estimates of C_M involve only well-defined constants and should be good. (6) Estimates of t_1 should be fair to good. (7) Estimates of i_u, made using Mair's formulae [168], involve $(F^E)^{-3}$ and could be too high (though comparisons [168] with experiment show little evidence of this).

APPENDIX B: ACRONYMS

ALIS™	Atomic level ion source
ASIS	Atomic size ion source
BSF	Best source field
BSV	Best source voltage
CFE	Cold field electron emission
CP	Charged particle
ECD	Emission current density
EHD	Electrohydrodynamic
FE	Field electron emission, *or* field emission, *or* electron field emission, *or* field electron.
FEM	field electron microscope (*or* microscopy), *or* field emission microscope (*or* microscopy)
FEVSC	Field-emitted vacuum space-charge
FI	Field ionization, *or* field ion, *or* field ion emission
FIM	Field ion microscope (*or* microscopy)
FN	Fowler-Nordheim

FOHA	Fall-off half-angle
FORD	Fall-off radius
FSFI	Free-space field ionization
FWHM	Full width at half maximum
FW50	Full width for 50% of the current
GFIS	Gas field ionization source, *or* gas field ion source
H&K	Hawkes and Kasper
He	Helium
HeSIM	Helium scanning ion microscope
ILIS	Ionic liquid ion source
ISQ	International System of Quantities
ITE	Intermediate temperature electron emission
KE	Kinetic energy
LMEHD	Liquid metal electrohydrodynamic
LMIS	Liquid metal ion source
NED	Normal energy distribution
PE	Potential energy
PED	Parallel energy distribution
Plane M	The plane of the extractor cap
PPE	Polarization potential energy
SAC	Supply-and-capture
SCPE	Spherical charged-particle emitter
SCPPI	Statistical Coulomb particle-particle interaction(s)
sd	Standard deviation
SEM	Scanning electron microscope (*or* microscopy)
SI	International System (of units)
SIM	Scanning ion microscope (*or* microscopy)

SL	Schottky limit
SN	Schottky-Nordheim
TE	Thermal electron emission
TF	Thermal-field
TED	Total energy distribution
W	Tungsten

NOTES

a. Or, in some cases, individual gas molecules. The word "atom" is used here to cover both cases.
b. Strictly, the V that appears in Equation 2.2 and other formulae is the magnitude of a mean *electrostatic potential difference* between points just outside the surfaces of the emitter and the surrounding electrodes. If these surfaces have different work-functions, then V differs slightly from the voltage difference between the emitter and surroundings. The effect is small (at most a few volts, usually less), and can usually be neglected. This chapter follows the literature in referring to V as "voltage."
c. The presence of a high electric field can change the chemistry of surface processes, essentially because (in some circumstances) it can significantly "lift" or "lower" orbitals in energy.
d. The usual machine split is into a "gun" and a "column." For this chapter, the split into a "source-module" and "post-source optics" is more convenient.
e. The absence of a widely recognized single-letter symbol for angular intensity is a nuisance. The symbol K is chosen here, to allow current i and its two relevant derivatives to be represented by the triplet $\{i, J, K\}$.
f. 1 Torr $\cong$ 133 Pa $= 1.33$ hPa ≈ 1.33 mbar.
g. Knauer's calculations are performed in the context of spherical symmetry. In principle, more than one way exists of adapting these calculations to the context of a conical beam from a field emitter.
h. The manufacturers' name for their particular machine is "helium ion microscope." The name used here is intended as a generic name for these machines. Reference 61 suggests that these machines could be known more generally as *picoprobers*.
i. Also called a "gas field ion source."
j. If the pressure gauge is measuring gas with temperature different from T_{bk}, but in equilibrium with the gas in the chamber, via a path with a radius small in comparison with the mean free path of a gas molecule, then the thermomolecular effect may operate. It can then be argued that T_{bk} should be replaced in Equations 2.60 and 2.61 by the gas temperature in the vicinity of the pressure gauge, and p_{bk} by the gauge reading.

REFERENCES

[1] International Standard ISO 80000–1:2009. Quantities and units—Part 1: General. (ISO, Geneva) (2009).
[2] Winkler, J. H. *Gedanken von den Eigenschaften, Wirkungen und Ursachen der Electricität nebst Beschreibung zweier elektrischer Maschinen.* Leipzig: Verlag B. Ch. Breitkopf, 1744.

[3]   Lilienfeld, J. E. *Am. J. Roentgenol.*, 9 (1922): 192.

[4]   Millikan, R. A., and C. C. Lauritsen. *Proc. Natl. Acad. Sci. U.S.A.* 14 (1928): 45.

[5]   Fowler, R. H., and L. Nordheim. *Proc. R. Soc. Lond. A* 119 (1928): 173.

[6]   Stern, T. E., B. S. Gossling, and R. H. Fowler. *Proc. R. Soc. Lond. A* 280 (1929): 383.

[7]   Kleint, C. *Prog. Surf. Sci.* 42 (1993): 101.

[8]   Kleint, C. *Surf. Interface Anal.*, 36 (2004) 387.

[9]   Müller, E. W. *Z. Phys.* 106 (1937): 541.

[10] Müller, E. W. *Z. Phys.* 131 (1951): 136.

[11] Müller, E. W., J. A. Panitz, and S. B. McLane. *Rev. Sci. Instr.* 39 (1968): 83.

[12] Müller, E. W., and K. Bahadur. *Phys. Rev.* 102 (1956): 618.

[13] Feynmann, R. P. APS meeting, Caltech, Pasadena, December 1959.

[14] Melmed, A. J. *Biographical Memoirs* 82 (2002): 3.

[15] Sommerfeld, A., and H. Bethe. In: *Handbuch der Physik*, eds. H. Geiger and K. Scheel. Berlin: Springer, 1933, Vol. 24/2, p. 133.

[16] Schottky, W. *Physik. Zeitschr.* 15 (1914): 872.

[17] Nordheim, L. W. *Proc. R. Soc. Lond. A* 121 (1928): 626.

[18] Schottky, W. *Z. Phys.* 14 (1923): 630.

[19] Miller, M. K., A. Cerezo, M. G. Heatherington, and G. D. W. Smith. *Atom probe field ion microscopy.* Oxford: Clarendon, 1996.

[20] Gadzuk, J. W., and E. W. Plummer. *Rev. Mod. Phys.* 45 (1973): 487.

[21] Modinos, A. *Field, thermionic and secondary electron emission spectroscopy.* New York: Plenum, 1984.

[22] Zhu, W. (ed.), *Vacuum microelectronics.* New York: Wiley, 2001.

[23] Fursey, G. *Field emission in vacuum microelectronics.* New York: Kluwer, 2005.

[24] Jensen, K. L. Electron emission physics. *Adv. Imaging Electron Phys.* 149 (2007): 1.

[25] Forbes, R. G., and J. H. B. Deane. *Proc. R. Soc. Lond. A* 463 (2007): 2907.

[26] Forbes, R. G. *J. Phys. D: Appl. Phys.* 18 (1985): 973.

[27] Sakurai. T., A. Sakai, and H. W. Pickering. Atom-probe field ion microscopy and its applications. In *Adv. Electronics Electron Phys., Suppl. 20.* Boston: Academic, 1989.

[28] Tsong, T. T. *Atom-probe field ion microscopy.* Cambridge, UK: Cambridge University Press, 1990.

[29] Dyke, W. P., and J. K. Trolan. *Phys. Rev.* 8 (1956): 89.

[30] Orloff, J. (ed.), *Handbook of charged particle optics*, 2nd ed. Boca Raton: CRC Press, 2009.

[31] Swanson, L. W., and G. A. Schwind. Review of ZrO/W Schottky cathode, Chap. 1 in Ref. 30.

[32] Method for reproducibly fabricating and using stable thermal-field emission cathodes, U.S. Patent No. 3817592, issued 18 June 1974.

[33] High angular intensity Schottky electron point source, U.S. Patent No. 6798126, issued 28 Sept. 2004.

[34] Atomic level ion source and method of manufacture and operation, U.S. Patent No. 7368727, issued 6 May 2008.

[35] Ion sources, systems and methods, U.S. Patent No. 7521693, issued 21 April 2009.

[36] Forbes, R. G., and N. N. Ljepojevic. *J. de Physique* 45 (Colloque C8) (1989): 3.

[37] Ljepojevic, N. N., and R. G. Forbes. *Proc. R. Soc. Lond. A* 450 (1995): 177.

[38] Leontovich, M. A. *Introduction to thermodynamics.* Moscow: Nauka, 1951 (in Russian).

[39] Maxwell, J. C. *A treatise on electricity and magnetism,* Vol. I, 1st ed., 1873, 3rd ed., 1891. (3rd ed. reprinted as an Oxford Classic Text, 1998, by Clarendon Press, Oxford.)

[40] Lord Raleigh. *Phil. Mag.* 14 (1882): 184.

[41] Lord Raleigh. *Phil. Mag.* 31 (1916): 177.

[42] Hendricks, C. D., and J. M. Schneider. *Amer. J. Phys.* 31 (1962): 450.

[43] Peters, J. M. H. *Eur. J. Phys.* 1 (1980): 143.

[44] Taylor, G. I. *Proc. R. Soc. Lond. A* 280 (1964): 383.

[45] Herring, C. *Phys. Rev.* 82 (1951): 87.

[46] Forbes, R. G., and G. L. R. Mair. Liquid metal ion sources, Chap. 2 in Ref. 30.

[47] Brandon, D. G. *Surf. Sci.* 3 (1964): 1.

[48] Forbes, R. G. *Appl. Surf. Sci.* 87/88 (1995): 1.

[49] Mair, G. L. R. *J. Phys. D: Appl. Phys.* 15 (1982): 2523.

[50] Hawkes, P. W., and E. Kasper. *Principles of electron optics,* Vol. 2. London: Academic, 1996.

[51] Jansen, G. H. Coulomb interactions in particle beams. In *Adv. Electronics Electron Phys., Suppl. 21.* Boston: Academic, 1990.

[52] Kruit, P., and X. R. Jiang. *J. Vac. Sci. Technol. B* 14 (1996): 1645.

[53] Jiang, X. R., J. E. Barth, and P. Kruit. *J. Vac. Sci. Technol B.* 14 (1996): 3747.

[54] Boersch, H. *Z. Phys.* 139 (1954): 115.

[55] Loeffler, K. H. PhD thesis, University of Berlin, 1964.

[56] Hamish, H., K. H. Loeffler, and H. J. Kaiser. *Proc. European Regional Conf. Electron Micros.* 3 (1964): 11.

[57] Orloff, J. Computational resources for electron microscopy, Appendix in Ref. 30.

[58] Barbour, J. P., W. W. Dolan, J. K. Trolan, E. E. Martin, and W. P. Dyke. *Phys. Rev.* 92 (1953): 45.

[59] Forbes, R. G. *J. Appl. Phys.* 104 (2008): 084303.

[60] Orloff, J., M. Utlaut, and L. W. Swanson. *High resolution focused ion beams: FIB and its applications.* New York: Kluwer, 2003.

[61] Forbes, R. G. Gas field ionization sources, Chap. 3 in Ref. 30.

[62] Ward, J. W., and R. L. Kubena. *J. Vac. Sci. Technol. B* 8 (1990): 1923.

[63] Born, M., and E. Wolf. *Principles of optics,* 3rd (revised) ed. Oxford: Pergamon, 1965.

[64] Wells, O. C. *Scanning electron microscopy.* New York: McGraw-Hill, 1974.

[65] Crewe, A. V. The scanning transmission electron microscope, Chap. 10 in Ref. 30.

[66] Hawkes, P. W. Aberrations, Chap. 6 in Ref. 30.

[67] Kruit, P., and G. H. Jansen, Space charge and statistical coulomb effects, Chap. 7 in Ref. 30.

[68] Krivanek, O. L., N. Dellby, and M. F. Murfitt. Aberration correction in electron microscopy, Chap. 12 in Ref. 30.

[69] Freitag, B., J. R. Jinschek, and A. Steinbach. *Microscopy and Analysis Nanotechnology Supplement* (November 2009), S5.

[70] Ishitani, T., K. Umemura, and Y. Kawanami, *Jap. J. Appl. Phys.* 33 (1994): L479.

[71] Bronsgeest, M. S., J. E. Barth, L. W. Swanson, and P. Kruit. *J. Vac. Sci. Technol. B* 26 (2008): 949.

[72] Swanson, L. W. *Nucl. Instr. Methods. Phys. Research* 218 (1983): 347.

[73] Prewett, P. D., and G. L. R. Mair. *Focused ion beams from liquid metal ion sources.* Taunton, UK: Research Studies Press, 1991.

[74] Bronsgeest, M. S., J. E. Barth, G. A. Schwind, L. W. Swanson, and P. Kruit. *J. Vac. Sci. Technol. B* 25 (2007): 2049.

[75] Swanson, L. W., G. A. Schwind, and A. E. Bell. *J. Appl. Phys.* 51 (1980): 3453.

[76] Komuro, M., T. Kanayama, H. Hiroshima, and H. Tanoue, *Appl. Phys. Lett.* 42 (1982): 908.

[77] Knauer, W. *Optik* 54 (1979): 211.

[78] Knauer, W. *Optik* 59 (1981): 335.

[79] Swanson, L. W. *J. Vac. Sci. Technol.* 12 (1975): 1228.

[80] Bell, A. E., and L. W. Swanson. *Phys. Rev. B* 19 (1979): 3353.

[81] Schwind, G. A., G. Magera, and L. W. Swanson. *J. Vac. Sci. Technol. B* 24 (2006): 2897.

[82] van Veen, A. H. V., C. W. Hagen, J. E. Barth, and P. Kruit. *J. Vac. Sci. Technol. B* 19 (2001): 2038.

[83] Wiesner, J. C., and T. E. Everhart. *J. Appl. Phys.* 44 (1973): 2140.

[84] Wiesner, J. C., and T. E. Everhart. *J. Appl. Phys.* 45 (1974): 2797.

[85] Kern, D. Dissertation, Univ. of Tübingen, 1978.

[86] Eupper, M. Diplomarbeit, Univ. of Tübingen, 1980.

[87] Ruska, E. *Z. Physik.* 83 (1933): 684.

[88] Gomer, R. *J. Chem. Phys.* 20 (1952): 1772.

[89] Southworth, H. N., and J. M. Walls. *Surface Sci.* 76 (1978): 129.

[90] Wilkes, T. J., G. D. W. Smith, and D. A. Smith. *Metallography* 7 (1974): 403.

[91] Smith, R., and J. M. Walls. *J. Phys. D: Appl. Phys.* 11 (1978): 409.

[92] Forbes, R. G. PhD Thesis, University of Cambridge, 1971.

[93] Jensen, K. L. *J. Appl. Phys.* 102 (2007): 024911.

[94] Paulini, J., T. Klein, and G. Simon. *J. Phys. D: Appl. Phys.* 26 (1993): 1310.

[95] Murphy, E. L., and R. H. Good. *Phys. Rev.* 102 (1956): 1464.

[96]   Deane, J. H. B., and R. G. Forbes. *J. Phys. A: Math. Theor.* 41 (2008): 395301.

[97]   Forbes, R. G. *J. Vac. Sci. Technol. B* 26 (2008): 788.

[98]   Modinos, A. *Solid-State Electronics* 45 (2001): 809.

[99]   Modinos, A. Private communication.

[100] de Jonge. N. *J. Appl. Phys.* 95 (2004): 673.

[101] Kruit, P., M. Bezuijen, and J. E. Barth. *J. Appl. Phys.* 99 (2006): 024315.

[102] Swanson, L. W., and A. E. Bell. *Adv. Electronics Electron Phys.* 32 (1973): 193.

[103] Bahm, A. S., G. A. Schwind, and L. W. Swanson. *J. Vac. Sci. Technol. B* 26 (2008): 2080.

[104] Dolan, W. W., and W. P. Dyke. *Phys. Rev.* 95 (1954).

[105] Barth, J. E., and P. Kruit. *Optik (Jena)* 101 (1996): 101.

[106] Barth, J. E., and M. D. Nykerk. *Nucl. Instr. Methods Phys. Res. A* 427 (1999): 86.

[107] A formula sheet is available from the author (<r.forbes@trinity.cantab.net>).

[108] Goldstein, J. L., et al. *Scanning electron microscopy and x-ray microanalysis*, 3rd (revised) ed. Amsterdam: Kluwer, 2003.

[109] Knoll, M., and E. Ruska. *Z. Phys.* 78 (1932): 318.

[110] Crewe, A. V., D. N. Eggenburger, J. Wall, and L. M. Welter. *Rev. Sci. Instr.* 39 (1968): 576.

[111] Crewe, A. V. *Science* 168 (1970): 1338.

[112] Good, R. H., and E. W. Müller, Field emission. In *Encyclopedia of physics,* Vol. 21, ed. S. Flugge. Berlin: Springer-Verlag, 1956, p. 176.

[113] Shrednik, V. N. *Sov. Phys. Solid State* 3 (1961): 1268.

[114] Swanson, L. W., and L. C. Crouser. *J. Appl. Phys.* 40 (1969): 4741.

[115] Swanson, L. W., and N. A. Martin. *J. Appl. Phys.* 46 (1975): 2029.

[116] Nottingham, W. B. *Phys. Rev.* 58 (1940): 906.

[117] Swanson, L. W., L. C. Crouser, and F. V. Charbonnier. *Phys. Rev.* 151 (1966): 327.

[118] Dionne, M., S. Coulombe, and J.-L. Meunier. *Phys. Rev. B* 80 (2009): 085429.

[119] Vladar, A. E., and M. T. Postek. Scanning electron microscopy, Chap. 9 in Ref. 30.

[120] Swanson, L. W., and G. A. Schwind. *Adv. Imaging Electron Phys.* 159 (2009): 63.

[121] Mackie, W. A., T. B. Xie, and P. R. Davis. *J. Vac. Sci. Technol. B* 17 (1999): 613.

[122] Kagarice, K. J., G. G. Magera, S. D. Pollard, and W. A. Mackie. *J. Vac. Sci. Technol. B* 26 (2008): 868.

[123] Fink, H.-W. *IBM J. Res. Develop.* 30 (1986): 460.

[124] Binh, V. T., N. Garcia, and S. T. Purcell. *Adv. Imaging Electron Phys.* 95 (1996): 63.

[125] Yu, M. L., and T. H. P. Chang. *Appl. Surf. Sci.* 146 (1999): 334.

[126] Rokuta, E., T. Itagaki, T. Ishikawa, B.-L. Cho, H.-S. Kuo, T. T. Tsong, and C. Oshima. *Appl. Surf. Sci.* 252 (2006): 3686.

[127] Hata, K., T. Yasuda, Y. Saito, and A. Ohshita. *Nucl. Instr. Methods Phys. Res. A.* 363 (1995): 239.

[128] Cloupeau, M., and B. Prunet-Foch. *J. Aerosol Sci.* 25 (1994): 1821.

[129] Gilbert, W. *De magnete.* London: Petrus Short, 1600 (in Latin). Translation by P. F. Mottelay. New York: Dover, 1958, see p. 89.

[130] Forbes, R. G. *Ultramicroscopy* 89 (2001): 1.

[131] Gray, S. *Phil. Trans. (1683–1775)* (of the Royal Society of London), 37 (1731/2): 227, 260. [Available online via JSTOR.]

[132] Priestly, J. *The history and present state of electricity, with original experiments,* printed for J. Dodsley, J. Johnson, and T. Cadell. London, 1768. (See pp. 46–47 in 3rd ed., 1775.)

[133] Krohn, V. E. *Prog. Astronautics Rocketry* 5 (1961): 73.

[134] Krohn, V. E. *J. Appl. Phys.* 45 (1974): 1144.

[135] Aitken, K. L. *Proc. Field Emission Day at ESTEC, Nordwijk, 9th April 1976,* European Space Agency (Paris) Report SP-119 (1977), p. 232.

[136] Wagner, A., and T. M. Hall. *J. Vac. Sci. Technol.* 16 (1979): 1871.

[137] Bell, A. E., and L. W. Swanson. *Appl. Phys. A* 41 (1986): 335.

[138] Rudenauer, F. G. *Surf. Interface Anal.* 39 (2007): 116.

[139] Jaworek, A. *J. Mater. Sci.* 42 (2007): 266.

[140] Giannuzzi, L. A., and F. A. Stevie (eds). *Introduction to focused ion beams.* New York: Springer, 2005.

[141] Yao, N. (ed.) *Focused ion beam systems: Basics and applications*. Cambridge, UK: Cambridge University Press, 2007.

[142] Niedrig, H. *Scanning Microscopy* 10 (1996): 919.

[143] Forbes, R. G . *Vacuum* 48 (1997): 85.

[144] Kingham, D. R., and L. W. Swanson. *Appl. Phys. A* 34 (1984): 123.

[145] van Dyke, M. D. (1965). Appendix in Ref. 147.

[146] van Dyke, M. D. (1969). Appendix in Ref. 148.

[147] Taylor, G. I. *Proc. R. Soc. Lond. A* 291 (1965): 145.

[148] Tayor, G. I. *Proc. R. Soc. Lond. A* 313 (1969): 453.

[149] Mair, G. L. R. *Nucl. Instr. Methods Phys. Res. B* 43 (1989): 240.

[150] Beckman, J. C. PhD thesis, Stanford University, 1997.

[151] Beckman, J. C., T. H. P. Chang, A. Wagner, and R. F. W. Pease. *J. Vac. Sci. Technol. B* 15 (1997): 2332.

[152] Kovalenko, V. P., and A. L. Shabalin. *Pis'ma Zh. Tekh. Fiz.* 15.3 (1989): 62 (in Russian). Translated as: *Sov. Tech. Phys. Lett.* 15 (1989): 232.

[153] Shabalin, A. L. *Pis'ma Zh.Tekh. Fiz.* 15.12 (1989): 27 (in Russian). Translated as: *Sov. Tech. Phys. Lett.* 15 (1989): 924.

[154] Mair, G. L. R. *J. Phys. D: Appl. Phys.* 17 (1984): 2323.

[155] Mair, G. L. R. *Vacuum* 36 (1986): 847.

[156] Mair, G. L. R. *J. Phys. D: Appl. Phys.* 17 (1997): 1945.

[157] Jousten, K., J. F. Homes, and J. Orloff. *J. Phys. D: Appl. Phys.* 24 (1991): 458.

[158] Mair, G. L. R. *J. Phys. D: Appl. Phys.* 25 (1992): 1284.

[159] Forbes, R. G. *Surf. Sci.* 61 (1976): 221.

[160] Mair, G. L. R., R. G. Forbes, R. V. Latham, and T. Mulvey. In *Microcircuit engineering '83*. London: Academic, 1983.

[161] Culbertson, R. J., G. H. Robertson, Y. Kuk, and T. Sakurai. *J. Vac. Sci. Technol.* 17 (1980): 203.

[162] Mair, G. L. R., D. C. Grindrod, M. S. Mousa, and R. V. Latham. *J. Phys. D: Appl. Phys.* 16 (1983): L209.

[163] Gesley, M. A., and L. W. Swanson. *J. de Phys.* 45 (1984): Colloque C9, 168.

[164] Puretz, J. *J. Phys. D: Appl. Phys.* 19 (1986): L237.

[165] Ward, J. W., and R. L. Kubena. *J. Vac. Sci. Technol. B* 8 (1990): 1923.

[166] Radlicka, T., and B. Lencová. *Ultramicroscopy* 108 (2008): 445.

[167] Mair, G. L. R. *J. Phys. D: Appl. Phys.* 20 (1987): 1657.

[168] Mair, G. L. R. *J. Phys. D: Appl. Phys.* 29 (1996): 2186.

[169] Ward, J. W., R. L. Kubena, and M. W. Utlaut. *J. Vac. Sci. Technol. B* 6 (1988): 2090.

[170] Kubena, R. L., and J. W. Ward. *Appl. Phys. Lett.* 51 (1987): 1960.

[171] de Castilho, C. M. C., and D. R. Kingham. *J. Phys. D: Appl. Phys.* 19 (1986): 147.

[172] Weinstein, B. W. PhD thesis, University of Illinois at Urbana-Champaign, 1975.

[173] Mair, G. L. R. *Nucl. Instr. Meth.* 172 (1980): 567.

[174] Komuro, M., H. Arimoto, and T. Kato. *J. Vac. Sci. Technol. B* 6 (1988): 923.

[175] Kosuge, T., H. Tanitsu, and T. Makabe. In T. Takagi (ed), *Proc. 12th Symp. on Ion Sources and Ion-Assisted Technology* (ISIAT '89), (1989) p. 43.

[176] Praprotnik, B., W. Driesel, Ch. Dietzsch, and H. Niedrig. *Surf. Sci.* 314 (1994): 353.

[177] Mühle, R., M. Döbeli, and A. Melnikov. *J. Phys. D: Appl. Phys.* 40 (2007): 2594.

[178] Ishitani, T., S. Umemura, S. Hosoki, S. Tayama, and H. Tamura. *J. Vac. Sci. Technol. A* 2 (1984): 1365.

[179] Ishitani, T., K. Umemura, and H. Tamura. *Jpn. J. Appl. Phys.* 23 (1984): L330.

[180] Clark, W. M., R. L. Seliger, M. W. Utlaut, A. E. Bell, L. W. Swanson, G. A. Schwind, and J. B. Jergenson. *J. Vac. Sci. Technol. B* 5 (1987): 197.

[181] Benassayag, G., P. Sudraud, and B. Jouffrey. *Ultramicroscopy* 16 (1985): 1.

[182] Suvorov, V. G. Private communication.

[183] Private communication to the author.

[184] Ward, J. W. *J. Vac. Sci. Technol. B* 3 (1985): 207.

[185] Georgieva, S., R. G. Vichev, and N. Drandarov. *Vacuum* 44 (1993): 1109.

[186] Galovich, C. S. *J. Appl. Phys.* 63 (1987): 4811.

[187] Galovich, C. S., and A. Wagner, *J. Vac. Sci. Technol. B* 6 (1988): 1186.

[188] Czarczynski, W. *J. Vac. Sci. Technol. A* 13 (1995): 113.

[189] Mair, G. L. R. *J. Phys. D: Appl. Phys.* 30 (1997): 921.

[190] Czarczynski, W. *J. Phys. D: Appl. Phys.* 30 (1997): 925.

[191] Ward, B. W., J. A. Notte, and N. P. Economu. *J. Vac. Sci. Technol. B* 24 (2006): 2871.

[192] Hill, R., J. Notte, and B. Ward. *Phys. Procedia* 1 (2008): 135.

[193] Fink, H.-W. *Phys. Scripta* 38 (1988): 260.

[194] Postek, M. T., and A. E. Vladar. *Scanning* 30 (2008): 457.

[195] Jousten, K., K. Böhringer, and S. Kalbitzer. *Appl. Phys. B* 46 (1988): 313.

[196] Southon, M. J. PhD thesis, Cambridge University, 1963.

[197] Southon, M. J., and D. G. Brandon. *Phil. Mag.* 8 (1963): 579.

[198] Forbes, R. G. *J. Microscopy* 96 (1971): 63.

[199] Forbes, R. G. *Appl. Surf. Sci.* 94/95 (1996): 1.

[200] Feldman, U., and R. Gomer. *J. Appl. Phys.* 6 (1966): 2380.

[201] ORION® Helium Ion Microscope Product Brochure, Carl Zeiss SMT (downloaded November 2009).

[202] Ernst, N., G. Bozdech, H. Schmidt, W. A. Schmidt, and G. L. Larkins. *Appl. Surface Sci.* 67 (1993): 111.

[203] Witt, J., and K. Müller. *J. de Phys.* 47 (1986): Colloque C2, 465.

[204] Lui, X., and J. Orloff. *Adv. Imaging Electron Phys.* 138 (2005): 147.

[205] Dyke, W. P., J. K. Trolan, W. W. Dolan, and G. Barnes. *J. Appl. Phys.* 24 (1953): 570.

[206] Carl Zeiss SMT Inc. (Nano Technology Systems Division) Press Release, 21 November 2008. (Downloaded from <http://www.zeiss.com/nts>, November 2009.)

[207] Purcell, S. T., V. T. Binh, and P. Thevenard. *Nanotechology* 12 (2001): 168.

[208] Lozano, P., and M. Martínez-Sánchez. *J. Colloid. Interface Sci.* 282 (2005): 415.

[209] Zoros, A. N., and P. C. Lozano. *J. Vac. Sci. Technol. B* 26 (2008): 2097.

[210] Forbes, R. G. *J. Aerosol Sci.* 31 (2000): 97.

[211] Tondare, V. N. *J. Vac. Sci. Technol. A* 23 (2005): 1498.

HOW TO SELECT COMPOUNDS FOR FOCUSED CHARGED PARTICLE BEAM ASSISTED ETCHING AND DEPOSITION

Tristan Bret and Patrik Hoffmann

1 INTRODUCTION

How can one design an experiment based on focused charged particle beam induced surface processing? We discuss here some of the underlying principles involved. Rather than a detailed state-of-the-art review (such as, for instance, the one recently compiled [1]), this chapter aims at giving a practical overview of the beam-induced events on the surface. These beam-induced events lead either to the fixation of the volatile chemical compounds on a substrate (as a nano-sized deposit) or to their combination with atoms from the substrate (forming volatile species and resulting in substrate etching). A practical method for the choice of precursors is illustrated in an example for both deposition of tin-containing compounds or etching of tin. The following discussion is presented in four sections (followed by a list of references):

1. Chemical Phenomena Occurring under an Electron Beam
2. How to Choose a Precursor?
3. Comparison with Published Results
4. Conclusion.

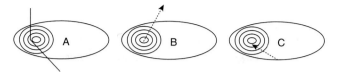

FIGURE 2.1: Schematic presentation of the excitation mechanisms of one highly energetic electron impinging on an atom (or atom in a molecule) or in atoms of the substrate or adsorbed molecules. (A) The atom with inner electron orbitals and one molecular orbital represented by the outer elliptical curve. The line represents the comparatively energetic impinging electron. (B) The most probable result of the electromagnetic interaction is the ionization of an electron from an atomic or molecular orbital (such as, in the example shown here, the most inner shell). (C) The excited and ionized system relaxes internally by filling the hole with an electron from an outer shell. The energy released by these processes can be transferred as photons or (in case the atom is part of a molecular structure or solid phase) as vibrations to the neighboring nuclei.

2 CHEMICAL PHENOMENA OCCURRING UNDER AN ELECTRON BEAM

What happens when bulk materials are irradiated by electrons? Excellent studies are available from technologically important fields such as electron lithography [2], analytical electron microscopy [3], and beam-related damage to materials [4]. Reviewing these fields is beyond the scope of this chapter, but we can summarize the fundamental interaction phenomena between the electron and the solid-state as follows:

- Elastic scattering processes (beam broadening, backscattering, diffraction . . .),
- Inelastic scattering processes (energy transfer, ionization, electron capture . . .),
- Secondary processes (secondary electron emission, Auger emission, X-rays . . .),
- Tertiary processes (beam induced desorption of molecules, atoms or ions . . .).

As a result of the inelastic interactions of the electrons penetrating into the substrate, excited ions are formed in the lattice as schematically presented in Figure 2.1 As the ions relax toward more stable electronic configurations, chemical bonds might break. If the atoms are confined, new bonds can form, which can lead to new material phases. If a gas is set free, it may gather as bubbles or diffuse away. Other volatile fragments might desorb from the surface as ions, neutral species, or recombined to form gases. These result in the loss of material from the substrate. The remaining fragments "reshuffle" their chemical bonds with the surrounding network, so that long-range crystalline order disappears in the irradiation region. Crystalline materials become partially amorphous[5]. Amorphous materials can crystallize locally. Cross-linking may occur as long as the effects of electron-induced amorphization or bond scission are limited. In a sense, the electron beam brings chemical mobility to the target.

Before spreading due to scattering effects, an impinging focused electron beam (FEB) can have a diameter of a few nanometers (10^{-9} m) in a modern electron microscope[3], while a focused ion beam (FIB) is typically at least one order of magnitude larger since ions are heavier

Table 2.1 The effects of electron irradiation on various organic compounds.

Compound	Electron energy & dose	Analysis method	Phenomena observed	Ref.
Alcohols, Ethers, Functional groups	40–100 keV 10^{-2} C/cm^2	Optical, TEM[a], Weight	Loss of H, O, and N; Residue: amorphous Carbon	[6]
Diazoketone (at 10°K)	6 keV 10^{-5} C/cm^2	FTIR[b]	"Wolff rearrangement," after loss of N_2	[7]
Nucleic acids (DNA, RNA), amino-acids	25–70 keV ~ 0.1 C/cm^2	TEM-EELS[c], Weight	"Loss of the low-energy-loss features associated to chemical functions" (i.e., amorphization); Reaction cross-sections correlate with the K-shell ionisation cross-sections.	[8–10]
Perfluoro-ethers	25 keV 10^{-5} C/cm^2	FTIR[b]	Atomic loss of O and F. Formation of COF_2 (gas) and acyl fluorides R-COF.	[11]
Partially fluorinated hydrocarbon films	10–90 eV	MS[d]	Dehydrofluorination (atomic loss of H and F), above an energy threshold of $\sim$30 eV.	[12]
Polyvinyl-chloride	100 keV 0.1 C/cm^2	FTIR[b]	Dehydrochlorination (atomic loss of H and Cl), chain scission, cross-linking, mass loss.	[13]

[a]Transmission Electron Microscopy; [b]Fourier Transform Infrared Spectroscopy; [c]Electron Energy Loss Spectroscopy; [d]Mass Spectrometry.

and slower than electrons[1]. The effects described above can thus be obtained on very small volumes, making them useful tools for nanotechnology.

For a more detailed understanding, effects of electron irradiation on many organic materials are listed in Table 2.1. For experimental and analysis reasons, electron irradiation was conducted over comparatively large solid volumes. That is, by scanning wide beams over several-cm^2 surfaces at energies above 10 keV, several-μm-thick films were exposed. This allows for the use of a wide range of analytical methods.

As can be seen from Table 2.1, the phenomena observed under electron exposure were similar, on many different compounds and over a wide energy range. A common feature was the loss upon irradiation of most of the H, N, O, F, and Cl content. Once set free from their original neighbors by the electron-induced liberation, two of these atoms might recombine and desorb from the sample. These "volatile elements" (H_2, N_2, etc.) are gases at room temperature. From the other elements, only C was lost, as recorded by "mass loss": this must be due to desorption of small C-containing volatile molecules such as CO, HCN, CH_4. Even a reactive

Table 2.2 The effects of electron irradiation on various inorganic compounds.

Compound	Electron energy & dose	Analysis method	Phenomena observed	Ref.
$NaBH_4$	80 keV, 0.1 C/cm^2	EELS[a]	H loss, formation of metallic Na platelets; Same behaviour from other hydrides (Li, Al. . .)	[16]
MgO	100 keV, ~10^5 A/cm^2	TEM[b]	O desorption from the surface, metal migration; "nm-scale e-beam ablation" on many other oxides	[17, 18]
AlF_3	20–100 keV, 0.5–20 μC/cm	TEM[b]	F desorption, Al surface clustering; Many other metal fluorides show the same effect	[19, 20]
NaCl	100 keV, ~1C/cm^2	AES[c], EELS[a]	Cl removal, formation of metallic Na clusters; 2-nm diameter holes in 50-nm thick NaCl films	[21]
Colloids (Au, Pd, Pt, Ru)	60–200 keV, 10–10^5 C/cm^2	TEM[b], electric	Organic surfactant "destruction," metal particle coalescence, increase in conductivity	[22, 23]

[a]Electron Energy Loss Spectroscopy; [b]Transmission Electron Microscopy; [c]Auger Electron Spectroscopy.

gas such as COF_2 could be identified as a gaseous irradiation product. The remaining material result was always a partially substituted amorphous carbon. Interestingly, this is similar to the chemical outcome of FEB-induced deposits from organic materials. In a more limited range of irradiation conditions, the same material (with composition: $C_9H_2O_1$, 90% sp^2 hybridization of the C phase and typically 2-nm-sized clusters) was obtained from four different organic precursors [14].

Every electron of the target can get ejected by the incoming beam, including the non-bonding core electrons if the energy is high enough (for instance, 280 eV to ionize the C K-shell). Due to the large number of ionization pathways, the process is selective only at low temperature, and only toward individual chemical bonds that are much less stable than all other bonds. The remarkable electron-induced Wolff rearrangement (a classical thermal reaction in organic synthesis), cited in Table 2.1, is one of few reactions where such a clear reaction channel can be written. Recently, other low energy induced electron reactions were studied, and the fundamental understanding in such processes constantly improves [15].

As can be seen in Table 2.2, the "organic" or "inorganic" nature of the target plays little role on electron-induced desorption of the volatile elements. The "volatile elements" can also originate from inorganic compounds. Many other electron irradiation results, compiled elsewhere ([24], p. 24), confirm this trend. The non-volatile residues reorganize under the beam into nanometer-sized clusters, or get slowly removed from the beam path, thanks to the electron-induced mobility.

It can be safely assumed that primary ionization is the underlying mechanism for these phenomena. Of course, other secondary phenomena can occur under the beam, such as electron-induced heating, when using high current density. Auxiliary effects due to electron-beam-induced heating were observed in several cases [25; 26]. The absolute temperature also has significant influence on the electron-induced effects [27; 28]. Direct momentum transfer to atomic nuclei can also occur at high electron energies, or under ion irradiation. Usual effects with ion beams (such as beam-induced "milling") are not possible with low-energy electron beams. But ionization takes place irrespective of the selected current density and beam energy, which makes it a ubiquitous effect of electron-induced processes.

What is the minimal resist thickness, for such an ionization process to take place? In principle, inelastic electron scattering can be observed from a single atom. Collective behaviors from several atoms (such as electron delocalization or plasmon resonances) only increase the collision cross-sections a little further. And indeed, electron patterning was demonstrated on self-assembled-monolayers [29] and was even claimed to be possible on the thinnest single-atom-layer one can imagine; such as the hydrogen terminated silicon surface! [30; 31]

We conclude this section by bringing the pieces of the puzzle together. If a monomolecular-thick layer of a volatile compound is reversibly adsorbed on a surface, the effects of electron or ion irradiation would probably be similar to those observed on thicker layers. As described above, most of the "volatile elements" (such as H, N, O, F, and Cl) and some volatile fragments (CO, COF_2, HCN...) are removed; while remaining elements cross-link or crystallize to form clusters. We consider here that the main reason for this transformation is electron-induced ionization (which, at the surface, also leads to secondary electron emission). Based on this empirical knowledge, we may choose an appropriate precursor for gas-phase transport.

3 HOW TO CHOOSE A PRECURSOR?

A major advantage of FEB-induced techniques is their versatility. In principle, the electron or ion microscope is by far the most complex, fragile, and expensive part of the setup. But an immense freedom is offered to the researcher for the choice of the gas to be injected into the system.

Thanks to the results gathered in Section 1, we know which elements of a molecule remain under the electron beam, and which elements leave. We can choose precursors for deposition and etching accordingly. However, not all ligands "fit" a chemical element. As a result, no general rule can be given as to which "packaging" makes an element volatile *and* whether it is adapted to electron-beam-induced processes. As an example, consider the combination of the first two group IV elements, carbon and silicon, with oxygen. Both form compounds with oxygen, but the volatility of stable room temperature oxides, CO_2 (gaseous carbon dioxide) and SiO_2 (silica glass), are quite different.

Here, another group IV element, tin (Sn), is selected in order to highlight general guidelines for local deposition and etching using charged particles (electrons or ions). Tin is an interesting choice. In its pure form, tin has a low melting point and high electrical and thermal conductivity, which offers a "nano-analogy" to classical soldering. Tin also has technologically important oxides (think, for instance, of the electrically conductive and optically transparent SnO_2 films used in liquid crystal displays), and a similar chemistry to that of silicon and germanium

(C-group). How can we make an "educated guess" of a few suitable precursors to deposit and etch tin?

3.1 TIN DEPOSITION

A quick look in the catalogue of a commercial chemical supplier (such as Aldrich, Avocado, ABCR, or STREM, to name a few) seems encouraging by the large numbers of volatile Sn-containing compounds available. Table 2.3 presents a few potential compounds for deposition of tin-containing materials:

Besides the molecular formula, several properties are highlighted in Table 2.3: melting point, boiling point, toxicity, some of the applications reported in literature, and pricing. Let us review their influences:

a. **Boiling point**: Too high boiling points (above 200°C at 1 bar) are impractical, since the precursor flow at room temperature is too low to compete with residual contamination. This disqualifies $SnBr_4$. Heating the precursor supply (reservoir, tubes) may lead to condensation on any cold wall. Hence, $Sn(hfac)_2$ is also disqualified, despite its analogy to other useful CVD precursors. On the other hand, too low boiling points (below 50°C, or gaseous compounds) imply flow control (such as manual needle valves, mass-flow controllers, or cooled reservoirs), since most electron microscopes are incapable of working at chamber pressures above 10^{-4} to 10^{-3} mBar. Large precursor flows may also lead to unwanted condensation on the substrate (due to adiabatic cooling of the gas entering the chamber).

b. **Toxicity/flammability**: For health and safety reasons, this point must not be underestimated. More stable, user-friendly and well-documented compounds are preferred (it is imperative to read corresponding Materials Safety Data Sheets [MSDS], before ordering any compound). Exposure to precursor vapors must be reduced by using tight and closed reservoirs, filled under N_2 or Argon in a glove box. The pump exhausts must be equipped with specific filter or trap units, to avoid releasing chemicals to the atmosphere. Most compounds can be handled safely if care is taken about their whole pumping and disposal cycle. With these precautions, tetramethyltin is a valid candidate, which the toxic $Sn(C_2H_5)_4$ is not. Despite the lower $Sn/(C+H)$ ratio, tetra-vinyltin is also worth a try. But most interesting is the "dimer" $Sn_2(CH_3)_6$, in which the $Sn/(C+H)$ ratio is higher than in $Sn(CH_3)_4$. Although this compound is less flammable and easier to handle than $Sn(CH_3)_4$, it is expensive and more toxic. But the Sn-Sn bond might lead to original material effects, making it definitely worth screening.

c. **Former uses in the literature**: References to Chemical Vapor Deposition (CVD) articles are usually a good sign. On the other hand, compounds optimized for catalysis, sol-gel deposition, or polymerization, among others, do usually not exhibit the correct set of properties for FEB-induced applications.

d. **Price**: A reasonable test compound, before being selected for a complete study, should not be so expensive that its delivery risks being interrupted by the supplier. 100 € is usually a good upper limit for 5 g of a compound. Many non-commercial compounds can also be obtained from smaller chemistry labs if their synthesis is simple.

e. **Molecular formula**: Simple ligands are preferred. From our experience, complex structures, with several different types of ligands, offer no particular advantage [32]. Since ligands get dissociated under irradiation (by electrons or ions) and can be integrated in the deposit, unnecessary elements must be avoided when a

Table 2.3 Sn compounds selected from commercial suppliers' catalogues (ABCR, Strem, Sigma-Aldrich).

Name	Formula	b.p. °C[1]; m.p. °C[2]	Comments	Price
Tetramethyltin, $Sn(CH_3)_4$	H_3C Sn CH_3 H_3C CH_3	75; −54	−$P_{vap}(20°) = 90$ mm Hg [3] −"For CVD of SnO_2," **−flashpoint: −12°C (9°F)** −toxicity, LD50: 195 mg/kg	25 g/87.5 €
Tin (IV) chloride	$SnCl_4$	114; −33	−$P_{vap}(20°)=19$ mm Hg [3] −"fumes in air" −toxicity, LD50: 99 mg/kg	25 g/27 €
Tin (IV) bromide	$SnBr_4$	204; 31	−"exothermic solution in water" −toxicity, LD50: 700 mg/kg	50 g/54.5 €
Dimethyl dichlorotin, $SnCl_2(CH_3)_2$	Cl CH_3 Sn Cl CH_3	190; 103	"Intermediate for SnO deposition on glass" (1 reference from 1969)	10g/87.5 €
Dimethylamino-trimethyltin, $(CH_3)_2N\text{-}Sn(CH_3)_3$	H_3C CH_3 $N{-}Sn{-}CH_3$ H_3C CH_3	126	−"useful dechlorinating agent" −flashpoint: 1°C (35°F) −HIGHLY TOXIC	5 g/71 €
Tetraethyltin, $Sn(C_2H_5)_4$	C_2H_5 C_2H_5 Sn C_2H_5 C_2H_5	181	−$P_{vap}(33°)=1$ mm Hg [3] −flashpoint: 107°C (225°F) −**HIGHLY TOXIC**; LD50: 10 mg/kg	5 g/148 €
Tetravinyltin, $Sn(C_2H_3)_4$	$H_2C{=}CH$ CH_2 $HC{-}Sn{-}CH$ H_2C $HC{=}CH_2$	163	−$P_{vap}(56°)=17$ mm Hg [3] −"Polymerization catalyst"	5 g/98.5 €
Hexamethylditin, $Sn_2(CH_3)_6$	CH_3 CH_3 $H_3C{-}Sn{-}Sn{-}CH_3$ CH_3 CH_3	88/45;[4] 24	−MSDS mentions no P_{vap}(RT) but an "irritating odor" (i.e., $P_{vap}(RT)\neq0$)[3] −flashpoint: 61°C (142°F) −toxicity: 25 mg/kg	5 g/120 €
Tin II hexafluoro-acetylacetonate, "Sn(hfac)₂,"	$\begin{bmatrix} F_3C & \\ & {=}O \\ & Sn \\ & O \\ F_3C & \end{bmatrix}_2$	125/2;[4] 71	Sn (II) analog to the well-known Cu(hfac)₂ used in Copper Chemical Vapor Deposition (CVD)	5 g/190.5 €

[1] Boiling point; [2] Melting point; [3] Vapor pressure; [4] Under the indicated pressure (in mm Hg).

pure compound is desired. There is no need for carbon (C), hydrogen (H), or fluorine (F) moieties, for instance, when the coordination sphere can consist only of Chlorine (Cl) in order to make the tin-containing molecule volatile. Hence, besides the other reasons named above, we let the "mixed" dimethyl-dichlorotin and dimethylamino-trimethyltin compounds aside. If one wants to investigate the possibility of incorporating nitrogen in the deposit (to obtain tin nitride Sn_3N_4 [33]), a nitrogen-containing precursor can be selected.

After screening these properties, only a few reasonable compounds remain, the tin halides and alkyls (namely $SnCl_4$, $Sn(CH_3)_4$, $Sn(C_2H_3)_4$, $Sn_2(CH_3)_6$). Checking the FIB/FEB literature yields a few hits with these compounds [34–36].

A new experimental study can start. *En route*, we will certainly meet other volatile Sn compounds. Synthetic chemists and the literature know many more interesting compounds than the commercially available molecules. To name a few, consider, for instance, the volatile trimethylcyclopentadienylstannane Me_3SnCp (a "pale yellow liquid with $P_{vap}(54°C) = 2.9$ mm Hg" [37], which is the Sn analog of a Pt precursor widely used in FIB/FEB studies [38]). C-free compounds were also synthesized, such as $Sn(NO_3)_4$ (which "sublimes under vacuum at 40°C" [39]), the "basic" $N(SnH_3)_3$ [40], the unstable Cl_3SnN_3 [41], or volatile ammonia adducts such as $SnCl(NH_2)_3$ [42].

FEB-deposits from hexamethylditin $Sn_2(CH_3)_6$ probably result in smooth nano-deposits with an average composition Sn:C:H:O with Sn or SnO_2 nano-clusters segregated in a C:H amorphous matrix (see Tables 2.1 and 2.2 and the discussion in Section 1). All the elements from the precursor are present. Oxygen is often incorporated in deposits due to its presence in water (which is the most pronounced background gas in electron microscopes). Integrated hydrogen in deposits can be observed using FTIR spectroscopy ([24], p. 147) or Rutherford Back-scattering (RBS) ([24], p. 150). The heavier elements can be quantified by Electron-Dispersive X-Ray Spectroscopy (EDX). Removal of residual C:H matrix may require addition of other gases, such as oxygen or water vapor [43, 44]. These O-containing gases will oxidize C and H into volatile compounds (CO_2, H_2O...), but will also oxidise Sn into SnO_2. Depositions at various substrate temperatures might also vary the atomic ratios. Increasing temperature would probably drive the process toward deposition of higher tin content, especially if the precursor molecules auto-catalytically decompose on the newly formed surface. In any case, a pure metal, a pure oxide or a metal-embedded-in-carbonaceous-matrix nano-material is obtained. This may be interesting for further applications. In any case, this opens a wide playing field for scientific studies. Nevertheless, there are no guaranties for obtaining a pure tin or a pure tin oxide as the deposit.

In contrast with FEB-assisted deposition, the FIB-assisted deposition is much faster due to larger collision cross-sections of ions (usually Ga^+) with respect to that of electrons. However, mechanical sputtering is unavoidable. Generally, lighter elements are sputtered more efficiently than heavier elements. So, the H, C, and O content in FIB-induced deposits is usually lower compared to FEB-induced deposits obtained with similar precursor, substrate, and assist gas. As a result, the metal content is higher in FIB-induced than in FEB-induced deposits, which allows for increased electrical conductivity. Note that in FIB processes, Ga atoms are co-implanted. This may be problematic in applications where Ga doping degrades the functionality of the deposited material. For example, in a hypothetical example where the main goal is the optical properties of the SnO_2 deposit, the Ga ion implantation must be taken into account, as it will stain the resulting material.

3.2 TIN ETCHING

We think that an electron-induced etching process basically consists in a surface chemical reaction. In the reaction, the elements of the substrate to be etched get combined with other elements brought by the precursor gas. If a volatile compound results, it can spontaneously desorb. This material transport away from the substrate is the root of the etch process. The products are stable (i.e., the etch process does not rely on the generation of ions or transient

Table 2.4 "Volatile" Sn compounds (from the CRC *Handbook of Chemistry and Physics*, pp. 4–91).

Formula	Name	Physical form	Melting point (°C)	Boiling point (°C)
SnH$_4$	**Stannane**	**Unstable colorless toxic gas**	**−146**	**−51.8**
SnH$_3$CH$_3$	Methylstannane	Colorless gas	N.A.	0
Sn(C$_2$H$_3$O$_2$)$_2$	**Tin(II) acetate**	**White crystals**	**183 (sublimates)**	**N.A.**
SnF$_2$	Tin(II) fluoride	White crystals, hygroscopic	213	850
SnF$_4$	Tin(IV) fluoride	White crystals	N.A.	705 (sublimates)
SnCl$_2$	Tin(II) chloride	White crystals	247	623
SnCl$_4$	**Tin(IV) chloride**	**Colorless fuming liquid**	**−33**	**114**
SnBr$_2$	Tin(II) bromide	Yellow powder	215	639
SnBr$_4$	**Tin(IV) bromide**	**White crystals**	**31**	**205**
SnI$_2$	Tin(II) iodide	Red-orange powder	320	714
SnI$_4$	Tin(IV) iodide	Yellow-brown crystals	143	364

species; see COF$_2$ formation in Table 2.1). They could in principle be isolated and analyzed, if the amounts of precursor gas and residual gases were not so comparatively large. Secondary processes, such as electron-induced desorption or milling by nuclear momentum transfer, are much less efficient: without the correct chemistry, nothing seems to happen. In the case of ion-beam-induced etching, the physical sputtering cannot be neglected. The chemical effects result in an enhancement of the FIB-induced etch rate, usually up to a factor 30 (which is considerable, and usually offers better material selectivity than physical sputtering).

Here, the substrate to be etched is tin. First of all, we have to look for a simple volatile Sn compound to be formed at the surface. Table 2.3 already lists a few, but let us check in the *Handbook of Chemistry and Physics* [45], in the section "Physical Constants of Inorganic Compounds," whether other volatile Sn compounds exist (which are not available commercially). Selected lines are reproduced in Table 2.4.

Several compounds are available. Stannane, Sn acetate, SnCl$_4$, and SnBr$_4$ have boiling points below ~200°C (as highlighted in Table 2.4). We can immediately discard Sn acetate, as an FEB process would lack the required selectivity to yield, from C-containing compounds, anything else than a solid C crust. Etching requires a clean vacuum, in which the contamination rate due to residual C-containing species is low enough to avoid the formation of such a crust. To generate SnH$_4$, SnCl$_4$, or SnBr$_4$, we need an H−, Cl−, or Br-containing gas. The gas decomposes on the surface under the electron beam irradiation, and the fragments must recombine with Sn into the corresponding volatile compound. Is the recombination reaction possible? In principle yes, but only if the system gains energy, i.e., if the products are more stable than the reactants. Free enthalpies of formation of the compounds (from CRC Handbook, *op. cit.*, "Standard thermodynamic properties of chemical substances") are reported in Table 2.5.

Stannane has a positive enthalpy of formation, which means that forming a Sn-H bond requires energy. This energy would have to be provided entirely by the impinging

Table 2.5 Free enthalpies of formation of the volatile Sn compounds
(CRC *Handbook*, *op. cit.*, pp. 5–8...18)

Molecular formula	Name	State	Standard free (Gibbs') enthalpy of formation $\Delta_f G^0$ (kJ.mol^{-1})
SnH_4	Stannane	gas	188.3
$SnCl_4$	Tin(IV) chloride	gas	−440.1
$SnBr_4$	Tin(IV) bromide	crystal	−350.2

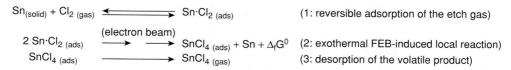

FIGURE 2.2: Simplified mechanism of electron beam induced Tin etching.

electron beam. But 188 kJ.mol^{-1} represents only ∼2 eV.atom^{-1}: this is deposited in the target in a few elastic collisions. As a result, a hydrogen source such as H_2 can be tested as an etch gas. Rapid desorption of gaseous stannane (b.p. −52°C, see Table 2.4) would displace a surface equilibrium toward the slow consumption of tin and H_2, i.e., result in metal etching.

As shown in Table 2.5, tin chloride and bromide form exothermically from the elements. The energy release is the driving force of the etching process. Luckily, the formation of the volatile Sn(IV) halides is more exothermic than that of the higher-boiling Sn(II) halides (not reported in Table 2.5), so the system will preferably form the more volatile compounds.

Based upon these data, we can design the etch experiment depicted in Figure 2.2.

Bringing a halogen gas into the vacuum chamber is, in principle, easy. Cl_2 is gaseous at room temperature (its boiling point is −34°C), and Br_2 is liquid (b.p. 59°C). Apart from acute toxicity considerations, the problem with these gases is that they have no permanent dipole moment. Their interactions with the substrate surface are limited to non-polar Van der Waals (London) interactions, so that adsorption (reaction 1) is not efficient. More specifically, they have short residence times, so the probability of their decomposition by the electron beam is low [46]. This is also valid for H_2. The effects of permanent dipole moments on surface residence times are compensated in dipolar species, such as HCl or HBr, by the lower molecular masses (as reflected by the boiling points, $T_b(HCl) = -85°C$, $T_b(HBr) = -67°C$). Providing simultaneously H and Cl to the tin could lead to the formation of compounds such as SnH_2Cl_2, in which the energy required to form each Sn-H bond is provided through the formation of a more energetic Sn-Cl bond, making the whole reaction exothermic. Such a chloro-stannane is stable, volatile (b.p. ∼ 59°C), and would desorb from the surface, resulting in the net removal of tin atoms from the substrate.

Using more "exotic" interhalogen compounds such as ClF (b.p. −100°C) or ClF_3 (b.p. 11.8°C) [47] could also lead to efficient surface formation and removal of volatile tin halides. In any case, after a gas-assisted etch test based on a halogen-containing gas the tin surface will most probably be chlorinated or fluorinated. Exposure of the sample to atmospheric air might have unwanted surface effects (such as oxidation, roughening, or peeling off of outmost layers).

Oxygen-containing gases (such as gaseous chlorine or bromine oxides [48]) would probably hamper the etching process, since the formation of SnO_2 is more exothermic ($\Delta_f G^0 = -515$ kJ.mol^{-1}) than that of $SnCl_4$. The presence of native SnO_2 at the surface, and that of O_2 or H_2O in the residual vacuum, might be an obstacle to the initiation of the etch process. However, since oxygen is also removed under the FEB (by electron-induced desorption, as mentioned in Section 1), a local delivery of the etch gas in large excess compared to the pressure of residual gases should allow for a FEB-induced halogenation of the surface (following reaction 2). The reaction might consist of several successive steps: first, oxygen must be removed, and then two to four adsorbed halogen atoms must be available in the vicinity of a tin atom to be etched. This makes surface refill a rate-limiting parameter. Detailed investigations of the surface reaction (*in situ* with electron spectroscopy or electron diffraction, or far away with mass spectroscopy) might show whether the volatile product is indeed $SnCl_4$ or further products thereof, but in any case desorption reaction (3) represents the atomic level of the etching process.

Interestingly, in contrast to CF_4 (b.p. $-128°$C), SiF_4 (b.p. $-86°$C), and GeF_4 (b.p. $-36.5°$C), SnF_4 is not volatile (with a sublimation point of $705°$C; see Table 2.4). Nevertheless, the popular fluorine precursor, xenon difluoride XeF_2, could also be tested as an etch gas, since it allows for the etching of many other elements [49]. XeF_2 is a volatile crystalline compound (sublimation point $114°$C), which yields 2 F atoms per molecule and 1 atom of the inert gas Xe, leaving no solid residue (which could act as an etch-stopping "crust" on the surface). The volatility of other less-well-documented tin fluorides (such as Sn_2F_6, $SnOF_2$, SnF_2Cl_2 . . .) is possibly sufficient to allow for an efficient etch process.

To complete the discussion, let us also briefly mention FIB-assisted etching. The interaction cross-sections are larger as compared to the FEB-induced process, and since the mechanical sputtering adds to the chemical-assisted etching, the process is even faster. Selecting the right precursor chemistry helps in increasing the selectivity of a material to the next, and can increase the etch rate up to roughly 30 times as compared to the milling without gas [50], but even non-optimized chemistry might enhance the etch rate by avoiding redeposition of sputtered atoms. Ga atoms are co-implanted in the process down to ~5 nm below the surface, which might cause amorphization or optical staining.

3.3 SUMMARY

The model case of tin treated here is no "real" example from the literature and, to our knowledge, it has not been experimentally studied in detail. A published article reports some side effects observed when two tin precursors were condensed to thick films on substrates and irradiated with 2 keV electrons. Tin chloride ($SnCl_4$) condensed on silicon resulted in films that contained up to 52 at.% Sn and 48 at.% Cl. Unfortunately, the relatively well-conducting layers peeled off the substrate and, when exposed to air, changed strongly due to hydrolysis of the chloride, probably leading to $SnOCl_2$. The other precursor, tin tetramethyl $Sn(CH_3)_4$, seems to have resulted in deposits with some of the hydrogen lost, but most of the carbon was still present, and the results of electron irradiation were not discussed in the article [36].

In summary, the chemical design steps in the choice of a deposition or etching precursor are the following:

- For deposition: look for a series of chemically "simple" volatile precursors, by screening both the commercial suppliers and the chemical literature.
- In the case studied here, the most interesting compounds for the deposition of tin-containing materials are: $SnCl_4$, $Sn_2(CH_3)_6$, and $Sn(NO_3)_4$.

- For etching: look for the volatile compounds of the element to be etched, which are formed exothermally (and selectively, in the presence of other elements).
- In the case studied here, several gases look promising for tin etching and would require experimental assessment: H_2, Cl_2, HCl, ClF_3, and XeF_2.

The same approach would be valid for any other pure element or compound. The interest of an in-depth study of a particular element lies in the original material structures and remarkable resolutions which can be obtained by the FEB-induced approach. Good knowledge of a particular chemical system (as shown here for tin) can also more easily lead to new approaches.

4 COMPARISON WITH PUBLISHED RESULTS

In Section 2.2, we defined a thermodynamic approach to try to design an FEB-induced etching process. The validity of this approach is checked by comparing with other elements than tin. In the following, assessment is done on a few published systems (as reviewed under [1], p. 1250).

All the reactions proposed in this section do not intend to describe the processes exactly: they aim at giving estimates of the involved chemical energies for thermodynamically possible pathways, to estimate whether a process is possible (negative ΔG^0) or not. When available, the standard free (Gibbs') enthalpies of formation ($\Delta_f G^0$) values of the compounds are used [45]. Where no $\Delta_f G^0$ are available, the exothermic nature of the reaction is approximated by using the standard enthalpies of formation ($\Delta_f H^0$), which neglects the entropic contributions $\Delta_f S^0$ to the reactions (let us recall that the three values are linked through the relation: $\Delta_f G^0 = \Delta_f H^0 - T\Delta_f S^0$, where T is the temperature). At room temperature, the entropic factors are usually small as compared to the large exchanges in bond energies involved. Qualitatively, $\Delta_f S^0$ tends to reduce the $\Delta_f G^0$ (i.e., the reaction gets more exothermic) when more gaseous species are present among the products than among the reactants.

Although all reactions take place under high vacuum, we mention in the discussion the boiling points at atmospheric pressure T_b of the compounds involved. This aims at giving a rough idea of the interactions of the molecular compound with a surface: the higher T_b, the stronger interactions a molecule has with its environment at a given temperature, and the more energy it requires to escape a surface.

   a. **Example 1**: FEB-induced, WF_6-assisted etching of SiO_2, as described in 1986 [51].

$$\text{Proposed reaction: } 2\,WF_6 + 3\,SiO_2 \rightarrow 2WO_3 + 3SiF_4$$

The free enthalpy of the reaction ΔG^0 is the sum of the standard enthalpies of formation $\Delta_f G^0$ of the products, subtracted from the $\Delta_f G^0$ of the reactants:

$$\Delta G^0 = -2\Delta_f G^0(WF_6) - 3\Delta_f G^0(SiO_2) + 2\Delta_f G^0(WO_3) + 3\Delta_f G^0(SiF_4)$$

i.e. $\Delta G^0 = -2^*(-1631.4) - 3^*(-856.3) + 2^*(-764) + 3^*(-1572.8)$

i.e. $\Delta G^0 = -414.7 \text{ kJ.mol}^{-1}$

The negative ΔG^0 value hints at a possible (i.e., exothermic) reaction. Forming the products releases chemical energy. However, ΔG^0 is not so negative that the reaction is spontaneous (the activation energy must be around the formation energy of the single reactants), so that the reaction has to be triggered locally by the electron beam.

One of the products has a low boiling point ($T_b(SiF_4) = -86°C$), which allows for spontaneous desorption at room temperature, as long as the deposition rate of the W-rich crust, which takes place competitively, is not in the same order of magnitude. Indeed, a temperature dependence of the process is reported: deposition takes place below room temperature (nota, $T_b(WF_6) = 17°C$), while the substrate gets etched above 50°C.

Other reactions take place, especially the electron-induced reduction of W^{+VI}: de-fluorination of WF_6 (see Table 2.2) or de-oxygenation of WO_3 (as reported later)[52]. They contribute to the reported formation of the metallic W crystallites.

b. **Example 2**: FEB-induced, Cl_2-assisted etching of GaAs, as described in 1989 [47].

> Proposed reaction: $GaAs + 3Cl_2 \rightarrow GaCl_3 + AsCl_3$
>
> Reaction free enthalpy: $\Delta G^0 = -\Delta_f G^0(GaAs) + \Delta_f G^0(GaCl_3) + \Delta_f G^0(AsCl_3)$
>
> i.e. $\Delta G^0 = -(-67.8) + (-454.8) + (-259.4)$
>
> i.e. $\Delta G^0 = -646.4 \text{ kJ.mol}^{-1}$.

Even though this ΔG^0 is more negative than the one in example 1, sample heating to 90°C was required to observe measurable etching. This is due to the relatively high boiling points of the products ($T_b(AsCl_3) = 130°C$, $T_b(GaCl_3) = 201°C$). Desorption of $GaCl_3$ produced at the surface is the rate-limiting step.

Although the same article reports the successful use of XeF_2 and ClF_3 to etch respectively SiO_2 and PMMA, nothing is said about the ability of these precursors to FEB-etch GaAs. Although the free enthalpies of formation of the fluorinated products are considerably lower ($\Delta_f G^0(GaF_3) = -1085.3 \text{ kJ.mol}^{-1}$, $\Delta_f G^0(AsF_3) = -774.2 \text{ kJ.mol}^{-1}$), the melting point of one of them is so high ($T_b(AsF_3) = 57.8°C$ but $T_m(GaF_3) > 1000°C!$) that most probably no practical electron-beam induced process can be set up. The FIB-induced, XeF_2-assisted etch process of GaAs seems not to be efficient either, since all further reports used Cl_2 ([1], p. 1252).

c. **Example 3**: FEB-induced, H_2O, O_2 or H_2-assisted etching of C, as described in 1964 [53].

> *Proposed reaction 1*: $2H_2O + 2C \rightarrow CH_4 + CO_2$
>
> Reaction free enthalpy: $\Delta G^0 = -2\Delta_f G^0(H_2O) + \Delta_f G^0(CH_4) + \Delta_f G^0(CO_2)$
>
> i.e. $\Delta G^0 = -2*(-228.6) + (-43.4) + (-394.4)$
>
> i.e. $\Delta G^0 = +19.4 \text{ kJ.mol}^{-1}$
>
> *Proposed reaction 2*: $H_2O + C \rightarrow H_2 + CO$
>
> Reaction free enthalpy: $\Delta G^0 = -\Delta_f G^0(H_2O) + \Delta_f G^0(CO)$
>
> i.e. $\Delta G^0 = 228.6 + (-137.2)$
>
> i.e. $\Delta G^0 = +91.4 \text{ kJ.mol}^{-1}$

Surprisingly, the thermodynamic formation of CO_2 from water at room temperature, even together with CH_4 which liberates some more energy, is not favored but slightly endothermic! However, the ΔG^0 is only slightly positive, so the reaction is balanced. At room temperature, the faster desorption of the volatile products ($T_b(CH_4) = -161°C$, $T_b(CO_2) = -78°C$), as compared to the reactants ($T_b(H_2O) = 100°C$), displaces the equilibrium toward the formation of the products. Moreover, 20 kJ.mol^{-1} is roughly 0.2 eV, which is in the same order of magnitude as the energy transferred by the elastic collision between one electron and the substrate. In other words, the reaction can start from an electronically excited system (through a C radical or an excited water ion), which tips the energy balance. Etching C through formation of CO is thermodynamically less probable. But the very low boiling point ($T_b(CO) = -191°C$) makes it an efficiently leaving fragment, and the energy (~ 1 eV) can be brought by the beam. The relative simplicity of the reaction (only 1 water molecule involved, against 2 molecules for the formation of CO_2) makes it a competitive pathway.

From a thermodynamic point of view, the C etch reactions from H_2 or O_2 are more probable compared to C etching from H_2O, since CH_4 or CO_2 are produced from the elements in their standard state (the reaction ΔG^0 are $\Delta_f G^0(CH_4)$ or $\Delta_f G^0(CO_2)$, respectively; the values are given above). However, although both reactions were also demonstrated, they are not as efficient as the C etch reaction using water[54]. The high polarity of water makes it a very "sticky" molecule, with a long residence time at surfaces, which leads to higher activation probabilities under electron irradiation.

d. **Example 4**: FIB-induced, Cl_2-assisted InP etching, as described in 1986 [50].

Proposed reaction: $InP + 2Cl_2 \rightarrow InCl + PCl_3$

Reaction enthalpy: $\Delta H^0 = -\Delta H^0(InP) + \Delta H^0(InCl) + \Delta H^0(PCl_3)$

i.e. $\Delta H^0 = -(-88.7) + (-186.2) + (-319.7)$

i.e. $\Delta H^0 = -417.2$ kJ.mol^{-1}

The reaction is clearly exothermic, even without considering the small entropic contributions. But one of the boiling points is very high ($T_b(PCl_3) = 76°C$, $T_b(InCl) = 608°C$). Other compounds can be formed from the elements, with even more negative ΔH^0 values (down to ~ -540 kJ.mol^{-1} for the reaction $InP + 3\,Cl_2 \rightarrow InCl_3 + PCl_5$), but they require more adsorbed Cl_2 and are less volatile ($T_b(PCl_5) = 160°C$, $T_m(InCl_2) = 235°C$, $T_m(InCl_3) = 586°C$, no listed T_b for both last compounds), so we concentrate on the proposed reaction. The high boiling point of InCl makes the desorption of this product the rate-limiting step in ion beam induced etching and would make the electron-induced process practically impossible to implement. Mechanical sputtering of the unvolatile fragment InCl by the FIB is necessary, and even so, a sample temperature above 140°C is required for efficient gas-enhancement of the etch process. The low melting temperature of In ($T_m(In) = 156°C$) makes it a mobile atom in these conditions: after P depletion due to PCl_3 desorption, metallic In droplets migrate outside the irradiation area.

In a later test study, Cl_2 was replaced by I_2 [55]. In this case, PI_3 can be formed ($\Delta_f H^0(PI_3) = -45.6$ kJ.mol^{-1}) instead of PCl_3, and InI ($\Delta_f H^0(InI) = -116.3$ kJ.mol^{-1}) instead of InCl, even if the reaction is not as exothermic ($\Delta H^0 = -73.2$ kJ.mol^{-1}) as the Cl_2-assisted InP etching. However, InI or InI_3 are not volatile either ($T_b(InI) = 712°C$, $T_m(InI_3) = 210°C$, no T_b), and

PI_3 decomposes upon evaporation at 227°C. This leads to iodination of the remaining fragments (formation of InI_3 from InI, as detected at the surface), and formation of In droplets, mixed with Ga from the ion beam.

Focused particle beam induced etching of InP is certainly hampered by the lack of volatile In compounds at ambient temperatures (e.g., $T_b(InF_3) > 1200°C$, $T_b(InBr) = 656°C$), so that neither an FEB-induced, nor a practical FIB-induced gas-enhanced etch process seems available for this technologically important compound.

The examples chosen in Section 3 show that the thermodynamic approach proposed helps to correctly interpret published processes, while justifying some of the observed physical and chemical effects. The approach can be summarized as follows:

A. For a beam-induced etch reaction to take place, usually at least one exothermic reaction can be written.
B. If all envisioned reactions are endothermic, the beam-induced etch process can still be possible, but the required energy must be small (a few eV.atom^{-1} at most), and the products must have low boiling points.
C. For an electron-beam-induced etch reaction to take place, all products must be volatile (boiling points below 200°C). Low product volatility can be partly compensated by heating the sample.
D. If one of the products is really not volatile (boiling point above ~300°C), no electron-beam-induced process will be possible, but an ion-beam-induced process might show good enhancement rates due to mechanical sputtering of this species.

This gives us a good confidence in the validity of the approach, when designing FEB- or FIB-induced etching experiments of new materials. A few more important aspects were highlighted: the importance of the relative boiling points, the concept of rate-limiting steps, and orders of magnitudes for the free enthalpy of reactions of thermodynamically possible reactions, as compared to the limited energy input from the beam.

5 CONCLUSIONS

In this chapter, we first highlighted some general principles of electron-induced chemistry on bulk materials and thin surface films. This showed that all the "volatile elements" (H, N, O, F, Cl, Br, I) can be desorbed easily after dissociation from the materials which contain them, and that some molecular fragments (such as COF_2) can also escape when their boiling points are low enough. Based upon this knowledge, we explained how to select molecular precursors to deposit the scarcely studied model element tin (Sn), and justified why pure metal deposits will be more difficult to obtain than alloys of Sn with C, H, O, and Cl. The etching of tin was addressed as a *Gedankenexperiment* based upon a thermodynamic approach, the available literature, and a list of possible gases conceivable as etch precursors. Finally, several etch processes were selected from the literature, to analyze the reasons of their success, especially in the light of the thermodynamic approach developed here: by comparing boiling points, standard free enthalpies of formation, and energy transfer from the particle beam, a solid analysis framework was built. The global coherence of this theory will hopefully help in the design of future beam-induced deposition and etch processes at the nanometer level.

ACKNOWLEDGMENTS

The present contribution was initiated by PH, composed and written by TB and extensively discussed and disputed by both. We thank Bamdad Afra (EPFL) and Laurent Bernau (Empa) for critical reading of the manuscript and their helpful comments.

REFERENCES

[1] Utke, I., P. Hoffmann, and J. Melngailis. Gas-assisted focused electron beam and ion beam processing and fabrication. *Journal of Vacuum Science & Technology* B 26 (2008): 1197–1276.

[2] Broers, A. N. Resolution limits for electron-beam lithography. *IBM Journal of Research and Development* 32 (1988): 502–513.

[3] Reimer, L. *Scanning electron microscopy*, ed. S. S. i. O. Sciences. Berlin: Springer, 1985.

[4] Pantano, C. G., A. S. D'Souza, and A. M. Then. Electron beam damage at solid surfaces. *Beam Effects, Surface Topography, and Depth Profiling in Surface Analysis* 5 (1998): 39–96.

[5] Hobbs, L. W. Topology and geometry in the irradiation-induced amorphization of insulators. *Nuclear Instruments & Methods in Physics Research Section B-Beam Interactions with Materials and Atoms* 91 (1994): 30–42.

[6] Reimer, L. Irradiation changes in organic and inorganic objects. *Laboratory Investigation* 14 (1965): 1082.

[7] Pacansky, J., and H. Coufal. Electron-beam-induced wolff rearrangement. *Journal of the American Chemical Society* 102 (1980): 410–412.

[8] Isaacson, M. Interaction of 25 keV electrons with nucleic-acid bases, adenine, thymine, and uracil. 1. Outer shell excitation. *Journal of Chemical Physics* 56 (1972): 1803.

[9] Isaacson, M. Interaction of 25 keV electrons with nucleic-acid bases, adenine, thymine, and uracil. 2. Inner shell excitation and inelastic-scattering cross-sections. *Journal of Chemical Physics* 56 (1972): 1813.

[10] Stenn, K. S., and G. F. Bahr. A study of mass loss and product formation after irradiation of some dry amino acids, peptides, polypeptides and proteins with an electron beam of low current density. *Journal of Histochemistry & Cytochemistry* 18 (1970): 574–580.

[11] Pacansky, J., and R. J. Waltman. Electron-beam irradiation of poly(perfluoro ethers) - identification of gaseous products as a result of main chain scission. *Journal of Physical Chemistry* 95 (1991): 1512–1518.

[12] Kelber, J. A., and M. L. Knotek. Electron-stimulated desorption from partially fluorinated hydrocarbon thin-films – molecules with common versus separate hydrogen and fluorine bonding sites, *Physical Review* B 30 (1984): 400–403.

[13] Vesely, D., and D. S. Finch. Chemical-changes in electron-beam-irradiated polymers. *Ultramicroscopy* 23 (1987): 329–337.

[14] Bret, T., S. Mauron, I. Utke, and P. Hoffmann. Characterization of focused electron beam induced carbon deposits from organic precursors. *Microelectronic Engineering* 78–79 (2005): 300–306.

[15] Allan, M. Study of triplet-states and short-lived negative-ions by means of electron-impact spectroscopy. *Journal of Electron Spectroscopy and Related Phenomena* 48 (1989): 219–351.

[16] Herley, P. J., and W. Jones. WOS:A1986D205100013, Transmission Electron-Microscopy of Beam-Sensitive Metal-Hydrides, 1–2 (1986).

[17] Hollenbeck, J. L., and R. C. Buchanan. Oxide thin-films for nanometer scale electron-beam lithography. *Journal of Materials Research* 5 (1990): 1058–1072.

[18] Salisbury, I. G., R. S. Timsit, S. D. Berger, and C. J. Humphreys. Nanometer scale electron-beam lithography in inorganic materials. *Applied Physics Letters* 45 (1984): 1289–1291.

[19] Langheinrich, W., and H. Beneking. Fabrication of metallic structures in the 10nm region using an inorganic electron-beam resist. *Japanese Journal of Applied Physics Part 1-Regular Papers Short Notes & Review Papers* 32 (1993): 6218–6223.

[20] Kratschmer, E., and M. Isaacson. Progress in self-developing metal fluoride resists. *Journal of Vacuum Science & Technology* B 5 (1987): 369–373.

[21] Muray, A., M. Scheinfein, M. Isaacson, and I. Adesida. Radiolysis and resolution limits of inorganic halide resists. *Journal of Vacuum Science & Technology* B 3 (1985): 367–372.

[22] Reetz, M. T., M. Winter, G. Dumpich, J. Lohau, and S. Friedrichowski. Fabrication of metallic and bimetallic nanostructures by electron beam induced metallization of surfactant stabilized Pd and Pd/Pt clusters. *Journal of the American Chemical Society* 119 (1997): 4539–4540.

[23] Hoffmann, P., G. Ben Assayag, J. Gierak, J. Flicstein, M. Maar-Stumm, and H. van den Bergh. Direct writing of gold nanostructures using a gold-cluster compound and a focused ion beam. *J. Appl. Phys.* 74 (1993): 7588–7591.

[24] Bret, T. Thesis, 2005, EPFL, Physico-chemical study of the focused electron beam induced deposition process.

[25] Kanaya, K. The distribution of temperature along a rod-specimen in the electron microscope. *Journal of Electron Microscopy* 4 (1956): 1–4.

[26] Utke, I., A. Luisier, P. Hoffmann, D. Laub, and P. A. Buffat. Focused-electron-beam-induced deposition of freestanding three-dimensional nanostructures of pure coalesced copper crystals. *Applied Physics Letters* 81 (2002): 3245–3247.

[27] Kunz, R. R., and T. M. Mayer. Catalytic growth rate enhancement of electron beam deposited iron films. *Appl. Phys. Lett.* 50 (1987): 962–964.

[28] I. Utke, D. Laub, T. Bret, P. A. Buffat, and P. Hoffmann. Nano-structure analysis of magnetic Co-containing tips obtained by focused electron beam induced deposition. *Microelectron. Eng.* 73–74 (2004): 553–558.

[29] Barraud, A., C. Rosilio, and A. Ruaudelteixier. Mono-molecular resists: A new approach to high-resolution electron-beam microlithography. *Journal of Vacuum Science & Technology* 16 (1979): 2003–2007.

[30] Tsubouchi, K., and K. Masu. Area-selective cvd of metals. *Thin Solid Films* 228 (1993): 312–318.

[31] Hui, F. Y. C., G. Eres, and D. C. Joy. Factors affecting resolution in scanning electron beam induced patterning of surface adsorption layers. *Applied Physics Letters* 72 (1998): 341–343.

[32] Luisier, A., I. Utke, T. Bret, F. Cicoira, R. Hauert, S.-W. Rhee, P. Doppelt, and P. Hoffmann. Comparative study of Cu-precursors for 3-D focused electron beam induced deposition. *J. Electrochem. Soc.* 151 (2004): C535–C537.

[33] Scotti, N., W. Kockelmann, J. Senker, S. Trassel, and H. Jacobs. Sn3N4, a tin(IV) nitride: Syntheses and the first crystal structure determination of a binary tin-nitrogen compound. *Zeitschrift Fur Anorganische Und Allgemeine Chemie* [in German] 625 (1999): 1435–1439.

[34] Christy, R. W. Conducting thin films formed by electron bombardment of substrate. *Journal of Applied Physics* 33 (1962): 1884.

[35] Vishnyakov, B. A., and K. A. Osipov. *Production of thin films of chemical compounds by electron beam bombardment*. Tel Aviv: Freund Publishing House Ltd., 1972.

[36] Funsten, H. O., J. W. Boring, R. E. Johnson, and W. L. Brown. Low-temperature beam-induced deposition of thin tin films. *Journal of Applied Physics* 71 (1992): 1475–1484.

[37] Davison, A., and P. E. Rakita. Fluxional behavior of cyclopentadienyl, methylcyclopentadienyl, and pentamethylcyclopentadienyl compounds of silicon, germanium, and tin. *Inorganic Chemistry* 9 (1970): 289–294.

[38] Koops, H. W. P., A. Kaya, and M. Weber. Fabrication and characterization of platinum nanocrystalline material grown by electron-beam induced deposition. *Journal of Vacuum Science & Technology* B 13 (1995): 2400–2403.

[39] Addison, C. C., and W. B. Simpson. Tin(4) nitrate: Relation between structure and reactivity of metal nitrates. *Journal of the Chemical Society* (1965): 598–602.

[40] Holleman, A. F., and E. Wiberg. *Lehrbuch der Anorganischen Chemie*, Vol. p. 736. Berlin: Walter de Gruyter, 1985.

[41] Dehnicke, K. Darstellung und Eigenschaften der Azidchloride $SnCl_3N_3$ $TiCl_3N_3$, $VOCl_2N_3$. *Journal of Inorganic & Nuclear Chemistry* 27 (1965): 809–815.

[42] Bannister, E., and G. W. A. Fowles. Reactions of tin(IV) halides with ammonia derivatives. 1. The reaction of tin(IV) chloride with liquid ammonia. *Journal of the Chemical Society* (1958): 751–755.

[43] Folch, A., J. Tejada, C. H. Peters, and M. S. Wrighton. Electron-beam deposition of gold nanostructures in a reactive environment. *Appl. Phys. Lett.* 66 (1995): 2080–2082.

[44] Perentes, A. PhD thesis, 2007, EPFL, Deposition of SiO2 by focused electron beam induced deposition.

[45] Lide, D. R., ed. *Handbook of chemistry and physics.* Boca Raton: CRC Press, 1995–1996.

[46] Christy, R. W. Formation of thin polymer films by electron bombardment. *Journal of Applied Physics* 31 (1960): 1680–1683.

[47] Matsui, S., T. Ichihasi, and M. Mito. Electron beam induced selective etching and deposition technology. *J. Vac. Sci. Technol.* B, **7** (1989): 1182–1190.

[48] Ruhl, G. DIIDW:2005402383, Etching of portion of chromium layer applied to substrate involves exposing substrate having chromium layer to gas atmosphere containing halogen and oxygen compounds, and directing electron beam onto portion of substrate, US2005103747-A1; DE10353591-A1, (2005).

[49] Coburn, J. W., and H. F. Winters. Ion-assisted and electron-assisted gas-surface chemistry: Important effect in plasma-etching. *Journal of Applied Physics* 50 (1979): 3189–3196.

[50] Ochiai, Y., K. Gamo, S. Namba, K. Shihoyama, A. Masuyama, T. Shiokawa, and K. Toyoda. Temperature-dependence of maskless ion-beam assisted etching of InP and Si using focused ion-beam. *Journal of Vacuum Science & Technology* B 5 (1987): 423–426.

[51] Matsui, S., and K. Mori. New selective deposition technology by electron-beam induced surface-reaction. *Journal of Vacuum Science & Technology* B 4 (1986): 299–304.

[52] Carcenac, F., C. Vieu, A. M. HaghiriGosnet, G. Simon, M. Mejias, and H. Launois. High voltage electron beam nanolithography on WO3. *Journal of Vacuum Science & Technology* B 14 (1996): 4283–4287.

[53] Löffler, K.-H. Ph.D. thesis (1964), Erzeugung freitragender Mikroobjekte durch elektronenstrahlaktivierten Kohlefolienabbau.

[54] Taniguchi, J., I. Miyamoto, N. Ohno, K. Kantani, M. Komuro, and H. Hiroshima. Electron beam assisted chemical etching of single-crystal diamond substrates with hydrogen gas. *Japanese Journal of Applied Physics Part 1- Regular Papers Short Notes & Review Papers* 36 (1997): 7691–7695.

[55] Callegari, V., and P. M. Nellen. Spontaneous growth of uniformly distributed In nanodots and InI3 nanowires on InP induced by a focused ion beam. *Physica Status Solidi a-Applications and Materials Science* 204 (2007): 1665–1671.

CHAPTER 3

GAS INJECTION SYSTEMS FOR FEB AND FIB PROCESSING THEORY AND EXPERIMENT

Vinzenz Friedli, Heinz D. Wanzenböck, and Ivo Utke

1 INTRODUCTION

The high number of applications using FEB- and FIB-induced processing (etching and deposition) is due to the flexibility imparted by the electron- and ion-beam resolution in combination with the many choices of volatile molecule species that can be used to deposit many different material compositions. These molecules reversibly adsorb (via physisorption) on the substrate (planar or non-planar) and are dissociated by the very local electron or ion impact. The non-volatile fragments are deposited, whereas volatile fragments are pumped away upon desorption.

Gas injection systems (GIS) with a straight cylindrical-tube nozzle design are frequently used for gas-assisted deposition and etching with FEB and FIB. They can supply a highly localized molecule flux to the electron- or ion-beam impact-area on the substrate while keeping the overall chamber pressure of the microscope low enough for FEB (FIB) operation. Moreover, substrates can be moved freely below such a GIS, in contrast to environmental subchambers [31; 67; 72]. It is crucial to know how efficiently dissociated molecules can be replenished ("refreshed") inside the irradiated area by new molecules from a GIS arrangement, since this will determine the regime in which deposition or etching will proceed; the process rate and the spatial resolution of the deposit or etch hole strongly depend on whether the process is molecule-limited or electron (ion)-limited [61, 81; 94; 93; 105; 107]. Furthermore, the chemical composition of the deposit material [46; 103] and its internal nanocrystal/matrix structure [101; 106] both depend, to a certain extent, on the process regime.

A limited number of measurements of gas flux near-field distributions in rarefied flow conditions can be found in the literature. In these studies, diatomic (H_2, N_2, CO) and noble atoms (He, Ne, Ar, Kr, and Xe) were used. The experiments were based on a pressure measurement using an ionization gauge [1; 11; 86] or a mass spectrometer [42] behind a sub-millimeter-sized entrance aperture, which spatially samples the flux distribution. However, the experimental complexity is large and the spatial resolution is limited to the aperture size, which must allow for a measurable molecule flux.

GIS-related total molecule flux measurements [7; 55; 88] and models [54] have also been studied. Attempts to determine and model the molecule flux distribution have been reported recently; however, they only consider specific GIS-substrate arrangements and specific pressure ranges (flow regimes) and molecules [28; 35; 102]. Only very recently [36], a comparative study of experiments and simulations for differing nozzle geometries was published.

Some attempts were made to optimize the nozzle shape, leading to the proposition of alternative nozzle geometries [14; 56; 83]. However, no experimental characterization can be found in the literature. It was further proposed that surfaces which are intercepting the gas flow can act as reflectors and serve to increase the local gas flux [37]. In some setups, the precursor flux has been increased and homogenized by arrangements of two symmetrically arranged tube nozzles [75].

The scope of this chapter mainly focuses on the gas injection system and does not deal with dissociation issues by FEB (FIB) impact on molecules. For a discussion of the properties of precursor molecules and their suitability for FEB- (FIB-) induced deposition and etching, as well as a vast review on precursor chemistry, the authors refer to a recent review article from Utke et al. [105].

In Section 2 we introduce the fundamentals of rarefied gas flows in GISs. The reviewed formulas of rarefied gas flow can be used as a benchmark for comparison to experimental and simulated results. In Section 3 we elaborate in depth how molecular and transient molecule flux through nozzles can be exactly simulated (direct simulation Monte Carlo), and introduce a simplified yet easy to implement simulation algorithm (test-particle Monte Carlo) for gas-assisted FEB (FIB) flux studies. We show further in Sect. 4 how well both codes agree with experimental results obtained from varying nozzle geometries, and we also address flux shadowing effects and electron (ion) gas-phase scattering. The second part of the chapter describes design aspects (Section 5), hardware components (Section 6), lab-made and commercial systems (Section 7), which supply the precursor molecule flux to the nozzle exit.

2 MOLECULE FLOW IN GAS INJECTION SYSTEMS

2.1 FUNDAMENTALS

A GIS consists of the following configuration (see Figure 3.1):

- Reservoir
- Supply system
- Nozzle.

The reservoir stores a solid, liquid, or gaseous precursor. From this source, volatile precursor molecules are delivered to the nozzle by the gas supply system. The fundamental parameters to characterize a GIS for FEB- (FIB-) induced processing are the gas flow rate (throughput) Q in the system, expressed in molecules per unit time or mbar l s^{-1}, and the impinging precursor molecule flux at the process area on the substrate J, expressed in molecules per unit area and unit time or cm^{-2} s^{-1}. The flow rate of the precursor in the GIS can be controlled in 3 ways:

1. with flow regulators (typically for gaseous and for liquid precursors),
2. by the GIS flow conductance, in most systems determined by the final nozzle, and
3. by controlling the temperature of the GIS (typically for solid precursors).

The flow rate determines the total amount of gas that is supplied to the vacuum chamber. The restriction on the maximum value of Q is imposed by the tolerated pressure inside the scanning electron (ion) microscope, allowing for safe working conditions of the electron (ion) source and optics. Higher chamber pressures are tolerated if the entire tool is designed as an environmental or a variable pressure microscope [31; 70]. In a nozzle-based GIS, the impinging precursor flux onto the substrate is controlled by the flow rate, the nozzle geometry, and position relative to the substrate. It is important to be aware of the flux pattern (the spatial distribution) that is produced by a nozzle-injected gas flow onto a substrate surface. The impinging flux to the processing area can be strongly affected by shadowing effects, e.g., at micron-sized obstacles on the substrate or at the objects fabricated by FEB- (FIB-) induced processing themselves. Scattering of electrons (ions) in the gas phase needs to be taken into account when optimizing the GIS throughput and impinging flux for FEB- (FIB-) induced processing.

For solid and liquid precursors, the pressure in the reservoir corresponds to the vapor pressure of the molecule, and for gaseous precursors, it corresponds to the gas pressure at a given temperature. Strictly speaking, this is an assumption, since the net flow of molecules through the nozzle exit does not allow for equilibrium between the condensed and gaseous phases of the molecule; however, this deviation is often negligible. At equilibrium conditions inside a closed precursor reservoir, the surface area of the solid or liquid in contact with the gas has no effect on the vapor pressure. However, a large surface may be advisable to replenish precursor that is consumed by the precursor flow eventually leaving the reservoir. The vapor pressure can be controlled by the reservoir temperature. The Clausius-Clapeyron relation [3] predicts an exponential temperature dependence of the vapor pressure. Experimentally, the total precursor flux emitted from a nozzle was revealed to have an exponential dependence on the precursor reservoir temperature [88].

With precursors that are gaseous at the storage conditions in the reservoir, the supply of a tolerable pre-nozzle pressure range requires a gas flux restriction system such as a valve or a mass flow controller.

Generally it should be noted that possible condensation of molecules on tubing surfaces that have lower temperatures than the precursor reservoir can change the flux distribution and lead to non-reproducible results. Condensed precursor inside the nozzle will produce a flow into the vacuum chamber long after shutting off the flow. This hampers clean post-imaging. Care should further be taken to minimize molecule reactions with the walls of the precursor reservoir and the tube system. In the specific case of Fe(CO)$_5$ under ultrahigh vacuum conditions, such wall reactions were observed and led to a high carbon monoxide concentration in the flux, since the iron was bound to the walls [43]. Evidently, any deposit composition will depend on the impinging flux composition and, furthermore, the uptake of precursor molecules at the tube walls will change the impinging flux distribution.

If no mass flow controller is used in the GIS arrangement, the total throughput Q through the GIS can be determined by measuring the (solid or liquid) precursor mass loss Δm using

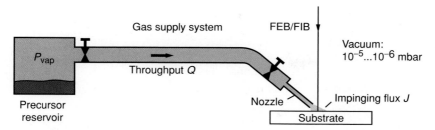

FIGURE 3.1: Schematics of a nozzle-based GIS for FEB- (FIB-) induced processing. The precursor reservoir contains the solid, liquid or gaseous precursor. High vapor pressure precursors are dosed by mass flow controllers or regulating valves which are not shown here. Reprinted with permission from [36]. Copyright (2009), IOP.

a precision balance with sub-milligram resolution. After an injection period Δt, the total throughput becomes $Q = (\Delta m / \Delta t) \cdot (N_A / M)$, where N_A is Avogadro's constant and M the molar mass of the precursor molecule. The corresponding total flux J_{tot} at the tube nozzle exit is then derived from the inner diameter d of its exit aperture:

$$J_{tot} = \frac{4Q}{\pi d^2}. \tag{1}$$

2.2 RAREFIED GAS FLOW

Low precursor gas flow rates injected into the vacuum chamber of scanning electron or ion microscopes are tolerated in order to keep the background pressure at high vacuum level. In these conditions, a low-pressure (or rarefied) molecule flow is present in the nozzle of the GIS. The rarefication of the gas is a quantity of fundamental importance since it determines the transitions between flow regimes obeying distinct physics. The degree of rarefication can be expressed through the dimensionless Knudsen number Kn: the ratio between the mean free path λ, which is the mean distance traveled by a particle without intermolecular collision, and a characteristic length of the containing region. This length can be some overall dimension of the flow, but a more precise choice is the scale of the gradient of a macroscopic quantity, as for example pressure $P/|\partial P/\partial z|$. For the gas flow in a circular tube an obvious choice of Knudsen number is the ratio of the mean free path to the tube diameter:

$$Kn \equiv \frac{\lambda}{d}. \tag{2}$$

Several flow regimes are empirically distinguished with varying Kn as illustrated in Figure 3.2. The different regimes have smooth transitions with Kn and require different mathematical frameworks. In rarefied flow conditions the Knudsen number describes if the molecule gas flow proceeds only with tube wall collisions (molecular flow) or in conjunction with collisions between molecules (transient flow) [4].

For an ideal homogeneous gas, modelled as hard (rigid) spheres, at pressure P (in Pa) the mean free path is: [59]

$$\lambda = \frac{kT}{\sqrt{2}\pi \delta_m^2 P}, \tag{3}$$

where k is the Boltzmann constant, T the temperature, and δ_m the molecule diameter.

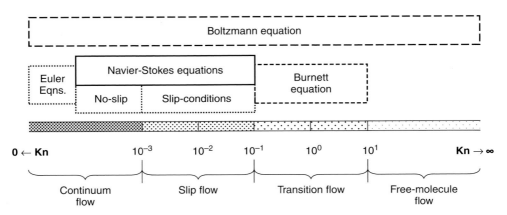

FIGURE 3.2: Generally accepted Knudsen number regimes and the corresponding mathematical framework (Reprinted with permission from [86]. Copyright (2000), IOP). As Kn increases, rarefaction effects become more important, and eventually the continuum approach breaks down altogether. Slip, transition and free-molecule (or molecular) flow describe the regime of rarefied flow.

For the description of gas flow through a GIS nozzle we consider a tube where a large pressure ratio between entrance and exit is preserved and the downstream condition is vacuum. As the precursor gas flows down the tube, the pressure and density decrease while the mean free path increases and the Knudsen number correspondingly increases. Conservation of mass requires the flow to accelerate down the constant-area duct, reaching speed of sound (Mach number $Ma = 1$) in the limiting case of choked-flow conditions (see Sect. 2.3.3). For the flow of some precursor gases with high source pressure the slip-flow, transition, and molecular regimes occur in the same tube, and thus, the full description of simple tube flow may in fact manifest a high complexity.

The Knudsen number inside the GIS reservoir is approximately obtained by inserting the static molecule vapor pressure for P. In contrast, the Knudsen number at the nozzle exit is mainly determined by the *dynamic* pressure, P_{dyn}, of the flowing gas. The static pressure at the nozzle exit, being approximately equivalent to the chamber background pressure, $P_{\mathrm{b}} < 10^{-4} - 10^{-5}$ mbar, can be neglected, since typically $P_{\mathrm{b}} \ll P_{\mathrm{dyn}}$. The dynamic pressure at the nozzle exit is straightforwardly obtained from the measured total molecule flux (see Equation 1):

$$P_{\mathrm{dyn}} = \frac{J_{\mathrm{tot}} \sqrt{2\pi M R T}}{N_{\mathrm{A}}}, \tag{4}$$

where $R = 8.314 \, \mathrm{J mol^{-1} K^{-1}}$ and is the universal gas constant. Approximate Knudsen numbers at the nozzle exit can be determined from the measured total throughput derived from Equations 1, 2, and 4:

$$Kn = \frac{\sqrt{RT/\pi M}}{8\delta_{\mathrm{m}}^{2}\left(Q/d\right)}. \tag{5}$$

2.3 THROUGHPUT

The throughput Q in terms of flow conductance C (in volume per second) follows the relation

$$Q = C\frac{P_u - P_d}{kT}, \tag{6}$$

where $P_u - P_d$ is the pressure decay between the upstream and downstream pressure of the bounded flow system. For the description of a GIS, the upstream condition is given by the reservoir pressure and the downstream condition given by the pressure in the microscope vacuum chamber. The throughput is constant along the entire duct, and it is related to the flux according to $J = Q/A$, where A is the duct cross-section. In the case of a flow across a large pressure ratio between the inlet and outlet, the throughput is solely controlled by the upstream pressure $Q \cong CP_u/(kT)$ and is independent of the downstream pressure, i.e., corresponding to choking of the flow. These conditions are typically realized in GIS nozzles. The downstream pressure P_d is depending on the pumping speed (in volume per unit time) in the vacuum system. Equation 6 can be used to determine the pressure decay within the precursor supply tubing based on measurable quantities: the geometry which determines the conductance and the net flux effusing into vacuum using Equation 1. However, the evaluation of the conductance of tubing components requires the knowledge of the prevailing flow regime.

The conductance is conveniently expressed in terms of the transmission probability W of a molecule entering the tube as

$$C = WC_a, \quad C_a = A\sqrt{\frac{RT}{2\pi M}}, \tag{7}$$

where C_a is the conductance of the entrance aperture with area A.

Several tubing components can be regarded as circuit elements connected in series, and accordingly, their conductances add up as $1/C_{tot} = 1/C_1 + 1/C_2 + \ldots + 1/C_n$. This expression is valid only if the flow in one component is not affected by the previous or succeeding component, i.e., if the molecular motion is disordered after passing through a large volume between two successive components. If this condition is not fulfilled, an approximate correction term can be introduced [95]. For improved accuracy, empirical correction factors [38] or simulations must be applied. Well-established formulations of the flow conductance of a variety of simple duct geometries can be found in textbooks, for example in [5].

2.3.1 Molecular flow regime

Molecular or free-molecular flow is the limiting case where the flow is controlled solely by wall collisions. In practical situations, a flow preserving $Kn > 10$ can be considered molecular and modeled without taking into account intermolecular collisions. Molecular flow can be described by a number of rigorously derived analytical formulations [59, 91]. The molecules, which are transmitted through a tube with length L, will be made up of two groups: a fraction of molecules that pass directly through the tube without striking the tube inner surface, and a fraction suffering wall collisions emitted from each element of the inner surface. The total flux through the tube may therefore be written:

$$J_m = J_{direct} + \int_0^L J_e(x) P_e(x) \, dx, \tag{8}$$

where J_{direct} is the direct number flux, $P_e(x)$ is the probability that a molecule reflected from the tube element of length dx at x leaves the tube without striking the tube inner surface, and $J_e(x)$ is the total number flux emitted from this element. The functions J_{direct} and $J_e(x)$ may be found for the cylindrical geometry by straightforward analysis. The solution for J_m can be found by numerical methods, assuming that $J_e(x)$ is a linear function of x. This problem was first solved by Clausing [17]. For circular tubes with radius r of any length, a formulation of the transmission probability in molecular flow conditions was phenomenologically derived by Santeler [87],

$$W_m = \frac{1}{1 + \frac{3}{8}\frac{L_e}{r}}, \tag{9}$$

where $L_e = L + L/(3 + 3L/(7r))$. For long tubes ($L \gg r$) Equation 9 reduces to $W_m = \frac{8r}{3L}$, which is in agreement to other models [59].

2.3.2 Transient flow regime

For Knudsen numbers $10 > Kn > 0.1$, boundary collisions *and* intermolecular collisions are important. These conditions are generally referred to as the transient flow regime or as the near-free molecular regime. The gas flow dynamics in the transient regime are intermediate between molecular flow and viscous flow. The transient flux can be described qualitatively by two counteractive effects [79]; on the one hand, the transmission probability is reduced for molecules that tend to gain non-axial velocity components by intermolecular collisions, and on the other hand, the transmission probability is increased by the development of continuum flow properties, known as the inset of drift velocity. With increasing pressure, and thus, decreasing Knudsen number, transient flux conditions may reveal a flux minimum below the molecular level before the transition to continuum fluxes. For short tubes, when L is comparable to d, the two effects may counterbalance each other and a minimum in conductance would then not be observed, as was confirmed experimentally [64, 65]. Several attempts to empirically model the transmission probability over the extended pressure range covering molecular and transition flow are found in the literature. Knudsen [52] deduced a formula for the transmission probability of long circular tubes in transient flow conditions from a series of measurements that can be written in the form given by Dushman [26]:

$$W_{tr} = \frac{8r}{3L}\left(0.0736/\overline{Kn} + \frac{1 + 1.254/\overline{Kn}}{1 + 1.548/\overline{Kn}}\right), \tag{10}$$

where $\overline{Kn}$ corresponds to the Knudsen number at the average pressure $\overline{P}$ in the tube. At low and high pressures the corresponding result takes values of the molecular (Equation 9) and the viscous transmission probability, respectively. Equation 10 was found to predict W_{tr} within 5% over the entire transition range for the flow of air in copper tubes [26]. For short tubes Equation 10 does not hold. Its range of validity is discussed in the literature [59], but due to the lack of experimental data, no general statement can be formulated. It should be noted that for short tube nozzles, Equation 10 overestimates the transmission probability.

There appear to be several approaches to model transitional flow, i.e., throughput in tubes with arbitrary length based on simple combinations of molecular and viscous models [59; 62].

The lack of a well-established and generally recognized model covering the whole pressure regime, however, is probably due to the dearth of experimental work in these conditions.

2.3.3 Choked flow

Precursor gas injection through a nozzle into vacuum proceeds in most cases in choked flow conditions resulting from the large pressure difference between the nozzle upstream and downstream pressure (in the vacuum chamber). In choked flow, the flow reaches the speed of sound at the nozzle exit. It is important to note that when the gas velocity becomes choked, the mass flow of the gas can still be increased by increasing the upstream pressure, e.g., by heating the precursor reservoir.

Choking is established if the pressure ratio of the upstream and downstream pressure fulfills the condition [59]:

$$\frac{P_{u}}{P_{d}} > \sqrt{\frac{\gamma + 1}{2}}^{\gamma/(\gamma-1)}, \qquad (11)$$

where γ is the ratio of the specific heats at constant pressure and volume, which takes typically a value of 1.1 for polyatomic molecules. Accordingly, for these molecules an absolute pressure difference across the tube of >1.7 is sufficient for choking. A description of the transmission probability W_{c} in choked tube-flow conditions covering the molecular and transient regime weighs the molecular transmission probability W_{m} (Equation 9) and the viscous transmission probability W_{v} according to [59]

$$W_{c} = \theta W_{m} + (1 - \theta) W_{v}, \quad \theta = k_{s}Kn/(k_{s}Kn + 1). \qquad (12)$$

The fitting parameter $k_{s} = 13.6$ is related to boundary conditions of a short tube. The laminar viscous transmission probability in choked flow conditions can be found in terms of the ratio of the upstream pressure P_{u} and the pressure P_{Ma} at Mach number $Ma = 1$ according to $W_{v} = \sqrt{\pi \gamma (\gamma + 1)} (P_{u}/P_{Ma})^{-1}$. The pressure ratio P_{u}/P_{Ma} can be further expressed in terms of the Mach number at the tube entrance [62]. The following section will exemplify the above theoretical framework.

2.4 CASE STUDY OF RAREFIED GAS FLOW IN A GAS INJECTION SYSTEM

2.4.1 Gas supply system

Figure 3.3 compares the Knudsen numbers for different solid precursors which span a typical range from low to high vapor pressure compounds used for FEB- (FIB-) induced processing. The pressure decay along the tube corresponds to a vertical displacement of the lines to higher Knudsen numbers. In Table 3.1 the vapor pressure at room temperature and the molecule size for these precursors are listed.

2.4.2 Nozzle

In the following, a tube nozzle of length 6 mm with an inner diameter of 0.4 mm is considered. The flow properties of this nozzle will be exemplified for the precursors $Cu(hfa)_2$ and $Co_2(CO)_8$.

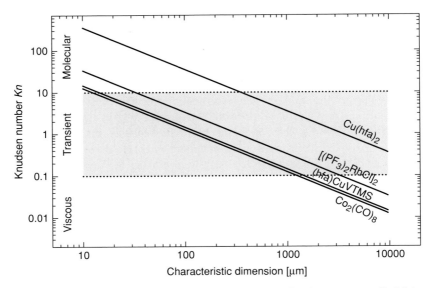

FIGURE 3.3: Illustration of the predominant flow regimes for the precursors $Cu(hfa)_2$ (low vapor pressure compound), $[(PF_3)_2RhCl]_2$, (hfa)CuVTMS, and $Co_2(CO)_8$ (high vapor pressure compound). The Knudsen number was calculated from the mean free path (Equation 3) at the molecule vapor pressure and the characteristic dimension in the flow, i.e., the tube inner diameter which supplies the precursor gas to the nozzle. $Kn = 0.1$ and 10 indicates the flow regime transitions.

Table 3.1 Summary of precursor properties.

Molecule	Vapor pressure [mbar] at room temperature	Molecule size [Å]	Reference
$Cu(hfa)_2$	4×10^{-3}	8.1[a]	[41]
$[(PF_3)_2RhCl]_2$	0.086	5.7[b]	[74]
(hfa)CuVTMS	0.10	8.6[a]	[15]
$Co_2(CO)_8$	0.134	7[b]	[40]

[a]The molecule size is determined from the compound density ρ according to $\delta_m = 1.122(M/(\rho N_A))^{1/3}$ [59].
[b]The molecule size is determined from the longest dimension of the molecule.

According to Equation 11, in the case of vacuum $P \leq 10^{-3}$ mbar on the downstream side and a pressure P_u in the upstream of the tube nozzle close to P_{vap}, the flow is always choked. These boundary conditions are present across the final tube in the GIS, and thus, the transmission probability in the pressure regime from molecular to continuum (viscous) flow can be found from Equation 12. The Mach number at the entrance of the tube in choked flow conditions was ~0.04 for $Co_2(CO)_8$, evaluated from flow simulations (see Section 3.3.). Generally, these simulations further revealed that in choked flow conditions the entrance Mach number is relatively insensitive to the gas species and the pressure, and is mainly determined by the tube geometry. Knowing the entrance Mach number for a given GIS tube geometry allows, thus, to estimate the transmission probability based on Equation 12 in a wide pressure

FIGURE 3.4: Model of the capillary showing the upstream side (pressure P_u) and the vacuum side (local pressure at the exit plane P_e). The regions with transient flow and molecular flow are indicated. At the pressure P_{t-m} the flow changes from transient to molecular.

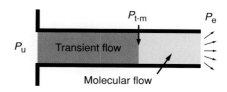

regime for any precursor. Using Equation 12 a flow conductance for the considered tube of $C_{Cu(hfa)_2} = 2.7 \times 10^{-7}$ m^{3}s^{-1} and $C_{Co_2(CO)_8} = 3.4 \times 10^{-7}$ m^{3}s^{-1} is found. The low conductance of the nozzle in most arrangements without flow restrictors is the flux-limiting element in the GIS.

The model in Figure 3.4 describes the pressure drop and the corresponding change in flow regime established in a tube nozzle. It exemplifies the typical transient-flow upstream condition for precursor gas at vapor pressure (see Figure 3.3). The angular velocity distribution of molecules leaving the nozzle is controlled by a molecular, transient, or mixed regime in the last portion of the tube depending on the location of the transition between the flow regimes.

The pressure P_{t-m} at which the flow changes from transient to molecular flow at $Kn = 10$ and at room temperature is found by combining Equations 2 and 3, leading to:

$$P_{t-m} \text{ [Pa]} = \frac{0.093}{\delta_m \text{ [nm]}^2 \, d \text{ [mm]}}. \tag{13}$$

For example, for the precursors Cu(hfa)$_2$ and Co$_2$(CO)$_8$ flowing through a tube nozzle with $d = 0.4$ mm, the transition pressure predicted by Equation 13 is 3.5×10^{-3} mbar and 4.7×10^{-3} mbar, respectively. For both precursors at their vapor pressure (see Table 3.1), the flow in this tube is in a transient regime. Using a first order approach, it is reasonable to assume a linear pressure decay along the tube for molecular *and* transient flow from P_{vap} (upstream) to $P_{dyn} = P_e$ (at the nozzle exit) [58]. In typical experimental conditions the total flux of $J_{Cu(hfa)2} = 1 \times 10^{16}$ cm^{-2} s^{-1} and Equation 4 predicts that the transition occurs at 12% of the tube length. For Co$_2$(CO)$_8$, $P_{dyn} > P_{t-m}$, and thus, it can be concluded that the transition regime will extend throughout the whole tube. After the expansion of the flow into the vacuum chamber, however, the pressure suddenly decreases and molecular conditions are established close to the nozzle exit.

2.5 IMPINGING FLUX

The issue of the locally impinging precursor flux on the substrate supplied by a tube nozzle (see Figure 3.5) has only marginally been addressed in the literature on FEB- and FIB-induced processing. The simulation of the spatial flux distribution onto the substrate $J(x, y)$ provides two important values: the *absolute local flux* determining the process regime and the *spatial flux uniformity* within the processing field. Figure 3.5 shows a schematic of a straight GIS tube nozzle and the relevant parameters that are important to the gas delivery to the process area.

2.5.1 Point sources: Far-field distribution

The molecular flow through an aperture in an infinitesimally thin wall separating a gas source and a vacuum chamber, referred to as a cosine emitter, is described by the cosine law of effusion.

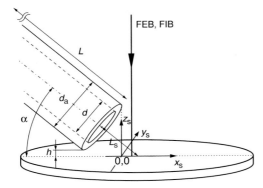

FIGURE 3.5: Scheme of a straight tube nozzle. The origin of the substrate frame of reference (x_s, y_s, z_s) is defined by the tube edge projection onto the substrate which is accessible to the normally incident FEB (FIB). The nozzle-substrate arrangement is defined by the incidence angle α and the clearance h. L_s is the distance from the tube exit plane to the substrate along the axis (Reprinted with permission from [36]. Copyright (2009), IOP).

It can be written in the form of the differential angular throughput with the polar angle θ as [44]:

$$dQ_\theta = \frac{Q}{\pi} \cos\theta \, d\Theta, \tag{14}$$

where Q is the total throughput and the solid angle element $d\Theta = \sin\theta \, d\theta \, d\phi$. We consider the local flux generated by this aperture impinging on a planar substrate placed at a large distance in comparison to the aperture diameter, i.e., the far-field distribution. Such a point source, which is incident under an angle α on a substrate plane (normal incidence: $\alpha = 90°$), produces a spatial flux profile along x_s calculated by [44]:

$$J(x_s) = J_0 \left(\frac{\sin(\alpha + \theta(x_s))}{\sin(\alpha)} \right)^2 \cos(\theta(x_s)) \sin(\alpha + \theta(x_s)),$$

$$\theta(x_s) = \arctan\left(\frac{\sin\alpha}{L_s/x_s - \cos\alpha} \right). \tag{15}$$

The local impinging flux $J(x_s)$ thus depends on the flux J_0 at the intersection point of the aperture symmetry axis with the substrate, the tilt angle α, and the aperture-substrate distance L_s along the tube axis. For normal incidence, it follows from Equation 15 that the cosine emitter produces a polar angular probability distribution $J(\theta) = J_0 \cos^4\theta$ and an impinging flux profile of $J(x_s) = J_0 \cos^4\theta(x_s)$ on the substrate surface. The distribution $J(\theta)$ is obtained by scanning a "point" flux detector radially around the tube exit surface center at a constant azimuth.

For molecular gas flow through tubes, Dayton [24] developed an analytical/numerical solution based on the work of Clausing [17], which predicts the angular probability distribution $J(\theta)$. This solution integrates the distribution across the tube exit surface.

2.5.2 Finite sources: Near-field distribution

Since the GIS nozzle in FEB (FIB) processing systems is located close to the substrate, the near-field impinging flux distribution must be considered. It has been demonstrated that the angular probability distribution is highly non-uniform across the finite tube surface by extending Dayton's model [57]. This result indicates that an analytical description of the spatial flux distribution becomes very complex in the near-field, and for this important reason, simulation approaches have been adopted.

The spatial flux distribution on the substrate produced by a nozzle is composed of the contribution from molecules which are scattered from a surface or from an intermolecular collision (in transient flow conditions) before leaving the tube (see Figure 3.4).

3 MONTE CARLO SIMULATIONS

3.1 GENERAL

The simulation of the flow of rarefied gases leads to the necessity to take into account that the gas is a discrete system composed of individual molecules in constant, random motion. The flow of gas is a superimposed ordered motion of molecules. However, at the molecular level, there is no distinction between the random and ordered motion. Probabilistic Monte Carlo (MC) simulations are the frameworks that allow the mathematical modeling of molecular and transient flows. In the following, fundamental MC simulation algorithms used for rarefied tube flow simulations are summarized.

3.2 ALGORITHMS

The simulation of a molecule trajectory starts at the entrance surface of the nozzle tube (see Figure 3.6). Molecules are uniformly generated over this surface and molecules "start" with a cosine angular velocity distribution [44]. The trajectory of the molecule is then determined by intramolecular collisions and collisions with the inner tube wall. The trajectory is finished when the molecule escapes from the tube either through the entrance surface, (physically this means it is backscattered into the source reservoir), or when it transmits through the exit surface. Transmitted molecules are spatially mapped on the substrate surface and the locally impinging flux J is normalized with respect to J_{tot}.

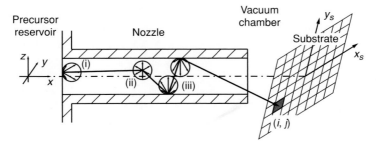

FIGURE 3.6: The molecule trajectories through the nozzle tube are traced by the following stochastic concepts: (i) Cosine point-source at tube entry. (ii) Hard sphere scattering after having travelled a free path s. (iii) Diffuse (cosine) nozzle wall scattering. The substrate is subdivided into a sampling grid for counting impinging molecules (Reprinted with permission from [36]. Copyright (2009), IOP).

In the following paragraphs the fundamental assumptions involved in the collision algorithms and the pressure distribution inside the flow are presented.

3.2.1 Wall scattering

Typically, diffuse scattering of molecules from the inner tube walls is assumed. This means that the desorption direction of the molecule is independent of the direction of incidence. In other words, the molecule desorption from the surface occurs randomly after negligible surface diffusion. The differential angular velocity probability, p, written in terms of the polar and azimuthal angles θ and ϕ, respectively, is described by the well-known cosine law:

$$\mathrm{d}p = \frac{1}{\pi} \cos\theta \sin\theta \, \mathrm{d}\theta \mathrm{d}\phi. \tag{16}$$

For microscopically rough surfaces, the diffuse scattering law is in many cases an adequate description that is well-established theoretically and experimentally (for references see the textbook by Lafferty [59]). Thomson and Owens pointed out that a combination of diffuse and specular reflections of the gas from the tube surface in some conditions matches better with the experimental results [99]. However, it was found that the agreement with experiments of gas flow in a typical nozzle of a GIS is reduced when adding a specular term [36].

3.2.2 Uptake

Chemisorption of molecules at the inner tube walls can be included through an uptake coefficient defined as the probability of sticking "forever" to the tube upon a collision with its wall. MC simulations show that the molecule flux was reduced and collimated by chemisorption, since the molecules transmitting without wall collisions now dominate the angular flux distribution, while the molecules having wall collisions are "lost" by uptake [92]. So far the practical importance of chemisorption inside the GIS was outlined only for the molecule $Fe(CO)_5$, where, as a result, the exit flux composition was enriched with CO molecules that dissociated from the parent molecule, leaving Fe on the tube wall [43]. Of note, chemisorption can change with increasing coverage by chemisorbed molecules, simply due to the fact that the chemisorbed material has different "interaction" properties with the precursor molecule than the initial tube material. It can favor physisorption (and reduce the initial uptake) or lead to increased auto-catalytic dissociation (and increase the initial uptake). Which of these processes will dominate for a specific molecule must be investigated experimentally, since no data is available in the literature. Uptake factors manifest as mass and color changes in the tubing system.

3.2.3 Intermolecular scattering

In MC simulations, transient molecule flow is included by simulating an intermolecular collision after the molecule reached its actual free path $s = \lambda \cdot \ln\left(Rn^{-1}\right)$ [59], with λ the mean free path and Rn a uniformly distributed random number between 0 and 1.

Scattering during intermolecular collisions can be simulated according to the hard sphere model. It is assumed that molecules are hard elastic spheres of diameter δ with a total collision cross-section of $\sigma = \delta^2 \pi$. In view of the random orientation and the large number of collisions, it is reasonable to assume a spherically symmetric geometry even for polyatomic molecules.

In the literature, modeling with more realistic collision cross-sections is well documented [6], but beyond the scope of this chapter. The differential angular velocity probability upon an intermolecular hard sphere collision is isotropic in all directions, as seen from the molecule's center of mass. This is a good approximation, and it has been found that the observable consequences of changes in the scattering law are generally very small [6].

3.3 DIRECT SIMULATION MONTE CARLO

In the direct simulation MC method, the trajectories of a very large number of simulated molecules are followed simultaneously. This approach allows the accurate simulation of flows of rarefied gases from the molecular regime far into the transition regime. It is currently the most reliable way to compute molecular flowfields in transient conditions, and it is more generally applicable than any analytical method [39]. A detailed exposition of the method is available in a textbook written by Bird, the pioneer of direct simulation MC [6].

The essential approximation in the direct simulation MC method is the uncoupling of the molecular motion and the intermolecular collisions over a small time interval or "step." A number of representative molecules are displaced (including the computation of the resulting boundary interactions) over the distances appropriate to this time step, followed by the calculation of a representative set of intermolecular collisions that are appropriate to this interval. The time step should be small in comparison with the mean collision time and, as long as this condition is met, the results are independent of this actual value. The initially imposed flowfield develops with time, and after a sufficiently large number of simulation steps, steady-state conditions are reached if the flow is stationary.

The simulation of the complete flowfield, i.e., the pressure gradients, has an enormous computational cost, which currently renders the usage of the direct simulation MC method unfavorable for GIS optimization and evaluation. While such computations converge reasonably fast for two-dimensional or axially symmetric simulations, the extension to three-dimensional flow problems is not practical on today's best-performing personal computers in a reasonable time.

The gas injection flowfield of the precursor molecules $Cu(hfa)_2$ and $Co_2(CO)_8$ simulated by direct simulation MC (DS3V [108]) is shown in Figure 3.7. These simulations demonstrate that intramolecular scattering outside of the tube nozzle is rare, since $Kn > 10$ close to the tube exit aperture in both cases. The flow of $Cu(hfa)_2$ (see Figure 3.7(a)) in the nozzle predicts the impinging substrate distribution to be entirely governed by the molecular flow regime in the last portion of the tube; while for the flow of $Co_2(CO)_8$ (see Figure 3.7(b)), the transient regime prevails up to the tube end. These findings are confirmed by the test-particle MC implementation described in the next section and compare very well to the experimentally determined flux distributions (see Section 4.1.1.).

3.4 TEST-PARTICLE MONTE CARLO SIMULATION

The simulation of molecular and, to some extend, of transient flow through arbitrary shaped tubing components is ideally suited for a probabilistic simulation within the framework of the test-particle MC method, which was first introduced by Davis [23]. This method relies on a probabilistic model similar to direct simulation MC. In contrast, the mean free path distribution, i.e., the pressure gradient, is not simulated in most test-particle MC implementations.

(a)

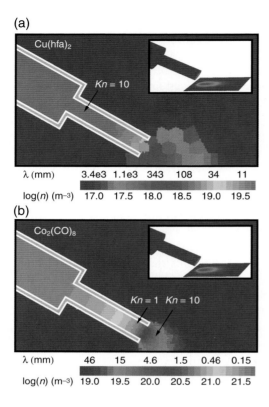

λ (mm)	3.4e3	1.1e3	343	108	34	11
log(n) (m-3)	17.0	17.5	18.0	18.5	19.0	19.5

(b)

λ (mm)	46	15	4.6	1.5	0.46	0.15
log(n) (m-3)	19.0	19.5	20.0	20.5	21.0	21.5

FIGURE 3.7: Three-dimensional direct simulation MC computation of the flowfield produced by a straight tube nozzle with diameter $d = 0.6$ mm ($d_a = 0.9$ mm) and length $L = 3$ mm at an incidence angle $\alpha = 30°$ and a clearance to the substrate $h = 360$ μm. Flow of **(a)** $Cu(hfa)_2$ at a total flux $J_{tot} = 6.4 \times 10^{16}$ cm^{-2} s^{-1} and **(b)** $Co_2(CO)_2$ at a total flux $J_{tot} = 2.1 \times 10^{18}$ cm^{-2} s^{-1} effusing from the tube into the vacuum chamber. The molecule density distribution n, the mean free path λ, and the corresponding Knudsen numbers $Kn = 1$ and 10 are indicated. Note the different range of λ in **(a)** and **(b)**. Outside of the tube, vacuum boundary conditions are employed. The tube upstream-pressure is maintained constant to adjust J_{tot} to the measured value. The computations were performed with diffuse scattering from the tube and substrate surfaces and full temperature accommodation to 27°C.

Recently, the authors presented a simulator for gas injection (GIS simulator [34]) implementing the test-particle MC method, which works in the molecular and transient flow regime. As an output, the impinging molecule flux distribution is obtained as a function of the nozzle geometry and the mutual arrangement between nozzle and substrate. Basically, molecule trajectories are computed consecutively in a large number, typically $10^6–10^7$ molecules which travel through the nozzle into the chamber, to predict the macroscopic flow distribution on a substrate. This approach is physically rigid only for molecular flow conditions where molecule trajectories are indeed independent of one another, since collisions occur only with the inner tube wall. However, it was shown experimentally [36] and by comparison with direct simulation MC [33] that test-particle MC simulations including intermolecular collisions give excellent results in the following way:

1. Transient flow simulations can be performed by setting the mean free path $\lambda = d \cdot Kn$ inside the entire nozzle. If Kn is defined by inserting Equation 5, it thus considers the flow regime prevailing near the end of the nozzle exit.
2. Outside of the nozzle the Knudsen number is set to infinity, i.e., molecules followed a straight trajectory after their last collision inside the tube nozzle until they hit the substrate (no molecular collisions outside the tube).
3. Consecutive trajectories inside the vacuum chamber are not taken into account.

It is the proportion between wall- and intermolecular-collisions near the nozzle exit that finally governs the angular distribution of molecules exiting the nozzle (see Section 2.4.2.).

This leads to the computational concept of a reduced nozzle (tube) length to decrease the computation time. A tube with reduced length L_r yields almost identical simulation results for impinging molecule distributions on a substrate compared to a long full-length nozzle. This concept results in a remarkable reduction in computational effort, since the probability of molecule transmission through a nozzle of reduced length is drastically increased compared to the nozzles of 100 mm length used in today's GISs. We determined the critical reduced nozzle length L_r by a series of simulations where the length of the nozzle was varied while its diameter was fixed. For molecular flow conditions, the spatial distribution of impinging molecules on the substrate is insignificantly altered for nozzle lengths $L_r > 15d$. For transient flow conditions, the critical reduced nozzle length depends on λ and $L_r = 3...5\lambda$. According to kinetic gas theory, this corresponds to a fraction $1 - e^{-L_r/\lambda} = 95...99\%$ of molecules that have suffered an intermolecular collision within the distance L_r. This illustrates that intermolecular collisions have a screening effect for molecules moving toward the nozzle exit. The consequence from the above simulation series is that the pressure gradient inside the nozzle between the reservoir and vacuum only *marginally* influences the molecule distribution on the substrate.

4 CASE STUDIES

4.1 STRAIGHT TUBE NOZZLE

4.1.1 Experiment and simulations

One experimental approach for characterization of the precursor flux distribution relies on the thermal decomposition of impinging precursor molecules on a heated substrate. In the precursor mass-transport-limited deposition regime, the shape of the deposited material represents the locally impinging flux on the substrate. In the following we refer to the method as local chemical vapor deposition (CVD) [36].

The straight tube nozzle is the conventional design used in most GISs. Figure 3.8(a) shows an optical image of the tube-substrate configuration in the reactor chamber.

Figure 3.8(b) shows that transient flow simulations with $Kn = 2$, obtained from Equation 5 and total flux measurements ($J_{tot} = 2 \times 10^{18}$ cm^{-2}s^{-1}), match the three-dimensional deposit topography very well. The good match between this simulation and the experiment is also obvious in the corresponding profile shown in Figure 3.8(c): the simulated full-width at half-maximum (FWHM) matches the experiment to better than 20%, and the deviation in the simulated flux value is <8% in the FEB- (FIB-) accessible region, $x_s > 0$. In contrast, the molecular flow simulated profile deviates considerably. That Equation 5 is a very good estimate for input in our MC simulations is also supported by experiments shown in Figure 3.8(d) using the precursor $Rh_2Cl_2(PF_3)_4$. The calculated corresponding Knudsen number of 10 (molecular flow), used as an input into the simulations, also showed a very good match with the deposit topography. Generally, the shape of the (simulated) impinging flux profiles is strongly sensitive to variations of Kn between 1 and 10, i.e., while changing from transient to molecular flow.

We would like to point out that the agreement, i.e., peak position, FWHM, and accuracy of the molecule distribution, between our simulations and the related experiments is very good compared to previous published attempts [28, 55]. For example, the deposit topography of condensed impinging water molecules, representing the impinging flux distribution, was

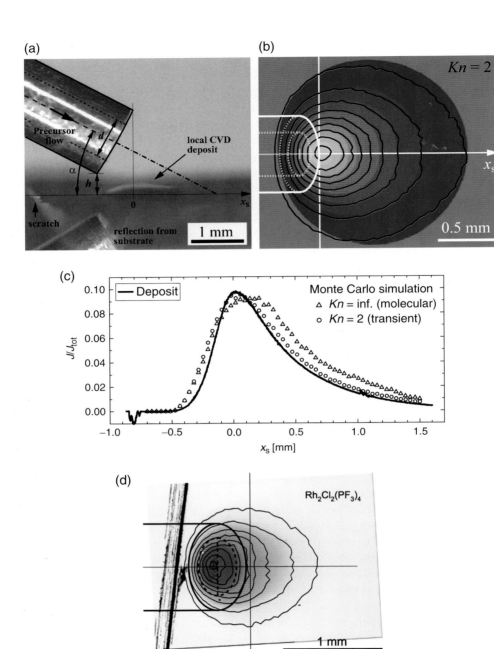

FIGURE 3.8: (a) Microscope image of the nozzle-substrate configuration during CVD. The inner tube diameter d and the tube axis are indicated ($d = 0.4$ mm, $d_a = 0.7$ mm, $\alpha = 30°$, $h = 220\,\mu$m) (Reprinted with permission from [36]. Copyright (2009), IOP). (b) Measured topography (filled contours) of the deposit superposed with simulated isoflux contours (dark lines) for transient flow with $Kn = 2$. The tube position and the reference system are also indicated (Reprinted with permission from [36]. Copyright (2009), IOP). (c) Simulations of impinging molecule flux along x_s (at $y_s = 0$) for molecular ($Kn = \infty$) and transient flow ($Kn = 2$). The measured height profile (solid line) from the deposit shown in **(b)** is superposed (Reprinted with permission from [36]. Copyright (2009), IOP). (d) Comparison of the impinging precursor distribution between experiment (interference colors indicate height) and MC simulation (isodensity contours) for molecular flow conditions ($Kn = 10$) (Reprinted with permission from [102]. Copyright (2006), Elsevier).

measured on a cryo-cooled substrate and modeled solving numerically the continuum Navier-Stokes equation [28]. Models and experiments between the distribution width deviated by nearly a factor of two.

4.1.2 Optimum straight nozzle configuration

This section aims to give guidelines for the optimum position of the straight tube nozzle relative to the substrate in terms of the maximum accessible precursor flux and uniformity within the FEB (FIB) processing field. All the results have been produced by the GIS simulator software. We considered a long tube with $L = 15d$, exceeding the minimum length at which the impinging distribution is influenced by changing the tube length.

In Figure 3.9, two-dimensional isoflux contour patterns are presented for a vertical tube-substrate distance of $h = 0.25d$ and $0.75d$ and a tube incidence of $\alpha = 60, 45, 30, 15, 0°$. The reference origin was chosen as shown in Figure 3.5, i.e., the obscured substrate region for the FEB (FIB) (due to the nozzle) always falls together with $x_s \leq 0$. It can readily be deduced that at large tube incidence angles, the flux peak is inaccessible for the FEB (FIB). The "accessible" flux profiles at $0 < x_s < d$ along the tube axis are summarized in Figure 3.10.

Close to the nozzle, a maximum flux is produced for large incidence angles, whereas at low incidence angles the flux level is strongly reduced even for the flux peaks situated at $x_s > 0$. Within the presented range for $h = 0.25d\ldots0.75d$, the accessible peak flux is maximized for 60° incidence, almost 20% of the total effusing flux, but has the strongest non-uniformity. Optimum uniformity at high flux levels, around 7% to 9%, is predicted at 15...30° incidence depending on h. At incidence $\alpha \rightarrow 0°$, a lower flux level results and is, again, strongly non-uniform within the simulated range.

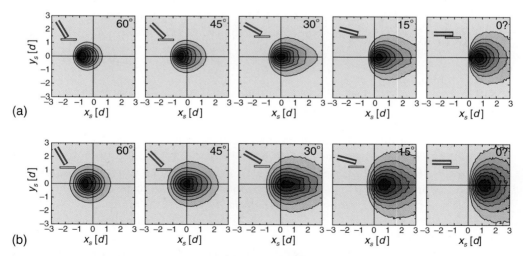

FIGURE 3.9: Simulated isoflux contours (molecular flow) at varying incidence angles and vertical distances (see insets for configuration). The nozzle obscures the substrate at $x_s \leq 0$ for the FEB (FIB), normally incident on the substrate plane. Nozzle dimensions: $L = 15d$, $d_a = 1.75d$. Nozzle-substrate clearance: **(a)** $h = 0.25d$, **(b)** $h = 0.75d$. For a tube nozzle with $d = 0.4$ mm ($d_a = 0.7$ mm), this results in a length $L = 6$ mm and $h = 100\,\mu$m and $300\,\mu$m, respectively. For the absolute flux scale see Figure 3.10 (Reprinted with permission from [36]. Copyright (2009), IOP).

(a)

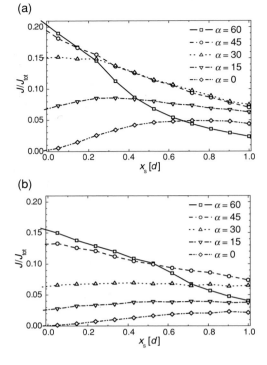

(b)

FIGURE 3.10: Normalized impinging flux profiles (*molecular* flow) on the substrate along x_s (at $y_s = 0$) corresponding to simulations in Figure 3.9. J_{tot} is the total flux effusing from the tube exit surface. Values at $x_s < 0$ are not shown, since these locations are not accessible by the FEB (FIB). Nozzle dimensions: $L = 15d$, $d_a = 1.75d$. Vertical nozzle-substrate distance: **(a)** $h = 0.25d$, **(b)** $h = 0.75d$ (Reprinted with permission from [36]. Copyright (2009), IOP).

In conclusion, if a uniform flux within the accessible region is to be assured, a tube incidence angle of between 15° to 30° is favorable in the molecular flow regime at $h = 0.25d\dots0.75d$. These results are scalable with tube dimensions as long as molecular conditions prevail. In transient conditions, the same qualitative trend is predicted by our simulations; however, simulations are case specific due to the corresponding Knudsen number.

4.1.3 Shadow effects

The strongly directed precursor flow produces shadow patterns behind substrate features, i.e., high aspect-ratio deposits and etched pits. Variations in impinging precursor flux due to shadowing have a strong impact on the process rates. It has been shown experimentally that the deposition rate is higher when the electron beam is scanned toward the flux and lower when scanned with the flux, as illustrated in Figure 3.11. In an attempt to span micron-wide slots by FIB deposited bridges, it was found that the horizontal growth rate is controlled by the direction of the gas flow [25]. When gas flow is perpendicular to the slot, the bridge grows much faster toward the nozzle than away from the nozzle. Gas flow parallel to the slot results in uniform growth from both sides.

Shadow effects due to a three-dimensional obstacle located on the substrate have been simulated by the test-particle MC method yielding quantitative results, as shown in Figure 3.12. The shaded areas with zero molecules impinging are strongly dependent on the obstacle geometry and mutual position of obstacle and nozzle. Here, a relatively large obstacle of 20 μm height and 10×10 μm² area was chosen to highlight the shadow effects. FEB deposits are one to two orders of magnitude smaller in lateral dimensions.

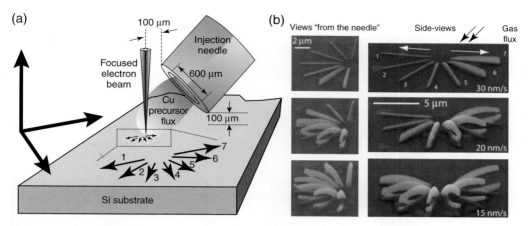

FIGURE 3.11: Lines deposited with (hfa)Cu-VTMS precursor and varying scan direction with respect to the gas flux. Note that the variation of gas pressure over the scanned field is negligible. Lines 1–7 are scanned with the same speed. Line 1 is scanned away from the gas feed tube and line 7 is scanned toward the gas feed tube. **(a)** Schematic of the experimental setup and example of writing sequence. **(b)** SEM micrographs, tilted (70°), of the resulting structures at three different scan speeds. Left: front views; right: side views (Reprinted with permission from [9]. Copyright (2005), Elsevier).

In practice, surface diffusion can supply additional molecules to the area which is shaded from directly impinging flux due to the concentration gradient. The critical dimension of surface diffusion is the molecular diffusion path $(D\tau)^{1/2}$, where D is the diffusion coefficient and τ the residence time of adsorption of a molecule on the surface. Inserting typical values, $\tau = 1$ ms and $D = 3 \times 10^{-8} \ldots 4 \times 10^{-7}$ cm^2 s^{-1} for Cu(hfa)$_2$ (copper hexafluoroacetylacetonate) molecules [104], yields an estimation of the average diffusion path length of 50. . .200 nm. Generally, the shadowed regions will be replenished by surface diffusion when their sizes are comparable to the average diffusion path leading to a homogenized coverage. This is probably the reason that homogeneous small-diameter pillars can be deposited with rotational symmetry and high aspect ratios even when shadow effects exist. When deposited pillar diameters came close to one micrometer, non-rotational symmetries, as well as internal structural inhomogeneities, were observed [101; 106].

The molecule contribution from the operating background pressure is another source of molecule replenishment of shadowed regions. For example, a straight tube nozzle of 6 mm length and 0.6 mm inner diameter typically injects a total flux $J_{\text{tot}} = 9 \times 10^{16}$ cm^{-2} s^{-1}, which raises the background chamber pressure to $P_{\text{b}} = 10^{-5}$ mbar during gas injection of the molecule Cu(hfa)$_2$ at room temperature. Rearranging Equation 4 and inserting P_{b} results in a background flux impinging on the substrate of $J_{\text{b}} = 7 \times 10^{14}$ cm^{-2} s$^{-1} \approx 0.4$ monolayers/s of Cu(hfa)$_2$ molecules. The injected molecule flux impinging on the substrate for various nozzle-substrate arrangements can be found from Figure 3.10. Choosing a tube position of 450 μm above the substrate at an angle of 30° gives a local impinging flux $J = 0.07 \cdot J_{\text{tot}} \approx 4$ monolayers/s from the nozzle. Thus, the background contribution is 10% in this arrangement. For precursors with higher vapor pressures, and correspondingly higher total injected flux, the background contribution would tend to be of less importance, provided the same GIS arrangement and background pressure. It should be noted that for a 60 mm long tube, the total injected flux would decrease by a factor of 10 due to the inverse scaling of the

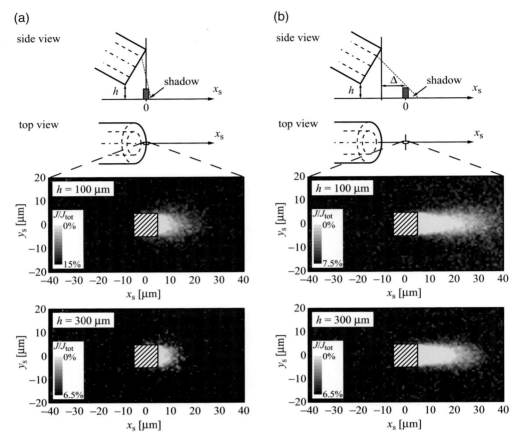

FIGURE 3.12: Side- and top-view illustration of nozzle-substrate arrangement with the corresponding simulated isoflux contours of impinging molecule flux J/J_{tot} around an obstacle (hatched box) of $20\,\mu m$ height and $10 \times 10\,\mu m^2$ footprint (the obstacle is scaled 10:1 in the side-view illustration). The simulation was performed in molecular flow conditions, for nozzle-substrate heights $h = 100\,\mu m$ and $h = 300\,\mu m$, and a tube incidence angle $\alpha = 30°$. The nozzle dimensions were $d = 0.4$ mm ($d_a = 0.7$ mm) with a length $L = 4$ mm. **(a)** The obstacle is located closest to the nozzle. **(b)** The obstacle is located at a distance $\Delta = 400\,\mu m$ away from the nozzle (Reprinted with permission from [36]. Copyright (2009), IOP).

throughput with the tube length [59]. Then background contribution and injected flux can become comparable.

A third source of molecule contribution arising from the intermolecular gas-phase scattering above the substrate surface can be quantified by the direct simulation MC method (see Section 3.3.). Simulating a typical (high) total flux of 2×10^{18} cm^{-2} s^{-1} of $Co_2(CO)_8$ molecules leaving the nozzle tube ($d = 0.6$ mm, $d_a = 0.9$ mm, and $L = 3$ mm) and impinging on an $h = 350\,\mu m$ distant substrate under an incidence angle $\alpha = 30°$ results in a maximum local impinging flux of $J = 3 \times 10^{17}$ cm^{-2} s$^{-1} \approx 1200$ monolayers/s (molecule size 0.4 nm^2), a mean free path between molecule collisions of $\lambda > 1$ mm at the tube exit, and $\lambda > 8$ mm near the substrate. Consequently, gas-phase scattering outside the tube and above the substrate is negligible for the GIS arrangement considered.

4.2 NOZZLE DESIGNS

Alternative nozzle geometries are inspired by the search for high *and* homogenously impinging local flux at the lowest total throughput, i.e., at the lowest pressure load for the microscope chamber. An obvious choice to increase the local accessible flux from a straight tube is to shape the tube end conically, allowing the positioning of the tube exit aperture closer to the substrate [54]. Nozzle designs creating a concentrator chamber above the substrate with a FEB (FIB) access hole can guide the molecule flux much closer to the substrate and consequently minimize molecule losses due to the divergent nature of effusion. Furthermore, shadow effects are reduced since molecules are supplied from the concentrator chamber walls. Two examples of beehive concentrators are illustrated in Figure 3.13. These nozzles need to be aligned relative to the FEB (FIB) in such a way that, while the nozzle is inserted, the beam would pass through the centre of the aperture at the top of the nozzle. The bottom opening of the nozzles should be parallel to the sample plane. The semi-spherical nozzle shape in Figure 3.13(b) is optimized to concentrate the molecules to the FEB (FIB) processing area. This is achieved since adsorbed molecules inside the virtual chamber preferentially desorb toward the sphere center, which is concentric with the working point on the substrate. Additionally, the gas delivery tube is directed toward the working point to take advantage of the directly substrate-impinging flux.

Two alternative GIS nozzle designs depicted in Figures 3.14(a) and 3.15(a) were analyzed experimentally and theoretically in a recent article [36]. Figures 3.14 and 3.15 summarize local CVD experiments with the precursor $Co_2(CO)_8$. The height profiles of the deposits along x_s (at $y_s = 0$) are superimposed with flux simulations in transient conditions, as

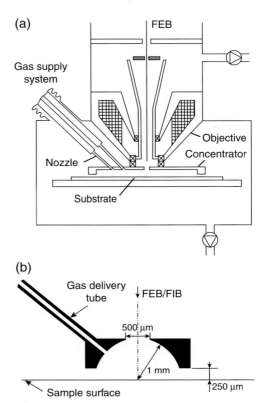

FIGURE 3.13: Different concentrator concepts for obtaining a high local precursor pressure. **(a)** Beehive concentrator design by NaWoTec (from [56]), and **(b)** an optimized design with a semi-spherical shape (Reprinted with permission from [83]. Copyright (2004), AIP).

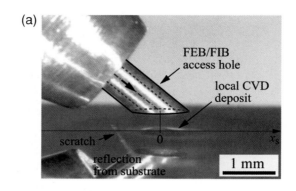

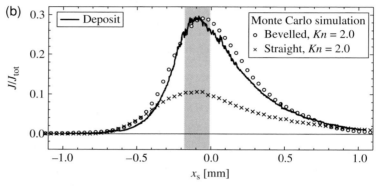

FIGURE 3.14: Local CVD experiments with bevelled nozzle. **(a)** Optical image of the nozzle-substrate configuration. The nozzle geometry is pointed out by an overlay. **(b)** Experimental and simulated profiles along x_s (at $y_s = 0$). The shaded area indicates the access hole position as seen by a zero-tilt impinging electron beam. Simulations were performed with $Kn = 2$ and nozzle dimensions of $L = 4.16$ mm, $d = 0.4$ mm, $d_a = 0.7$ mm, $d_h = 170\,\mu$m. Nozzle-substrate configuration: $\alpha = 40°$, $h = 287\,\mu$m. For comparison, the simulated profile of a straight nozzle with equivalent tube axis, dimensions and nozzle-substrate clearance is plotted (Reprinted with permission from [36]. Copyright (2009), IOP).

predicted by Equation 5. A very good agreement with the experiment can be seen for the bevelled nozzle design in Figure 3.14(b); the simulated FWHM corresponds within 12% to the FWHM of the deposit and deviations of the simulated flux values are <10% inside the FEB- (FIB-) accessible region (access hole). With the present conditions, the FEB (FIB) access hole had no significant influence on the flux distribution, as was confirmed by both the simulation and the experiment. The placement of the FEB (FIB) permitted access to the maximum molecule flux. Compared to a straight tube nozzle, approximately three times higher molecule flux can be achieved and accessed with a bevelled nozzle at the same throughput (see Figure 3.14(b)).

Experiments and simulations with the doubly perforated tube, closed at one end, are summarized in Figure 3.15. The simulated impinging local flux again matches well with the experimental profile (see Figure 3.15(b)). The peak position is perfectly matched and the simulated FWHM deviates <16% from the experiment.

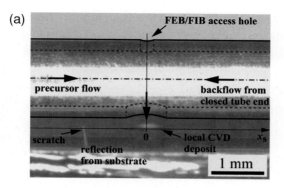

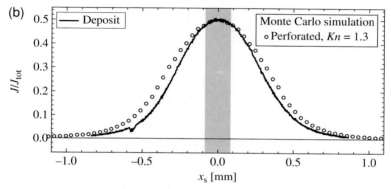

FIGURE 3.15: Local CVD experiments with perforated nozzle. **(a)** Optical image of the nozzle-substrate configuration. The nozzle geometry is pointed out by an overlay. **(b)** Experimental and simulated profiles along x_s (at $y_s = 0$). The shaded area indicates the access hole position as seen by a zero-tilt impinging electron beam. Simulations were performed with $Kn = 1.3$ (within the tube) and a nozzle geometry of $L = 10$ mm, $d_i = 1.07$ mm, $d_a = 1.5$ mm, $d = 0.64$ mm, $d_h = 160\,\mu$m. Nozzle-substrate configuration: parallel, $h = 330\,\mu$m. Note the J/J_{tot} axis scale with respect to Figures 9(c) and 14(b) (Reprinted with permission from [36]. Copyright (2009), IOP).

Table 3.2 compares simulations of the maximum accessible impinging flux for the three nozzle geometries. The inner diameter was kept constant for the three nozzles. The figure of merit is given in percentage of J_{max}/J_{tot} of a straight tube nozzle. It is evident that the highest accessible molecule flux is achieved with the doubly-perforated nozzle design. This becomes especially important when low vapor pressure precursors are to be supplied. Furthermore, the rotational symmetry of such a nozzle design prevents shadow effects and results in more homogeneous three-dimensional deposits.

4.3 BACKGROUND IRRADIATION

Within the gas phase volume above the substrate, collisions between molecules with incident electrons and ions, as well as emitted electrons and sputtered atoms, occur, leading to scattering, ionization, and dissociation (see Figure 3.16(a)). The scattered electrons (ions) form a beam skirt

Table 3.2 Comparison of simulated local impinging flux at a fixed nozzle-substrate clearance h for three nozzle geometries. The nozzle diameter was set to $d = 0.4$ mm for all geometries, the remaining dimensions and Knudsen numbers are identical to Figures 3.8(a-c), 3.14 and 3.15. J_{max} is the maximum accessible flux with a zero-tilt incident electron beam. The percentages give the enhancement of maximum accessible flux with respect to the straight nozzle design.

	J_{max}/J_{tot}		
H [μm]	Straight	Bevelled	Perforated
100	0.15 (100%)	0.70 (460%)	0.81 (540%)
300	0.07 (100%)	0.27 (390%)	0.32 (460%)

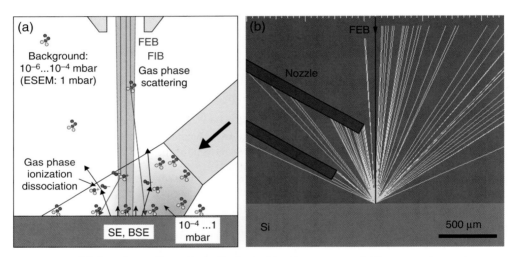

FIGURE 3.16: **(a)** Gas phase effects. Typically, the incident beam passes 5–10 mm gas phase at background pressure and about 1 mm at locally increased pressure before impinging on the specimen (Reprinted with permission from [105]. Copyright (2008), AIP). **(b)** MC-simulated backscattered electron trajectories interfering with a straight tube nozzle. The primary electron energy was 20 keV. Note that secondary electrons are generated at all impact points (not shown).

which manifests as a broad low intensity background irradiation onto the substrate. With the invention of environmental scanning electron microscopes, the electron beam skirt formation and its consequences have been studied in detail [22; 21]. Less studied is the interference of the nozzle with backscattered electrons emitted from the focusing spot of the FEB, as illustrated in Figure 3.16(b) by an MC simulation. Electrons which hit the nozzle produce an additional background of backscattered and secondary electrons on the substrate, which can lead to precursor decomposition in a broad area around the FEB impact. This background irradiation can be minimized by biasing the nozzle with an electrostatic potential [83]. Alternatively, a larger distance between nozzle and substrate would also decrease the effects of secondary electron emission from the nozzle.

4.3.1 Gas phase scattering

Scattering in the gas phase above the substrate becomes significant when the mean free path of electrons or ions $\lambda_{e,i}$ in a gas with density $\rho = MP/(RT)$,

$$\lambda_e = \frac{M}{N_A \rho \sigma}, \tag{17}$$

becomes smaller than the distance traversed by the primary beam in the gas phase. The MC trajectory simulation is the same as for solids [84].

The electron mean free path scales inversely with gas pressure, which can range from 10^{-6} mbar (background pressure) to about 1 mbar (local injection gas pressure or background pressure in environmental and variable pressure microscopes). The fraction of gas phase scattered incident electrons or ions becomes:

$$N_s = 1 - \exp\left(-L/\lambda_e\right), \tag{18}$$

where L is the length traversed by the beam in the gas phase. For tube-nozzle-based GISs, the local pressure must be taken into account.

Table 3.3 shows that for gas phase scattered incident electrons, the skirt full-widths, comprising 50% of all electrons, are in the micrometer range. Since the flux distribution of the scattered electrons and ions scales with the inverse square of the full-widths, a very low background flux is obtained, meaning that the resolution of the primary beam is not lost, even at relatively high collision percentages. Part of the gas phase collisions can produce ionized molecules at an amount given by the ionization cross-section and its energy dependence. At keV incident energies, the efficiency for all these mechanisms is low. This is why gas phase initiated reactions within the incident charged-particle beam volume and within the backscattered electron "cloud" might be mostly neglected. However, it is the "fast" emitted

Table 3.3 Skirt MC simulations (Mocasim [85]) of incident electrons (zero diameter beam) passing 1 mm $Co_2(CO)_8$ in the gas phase $P = 0.023$ mbar ($\rho = 3.2 \times 10^{-7}$ g cm^{-3}). Mean free electron paths (λ_e), the fraction of scattered incident primary beam electrons (N_s), and skirt radius comprising 10% and 50% of all scattered electrons are given.

Molecule	$Co_2(CO)_8$					
Energy	1 keV	3 keV	5 keV	10 keV	20 keV	30 keV
λ_e [mm]	6.2	14.6	22.3	40.1	72.9	103.5
N_s [%]	15.1	6.5	4.4	2.5	1.3	1.0
$r_{10\%}$ [μm]	3.9	2.1	1.4	0.8	0.5	0.4
$r_{50\%}$ [μm]	28.4	17.9	13.6	9.2	6.2	5.0

secondary electrons that can ionize molecules with the highest efficiency in the gas phase just above the substrate, since the ionization cross-section peak is at around 100 eV. The same holds for other low-energy dissociation mechanisms.

The skirt formation has been studied quantitatively by MC simulations of electron (ion) scattering in the gas phase for the case of a typical tube nozzle (see Figure 3.17). These simulations involved an electron (ion) beam passing through 1 mm of precursor gas before impinging onto the substrate. The flux distribution $f(r)$ of scattered and non-scattered electrons (ions) at the substrate plane is a very sharp peak function with a maximum value at $r = 0$. In Figure 3.17(b) and (d), this distribution is visualized in a radial plot of the integrated distribution $S(r) = \int_0^r f(r)\, 2\pi r\, dr$, $S(\infty) = 1$.

However, the role of secondary electron emission on gas phase ionization, or dissociation in general, has not yet been studied systematically in this context to our knowledge. A better resolution in focused electron beam induced deposition of dots with reduced or even closed precursor supply was noticed [20, 97], which might hint to an additional supply mechanism mediated by a secondary electron-gas phase reaction. A way to distinguish between surface or gas phase controlled deposition (etching) is to change the substrate temperature. For surface-controlled dissociation, a decrease in process rate with increasing temperature is detected, which allows us to determine the desorption energy. In the case of gas-phase-determined molecule dissociation, substrate heating should result in negligible changes of process rates.

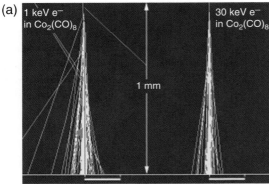

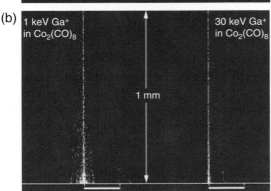

FIGURE 3.17: (a) MC-simulated electron trajectories (MOCASIM [85]) and **(b)** Ga^+ ion trajectories (SRIM [111]) from a zero-diameter beam passing 1mm of $Co_2(CO)_8$ in the gas phase at $P = 0.023$ mbar ($\rho = 3.2\ 10^{-7}$ g cm^{-3}). 100 scatter-trajectories are plotted. The scale bars are 200 μm. **(c), (d)** The corresponding integrated spatial distribution S of scattered *and* non-scattered electrons (ions) at the specimen plane. Note that the intersection with the 10% S and 50% S lines give the radii r containing 10% and 50% of the electrons (ions). The fraction of incident electrons which undergo scattering is given in Table 3.3.

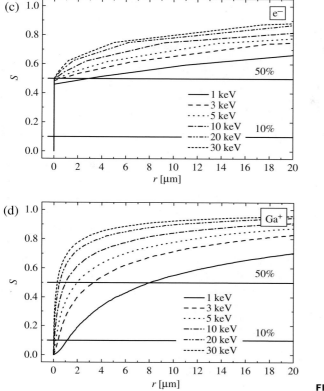

FIGURE 3.17: (Continued)

It should be noted that the beam skirt can lead to reduced surface diffusion because it irradiates the surroundings and decomposes the precursor before a substantial amount can diffuse toward the main beam region. Pre-scanning of a large micron-size region is one of the techniques commonly used in SEM to reduce contamination by surface diffusion [47].

5 DESIGN ASPECTS OF GAS SUPPLY SYSTEMS

While the nozzle is the key part determining the molecule distribution delivered to the process area, a GIS also includes a gas supply system that provides a defined pre-nozzle pressure and throughput of the volatile precursor. The flow conductance and the corresponding pressure gradients of simple gas supply systems, or of single components, can be analyzed using flow theory [59]. However, in most cases it is more efficient to experimentally determine the relevant parameters of the gas supply systems (see Section 2.1.).

In this section, the possibilities for the design of gas supply systems are summarized. Based on the requirements defined by the physical and chemical properties of the precursor, the methods to regulate and to measure the precursor throughput will be presented. Finally, several state-of-the-art systems integrating the described components will be discussed.

5.1 GENERAL ASPECTS

The gas supply system may be enclosed entirely in the process chamber or transfer the gas from an external reservoir into the chamber via vacuum feed-throughs. An important issue in the gas supply design is the minimization of the dead volume in the system. On one hand, the dead volume holds precursor gas that is not actively used for the process, which may be a significant cost factor with expensive precursors. On the other hand, the dead volume must be emptied in an evacuation step, which extends the pumping time required. This can be achieved in compact designs with short and thin tubing interconnects and with an appropriate valve selection. Furthermore, for fast response time upon switching on or off the gas flow, a shutdown-valve should be inserted close to the nozzle.

5.2 CONSTRUCTION MATERIALS FOR GAS SUPPLY SYSTEMS

The materials used for the construction of a gas supply system must fulfill the following general requirements: the material must be gas-tight and not outgas in order to maintain a sufficient vacuum level (for outgassing rates of construction materials in vacuum see, for example, [53; 71; 110]), be non-reactive with the precursor gas, and be mechanically firm enough to establish vacuum tight fittings. Furthermore, GIS components which are located in the vicinity of the charged particle beam, e.g., the nozzle, must be non-magnetic.

5.2.1 Metals

Stainless steel is popular as constructive material in vacuum components. Especially 316L molybdenum-bearing-grade steel is used. This austenitic steel contains additions of <0.03% C, 16–18.5% Cr, 10–14% Ni, 2–3% Mo, <2% Mn, <1% Si, <0.045% P, <0.03% S. The molybdenum gives 316 good overall corrosion resistant properties in chloride environments [89]. It has excellent forming and welding characteristics. The austenitic structure also gives this grade excellent toughness, even down to cryogenic temperatures. For less reactive precursors, austenitic carbon steel or stainless steel 304-grade can be used. Thermal pre-treatment of steel can provide non-corrosive surfaces [73].

For highly reactive precursors, e.g., halogenides, which react with residual moisture forming hydrofluoric, hydrochloric, or hydrobromic acid, special Ni-based superalloys such as Monel™, Inconel™, and Hastelloy™ can be used [60]. Superalloys are mechanically stronger than pure nickel, but are extremely corrosion resistant and non-ferromagnetic.

Monel™ designates a group of nickel-copper alloys containing about 66% nickel and 31.5% copper, with small amounts of iron, manganese, carbon, and silicon. Monel™ alloys are resistant to corrosion by many agents, including rapidly flowing sea water. They can be fabricated readily by hot- and cold-working and welding. Monel™ 400 contains more than 63% Ni/Co-alloy, 27–33 Cu, 2% Fe, 2.3–3.2% Al, 1.5% Mn, 0.3–0.9% Ti and 0.5% Si. Monel™ has an excellent corrosion resistance against hydrofluoric acid, sulfuric acid, phosphoric acid, organic acids, as well as alkalies and chlorine [89]. It has also a good resistivity against hydrochloric acid; only insufficient resistivity exists against oxidizing acids such as nitric acids.

Inconel™ 625 is a nickel-chromium-molybdenum alloy, being especially resistant to pitting and crevice corrosion in a wide range of corrosive media. Inconel™ 625 consists of more than 58% Ni, 20–23% Cr, 8–10% Mo, 3.1–4.1% Nb+Ta and max. 5% Fe. It typically finds applications in marine, aerospace industries, and chemical processing.

Hastelloy™ X is a nickel-based superalloy that possesses excellent resistance to corrosion. Additionally, it delivers strong strength characteristics at elevated temperatures and is well suited for forming and welding. Hastelloy™ X contains 45–55% Ni, 20.5–23% Cr, 17–20% Fe, 8–10% Mo, <1% W, Si, Mn, Al, Ti, Cu.

For less corrosive precursor gases, brass may also be used as constructive material. Brass contains zinc as the principal alloying element, with or without other designated alloying elements such as iron, aluminium, or manganese. The high outgassing rate of brass, due to the high vapor pressure of zinc, limits its application to moderate vacuum levels. Brass alloys containing aluminium as an alloying element are mechanically stronger and more corrosion resistant [90]. The aluminium causes a hard layer of thin, transparent and self-healing aluminium oxide (Al_2O_3) on the surface. Especially tubes, tube fittings and valves made of brass are widely used. Brasses have good resistance to atmospheric corrosion, but corrosion results in the formation of a thin protective green "patina." However, brass is not suitable for reactive chemicals, such as reactive halogenides or ammonia. In ammonia, corrosion cracking of brass is a known problem, and with reactive chlorine species a zinc chloride is formed easily [19].

Aluminium as a construction material is rarely used for tubing or valves, but it is widely used for vacuum chambers and for vacuum flanges, especially Klein flanges (KF). Conflat flanges (CF) typically are made of stainless steel with copper gaskets. Aluminium can be easily machined and allows a low background pressure. It has a corrosion resistance only for inert gases or oxidizing media [90]. The corrosion resistance, as well as surface hardness, can be increased by anodic coatings of the aluminium surface and also aluminium alloys with improved corrosion resistance exist. Aluminium is especially sensitive to halides, and pitting corrosion by chloride solutions is frequently encountered. Due to the low hydrogen diffusion rate of aluminium (as well as carbon-free iron), it is suitable for hydrogen gas supply systems.

5.2.2 Ceramics

For precursor reservoirs and gas supply systems, silicon oxide (quartz glass) and borosilicate glass have been used. Among the benefits are optical transparency, the smooth surface, and the excellent corrosion resistance against most acids, alkaline as well as all oxidizing and most reducing agents. Among the downsides of glass are the mechanical fragility and the adsorption of water vapor on the glass surface resulting in outgassing problems. Sodium borosilicate glasses are preferred over quartz glass for their better chemical inertness.

5.2.3 Polymers

Apart from metallic and ceramic materials, elastomers are used for sealings or as flexible tubing components. A problem for all hydrocarbon-based materials is their outgassing rate of 10^{-7} to 10^{-6} mbar l cm^{-2} sec^{-1}. Also mechanical stability at temperatures above 100°C is an issue that has to be considered.

Polytetrafluoroethylene (PTFE), better known by the brand name Teflon™, is a fluorocarbon polymer with an extraordinary chemical inertness. PTFE can be in continuous contact with most aggressive organic and inorganic chemicals, including halogenides, oxidizing and reducing acids, alkaline lyes, and ammonia with no detectable chemical reaction taking place. Also absorption of chemicals in the PTFE material is unusually low. Because of the high electronegativity of fluorine, PTFE is not wet by polar substances, e.g., water, and apolar substances, e.g., oil. PTFE can also be used over a wider temperature range from 200°C (the melting point is 327°C) down to cryogenic temperatures.

Polysiloxane, also referred to as silicone rubber, is a material class that provides a high elasticity in comparison to most organic polymers. The backbone of polysiloxanes consists of Si-O-Si units, resulting in large bond angles in comparison to C-C-C units, yielding a much more flexible polymer. Polysiloxanes are temperature-stable up to 250°C [49], chemically inert, and resistant to oxidation, hydrolysis, and degradation. Due to the high gas diffusion rate in polysiloxanes [68], this material cannot be used in the reservoir part, but may be suitable as flexible tubing inside the chamber at moderate vacuum levels.

Other organic polymers such as polyimide (PI), polyvinyl chloride (PVC) or polymethyl-methacrylate (PMMA) may be used for gas supply systems, but all suffer from a content of residual volatile oligomers, as well as the absorption of several percent of water. Both effects result in significant outgassing problems and incompatibility with water-sensitive precursors. Above 80°C (PVC, PMMA) to 150°C (PI) the mechanical stability also degrades. Therefore, organic polymers are avoided whenever an alternative is available.

6 COMPONENTS OF GAS SUPPLY SYSTEMS

6.1 PRECURSOR RESERVOIR

Both research and commercial precursor reservoirs are typically fabricated of stainless steel (see Figure 3.18(a)). Steel containers offer a mechanically stable containment fulfilling applicable safety regulations and can be adapted to the desired design by mechanical processing. In research, glass reservoirs are also used for precursors not sensitive to residual adsorbed water (see Figure 3.18(b)). The advantage of glass containers is that the consumption of the precursor and color changes due to decomposition in the reservoir may be conveniently monitored. With any reservoir material, it has to be considered that the reservoir material may react or catalytically initiate a decomposition of the precursor, so that it may be advisable to apply a coating on the reservoir walls. For degassing liquid precursors, it is advisable to design the reservoir such that it can be immersed into a cooling media, e.g. liquid nitrogen, to perform repeated freeze-pump-thaw cycles.

6.2 SHUTDOWN-VALVES

A system to shut down the precursor gas flow is required with every gas supply system. The shutdown-valve is essential to stop introduction of the gas flux to the nozzle, in order to enable the imaging operation of the electron microscope or the focused ion beam system. Furthermore, precursor consumption is a significant cost factor and shall be limited to processes requiring the gas in the vacuum chamber. Different technical concepts are at hand to realize this function. An important selection criterion concerning the chemical inertia against precursors is the sealing, which may be a metal-metal-contact, but may also consist of polymers such as Viton™, Kalrez™, Teflon™, or other fluoroplastomers.

FIGURE 3.18: (a) Glass reservoir with flange fitting [12]
(b) Precursor reservoir made of stainless steel [96].

6.2.1 Manually operated shutdown-valves

Various types of mechanical valves exist, including ball, plug, bellow-sealed plug, diaphragm, and needle valves. Ball and plug valves are commonly used as shutdown-valves. These valves do not offer the fine control that is necessary for applications requiring regulation, but they are durable, easy to maintain, and achieve perfect shut-off even after years of use. The isolation to the outside environment is usually accomplished by polymer sealings (Viton™ or Kalrez™), hence, the compatibility with the used precursor should be checked. For toxic or sensitive precursors, a physical separation to the outside environment is advisable. This can be realized by bellow-sealed valves or diaphragm valves. Due to the complex setup, these valves are comparatively large. Shut-off needle valves are compact, but have a low maximum flow coefficient. Alternatively, gate or butterfly valves can be used. These valves are mainly used for tube diameters larger than 1 inch, and are popular for controlling the pumping flux of vacuum pumps. For gas supply systems, these valves are of inferior relevance. A detailed description of the various shutdown-valves types can be found for example in [48].

6.2.2 Actuated shutdown-valves

All shutdown-valves described above can be actuated by pneumatic pressure or by electrome-chanical operation. This is a common feature within automated gas supply systems.

Pneumatic valves are operated by a pilot medium under pressure, usually compressed air. Pneumatic valves allow a fast opening and closing via remote control, which is often desirable for safety of the gas supply system. Pneumatic operation is even feasible within the vacuum chamber, if vacuum-tight connectors are used. However, commercial systems typically apply the pneumatic actuators outside of the vacuum chamber.

Direct electric actuation is usually performed with solenoid valves. They operate with an electric current through a solenoid—a coil of wire—that induces magnetic fields. The magnetic force moves a piston, which opens or closes the valve. The current to actuate the valve is typically higher than the current needed to maintain the valve in the actuated position. The magnetic fields produced in solenoids must be properly shielded to avoid interference with the FEB (FIB). Typically, solenoid valves are monostable and they are offered as normally-closed or normally-open types. Because of electrical operation, the use of flammable precursors is not safe due to the risk of a spark inflammation. The thermal input from the energy dissipation in the activated solenoid increases the temperature in an uncontrolled way, which is undesirable for GISs. This effect may lead to a premature decomposition of the precursor already within the hot valve and, eventually, to clogging of the valve. For thermally sensitive precursors, solenoid valves should therefore be used with care in gas supply systems. However, some solenoid valves are available with a latching mechanism. These valves are actuated by a short current-pulse, preventing the system from heating up.

Actuated valves also allow automated pre-evacuation cycles and purge cycles (see Section 6.6.) to be implemented in the operation of the GIS. This may be done for preconditioning of the system. Repeated purging with inert gas in combination with evacuation of the gas supply manifold is used to remove residual gases, water vapor originating from leaks to the outside environment, and potentially decomposed precursor. For toxic precursors, this evacuation and purging operation is essential for replacement or refill of the precursor without being exposed to the toxic vapor stored within the gas manifold.

For several commercial GISs, an actuated shutdown-valve is implemented directly within the precursor reservoir (see Figure 3.19(a)). The central part of this compactly designed valve system is a cylindrical passage with a solid actuator piston inserted co-axially. The outer diameter of the actuator piston (plunger) is smaller than the inner diameter of the cylindrical passage in order to form a channel for gas delivery. The actuator piston can be pushed toward an O-ring seal, closing the passage, or can be pulled away from the O-ring, opening the passage by a circular aperture. The actuator piston is mounted on a vacuum-sealed, flexible membrane, which can be operated by an actuator located outside of the vacuum chamber. The actuated piston is frequently driven by a pneumatic cylinder (see Figure 3.19(b)). The pressure of the pilot medium enters the actuator cylinder and acts on the piston, which allows the seal to open or to close through the stem. The actuation of the seal into its rest position is either driven by a return spring or by a double-acting pneumatic configuration.

(a)

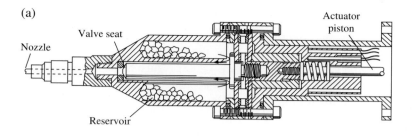

(b)

FIGURE 3.19: **(a)** Compact GIS with integrated shutdown-valve as patented by FEI [82]. **(b)** Pneumatic Cylinders [2].

6.3 FLUX REGULATING SYSTEMS

While shutdown-valves solely block the precursor flux into the vacuum chamber, for instance while operating in imaging mode, a regulation system is required to adjust the flux to the optimum throughput. This is typically achieved by introducing a regulating valve (see Section 6.3.1) or—for highest precision—a mass flow controller (see Section 6.3.2.). Another option to lower the precursor flux is by cooling the precursor reservoir with a Peltier element or a cooling medium. Cooling results in a reduction of the flux due to the temperature dependence of the precursor vapor pressure. Note that it is sufficient to cool the precursor reservoir, while all other parts of the gas manifold may remain at ambient temperature. Alternatively, the molecule flux can be controlled via the nozzle diameter and length or by introducing a fixed aperture. If the precursor flux is too low, the precursor reservoir can be heated, or the precursor can be transported with a carrier gas flux. Heating of the precursor in the reservoir will result in a larger amount of gas evaporating. As the vapor will condense or crystallize on colder spots, it is necessary to heat the entire gas supply line from the reservoir to the nozzle. Optionally, the delivery of sufficient amounts of precursor at room temperature is achieved by using a carrier gas saturated with the precursor. The flux of the carrier gas will transport and drag precursor molecules through the nozzle. However, it has to be considered, that the total flux through the nozzle is limited by the tolerated maximum chamber pressure. Furthermore, the carrier gas may also co-adsorb on the surface leading to lower precursor coverage. To reduce this undesired side effect, inert carrier gases such as Ar or N_2 are used. Alternatively, the gas flux of low vapor pressure precursors is increased by a supply with large flow conductivity connecting the reservoir to the nozzle, i.e., implemented in compact systems which incorporate the entire GIS inside the vacuum chamber.

6.3.1 Regulating valves

Regulating valves operate as variable apertures to control the precursor flux through the gas-supply manifold. The flux is determined by the size of an aperture cross-section at the valve seating. The needle valve shown in Figure 3.20(a) regulates the throughput by adjusting the stem position, which increases or reduces the aperture cross-section. The stem may be operated manually or motor-driven.

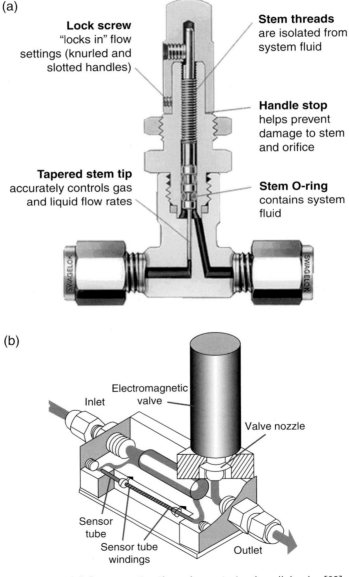

FIGURE 3.20: (a) Cross-section through a metering (needle) valve [96].
(b) Schematic illustration of a mass flow controller [18].

For selecting the proper valve, technical data provided by the valve manufacturer should be used as a basis. For vacuum-designed valves the minimum and maximum controllable gas flow, as well as tightness, is given in mbar l s^{-1}. Sometimes flow values are given in standard cubic centimetres per minute (sccm) which can be converted to the desired flow units as:

$$1 \text{ sccm} = 0.017 \text{ mbar l s}^{-1} = 4.48 \times 10^{17} \text{ molecules s}^{-1}.$$

A gas flow of 0.0003 sccm ($= 5 \times 10^{-6}$ mbar l s^{-1} $= 1 \times 10^{14}$ molecules s^{-1}) to 600 sccm ($= 10$ mbar l s^{-1} $= 3 \times 10^{20}$ molecules s^{-1}) is the useful range for gas-assisted FEB and FIB processing. For valves designed for ambient pressure operation, the gas flux is typically stated as the C_V-value, experimentally determined by the manufacturer. This flow coefficient offers a standard method of comparing valve capacities and sizing valves for specific applications widely accepted by industry. By definition, the C_V-value expresses the flow capacity in imperial units: U.S. gallons per minute of water that a valve will pass for a pressure drop of 1 psi (pounds per square inch). However, the C_V-value is not a relevant indicator for the molecular flow regime; it can be used as a qualitative selection criterion only. A small C_V-value, as found with ultrafine dosing valves, should be preferred.

The maximum recommended long-time operating temperature needs to be considered if the valve is used within a heated gas supply line. This is the highest temperature that must never be exceeded in order to not damage the valve. This temperature limit must be respected when baking-out the valve to desorb less-volatile residual components which have been accumulated within the valve.

6.3.2 Mass flow controllers

A mass flow controller (MFC) regulates the gas flow based on the feedback from an integrated flow sensor. The channel for the main gas flow and all other components of the MFC are mounted on one base (see Figure 3.20(b)). A flow splitter divides the gas stream precisely in two portions at a ratio that is constant over the entire flow regime. One portion flows through the main flow path with a larger diameter, the second portion flows over the bypass containing the flow sensor. Integrated electronics adjust the control valve in closed-loop operation to maintain the set flow rate. The valves may be actuated either by a solenoid, a piezoelectric, or a thermal actuator.

The sensor is built from two resistive temperature-sensing elements. The gas flow through the evenly heated bypass tube transports heat from the upstream section to the downstream section. This creates a temperature difference between the two temperature sensors yielding the output signal. The heat uptake by the gas flow passing by the thermal sensors is proportional to the gas mass. Since molecules have a specific heat capacity, the total heat transport is gas-type dependent. Therefore, it is essential to calibrate the MFC for the precursor gas used.

The maximum MFC throughput is typically specified in standard cubic centimetres per minute. For injecting gas into vacuum chambers, typically, the models with the lowest maximum flow values are useful. A reliable control of the flow with 1% accuracy is only feasible for flow rates between 10% and 100% of the maximum specified flow range. Below 5% of the maximum specified flow range, the accuracy of the flow is only 5%. Hence, for a 1 sccm MFC, the minimum controllable flow is 0.05 sccm–this corresponds to a flow rate of 8.5×10^{-4} mbar l s^{-1} ($= 2.2 \times 10^{16}$ molecules s^{-1}). MFCs are only suited for the injection of gaseous precursors or the vapor phase of highly volatile precursors. For less volatile substances, other regulating concepts are used.

6.3.3 Pressure-controlled dosing valve

For controlled dosing of precursors, a pressure-controlled feedback system may also be used. The pressure in a gas supply manifold is measured after the dosing valve with a pressure gauge, such as a capacitance manometer. The pressure reading is used as a feed-back for closed-loop operation of a needle valve, for instance. The advantage of this approach is that a specific pressure in the gas manifold can be adjusted. This gives information on the actual flow regime, allows regulating small precursor flows of less-volatile substances with high precision, and allows molecule-independent operation of the gas supply system when absolute pressure gauges are used. However, the precursor flow cannot be directly measured as compared to MFCs. Pressure-controlled dosing valves are not often used due to the high costs and the high constructive volume required for these solutions.

6.4 HEATING SYSTEM

Heating systems for GISs are widely used to increase the vapor pressure of low-volatile solid and liquid precursors in the reservoir in order to produce a sufficient source pressure. In many commercial systems, thermal management is also used to control the flow rate of precursor. The thermal management of the entire GIS is a critical issue. The heat supplied by an external heating system must be low enough to avoid premature thermal decomposition of the precursor molecules. On surfaces with a lower temperature in the gas supply system than the temperature of the reservoir precursor, molecules will condense. This may especially occur at thermal bridges of the supply manifold to colder parts, e.g., at the wall mounting to the vacuum chamber. In well-designed GISs, the heating systems of the precursor reservoir and of the supply manifold and nozzle are separated. Typically, the source pressure is defined by the temperature of the reservoir, and the supply manifold is heated to a temperature several degrees above as a precaution against condensation of the precursor vapor within the supply manifold. Whenever heating systems are used, longer evacuation time may be required, as the precursor adsorbs on the cold walls of the vacuum chamber and develops a lower partial pressure than in the heated GIS.

Heating systems are also used to adjust the throughput as an alternative to regulating valves. By increasing the temperature of the precursor reservoir, a higher nozzle throughput can be set, while decreasing the reservoir temperature will lead to a lower throughput. Using the temperature to regulate the flow rate is considered an accurate adjustment method and the only practical way to increase the throughput of low vapor-pressure precursors in a given supply system. Flow regulation by thermal management is preferred with many commercial suppliers, since the maintenance otherwise required for valves is significantly reduced and it allows for a compact design.

With all currently known systems, electrically-resistive heating is used. Usually, a thin metal wire or a thin metal film is heated by the electrical current passing through the element. Simple low-cost thermostats use a pulse control for heating to switch the heating elements on or off. With these systems, overshoot and undershoot must be expected, and pulse current signals within the vacuum chamber may compromise the beam stability.

For a stable temperature, thermo-controllers with a variable analog output are advisable.

Heating cables, heating jackets, and alternative heating concepts may be used. These options will be addressed in the following.

6.4.1 Heating cables

Individual heating cables are usually wrapped around the components, e.g. a gas supply tube (see Figure 3.21(a)). Heating wires are typically made of special alloys, such as constantan or manganin, with a low resistive temperature coefficient. Thus, the electrical resistance stays constant for a wide temperature range. The heating efficiency depends on the wire cross-section.

The metal wires are embedded in an insulating cladding. Outside of the vacuum chamber, glass-fiber fabric or silicone elastomer may be used. Within the vacuum chamber, only materials with low outgassing rate at elevated temperatures may be used. Due to its suitability in high- and ultra-high-vacuum environments, often Kapton™, poly(4,4'-oxydiphenylene-pyromellitimide), is used as insulator material. This polyimide film remains thermally stable from −273°C to +400°C and has a comparatively high thermal conductivity and good dielectric qualities. Alternatively, metal wires embedded in vacuum-tight metal tubings can be used. The electrical insulation between the heating wire and the outside metal tube is formed by compressed inorganic oxide powders. The outer metal tubing is highly corrosion resistive to etch gases, but difficult to structure due to their low flexibility. For GISs, often factory-configured heating elements with mineral insulation and metal shielding are used [16; 98].

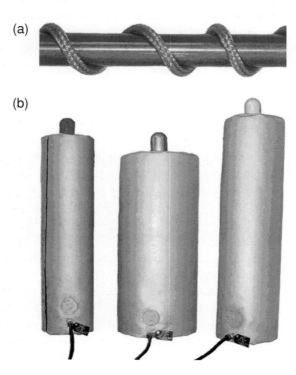

(a)

(b)

FIGURE 3.21: Heating elements showing **(a)** a heating cable and **(b)** a heating jacket [10].

Also, structured metal thin films on a vacuum-suitable support material, like ceramic or Kapton™, may be used. Meander-shaped layouts allow electrical heating of large areas. Useful applications are heating of baseplates or cylindrical reservoirs.

6.4.2 Heating jackets

For gas supply components outside of the chamber, heating jackets are also available (see Figure 3.21(b)). Heating jackets consist of electrical heating wires and thermal sensors integrated into a thermal insulation jacket. Customized options can offer turnkey solutions for complex gas manifold geometries. Due to the high outgassing rate, this option is not available for components in the vacuum chamber.

6.4.3 Alternative heating concepts

Efficient heating is also achieved via a heating medium. A closed tube system of narrow-bore tubes with a fluid, such as water, oil, or a gas, as medium for heat transfer has been acknowledged as successful approach for maintaining components at a specific temperature. Heating or cooling of the transfer medium can be performed at a remote spot using a heat exchanger or a remote electrical heating system. Alternatively, a thermal bridge constructed from a material with a high thermal conductivity may be used. One end of the thermal bridge is heated or cooled on a remote spot and the other end is connected to the GIS component. Typically such a system requires a thermally-isolated feed-through into the vacuum chamber. Currently, none of these concepts has been applied in GISs, but may be useful in special cases [32; 100].

6.5 COOLING SYSTEMS

For volatile substances with a high vapor pressure, often the gas flow needs to be reduced. While the conventional approach would include a regulating valve, cooling of the precursor reservoir may also be applied to reduce the vapor pressure by thermodynamic means. Cooling of the precursor also allows accurate adjusting of the flow of the precursor vapor while omitting mechanically actuated components. Cooling of other parts of the GIS, such as the gas tubing manifold and the nozzle head, is not required. For cooling of the reservoir, the use of Peltier elements has proven to be convenient. Peltier elements are solid-state devices using the thermoelectric effect to create a temperature difference between the front and rear face of the device. Typically, Peltier elements are designed to cool by 20–50 K with respect to the heat-sink temperature. However, the excess heat generated on the rear side must be dissipated, which requires sophisticated engineering for applications inside vacuum chambers.

6.6 PURGING SYSTEMS

For decontamination of unwanted (toxic) substances, a purging system must be included in the gas supply system. Purging is typically performed by repeating the cycle of (1) filling the gas lines with an inert gas and (2) subsequently evacuating the gas manifold. This way the unwanted

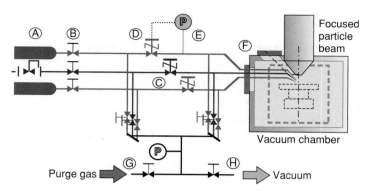

FIGURE 3.22: Schematic setup of a purging system: (A) Precursor reservoir, (B) shut-down valve, (C) manual regulating valve, (D) pressure controlled regulating valve, (E) pressure gauge, (F) gas feed-through into vacuum chamber, (G) purge gas inlet with shutdown-valve, (H) vacuum port with shutdown-valve.

substance is diluted and pumped off with the excess of inert gas. A schematic setup of a purging system with a GIS is illustrated in Figure 3.22.

A purging system consists of an inlet for inert purge gas and a vacuum port for quickly evacuating the entire gas line to remove toxic or corrosive residues. With full-fledged purging systems, even two purge gas inlets and two vacuum ports may be present—one pair before and one pair after the dosing unit. This setup allows separate purging or evacuating of either the "high-pressure" branch before the dosing valve, including the precursor reservoir, or the "low-pressure" branch after the dosing valve with a direct connection to the vacuum chamber.

In the case of a replacement of an empty reservoir of a toxic precursor, the entire gas line must be purged from toxic residues. With the reservoir valve closed, the empty precursor container can then be removed safely. Also, to precondition certain precursor compounds before their usage for gas-induced processing, a purging system is required. For example, the molecule $Co_2(CO)_8$ is sometimes stabilized by hexane, which needs to be first pumped away by a purging cycle to allow for reproducible chemistry supplied for FEB- (FIB-) induced deposition.

For corrosive precursors, purging systems are also used for reconditioning the gas supply manifold in order to avoid corrosion or to prevent re-crystallization of the solid precursor inside the gas supply line during extended standby periods.

6.7 NOZZLE POSITIONING SYSTEMS

It is essential to position the sample in close vicinity to the substrate to ensure a sufficiently high precursor concentration on the surface. Applications that do not require sample tilt—such as FEB-induced processing for photomask repair [27]—allow mounting the nozzles in a fixed geometry. For retaining the precursor gas at the sample surface, permanently mounted concentrators may be used also.

The majority of focused-particle-beam systems are equipped with stages that can be tilted for inspection and processing of the sample. These systems require retractable nozzle systems.

During operation of the GIS, the nozzle opening is typically positioned <0.5 mm above the sample surface. Retracting the nozzle from the sample surface should allow for free maneuverability of the sample stage. Depending on the requirements, the nozzle can be inserted and retracted one-dimensionally along an axis intersecting to the processing area, or it can be positioned by a three-dimensional positioning system. Either the entire GIS, including reservoir and supply system, is moved, or only the nozzle head, which is connected by flexible tube elements to the GIS mounted to the vacuum chamber.

6.7.1 One-dimensional positioning systems

In such a system, the nozzle is mounted on a movable component that can be positioned along a path that is coincident to the path of the particle beam striking the substrate (see Figure 3.23(a)). The advantage of such a one-dimensional positioning system is its potentially compact design and that the end points at the inserted and retracted positions are well defined. On the other hand, a one-dimensional positioning system requires the entire GIS to be mounted in a specific orientation with high precision so that the nozzle head directly aims at the focus spot of the particle beam. Inserting and retracting the nozzle can be performed with a pneumatic or an electromechanic actuator. With lab-made systems, manual operation has also

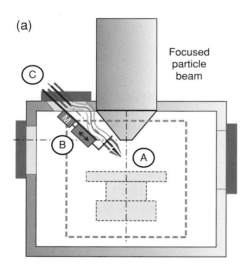

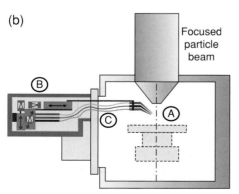

FIGURE 3.23: Schematic illustrations of a **(a)** one-dimensional nozzle insertion/retraction and a **(b)** three-dimensional nozzle positioning system with flexible gas supply connecting the precursor reservoir to the nozzle. (A) The nozzle head with 3 individual gas nozzles, (B) positioning unit consisting of motorized translational stages, (C) port to vacuum chamber.

been successfully performed. This method allows an easy positioning of the nozzle, but limits the construction options regarding mounting the GIS to the process chamber, i.e., appropriate access flanges need to be available.

6.7.2 Three-dimensional positioning systems

Three-dimensional positioning is achieved by mounting the nozzle head to an x-y-z stage (see Figure 3.23(b)). Usually, motor-actuated stages are used, enabling remote control of nozzle positioning. The x-y-z stage gives full flexibility in positioning the nozzle and does not place stringent requirements on the accurate mounting of the GIS. Hence, these GISs are often flanged conveniently to the side walls of the vacuum chamber.

7 STATE-OF-THE-ART GAS INJECTION SYSTEMS

With all the different components of a GIS discussed above, this section will focus on the GIS as a functional entity. An overview on the different types of GISs available today in research and industry is presented.

7.1 LAB-MADE GAS INJECTION SYSTEMS

In numerous research labs the upgrade of scanning electron (ion) microscopes toward gas-induced deposition and etching has been achieved by custom-built GIS solutions. In most cases the gas supply system has been designed for a specific precursor class. A convenient precursor class is metal carbonyls, as several representatives are liquid or solid substances with a vapor pressure in the mbar range. In many cases the functionality of the GIS was intentionally restricted, so that some components of the GIS discussed above were omitted, without compromising the targeted research goals. Typically, lab-made systems are built using commercially available components that are reliable and easy to maintain.

GISs with fixed nozzle positions allow one to easily mount the nozzle holder either on the chamber wall or directly on the sample holder. Non-retractable nozzles have the benefit that they are in the same position for the entire research experiment guaranteeing the best reproducibility. However, tilting functions of the sample stage may be restricted.

For low-vapor-pressure precursors (in the 0.1 mbar-range and below), space- and cost-saving GIS realizations are built without a shutdown-valve. They are often preferred, if the entire GIS is implemented within the vacuum chamber. In this case the omission of the shutdown-valve is less critical, as precursor outflow is negligible during sample exchange and evacuation of the chamber. No electrical vacuum feed-throughs are required for the valve control. Since the flow resistance of the entire GIS remains constant, a very homogeneous precursor flow is ensured, which is advantageous for the reproducibility of focused-beam-induced deposition experiments. On the other hand, no shutdown-valve also means that there is a permanent flow of the precursor molecules into the vacuum chamber. The precursors should be non-toxic, non-pyrophorous, and not sensitive to oxygen or water-moisture during venting of the vacuum chamber. Inspection of the processed samples should be performed with the nozzle retracted from the sample or after removal of the compact GIS unit to avoid additional deposition or etching.

Compact lab-made GISs are sometimes equipped with a heating system, which allows increasing the gas flow during gas-assisted experiments [51; 66]. For gaseous and highly volatile precursors, often no temperature control is implemented. This configuration has proven successful for many gases, including oxygen, hydrogen, chlorine and inert gases [78; 30]. Also, precursors with a high vapor pressure, such as water, alkoxy-siloxanes, and selected metalorganic precursors, (i.e., iron pentacarbonyl or Cu(I) trimethylvinylsilane), were successfully accomplished without heating [8; 63; 45].

Lab-made GISs have been built for operation with a single precursor or for multiple precursors. These systems are equipped with a single nozzle or a multi-nozzle head. For multiple nozzles the radial mounting to a joint focal point on the sample plane is aspired to deliver simultaneously precursor and assistance gases to the process area. Furthermore, switching between different precursors is available without repositioning the nozzle. Multiple tubes injecting the same precursor can strongly increase the flux to the process area [75]. The use of a single needle GIS for several precursors is only advisable if the used precursors do not interact with each other or if the precursor supply can be purged before switching precursors.

As an example, a version of a lab-made GIS with manual positioning capabilities is presented in Figure 3.24(a). This GIS allows the introduction of liquid and solid precursors with a vapor pressure between 0.05 to 200 mbar. The precursor reservoirs and the gas supply manifold are built from stainless steel each with an individual shutdown-valve. A precision needle-valve is implemented for regulating the precursor flow through the nozzle. A flange with gas feed-throughs is transferring the precursor gas into the chamber. Within the chamber, the precursors are guided to the nozzle head through individual flexible tubings of vacuum-tested silicone rubber. The nozzle head consists of three circular straight metal tubes, each with a 0.8 mm inner diameter. The nozzles are arranged radially directed to a joint spot. The entire nozzle head may be retracted to park the nozzles and free the sample stage. It is also adjustable in height and side-to-side, utilizing a miniaturized dual-axis stage. The adjustment is necessary only with the first installation of the GIS and with a change in the working distance of the focused beam. This system has been successfully employed with iron pentacarbonyl $(Fe(CO)_5)$ for iron deposition as well as with siloxane and water for deposition of dielectric material [109].

Another lab-made system is illustrated in Figure 3.24(b). The gas supply system features a pressure controlled needle valve and a purging system. This GIS is designed for injecting precursors with a vapor pressure of 0.5 mbar or higher–including gaseous precursors. The core part is the controller-operated needle valve with two pressure gauges (see Section 6.3.3.). The pressure is measured with capacitive pressure controllers that are operating independent of the used precursor type. Hence, no calibration for different gases is necessary, which makes this system applicable for a wide range of precursors. Due to the low conductivity of the valve and the long flow supply path, the system is not suited for injection of low-vapor-pressure precursors. The purging system has two inlets: one between the reservoir and the valve, and the second one between the valve and the nozzle head. Both sections can be evacuated independently using an oil-free vacuum pump. Typically, nitrogen or argon is used as a purge gas. Besides removing residues of toxic or corrosive precursors, the purging system allows the evacuation of the dead volume of the supply tubes from excess precursor, which reduces the necessary evacuation time before the system can be used for high-resolution imaging again.

The nozzle head consists of maximally four independent gas nozzles. In Figure 3.24(b) three different nozzle heads with different nozzle diameters are shown. It should be noted that all nozzles are bent to direct the precursor flow to the focused-beam processing area. The precursor from the external supply manifold is introduced to the chamber wall via gas feed-throughs mounted to a joint flange. The same flange also holds the electrical feed-through for controlling the stepper motor, which inserts and retracts the nozzle head mounted on a rotating pivot arm. In the park position (shown in Figure 3.24(b)), the pivot arm is rotated

(a)

(b)

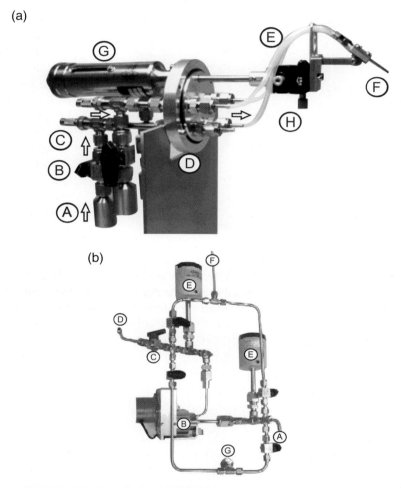

FIGURE 3.24: **(a)** Manually operated GIS for four precursors: (A) Stainless steel precursor reservoir, (B) shutdown-valve, (C) needle-valve for dosing, (D) vacuum feed-through, (E) flexible precursor tubes, (F) nozzles, (G) linear manipulator, (H) manually operated two-axis stage. [45] **(b)** Lab-made GIS showing the gas supply system with a pressure-controlled dosing valve: (A) Connection to precursor reservoir with shutdown-valve, (B) pressure controlled needle-valve for dosing, (C) shutdown-valve, (D) connection to vacuum feed-through, (E) pressure gauges, (F) purge gas inlet, (G) vacuum port. **(c)** Lab-made motorized GIS with the nozzles mounted on a pivot arm: (A) Nozzle heads with different nozzle diameters, (B) positioning unit with motorized pivot arm and two-axis stage, (C) stepper motor, (D) vacuum feed-through, (E) connection to gas supply manifold [109].

(c)

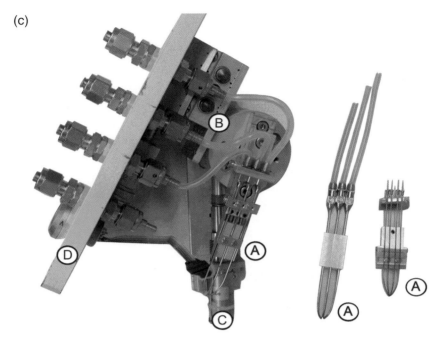

FIGURE 3.24: (Continued)

to a position parallel to the chamber wall in order to not obstruct the movement of the sample stage. The nozzle head can be adjusted in height and sideways by an x-y table that can be adjusted manually at first installation of the system. This way, a full three-dimensional positioning of the nozzle openings within the vacuum chamber is feasible. This system has been successfully applied for addition of oxygen, hydrogen, various siloxanes, and metalorganic copper precursors [109].

7.2 COMMERCIAL GAS INJECTION SYSTEMS

Commercial GISs have first been introduced with focused ion beam systems. Since the emergence of DualBeam™ and CrossBeam™ tools combining FEB and FIB within one microscope, GISs are available from all major suppliers of focused beam systems. Furthermore, several companies specializing in focused-beam accessories produce and offer proprietary GISs to be combined with different tools. In commercial tools, each precursor line of the GIS is designed and optimized for a specific precursor. Table 3.4 summarizes the precursor chemistry that is offered today in commercial GISs. New applications will constantly drive the extension of this list. As compared to lab-made systems, focus of the commercial GISs is put on reliability and safety. The precursor reservoirs are in a safety casing, so that national safety regulations are met. Hence, precursor reservoirs are not easily accessible to the operator and are intended to be changed or refilled only by trained, authorized personnel. Commercial systems are all fully automated or at least partially automated, giving users a tool that is easily and convenient to operate. The control software implements several safety interlocks. Although many features are common to all commercial GISs, the technical realisation can be

Table 3.4 The precursors and their functions offered with commercial GISs.

Precursor	Formula	Function
Pentacene	$C_{22}H_{14}$	carbon deposition
Tungsten carbonyl	$W(CO)_6$	tungsten deposition
Methyl-cyclopentadienyl-(trimethyl) platinum	$(CH_3)_3(CH_3C_5H_4)Pt$	platinum deposition
Dimethylgold-acetylacetonate	$C_7H_{13}AuO_2$	gold deposition
Tetramethoxysilane	$(CH_3O)_4Si$	silicon oxide deposition
Tetraethoxysilane	$(C_2H_5O)_4Si$	silicon oxide deposition
Tetramethyl-cyclo-tetra-siloxane	$(CH_4SiO)_4$	silicon oxide deposition
Pentamethyl-cyclo-penta-siloxane	$(CH_4SiO)_5$	silicon oxide deposition
Water	H_2O	carbon, photoresist etching
Xenon difluoride	XeF_2	silicon etching
Chlorine	Cl_2	aluminium, silicon etching
Nitryl chloride	NO_2Cl	aluminium, silicon etching
Bromine	Br_2	copper, titanium etching
Iodine	I_2	silicon, metal etching

very different. The main difference between the GIS types is the use of a single nozzle GIS capable of injecting one specific precursor versus a multi-nozzle GIS capable of introducing several different precursors through a single nozzle head.

7.2.1 Single-nozzle

The advantage of single-nozzle GISs is their reliability, fast operation, reproducible positioning, compactness, and low cost. All single-nozzle GISs are intended for mounting to the chamber such that they can be inserted and retracted with one single-axis motion. Hence, only a one-dimensionally-actuated system with two end positions to insert and retract the nozzle from the sample is required.

A commercial realisation by FEI [29] of a single-nozzle GIS is illustrated in Figure 3.25(a). This system has a shutdown-valve integrated in the precursor crucible (see Figure 3.19(a)). The valve actuator is utilized to both open the valve, and at the same time, to insert the gas nozzle to a position above the sample. For low-vapor-pressure precursors, the entire flow path including the nozzle can be heated.

A single-nozzle GIS from Orsay Physics [77] is shown in Figure 3.25(b). The precursor container is enclosed in the metal casing mounted perpendicular to the nozzle direction. The GIS is equipped with a heating system optimized for injection of low-vapor-pressure precursors. Also, a cooling version for high vapor pressure precursors is available. The reservoir inside of the container is opened by a pneumatic actuator. The nozzle can be positioned from the outside either manually or with an actuator. In the inserted position, the nozzle is positioned at a few tens of microns from the sample surface. In its park position, the nozzle is 25 mm retracted from the sample. The movement from the park position to the working position is done within only 1 second.

(a)

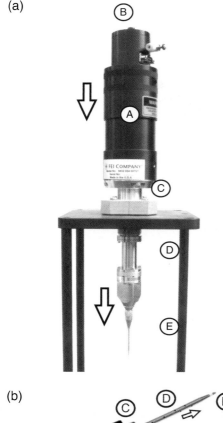

(b)

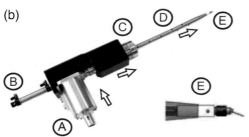

FIGURE 3.25: Single-nozzle based GISs.
(a) FEI-GIS [29]: (A) Precursor container,
(B) pneumatic actuator for insertion/retraction,
(C) vacuum flange, (D) injection line (E) nozzle.
(b) MonoGIS™ from Orsay Physics [77]:
(A) Precursor reservoir with pneumatic valve,
(B) pneumatic actuator for insertion/retraction,
(C) position adjustment system with vacuum
flange, (D) injection line, (E) nozzle.

7.2.2 Single-nozzle for multiple precursors

If precursor supply tubes can be reconditioned by purging from the previous precursor, a single nozzle may also be used for multiple precursors. This concept is offered by Omniprobe [76] (see Figure 3.26).

The OmniGIS™ is designed as an injection system for three different solid or liquid precursors and two inert gases. Adapters with dosing valves can be used as add-on components for gaseous precursors. The system uses vacuum-sealed precursor reservoir cartridges, which guarantee safe handling of precursors and eliminate the need for breaking the vacuum to change the gas chemistry to be utilized. The precursor containers can be heated or cooled in a temperature range between 0–40° C to achieve the appropriate working vapor pressure. The inert gases are supplied from external sources and can be used as purge or as carrier gases to assist transport of the precursor to the delivery site. This system

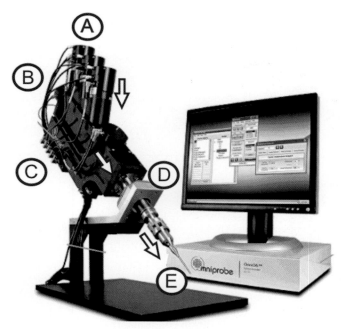

FIGURE 3.26: Single-nozzle GIS for multiple precursor from Omniprobe (OmniGIS™) [76]: (A) Three reservoirs for solid precursors, (B) linear translation drive in casing, (C) two ports for gaseous precursors, (D) vacuum flange, (E) nozzle.

has also incorporated, as a unique feature, a closed-loop mass flow control. Inserting and positioning the nozzle to the working position is performed with stepper motors equipped with encoder feedback. Inserting and retracting the nozzle is slower than with pneumatic actuation, but nozzle height above the sample can be adjusted and vibrations are avoided. All gases are delivered with a single nozzle. With non-reactive precursors this allows mixing gases prior to the injection to the substrate so that a homogeneous gas mixture reaches the surface.

The OmniGIS™ is capable of pulse delivery of precursor gases which reduces overspray at the valves minimum duty cycle (see Figure 3.27). Sequential cycling of precursors with precise flow control and carrier gas mixing allows operation modes similar to atomic layer deposition (ALD) [80]. Pulsing is claimed to be beneficial for faster deposition and sharper edges.

7.2.3 Multi-nozzle

For sophisticated focused-beam processing, multi-nozzle GISs are preferably used, allowing mixing of multiple precursor gases locally in the vacuum chamber above the processing area. This avoids mixing of reactive gases at higher densities inside the supply tubes. Typically, such systems are intended for complex gas chemistries for beam-induced processing.

The nozzle heads of multi-nozzle GISs typically have the capability to position the nozzles to the processing area by motorized stages in three dimensions within a few hundred microns. These systems can be flanged to the vacuum chamber with more flexibility than GISs with

174   FUNDAMENTALS AND MODELS

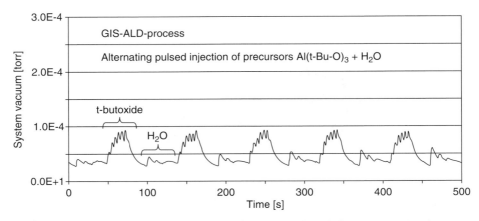

FIGURE 3.27: Sequential cycling of two precursors with the OmniGIS™ (Reprinted with permission from [80]).

one-dimensional positioning. This is especially important for densely packed focused beam systems with multiple detectors and add-on components.

The GIS-5 from Orsay Physics (Figure 3.28(a)) features a very compact design with the heated precursor cartridges directly flanged to the chamber. This system does not allow for complex gas dosing units and only shutdown-valves are implemented in the supply manifold. The Micrion-GIS nozzle head shown in Figure 3.28(b) is entirely enclosed inside the vacuum chamber. Precursors that do not require heating are supplied from an external gas manifold. Flexible tubes connect the gas feed-throughs to the nozzle head. Additionally, a heated reservoir for a low-volatile precursor is directly incorporated in the nozzle head featuring a heated nozzle. In Figure 3.28(c) the multi-nozzle head of a GIS realization by Carl Zeiss is shown. A bundle of microtubes, which can be heated, leads the gas to the nozzle head. It contains five individual nozzles, one for each precursor gas supplied. Although the nozzles are directed toward a center point, an individual positioning of the nozzles relative to the FEB (FIB) by a motor-driven micro stage with three degrees of freedom allows for optimization of the impinging flux. The NanoChemix™ system from FEI (see Figure 3.28(d)) features a tri-nozzle system with two opposing nozzles, which reduce gas shadowing effects on uneven surfaces and provides the ability to mix gases. The dual-nozzle-based delivery can also maintain more homogeneous gas coverage leading to a better floor uniformity or planarity of the processed structure. With the NanoChemix™ gas delivery system, a third central nozzle delivers the precursor for metal depositions.

Generally, precursor reservoirs may be either mounted directly to the chamber wall on a vacuum-sealed flange (see Figure 3.28(a)) or located outside of the vacuum chamber in spacious gas-tight casings (see Figure 3.29).

External gas manifolds are detached units not directly mounted to the focused beam tool, but placed in close vicinity to reduce the length of the supply tubes. The precursor reservoirs are enclosed in gas-tight safety casings, which are vented to a filter. With commercial systems the gas tubes are orbital welded to ensure absolute tightness of the system. Due to the length of the supply lines and the necessity of heating the entire gas supply manifold, precursor heating can be hardly realized with external gas supply systems.

(a)

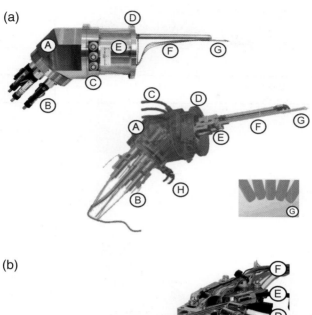

(b)

FIGURE 3.28: Multi-nozzle GISs. **(a)** GIS-5 from Orsay Physics [77]: (A) Precursor reservoirs, (B) pneumatic actuators, (C) electrical feed-through for motorized stages, (D) vacuum flange, (E) motorized stage, (F) heated-gas transfer lines, (G) nozzle head consisting of five individual nozzles, (H) cooling circuit. **(b)** Nozzle head of the Micrion-GIS [69]. A heated reservoir for tungsten carbonyl $W(CO)_6$ is directly attached. Other precursors are supplied by flexible tubes. The system features four individual nozzles bent to approximately focus on the sample spot. The motorized stage allows optimizing the position of the nozzles relative to the FEB (FIB). (A) Nozzle head ($W(CO)_6$-line is heated) (B) heated $W(CO)_6$ reservoir (C), (D), (E) ports for Cl_2 or XeF_2, for water and for siloxane (F) electrical connections to valve and reservoir heating. **(c)** Multi-nozzle GIS mounted on a Carl Zeiss [13] CrossBeam™ workstation: (A) Nozzle head with five individual gas nozzles, (B) heated precursor tubing, (C) focused ion beam optics, (D) focused electron beam optics. **(d)** Dual-nozzle design of the FEI NanoChemix™ [29]: (A) Nozzle head with two opposing nozzles, (B) vacuum flange, (C) precursor tubings in casing.

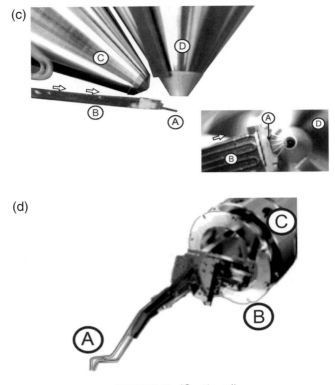

(c)

(d)

FIGURE 3.28: (Continued)

7.2.4 Miniaturized nozzle head

A miniaturized nozzle head incorporating a micro-valve mounted on a micromanipulator is shown in Figure 3.30. The GIS consists of external, temperature-controlled reservoirs, a gas supply equipped with regulating valves and vacuum feed-throughs mounted to the vacuum chamber. Inside the chamber, the gas is supplied by a flexible tube that connects to the nozzle head. The nozzle can be positioned relative to the processing area in three dimensions by a micromanipulator from Kleindiek Nanotechnik actuated by piezo stick-slip motors.

8 CONCLUSIONS AND OUTLOOK

The first part of this chapter introduced the theoretical and experimental background for the analysis and design of nozzle-based gas injection systems for gas-assisted FEB and FIB deposition and etching.

The relevant precursor flow can be characterized by the molecule throughput and the molecule flux impinging at the process area. For the typical case of rarefied flow conditions (molecular and/or transient flow) inside the nozzle of a GIS, this parameter and distribution can be determined using the analytical and simulation framework. The available analytical theory is capable to model the molecule throughput; however, it fails to predict the precursor flux distribution impinging on a substrate produced by a nozzle. A solution to this problem is the Monte Carlo simulation modeling approach. The Monte Carlo simulations presented for several nozzle designs relate very well to experimentally determined molecule flux distributions impinging on the substrate. Consequently, the Monte Carlo simulations can serve as a design

(a)

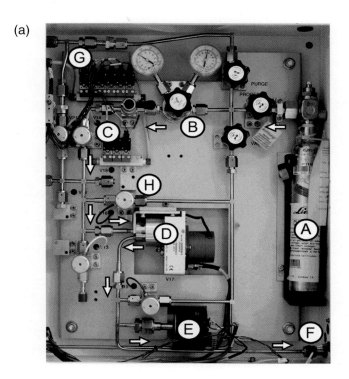

(b)

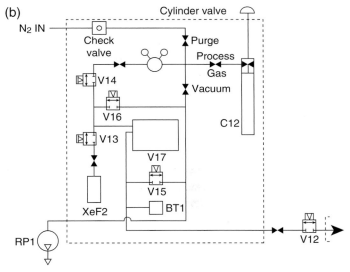

FIGURE 3.29: (a) Gas supply manifold and **(b)** corresponding scheme for a gaseous precursor (chlorine): (A) 500 cl gas cylinder, (B) reducing valve, (C) shutdown-valve, (D) pressure-controlled dosing valve (needle valve actuated by a stepper motor), (E) capacitive pressure gauge, (F) port to vacuum chamber, (G) inert gas port with pneumatic valve for purging, (H) vacuum port with pneumatic valves for evacuation. **(c)** Gas-supply manifold and **(d)** corresponding scheme for a liquid precursor (Tetraethoxysilane) in an external safety casing: (A) Precursor reservoir cooled by Peltier element, (B) manual shutdown-valve, (C) pneumatic shutdown-valve, (D) pressure-controlled dosing valve, (E) capacitive pressure gauge, (F) port to vacuum chamber, (G) inert gas port with pneumatic valve for purging, (H) vacuum port with pneumatic valves for evacuation.

(c)

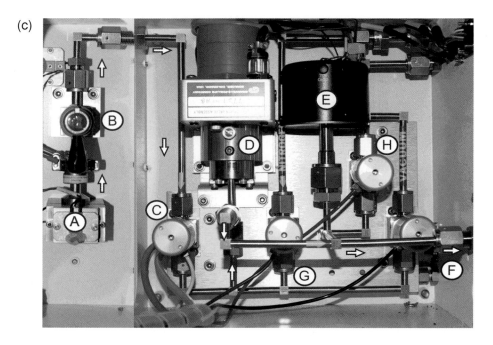

(d)

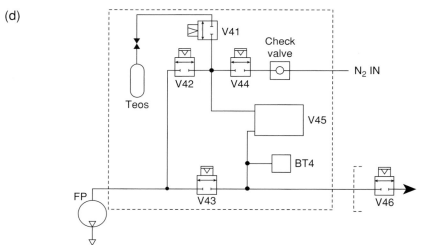

FIGURE 3.29: (Continued)

tool for the nozzle shape and mutual arrangement relative to the substrate in order to optimize the local impinging flux at a minimum total throughput. Nozzle designs can further be analyzed in terms of other aspects involved, such as the electron (ion) interactions within the gas phase leading to beam skirt formation, the generation of secondary electrons at the nozzle, and the directional nozzle-flow leading to shadowing effects in FEB (FIB) deposits and etch pits.

In the second part of this chapter, an overview of the practical aspects of GIS designs was given. Suitable materials and vacuum components are illustrated specifically for their usage

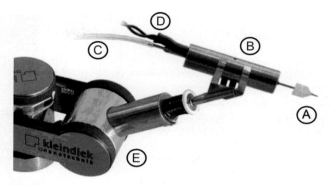

FIGURE 3.30: Miniaturized nozzle-head positioned by a micromanipulator from Kleindiek Nanotechnik [50]: (A) Nozzle, (B) miniaturized precursor valve, (C) flexible tubing, (D) electrical connector to the valve, (E) multi-axis positioning unit.

in GISs. The description of the various concepts can serve as a support for the selection of components for the design of a GIS. Today, already many lab-made GIS implementations are used in research institutes. Commercial solutions are available either integrated in FEB (FIB) processing tools or as add-on systems for usage in standard electron (ion) microscopes. The current strong development of (commercial) GISs proceeds in parallel with the growing field of applications made possible using the gas-assisted focused electron (ion) beam induced processing technique.

The determination and control of the locally impinging precursor flux to the FEB (FIB) impact site is of major importance in the future to gain better control over the deposition and etch process. Accordingly, sophisticated and flexible precursor GIS designs with the ability to control the precursor flux to the substrate are of great interest to tailor the processing throughput and the properties of deposited materials. Further optimization of gas injection of mixed gas flows evolving from multi-nozzle systems must be investigated to allow gas-assisted FEB- (FIB-) induced processing to enter into the field of local multilayer deposition. The simulation and experimental characterization approach proposed in this chapter is appropriate for conducting such studies.

REFERENCES

[1] Adamson, S., and J. R. McGilp. Measurement of gas flux distributions from single capillaries using a modified, UHV-compatible ion gauge, and comparison with theory. *Vacuum* 36 (1986): 227–232.
[2] Airtac, http://www.airtacworld.com/(2009)
[3] Atkins, P. W., and J. De Paula. *Atkins' physical chemistry.* Oxford: Oxford University Press, 2006.
[4] Barber, R. W., and D. R. Emerson. Challenges in modeling gas-phase flow in microchannels: From slip to transition. *Heat Transfer Engineering* 27 (2006): 3–12.
[5] Berman, A. *Vacuum engineering calculations, formulas, and solved exercises/* San Diego: Academic Press, 1992.
[6] Bird, G. A. *Molecular gas dynamics and the direct simulation of gas flows.* Oxford: Clarendon, 1994.
[7] Blauner, P. G., J. S. Ro, Y. Butt, and J. Melngailis. Focused ion-beam fabrication of sub-micron gold structures. *Journal of Vacuum Science & Technology B* 7 (1989): 609–617.
[8] Bret, T., I. Utke, A. Bachmann, and P. Hoffmann. In situ control of the focused-electron-beam-induced deposition process. *Applied Physics Letters* 83 (2003): 4005–4007.

[9] Bret, T., I. Utke, and P. Hoffmann. Influence of the beam scan direction during focused electron beam induced deposition of 3D nanostructures. *Microelectron Eng* 78–79 (2005): 307–313.

[10] BriskHeat Corporation, http://www.briskheat.com/(2009).

[11] Buckman, S. J., R. J. Gulley, M. Moghbelalhossein, and S. J. Bennett. Spatial profiles of effusive molecular-beams and their dependence on gas species. *Meas. Sci. Technol.* 4 (1993): 1143–1153.

[12] Caburn MDC, http://www.caburn.com/(2009).

[13] Carl Zeiss, http://www.smt.zeiss.com/(2009).

[14] Casella, R. A. Patent No. WO9749116. (1999).

[15] Choi, K. K., and S. W. Rhee. Effect of the neutral ligand (L) on the characteristics of hexafluoroacetylacetonate (hfac)Cu(I)-L precursor and on the copper deposition process. *Thin Solid Films* 409 (2002): 147–152.

[16] Chromalox, http://www.chromalox.com/(2009).

[17] Clausing, P. Flow of highly rarefied gases through tubes of arbitrary length. *Journal of Vacuum Science & Technology* 8 (1971): 636.

[18] Cole-Parmer, http://www.coleparmer.co.uk/(2009).

[19] Craig, B. D. *Handbook of corrosion data.* Materials Park, Ohio: ASM International, 1995.

[20] Crozier, P. A., J. Tolle, J. Kouvetakis, and C. Ritter. Synthesis of uniform GaN quantum dot arrays via electron nanolithography of D/sub 2/GaN/sub 3. *Applied Physics Letters* 84 (2004): 3441–3443.

[21] Danilatos, G. D. Equations of charge-distribution in the environmental scanning electron-microscope (ESEM). *Scanning Microscopy* 4 (1990): 799–823.

[22] Danilatos, G. D. Foundations of environmental scanning electron-microscopy. *Advances in Electronics and Electron Physics* 71 (1988): 109–250.

[23] Davis, D. H. Monte-Carlo calculation of molecular flow rates through a cylindrical elbow and pipes of other shapes. *Journal of Applied Physics* 31 (1960): 1169–1176.

[24] Dayton, B. B. *Gases and vacua*, ed. A. H. Beck. Oxford: Pergamon, 1965.

[25] DeMarco, A. J., and J. Melngailis. Lateral growth of focused ion beam deposited platinum for stencil mask repair. *Journal of Vacuum Science & Technology B* 17 (1999): 3154–3157.

[26] Dushman, S. *Scientific foundations of vacuum technique.* New York: Wiley, 1962.

[27] Edinger, K., H. Becht, J. Bihr, V. Boegli, M. Budach, T. Hofmann, H. W. P. Koops, P. Kuschnerus, J. Oster, P. Spies, and B. Weyrauch. Electron-beam-based photomask repair. *Journal of Vacuum Science & Technology B* 22 (2004): 2902–2906.

[28] El-Morsi, M. S., A. C. Wei, G. F. Nellis, R. L. Engelstad, S. Sijbrandij, D. Stewart, and H. Mulders. Gas flow modeling for focused ion beam (FIB) repair processes. In *Proceedings of SPIE – The International Society for Optical Engineering*, ed. W. Staud and J. T. Weed. Monterey, CA (2004), pp. 467–476.

[29] FEI Company, http://www.fei.com/(2009).

[30] Fischer, M., H. E. Wanzenboeck, J. Gottsbachner, S. Muller, W. Brezna, M. Schramboeck, and E. Bertagnolli. Direct-write deposition with a focused electron beam *Microelectron Eng* 83 (2006): 784–787.

[31] Folch, A., J. Servat, J. Esteve, J. Tejada, and M. Seco. High-vacuum versus "environmental" electron beam deposition. *Journal of Vacuum Science & Technology B* 14 (1996): 2609–2614.

[32] Frankel, D. J., B. Fruhberger, R. H. Jackson, and D. J. Dwyer. Ultrahigh-vacuum cold finger for surface-reactions studies. *Rev. Sci. Instrum.* 64 (1993): 2368–2370.

[33] Friedli, V. *Focused electron- and ion-beam induced processes: In situ monitoring, analysis and modeling.* Lausanne: Ecole Polytechnique Fédérale de Lausanne, 2008, p. 157.

[34] Friedli, V. http://www.empa.ch/GISsimulator (2009).

[35] Friedli, V., C. Santschi, J. Michler, P. Hoffmann, and I. Utke. Mass sensor for in situ monitoring of focused ion and electron beam induced processes. *Applied Physics Letters* 90 (2007): 053106.

[36] Friedli, V., and I. Utke. Optimized molecule supply from nozzle-based gas injection systems for focused electron- and ion-beam induced deposition and etching: Simulation and experiment. *Journal of Physics D-Applied Physics* 42 (2009): 125305.

[37] Fujita, J., M. Ishida, T. Sakamoto, Y. Ochiai, T. Kaito, and S. Matsui. Observation and characteristics of mechanical vibration in three-dimensional nanostructures and pillars grown

by focused ion beam chemical vapor deposition. *Journal of Vacuum Science & Technology B* 19 (2001): 2834–2837.

[38] Fustoss, L., and G. Toth. Problem of compounding of transmission probabilities for composite systems. *Journal of Vacuum Science & Technology* 9 (1972): 1214.

[39] Gad-el-Hak, M. The fluid mechanics of microdevices: The Freeman Scholar Lecture. *J. Fluids Eng.-Trans. ASME* 121 (1999): 5–33.

[40] Garner, M. L., D. Chandra, and K. H. Lau. Low-temperature vapor pressures of W-, Cr-, and Co-carbonyls. *Journal of Phase Equilibria* 16 (1995): 24–29.

[41] Gromilov, S. A., I. A. Baidina, P. A. Stabnikov, and G. V. Romanenko. Crystal structure of copper(II) bis-hexafluoroacetylacetonate. *Journal of Structural Chemistry* 45 (2004): 476–481.

[42] Guevremont, J. M., S. Sheldon, and F. Zaera. Design and characterization of collimated effusive gas beam sources: Effect of source dimensions and backing pressure on total flow and beam profile. *Rev. Sci. Instrum.* 71 (2000): 3869–3881.

[43] Henderson, M. A., R. D. Ramsier, and J. T. Yates. Minimizing ultrahigh-vacuum wall reactions of $Fe(CO)_5$ by chemical pretreatment of the dosing system. *Journal of Vacuum Science & Technology A-Vacuum Surfaces and Films* 9 (1991): 2785–2787.

[44] Herman, M. A., and H. Sitter. *Molecular beam epitaxy fundamentals and current status.* Berlin: Springer, 1996.

[45] Hochleitner, G., H. D. Wanzenboeck, and E. Bertagnolli. Electron beam induced deposition of iron nanostructures. *Journal of Vacuum Science & Technology B* 26 (2008): 939–944.

[46] Hoyle, P. C., M. Ogasawara, J. R. A. Cleaver, and H. Ahmed. Electrical-resistance of electron-beam-induced deposits from tungsten hexacarbonyl. *Applied Physics Letters* 62 (1993): 3043–3045.

[47] Hren, J. J. *Introduction to analytical electron microscopy*, ed. J. J. Hren, et al. New York: Plenum Press, 1979.

[48] Jousten, K. *Handbook of vacuum technology.* Weinheim: Wiley-VCH, 2008.

[49] Kim, E. S., H. S. Kim, S. H. Jung, and J. S. Yoon. Adhesion properties and thermal degradation of silicone rubber. *Journal of Applied Polymer Science* 103 (2007): 2782–2787.

[50] Kleindiek Nanotechnik, http://www.nanotechnik.com/(2009).

[51] Klingenberger, D., and M. Huth. Modular ultrahigh vacuum-compatible gas-injection system with an adjustable gas flow for focused particle beam-induced deposition. *Journal of Vacuum Science & Technology A* 27 (2009): 1204–1210.

[52] Knudsen, M. Molecular current of hydrogen through channels and the heat conductor manometer. *Ann. Phys.-Berlin* 35 (1911): 389–396.

[53] Kohl, W. H. *Handbook of materials and techniques for vacuum devices.* New York: Reinhold, 1967.

[54] Kohlmann, K. T., M. Thiemann, and W. H. Brunger E-Beam induced X-ray mask repair with optimized gas nozzle geometry. *Microelectron. Eng.* 13 (1991): 279–282.

[55] Komuro, M., N. Watanabe, and H. Hiroshima. Focused Ga ion-beam etching of Si in chlorine gas. *Japanese Journal of Applied Physics Part 1-Regular Papers Short Notes & Review Papers* 29 (1990): 2288–2291.

[56] Koops, H. Patent No. WO03071578 (2003).

[57] Krasuski, P. Angular-distribution of flux at the exit of cylindrical-tubes. *Journal of Vacuum Science & Technology a-Vacuum Surfaces and Films* 5 (1987): 2488–2492.

[58] Kurepa, M. V., and C. B. Lucas. The density gradient of molecules flowing along a tube. *Journal of Applied Physics* 52 (1981): 664–669.

[59] Lafferty, J. M. *Foundations of vacuum science and technology.* New York: Wiley, 1998.

[60] Landolt, D. *Corrosion and surface chemistry of metals.* Lausanne: EPFL Press, 2007.

[61] Lassiter, M. G., and P. D. Rack. Nanoscale electron beam induced etching: A continuum model that correlates the etch profile to the experimental parameters. *Nanotechnology* 19 (2008): 455306.

[62] Livesey, R. G. Solution methods for gas flow in ducts through the whole pressure regime. *Vacuum* 76 (2004): 101–107.

[63] Luisier, A., I. Utke, T. Bret, F. Cicoira, R. Hauert, S. W. Rhee, P. Doppelt, and P. Hoffmann. Comparative study of Cu precursors for 3D focused electron beam induced deposition. *Journal of the Electrochemical Society* 151 (2004): C535–C537.

[64] Lund, L. M., and A. S. Berman. Flow and self-diffusion of gases in capillaries I. *Journal of Applied Physics* 37 (1966): 2489.

[65] Lund, L. M., and A. S. Berman. Flow and self-diffusion of gases in capillaries II. *Journal of Applied Physics* 37 (1996): 2496.

[66] Matsui, S., T. Kaito, J. Fujita, M. Komuro, K. Kanda, and Y. Haruyama. Three-dimensional nanostructure fabrication by focused-ion-beam chemical vapor deposition. *Journal of Vacuum Science & Technology B* 18 (2000): 3181–3184.

[67] Matsui, S., and K. Mori. New selective deposition technology by electron-beam induced surface-reaction. *Jpn. J. Appl. Phys. Part 2 – Lett.* 23 (1984): L706–L708.

[68] Merkel, T. C., V. I. Bondar, K. Nagai, B. D. Freeman, and I. Pinnau. Gas sorption, diffusion, and permeation in poly(dimethylsiloxane). *Journal of Polymer Science Part B-Polymer Physics* 38 (2000): 415–434.

[69] Micrion Corporation merged with FEI Company in 1998, http://www.fei.com/(2009).

[70] Molhave, K., D. N. Madsen, S. Dohn, and P. Boggild. Constructing, connecting and soldering nanostructures by environmental electron beam deposition. *Nanotechnology* 15 (2004): 1047–1053.

[71] O'Hanlon, J. F. *A user's guide to vacuum technology.* Hoboken: Wiley, 2003.

[72] Ochiai, Y., K. Shihoyama, T. Shiokawa, K. Toyoda, A. Masuyama, K. Gamo, and S. Namba. Characteristics of maskless ion-beam assisted etching of silicon using focused ion-beams. *Journal of Vacuum Science & Technology B* 4 (1986): 333–336.

[73] Ohmi, T., M. Yoshida, Y. Matudaira, Y. Shirai, O. Nakamura, M. Gozyuki, and Y. Hashimoto. Development of a stainless steel tube resistant to corrosive Cl-2 gas for use in semiconductor manufacturing. *Journal of Vacuum Science & Technology B* 16 (1998): 2789–2795.

[74] Ohta, T., F. Cicoira, P. Doppelt, L. Beitone, and P. Hofmann. Static vapor pressure measurement of low volatility precursors for molecular vapor deposition below ambient temperature. *Chemical Vapor Deposition* 7 (2001): 33–37.

[75] Okada, S., T. Mukawa, R. Kobayashi, J. Fujita, M. Ishida, T. Ichihashi, Y. Ochiai, T. Kaito, and S. Matsui. Growth manner and mechanical characteristics of amorphous carbon nanopillars grown by electron-beam-induced chemical vapor deposition. *Japanese Journal of Applied Physics Part 1-Regular Papers Brief Communications & Review Papers* 44 (2005): 5646–5650.

[76] Omniprobe, Inc., http://www.omniprobe.com/(2009).

[77] Orsay Physics, http://www.orsayphysics.com/(2009).

[78] Perentes, A., and P. Hoffmann. Focused electron beam induced deposition of Si-based materials from SiOxCy to stoichiometric SiO2: Chemical compositions, chemical-etch rates, and deep ultraviolet optical transmissions. *Chemical Vapor Deposition* 13 (2007): 176–184.

[79] Pollard, W. G., and R. D. Present. On gaseous self-diffusion in long capillary tubes. *Physical Review* 73 (1948): 762–774.

[80] Principe, E. L., C. Hartfield, R. Kruger, A. Smith, R. Dubois, K. Scammon, and B. Kempshall. Atomic layer deposition an vapor deposition SAMS in a crossbeam FIB-SEM platform: A path to advanced material synthesis. *Microscopy Today* 17 (2009): 18–25.

[81] Randolph, S. J., J. D. Fowlkes, and P. D. Rack. Focused, nanoscale electron-beam-induced deposition and etching. *Critical Reviews in Solid State and Materials Sciences* 31 (2006): 55–89.

[82] Rasmussen, J. Patent No. US5435850. (1993)

[83] Ray, V. Gas delivery and virtual process chamber concept for gas-assisted material processing in a focused ion beam system. *Journal of Vacuum Science & Technology B* 22 (2004): 3008–3011.

[84] Reimer, L. *Scanning electron microscopy physics of image formation and microanalysis.* Berlin: Springer, 1998.

[85] Reimer, L., M. Kassens, and L. Wiese. Monte Carlo simulation program with a free configuration of specimen and detector geometries. *Mikrochimica Acta* 13 (1996): 485–492.

[86] Rugamas, F., D. Roundy, G. Mikaelian, G. Vitug, M. Rudner, J. Shih, D. Smith, J. Segura, and M. A. Khakoo. Angular profiles of molecular beams from effusive tube sources: I. Experiment. *Meas. Sci. Technol.* 11 (2000): 1750–1765.

[87] Santeler, D. J. New concepts in molecular gas-flow. *Journal of Vacuum Science & Technology A-Vacuum Surfaces and Films* 4 (1986): 338–343.

[88] Scheuer, V., H. Koops, and T. Tschudi. Electron beam decomposition of carbonyls on silicon. *Microelectron. Eng.* 5 (1986): 423–430.

[89] Schweitzer, P. A. *Corrosion engineering handbook,* vol. 1. Boca Raton, FL: CRC Taylor & Francis, 2007.

[90] Schweitzer, P. A. *Corrosion resistance tables: Metals, nonmetals, coatings, mortars, plastics, elastomers and linings, and fabrics.* New York: Marcel Dekker, 2004.

[91] Shen, C. *Rarefied gas dynamics fundamentals, simulations and micro flows.* Berlin: Springer, 2005.

[92] Smith, C. G., and G. Lewin. Free molecular conductance of a cylindrical tube with wall sorption. *Journal of Vacuum Science & Technology* 3 (1966): 92.

[93] Smith, D. A., J. D. Fowlkes, and P. D. Rack. Simulating the effects of surface diffusion on electron beam induced deposition via a three-dimensional Monte Carlo simulation. *Nanotechnology* 19 (2008): 415704.

[94] Smith, D. A., J. D. Fowlkes, and P. D. Rack. Understanding the kinetics and nanoscale morphology of electron-beam-induced deposition via a three-dimensional Monte Carlo simulation: The effects of the precursor molecule and the deposited material. *Small* 4 (2008): 1382–1389.

[95] Stubblefield, V. E. Net molecular-flow conductance of series elements. *Journal of Vacuum Science & Technology A-Vacuum Surfaces and Films* 1 (1983): 1549–1552.

[96] Swagelok, http://www.swagelok.com/(2009).

[97] Tanaka, M., M. Shimojo, K. Mitsuishi, and K. Furuya. The size dependence of the nano-dots formed by electron-beam-induced deposition on the partial pressure of the precursor. *Applied Physics A-Materials Science & Processing* 78 (2004): 543–546.

[98] Thermon, http://www.thermon.com/(2009).

[99] Thomson, S. L., and W. R. Owens. Survey of flow at low-pressures. *Vacuum* 25 (1975): 151–156.

[100] Underwood, J. M., and J. C. Price. A surface-sensitive UHV dielectric spectrometer for studies of nanoscale molecular systems on a planar surface. *Rev. Sci. Instrum.* 79 (2008): 093905.

[101] Utke, I., T. Bret, D. Laub, P. Buffat, L. Scandella, and P. Hoffmann. Thermal effects during focused electron beam induced deposition of nanocomposite magnetic-cobalt-containing tips. *Microelectron. Eng.* 73–74 (2004): 553–558.

[102] Utke, I., V. Friedli, S. Amorosi, J. Michler, and P. Hoffmann. Measurement and simulation of impinging precursor molecule distribution in focused particle beam deposition/etch systems. *Microelectron. Eng.* 83 (2006): 1499–1502.

[103] Utke, I., V. Friedli, J. Michler, T. Bret, X. Multone, and P. Hoffmann. Density determination of focused-electron-beam-induced deposits with simple cantilever-based method. *Applied Physics Letters* 88 (2006): 031906.

[104] Utke, I., V. Friedli, M. Purrucker, and J. Michler. Resolution in focused electron- and ion-beam induced processing. *Journal of Vacuum Science & Technology B* 25 (2007): 2219–2223.

[105] Utke, I., P. Hoffmann, and J. Melngailis. Gas-assisted focused electron beam and ion beam processing and fabrication. *Journal of Vacuum Science & Technology B* 26 (2008): 1197–1276.

[106] Utke, I., J. Michler, P. Gasser, C. Santschi, D. Laub, M. Cantoni, P. A. Buffat, C. Jiao, and P. Hoffmann. Cross section investigations of compositions and sub-structures of tips obtained by focused electron beam induced deposition. *Adv. Eng. Mater.* 7 (2005): 323–331.

[107] van Dorp, W. F., and C. W. Hagen. A critical literature review of focused electron beam induced deposition. *Journal of Applied Physics* 104 (2008): 081301.

[108] Visual Simulation Programs created by Bird GA, http://www.gab.com.au/(2009).

[109] Wanzenboeck, H. D., M. Fischer, R. Svagera, J. Wernisch, and E. Bertagnolli. Custom design of optical-grade thin films of silicon oxide by direct-write electron-beam-induced deposition. *Journal of Vacuum Science & Technology B* 24 (2006): 2755–2760.

[110] Yoshimura, N. *Vacuum technology practice for scientific instruments.* Berlin: Springer, 2007.

[111] Ziegler, J. F. http://www.srim.org/(2009).

CHAPTER 4

FUNDAMENTALS OF INTERACTIONS OF ELECTRONS WITH MOLECULES

John H. Moore,[†] Petra Swiderek, Stefan Matejcik, and Michael Allan

1 INTRODUCTION

The electron-molecule processes considered here are:

$$e^-(E_i) + AB \rightarrow AB + e^-(E_i) \qquad \text{elastic scattering} \qquad (1)$$

$$e^-(E_i) + AB \rightarrow AB(v) + e^-(E_r) \qquad \text{vibrational excitation (VE)} \qquad (2)$$

$$e^-(E_i) + AB \rightarrow AB^* + e^-(E_r) \qquad \text{electronic excitation (EE)} \qquad (3)$$

$$e^- + AB \rightarrow A^\bullet + B^- \qquad \text{dissociative electron attachment (DEA)} \qquad (4)$$

$$e^- + AB \rightarrow A^\bullet + B^\bullet + e^- \qquad \text{neutral dissociation (ND)} \qquad (5)$$

$$e^- + AB \rightarrow A^\bullet + B^+ + 2e^- \qquad \text{dissociative ionization (DI)} \qquad (6)$$

$$e^- + AB \rightarrow A^- + B^+ + e^- \qquad \text{bipolar dissociation (ion pair formation) (BD)} \qquad (7)$$

Rotational excitation is not explicitly listed; the elastic and inelastic cross-sections are meant to be integrated over rotational transitions.

The knowledge of the absolute cross-sections for these processes in the gas phase is useful not only for the understanding of gaseous plasmas [1], but also as a starting point for the understanding of dense media. An example of the latter are the Monte-Carlo simulations of electron interactions with dense CH_4 and H_2O, motivated by the need to understand and to optimize radiotherapy [2–4]. The input of these simulations are the absolute cross-sections for the above processes, both as a function of electron energy and of scattering angle. Similar simulations have been also performed for FEBIP [5–7].

Measurement of each of the various processes requires specialized instruments, which are generally not all present in one laboratory. Different instruments are further often needed to cover different energy ranges.

A great body of existing measurements were performed with the aim of better understanding the resonant phenomena in the collisions. They often cover the energy range of about 0.1-30 eV and emphasize high resolution. They are carried out with instruments using thermionic electron sources and hemispherical or trochoidal electron energy selectors.

A second class of measurements covers the very low energy region, 1-200 meV, with an extremely high resolution, and relies on photoelectron sources [8]. This regime is important for the application because every electron in the dense media, including the many secondary electrons, will finally be slowed down to these energies, and because the cross-sections, in particular for DEA (eq. 4), can be extremely large at these low energies.

Finally, both for the medical applications and for FEBIP, cross-sections are needed also at high energies, about 30–1000 eV. These measurements generally require different instruments, which do not need high resolution but emphasize high sensitivity, required by the low values of the cross-sections at high energies, and also addresses other specific problems, like the danger of "polluting" the incident high energy electron beam with slower electrons resulting from inelastic collisions with the metallic apertures. Such slower electrons can seriously distort the data because of the much larger cross-sections at low energies. The emerging medical applications led to the construction of new instruments and recent measurements of data in this high energy regime, particularly by García and coworkers [9].

An important issue is that it is much easier to measure relative cross-sections, the shapes of the spectra; a substantial body of literature on such spectra exists. While the relative data are useful for unraveling the (resonant) mechanism of the processes, they are not useful for simulations and will generally not be covered here.

Another important issue is that it is experimentally much easier to detect charged particles, electrons and positive or negative ions, than to detect neutral products from the electron-molecule collisions. Whereas in the cases of DEA and DI (Equations (4) and (6) above) the detected charged particle allows conclusions on the cross-section for the complementary neutral fragment, in the very important class of ND (Equation (5)) the detection of the neutral particle is a prerequisite for measuring the cross-section. Such measurements are consequently generally neglected, very rare and valuable, and we shall devote the Section 4 to them. Neutral products are also detected, by thermal desorption, in the condensed phase experiments described in Section 6.

As a result, complete sets of cross-sections, covering all the above processes and the entire energy range, are extremely rare. Moreover, the choice of the targets was influenced in the past by the prospective applications in plasmas for electronics manufacture and does not include metalorganic compounds relevant for FEBIP. An example are the cross-sections for CF_4 given in the book of Christophorou and Olthoff, ref. [1], and reproduced in Figure 4.1. Even this set is not "full" in the sense that it does not contain the angular distributions, required for detailed simulations.

Apart from the experiments, the progress of theory is very important because many relevant cross-sections, those involving transient molecules (like CF_2), and vibrationally and electronically excited molecules, are very hard or impossible to measure, and we depend on theory to obtain them. The present chapter will therefore present comparisons between experiment and theory whenever available.

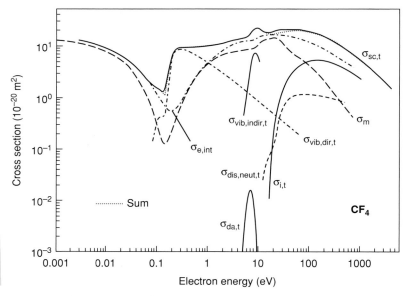

FIGURE 4.1: A complete set of integral cross-sections for CF_4. Total scattering: $\sigma_{sc,t}$; elastic cross-section: $\sigma_{e,int}$; vibrational excitation: $\sigma_{vib,dir,t}$ and $\sigma_{vib,indir,t}$; ionization: $\sigma_{i,t}$; neutral dissociation: $\sigma_{dis,neut,t}$; dissociative electron attachment: $\sigma_{da,t}$; momentum transfer: σ_m. Reproduced from Reference [1] (page 6, Figure 1.2), with kind permission of Springer Science and Business Media.

This chapter provides a brief overview of the subjects mentioned above, with few illustrative examples. It is organized according to the fundamental processes, approximately in the order given by the Equations (1–7). It is, necessarily, to a certain degree divided according to the techniques used to measure the cross-sections. There is a certain unavoidable "cross-linking" between the sections, given by the fact that certain processes, in particular DEA and DI, yield both a neutral and a charged fragments.

Section 2 describes electron scattering processes which do not immediately lead to a chemical change of the target molecule, that is, elastic scattering, and vibrational and electronic excitation. All are important—elastic scattering changes the direction of the electron and is thus responsible for the widening of the incident beam by repeated collisions in dense media. Vibrational excitation slows the electrons down and heats the target. Electronic excitation is an important initial step leading to neutral dissociation (5) and also a means of energy deposition.

Section 3 describes the first (in the sense that it has the lowest threshold energy) process leading to chemical change, the dissociative electron attachment (4).

Section 4 describes the measurements which detect neutral dissociation products, which pose a particular challenge experimentally and are the only means to obtain data on neutral dissociation (5).

Section 5 concentrates on experiments where positive ions are detected.

Section 6 provides a bridge between the gas-phase cross-sections and the applications in the condensed phase. An experiment which provides absolute cross-sections for chemical changes in the condensed phase is described, which yields information not only on unimolecular primary processes, but also on the subsequent reactions of the transient species formed initially.

The ways in which the resonances and cross-sections are influenced by the condensed media are discussed.

Section 7 provides a brief summary, conclusions and outlook.

2 ELECTRON SCATTERING

This section will start with a brief description of resonances and then present illustrative examples of measured and theoretical cross-sections for elastic scattering and for vibrational and electronic excitation.

2.1 THE ROLE OF RESONANCES

At suitable incident energies the electron is often temporarily captured by the target molecule to form a negative ion $\{AB^-\}_j$, called a resonance, with a lifetime typically in the ps time domain. Despite their short lifetime, resonances often dramatically increase the cross-sections for the inelastic processes and for DEA. The processes of vibrational and electronic excitation, and of DEA, are generally dominated by resonances:

$$
\begin{array}{lll}
e^- + AB & \rightarrow \quad \{AB^-\}_j \quad \rightarrow AB + e^- & \text{elastic scattering} \\
& \rightarrow AB(v) + e^- & \text{VE} \\
& \rightarrow AB^* + e^-(E_r) & \text{EE} \\
& \rightarrow A^\cdot + B^- & \text{DEA}
\end{array}
$$

The process is schematically illustrated in Figure 4.2. The attachment of an electron transfers the initial wave packet of the nuclei to the resonant potential surface. The surface is usually repulsive in the Franck-Condon region and the nuclei start to separate, to relax. A loss of the electron by fast autodetachment occurs as the nuclei move; the wave packet "rains down" back onto the potential surface of the neutral molecule. The fast autodetachment leads, by the

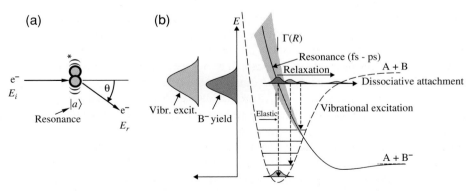

FIGURE 4.2: (a) A schematic diagram of an electron-molecule collision. (b) A schematic diagram of the role of a resonance in the electron collision.

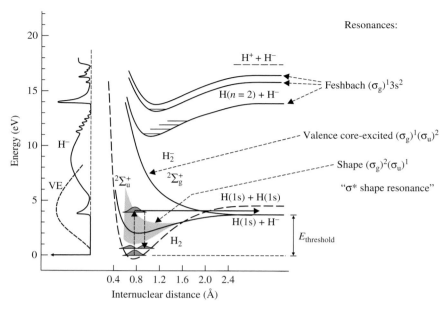

FIGURE 4.3: Schematic diagram of selected resonances in H_2. Schematic VE and DEA (yield of H^-) cross-sections are shown (rotated by 90°) on the left of the potential curves.

uncertainty principle, to an energy width Γ of the resonance, which is typically in the 1 meV – 5 eV range.

The system may fall back into the ground vibrational level of the target molecule (elastic scattering); with more relaxation it falls into vibrationally excited levels. DEA results when it "survives" beyond the "stabilization point" (curve crossing), R_c. The resulting VE and DEA cross-sections are shown (rotated by 90°) on left of the potential curve.

Figure 4.3 illustrates, on the example of H_2 [10], several frequently encountered types of resonances. The "shape resonance" results when an electron is temporarily captured into an normally unoccupied (virtual) orbital and resides there for a short time because it has to tunnel through a centrifugal barrier in order to leave. In this case it is the σ_u LUMO (lowest unoccupied molecular orbital), the resulting $^2\Sigma_u$ resonance has the configuration $(\sigma_g^2)(\sigma_u)$ and may be called a "σ^* shape resonance" for short. Its "parent state" (the state of the neutral target obtained when the "extra" electron is formally removed) is the electronic ground state of H_2. The width of this resonance in H_2 is more than 2 eV [11] and it consequently gives rise to a very broad band in the VE cross-section [12; 13].

Interestingly, the same resonance gives rise to a DEA band which is much narrower than the band in VE. This is because the DEA band has a vertical onset at the DEA threshold, and, on the high energy side, the DEA cross-section falls rapidly because the nuclear wave packet has very little chance to survive when the attachment occurs at low internuclear distance, that is, at higher energy. This illustrates that both the VE and DEA are complementary means of detecting resonances, but one and the same resonance may appear very differently in both channels.

Another noteworthy aspect is that, because of the large autodetachment width and thus very fast autodetachment, the elastic and VE channels "win" by far (about a factor of 10^5) in the competition with DEA. This has two important consequences: The DEA cross-section is small,

0.16×10^{-24} m^2 [14], and the isotope effect is large—the cross-section for D$^-$/D$_2$ is about $200\times$ smaller than that for H$^-$/H$_2$. This can be easily rationalized qualitatively, the heavier deuterium moves, at the same energy, slower than hydrogen, and thus needs a longer time to reach the stabilization point at R_c, leaving more time for autodetachment. Large isotope effects are not uncommon in DEA by low-lying shape resonances, further examples are CH$_3$OH, C$_2$H$_5$OH [15], and C$_2$H$_2$ [16]. This suggests that one may gain insight into the role of DEA in certain application cases by using deuterated precursors.

A repulsive resonance with the configuration $(\sigma_g)(\sigma_u)^2$, called a valence core-excited resonance, results when the incoming electron excites an electron of the target molecule before being captured in the same valence orbital. The parent state is a valence-excited state of H$_2$. The term symbol is $^2\Sigma_g^+$ and it can conveniently be written as $^2(\sigma_g, \sigma_u^2)$, meaning that, with respect to the target H$_2$, the resonance has a hole in the σ_g orbital and an additional double occupation of the σ_u orbital. This resonance causes only little VE because the probability of its formation, being a two-electron process, is quite low in comparison with a shape resonance. But it causes a broad DEA band, because it has a much narrower autodetachment width than the shape resonance and the nuclei thus have a larger probability to reach R_c.

At still higher energies a large number of core-excited resonances with double occupation of Rydberg-like orbitals is found. The lowest can be written as $^2(\sigma_g, 3s^2)$. Its parent state is the $^3(\sigma_g, 3s)$ Rydberg state of H$_2$, and its "grandparent state" is the ground state of H$_2^+$, $^2(\sigma_g)$. The resonance lies about 0.4 eV below its parent Rydberg state and is called a Feshbach resonance because it cannot decay into its parent state. Since the two excited electrons reside in a Rydberg-like orbital, which is spatially diffuse and have little density between the nuclei, they do not strongly contribute to binding and the potential curve of the Feshbach resonance is *a priori* similar to that of the cation, that is, not dissociative. In reality, however, the Feshbach resonances are often predissociated by repulsive (valence) states, and they are responsible for the sharp structures and bands in the 11–18 eV range of the DEA spectrum in Figure 4.3. In fact, such predissociations are common among many molecules and Feshbach resonances are a major and very frequent cause of DEA in the 6–15 eV range of electron energies.

Sometimes large cross-sections and sharp structure are found at low energies which cannot be assigned to any of the above resonance types. Examples are HF ([17] and references therein), where a narrow threshold peak followed by sharp structures is found in VE, although only a broad shape resonance similar to that of H$_2$ would be expected. The threshold peak and the structures are assigned to Vibrational Feshbach Resonances (VFR) and are due do dipole and polarizability binding of the incoming electron at elongated interatomic distances. A similar threshold peak and structures are found, for example, in CO$_2$ [18; 19], where it is ascribed to a "virtual state". These "exotic resonances" near threshold are the subject of considerable interest [20], but it is uncertain whether they are also found in condensed state.

2.2 ELASTIC SCATTERING

As already mentioned, elastic scattering changes the direction of the electrons and is thus important in the simulations like those of García and coworkers—it influences how strongly the incident beam widens in condensed media. The results for tetrahydrofuran (THF, Ref. [21] and references therein) are shown in Figure 4.4 as an illustrative example of measured and calculated elastic cross-section. Both calculated results were obtained by *ab initio* calculations and the agreement is seen to be satisfactory. Critical are low energies, where target polarization

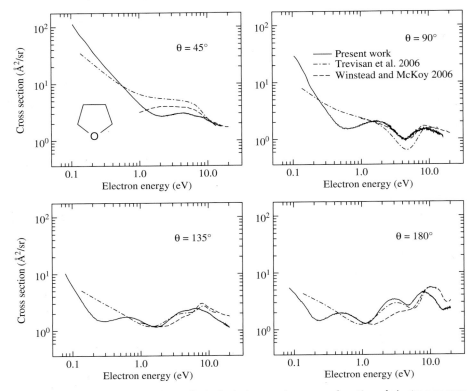

FIGURE 4.4: Elastic cross-sections of tetrahydrofurane shown as a function of electron energy at four representative scattering angles [21]. The calculated data of Trevisan et al. [22] and of Winstead and McKoy [23] are shown for comparison.

becomes important, and low energies combined with low scattering angles, where the cross-section becomes very large due to long-range dipole ($\mu = 1.75$ D for THF) interaction. It should be pointed out, however, that the numerical requirements of this type of calculations are very large and the method is consequently not easily scalable to much larger molecules. An alternative in this respect is the independent-atom method (IAM), employing a quasi-free nonempirical model, which has recently been revised to improve its foundation and accuracy and yields satisfactory results at higher energies, from about 30 eV to a few keV [9].

Preliminary elastic cross-sections of a FEBIP-relevant compound, $Pt(PF_3)_4$, are shown in Figure 4.5. Additional measurements will be required to obtain a more complete set of cross-sections, but already at this stage it is clear that this nearly spherical molecule with many electrons leads to interesting features. The angular distribution is unusual in the sense that it has a narrow minimum around 40°. Deep Ramsauer-Townsend minima appear in the energy-dependence of the cross-section.

2.3 VIBRATIONAL EXCITATION

The role of vibrational excitation is primarily slowing-down of the electrons and heating the target. The cross-sections of methane (Ref. [24] and references therein) are shown in

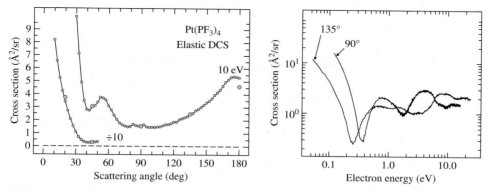

FIGURE 4.5: Elastic cross-sections of $Pt(PF_3)_4$ shown as a function of scattering angle at 10 eV on the left, and as a function of electron energy at the scattering angles $\theta = 90°$ and $135°$ on the right.

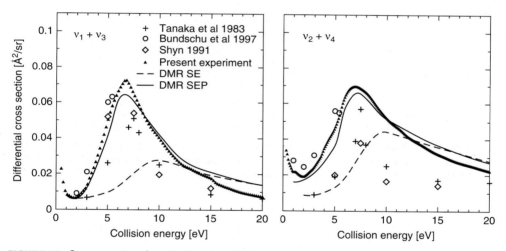

FIGURE 4.6: Cross-sections for vibrational excitation of methane shown as a function of electron energy at the scattering angle $\theta = 90°$ (from ref. [24]). The experimental results from Fribourg are shown by triangles, the remaining symbols show earlier experimental data of Tanaka et al., Buntschu et al. and Shyn. The solid and dashed lines show the results of two levels of theory as explained in the text.

Figure 4.6 as an illustrative example. The closely-spaced individual modes can not be resolved, primarily because of the rotational broadening of the vibrational bands, and the sums for $v_1 + v_3$ (both are C-H stretch vibrations) and $v_2 + v_4$ (both are H-C-H deformation vibrations) are shown. Methane is also illustrative of the present capacity of theory, as Figure 4.6 compares the experiment with the results of calculations carried out using the discrete momentum representation (DMR) method of Čársky and Čurík. This method is also fully *ab initio* and has the advantage of being applicable even to larger, many-modes molecules. It has recently been improved by including the target polarizability (results labeled as SEP, static exchange with polarizability, in Figure 4.6), which brings a substantial improvement over the older version without polarizability (results labeled as SE in the Figure) at energies below 10 eV. The agreement of experiment and the SEP theory is very satisfactory.

2.4 ELECTRONIC EXCITATION

Electronic excitation by electron impact has been extensively studied both experimentally and theoretically in atoms, where it is important for lighting applications (for an illustrative example, see Reference [25] and references therein). Absolute measurements in polyatomic molecules are much more rare, and the corresponding theory is much less advanced.

The electronic excitation of the lowest electronic state in ethene, a prototype of π electronic systems, will be presented here as an illustrative example of the state of experiment and of *ab initio* theory in polyatomic molecules [26].

The results are shown in Figure 4.7. For the comparison with theory it is important that the experiment covers the entire angular range from 0° to 180°. The figure also illustrates the problems encountered with *ab initio* calculations. The older version of the theory reproduced well the shapes and overall trends of the cross-sections, but overestimated its magnitude, by about a factor of two within the first about 2 eV above threshold. Later it was realized that this discrepancy is due to the neglect of the target polarization, and its inclusion resulted in a great improvement of the magnitude near threshold. But the theoretical effort is substantial, it cannot be scaled to much larger molecules, and the theory is useful only within the first about 4 eV above threshold; it fails as more "final channels", possibilities of the resonances to decay into higher-lying electronic states, open up. It is clear that *ab initio* theory is very useful in the near-threshold energies where more approximate theories fail, but cannot provide all the data required for simulations.

For the applications, it is important to know also the cross-sections for the excitation of the higher excited states, and to know the cross-sections at higher energies (References [2; 4] and references therein). At higher energies, the direct excitation of dipole-allowed transitions becomes dominant and the subsequent fragmentation of the excited states may be the primary mechanism of neutral dissociation.

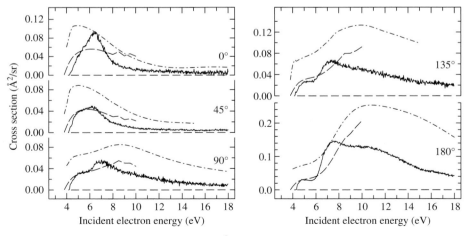

FIGURE 4.7: Cross-sections for exciting the $\tilde{a}\,{}^3B_{1u}$ triplet state of ethene, summed over all ro-vibrational transitions [26]. Dash-dotted lines show the older calculated results of Sun et al. [27], dashed lines the more recent theoretical data [26], which includes target polarization and reproduces better the near threshold region.

3 DISSOCIATIVE ELECTRON ATTACHMENT

Dissociative electron attachment cross-sections span many orders of magnitude, as can be seen in the overview Figure 4.8. In some cases, like the hydro- and fluoro-carbons discussed in Section 4, including CF_4 shown in Figure 4.1, DEA cross-sections are negligible as means of producing reactive intermediates, the radicals. In other cases, in particular at low energies, DEA cross-sections may become very large, up to 10^{-15} cm^2. The size of the cross-section is to a large degree given by the competition of dissociation with autodetachment. Since autodetachment tends to be slower at low energies, when the electron leaves with little energy, DEA cross-sections are larger there than at higher energies, when the autodetachment is faster. This argument is valid for the shape resonances. The Feshbach resonances may have slow autodetachment even at higher energies, 6-15 eV, making the cross-sections larger, but not as high as at very low energies, because the probability of forming the Feshbach resonances, a two-electron process, is lower than that for forming the shape resonances.

This section will present few illustrative examples of DEA, intended to show also the progress of theory, and various possibilities of "control", ways to influence the outcome of the reaction.

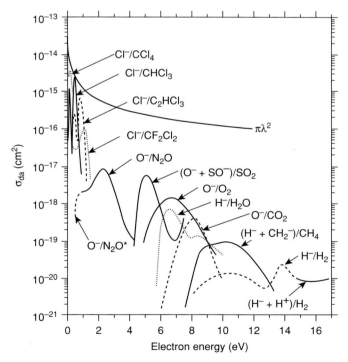

FIGURE 4.8: Dissociative electron attachment cross-sections for a number of molecules. Reproduced from Reference [1] (page 11, Figure 1.5), with kind permission of Springer Science and Business Media.

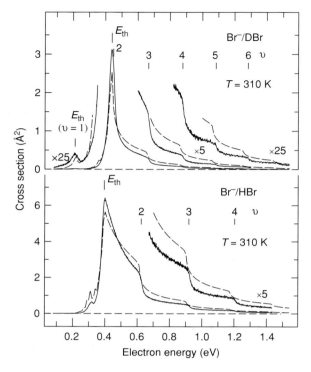

FIGURE 4.9: Solid lines: high resolution (10 meV) absolute cross-sections for Br$^-$/HBr and Br$^-$/DBr, obtained by normalizing earlier relative spectra [28] to the absolute values measured later [29]. Dashed lines: predictions of the nonlocal resonance theory [28; 30], with the final temperature taken into account, but without convolution with an apparatus function. Note that the experimental and theoretical data sets are independently on absolute scales, without any mutual normalization or scaling! The vibrational thresholds (v) and the DEA thresholds E_{th} are indicated. Reproduced from Reference [29].

3.1 DIATOMIC MOLECULES

Measured and calculated DEA spectra of HBr are shown in Figure 4.9 as an example where detailed experiment and a very successful calculation are available.

The cross-section is fairly large, although one would expect a σ^* shape resonance with a very fast autodetachment like that in H_2, described in Section 2.1, and consequently a small cross-section. The large size of the cross-section is due to dipole binding of the incoming electron and the ensuing vibrational Feshbach resonances. The larger cross-section reveals a more favorable competition with autodetachment, confirmed also by the isotope effect, which is seen in Figure 4.9 to be about a factor of two, sizable, but much less than in H_2.

Interesting are the downward steps at the energies of the vibrational levels of HBr. They were first discovered in HCl [31] and are due to "interchannel coupling": the channel of DEA becomes less populated when the channel of vibrational excitation into a given v opens up.

The threshold phenomena which obviously dominate DEA in this case cannot be described by the "local" resonance model, where the resonance is described only by ist potential curve and R-dependent width Γ, but a "nonlocal" resonance model devised by Domcke and coworkers [30], where Γ is also a function of energy, becomes successful. Unfortunately, this model can at present not be extended to molecules with more than two atoms.

3.2 POLYATOMIC MOLECULES: ACETYLENE

Acetylene is mentioned here because it is perhaps the only example where absolute DEA cross-section for a more than three-atomic molecule was calculated *ab initio* and the result was validated experimentally. It thus shows the direction for the future.

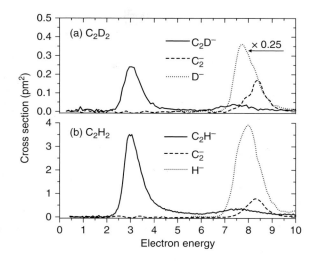

FIGURE 4.10: Dissociative electron attachment cross-sections for (a) C_2D_2 and (b) C_2H_2.

The absolute cross-sections are shown in Figure 4.10 (see also Azria and Fiquet-Fayard [32] for an earlier measurement). There is a certain similarity with the spectra of H_2 in Figure 4.3. The band at 3 eV has an onset at the thermodynamic threshold and is due to a shape resonance with a temporary electron capture in the π_g orbital. The cross-section at this band was successfully calculated by Chourou and Orel [33].

The isotope effect was measured later [16]. It was then realized that the cross-section rises rapidly with initial vibrational excitation of the target, that is, with the temperature, and a substantial rise is found already at room temperature. It was necessary to include the temperature dependence to correctly reproduce the isotope effect [34].

The theory provided important insight into the mechanism of the dissociation, which is symmetry-forbidden in the linear geometry. The π^* resonance of acetylene has to bend before it can dissociate. In the bent state the dissociation proceeds without activation barrier, but the necessity to bend makes the dissociation time longer, the competition with autodetachment less favorable, and the DEA cross-section small. In fact, the role of DEA in the overall production of reactive intermediates in acetylene is, like in the case of CF_4 described in Section 4 below and shown in Figure 4.1, presumably very small.

Calculating the cross-section for the 8 eV Feshbach resonance remains, however, beyond the capacity of the current theory, although Feshbach resonance-mediated DEA was calculated in a pioneering work for the triatomic molecule H_2O [35; 36].

3.3 CONTROLLING THE OUTCOME OF DEA

In DEA it is common that for a given target different resonances, formed at different incident energies, dissociate into different fragments. An example has already been presented in Figure 4.10, where C_2H^- was produced at 3 eV and H^- and C_2^- were produced at 8 eV. DEA is in this respect fundamentally different from photochemistry, ruled in most cases by the Kasha's rule, which says that radiationless transitions from higher excited states to the lowest excited state are generally faster than chemical reactions. That means that photochemistry from higher excited states is not different from that of the lowest excited state.

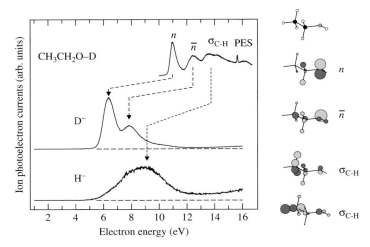

FIGURE 4.11: Dissociative electron attachment (H^- and D^- yields) spectra of partially deuterated ethanol, C_2H_5OD. The photoelectron spectrum shown on the top aids the assignment of the Feshbach resonance bands. Shown on the right are schematic diagrams of the orbitals involved in the cationic states in the photoelectron spectrum and which are only singly occupied in the Feshbach resonances.

Other examples of selective DEA have recently been observed in molecules of biological relevance, the nucleobases [37]. Another example was reported by Prabhudesai et al. who observed, using selective deuteration, that for methanol, ethanol, acetic acid, and *n*-propyl amine H^- is lost from the heteroatom around 6.5 and 7.7 eV, and from the alkyl group around 10 eV [38].

A number of selectivities were observed for alcohols and ethers, for example that Feshbach resonances with core hole on the oxygen lone pair orbitals n or $\bar{n}$ break the O-H bond but not the O-C bond, and this observation was rationalized using potential curves of the parent Rydberg states [39]. A selectivity was even found in cleaving various C-O bonds in asymmetric ethers [40].

Figure 4.11 shows an illustrative example of selectivity of the type reported by Prabhudesai et al. [38]. The figure is based on the data of Reference [15]. The assignment of the DEA bands is aided by comparison with the grandparent states of the cation, revealed by the photoelectron spectrum (PES) on the top of the figure. Since the binding energy of the two $3s$ electrons of the Feshbach resonance with respect to the cation is always about 4.5 eV, the two bands in the D^- yield must be the $^2(n, 3s^2)$ and $^2(\bar{n}, 3s^2)$ Feshbach resonances, where n and $\bar{n}$ are the out-of-plane and the in-plane lone pair orbitals localized predominantly on the oxygen atom, as shown by the orbital diagrams in the figure. D, bound to the O-atom, is thus ejected (as a negative ion) exclusively by resonances where the hole is localized predominantly on the O-atom. The signal around 9 eV in the H^- yield must be due to Feshbach resonances of the type $^2(\sigma, 3s^2)$, with a hole in one of the σ orbitals. H, bound to the C-atom, is thus ejected predominantly by resonances where the hole is localized primarily on the alkyl group.

These selectivities open up, in principle, the possibility of controlling the chemistry by tuning the electron energy. This possibility can presumably not be used in practice, however, because there is not sufficient control over the energies of the secondary electrons in FEBIP, and

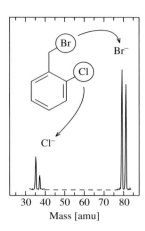

FIGURE 4.12: Dissociative electron attachment cross-sections for two halosubstituted toluenes, at an incident energy of about 0.4 eV. The signal of the halogen attached directly to the ring is always weaker.

because this selectivity concerns only DEA, whereas a number of other processes contribute to the production of reactive intermediates at the same time.

There is a second kind of selectivity whereby fragments in different chemical surroundings have different sensitivities to being dissociated by electrons. An interesting example are halogen atoms, connected to an aromatic ring or to a double bond, either directly or *via* a methylene ($-CH_2-$) group.

The result may seem surprising at first: although the electron is captured into a π^* orbital, the halogen situated further away from the aromatic ring (or a double bond) is removed preferentially [41;42].

This opened up the possibility to synthesize dihalo substituted toluenes which lost preferentially Cl^- or Br^- upon the attachment of an electron into the π^* orbital of the benzene ring, as shown in Figure 4.12 [43]. The principle is operative also for substituents other than halogen, namely alkoxy [44].

Another way to influence DEA is to choose compounds with "good leaving groups". These are generally halogens, where DEA is driven by their large electron affinity. But compounds with very stable neutral fragments, for example phenyl azide, which loses N_2 upon attachment of 0-0.5 eV electrons, also have large DEA cross-sections [45].

4 REACTIVE NEUTRAL FRAGMENTS FROM ELECTRON IMPACT FRAGMENTATION

Electron-impact fragmentation of molecules in a gas yields ions and neutrals. The neutrals are in the majority; many of these are chemically reactive. The sticking probability for the neutrals at nearby surfaces is much less than for the ions. In practical situations, in the absence of confining fields, ions may be quickly lost to the walls of an apparatus so that the concentration of neutrals, even highly reactive radicals, can increase by orders of magnitude over that of the ions (see for example Reference [46]).

Three electron-impact fragmentation processes yield neutral species:

neutral dissociation (ND)

$$e^- + AB \rightarrow A + B + e^-,$$

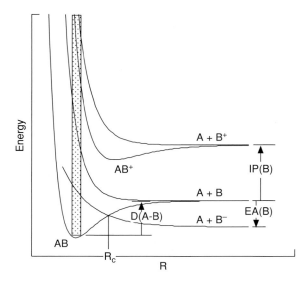

FIGURE 4.13: A potential energy diagram for the polyatomic molecule AB representing a cut through electronic potential energy surfaces along the A-B bond direction. Shown are typical potential energy curves for bound and unbound states of the neutral AB molecule and the AB^+ ion, as well as an unbound state of the negative ion, AB^-. IP(B) is the ionization potential of the B fragment, EA(B) the electron affinity of B, and D(A-B) is the A-B bond energy.

dissociative ionization (DI)

$$e^- + AB \rightarrow A + B^+ + 2e^-,$$

dissociative attachment (DEA)

$$e^- + AB \rightarrow A + B^-,$$

(It is understood in that AB represents in general a polyatomic molecule, and A or B or both represent polyatomic fragments.)

A few broad generalizations will help in understanding the nature of the products of electron-impact fragmentation. From elementary chemistry, recall that, with very few exceptions, stable molecules possess an even number of electrons; stability is derived from having all electrons paired.[1] Neutral dissociation requires the rupture of a chemical bond and the separation of a bonding pair of electrons. It follows that both fragments are odd-electron species. They are *radicals* and tend both to be very reactive. In chemical notation the unpaired electron is explicitly notated. For example, neutral dissociation of CF_4:

$$e^- + CF_4 \rightarrow {}^\bullet CF_4 + {}^\bullet F + e^-.$$

Dissociative ionization and dissociative attachment rupture a bond and ionize one of the fragments. Stable ions tend to be even-electron species (for example, F^+, F^-, CF_3^+). The neutral fragment of dissociative ionization or dissociative attachment is thus usually a radical.

Much can be learned about the process of electron-impact fragmentation from an examination of a generalized potential-energy diagram for the target molecule (Figure 4.13). (In fact, a many-dimensional energy surface is required to describe each electronic state of a polyatomic molecule. Figure 4.13 describes a cut through some of these surfaces along the A-B bond direction.) The initial step in electron-impact-induced chemistry involves excitation of

---

[1] "Nature abhors a vacuum ... and She's none too fond of an unpaired electron either." Anonymous.

the parent to a dissociative electronic state or to a bound state with sufficient excess energy to place the molecule above the dissociation limit. An electronic transition occurs in a brief time compared to the period of vibrational motion in a molecule; the positions of the atomic nuclei are essentially frozen. This is the Franck-Condon approximation. On the potential-energy diagram, electronic excitation is represented as a vertical transition from near the equilibrium confirmation of the parent on the ground electronic state surface to an excited state surface in an unchanged geometry. This so-called "Franck-Condon region" is represented by the shaded area on the figure. Subsequent fragmentation may occur as the molecule relaxes along the excited state surface. The *threshold* energy for dissociation is the asymptotic limit of each potential curve at large A-B separation. With regard to the production of neutral fragments, a number of generalizations can be drawn from Figure 4.13: First, it is obvious that the (Franck-Condon) transition energy to a dissociative state is significantly greater than the threshold energy. Second, the lowest threshold for neutral dissociation is simply the A-B bond energy ...typically about 4 eV. Finally, the threshold for dissociative ionization is the sum of the A-B bond energy and the ionization energy of A or B. Ionization energies are typically of the order of 10 eV. The threshold for dissociative ionization may be two to three times that for neutral dissociation. Similarly, the Franck-Condon transition energy for dissociative ionization may be tens of eV greater than for neutral dissociation.

Dissociative attachment involves the essentially instantaneous excitation to an electronic state of the parent negative ion (AB^-) followed a relatively leisurely relaxation to fragments [47]. The relaxation, however, must compete with autodetachment of the electron. Dissociative attachment can only proceed if the electron-molecule complex persists for a sufficiently long time for the parent negative ion to relax past a critical crossing point at R_c. The lifetime of the parent temporary negative ion is roughly inversely related to the energy of the attaching electron. Persistent negative ion states in some saturated molecules (alkanes, for example) typically involve the (resonant) capture of electrons with energies less than 1 eV. For unsaturated molecules (alkenes, alkynes) there may be long-lived negative ion states formed in the capture of electrons with energies up to 2 or 3 eV. Dissociative electron attachment has not been found to contribute to fragmentation in saturated hydrocarbons and fluorocarbons. Dissociative attachment is significant for many unsaturated hydrocarbons and unsaturated fluorocarbons as well as saturated chloro- and bromocarbons.

Electronic excitation to ground state of the parent ion deserves special attention in the present context. The ion, AB^+, having lost a strongly-bonding electron, tends to assume a geometry very different from the parent neutral, as suggested by the displacement of the minimum of the ground state ion potential energy curve relative to that of the ground state neutral. Vertical excitation of the neutral to the bound state of the ion frequently leaves the ion with energy in excess of its dissociation limit. As a consequence, there are many chemical species for which the parent ion cannot be produced by electron impact.

The likelihood of any particular electron-impact-induced process is specified as a cross-section [48]. In general, the cross-section for electronic excitation as a function of the energy of the impacting electron rises from zero at the threshold to a broad maximum at the energy corresponding to the Franck-Condon region and then slowly decreases with increasing electron energy. The maximum cross-section for the sum total of all electronic excitation leading to fragmentation (exclusive of dissociative attachment) typically reaches a maximum for an impacting electron energy of about 100 eV. The magnitude of the sum total cross-section roughly amounts to the gas kinetic cross-section of the target molecule, of the order of 10×10^{-20} m^2. As suggested by the potential-energy diagram, the cross-section for neutral dissociation peaks at lower energy (30–70 eV) than that for dissociative ionization (50–150 eV). Below 50 eV, neutral dissociation is usually the main source of radicals. Above 100 eV, both neutral dissociation and dissociative ionization are important sources of radicals from electron-impact fragmentation of polyatomic molecules.

To determine the cross-section for dissociation, one must detect one or the other of the fragments. Experimentally it is much easier to detect a charged particle, so, if the choice exists, a charged fragment of dissociation is detected rather than a neutral fragment. This is obvious from the literature describing radical production by electron-impact; there are many more reports of measurements of cross-sections for dissociative ionization than for neutral dissociation since a measurement of a neutral dissociation cross-section requires the detection of one or the other of the neutral products.

Cross-section measurements should be carried out under single-collision conditions. Real-time measurements of neutral dissociation cross-sections require essentially single-radical sensitivity. Mass spectrometric and optical techniques are the obvious choices [49]. Optical methods can be very sensitive in the detection of atoms. For molecular species, however, transition intensity is spread over many rovibrational components of a transition, effectively reducing the sensitivity of optical detection of either emitter or absorber. In addition, optical techniques, such as laser-induced fluorescence, suffer a lack of generality; one must know in detail the spectrum of each species to be detected. On the other hand, a mass spectrometer can be employed to detect almost any volatile species at concentrations below $10^2\,\mathrm{cm}^{-3}$ owing to the fact that the typical electron-impact-ionization source ionizes all species with approximately the same high efficiency. This lack of specificity, however, may be a curse rather than a blessing. To measure a cross-section for radical production, a beam of electrons is passed through a target gas at a sufficiently low density that only a small fraction of the target is dissociated; the radical precursor is present in much higher concentration than the radical. The problem that arises with mass spectrometric detection is that electron-impact ionization in the mass spectrometer source of both the radical products and the target gas yields the same ionic species. For example, the 69 amu CF_3^+ peak is the most prominent feature in the mass spectrum of both the parent CF_4 and the radical fragment CF_3. As discussed above, the parent ion of the stable target molecule (i.e., CF_4^+) is not produced in the electron-impact source of a mass spectrometer. A degree of discrimination between parent and radical product has been achieved by adjusting the ionizer electron energy to the threshold for the species of interest (*vide infra*). Some specificity has also been achieved with a multi-photon ionization source.

Alkyl radicals react with many main group metals to produce volatile polyalkyl metal complexes. In 1929 Paneth and Hofeditz first demonstrated the existence of radicals in an experiment in which the photodecomposition products of an organic compound were exposed to a mirror of lead [50]. The disappearance of the mirror along with their detection and analysis for tetramethyl lead were essentially conclusive. Subsequently, Rice and Dooley [51] and Belchetz and Rideal [52] showed that methyl radicals reacted at a tellurium mirror to yield appreciable quantities of dimethyl ditelluride, dimethyl telluride, and a small quantity of hydrogen telluride.

Corrigan [53] and then Winters and Inokuti [54] took essentially the opposite tack in making the first quantitative measurements of the total dissociation cross-section. Electron irradiation of a gas was carried out in a closed vessel whose walls were coated with a titanium getter that permanently sequestered radical fragments of electron impact dissociation. The total dissociation cross-section was obtained from the pressure drop in the container, the irradiating beam current, and time of exposure. Employing these data, total neutral dissociation cross-sections have been calculated for CH_4, CF_4, and C_2F_6 as the difference between the total dissociation and total ionization cross-sections [55–61].

Motlagh and Moore developed a specific and nearly universal technique for the quantitative analysis of radicals that has been employed in the measurement of *partial* cross-sections for the production of neutrals by electron impact on CH_4, CH_3F, CH_2F_2, CHF_3, CF_4, C_2F_6, and C_3F_8 [62]. (A partial cross-section is a cross-section for a process such as ionization or dissociation with the production of one specific product.) The technique, based on the method by which radicals in the gas phase were first identified, relies upon the efficient reaction of radicals with tellurium to yield volatile and stable organotellurides. A beam of electrons passes through a target gas in a collision cell that has a tellurium mirror on its inner surface. Radicals from electron-impact fragmentation react with tellurium within their first few encounters with the wall to produce volatile tellurides. (An electrical bias prevents ions from reaching the tellurium surface.) The telluride partial pressure is measured mass spectrometrically and related to the radical production rate. The technique is specific for radicals since a target gas of stable (even-electron) molecules does not react at the tellurium surface. In addition, the portion of the mass spectrum under observation is displaced by more than 128 amu (the nominal tellurium mass) from the region displaying peaks characteristic of the parent gas.

It can reasonably be argued that above the dissociation threshold the total dissociation cross-section is equal to the sum of the cross-sections for excitation to all available electronic and ionic states [54]. Following this line of reasoning, Mi and Bonham have carefully measured the total inelastic electron-scattering cross-section (equivalent to the total excitation cross-section) for N_2 and CF_4 and obtained the neutral dissociation cross-section by subtracting the total ionization cross-section [63].

Sugai and collaborators have worked to develop threshold-energy ionization/mass spectroscopy as a technique for observing radicals from electron-impact fragmentation (Reference [64] and references therein). The technique relies on the fact that the threshold for ionization of radical fragments is invariably 3 to 5 eV lower than that for the stable parent from which the radical is derived. To obtain radical specificity, the electron energy in the mass spectrometer ion source is set in the range below the ionization energy of the parent. Great care is required in setting the energy and controlling the energy width of electrons in the ion source since the concentration of radicals to be ionized and detected is typically orders of magnitude less than that of the parent. Sensitivity suffers; the ionization efficiency is low near threshold. This technique has permitted the determination of relative partial dissociation cross-sections for the production of a range of neutral fragments for each parent that has been investigated. Placing the cross-sections on an absolute scale has involved a difficult determination of a number of instrument variables and normalization to other measurements.

Perrin, Schmitt, De Rosny, Drevillon, Huc, and Lloret derive dissociation cross-sections from a kinetic analysis of molecular dissociation in a constant-flow multipole dc plasma reactor [65].

Including the measurements mentioned above, total dissociation or neutral dissociation cross-sections have been reported for methane [67] (*vide infra*); all the fluorinated methanes [54; 62; 63; 68; 69]; perfluorinated ethane [54; 62], propane [62; 69] and cyclobutane [70]; C_3HF_7O [64]; silane [65] and disilane [65]. Several authors, notably Christophorou, Olthoff, and Rao [1; 71–73]; Shirai and coworkers [74]; and Morgan [57] have undertaken reviews and evaluations of these data attempting to justify interrelated measurements such as total dissociation cross-sections and cross-sections for neutral dissociation and dissociative ionization, as well as partial cross-sections for production of specific product fragments. Experimental data on electron-impact fragmentation of methane, CH_4, and perfluoromethane, CF_4, represent a significant proportion of what is available. By way of example we show a sample of data for the production of methyl radical (CH_3) from methane (CH_4), and perfluoromethyl radical (CF_3) from perfluoromethane (CF_4). Neutral dissociation and dissociative ionization are the primary processes leading to electron-impact fragmentation for both CH_4 and CF_4; dissociative attachment is not significant in either case.

The two processes contributing to methyl radical production from methane are:

$$e^- + CH_4 \rightarrow CH_3 + H + e^- \quad \text{(neutral dissociation)} \tag{8}$$

$$e^- + CH_4 \rightarrow CH_3 + H^+ + 2e^- \quad \text{(dissociative ionization).} \tag{9}$$

A collection of data bearing upon the production of CH_3 from electron impact on CH_4 is shown in Figure 4.14. The total dissociation cross-section ("total dis." in the figure) was obtained by Winters [65] using the gettering technique. The cross-section for methyl radical production by reactions (8) and (9) ('CH_3 [n.d. + d.i.]') was obtained with the telluride-conversion technique

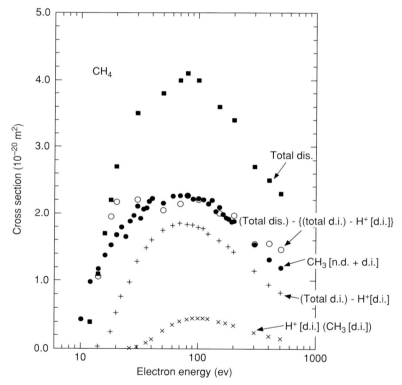

FIGURE 4.14: Cross-sections for the production of CH_3 resulting from electron impact on CH_4. The cross-section for the production of CH_3 by both neutral dissociation and dissociative ionization [●, ('CH_3 [n.d. + d.i.]', Reference [62]] has been normalized to the difference (○) between the total dissociation cross-section (■, 'total dis.', Reference [65]) and the total dissociative ionization cross-section apart from the contribution of dissociative ionization to CH_3 production (+, '(total d.i.) – H^+ [d.i.]', Reference [66]). The cross-section for production of CH_3 by dissociative ionization is taken equal to the cross-section for the production of H^+ by dissociative ionization (✕ , 'H^+ [d.i.] (= CH_3 [d.i.])').

by Motlagh and Moore [62]. The cross-section for reaction (9), the partial ionization cross-section for the production of H$^+$, has been measured by Straub, Lin, Lindsay, Smith, and Stebbings [66], who also measured the total ionization cross-sections ("total d.i.") as well as the partial ionization cross-sections for the other jor dissociative ionization channels:

$$e^- + CH_4 \rightarrow CH_3^+ + H + 2e^- \tag{10}$$

$$e^- + CH_4 \rightarrow CH_2^+ + 2H \text{ (or } H_2) + 2e^- \tag{11}$$

$$e^- + CH_4 \rightarrow CH^+ + 3H \text{ (or } H_2 + H) + 2e^-. \tag{12}$$

As implied by reaction (9) and indicated on the figure, the cross-section for the production of H$^+$ by dissociative ionization is identical to that for the production of CH$_3$ by dissociative ionization ("H$^+$ [d.i.] (= CH$_3$ [d.i.])"). It follows that subtracting all the dissociative ionization, apart from this channel, from total dissociation leaves the cross-section for CH$_3$ production. This subtraction is shown on the figure ("(total dis.) – (total d.i.) – H$^+$ [d.i.]"). The agreement with the direct measurement of total CH$_3$ production reflects the internal consistency of the various measurements.

With regard to generalizations about cross-sections made in the introduction above, it should be noted that, for methyl radical production from electron-impact on methane, virtually all of the radical production is attributable to neutral dissociation below about 50 eV. The maximum cross-section for radical production by neutral dissociation is at lower energy (~70 eV) than for production by dissociative ionization (~100 eV). The threshold energy for neutral dissociation (~12 eV) is distinctly below that for dissociative ionization (~18 eV).

A collection of data bearing upon the production of CF$_3$ from electron impact on CF$_4$ is shown in Figure 4.15. The two processes contributing to CF$_3$ production are:

$$e^- + CF_4 \rightarrow CF_3 + F + e^- \quad \text{(neutral dissociation)} \tag{13}$$

$$e^- + CF_4 \rightarrow CF_3 + F^+ + 2e^- \quad \text{(dissociative ionization).} \tag{14}$$

Absolute neutral dissociation cross-sections at three energies as reported by Mi and Bonham are shown on the figure [63]. These data represent the difference between their measurements of the total inelastic electron-scattering cross-section and the total ionization cross-section. Relative cross-section measurements for CF$_3$ radical production by reactions (13) and (14) ("CF$_3$ [n.d. + d.i.]") were carried out employing the telluride-conversion technique by Motlagh and Moore [62]. These measurements are placed on an absolute scale by normalization to the Mi and Bonham cross-sections under the assumption that neutral dissociation is the sole source of CF$_3$ radicals below about 40 eV. Cross sections for the production of F$^+$ have been measured by Ma, Bruce, and Bonham [75–77], and by Poll, Winkler, Margreiter, Grill, and Mark [78]. These data are in excellent agreement with one another. The recommended average [71] of these two sets of data (following various corrections [77; 79]) is shown in Figure 4.15. As implied by reaction (14) and indicated on the figure, the cross-section for the production of F$^+$ by dissociative ionization is identical to that for the production of CF$_3$ by dissociative ionization ("F$^+$ [d.i.] (= CF$_3$ [d.i.])"). The cross-section for the production of CF$_3$ by neutral dissociation ("CF$_3$ [n.d.]") is obtained as the difference between that for production by both neutral dissociation and dissociative ionization and that for dissociative ionization alone.

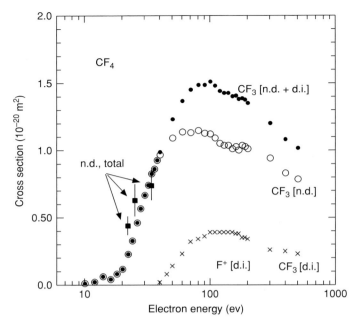

FIGURE 4.15: Cross-sections for the production of CF_3 resulting from electron impact on CF_4. The cross-section for the production of CF_3 by both neutral dissociation and dissociative ionization (●, "CF_3 [n.d. + d.i.]", Reference [62]) is normalized to the total neutral dissociation cross-section (■, "n.d., total", Reference [63]) below 40 eV where only neutral dissociation contributes. The cross-section for production of CF_3 by dissociative ionization is taken as equal to the cross-section for production of F^+ by dissociative ionization (✕, "F^+ [d.i.] (= CF_3 [d.i.])", References [71; 75–79]). The cross-section for the production of CF_3 by neutral dissociation (○, "CF_3[n.d.]") is taken as the difference between the normalized cross-section for production of CF_3 by both neutral dissociation and dissociative ionization and the cross-section for production of CF_3 by dissociative ionization.

The data for electron-impact fragmentation of CF_4 support the generalizations above. For CF_3 radical production from electron-impact on CF_4, neutral dissociation accounts for virtually all of the radical production below about 50 eV. The maximum cross-section for radical production by neutral dissociation is at lower energy ($\sim$70 eV) than for production by dissociative ionization ($\sim$120 eV). The threshold energy for neutral dissociation ($\sim$18 eV) is distinctly below that for dissociative ionization ($\sim$40 eV).

5 DISSOCIATIVE IONIZATION

5.1 INTRODUCTION

The processes that result in formation of the positive ions are known under the common name "electron impact ionization" (EII). The description of the EII reactions is a complex task that includes many problems such as the kinetics, energetics of the reactions, the mechanism of ion formation, the dissociation and the distribution of products. Several reviews and books have been written on this topic over the years. We note in particular the review by Märk [80], Märk and Dunn [81], as well McDaniel [82], Illeneberger and Momigny [83], Christophorou and Olthoff [1], and reviews by Becker and Tarnovsky [84].

According to [7] the dissociative ionization (DI) is the most relevant EII reaction channel to the FEBIP technique. We will try to present some essential aspects of EII and DI processes which may be interesting for the FEBIP user. The dissociative ionization depends on the size of the molecule, its chemical composition, structure, initial state, and the energy of the interacting electrons. In the case of organometallic molecules we deal with large polyatomic molecules and the number of the dissociative channels may be very large. For the FEBIP technique, those DI channels are important, which lead to formation of small fragments (ionic or neutral) with metal atom inside. In an ideal case, "naked" metallic ions or "naked" metal atoms are formed, which can be deposited on the surface and form metallic layers. Detailed knowledge of the dissociation processes and their cross-sections at electron energies relevant for FEBIP technique would enable one to select suitable precursor molecules and tune the electron energy, and thus finally improve the performance of this technique. Unfortunately, so far only very little has been done on the field DI to organometallic compounds, especially concerning the kinetics of these reactions.

5.2 ELECTRON-IMPACT IONIZATION

The EII, generally, is the interaction of the electrons with a neutral targets (molecules or atoms), which results in formation of a positively charged particles (molecular ions, fragment ions, metastable ions, multiply charged ions . . .), two or more electrons and also neutral fragments or radicals. In the case of EII to the atomic and molecular targets, single and multiple ionizations are possible reaction pathways:

$$e^- + M \rightarrow M^+ + e_s^- + e_e^- \qquad \text{single ionization} \qquad (15)$$

$$\rightarrow M^{n+} + e_s^- + ne_e^- \qquad \text{multiple ionization} \qquad (16)$$

$$\rightarrow M\text{-}R^+ + R^+ \qquad \text{Coulomb explosion} \qquad (17)$$

$$\rightarrow M^{(K)+} + e_s^- + ne_e^- \qquad \text{innershell ionization} \qquad (18)$$

$$\rightarrow M^{**} \rightarrow M^{*+} + 2e^- \qquad \text{auto-ionization} \qquad (19)$$

$$\rightarrow M^{*+} \rightarrow (M\text{-}R)^+ + R \qquad (20)$$

In addition, in the case of molecular targets (as it is in case of organometallic molecules), dissociative channels – dissociative ionization (DI) occurs:

$$e^- + M \rightarrow M^{*+} \rightarrow (M\text{-}R)^+ + R + e_s^- + e_e^- \qquad \text{dissociative ionization} \qquad (21)$$

$$\rightarrow (M\text{-}R)^+ + R^- + e_e^- \qquad \text{ion pair formation} \qquad (22)$$

M – denotes the neutral target, R – the neutral fragment or radical, M^+ – ions, e_s^-, e_e^- – "scattered" and "ejected" electrons. The reaction (22), called "ion pair formation," is often reported along with dissociative ionization; however, in contrast to the dissociative ionization

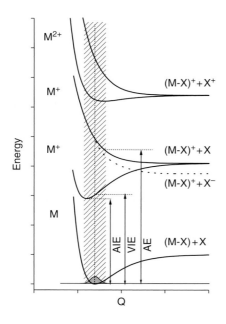

FIGURE 4.16: Potential energy curves of molecule M and the different ionic states M^+, M^{2+}, fragment ions $(M-X)^+$, X^+, X^-, and the neutral fragment, or radical, X. Q is one of the molecular reaction coordinates. AIE – adiabatic ionization energy, VIE – vertical ionization energy, AE – appearance energy of the ion $(M-X)^+$.

the fragment R is negatively charged. The process of dissociation may be fast if it occurs on the timescale of pico-seconds or metastable if the lifetime of the unstable ion spans from nanoseconds up to microsecond scale.

High energy electrons have the ability to induce multiple ionizations (16). The stability of the multiply charged ions depends on the electronic structure and geometry of the molecular ions and on the elemental composition of the multiply charged ions. The multiply charged ions may further decompose *via* a process called Coulomb explosion (17). This process occurs if the repulsive Coulomb force of the two (or more positive charges) in the multiply charged ion exceeds the intra-molecular forces which bind the molecule together.

The electron impact ionization is a fast process for which the Born-Openheimer approximation is valid (Figure 4.16). Typical time scale for direct ionization event is 10^{-16} s, which is much shorter then a typical vibrational period of a molecule. This means that during the ionization event we neglect the movement of the molecular nuclei. This fast process we call direct ionization. By varying the incident electron energy, we are able to reach different bound states of the molecular ion and also the states of the continuum.

Positive ions may be formed also *via* an indirect mechanism, e.g., auto-ionization (19 and 20), where the molecules are excited to doubly excited states, which decay radiationlessly into an ion $(M^+)^*$, which may further dissociate. The indirect processes can be recognized by the structures in the measured ionization cross-sections as the process has resonant character. In some atoms and molecules the indirect ionization may play important role [81; 84].

The EII and DI are generally endothermic reactions, i.e., there exists for each of the reactions a particular threshold, and only electrons with an kinetic energy above this threshold, may initiate the reaction. In the case of the reaction EII, the threshold is called ionization energy (IE) of the particle (atom or molecule). Typical values of the ionization energies of the molecules are in the range of 10 eV [85]. The values of the ionization energies reflect the chemical composition, the electronic structure of the molecule, the geometry and the internal energy of the molecules prior to the reaction.

If the incident energy of the electron exceeds the dissociation limit of the molecular ion state (Figure 4.16), dissociative ionization may occur (reaction channel (21)). The dissociative ionization proceeds *via* the formation of the excited molecular ion M^{*+}, which decays further into fragment ions—primary ions and neutral fragments. The primary ions with excess of energy may further decompose into secondary ions. The molecules fragment into (1) ions which contain functional group, (2) ions formed by cleavage of a functional group, and (3) ions formed *via* rearrangement of the bonds within the ion [86]. The fragmentation of the molecule can be theoretically treated quantitatively only for small molecules (diatomic and small polyatomic). The potential-energy surfaces of the neutral and ionic states can be calculated using *ab initio* quantum chemical methods and the dissociation can be treated in terms of molecular dynamics. For larger polyatomic systems (inclusive majority of the organometallic molecules) the quasiequilibrium theory provides a tool to study the dissociation of the ions [86]. Lango et al. [87] applied quasi-equilibrium theory to study unimolecular decay of the organometallic ions.

The cross-section is generally a quantity which describes the probability of the ionization of electrons with given kinetic energy interacting with molecules; it has the dimension of an area (usually the units are m^2, cm^2). The total ionization cross-section gives the probability of the formation of any positive ion without detailed information about the nature of the ions formed. The partial ionization cross-section reflects the efficiency of formation of a specific ion (molecular ion, fragment ions, multiply charged ion).

The theory of the ionization reaction is a complex task, because the ionization is a many body problem, which is not trivial to solve rigorously. Therefore mainly semi-empirical methods are employed to calculate total cross-sections for EII. There exist several semi-empirical methods based on the additive rule [88], which express the total single ionization cross-section as the sum of the ionization cross-sections of the constituent atoms. These methods have limitations, since molecular bonding is not taken into account. Therefore new concepts (modified additive rule) have been proposed which attempt to account for the molecular bonding [89], and which give reasonable agreement in total ionization cross-section for whole range of molecules. A different semiempirical method is the so called DM formalism introduced by Deutsch and Märk [90]. In recent years the Binary Encounter Bethe (BEB) of Kim [91] gained recognition with very good results for atmospheric and hydrocarbon molecules. The usability of this method should be proofed for organometallic compounds.

The situation on the partial cross-sections for DI is even more complicated. Generally the most reliable information about total DI cross-section can be obtained at the present time using experimental methods. The overview of various experimental methods employed to study EII and DI can be found in [80; 81; 84]. Dissociative ionization has been intensively studied experimentally already for several decades. Dedicated experimental techniques were developed to study EII [81; 82], which allow the measurement of absolute or relative double and triple differential ionization cross-sections.

Absolute partial and absolute total ionization cross-sections, on the other hand, are the result of experiments which are not differential in electron scattering angle, energy of the outgoing electrons, and electron spin. For many areas of applications (plasma physics, electric discharges, and also FEBIP) angle-integrated absolute total cross-sections and absolute partial cross-sections (mass selected) are of high importance, as these cross-sections govern the production rate of a variety of reactive species and secondary electrons. In recent decades, advances in measurement of absolute DI cross-sections have been achieved; however, some problems still persist: (1) achievement of reliable detection efficiency of the mass spectrometers

Table 4.1 The ionization energies (IE) of selected organometallic molecules, the appearance energies (AE) of the metal atom ions Me^+, and the relative abundances of Me^+ in mass spectra at electron energy of 70 eV (ratio (Me^+/Total ion intensity)).

Molecule	IE (eV)	AE(Me^+) (eV)	Method	Relative Me^+ intensity
$Al(CH_3)_3$	9.09 ± 0.26	14.6 ± 0.2	EI [97]	16% [98]
$Ga(CH_3)_3$	9.87 ± 0.02	13.24 ± 0.03	EI [99]	27% [98]
$Co(CO)_3NO$	7.89 ± 0.3	14.03 ± 0.3	EI [100]	21% [100]
$CpCo(CO)_2$	7.08 ± 0.3	13.47 ± 0.3	EI [100]	11% [100]
Cp_2Co	5.35 ± 0.3	14.19 ± 0.3	EI [100]	12% [100]
$(RhCl(CO)_2)_2$	9.01	–	PE [101]	10% [85]
$Re_2(CO)_{10}$	8.49 ± 0.02	28.96	PE [102]	<1% [85]
$(CH_3C_5H_4)_2Fe$	6.6 ± 0.2	14.9 ± 0.3	EI [103]	1.5% [103]

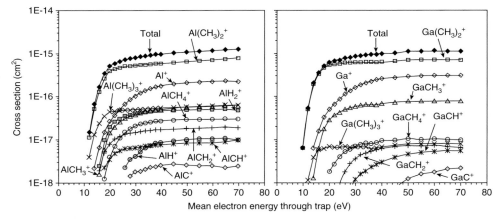

FIGURE 4.17: Partial cross-sections for dissociative ionization of a) TMA b) TMG. Reprinted from [98] with permission from Elsevier.

for ions with different masses; (2) discrimination effects at the ion source; (3) in the case of the dissociative ionization, problems with kinetic energy release. For a detailed discussion of the advantages and limitations of the various experimental setups see [82]. At the present time, several laboratories in the world have mastered the technique of measurement of absolute partial cross-sections for DI, e.g., see the following articles [92–96].

Besides the cross-section, the experiment may provide also other quantities like ionization energies of the molecules and the appearance energies of the fragment ion formed upon DI. From these quantities we may obtain information about the structure of the molecule, of the molecular and fragment ions, and the reaction enthalpies of the dissociation channels.

5.3 DISSOCIATIVE IONIZATION OF MOLECULES

The number of molecules for which absolute ionization cross-sections are available increases only slowly. Most of the molecules are simple di-atomics H_2, N_2, O_2, CO, NO, and HCl, and tri-atomic molecules such as H_2O, CO_2, N_2O, NO_2, and SO_2 [84]. The studies of polyatomic molecules were stimulated largely by the need for ionization cross-section data in various areas of applications (C_2H_2, NH_3, CH_4, CCl_4, CF_4, SF_6, Si_2H_6, C_2H_6, and C_3H_8 (see [81; 84] for a more complete list of references)). In the past decade the list grew only very slowly. For molecules the single ionization processes are the dominant mechanisms, whereas multiple ionization and ion pair formation tend to have much smaller cross-sections.

The importance of DI increases with the size of the molecules. In many molecules the cross-sections for the formation of fragment ions are larger than the parent ionization cross-section (C_2H_6, CF_4, NF_3 ...) [84]. The initial state of the molecules also has an influence in the DI processes [104; 105]; it may influence (decrease) the appearance energy of the fragment ions and also change the distribution of the fragments.

The needs for the cross-sections and other data concerning EII and DI for organometallic molecules relevant to FEBIP applications are very large. The kinetic data are of high importance for understanding processes involved in these techniques, for computer modeling and optimization of the processes. For the majority of the organometallic compounds, the cross-sections for EII and DI were not measured so far. In fact, we have found only a few publications with partial cross-sections for DI of organometallic molecules. Jaio et al. [98] measured total ionization cross-section to trimethyl-aluminum (TMA) and trimethyl-gallium (TMG) molecules. They applied Fourier Transform mass spectrometry with electron ionization source to measure partial cross-sections for DI to TMA and TMG molecules. The Figure 4.17a presents the measured cross-sections for TMA, TMG in Figure 4.17b shows very similar cross-sections and fragmentation pattern. The dominant product in the whole measured energy range was the $Al(CH_3)_2^+$ ion. The parent molecular ion was more than one order of magnitude weaker. This indicates that the molecular ion decays rapidly into other products. A very interesting result of this study is the fact that the metal ion Al^+ is the second most abundant species formed at electron energies above 20 eV (see Table 4.1). In the case of TMG (Figure 4.1b) there is a very similar picture with one important difference that the cross-section for metal ion Ga^+ has the second largest cross-section in the whole measured energy scale. The Ga^+ ion make, at 70 eV, almost 27% of all the ions formed.

A more reliable method of partial cross-section measurements was applied by Popovic [106], who measured partial cross-sections for DI to Barium derivatives (BaO, Ba, BaF_2 and BaI_2). Basner et al. [107] investigated EII to $CpPtMe_3$ – $(C_5H_5)Pt(CH_3)_3$, $(MeCp)_2Ru$ – $(CH_3C_5H_4)_2Ru$ and $(MeCp)_2Fe$ – $(CH_3C_5H_4)_2Fe$. They measured ionization energies, total ionization cross-sections, and also mass spectra of the molecules. The contribution of metal atom ions was, according to this work, small.

Another type of DI experiments is the study of the threshold behavior of the mass selected ion yields. Opitz [100] in his study to several organometallic compounds of cobalt measured the mass spectra of the molecules and in addition also ionization energies of the molecules and appearance energies of the fragment ions. From the appearance energies he determined some important quantities, like bond dissociation energies of various functional groups in the molecules and heats of formation of molecular ions and selected fragment ions. Similar studies were done also for molybdenum complexes [108], ferrocene and chloro-ferocene [109].

For many organometallic compounds, electron ionization mass spectra are available, ionized typically by 70 eV electrons ([110–113] ...) Electron ionization mass spectroscopy is with NMR and IR spectroscopy a standard tool that is used to characterize synthesized molecules. The mass spectrometric studies give some basic information about the structure of the molecules, the functional groups of the molecule, and fragment ions formed by DI. These data do not provide information about the cross-sections as a function of electron energy and about the thresholds of the DI channels.

Information about the ionization energies, threshold energies for DI processes can be obtained from other types of experiments, (photo ionization [114], or photoelectron spectrometry [101; 115; 116]) and quantum chemical calculations [117; 118]. Except for electron ionization mass spectrometry, several different ionization methods were applied to study mass spectra of the organometallic molecules (field ionization [119], VUV photoionization [114; 120], multi-photon ionization [121]).

The molecular targets relevant to FEBIP technique are various organometallic compounds which decompose under electron impact into fragments (neutral and ionic), which contain a metal atom and dissociate at some extent into metal atoms and metal atom ions. These particles then may deposit on the surface and form metallic structures. The products of DI reaction may also undergo ion-molecular reactions [122] and chemical reactions on the surface, which may modify the composition of the metallic structures.

The dissociative ionization of the organometallic molecules may be initiated by the primary electrons (with kinetic energies in the range of several tens of keV). However, we also should not underestimate the role of the secondary electrons to initiate DI reaction in FEBIP. The relative contribution of secondary electrons to DI events may be in some systems comparable with the contribution of primary electrons. There exist several reasons for this implication: (1) the cross-sections for DI and EII generally decrease with the increasing electron energy and thus for primary electrons (kinetic energy in the range of tens of keV) the cross-sections for DI could be relatively small, (2) the ionization energies of the organometallic compounds and the DI thresholds are usually low (often far below 10 eV see Table 4.1) even for metal atom ions, thus (3) the thresholds for DI are within the interval energies of the secondary electrons.

6 ELECTRON-INDUCED REACTIONS IN THE CONDENSED PHASE

The previous sections have shown that electron-molecule interactions in gas-phase induce chemical reactions, namely, dissociative electron attachment (DEA), dissociative ionization (DI), and neutral dissociation (ND). The same processes can also be induced in condensed phases or adsorbates on surfaces, the latter being the situation relevant to FEBIP. The dense environment in such systems, however, can modify the outcome of the initial electron-molecule interaction. Furthermore, it offers reaction partners to the initially formed reactive fragments. Consecutive reactions thus occur and have to be considered to understand the final products of electron-induced chemistry. Also, while gas-phase studies are capable of detecting the immediate reactive products of the initial electron-molecule collision, the consecutive reactions in the condensed phase may occur so rapidly that only stable, i.e., closed-shell, final products can in fact be monitored. Typically, because the chemistry of highly reactive fragments often is not very selective, electron-induced reactions in the condensed phase yield a complex mixture of products which can, as an additional complication, usually not all be detected by the same method. Different analytical techniques are therefore, in the ideal case, applied to the same system. An experimental procedure that separates different products after the electron exposure is useful but may by itself induce more consecutive reactions.

The meaning of cross-sections derived from condensed-phase measurements requires specific attention. For example, a cross-section for formation of a specific product may refer to a sequence of elementary reaction steps in which the electron-molecule interaction is only the initiating process. Equally, if the detection of a product requires its thermal desorption from a surface or a condensed phase, the thermal activation may lead to additional reactions so that the detected product is not necessarily the one that was initially formed by electron-induced chemistry. This section discusses selected examples of electron-induced reactions in a condensed phase. Typical methods applied to condensed-phase electron-induced reactions are presented first. The effect of the condensed phase on the initial electron-molecule interactions is then addressed and examples are presented that underline the importance of intermolecular reactions. Finally, examples of cross-section measurements are discussed.

6.1 METHODS FOR STUDYING ELECTRON-INDUCED REACTIONS IN CONDENSED PHASES

Due to the restricted penetration depth of electron beams [124], surface science methods are applied to the study of electron-induced reactions, the most frequently used being electron-stimulated desorption (ESD), both of ions [125; 126] and neutrals [127; 128], thermal desorption spectrometry (TDS) [123; 129], X-ray photoelectron spectroscopy (XPS) [128; 130], high-resolution electron energy loss spectroscopy (HREELS) [131; 132], reflection-absorption infrared spectroscopy (RAIRS) [128; 130; 133], and low-energy electron transmission (LEET) [134]. In these experiments the samples must be conductive to avoid excessive charging during exposure to the electron beam. This calls for thin multilayer films or monolayer adsorbates on conductive surfaces. In contrast to the actual FEBIP process, the experiments are performed under ultrahigh vacuum (UHV) to avoid adsorption of unknown amounts of residual gas and thus be able to control the composition of the sample. UHV requires that the experiments are performed at low temperatures to obtain a sufficient and well-defined surface coverage. On the other hand, small product molecules can be trapped at low temperature and thus identified following electron exposure. Nonetheless, experiments in which a room temperature surface is exposed to a stream of the FEBIP precursor molecules during electron exposure and which thus simulate the actual FEBID process are possible [135]. Also, the clean environment can be exploited to study the effect of impurities like the residual gases present in FEBIP by admixing them in a controlled manner.

Motivated by the wish to understand the effect on the radiation chemistry of low-energy secondary electron that are formed abundantly under exposure to high-energy beams, many experiments so far have focused on electrons with kinetic energies below 20 eV. Nonetheless, commercial electron guns that are frequently used in these experiments can be tuned to energies up to 500 eV. Typical current densities applied to the samples are of the order of a few $\mu A/cm^2$. To achieve sufficient sensitivity, sample areas of 1 cm^2 and more are used.

As mentioned in the outline, different methods used in the study of electron-induced reactions of adsorbates and condensed phases have specific advantages and disadvantages. For example, the frequently used XPS yields information on elemental composition of the sample and the oxidation state of the elements but care must be taken to avoid extended X-ray exposure as secondary electrons themselves are known to induce reactions [136]. ESD mass-spectrometrically detects fragments desorbing under electron exposure but has little sensitivity toward heavier species, which often do not have sufficient kinetic energy to overcome the barrier at the interface between molecular film and the vacuum that results from attractive forces within the condensed phase. LEET is complementary in the sense that it can be used to measure the charging of the sample, which can be traced back to electrons or ions trapped within the sample. Methods measuring vibrational excitation spectra (HREELS, RAIRS) provide information on species remaining in the film. HREELS, on the other hand, can be difficult to interpret when a mixture of products is formed and their bands overlap. RAIRS has a better resolution but suffers from lower sensitivity. Finally, different products can be separated from each other because of their different desorption temperatures by using TDS, which monitors the neutral species desorbing upon temperature increase of the sample.

Despite the fact that one cannot easily distinguish if products detected by TDS have been formed as an immediate consequence of electron exposure or only after activation due to the temperature increase, this method has recently provided valuable insight into the mechanisms of electron-induced reactions by comparing with TDS experiments on samples with known composition [137]. Based on this it is possible to reliably identify products [138; 139] and also to determine cross-sections for the formation of specific products [139]. The discussion of condensed-phase electron-induced reactions will specifically focus on some of these results but other important literature will be included as well.

The examples that will be discussed here to demonstrate the level of insight that can be obtained into condensed-phase electron-induced reactions concern simple organic compounds such as may be used for depositing carbonaceous deposits in FEBIP as well as hexamethyldisiloxane (HMDSO), a silicon-containing precursor. Nonetheless, a few UHV studies using the methods described above have been performed on the organometallics FEBIP precursors $Fe(CO)_5$ [151], $W(CO)_6$ [144], $Ni(CO)_4$ [145], ferrocene ($Fe(C_5H_5)_2$) [146], $Mo(CO)_6$ [147], $(C_6D_6)Cr(CO)_3$ [148], Cu(I)(hfac)(vmts) [149], and trimethyl(methylcyclopentadienyl)platinum(IV) (MeCpPtMe$_3$) [128; 135]. In addition, some FEBIP-related studies concern the organic compounds $CH_3CHNH_2CH_2NH_2$ [150] and C_2H_4 [143]. Several of these studies apply Auger electron spectroscopy (AES) as an alternative to XPS for monitoring the elemental composition and the metal deposition. The most comprehensive study so far concerns MeCpPtMe$_3$ and includes a comparison of cross-sections for its decomposition obtained from XPS, RAIRS, and ESD [128].

6.2 EFFECT OF THE CONDENSED PHASE OR SURFACE ON INTERMEDIATES AND REACTIVE PRODUCTS OF ELECTRON-MOLECULE INTERACTIONS

By embedding a molecule in a condensed phase, the formation and evolution of neutral or ionic excited species produced by electron-molecule interaction is modified with respect to the gas phase. These effects have been summarized recently [152–154] and need not be repeated in detail, although a few effects may be highlighted here.

In general, a condensed phase is a polarizable molecular environment that energetically stabilizes charged states or states with higher dipole moment with respect to the gas phase or to a neutral or less polar ground state. Charged states in a condensed phase are thus typically formed at slightly lower electron energy than in the gas phase. This modifies the relative energies of the various electronic potential energy curves and can, for example, change the position of the critical crossing point and thus the branching ratio between autodetachment and fragmentation. Also, a negative ion state may drop in energy below a neutral state that would, in the gas phase, be the product of autodetachment, thereby increasing the lifetime of the charged state and enhancing its fragmentation probability. Similarly, by energy transfer to neighboring molecules, a negative ion state with potential energy minimum below the neutral ground state and thus accessible to free electrons only via its vibrationally excited states may undergo vibrational relaxation and thus form a stable negative ion. This contributes to charging of a condensed phase under electron exposure. Distance-dependent charge transfer to an adjacent metal surface can deactivate negative ion states and thus suppress fragmentation. Furthermore, thermalization of the incoming electrons through multiple scattering in the condensed phase needs to be considered, as it lowers the average energy at which an individual electron-molecule interaction takes place. As a final example, dissociation processes can be hindered in a condensed phase by the so-called cage effect, which prevents the fragments from drifting apart and thus favors recombination.

The question relevant to FEBIP is if gas phase data on electron-induced fragmentation processes are in fact useful for understanding the corresponding condensed-phase chemistry. While DEA, DI, and ND are equally possible in the condensed phase, although at more or less modified electron energy, the fragmentation cross-section can be modified by orders of magnitude [153; 154]. Because of a delicate balance between the different effects listed above, the actual deviation from the gas phase behavior needs to be considered separately for each specific case. Furthermore, the fragments resulting from electron-molecule interaction may simply recombine with other fragments or may decay themselves if the excess energy gained by formation of the new bond is sufficient to dissociate other bonds.

An example where a known gas-phase resonance survives in the condensed phase and leads to the expected product is shown in Figure 4.18. In the gas phase, DEA to acetonitrile (CH_3CN) in the energy range between 6 and 8 eV leads to dissociation along the CC bond as detected by formation of a dominant negative ion CN^- and smaller quantities of CH_3^- [155]. In the

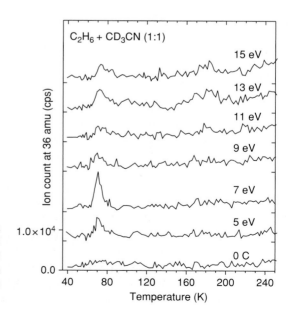

FIGURE 4.18: TDS recorded at 36 amu of a multilayer mixture of ethane (C_2H_6) and deuterated acetonitrile (CD_3CN) before and after electron exposure of 1800 $\mu C/cm^2$. The desorption peak between 70 and 75 K is ascribed to formation of C_2D_6 [123].

condensed phase, formation of C_2D_6 from deuterated acetonitrile is observed predominantly at 7 eV [123]. The same DEA process is thus active in the condensed phase and the resulting neutral CD_3 fragments recombine to yield the stable product C_2D_6. In contrast, recent results on ammonia (NH_3) in the condensed phase have shown that a DEA process in the 7–11 eV range detected as ESD of D^- from ND_3 [125] and consequently leaving behind ND_2 radicals rather produces N_2 most likely via disproportionation of the initially formed radicals [157]. This suggests that ammonia is particularly easy to decompose to volatile fragments by electron exposure and is thus probably a suitable ligand for FEBIP precursors.

6.3 MOLECULAR SYNTHESIS INDUCED BY ELECTRON-MOLECULE INTERACTIONS

In FEBIP the desired outcome of the electron-molecule interaction is complete fragmentation to the point where only the desired element remains on the surface and all other fragments have sufficient volatility to desorb. While the previous sections on electron-molecule interactions in the gas phase have also emphasized fragmentation, in the condensed phase these initiating reactions are very likely succeeded by formation of new bonds driven by the reactivity of fragments with unpaired electrons [123]. These reactions can lead to the synthesis of stable larger and thus less volatile species and, in the extreme case, large-area crosslinking of the irradiated material [130]. Such reactions may well be responsible for the often remaining impurities in the deposits formed under FEBIP. Therefore, a few examples shall illustrate the mechanisms of reactions taking place in electron-induced synthesis.

The first example concerns the reactions of acetaldehyde (CH_3CHO) [139], a molecule that releases important amounts of CO under exposure to electrons above roughly 10 eV. To form this product, two bonds must dissociate, initially yielding H and CH_3 radicals. It has been shown by use of TDS that these highly reactive species not only recombine to form CH_4 but can also react with neighboring intact molecules as summarized in Figure 4.19. These reactions synthesize a variety of products, some of which in fact are larger than the parent compound acetaldehyde. It must be noted that in addition to the radical mechanisms shown in Figure 4.19, recent unpublished quantum chemical calculations provide evidence that some of the products and more specifically 2-propanol ($(CH_3)_2CHOH$) may be formed via cation-driven chemistry.

Another example shows convincingly that ionization-driven chemistry can contribute to the formation of larger and more complex species [156]. Figure 4.20 shows that ionization of ethylene (C_2H_4) initiates the synthesis of a larger product because the cation interacts attractively with a neutral molecule, in this case ammonia (NH_3), that carries a partial negative charge. Above the ionization threshold, ethylene and ammonia thus react to form aminoethane ($CH_3CH_2NH_2$) as again deduced from TDS results. Alternatively, an equivalent reaction mechanism starting with ionization of ammonia can be conceived [156].

Other examples of the synthesis of larger species under low-energy electron exposure include the formation of deuterium peroxide (D_2O_2) from deuterated water (D_2O) [158], the production of ethylene glycol ($HOCH_2CH_2OH$), ethanol (CH_3CH_2OH), dimethyl ether (CH_3OCH_3) in methanol (CH_3OH) multilayers [129], the synthesis of the amino acid glycine (H_2NCH_2COOH) from mixtures of ammonia (NH_3) and acetic acid (CH_3COOH) [159], as well as the formation of various larger fluoroiodocarbons from trifluoroiodomethane (CF_3I) [160] and of different chlorocarbons from carbon tetrachloride (CCl_4) [161].

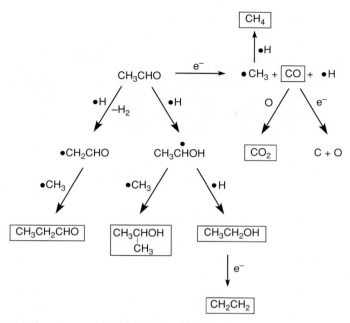

FIGURE 4.19: Products identified in TDS measurements of multilayer films of acetaldehyde (CH$_3$CHO) after electron exposure at 15 eV [139]. Boxes mark the products that have been identified by TDS.

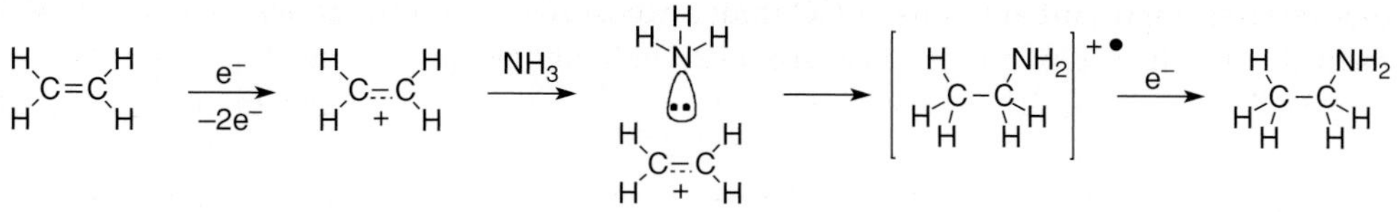

FIGURE 4.20: Synthesis of aminoethane (CH$_3$CH$_2$NH$_2$) in multilayer mixed films of ethylene (C$_2$H$_4$) and ammonia (NH$_3$) induced by electron impact ionisation [156]

6.4 CROSS SECTIONS FOR ELECTRON-INDUCED REACTIONS IN CONDENSED PHASES

Based on the described UHV surface analytical methods, cross-sections for electron-induced reactions in the condensed phase can be obtained. Due to the specific interest in DEA, these measurements were so far often restricted to the 0–20 eV range [131; 139; 153], although experiments aiming at FEBIP precursors presently extend this range to energies reaching up to 1 keV [128]. Reference [153] summarizes different approaches to cross-section measurements, including XPS experiments aiming at cross-sections for specific surface modifications, HREELS studies measuring the production of CO from an oxygen-containing organic compound, cross-sections for dissociation and desorption of molecules from a surface obtained by mass spectrometry, and LEET experiments from which cross-sections for charge trapping, both as solvated electron or in the form of stable anions or anionic fragments resulting from DEA, can be deduced.

The measurement of cross-sections for the decay of the initial sample is a relatively simple task because, under the condition of constant electron current density on the surface and assuming that reactions are initiated by a single electron-molecule collision, it follows simple first order kinetics [123; 128]. In contrast, it is more difficult to deduce values for the various products of electron-induced reactions. This is due to the fact that not all methods, especially those giving information on the chemical identity of a molecular product (RAIRS, HREELS, TDS), yield an absolute measure of the quantity of material. In such cases, signals must first be chosen that can unequivocally be ascribed to a specific product such as the lowest electronic excitation which is used for the quantification of CO in HREELS [131; 132; 153] or a specific desorption signal with a characteristic molecular mass and desorption temperature in TDS [139; 158; 163; 164]. Then the absolute intensity of these signals must be determined by comparing with reference samples containing a known amount of the product. An example of such a procedure using TDS is shown in Figures 4.21 and 4.22. Here, the production of CH_4 from hexamethyldisiloxane (HMDSO) is quantified from its characteristic desorption signal recorded by setting the mass spectrometer to 16 amu. Figure 4.21 also includes the desorption signal for the parent compound HMDSO recorded at 147 amu. The desorption of CH_4 starts with a sharp peak at 55 K but also extends up to the desorption temperature of HMDSO (140–165 K) because a part of the product quantity is trapped in the condensed HMDSO film. This is proven by reference samples, which show a similar behavior [157]. Integration of the CH_4 desorption signal over the complete temperature range up to 165 K thus yields a measure from which the produced amount of CH_4 can be quantified by comparison with the same value obtained from the reference samples as shown in Figure 4.22. From this plot, the cross-section for formation of the product can easily be calculated taking into account the number of electrons applied to the film and the surface area. This is explained in detail in [163].

A comprehensive UHV study on cross-sections for degradation under electron exposure has been dedicated to the FEBIP precursor $MeCpPtMe_3$ [128]. This example shows that

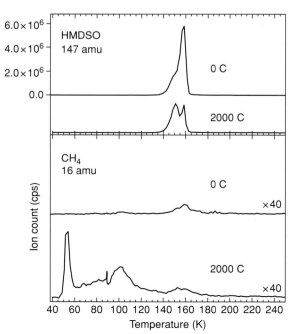

FIGURE 4.21: TDS showing the depletion of hexamethyldisiloxane (HMDSO, 147 amu) from a multilayer films and the formation of methane (CH_4, 16 amu) in the same film upon electron exposure at 15 eV [157]. The integrated desorption signal for CH_4 serves as measure of the produced amount in the determination of the formation cross-section.

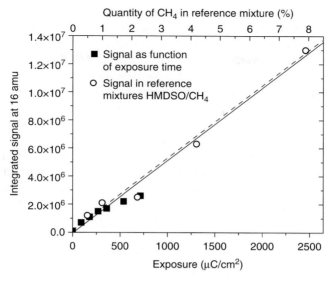

FIGURE 4.22: Plot of the integrated desorption signal recorded at 16 amu (CH_4) as function of electron exposure at 15 eV and plot of the same integrated intensity for reference mixtures as function of CH_4 content [157]. The data yield a cross-section of 5.1×10^{-18} cm^2 for the formation of CH_4.

different methods must be applied to both follow the deposition of the desired element (XPS) and the decomposition of the organic material (RAIRS, MS). An increasing use of TDS in the future is anticipated to yield more detailed information on the chemical nature of the organic material remaining on the surface. The method described above may thus be of interest for fundamental studies on the chemical processes related to FEBIP.

7 CONCLUSIONS AND OUTLOOK

Complete sets of cross-sections, covering all the above processes and the entire energy range, are very rare. The choice of the targets for the measurements in the past was influenced by potential applications in plasma processing for semiconductor manufacture. An example are the cross-sections for CF_4 and SF_6 given in the book of Christophorou and Olthoff, Reference [1], on pages 6 and 727, respectively. In view of the emerging FEBIP application it would be desirable to measure similar cross-section sets at least for several prototype FEBIP precursor molecules. Such an enterprise promises to be interesting also from the purely scientific point of view—by directing the attention of the electron-collision community toward metalorganic compounds.

Dissociative ionization may be the process most relevant to FEBIP. It may be initiated by both the primary electrons (with kinetic energies in the range of several tens of keV) and the secondary electrons. The relative contribution of secondary electrons may in some systems be comparable with the contribution of primary electrons because the cross-sections are larger at low energies, the ionization energies of the organometallic compounds and the DI thresholds are usually low, within the typical energies of the secondary electrons.

Measuring a complete set of cross-sections requires the collaboration of several laboratories and is work-intensive and time-consuming. A serious problem is that a crucial part of the equipment required for such complete cross-section measurements, the Maryland instrument for the detection of neutral radicals, existed only once in the world, and is no longer operational.

Many cross-sections, those on transient molecules and on excited molecules, are hard or impossible to measure and help of theory is indispensable. At the current state, theory made a substantial progress in calculating the elastic cross-sections, but calculating cross-sections for vibrational and electronic excitation is already much more difficult. Methods for calculating the "chemical" process of DEA *ab initio* for molecules with more than three atoms are just beginning to appear, with the case of acetylene mentioned in Section 3 being promising. Calculations of dissociative ionization and neutral dissociation of polyatomic molecules are generally not possible at the present time.

Porting the knowledge from the cross-sections to the FEBIP-like conditions poses a number of challenges. The reactive fragments produced in the primary collisions of electrons with the precursor molecule are likely to be involved in further chemical reactions which can only be infered indirectly from the final stable products identified by different surface analytical tools. All the transient chemical species produced are subject to further electron collisions and subsequent chemical change about which we do not have direct information. The primary electron beam is likely to cause local heating, invoking electron collisions with vibrationally excited molecules, the cross-sections for which often differ substantially from those for cold molecules, and which are hard to measure.

An important question is whether it is possible to control the outcome of the chemical changes induced by electrons. In DEA attachment there are many cases in which the outcome depends on electron energy, opening the possibility of "control" in principle. It will be difficult to invoke in reality, however, because it would require the use of quasi-monoenergetic electrons in the 1–15 eV range, which cannot be focussed to sufficiently small spots. The energy-distribution of the electrons is further "smeared-out" by inelastic collisions in the dense medium.

A second, perhaps more promising approach to "control" is to design custom precursors where certain cross-sections are enhanced and others suppressed. This is possible in some cases—the presence of stable "leaving groups" like halogens, CO_3^-, or N_2 can enhance DEA cross-sections. Local symmetry may also strongly affect DEA cross-sections as has been demonstrated with the dihalo-toluenes. The presence of π virtual orbitals leads to pronounced shape resonances and DEA patterns distinctly different from those of saturated compounds.

ACKNOWLEDGMENT

This research is part of the Swiss National Science Foundation project No. 200020-113599/1 and of COST Action CM0601.

REFERENCES

[1] L. G. Christophorou and J. K. Olthoff. *Fundamental electron interactions with plasma processing gases*. New York: Kluwer Academics/Plenum Publishers, 2004.

[2] A. Muñoz, J. C. Oller, F. Blanco, J. D. Gorfinkiel, P. Limão-Vieira, and G. García. Electron-scattering cross-sections and stopping powers in H_2O. *Phys. Rev. A*, vol. 76, no. 5, p. 052707, 2007.

[3] A. Muñoz, J. C. Oller, F. Blanco, J. D. Gorfinkiel, P. Limão-Vieira, and G. García. Erratum for: Electron-scattering cross-sections and stopping powers in H_2O. *Phys. Rev. A*, vol. 76, no. 5, p. 069901, 2007.

[4] A. Muñoz, J. C. Oiler, F. Blanco, J. D. Gorfinkiel, P. Limão-Vieira, A. Maira-Vidal, M. J. G. Borge, O. Tengblad, C. Huerga, M. Téllez, and G. García. Energy deposition model based on electron scattering cross section data from water molecules. *J. Phys.: Conf. Ser.*, vol. 133, p. 012002, 2008.

[5] N. Silvis-Cividjian and C. W. Hagen. Electron-beam–induced nanometer-scale deposition. vol. 143, p. 1, 2006.

[6] C. W. Hagen, N. Silvis-Cividjian, and P. Kruit. Resolution limit for electron beam-induced deposition on thick substrates. *SCANNING*, vol. 28, p. 204, 2006.

[7] C. W. Hagen, W. F. van Dorp, P. Crozier, and P. Kruit. Electronic pathways in nanostructure fabrication. *Surface Science*, vol. 602, p. 3212, 2008.

[8] M.-W. Ruf, M. Braun, S. Marienfeld, I. I. Fabrikant, and H. Hotop. High resolution studies of dissociative electron attachment to molecules: Dependence on electron and vibrational energy. *J. Phys. Conf. Ser.*, vol. 88, p. 012013, 2007.

[9] H. Kato, M. C. Garcia, T. Asahina, M. Hoshino, C. Makochekanwa, H. Tanaka, F. Blanco, and G. Garcia. Absolute elastic differential cross-sections for electron scattering by $C_6H_5CH_3$ and $C_6H_5CF_3$ at 1.5–200 eV: A comparative experimental and theoretical study with C_6H_6. *Phys. Rev. A*, vol. 79, p. 062703, 2009.

[10] G. J. Schulz. Resonances in electron impact on diatomic molecules. *Rev. Mod. Phys.*, vol. 45, p. 423, 1973.

[11] C. Mündel, M. Berman, and W. Domcke. Nuclear dynamics in resonant electron-molecule scattering beyond the local approximation: Vibrational excitation and dissociative attachment in H_2 and D_2. *Phys. Rev. A*, vol. 32, p. 181.

[12] J. Horáček, M. Čížek, K. Houfek, P. Kolorenč, and W. Domcke. Dissociative electron attachment and vibrational excitation of H_2 by low-energy electrons: Calculations based on an improved nonlocal resonance model. *Phys. Rev. A*, vol. 70, p. 052712, 2004.

[13] J. Horáček, M. Čížek, K. Houfek, P. Kolorenč, and W. Domcke. Dissociative electron attachment and vibrational excitation of H_2 by low-energy electrons: Calculations based on an improved nonlocal resonance model. II. vibrational excitation. *Phys. Rev. A*, vol. 73, p. 022701, 2006.

[14] G. J. Schulz and R. K. Asundi. Isotope effect in the dissociative attachment in H_2 at low energy. *Phys. Rev.*, vol. 158, p. 25, 1967.

[15] B. C. Ibănescu, O. May, A. Monney, and M. Allan. Electron-induced chemistry of alcohols. *Phys. Chem. Chem. Phys.*, vol. 9, p. 3163, 2007.

[16] O. May, J. Fedor, and M. Allan. Isotope effect in dissociative electron attachment to acetylene. *Phys. Rev. A*, vol. 80, p. 012706, 2009.

[17] M. Čížek, J. Horáček, M. Allan, I. I. Fabrikant, and W. Domcke, Vibrational excitation of hydrogen fluoride by low-energy electrons: Theory and experiment. *J. Phys. B*, vol. 36, p. 2837, 2003.

[18] M. Allan. Selectivity in the excitation of Fermi-coupled vibrations in CO_2 by impact of slow electrons. *Phys. Rev. Lett.*, vol. 87, p. 033201, 2001.

[19] M. Allan. Vibrational structures in electron-CO_2 scattering below the $^2\Pi_u$ shape resonance. *J. Phys. B: At. Mol. Opt. Phys.*, vol. 35, p. L387, 2002.

[20] H. Hotop, M.-W. Ruf, M. Allan, and I. I. Fabrikant. Resonance and threshold phenomena in low-energy electron collisions with molecules and clusters. *Adv. At. Mol. Opt. Phys.*, vol. 49, p. 85, 2003.

[21] M. Allan. Absolute angle-differential elastic and vibrational excitation cross-sections for electron collisions with tetrahydrofuran. *J. Phys. B: At. Mol. Opt. Phys.*, vol. 40, p. 3531, 2007.

[22] C. S. Trevisan, A. E. Orel, and T. N. Rescigno. Elastic scattering of low-energy electrons by tetrahydrofuran. *J. Phys. B: At. Mol. Opt. Phys.*, vol. 39, p. L255, 2006.

[23] C. Winstead and V. McKoy. Low-energy electron scattering by deoxyribose and related molecules. *J. Chem. Phys.*, vol. 125, p. 074302, 2006.

[24] R. Čurík, P. Čársky, and M. Allan. Vibrational excitation of methane by slow electrons revisited: Theoretical and experimental study. *J. Phys. B: At. Mol. Opt. Phys.*, vol. 41, p. 115203, 2008.

[25] M. Allan, K. Franz, H. Hotop, O. Zatsarinny, and K. Bartschat. Absolute angle-differential cross-sections for electron-impact excitation of neon within the first 3.5 eV above threshold. *J. Phys. B*, vol. 42, p. 044009, 2009.

[26] M. Allan, C. Winstead, and V. McKoy. Electron scattering in ethene: Excitation of the $\tilde{a}^3B_{1u}$ state, elastic scattering and vibrational excitation. *Phys. Rev. A*, vol. 77, p. 042715, 2008.

[27] Q. Sun, C. Winstead, V. McKoy, and M. A. P. Lima. Low-energy electron-impact excitation of the $\tilde{a}\,^3B_{1u}$ ($\pi \rightarrow \pi^*$) state of ethylene. *J. Chem. Phys.*, vol. 96, p. 3531, 1992.

[28] M. Čížek, J. Horáček, A.-C. Sergenton, D. B. Popović, M. Allan, W. Domcke, T. Leininger, and F. X. Gadea. Inelastic low-energy electron collisions with the HBr and DBr molecules: Experiment and theory. *Phys. Rev. A*, vol. 63, p. 062710, 2001.

[29] J. Fedor, O. May, and M. Allan. Absolute cross-sections for dissociative electron attachment to HCl, HBr, and their deuterated analogs. *Phys. Rev. A*, vol. 78, p. 032701, 2008.

[30] J. Horáček, M. Čížek, P. Kolorenč, and W. Domcke. Isotope effects in vibrational excitation and dissociative electron attachment of DCl and DBr. *Eur. Phys. J. D*, vol. 35, p. 255, 2005.

[31] R. Abouaf and D. Teillet-Billy. Fine structure in dissociative-attachment cross-sections for hydrogen chloride and deuterium chloride. *J. Phys. B: At. Mol. Phys.*, vol. 10, p. 2261, 1977.

[32] R. Azria and F. Fiquet-Fayard. Atachment électronique dissociatif sur C_2H_2 et C_2D_2. *J. Physique (Paris)*, vol. 33, p. 663, 1972.

[33] S. T. Chourou and A. E. Orel. Dissociative attachment to acetylene. *Phys. Rev. A*, vol. 77, no. 4, p. 042709, 2008.

[34] S. T. Chourou and A. E. Orel. Dissociative attachment to acetylene. *Phys. Rev. A*, p. to be published, 2009.

[35] D. J. Haxton, T. N. Rescigno, and C. W. McCurdy. Dissociative electron attachment to the H_2O molecule. II. nuclear dynamics on coupled electronic surfaces within the local complex potential model. *Phys. Rev. A*, vol. 75, p. 012711, 2007.

[36] D. J. Haxton, T. N. Rescigno, and C. W. McCurdy. Erratum. *Phys. Rev. A*, vol. 76, p. 049907, 2007.

[37] S. Ptasińska, S. Denifl, P. Scheier, E. Illenberger, and T. D. Märk. Bond- and site-selective loss of H atoms from nucleobases by very-low-energy electrons (<3 eV). *Angew. Chem. Int. Ed.*, vol. 44, p. 6941, 2005.

[38] V. S. Prabhudesai, A. H. Kelkar, D. Nandi, and E. Krishnakumar. Functional group dependent site specific fragmentation of molecules by low energy electrons. *Phys. Rev. Lett.*, vol. 95, p. 143202, 2005.

[39] B. C. Ibanescu and M. Allan. A dramatic difference between the electron-driven dissociation of alcohols and ethers and its relation to Rydberg states. *Phys. Chem. Chem. Phys.*, vol. 10, p. 5232, 2008.

[40] B. C. Ibanescu and M. Allan. Selective cleavage of the C-O bonds in alcohols and asymmetric ethers by dissociative electron attachment. *Phys. Chem. Chem. Phys.*, p. at press, 2009.

[41] R. Dressler, M. Allan, and E. Haselbach. Symmetry control in bond cleavage processes: Dissociative electron attachment to unsaturated halocarbons. *Chimia*, vol. 39, p. 385, 1985.

[42] K. L. Stricklett, S. C. Chu, and P. D. Burrow. Dissociative attachment in vinyl and allyl chloride, chlorobenzene and benzyl chloride. *Chem. Phys. Lett.*, vol. 131, p. 279, 1986.

[43] C. Bulliard, M. Allan, and E. Haselbach. Intramolecular competition of phenylic and benzylic CX bond breaking in dissociative electron attachment to dihalotoluenes. *J. Phys. Chem.*, vol. 98, p. 11040, 1994.

[44] C. Bulliard, M. Allan, and S. Grimme. Electron energy loss and dissociative electron attachment spectroscopy of methyl vinyl ether and related compounds. *Int. J. of Mass Spectr. and Ion Proc.*, vol. 205, p. 43, 2001.

[45] S. Živanov, B. C. Ibănescu, M. Paech, M. Poffet, P. Baettig, A.-C. Sergenton, S. Grimme, and M. Allan. Dissociative electron attachment and electron energy loss spectra of phenyl azide. *J. Phys. B: At. Mol. Opt. Phys.*, vol. 40, p. 101, 2007.

[46] M. A. Lieberman and A. J. Lichtenberg, *Principles of plasma discharges and materials processing*. New York: John Wiley & Sons, 1994.

[47] I. Shimamura and K. Takayanagi, eds., *Electron-molecule collisions*. New York: Plenum Press, 1984.

[48] A. Gilardini, *Low energy electron collisions in gases*. New York: John Wiley & Sons, 1972.

[49] W. Hack. Detection methods for atoms and radicals in the gas phase. *International Reviews in Physical Chemistry*, vol. 4, 1985.

[50] E. Paneth and W. Hofeditz. Preparation of free methyl. *Ber. Dt. Chem. Ges. B*, vol. 62, p. 1335, 1929.

[51] F. O. Rice and M. D. Dooley. Thermal decomposition of organic compounds from the standpoint of free radicals. XII. the decomposition of methane. *J. Am. Chem. Soc.*, vol. 56, p. 2747, 1934.

[52] L. Belchetz and E. Rideal. The primary decomposition of hydrocarbon vapors on carbon filaments. *J. Am. Chem. Soc.*, vol. 57, p. 1168, 1935.

[53] S. J. B. Corrigan. Dissociation of molecular hydrogen by electron impact. *J. Chem. Phys.*, vol. 43, p. 4381, 1965.

[54] H. F. Winters and M. Inokuti. Total dissociation cross-section of tetrafluoromethane and other fluoroalkanes for electron impact. *Phys. Rev. A*, vol. 25, p. 1420, 1982.

[55] Y. Ohmori, K. Kitamori, M. Shimozuma, and H. Tagashira. Boltzmann equation analysis of electron swarm behavior in methane. *J. Phys. D*, vol. 19, p. 437, 1986.

[56] Y. Nakamura. *Gaseous electronics and their applications*, p. 178. Tokyo, Japan: KTK Scientific, 1991.

[57] W. L. Morgan. A critical evaluation of low-energy electron impact cross sections for plasma processing modeling. II: Tetrafluoromethane, silane, and methane. *Plasma Chem. Plasma Proc.*, vol. 12, 1992.

[58] M. Hayashi. *Swarm studies and inelastic electron-molecule collisions*, p. 167. New York: Springer, 1987.

[59] R. A. Bonham. Electron impact cross-section data for carbon tetrafluoride. *Jpn. J. Appl. Phys. 1*, vol. 33, 1994.

[60] D. K. Davies, L. E. Kline, and W. E. Bies. Measurements of swarm parameters and derived electron collision cross-sections in methane. *J. Appl. Phys.*, vol. 65, 1989.

[61] F. J. de Heer. Electron excitation, dissociation and ionization of hydrogen, deuterium and tritium molecules, simple hydrocarbons and their ions. *Phys. Scr.*, vol. 23, p. 170, 1981.

[62] S. Motlagh and J. H. Moore. Cross sections for radicals from electron impact on methane and fluoroalkanes. *J. Chem. Phys.*, vol. 109, p. 432, 1998.

[63] L. Mi and R. A. Bonham. Electron-ion coincidence measurements: The neutral dissociation cross-section for CF$_4$. *J. Chem. Phys.*, vol. 108, p. 1910, 1998.

[64] H. Tanaka, H. Toyda, and H. Sugai. Cross-sections for electron-impact dissociation of alternative etching gas, C$_3$HF$_7$O. *Jpn. J. Appl. Phys.*, vol. 37, p. 5053, 1998.

[65] J. Perrin, J. P. M. Schmitt, G. D. Rosny, B. Drevillon, J. Huc, and A. Lloret. Dissociation cross-sections of silane and disilane by electron impact. *Chem. Phys.*, vol. 73, p. 383, 1982.

[66] H. C. Straub, D. Lin, B. G. Lindsay, K. A. Smith, and R. F. Stebbings. Absolute partial cross-sections for electron-impact ionization of CH$_4$ from threshold to 1000 eV. *J. Chem. Phys.*, vol. 106, 1997.

[67] H. F. Winters. Dissociation of methane by electron impact. *J. Chem. Phys.*, vol. 63, p. 3462, 1975.

[68] M. Goto, K. Nakamura, H. Toyoda, and H. Sugai. Cross section measurements for electron-impact dissociation of CHF$_3$ into neutral and ionic radicals. *Jpn. J. Appl. Phys., Pt. 1*, vol. 33, 1994.

[69] J. E. Baio, H. Yu, D. W. Flaherty, H. F. Winters, and D. B. Graves. Electron-impact dissociation cross-sections for CHF$_3$ and C$_3$F$_8$. *J. Phys, D: Appl. Phys.*, vol. 40, p. 6969, 2007.

[70] H. Toyoda, M. Ito, and H. Sugai. Cross section measurements for electron-impact dissociation of C$_4$F$_8$ into neutral and ionic radicals. *Jpn. J. Appl. Phys.*, vol. 36, p. 3730, 1997.

[71] L. G. Christophorou, J. K. Olthoff, and M. V. V. S. Rao. Electron interactions with CF$_4$. *J. Phys. Chem. Ref. Data*, vol. 25, 1996.

[72] L. G. Christophorou, J. K. Olthoff, and M. V. V. S. Rao. Electron interactions with CHF$_3$. *J. Phys. Chem. Ref. Data*, vol. 26, 1997.

[73] L. G. Christophorou and J. K. Olthoff. Electron interactions with C$_2$F$_6$. *J. Phys. Chem. Ref. Data*, vol. 27, 1998.

[74] T. Shirai, T. Tabata, H. Tawara, and Y. Itikawa. Analytic cross-sections for electron collisions with hydrocarbons: CH$_4$, C$_2$H$_6$, C$_2$H$_4$, C$_2$H$_2$, C$_3$H$_8$, and C$_3$H$_6$. *At. Data Nucl. Data Tables*, vol. 80, p. 147, 2002.

[75] C. Ma, M. R. Bruce, and R. A. Bonham. Absolute partial and total electron-impact-ionization cross-sections for tetrafluoromethane from threshold up to 500 eV. *Phys. Rev. A*, vol. 44, p. 2921, 1991.

[76] C. Ma, M. R. Bruce, and R. A. Bonham. Erratum. *Phys. Rev. A*, vol. 45, p. 6932, 1992.

[77] R. A. Bonham. Electron impact cross-section data for carbon tetrafluoride. *Jpn. J. Appl. Phys. 1*, vol. 33, p. 4157, 1994.

[78] H. U. Poll, C. Winkler, D. Margreiter, V. Grill, and T. D. Märk. Discrimination effects for ions with high initial kinetic energy in a Nier-type ion source and partial and total electron ionization cross-sections of carbon tetrafluoride. *Int. J. Mass Spectrom. Ion Processes*, vol. 112, p. 1, 1992.

[79] M. R. Bruce, C. Ma, and R. A. Bonham. Electron impact cross-section data for carbon tetrafluoride. *Chem. Phys. Lett.*, vol. 190, p. 285, 1992.

[80] T. D. Märk. *Electron-molecule interactions and their applications*. Orlando: Academic Press, 1984.

[81] T. D. Märk and G. H. Dunn. *Electron impact ionisation*. Springer-Verlag, 1985.

[82] E. W. McDaniel. *Atomic collisions, electron and photon projectiles*. New York: John Wiley & Sons, 1986.

[83] E. Illenberger and J. Momigny, eds. *Gaseous molecular ions*. New York: Steinkopf Verlag Darmstadt/Springer Verlag, 1992.

[84] K. H. Becker and V. Tarnovsky. Electron-impact ionization of atoms, molecules, ions and transient species. *Plasma Sources Sci. Technol.*, vol. 4, p. 307, 1995.

[85] in *NIST Chemistry WebBook, NIST Standard Reference Database Number 69* (P. J. Linstrom and W. G. Mallard, eds.), http://webbook.nist.gov, retrieved July 21, 2009.

[86] F. W. McLafferty and F. Tureček, *Interpretation of mass spectra*, 4th edition, p. 55. University Science books, 1993.

[87] J. Lango, L. Szepes, P. Csaszar, and G. Inorta. Studies on the unimolecular decomposition of organometallic ions. *J. Organometal. Chem.*, vol. 269, p. 133, 1984.

[88] H. Deutsch, D. Margreiter, and T. D. Märk. Total electron impact ionization cross-sections of free molecular radicals: a case of failure of the additivity rule? *Int. J. Mass Spectrom Ion Proc.*, vol. 93, p. 259, 1989.

[89] H. Deutsch, T. D. Märk, V. Tamovsky, K. Becker, C. Comelissen, L. Cespive, and V. Bonacic-Koutecky. Measured and calculated absolute total cross-sections for the single ionization of CF_x and NF_x by electron impact. *Int. J. Mass Spect. Ion Proc.*, vol. 137, p. 77, 1994.

[90] T. D. Märk. Ionization by electron impact. *Plasma Phys. Control. Fusion*, vol. 34, p. 2083, 1992.

[91] Y. K. Kim and M. E. Rudd. Binary-encouter-dipole model for electron-impact ionization. *Phys. Rev. A*, vol. 50, p. 3954, 1994.

[92] S. J. King and S. D. Price. Electron ionization of H_2O. *Int. J. Mass Spect.*, vol. 277, p. 84, 2008.

[93] C. Tian and C. R. Vidal. Electron impact dissociative ionization of CO_2: Measurements with a focusing time-of-flight mass spectrometer. *J. Chem. Phys.*, vol. 108, p. 927, 1998.

[94] R. Basner, M. Schmidt, and K. Becker. Absolute total and partial cross sections for the electron impact ionization of diborane (B_2H_6). *J. Chem. Phys.*, vol. 118, p. 2153, 2003.

[95] K. Gluch, P. Scheier, W. Schustereder, T. Tepnual, L. Feketeova, C. Mair, S. Matt-Leubner, A. Stamatovic, and T. D. Märk. Cross sections and ion kinetic energies for electron impact ionization of CH_4. *Int. J. Mass Spect.*, vol. 228, p. 307, 2003.

[96] M. A. Mangan, B. G. Lindsay, and R. F. Stebbings. Absolute partial cross sections for electron-impact ionization of CO from threshold to 1000 eV. *J. Phys. B: At. Mol. Opt. Phys.*, vol. 33, p. 3225, 2000.

[97] R. E. Winters and R. W. Kiser. Ionization and fragmentation of dimethylzinc, trimethylaluminum, and trimethylantimony. *J. Organometal. Chem.*, vol. 10, p. 7, 1967.

[98] C. Q. Jiao, C. A. DeJoseph Jr., P. Haaland, and A. Garscadden. Electron impact ionization and ion chemistry in trimethylaluminum and in trimethylgallium. *Int. J. Mass Spectr.*, vol. 202, p. 345, 2000.

[99] F. Glockling and R. G. Strafford. Electron impact studies on some group III metal alkyls. *J. Chem. Soc. A*, p. 1761, 1971.

[100] J. Opitz. Electron impact ionization of cobalt-tricarbonyl-nitrosyl, cyclopentadienyl-cobalt-dicarbonyl and biscyclopentadienyl-cobalt: Appearance energies, bond energies and enthalpies of formation. *Int. J. Mass Spectr.*, vol. 225, p. 115, 2003.

[101] J. F. Nixon, R. J. Suffolk, M. J. Taylor, J. G. Norman Jr., D. E. Hoskins, and D. J. Gmur. Photoelectron and electronic spectra of $Rh_2Cl_2(CO)_4$ and $Rh_2Cl_2(PF_3)_4$: Assignments from SCF-Xa-SW calculations. *Inorg. Chem.*, vol. 19, p. 810, 1980.

[102] G. D. Michels and H. J. Svec. Characterization of $MnTc(CO)_{10}$ and $TcRe(CO)_{10}$. *Inorg. Chem.*, vol. 20, p. 3445, 1981.

[103] S. Barfuss, K. H. Emrich, W. Hirschwald, P. A. Dowben, and N. M. Boag. A mass spectrometric investigation of chloro-, bromo- and methylferrocenes by electron and photon impact ionization. *J. Organomet. Chem.*, vol. 391, p. 209, 1990.

[104] E. Vasekova, M. Stano, S. Matejcik, J. D. Skalný, P. Mach, T. D. Märk, and J. Urban. Electron impact ionization of C_2H_6: ionization energies and temperature effects. *Int. J. of Mass Spect.*, vol. 235, p. 155, 2004.

[105] S. Denifl, S. Matejcik, J. D. Skalný, M. Stano, P. Mach, J. Urban, P. Scheier, T. D. Märk, and W. Barszczewska. Electron impact ionization of C_3H_8: appearance energies and temperature effects. *Chem. Phys. Lett.*, vol. 402, p. 80, 2005.

[106] A. Popovic. Mass spectrometric determination of the ionisation cross-sections of BaO, Ba, BaF_2 and BaI_2 by electron impact. *Int. J. Mass Spect.*, vol. 230, p. 99, 2003.

[107] R. Basner, M. Schmidt, and H. Deutsch. Electron impact ionization and fragmentation of metal-organic compounds used in plasma assisted thin film deposition techniques. *Contrib. Plasma Phys.*, vol. 35, p. 375, 1995.

[108] D. Bruch, J. Opitz, and G. von Bünau. Electron impact and multiphoton ionization and fragmentation of molybdenum-cyclopentadienyl-dicarbonyl-nitrosyl at 351, 248 and 193 nm. *Int. J. Mass Spect. Ion Proc.*, vol. 171, p. 147, 1997.

[109] S. Barfuss, M. Grade, W. Hirschwald, W. Rosinger, N. M. Boag, D. C. Driscoll, and P. A. Dowben. The stability and decomposition of gaseous chloroferrocenes. *J. Vac. Sci. Technol. A*, vol. 5, p. 1451, 1987.

[110] V. H. Dibeler. Mass spectra of the tetramethyl compounds of carbon silicon, germanium, tin, and lead. *J. Research NBS*, vol. 49, p. 235, 1952.

[111] R. B. King. Mass spectra of organometallic compounds. VI. some indenylmetal derivatives and related compounds. *Can. J. Chem.*, vol. 47, p. 559, 1969.

[112] A. Bjarnason. Fourier transform mass spectrometry of several organometallic complexes: Laser desorption versus electron impact ionization. *Oganometallics*, vol. 9, p. 657, 1990.

[113] M. N. Rocklein and D. P. Land. Mass spectra of $Fe(CO)_5$ using Fourier transform mass spectrometry (FTMS) and laser induced thermal desorption FTMS with electron ionization, charge exchange, and proton transfer. *Int. J.Mass Spect.*, vol. 177, p. 83, 1998.

[114] S. Georgiou, E. Mastoraki, E. Raptakis, and Z. Xenidi. The potential of VUV photoionisation mass spectrometry in monitoring photofragmentation of organometallics. *Laser Chem.*, vol. 13, p. 113, 1993.

[115] P. J. Bassett, B. R. Higginson, D. R. Lloyd, N. Lynaugh, and P. J. Roberts. Helium-I photoelectron spectra of tetrakis(trifluoro-phosphine)-nickel(0), -palladium(0), and -platinum(0). *J. Chem. Soc. Dalton Trans.*, vol. 21, p. 2316, 1974.

[116] D. S. Yang, G. M. Bancroft, R. J. Puddephatt, K. H. Tan, J. N. Cutler, and J. D. Bozek. Assignment of the valence molecular orbitals of $CpPtMe_3$ and $Me_2Pt(COD)$ using variable-energy photoelectron spectra. *Inorg. Chem.*, vol. 29, p. 4956, 1990.

[117] P. Seuret, F. Cicoira, T. Ohta, P. Doppelt, P. Hoffmann, J. Weber, and T. A. Wesolowski. An experimental and theoretical study of $[RhCl(PF_3)_2]_2$ fragmentation. *Phys. Chem. Chem. Phys.*, vol. 5, p. 268, 2003.

[118] M. Polášek and J. Kubišta. Bis(eta^5-cyclopentadienyl)titanium(II) in the gas phase: Mass spectrometric and computational study of the structure and reactivity. *J. Organomet. Chem.*, vol. 692, p. 4073, 2007.

[119] R. B. Sohnlein and D.-S. Yanga. Pulsed-field ionization electron spectroscopy of group 6 metal Cr, Mo, and W bis benzene sandwich complexes. *J. Chem. Phys.*, vol. 124, p. 134305, 2006.

[120] F. Qi, S. Yang, L. Sheng, H. Gao, Y. Zhang, and S. Yu. Vacuum ultraviolet photoionization and dissociative photoionization of $W(CO)_6$. *J. Chem. Phys.*, vol. 107, p. 10391, 1997.

[121] B. Samorski, J. M. Hosenlopp, D. Rooney, and J. Cheiken. The effect of deuteration on multiphoton dissociation of benzene chromium tricarbonyl. *J. Chem. Phys.*, vol. 85, p. 3326, 1986.

[122] P. B. Armentrout and J. L. Beauchamp. The chemistry of atomic transition-metal ions: Insight into fundamental aspects of organometallic chemistry. *Acc. Chem. Res.*, vol. 22, p. 315, 1989.

[123] I. Ipolyi, W. Michaelis, and P. Swiderek. Electron-induced reactions in condensed films of acetonitrile and ethane. *Phys. Chem. Chem. Phys.*, vol. 8, p. 180, 2007.

[124] D. R. Lide, *CRC Handbook of Chemistry and Physics*. 86th Edition. Taylor & Francis, 2005.

[125] M. Tronc, R. Azria, Y. L. Coat, and E.Illenberger. Threefold differential electron-stimulated desorption yields of D^- anions from multilayer films of D_2O and ND_3 condensed on platinum. *J. Phys. Chem.*, vol. 100, p. 14745, 1996.

[126] L. Sanche. Electron resonances in DIET. *Surf. Sci.*, vol. 451, p. 82, 2000.

[127] M. A. Huels, P.-C. Dugal, and L. Sanche. Degradation of functionalized alkanethiolate monolayers by 0-18 eV electrons. *J. Chem. Phys.*, vol. 118, p. 11168, 2003.

[128] J. D. Wnuk, J. M. Gorham, S. Rosenberg, W. F. van Dorp, T. E. Madey, C. W. Hagen, and D. H. Fairbrother. Electron induced surface reactions of the organometallic precursor trimethyl(methylcyclopentadienyl)platinum(IV). *J. Phys. Chem. C*, vol. 113, p. 2487, 2009.

[129] T. D. Harris, D. H. Lee, M. Q. Blumberg, and C. R. Arumainayagam. Electron-induced reactions in methanol ultrathin films studied by temperature-programmed desorption: A useful method to study radiation chemistry. *J. Phys. Chem.*, vol. 99, p. 9530, 1995.

[130] W. Eck, V. Stadler, W. Geyer, M. Zharnikov, A. Gölzhäuser, and M. Grunze. Generation of surface amino groups on aromatic self-assembled monolayers by low energy electron beams: A first step towards chemical lithography. *Adv. Mater.*, vol. 12, p. 805, 2000.

[131] M. Lepage, M. Michaud, and L. Sanche. Low-energy electron scattering cross section for the production of CO within condensed acetone. *J. Chem. Phys.*, vol. 113, p. 3602, 2000.

[132] C. Jäggle, P. Swiderek, S.-P. Breton, M. Michaud, and L. Sanche. Products and reaction sequences in tetrahydrofuran exposed to low-energy electrons. *J. Phys. Chem. B*, vol. 110, p. 12512, 2006.

[133] C. Olsen and A. Rowntree. Bond-selective dissociation of alkanethiol based self-assembled monolayers adsorbed on gold substrates, using low-energy electron beams. *J. Chem. Phys.*, vol. 108, p. 3750, 1998.

[134] L. Sanche. Transmission of 0–15 eV monoenergetic electrons through thin-film molecular solids. *J. Chem. Phys.*, vol. 71, p. 4860, 1979.

[135] M. N. Hedhili, J. H. Bredehöft, and P. Swiderek. Electron-induced reactions of MeCpPtMe$_3$ investigated by HREELS. *J. Phys. Chem. C*, vol. 113, p. 13282, 2009.

[136] R. L. Graham, C. D. Bain, H. A. Biebuyck, E. Laibinis, and G. M. Whitesides. Damage to CF$_3$CONH-terminated organic self-assembled monolayers (SAMs) on Al, Ti, Cu, and Au by Al Kα X-rays is due principally to electrons. *J. Phys. Chem.*, vol. 97, p. 9456, 1993.

[137] E. Burean, I. Ipolyi, T. Hamann, and P. Swiderek. Thermal desorption spectrometry for the study of electron-induced reactions. *Int. J. Mass Spectrom.*, vol. 277, p. 215, 2008.

[138] P. Swiderek, C. Jäggle, D. Bankmann, and E. Burean. Fate of reactive intermediates formed in acetaldehyde under exposure to low-energy electrons. *J. Phys. Chem. C*, vol. 111, p. 303, 2007.

[139] E. Burean and P. Swiderek. Electron-induced reactions in condensed acetaldehyde: Identification of products and energy dependent cross sections. *J. Phys. Chem. C*, vol. 112, p. 19456, 2008.

[140] J. S. Foord and R. B. Jackman. Studies of adsorption and electron-induced dissociation of Fe(CO)$_5$ on Si(110). *Surf. Sci.*, vol. 171, p. 197, 1986.

[141] M. A. Henderson, R. D. Ramsier, and J. T. Yates, Jr. Photon- versus electron-induced decomposition of Fe(CO)$_5$ adsorbed on Ag(111): Iron film deposition. *J. Vac. Sci. Techn. A*, vol. 9, p. 1563, 1991.

[142] M. A. Henderson, R. D. Ramsier, and J. T. Yates, Jr. Low-energy electron-induced decomposition of Fe(CO)$_5$ adsorbed on Ag(111). *Surf. Sci.*, vol. 259, p. 173, 1991.

[143] O. Guise, H. Marbach, J. Levy, J. Ahner, and J. T. Yates, Jr. Electron-beam-induced deposition of carbon films on Si(100) using chemisorbed ethylene as a precursor molecule. *Surf. Sci.*, vol. 571, p. 128, 2004.

[144] J. R. Swanson, F. A. Flitsch, and C. M. Friend. Low energy electron induced decomposition on surfaces: W(CO)$_6$ on Si(111)-(7 × 7). *Surf. Sci.*, vol. 215, p. L293, 1989.

[145] R. D. Ramsier, M. A. Henderson, and J. T. Yates, Jr. Electron induced decomposition of $Ni(CO)_4$, adsorbed on Ag(111). *Surf. Sci.*, vol. 257, p. 9, 1991.

[146] K. Svensson, T. R. Bedson, and R. E. Palmer. Dissociation and desorption of ferrocene on graphite by low energy electron impact. *Surf. Sci.*, vol. 451, p. 250, 2000.

[147] Y. Wang, F. Gao, M. Kaltchev, and W. T. Tysoe. The effect of electron beam irradiation on the chemistry of molybdenum hexacarbonyl on thin alumina films in ultrahigh vacuum. *J. Mol. Catal. A*, vol. 209, p. 135, 2004.

[148] R. D. Ramsier and J. T. Yates, Jr. Thermal and electron-induced behavior of d_6-benzene-chromium-tricarbonyl adsorbed on Ag(111). *Surf. Sci.*, vol. 289, p. 39, 1993.

[149] S. Mezhenny, I. Lyubinetsky, W. J. Choyke, and J. T. Yates, Jr. Electron stimulated decomposition of adsorbed hexafluoroacetylacetonate Cu(I) vinyltrimethylsilane, Cu(I)(hfac)(vmts). *J. Appl. Phys.*, vol. 85, p. 3368, 1999.

[150] J. D. Wnuk, J. M. Gorham, and D. H. Fairbrother. Growth and microstructure of nanoscale amorphous carbon nitride films deposited by electron beam irradiation of 1,2-diaminopropane. *J. Phys. Chem. C*, vol. 113, p. 12345, 2009.

[151] T. Lukasczyk, M. Schirmer, H. P. Steinrück, and H. Marbach. Electron-beam-induced deposition in ultrahigh vacuum: Lithographic fabrication of clean iron. *Small*, vol. 4, p. 841, 2008.

[152] L. Sanche. Effects of the solid phase on resonance stabilization, dissociative attachment and dipolar dissociation. in *Linking the gaseous and condensed phases of matter* (L. G. Christophorou, E. Illenberger, and W.-F. Schmidt, eds.), p. 377, Plenum Press, New York, 1994.

[153] A. Bass and L. Sanche. Absolute and effective cross-sections for low-energy electron-scattering processes within condensed matter. *Radiat. Environ. Biophysics*, vol. 37, p. 243, 1998.

[154] I. Bald, J. Langer, P. Tegeder, and O. Ingólfsson. From isolated molecules through clusters and condensates to the building blocks of life – A short tribute to Prof. Eugen Illenberger's work in the field of negative ion chemistry. *Int. J. Mass Spectrom.*, vol. 277, p. 4, 2008.

[155] W. Sailer, A. Pelc, P. Limão-Vieira, N. J. Mason, J. Limtrakul, and T. D. M. P. Scheier, M. Probst. Low energy electron attachment to CH_3CN. *Chem. Phys. Lett.*, vol. 381, p. 216, 2003.

[156] T. Hamann, E. Böhler, and P. Swiderek. Low-energy electron-induced hydroamination of an alkene. *Angew. Chem. Int. Ed.*, vol. 48, p. 4643, 2009.

[157] I. Ipolyi, E. Burean, T. Hamann, M. Cingel, S. Matejcik, and P. Swiderek. Low-energy electron-induced chemistry of condensed-phase hexamethyldisiloxane: Initiating dissociative process and subsequent reactions. *Int. J. Mass Spectrom.*, vol. 282, p. 133, 2009.

[158] X. Pan, A. D. Bass, J. Jay-Gerin, and L. Sanche. A mechanism for the production of hydrogen peroxide and the hydroperoxyl radical on icy satellites by low-energy electrons. *Icarus*, vol. 172, p. 521, 2004.

[159] A. Lafosse, M. Bertin, A. Domaracka, D. Pliszka, E. Illenberger, and R. Azria. Reactivity induced at 25 K by low-energy electron irradiation of condensed NH_3-CH_3COOD (1:1) mixture. *Phys. Chem. Chem. Phys.*, vol. 8, p. 5564, 2006.

[160] N. Nakayama, E. E. Ferrenz, D. R. Ostling, A. S. Nichols, J. F. Faulk, and C. R. Arumainayagam. Surface chemistry and radiation chemistry of trifluoroiodomethane (CF_3I) on Mo(110). *J. Phys. Chem. B*, vol. 108, p. 4080, 2004.

[161] L. D. Weeks, L. L. Zhu, M. Pellon, D. R. Haines, and C. R. Arumainayagam. Low-energy electron-induced oligomerization of condensed carbontetrachloride. *J. Phys. Chem. C*, vol. 111, p. 4815, 2007.

[162] E. Burean and P. Swiderek. Thermal desorption measurements of cross-sections for reactions in condensed acetaldehyde induced by low-energy electrons. *Surf. Sci.*, vol. 602, p. 3194, 2008.

[163] E. Burean, I. Ipolyi, T. Hamann, and P. Swiderek. Thermal desorption spectrometry for the identification of products formed by electron-induced reactions. *Int. J. Mass Spectrom.*, vol. 277, p. 215, 2008.

CHAPTER 5

SIMULATION OF FOCUSED ION BEAM MILLING

Heung-Bae Kim and Gerhard Hobler

1 ION-SOLID INTERACTION AND ANALYSIS

The quality of ion beam direct fabrication depends critically on the interactions between the impinging ion beam and the target. Thus, understanding the basics of ion beam–solid interactions may greatly enhance the ability to achieve optimum results using an ion beam system. Depending on process parameters, these interactions can result in swelling, deposition, sputtering, etching, redeposition, implantation, backscattering, or nuclear reaction [1]. When an energetic ion impinges on a surface, it interacts with the target nuclei (nuclear stopping) and the target electrons (electronic stopping). While electronic stopping mainly acts as a dragging force on the projectile, nuclear stopping leads to both energy transfer from the projectile to the target atoms and deflection of the projectile. The latter determines the spatial distribution of the interactions further down the projectile path and may also lead to backscattering of the incident ion from the sample. The energy exchange can be sufficient to dislodge a surface atom from a weakly bound position and cause its relocation on the surface to a more strongly bound position. Greater energy exchange leads to physical sputtering, when enough momentum is transferred to entirely free one or more atoms. Ions can penetrate into the lattice and become trapped as their energy is expended (ion implantation). In addition, chemical effects may occur. When the ions react with the surface atoms, new compounds can be formed on the sample surface, or the outermost layer of atoms or molecules may leave into the gas phase (chemical sputtering). Alternatively, when precursor molecules have been adsorbed to the surface as a result of gas injection, they may be decomposed by the impinging ion and undergo further

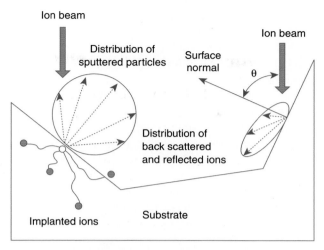

FIGURE 5.1: Schematic illustration of ion solid interactions: incidence angle (θ), sputtering, sputtered particles distribution, ion penetration (implantation), backscattering reflection of ions and their distribution.

reactions with the target material or with each other. Depending on whether the target is involved in these reactions and whether the reaction products are volatile, this process results in ion-beam induced deposition or etching. Finally, secondary electron emission occurs under suitable conditions of ion bombardment of metal surfaces. Secondary ion emission results when surface atoms are excited to ionized states and emitted from the sample [2–4]. The processes more relevant to physical sputtering are illustrated in Figure 5.1.

If the ion energy is adequate, the collision can transfer sufficient energy to the surface atom to overcome its surface binding energy (e.g., 4.7 eV for Si), and the atom is ejected as a result. This interaction is called sputtering and is the governing effect in ion beam induced direct fabrication. Because the interaction depends solely on momentum transfer to remove the atoms, sputtering is a purely physical process. The sputtering yield, defined as the number of atoms ejected per incident ion, is a measure of the efficiency of material removal and is defined by Equation (1).

$$Y(\theta) = \frac{\text{Number of sputtered atoms}}{\text{Number of incident ions}} \tag{1}$$

The sputtering yield is normally a function of many variables, including masses of ions and target atoms, ion energy, direction of incidence with respect to the target surface, target temperature, and ion flux. In particular, $Y(\theta)$ strongly depends on the angle θ between the surface normal and the incidence direction, which will be discussed in the next subsection. Initially, the sputtering yield increases as the ion energy increases, but it starts to decrease as the energy is increased past the level where the ions can penetrate deep into the substrate. At this stage, implantation or doping can take place, in which the ions become trapped in the substrate as their energy is expended. As a result, the proper energy for sputtering is between 10 and 100 keV for most of the ion species used for direct fabrication. During sputtering, a portion of the ejected atoms is redeposited into the sputtered region and this redeposition makes it difficult to control the amount of material removed by sputtering. In fact, the essence of ion beam induced direct fabrication is to carefully control both the material sputtering and the redeposition, so that a precise amount of material can be removed.

Sputtering yields and the angular distribution of sputtered atoms and reflected ions may be obtain either by simulation [5] or by experiment [6]. In the case of experiment, the sputtering yield is usually determined by measuring the depth after milling of a defined area (e.g., box milling in a focused ion beam tool). Angular distributions may be obtained by collecting the sputtered particles onto a thin foil (usually Al), which is bent semi-cylindrically along the curve of a circle centered at the eroded spot of the target [7]. After collecting the sputtered particles, the collector foil is flattened and the amount of deposits on the foil is then estimated using an electron probe analyzer. This method is widely used for the calculation of sputtered particle distributions and is called collector technique.

In the case of simulation, the ion-solid interaction is modeled by a sequence of instantaneous binary collisions with mean free paths over which the ion experiences a continuous inelastic energy loss. SRIM [8] is a well-known computer program that allows the calculation of sputtering and backscattering yields as well as implantation profiles in amorphous targets. SRIM does not take the compositional changes of the target into account that result from the implantation of the ions and from mixing. Dynamical computer programs such as TRIDYN [9] deal with these effects by dividing the target into a series of layers parallel to the surface and keeping track of the atoms added and removed from each layer and relaxing the local target density. As shown in Figure 5.2, the sputtering yields obtained from TRIDYN are much closer to the values obtained from experiments than those of SRIM. Several other Monte Carlo codes exist with similar capabilities as SRIM and TRIDYN; e.g., the implant simulator IMSIL [10] has recently been extended [11] to implement similar physics as TRIDYN, providing similar results as TRIDYN, as shown in Figure 5.2.

Some sputtering yields that were calculated by SRIM, IMSIL, and TRIDYN are compared to the values obtained by experiments as listed in Table 5.1.

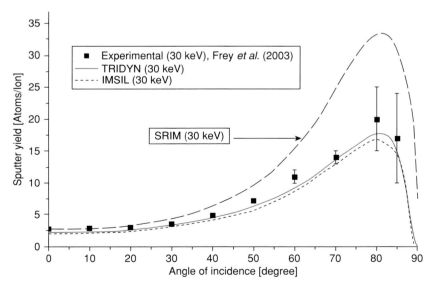

FIGURE 5.2: Comparison of sputtering yields of Si for Ga+ ions at 30 keV between SRIM, TRIDYN, IMSIL and experiments. Experimental data from Reference [12], with permission.

Table 5.1 List of silicon sputtering yields: Ga ion, normal incidence

Substrate (Energy)	SRIM	TRIDYN	IMSIL	Experimental
Si (30 keV)	2.6	2.2	1.96	2.4 [12,13]
Si (20 keV)	2.4	1.98	1.84	1.94 [14]
SiO_2 (30 keV)	0.74	0.68	0.73	0.84 [15]

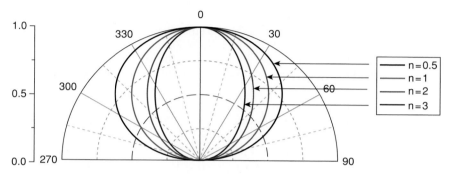

FIGURE 5.3: Calculated lobe-shaped distributions of sputtered atoms with various cosine exponents.

The sputtered atoms often exhibit an ideal, diffuse angular distribution, being sputtered away in a broad range of angles centered on the surface normal. When the incidence energy is high enough (above ~1 keV) and the incidence angle is not too large, the angular distribution of the emitted atoms may reasonably well be modeled by Equation (2) with $n = 1$, known as cosine emission law, as is illustrated in Figure 5.3.

$$f_{3D}(\alpha) \propto \cos^n(\alpha) \tag{2}$$

A large number of experimental observations [16] indicate that the emitted atom distribution cannot be modeled with the cosine emission rule but exhibits a heart-shaped distribution function when the incidence energy is low (below 1 keV). The distribution may in fact be more sideways, described by an "under cosine" shape (where n is less than 1 in Equation (2)). In other cases, though, it can exhibit asymmetric features like a peak just a few degrees from the perpendicular of the incidence direction [17]. In addition, studies of ion bombarded single crystals reveal that atom emission reflects the lattice symmetry (crystallographic effect). In case of FCC metals it has been observed that atoms are preferentially ejected along the (110) direction, but ejection in (100) and (111) directions also occurs to lesser extents. For BCC metals (111) is the usual direction for atom ejection [5].

When angular distributions or sputtering yields are calculated by binary collision simulations, attention has to be given to the surface binding energy, which is the potential barrier an atom has to overcome to leave the target. It may be assumed isotropic or planar. This potential causes energy loss during ejection of atoms. In the planar potential model, the sputtered atoms also experience refraction, so that the final emission angle is changed as a result of the planar potential. This effect is more pronounced for low energies and large emission angles of the sputtered atoms [18]. Figure 5.4 shows simulation results for 30 keV Ga impinging on a silicon target at two different incidence angles. Panels (a) and (c) correspond to normal

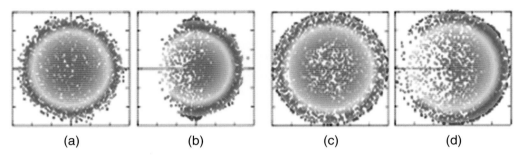

(a)   (b)   (c)   (d)

FIGURE 5.4: Top views of simulated sputtered atoms distributions gathered on a half sphere. 30 keV Ga ion incident into Si substrate. (a) and (c): normal incidence, (b) and (d): incidence at 70° from the left. (a) and (b) show the distribution without refraction, (c) and (d) after application of the planar surface potential (calculated with IMSIL). Color code represents ejection angle of sputtered atoms (from red (90) to violet (0)).

incidence, while panels (b) and (d) correspond to an incidence angle of 70°. The incidence direction is from the left to the center of the plots. At the incidence angle of 70° (panels (b) and (d)) the distribution is more preferential to the right. In cases (c) and (d), where the planar model for the surface binding energy has been applied, fewer particles are observed in the center region when compared to (a) and (b), respectively, and the periphery is more populated. This is a result of the refraction of the atoms by the planar potential away from the surface normal. While there are arguments in favor of the planar model [19], it is not completely clear which model should be favored.

When energetic ions are incident onto a substrate at a grazing angle, some of them are reflected at the surface, retaining much of their initial ion energy. Others penetrate into the substrate, but are scattered backward and eventually leave the target. These reflected and backscattered particles may hit opposing surfaces of the substrate and cause a secondary sputtering. Unlike the sputtered particles, which usually have low energies, the reflected and backscattered particles' energy and angular distribution are important. From simulations [16] it has been found that most of the reflected ions keep their energy and are reflected or backscattered by specular reflection. The reflection coefficient, the ratio between the number of backscattered projectiles and number of projectiles, is greatly increased, as the incidence angle is increased (Figure 5.5) and approaches almost 1 as the incidence angle approaches 90°. The energy distribution of backscattered particles changes dramatically when the angle of incidence is varied, as shown in Figure 5.6.

2 MODELING OF THE ION MILLING PROCESS

The surface movement can be described by optical wave propagation and thus can be modeled by optical theory [20]. In the case of sputtering, the erosion rate is anisotropic, i.e., the yield varies as a function of the incidence angle. The general form of first order erosion theory can be described by Equation (3) with the surface S given by Equation (4):

$$\frac{\partial S}{\partial t} + c \cdot \nabla S = 0 \qquad (3)$$

$$S(x, y, z, t) = 0, \qquad (4)$$

where c is the normal velocity of surface movement.

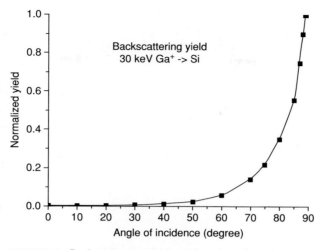

FIGURE 5.5: Backscattering yield as a function of incidence angles, 30 keV Ga+ ions incident onto a Si substrate (calculated by IMSIL).

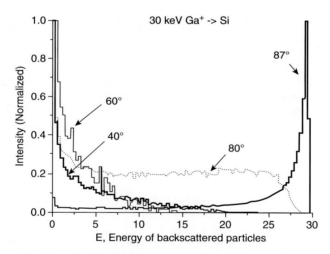

FIGURE 5.6: Calculated data of energy distributions under various incidence angles, 30 keV Ga+ ions incident onto a Si substrate (calculated by IMSIL).

Within the general form of erosion theory, several approaches to ion beam related topography simulation have been proposed in the literature. Neureuther et al. [21] have demonstrated the simulation of the ion milling process with a string segments motion algorithm. They used experimentally obtained angular dependencies of sputtering rates but did not consider redeposition. The redeposition flux was included in the simulation by Yamauchi et al. [22], but their approach was applied to a very simplified geometry only. Smith et al. [23] simulated ion beam induced erosion and deposition with the optical wave front method, which determines the erosion direction analogous to the way refracted light changes the direction at the interface of two media. Katardjiev et al. [24] developed the 3D simulation code DINESE, which is capable of topography simulation during ion milling and focused ion beam bombardment, chemical vapor deposition, and reactive ion etching. They used the T-DYN code [25] to obtain sputtering yields but did not consider redeposition fluxes. Ishitani et al. [26] modeled the sputtering and redeposition fluxes for the focused ion beam process under the assumption that the sputtered atoms are emitted according to a cosine distribution with regard to the surface normal and demonstrated the formation of sloped trench sidewalls due to redeposition. Even if the proposed sputtering and redeposition model was the most advanced at the time, they did not obtain the experimental values of the sputtering yields and sputtered atoms distributions for the combinations of ion species, ion energy, and substrate they investigated. Ximen et al. [27] used a simple computer simulation in 3D space without redeposition, and they simulated various complicated 3D structures, including arbitrary curved paths. More recently, a completely different approach has been implemented by Boxleitner et al. [28]. Their FIBSIM code combines cell based topography simulation and dynamic Monte Carlo simulation of the collision cascades. It has several features, like the prediction of the thickness of doped or damaged layers and intermixing of different layers. While physically rigorous, the approach appears to be computationally expensive. Platzgummer et al. [29] developed the 2D simulation code IonShaper® for ion beam induced erosion and deposition in focused ion beam processing, including secondary sputtering by backscattered and reflected ions. They demonstrated the 2D simulation of nanoimprint template fabrication with experimentally obtained sputtering yields. Recently, Kim et al. [30–32] have performed two- and three-dimensional string-based simulation of focused ion beam processing as well as full three-dimensional level set simulation. The simulation model and simulation method will be presented in the following.

In the simulation model, two kinds of fluxes are considered in the vacuum as shown in Figure 5.7, the incident ion flux $F_{incident}$ and the flux of sputtered atoms $F_{sputtered}$.

The incident ion flux, which is described by a Gaussian profile [1 and 2], causes sputtering and the flux of sputtered atoms causes redeposition. The total flux F_{total} at each position on the surface in the direction of the surface normal is calculated as the sum of the total flux of sputtered atoms (direct flux, F_{direct}) and the (always negative) total flux of redeposited atoms (indirect flux, $F_{indirect}$).

$$F_{total} = F_{direct} + F_{indirect} \qquad (5)$$

The direct flux, F_{direct}, is related to the incident ion current density $F_{incident}$ above the substrate that arrives without any interaction with the substrate.

$$F_{direct} = F_{incident} Y(\theta) \cos(\theta) \qquad (6)$$

The total flux of redeposited atoms may be calculated from the sputtered fluxes according to:

$$F_{indirect} = -S_c \int \frac{F_{direct} f_{3D}(\alpha) \cos(\beta)}{d^2} dA, \qquad (7)$$

where the integral extends over all visible nodes. $f_{3D}(\alpha)$ is the three-dimensional sputtered atoms distribution function which characterizes the angular dependence of the sputtered atom flux, α is the emission angle of the sputtered atoms that is measured between the surface normal and the direction toward the segment position under consideration, and β is the sputtered atoms incidence angle that is measured between the surface normal and the direction

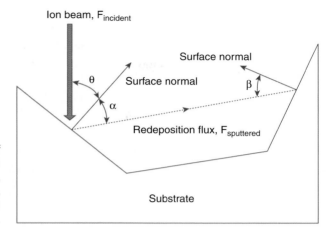

FIGURE 5.7: Schematic illustration of the ion beam simulation model (θ: incidence angle of the ion beam, α: emission angle of sputtered atoms, β: incidence angle of redeposited atoms).

of sputtered atoms. d denotes the distance traveled in the vacuum by the sputtered atoms. The sticking coefficient S_c here is defined as the ratio of the flux of species that properly react and stay on the surface to the flux of incident species. In other words, it is the redeposited fraction of the sputtered atom flux incident on the surface. It depends on the chemistry and the material being deposited, and on the substrate temperature. In general, for FIB simulations $S_c = 1$ can be used if no chemical reactions occur due to a reactive atmosphere.

In the case of 2D, the total flux of redeposited atoms may be calculated from the sputtered flux from all visible segments dl and is given by Equation (8):

$$F_{indirect} = -S_c \int \frac{F_{direct} f_{2D}(\alpha) \cos(\beta)}{d} dl, \qquad (8)$$

where f_{2D} is the two-dimensional normalized sputtered atoms distribution function which characterizes the angular dependence of the sputtered atom flux. It can be calculated by projecting the f_{3D} function into the two-dimensional plane. In addition, in the case of rotational symmetry, the indirect flux may be written as:

$$F_{indirect} = 2rS_c \int dl \int_0^{\pi} d\varphi \frac{F_{direct} f_{3D}(\alpha) \cos(\beta)}{d^2} \qquad (9)$$

where r represents the distance of the indexed point i from the symmetry axis, which is generally the center of the beam, f_{3D} is the three-dimensional normalized sputtered atoms distribution function, dl is the infinitesimal slant height of the truncated cone segment, d is the distance between the sputtering source and the redeposition target and φ is the rotational angle. The angles α, β and distance d are functions of the angle φ, while F_{direct} is independent of φ.

Finally, the ion erosion rate or surface advancement velocity $v_{\perp}$ (m s^{-1}) is calculated by the total flux divided by N, which is the number of atoms incorporated per unit volume in the substrate:

$$v_{\perp} = \frac{F_{total}}{N}. \qquad (10)$$

During simulation, then, the amount of each point movement is calculated by the speed multiplied by the time step Δt, which is usually chosen so that the maximum distance of point movement does not exceed the minimum segment length.

Deviations from the cosine rule can be described by the power n in:

$$f_{3D}(\alpha) = \frac{n+1}{2\pi} \cos^n(\alpha), \tag{11}$$

where $n < 1$ corresponds to more sputtering toward the sides, while $n > 1$ describes a preference of near-normal directions. It should be noted that the distribution Equation (11) is always symmetric about the surface normal. An attempt to model nonsymmetric angular sputtering distributions has been described in Reference [33].

3 TOPOGRAPHY SIMULATION OF FOCUSED ION BEAM MILLING

There are basically four algorithms for simulating surface propagation. Dill's cell method [34], introduced in 1975, is a volumetric approach in which the material is divided into a matrix of tiny cells, each characterized as being either removed or un-removed. The etch surface is tracked by noting the etch state of each cell in the material. This algorithm is robust and easy to implement but is also slow and inefficient. Jewett's string model [35] and Hagouel's ray model [36], on the other hand, are faster and more accurate etching algorithms. The two models are both surface-advancement algorithms, in which a mesh of connected points is used to represent the surface of the material as it is being changed. However, the surface-advancement algorithms are difficult to implement, and also present difficult algorithmic and geometric problems in the treatment of boundaries and in the elimination of loops. In the ray model, the erosion direction is determined analogous to the way refracted light changes the direction at the interface of two media.

In addition, the level set method introduced by Osher and Sethian [37] is a highly robust and accurate computational technique for tracking of moving interfaces and has been applied, among others, to etching, deposition, and photolithography processes in semiconductor manufacturing. It originates from the idea of viewing the moving front as a particular level set of a higher dimensional function, so that topological merging and breaking, sharp gradients and cusps can form naturally, and the effects of curvature can be easily incorporated. Figure 5.8 schematically illustrates string, cell, and level set method.

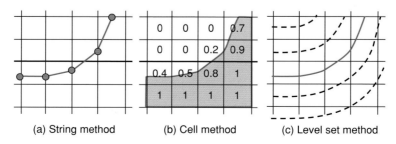

(a) String method   (b) Cell method   (c) Level set method

FIGURE 5.8: Schematic illustration of (a) string method, (b) cell method, and (c) level set method.

The string method is a surface-advancement algorithm in which the actual points on the surface of the material are calculated as the surface moves in time. The string method uses a "string" of connected points to approximate the boundary between the two substance regions (often one of them being vacuum). This surface is advanced in small discrete time-steps by moving each point along a vector normal to the local surface. This advancement or direction vector is the average of the normal vectors of the segments adjacent to the point. The string is inherently faster than the cell method, as it only keeps track of the surface instead of the entire volume of the material. However, the string algorithm is difficult to set up and requires careful attention to details in regularizing the surface. The greatest weakness of the string algorithm is that loops can and will form in the boundary surface, especially when the surface moving speed in the substance has a highly varying spatial distribution. Small loops initially develop in the surface, but as the simulation continues, these loops begin to expand and also to intersect each other. This consumes precious computation time and memory and can result in the program crashing if predefined storage limits exist. The destructive loop behavior observed in this set of simulations is actually caused by the surface-advancement method itself. The string algorithm moves the surface using normal vectors determined from the local surface. Errors in the local surface (e.g., loops) cause inaccurate calculation of subsequent normal vectors, which in turn results in incorrect etch surfaces. To avoid this problem, it is necessary to remove the loops before they become too complex. The string thus has to be "de-looped" every few time-steps; the intersection of all the segments in the string must be found and the loops deleted [38]. This string method will be discussed in detail later for two-dimensional simulation.

A different approach to front motion is provided by the cell method, introduced by Noh and Woodward [38] and based instead on an Eulerian view. These techniques have appeared in a variety of forms and have been introduced under many names, such as the "volume-of fluid" and the "method of partial fractions." The basic idea is as follows. Imagine a fixed grid on the computational domain, and assign values to each grid cell based on the fraction of that cell containing material inside the interface. Given a closed curve, we assign a value of unity to those cells completely inside this curve, a cell value of zero to those completely outside, and a fraction between 0 and 1 to cells that straddle the interface, based on the amount of the cell inside the front.

The idea, then, is to rely solely on these "cell fractions" shown in Figure 5.9 to characterize the interface location. Approximation techniques are then used to reconstruct the front from these cell fractions. The original Noh and Woodward algorithm was known as "SLIC" for "Simple Line Interface Calculation" and reconstructed the front as either a vertical line or a horizontal line.

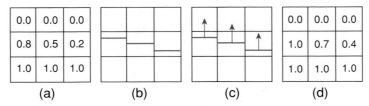

FIGURE 5.9: Reconstruction and advection of volume fractions:
(a) volume fractions, (b) reconstructed front, (c) front advection,
and (d) new volume fractions [38].

In order to evolve the interface, the cell fractions on this fixed grid are updated to reflect the progress of the front. Suppose that one wishes to advance a front under the transport velocity u. Noh and Woodward provide a methodology in which the value in each cell is updated under this transport velocity in each coordinate direction by locally reconstructing the front and then material in neighboring cells under this motion. After completing coordinate sweeps, one has produced new cell fractions at the next time step corresponding to the updated front [38].

The cell method can be powerful. However, there are some drawbacks: such techniques are inaccurate. Since the approximation to the front through volume fractions is relatively crude, a large number of cells are often required to obtain reasonable results. Evolution under complex speed functions is problematic. Calculation of intrinsic geometric properties of the front, such as curvature and normal direction, may be inaccurate. Considerable work may be required to develop higher order versions of such schemes.

The basic idea behind the level set method is to represent the surface in question at a certain time t as the zero level set of a certain function $\Phi(t, x)$, the so-called level set function. The evolution of the surface in time is caused by forces, or fluxes of particles reaching the surface in the case of the FIB processing. The surface at a certain time t is determined by solving time-dependent Hamilton-Jacobi and related equations on fixed structured Euclidean grids. The partial differential equation to be solved is of the form:

$$D_t \Phi(x, t) = H(x, t, \Phi, \nabla\Phi, D_x^2\Phi) = 0, \quad x \in R^n, t \geq 0, \tag{12}$$

where $D_t\Phi$ is the partial derivative of Φ with respect to the time variable t, $\nabla\Phi = D_x\Phi$ is the gradient of Φ, and $D_x^2\Phi$ is the Hessian matrix of the second partial derivatives with respect to the space variable. In the simulation of ion beam induced direct fabrication, the speed of the surface Γ is given from external physics, and in particular the speed function $v_\perp(x, t)$ of Γ in its normal direction is given as discussed in the previous subsection. In this case the level set equation reads:

$$D_t \Phi(x, t) + v_\perp(x, t) |\Phi(x, t)| = 0. \tag{13}$$

The initial condition $\Phi(x, t = 0)$ is given by the signed distance function, i.e., the distance of the point x from the surface Γ, with a positive (negative) sign if the point is outside (inside) the surface. The simplest numerical method to solve the hyperbolic level set equation is by a combination of forward Euler time discretization and upwind spatial differencing.

The advantages of level set methods include the following: (1) Shocks and rarefactions can develop in the slope, corresponding to corners and fans in the evolving interface, and numerical techniques designed for hyperbolic conservation laws can be exploited to construct upwind schemes that produce the correct, physically reasonable entropy solution. These are naturally captured in the above representations. (2) The front is free to change topology as it evolves; no special care is required. In the level set perspective, the front at time t is given by the set of all points (x, y) such that $\phi(x, y, t) = 0$. While the solution ϕ is a single-valued function, the lower-dimensional set of points corresponding to the front may break, merge, and consist of multiple regions. (3) Finite difference schemes may be employed to compute the solution to the PDEs in a relatively straightforward manner. (4) There are no differences in the above construction for hypersurfaces propagating in three or more space dimensions.

In the string model, discretized points on the evolution front are sequentially indexed by $i = 0, 1, 2, \ldots, n$. At each point, the normal direction is calculated as the direction of the bisecting line of the angle between the adjacent segments as shown in Figure 5.10.

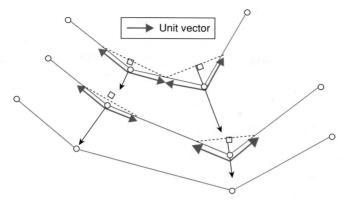

FIGURE 5.10: Illustration of normal direction in string-based simulation.

The method is therefore also called angle bisector algorithm. Then the points move along this direction at a speed that is proportional to the total flux of each point. The surface is represented by a string of points connected by straight line segments where the point velocity is taken as the average of the two segment etching rates. The angle bisector algorithm and its derivatives are popular because advancing the surface along the normal is consistent with popular intuitions about surface evolution, and because no a priori knowledge of the etching rate behavior is required. Although intuitively satisfying and easy to implement, the angle bisector method suffers from certain penalties when applied to ion etching systems. The time integration step must be small enough to ensure that the surface evolution is gradual. Generally the etching rate is a strong function of slope (or ion arrival angle); errors in the point locations at one time step result in errors in the etching rates at the next. If the time step is too large, the profile may become unstable. Several researchers have improved on the string method by choosing an optimal direction of point movement using the so-called method of characteristics [20 and 39]. If the calculation time step is small enough, there are no big differences between the angle bisector algorithm and the method of characteristics.

During string-based simulation, several artifacts would occur without additional measures. The most important are loop, rarefaction fan, and sharp edge formation. In addition, smoothening is necessary to guarantee long-time stability and accuracy. It keeps points far enough from each other and removes simulation noise. Loops are eliminated by detecting intersecting segments, insertion of a new point at the intersection point, and removal of the loop. Rarefaction points, which lead to topography distortions when they move outward, are prevented by replacing the point by two nearby points if the angle between the two adjacent segments is less than 90°. The new points are inserted on the old segments as to divide each of them at a certain fraction of their length. The fraction should be chosen between 0 and 1 so that the new segment lengths are greater than the minimum segment length. In addition, points are inserted or removed to keep segment lengths within user specified maximum and minimum values. By insertion and removal of points, smoothening is automatically done.

Katardjiev et al. [24] used a regular grid in the x-y plane and moved the nodes in z direction for the simple focused ion beam simulation without redeposition, so that the surface is always represented in the form $z = f(x, y)$ as illustrated in Figure 5.11.

The intersection points of the grid form the calculation nodes, and the patches with four corners represent the surface. Unfortunately, during the simulation, some of the patches

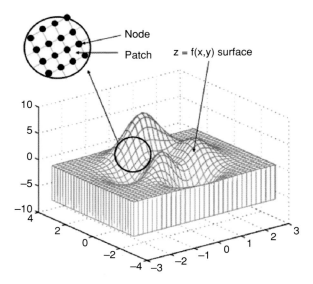

FIGURE 5.11: Illustration of surface representation with a rectangular grid.

can be excessively stretched (Figure 5.11), which leads to numerical errors at concave or convex areas of the surface. Therefore, some surface reconstruction is necessary after a certain number of time steps. The reconstruction concept is that the distorted and overstretched surface nodes are re-determined by bilinear interpolation with evenly spaced nodes in x- and y-direction. Because of the $z = f(x, y)$ surface representation (only one z value is possible), the bilinear interpolation based reconstruction is limited to non-overhanging (non-undercut) structures.

In an attempt to overcome this limitation, we have applied an initially regular grid to our algorithm, which advances the nodes perpendicular to the surface. During the simulation the initially rectangular patches are distorted. The more reconstructions, the more accuracy will be achieved, because the surface distortion causes errors during calculation of surface normal and area. Patches are kept at approximately equal size by insertion of grid lines in the middle of existing lines if the maximum inter-line distance exceeds a user-specified limit. In a structured grid, however, this leads to unnecessarily dense nodes in parts of the simulation area (Figure 5.12(a)) and can also lead to numerical errors during calculation of surface normal and area as seen in the lower part of Figure 5.12(b). The numerical errors can be reduced by a reconstruction technique (Figure 5.12(c)). Therefore the method is again restricted to non-overhanging structures. Nevertheless, it is well suited for simple geometries.

4 EXPERIMENTAL VALIDATION

Experiments were carried out [40] at an acceleration voltage of 50 kV with a selectable 50 μm beam-limiting aperture corresponding to a beam current of 45 pA with a silicon substrate and Ga^+ ions. The beam diameter (FWHM) was 68 nm and the current density at the center of the beam 0.8 A/cm^2.

Rectangular boxes were milled using an array of 30 × 6 pixels with the scan direction along the 30 pixels. One pass was performed in the serpentine scan mode as illustrated in Figure 5.13.

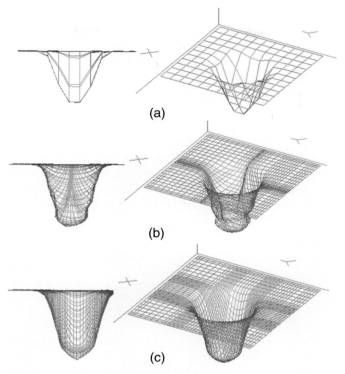

FIGURE 5.12: Illustration of point insertion schemes and sharp edge removal by interpolation. Simulation without insertion and interpolation (a), only with point insertion (b), and with both point insertion and interpolation (c).

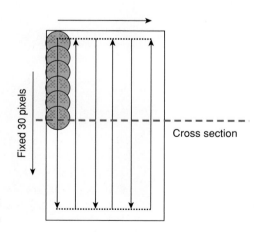

FIGURE 5.13: Schematic illustration of scan strategy and the position of the cross-section.

Dwell times of 9 ms and 11 ms were used, and the pixel overlap was 40%, 50% and 60%, resulting in different sizes of the exposed rectangular area. The resulting topographies were investigated with SEM. A 40 nm Au layer was deposited for protecting the structure during FIB sectioning in the middle of the traverse beam direction as depicted.

It is well known that under slow scan conditions the depth and shape of FIB-machined wells may depend on the beam dwell time even if the total dose is kept constant (fixed beam current and total machining time) [41]. This is usually explained by the increase in sputtering rates when oblique angles develop under the beam and by the accumulation of redeposited material. Generally, the fabrication in the micro/nano range by focused ion beams is determined not only by the basic process parameters such as beam shape and doses, but also scan strategy and the number of passes. Figure 5.14 shows our experimental results obtained with the 30 × 6 pixel array box milling. The fabricated structure shapes are asymmetric and the maximum depth is drastically increased as the overlap is increased (the pixel

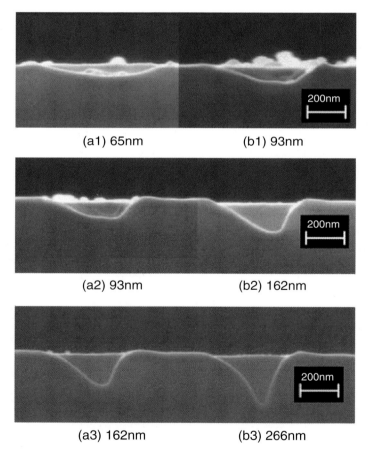

(a1) 65nm   (b1) 93nm

(a2) 93nm   (b2) 162nm

(a3) 162nm   (b3) 266nm

FIGURE 5.14: Cross-sectional SEM images milled on a Si substrate with Ga+ ion beam accelerated at 50 keV. One pass and a total of 6 scans were applied with 40, 50 and 60% beam overlap for the 1st, 2nd, and 3rd row, respectively. Dwell times of 9 and 11 ms were applied for column (a) and (b), respectively. The value below each SEM image is the measured maximum depth.

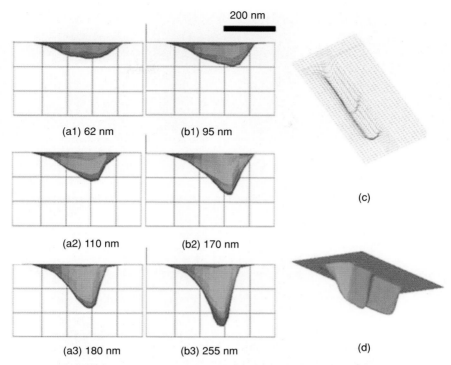

200 nm

(a1) 62 nm (b1) 95 nm

(c)

(a2) 110 nm (b2) 170 nm

(a3) 180 nm (b3) 255 nm (d)

FIGURE 5.15: Simulation results of experiments shown in Figure 14(a1)–(b3). (a), (b) cross sections. The values below each simulation are the maximum depths. (c) surface during milling seen from the top/left presented with grid lines and (d) from the bottom/right presented with surface shading for 60% beam overlap and 11 ms dwell time.

spacing decreased). In addition the ~20% larger dwell time in panels (b) results in a 50–70% increase in depth compared to panel (a). The simulation results well predict the topography shape and the non-linear dependence of the maximum depths on pixel overlap and dwell time.

As shown in Figures 5.15(c) and (d), a step-like contour develops under the beam both perpendicular and parallel to the scan direction, if pixel spacing and dwell time are large enough. These conditions are therefore well suited to test the 3D simulation code.

For the purpose of further investigating the redeposition mechanism we have performed another set of experiments, varying the number of scan lines from 8 to 12 while keeping the number of pixels per line constant at 30. Dwell time and pixel spacing were fixed to 9 ms and 60% overlap, respectively. The experimental results are shown in the upper part of Figure 5.16.

It can be observed that the trench depth is increased as the number of scans is increased while the trench becomes narrower near the bottom. This behavior is well reflected in the simulation results shown in the lower part of Figure 5.16. It also can be observed that the curvature of the left sidewall in the simulation well reflects the experiment, in particular the pronounced change in slope in Figure 5.16(c).

In order to demonstrate the importance of redeposition and sputtering yield enhancement, we repeated the simulation of Figure 5.16(c) without consideration of the redeposition flux and varied the number of passes (Figure 5.17). As shown in Figure 5.17(b), the depth is much larger and the slope of the left side wall is quite different when redeposition is neglected in the single-pass simulation. In addition, as the number of passes is increased while leaving

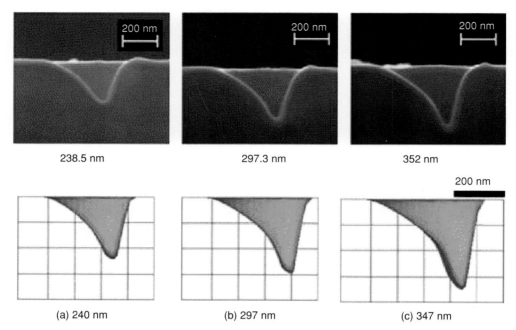

238.5 nm   297.3 nm   352 nm

(a) 240 nm   (b) 297 nm   (c) 347 nm

FIGURE 5.16: Cross-sectional SEM images milled on a Si substrate with Ga+ ion beam accelerated at 50 keV, 60% overlap, 9 ms dwell time and a total of 8, 10 and 12 passes for (a), (b) and (c), respectively. The values below the images are the maximum depth.

the total dose the same, the shape of the well changes drastically as shown in Figures 5.17(c) and (e). The reduction in the depth of the trench bottom is due to the flatter bottom, which results in a lower sputtering yield because of its angular dependence. From Figures 5.17(d) and (f) it can be seen that redeposition is of lower importance in multi-pass scans, which is also due to the fact that the aspect ratio of the structure is lower.

In addition to string-based two- and three-dimensional simulations, we have also investigated one of the sample lift-out techniques during *in situ* or *ex situ* TEM sample preparation [42], which uses two sample stage tilts to about $\sim 45°$ in two opposite directions and FIB box mills in each orientation. For the sake of reducing time-consuming and unnecessary operations, the TEM sample preparation can be optimized by simulation. This is a case where surface merge and separation occur, which are automatically handled by the level set method. The SEM images and the simulation results are shown in Figure 5.18. Although the same dose is delivered to each of the trenches, the second trench (right side) is slightly deeper than the first one (left side) because it crosses the already sputtered first trench. It is also broader because of redeposition into the first trench during the second trench formation, which is not subject to resputtering due to shadowing of the ions. Notice again the excellent agreement of the trench depths at the various stages of processing.

5 SIMULATION FOR THE DEVELOPMENT OF APPLICATIONS

Several reports have shown that focused ion beam simulation can be useful in realizing predefined structures. Nellen et al. [43] fabricated photonic structures using focused

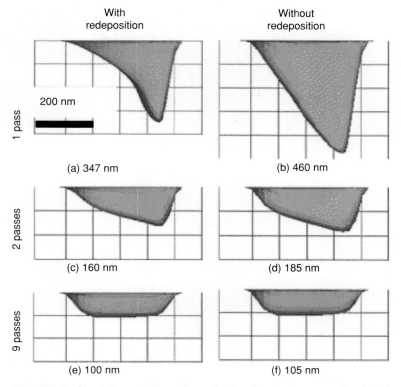

FIGURE 5.17: Simulation results performed using Ga+ ion beam accelerated at 50 keV, 60% overlap and 9 ms dwell time with a total of 12 scans. (a), (c), (e) with consideration of the redeposition flux. (b), (d), (f) neglecting the redeposition flux. (a), (b) single pass, (c), (d) 2 passes, (e), (f) 9 passes. The dwell time was adjusted as to leave the total dose the same. The value below each image is the maximum depth.

ion beam with the aid of some simulations. Waveguides and photonic crystals were fabricated by circular holes arrays. Different scan patterns were investigate, conventional and circular scan with constant dwell time, and circular scan with the dwell time linearly or parabolically increasing with the radius. For the fabrication of Fresnel microlenses, they simulated and demonstrated Fresnel lenses on silicon and on optical glass fiber tips.

For the fabrication of grating-like structures Kim et al. [40] have performed a series of simulations and demonstrated them on silicon substrate. A diffraction grating is an optical surface with grooves in it, shaped and spaced so as to disperse incident polychromatic light into a sharply defined spectrum. The conventional fabrication methods for the gratings are photolithography with a gray-scale mask, electron-beam lithography, and laser direct writing upon a photoresist to form a pattern that is then transferred from the photoresist to a substrate by dry or wet etching or by electroplating metal for further molding [44]. In addition, a focused ion beam direct milling method was reported in Reference [45]. The diffracted structures were generated on a substrate by scanning a focused ion beam in a specific area only. No patterned direct writing was required. But the fabricated structures were not controllable in

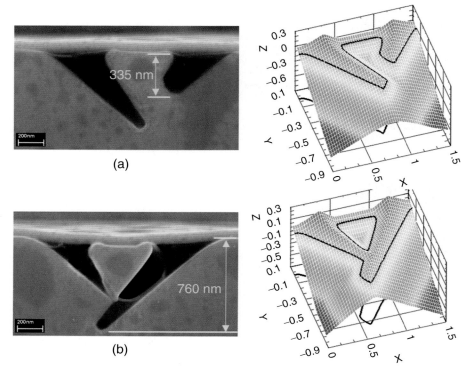

FIGURE 5.18: Comparison of simulated surface contours (right) and experimental results (left) of TEM sample preparation with two inclined trenches. The right trench has received 40% and 100% of the dose of the left trench in panels (a) and (b), respectively. The values at the z-axis and the gray-leveled surfaces in the 3-D graphs on the right are the level set function, which roughly corresponds to the signed distance from the surface. The surface is indicated by the bold lines, which can be compared to the experimental results.

shape and dimension, and the uniformity of groove depth was too poor to guarantee the grating performance. It may be expected that the shape and overall size would be controllable using a simulation technique.

For the fabrication of blazed diffractive grating with predefined geometry, three steps are followed: (1) define grating requirements such as wavelength, diffractive order and efficiency; (2) define grating overall dimensions and shape according to the predefined requirements by optical simulation; (3) obtain process parameters by FIB simulation. As step 2 is concerned, the width and depth of the structure are fundamental for the grating performance. In addition, the sidewall planarity is an important factor. We performed a series of simulations, varying the dwell time and pixel spacing [40] and determined the process conditions leading to the desired geometry. A series of SEM images of gratings fabricated with various dwell times and the optimum pixel spacing is shown in Figure 5.19(a)–(d). The fabricated grating which has a flat sidewall is shown in Figure 5.19(b), the corresponding simulation in Figure 5.20.

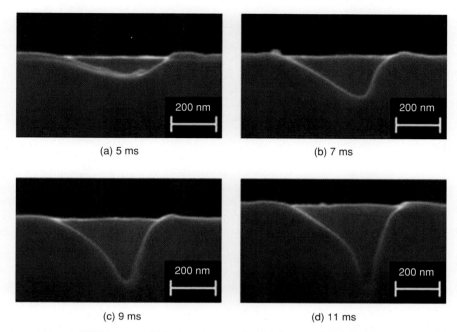

(a) 5 ms   (b) 7 ms

(c) 9 ms   (d) 11 ms

FIGURE 5.19: SEM images of fabricated gratings with various dwell times and optimum pixel spacing. Image (b) shows the planar sidewall grating.

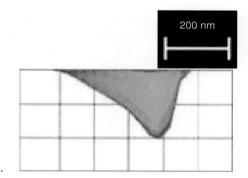

FIGURE 5.20: Simulation result of Figure 5.19(b).

REFERENCES

[1] Brodie, I., and J. J. Muray. *The physics of micro/nano-fabrication*. New York, London: Plenum Press, 1992.

[2] Tseng, A. A. Recent developments in micromilling using focused ion beam technology. *J. Micromech. Microeng.* 14 (2004): R15–R34.

[3] Tanemura, M., T. Aoyama, A. Otani, M. Ukita, F. Okuyama, and T. K. Chini. Angular distribution of In and P particles sputtered from InP by inert-gas ion bombardment. *Surf. Sci.* 376 (1997): 162–168.

[4] Plummer, D., D. Deal, and B. Griffin. *Silicon VLSI technology*. New Jersey: Prentice Hall, 2000.

[5] Eckstein, W. *Computer simulation of ion-solid interactions*. Berlin: Springer, 1991.

[6] De Winter, D. A. M., and J. J. L. Mulders. Redeposition characteristics of foucsed ion beam milling for nanofabrication. *J. Vac. Sci. Technol.* B 25 (2007): 2215–2218.

[7] Mueller, K. P., and H. C. Petzold. *Proc. SPIE* 1263 (1990): 12.

[8] Ziegler, F. *SRIM instruction manual.* http://www.srim.org (2003).

[9] Moeller, W., and M. Posselt. *TRYDYN_FZR user manual.* Dresden: Forschungszentrum Rossendorf, 2002.

[10] Hobler, G. Monte Carlo simulation of two-dimensional implanted dopant distributions at mask edges. *Nucl. Instr. Meth.* B 96 (1995): 155–162.

[11] Hobler, G. *IMSIL 2009-01 manual.* Vienna: TU, 2009.

[12] Frey, L., C. Lehrer, and H. Ryssel. Nanoscale effects in focused ion beam processing. *Appl. Phys. A* 76 (2003): 1017–1023.

[13] Lugstein, A., B. Basnar, J. Smoliner, and E. Bertagnolli. FIB processing of silicon in the nanoscale regime. *Appl. Phys. A* 76 (2003): 545–548.

[14] Adams, D. P., and M. J. Vasile. Accurate focused ion beam sculpting of silicon using variable pixel dwell time approach. *J. Vac. Sci. Technol.* B 24 (2006): 836–844.

[15] Xu, X., D. Della Ratta, S. Sosonkina, and J. Melngailis. Focused ion beam induced deposition and ion milling as a function of angle of incidence. *J. Vac. Sci. Technol.* B 10 (1992): 2675–2680.

[16] Yamamura, Y., T. Takiguchi, and T. Tawara. *Data compilation of angular distributions of sputtered atoms.* Nagoya: National Institute for Fusion Science, 1990.

[17] Ebm, C., and G. Hobler. Assessment of approximations for efficient topography simulation of ion beam processes: 10 keV Ar on Si. *Nucl. Instr. Meth.* B 267 (2009): 2987–2990.

[18] Smith, R. *Atomic & ion collisions in solids and at surfaces.* London: Cambridge University Press, 1997.

[19] Eckstein, W. Sputtering yields. In *Sputtering by particle bombardment: Experiments and computer calculations from threshold to MeV energies,* eds. R. Behrisch, W. Eckstein. Berlin, Heidelberg, New York: Springer, 2007.

[20] Katardjiev, I. V. Simulation of surface evolution during ion bombardment. *J. Vac. Sci. Technol.* A 6 (1988): 2434–2442.

[21] Neureuther, A. R., C. Y. Liu, and C. H. Ting. Modeling ion milling. *J. Vac. Sci. Technol.* 16 (1979): 1767–1771.

[22] Yamauchi, N., T. Yachi, and T. Wada. A pattern edge profile simulation for oblique ion milling. *J. Vac. Sci. Technol.* 2 (1984): 1552–1557.

[23] Smith, R., S. J. Wilde, G. Carter, I. V. Katardjiev, and M. J. Nobes. The simulation of twodimensional surface erosion and deposition process. *J. Vac. Sci. Technol.* 5 (1987): 579–585.

[24] Katardjiev, I. V., G. Carter, M. J. Nobes, S. Berg, and H. O. Blom. Three-dimensional simulation of surface evolution during growth and erosion. *J. Vac. Sci. Technol.* A 12 (1994): 61–67.

[25] Biersack, J. P., and S. Berg. T-DYN Monte Carlo simulations applied to ion assisted thin film processes. *Nucl. Instr. Meth.* B 59–60 (1991): 21–27.

[26] Ishitani, T., and T. Ohnishi. Modeling of sputtering and redeposition in focused-ion-beam trench milling. *J. Vac. Sci. Technol.* A 9 (1991): 3084–3089.

[27] Ximen, H., R. K. DeFreez, J. Orloff, R. A. Elliott, G. A. Evans, N. W. Carlson, W. Carlson, M. Lurie, and D. P. Bour. Focused ion beam micromachined three-dimensional features by means of a digital scan. *J. Vac. Sci. Technol.* 8 (1990): 1361–1365.

[28] Boxleitner, W., G. Hobler, V. Klueppel, and H. Cerva. Simulation of topography evolution and damage formation during TEM sample preparation using focused ion beams. *Nucl. Instr. Meth.* B 175 (2001): 102–107.

[29] Platzgummer, E., A. Biedermann, H. Langfischer, S. Eder-Kapl, M. Kuemmel, S. Cernusca, H. Loeschner, C. Lehrer, L. Frey, A. Lugstein, and E. Bertagnolli. Simulation of ion beam direct structuring for 3D nanoimprint template fabrication. *Microelectron. Eng.* 83 (2006): 936–939.

[30] Kim, H. B., G. Hobler, A. Lugstein, and E. Bertagnolli. Simulation of ion beam induced micro/nano fabrication. *J. Micromach. Microeng.* 17 (2007): 1178–1183.

[31] Kim, H. B., G. Hobler, A. Steiger, A. Lugstein, and E. Bertagnolli. Full three-dimensional simulation of focused ion beam micro/nanofabrication. *Nanotechnology* 18 (2007): 245303.

[32] Kim, H. B., G. Hobler, A. Steiger, A. Lugstein, and E. Bertagnolli. Level set approach for the simulation of focused ion beam processing on the micro/nano scale. *Nanotechnology* 18 (2007): 265307.

[33] Zhang, Z. L., and L. Zhang. Anisotropic angular distribution of sputtered atoms. *Radiat. Eff. Def. Sol.* 159 (2004): 301–307.

[34] Dill, F. H., A. R. Neureuther, J. A. Tutill, and E. J. Walker. Modeling projection printing of positive photoresists. *IEEE Trans. Electron. Devices* 22 (1975): 45–64.

[35] Jewett, E., P. I. Hagouel, A. R. Neureuther, and T. Van Duzer. Lineprofile resist development simulation technique. *Polymer Eng. Sci.* 17 (1977): 381–384.

[36] Hagouel, P. I. X-ray lithography fabrication of blazed diffracting gratings. PhD dissertation, University of California, Berkeley (1976).

[37] Osher, S., and J. A. Sethian. Fast propagating with curvature-dependent speed: Algorithm based on Hamilton-Jacobi formulation. *J. Comp. Phys.* 79 (1988): 12–49.

[38] Sethian, J. A. *Level set methods and fast marching methods.* Cambridge: Cambridge University, 1996.

[39] Shaqfeh, E. S. G., and C. W. Jurgensen. Simulation of reactive ion etching. *J. Appl. Phys.* 66 (1989): 4664.

[40] Kim, H. B., G. Hobler, A. Steiger, A. Lugstein, and E. Bertagonolli. Simulation-based approach for the accurate fabrication of blazed grating structures by FIB. *Opt. Express* 15 (2007): 9444–9449.

[41] Santamore, D., K. Edinger, J. Orloff, and J. Melngailis. Focused ion beam yield change as a function of scan speed. *J. Vac. Sic. Technol.* B 15 (1997): 2346–2349.

[42] Giannucci, L. A., and F. A. Stevie. *Introduction to focused ion beams.* Boston: Springer, 2005.

[43] Nellen, P. M., V. Callegari, and R. Broennimann. FIB-milling photonic structures and sputtering simulation. *Microelectron. Eng.* 83 (2006): 1805–1808.

[44] Hecht, E. *OPTICS.* Reading, MA: Addison Wesley Longman, 2002.

[45] Fisher, R. E., and B. T. Galeb. *Optical system design.* New York: McGraw-Hill, 2000.

CHAPTER 6

FEB AND FIB CONTINUUM MODELS FOR ONE ADSORBATE SPECIES

Ivo Utke

1 INTRODUCTION

The following introductory considerations will set the stage for gas assisted FEB (FIB) deposition and etch models based on the continuum approach. One might argue that one-adsorbate-species conditions seldom prevail in experiments: even in ultra-high vacuum, where it takes several hours before one monolayer of background molecules impinges on the substrate, several species of adsorbates can be created by electron induced dissociation of the functional "parent" adsorbate injected into the chamber. Not to mention the presence of hydrocarbons and water impinging at around one monolayer per second in high-vacuum chambers working at around 10^{-6} mbar. However, when injecting the functional adsorbate at comparatively high flux, often the one-adsorbate-species models represent a good approximation. Furthermore, they allow the governing principles of rates and resolution of FEB and FIB processing to be transparently discussed using analytical solutions to the governing equations.

1.1 SET OF EQUATIONS

The schematic view of gas-assisted FEB (FIB) processing is depicted in Figure 6.1. A gas-injection system supplies volatile molecules to the substrate where they can reversibly adsorb and desorb without spontaneously forming a chemical bond with the surface. The adsorbates are depleted via dissociation under local beam irradiation and replenished via surface diffusion and gas phase transport.

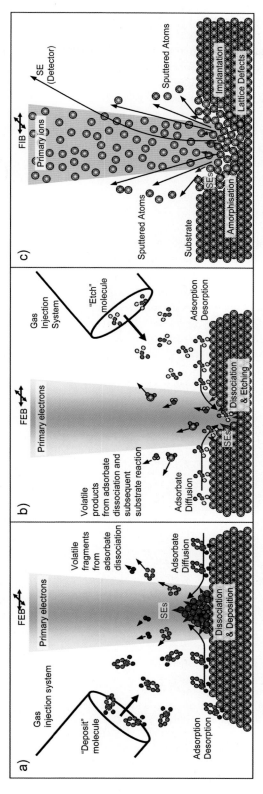

FIGURE 6.1: Gas-assisted FEB (FIB) processing involves a gas injection system that supplies functional volatile molecules to the substrate. (a) The adsorbed molecule is dissociated into a non-volatile fragment forming the functional deposit while the volatile fragments are pumped away. (b) The adsorbed molecules for etching should not spontaneously etch the substrate. Only the dissociation products obtained by FEB (FIB) irradiation should react with the substrate elements to volatile fragments. (c) FIB involves also physical sputtering in addition to the chemical contributions shown in (a) and (b) for FEB. Ion impact leads to amorphization of the near-surface layers by dislocating target atoms via collision cascades and to implantation of source ions. For clarity the gas kinetics was omitted.

Conceptually, gas-assisted FEB and FIB etching and deposition is considered to be a chemical reaction between electrons (ions) and adsorbates. Without any charged particle irradiation the adsorbates desorb spontaneously due to thermal energy fluctuations (this type of adsorption is sometimes called physisorption). Ideally, a clean surface will be eventually regained when the supply of molecules is removed. The necessary type of primary reaction leading to deposition or etching is generally agreed to be a dissociation reaction of the adsorbate with the electrons (ions). A simple route leading to deposition is when some of the dissociation products are non-volatile; the volatile dissociation products are pumped away. Deposition can also take place when volatile primary dissociation products (often highly reactive radicals) cross-link to other adsorbates or cross-link between each other in a secondary reaction (polymerization) on the surface. In contrast, etching requires the electron (ion) induced dissociation products to react with the substrate atoms in a secondary reaction to produce volatile compounds; for efficient etching, all the reaction products should be volatile to avoid competitive deposition.

To what extent electron (ion) induced dissociation of gas phase molecules within the beam volume is (not) involved in deposition or etching can be determined by heating or cooling the substrate. The majority of articles report that the deposition rate decreases with increasing temperature due to thermally enhanced adsorbate desorption [1; 2; 3] and thus the concept of electron induced dissociation of adsorbates on surfaces is now generally accepted.

In a system with rotational symmetry, as shown in Figure 6.2, the vertical *chemical* deposition or etch rate $R_{ch}(r)$ (in units of distance per unit time) as a function of the distance r from the center of the electron flux distribution $f(r)$ (in units of electrons per unit area and time) is for

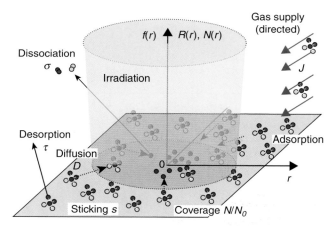

FIGURE 6.2: Reference system and processes involved in gas-assisted FEB and FIB deposition. Inside the irradiated area precursor adsorbates are depleted by electron (ion) dissociation. Replenishment occurs by gas phase transport and by surface diffusion. The symbols J, s, τ, D, and σ denote the molecule flux, the sticking probability, the adsorbate residence time, the diffusion coefficient, and the dissociation cross-section. $f(r)$ signifies the beam distribution, $R(r)$ the deposition (etch) rate, and $N(r)$ the adsorbate density.

steady state:

$$R_{ch}(r) = VN(r)\sigma f(r) = R_{FEB}(r), \tag{1a}$$

where V is the volume of the *deposited* fragment or the etched atom, $N(r)$ is the number of adsorbates per surface unit, and σ is the electron impact dissociation cross section. The chemical rate defined in Equation 1a is equivalent to the FEB deposition (etch) rate since electrons with energies smaller than 30 keV (typical maximum SEM energy) have negligible physical (sputter) impact on most atoms or molecules. In contrast, as shown in Figure 6.1c, for gas-assisted FIB deposition and etching, *physical* sputtering of the impinging ions with a yield Y_S must be taken into account since an ion is many thousand times heavier than an electron (for instance, the mass of a Ga-ion is about 128,000 times heavier than an electron):

$$R_{FIB}(r) = \underbrace{R_{ch}(r)}_{chemical} \pm \underbrace{Y_S Vf(r) \cdot (1 - N(r)/N_0)}_{physical}. \tag{1b}$$

The distribution $f(r)$ signifies now the ion flux distribution. The plus sign applies for gas-assisted FIB etching, whereas the minus sign holds for gas-assisted FIB deposition. Physical sputtering, also known as FIB milling, adds to the chemical etch rate but reduces the chemical deposition rate. The milling rate becomes *adsorbate-dependent* with gas assistance and is proportional to the relative number of substrate locations not covered by adsorbates $(1 - N(r)/N_0)$ [4; 5]. A physical argument is that adsorbate covered surface atoms have too heavy a mass to be sputtered with the kinetic energy provided by the collision cascade. Furthermore, in the excited surface atom model (see Figure 6.3), this energy is supposed to dissociate the adsorbates. Equation 1b would predict a reduced sputter rate when an adsorbate dissociates into "non-etching" fragments ($R_{ch} = 0$).

Sometimes the coverage is small $N/N_0 \ll 1$ and the approximation:

$$R_{FIB}(r) \approx R_{ch}(r) \pm Y_S Vf(r) \tag{1c}$$

can be used.

Often the "FIB notation" of equations 1b and 1c in terms of yields is found in literature, where the yields are defined via the relation $Y = (R/V)/f$:

$$Y_{net} = Y_{ch} \pm \left(1 - N/N_0\right) Y_S \approx Y_{ch} \pm Y_S, \tag{1d}$$

where Y_{net} is the net deposition or etch yield, $Y_{ch} = N(r)\sigma$ the chemical deposition or etch yield due to dissociation of the adsorbate, and Y_S the physical sputter yield. The right-hand term approximation in Equation 1d holds for low coverages. The dependence on radius is not explicitly noted anymore since the yields are given in units of dissociated or sputtered atoms per *incident* ions (at a given radius r). However, when using the notation of Equation 1d, one should keep in mind that the chemical yield depends on the adsorbate density inside the irradiated area $Y_{ch} = N\sigma$. Evidently, the *maximum* chemical (reaction) yield is $N_0\sigma$, given in units of dissociated atoms per ion. Noteworthy is that the chemical reaction yield for *deposition* must override the physical sputter yield in order to get a net deposit; otherwise material removal (sputtering) will occur.

The next conceptual point is a differential adsorption rate equation describing the surface kinetics and coverage of molecules. Four key processes, as shown in Figure 6.2, are generally considered to determine the surface density $N(r, t)$ of adsorbed molecules (adsorbates): (a) adsorption from the gas phase governed by the precursor flux J, the sticking probability s, and coverage N/N_0; (b) surface diffusion from the surrounding area to the irradiated area governed by the surface diffusion coefficient D and the surface concentration gradient; (c) spontaneous thermal desorption of adsorbates (physisorbed molecules) after a residence time τ; and (d) adsorbate dissociation by beam irradiation given by the rate σf. It follows that the molecule adsorption rate dN/dt is given by:

$$\frac{\partial N}{\partial t} = \underbrace{sJ\left(1 - \frac{N}{N_0}\right)}_{Adsorption} + \underbrace{D\left(\frac{\partial^2 N}{\partial r^2} + \frac{1}{r}\frac{\partial N}{\partial r}\right)}_{Diffusion} - \underbrace{\frac{N}{\tau}}_{Desorption} - \underbrace{\sigma f N}_{Dissociation} \qquad (2)$$

The adsorption term in Equation 2 describes a non-dissociative Langmuir adsorption, where N_0 is the maximum monolayer density given approximately by the inverse of the adsorbate size. This adsorption type accounts for surface sites already occupied by non-dissociated precursor adsorbates and limits the coverage to N_0. The parameters $N = N(r, t)$ and $f = f(r, t)$ are considered time and position dependent.

By solving equations 1 and 2, one obtains the spatial distributions of deposition and etch rates for FEB or FIB processing, or in other words, the shapes of deposits and etch holes.

1.2 SIMPLIFICATIONS INVOLVED

The above set of equations made several assumptions, the implications of which are outlined below.

1.2.1 Kinetics of electron (ion) impact reaction

The proportionality $R \sim Nf$ assumes that one electron (ion) induced dissociation reaction with the adsorbate governs the deposition (etch) rate. Any possible follow-up reactions leading to the final dissociation product (such as the formation of a volatile product between the molecule fragment and the substrate element in case of an etch reaction) will proceed at a time scale smaller than two successive electron (ion) induced dissociation events. Of note is that such secondary reactions are generally temperature dependent, while primary dissociation reactions will not. The proportionality $R \sim Nf$ is today generally agreed upon, but there is an example reported for the molecule butadiene C_4H_6 which shows that intermediate reactions can lead to other relations, such as a square root dependence of the deposition rate on current density $R \sim j^{1/2}$, where $j = f/e_0$ and e_0 is the electron charge. Here the rate-determining reaction was found to be the generation of active ions in the freshly deposited material, which were interacting with the adsorbed molecule [6].

Another simplification is that follow-up reactions in deposited material are not covered by the above model, i.e., it does not include irradiation chemistry processes in the bulk of deposits due to penetrating electrons.

1.2.2 Incident and emitted electron (ion) flux distributions

The incident primary electron (ion) beam will be assumed to be Gaussian distribution:

$$f(r) = f_0 \exp(-r^2/2a^2) \tag{3a}$$

with $f_0 = (I_P/e_0)/(2\pi a^2)$, where $f(r)$ is in electrons or ions per unit area and time, "a" is the standard deviation, I_P the beam current, and e_0 the electron charge. I_P can be measured in a Faraday cup. For the above Gaussian distribution, it follows that the full width at half maximum is:

$$FWHM_B = 2a\sqrt{2\ln 2} \ . \tag{3b}$$

The Gaussian distribution gives quite a realistic model for the central part of charged particle beams in scanning microscopes; for details see a recent review [7].

A more complex issue is the emitted electron flux generated due to interactions of the primary electrons (ions) incident on solid matter (substrate or deposit itself). These emitted fluxes will be equally, if not dominantly, important for adsorbate dissociation. The interactions and emitted fluxes are schematically shown in Figure 6.3. According to Figure 6.3a, for gas-assisted FEB processing the emitted electron flux consists of secondary electrons (generated both by primary and backscattered electrons), which can efficiently dissociate adsorbates. To what extent secondary or primary and backscattered electrons govern the dissociation of adsorbates is still an open question and seems to depend on the molecular species. According to Figure 6.3b, for gas-assisted FIB processing, the emitted secondary electron flux and the surface density of excited surface atoms (ESA) produced by the collision cascades are responsible for adsorbate dissociation. It has been shown experimentally that the chemical deposition and

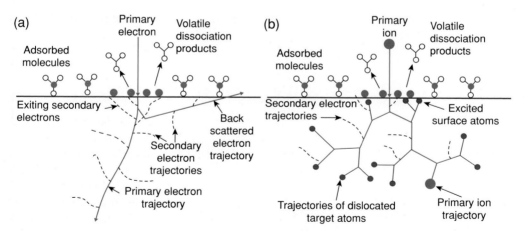

FIGURE 6.3: (a) Model of gas assisted FEB deposition and etching. Primary electrons generate an emitted flux of secondary electrons and back scattered electrons. All electrons can dissociate adsorbates via electronic excitation. All electrons in the bulk continue to interact and can cause further dissociation in a growing deposit volume. (b) Excited surface atom (ESA) model for gas-assisted FIB deposition and etching. The primary ions generate secondary electrons and a collision cascade of substrate atoms. The non-sputtered surface atoms remain excited and cause adsorbate dissociation together with exiting secondary electrons. Reprinted with permission from [7]. Copyright 2008, American Institute of Physics.

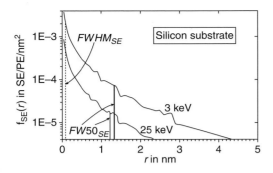

FIGURE 6.4: MC-simulated radial distributions of emitted secondary electrons (SE) per primary electron (PE): zero diameter electron beam impinging normal with 25 keV and 3 keV at $r = 0$ on planar bulk silicon. The SE exit depth was taken as 5.2 nm and the mean exit energy as 70 eV. The *FWHM* and *FW50* (full width containing 50% secondary electrons) are indicated. Replotted from Utke et al. [7].

etch yield are linearly proportional to the number of excited surface atoms [8; 9; 10]. The secondary electron emission induced by primary ions was found to play a minor role for the energy ranges (2–10 keV) and ion species (He^+, Ne^+, Ar^+, Kr^+, Xe^+) investigated.

For continuum modeling this means, basically, that the distribution of the excited surface atoms and of the emitted secondary and backscattered electrons should be considered. These distributions can presently only be estimated by Monte Carlo simulations. The SEs show a non-Gaussian distribution and a sharp peak with a very small full-width at half maximum for a zero diameter simulated incident beam, $FWHM_{SE} \ll 1$ nm (see Figure 6.4). Consequently, for non-zero diameter beams with $FWHM_B > FWHM_{SE}$, the SE-distribution can be taken as the incident Gaussian distribution. For the case $FWHM_B < FWHM_{SE}$, which basically only prevails in scanning transmission microscopes (STEM), we approximate the SE-distribution of Figure 6.4 as an exponential distribution $f(r) = f_0 \cdot \exp(-r/2a)$.

Continuum models can in principle implement more complex spatial distributions to cope with any distributions of emitted electron or excited surface atoms, see for an example Chapter 7 by Lobo and Toth in this book. However, at this stage this would add too much complexity just to be able to explain side effects, for instance halo deposition, at the expense of clarity for the common case of today's "standard" sized Gaussian distributed electron and ion beams. Using the above Gaussian and exponential distributions will illustrate the fundamental concepts we are facing in gas-assisted deposition at a fairly realistic level while keeping the mathematics simple.

Another simplification we make here concerns the planarity of the surface: As soon as the surface becomes non-planar, the above flux distributions will alter. Flux alterations can be expected for etch holes or pillar deposits with aspect ratios larger than one. Consequently, the calculated rates *throughout this chapter* hold strictly only for planar surfaces and approximately for aspect ratios of etch holes or deposits smaller than unity. Electron (ion) distributions on high aspect ratio curved surfaces can be approximately accessed by MC simulations, which are discussed for electrons in Chapter 8 by Mitsuishi in this book. Nevertheless, the general lines concerning lateral deposit or etch hole resolution derived in the following sections will still hold true at the apex of such high-aspect ratio structures.

1.2.3 Energy dependence of cross-sections, emitted electrons, and surface excited atoms

Another simplification involves the introduction of a "non-energy-dependent" interaction (dissociation) cross-section replacing the integral over the energy spectrum generated by the emitted electron flux and the excited surface atoms:

$$R_{ch}(r) = \underbrace{Vn(r) \int_0^{E_{PE}} \sigma(E)f(r,E)dE}_{\text{Monte Carlo}} \cong \underbrace{Vn(r)\sigma f(r)}_{\text{Continuum}}. \qquad (4)$$

Neglecting the energy dependence greatly simplifies the computation. Again, it would be desirable to perform comparative MC and continuum simulations once energy-dependent cross-sections are known for FEBIP/FIBIP relevant molecules to pin down the difference in both approaches. The simplification implies that cross-sections σ, derived from continuum models, should be considered as an *integral value over the entire energy spectrum which contains all channels responsible for deposition (etching)*. Depending on the domination of the one or other dissociation mechanism, the dependence $\sigma(E_{PE})$ of such an obtained cross-section might fall into one of the types described in Chapter 4 by Moore and colleagues in this book. Analogous arguments apply to gas-assisted FIB deposition and etching.

1.2.4 Adsorption isotherms

Apart from surface diffusion, molecule adsorption from the gas phase will determine the material transport for deposition and etch reactions. Gas-assisted FEB and FIB processing needs molecules that reversibly adsorb on surfaces, i.e., do not form permanent strong chemical bonds with the surface (chemisorption) or spontaneously etch without being excited and dissociated by electrons or ions. This type of adsorption is generally termed *physisorption*. It is mainly due to weak van der Waals forces and "permits" the adsorbates to leave the surface by spontaneous thermal desorption after an average residence time τ. Adsorption isotherms describe the average adsorbate coverage $\theta = N/N_0$ on a substrate surface in contact with gas molecules at a temperature T. Such isotherms can be derived from thermodynamic equilibrium potentials of the gas phase at pressure p and the adsorbed phase, usually treated as two-dimensional lattice gas. Including interactions between adsorbates and surface triggered dissociation of adsorbates, the Fowler-Frumkin isotherms are obtained [11]:

$$\left(\frac{\theta}{1-\theta}\right)^m \exp\left(\frac{W(\theta)}{kT}\right) = K_{eq}(T) \cdot p = s\tau(T) \cdot J/N_0, \qquad (5)$$

where $W(\theta) = w\theta$ describes the attractive/repulsive interaction potential between the adsorbates and $m = 1$ the non-dissociative molecule adsorption and $m = 2$ the dissociative molecule adsorption. $K_{eq}(T)$ is a temperature-dependent equilibrium constant. In the right term we introduce kinetic parameters generally used in FEB (FIB) notation: the pressure p is replaced by the impinging flux J (defining the local pressure above the substrate $p \sim J$) and the equilibrium constant assumes the value $s\tau(T)/N_0$. The mean residence time of the adsorbate on the surface depends on temperature via an Arrhenius relation $\tau = \tau_0 \cdot \exp(E_{des}kT)$, where E_{des} is the activation barrier for desorption and τ_0^{-1} is a characteristic desorption frequency of the adsorbate-substrate system (not to be confused with the so-called attempt frequency $kT/h = 8.3 \cdot 10^{12}$ s^{-1} at 25°C known from transition state theory). The sticking probability s is independent of adsorbate coverage and accounts for events of prompt scattering of impinging molecules on the free surface sites. Such an event can be pictured as an interaction where no van der Waals "bond" is established and where the molecule leaves the surface at a time scale much shorter than the residence time τ. Generally, it is assumed that the whole substrate surface is available for adsorbates and that no specific adsorption sites exist, i.e. $s = 1$. Of note is that the literature sometimes confuses coverage, N/N_0, with the sticking probability, s. Also, the mechanisms underlying "sticking" or prompt scattering are not uniquely defined.

The pragmatic approach in all FEB (FIB) models at present assumes a non-dissociative Langmuir adsorption isotherm. It assumes that the molecule stays intact upon adsorption ($m = 1$) and that adsorbates do not interact ($w = 0$). Introducing these values into Equation 5 gives the adsorbate coverage at equilibrium at a given temperature and at a given molecule flux J:

$$\theta = \frac{sJ/N_0}{sJ/N_0 + 1/\tau}. \tag{5a}$$

Since Langmuir adsorption was assumed in setting up the adsorption rate in Equation 2, the same result is obtained from the steady state solution $dN/dt = 0$ of Equation 2 for zero irradiation, i.e., $\sigma f N = 0$ (and consequently $D\nabla^2 N = 0$).

Adsorption isotherm measurements validating or dismissing interactions between FEBIP and FIBIP relevant adsorbates have not yet been performed. Dissociative adsorption of the molecules Me_2-Au(tfa) and (hfa)CuVTMS at room temperature was observed and led to one permanent chemisorbed monolayer on the substrate. However, successive adsorption onto the chemisorbed molecule monolayer was reversible [12; 13].

The different adsorbate coverages resulting from adsorbate-adsorbate and adsorbate-surface thermodynamics are shown in Figure 6.5 as a function of impinging molecule flux (pressure) at constant temperature. The isotherms with $w \neq 0$ are known as Frumkin-Fowler isotherms. At low molecule fluxes (pressures), Langmuir and Fowler-Frumkin isotherms do not differ considerably from each other for the same adsorption mechanism (dissociative or non-dissociative). This is simply due to the fact, that at low coverage the interaction with neighboring adsorbates is not an important contribution. For non-dissociative adsorption, the coverage increases linearly with the flux in the small coverage/flux region and levels off as the coverage increases. For dissociative adsorption, the coverage is larger (two adsorbate products due to dissociation) and changes with a square root dependence on the flux (pressure). When strong interactions in the order of 100 meV $\geq 4\,kT$ between adsorbates exist, then they can become very important and evident: while repulsive interaction keeps the coverage at high fluxes low, almost monolayer coverage is obtained for attractive interaction. The difference between these isotherms can take a value of about one order of magnitude within the flux (pressure) range considered in Figure 6.5.

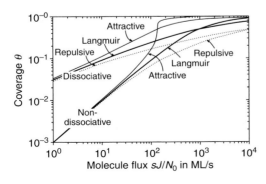

FIGURE 6.5: Non-dissociative and dissociative Langmuir adsorption isotherms and repulsive and attractive Fowler-Frumkin adsorption isotherms from Equation 5. The residence time of the adsorbate was set to $\tau = 10^{-3}$s, the repulsive interaction potential to $w = +5kT$, the attractive interaction to $w = -4kT$.

All of the isotherms discussed above limit coverage to a maximum of one monolayer, which makes sense with regard to the volatility of the molecules used in gas assisted FEB (FIB) processing. Multilayer condensation, as described by the Brunauer-Emett-Teller (BET) isotherm, can in principle occur when the substrate is cooled down below the condensation temperature.

1.2.5 Other simplifications

Here follows a list of other simplifications which apply throughout this chapter.

a) The residence time of the volatile dissociation products is considered small with respect to the residence time of the intact adsorbate; in other words, the volatile reaction products will not occupy surface sites and will not be dissociated by irradiation. The latter process can lead to unwanted effects in FEB etching: re-dissociation of the volatile "etch" product containing the substrate atom will counteract the wanted etch process. The corresponding continuum model was developed by Lassiter et al. [14].

b) Desorption is considered to be thermally activated. Electron-stimulated desorption, being a special case for insulating materials [15], is not accounted for.

c) Other transport mechanisms mediated by electric field gradients and appearing mostly on insulating samples are not considered.

d) Since this chapter considers FEB (FIB) models for one adsorbate species only, it will not enter into issues of competitive adsorption when multiple species are involved, described via the Langmuir-Hinshelwood formalism.

Diffusion of the physisorbed adsorbates is considered by Fick's diffusion laws. The involved surface diffusion coefficient is taken as coverage independent and constant at a given temperature.

2 GAS-ASSISTED FEB AND FIB DEPOSITION AND ETCHING

We will review the steady state solutions of Equations 1 and 2 with and without diffusion. Based on these results, we can proceed to:

a) define scaling laws for the resolution of low aspect ratio deposits (etch holes) in terms of fundamental dimensionless variables,

b) define three distinct process regimes: electron (ion) limited, adsorbate-limited, surface-diffusion enhanced, and

c) introduce concepts to extract adsorbate kinetics and adsorbate-charged particle interaction parameters from comparison with experiments.

2.1 SOLUTIONS WITHOUT SURFACE DIFFUSION

In this chapter we introduce the concepts of *irradiative depletion*, i.e., the "consumption" of adsorbates via dissociation by electron (ion) impact, and of *replenishment* with non-dissociated "fresh" adsorbates via gas phase transport. We will derive and discuss the related time scales of these processes and introduce the *electron- (ion-) limited* process regime and the *adsorbate-limited* process regime.

2.1.1 Continuous exposures

The first theory of charged particle deposition rates was developed for electron beams and, in fact, was not formulated for a focused beam but rather for a large beam of over 1 millimeter spot size. This had the advantage that surface diffusion of adsorbates, the diffusion paths of which are generally ≤ 1 µm, could be neglected as a substantial contribution to deposition or etching within this large area. Without the diffusion term (and ion sputtering) Equation 2 becomes:

$$\frac{\partial N(r)}{\partial t} = sJ\left(1 - \frac{N(r)}{N_0}\right) - \frac{N(r)}{\tau} - \sigma f(r)N(r). \tag{6}$$

This equation is identical to the expressions of Ennos [16] and Christy [1] derived in the 1950s and 1960s, except for the $-N/N_0$ term and the explicitly noted radial dependency. The $-N/N_0$ term confines the maximum coverage of adsorbates to one complete monolayer. This term can be only neglected in the low pressure part of the Langmuir isotherm. The molecule flux impinging on the substrate J is taken as constant along the deposition site, which generally holds true at a lateral scale of roughly 100 µm for today's gas injection systems (GIS), see Chapter 3 by Friedli and colleagues in this book.

Solving eqn. 6 for steady-state (continuous exposure), $dN/dt = 0$, we obtain:

$$N(r) = \frac{sJ}{sJ/N_0 + 1/\tau + \sigma f(r)} = sJ\tau_{eff}(r), \tag{7}$$

with the effective residence time of the adsorbates $\tau_{eff}(r) = (sJ/N_0 + 1/\tau + \sigma f(r))^{-1}$. The vertical chemical deposition or etch rate in units of distance per unit time from Equation 1a becomes:

$$R_{ch} = VsJ\tau_{eff}(r) \cdot \sigma f(r) = V\frac{sJ}{sJ/N_0 + 1/\tau + \sigma f(r)}\sigma f(r). \tag{8}$$

For any peak function $f(r)$ with a peak value $f_0 = f(r = 0)$, an effective residence time in the center of the electron or ion beam is defined:

$$\tau_{in} = \tau_{eff}(r = 0) = 1/(sJ/N_0 + 1/\tau + \sigma f_0). \tag{9a}$$

This is the time scale at which irradiative depletion (dissociation) of adsorbates occurs. The effective residence time far away from the electron beam center becomes:

$$\tau_{out} = \tau_{eff}(r \to \infty) = 1/(sJ/N_0 + 1/\tau). \tag{9b}$$

This is the time scale at which refreshment with intact adsorbates from the gas phase occurs. The adsorbate density outside the irradiated area is:

$$N_{out} = sJ\tau_{out} \tag{10a}$$

and at the beam center:

$$N_{in} = sJ\tau_{in}. \tag{10b}$$

Note that the solution found by Christy back in the 1960s is obtained by setting $f(r) = f_0$ and $sJ/N_0 = 0$ in Equation 8.

With Equation 9 one can define a useful dimensionless parameter, the *irradiative depletion* $\tilde{\tau}$ of adsorbates due to electron (ion) dissociation at the center of the focused beam. Using the effective residence times inside and outside the irradiated area:

$$\tilde{\tau} = \tau_{out}/\tau_{in} = 1 + \sigma f_0/(1/\tau + sJ/N_0). \tag{11}$$

Experimentally, $\tilde{\tau}$ can easily be varied by changing the electron (ion) beam current density or by varying the molecule flux. When $\tilde{\tau} = 1$, there is no irradiative depletion of adsorbates (they can be replenished via an intense molecule flux or the current density is very low). Conversely, when say, $\tilde{\tau} = 10$, then dissociation lowers the adsorbate coverage by a factor of 10 at the center of the beam with respect to the outside non-irradiated region. We now turn to the dependence of the chemical rate R on the electron (ion) flux f plotted in Figure 6.6. For a small electron (ion) flux, or, more precisely, a low dissociation rate, σf, compared to the molecule adsorption rate, $sJ/N_0 + 1/\tau$, R is proportional to f. In other words, proportionality is obtained when irradiative depletion is low, $\tilde{\tau} \cong 1$. A plateau is reached at high flux when $\sigma f \gg sJ/N_0 + 1/\tau$, or $\tilde{\tau} \gg 1$.

These inequalities define the *electron- (ion-) limited* process regime and the *adsorbate-limited* process regime, respectively. Within the electron- (ion-) limited regime there are always sufficient adsorbates available to be dissociated, and the process rate (dissociation reaction) is limited by the number of electrons (ions) and their efficiency to dissociate surface adsorbed molecules. Within the adsorbate-limited regime there are insufficient adsorbates available to be dissociated by the electrons (ions). The terms reaction-rate-limited (for the electron- (ion-) limited regime) and "mass-transport-limited" (for the adsorbate-limited regime) are also in use. These terms make an analogy to terms used in chemical vapor deposition (CVD) literature. Although CVD and gas-assisted FEB (FIB) deposition both involve chemical

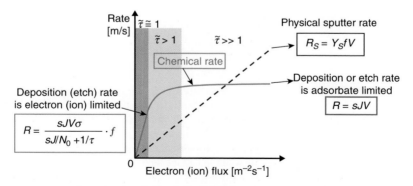

FIGURE 6.6: Generic plot of chemical and physical rate versus electron (ion) flux according to Equation 8. Process regimes are indicated. When the physical sputter rate is larger than the chemical deposition rate, no net deposition can be accomplished.

reactions, we would like to point out that the fundamental difference is in the dissociation mechanism. Gas-assisted FEB (FIB) processing relies on *non-thermal* electronically excited dissociation, in contrast to *thermally* excited molecule dissociation in CVD.

The generic plot in Figure 6.6 was confirmed in many experiments over the years. In gas-assisted FIB deposition, the transition from the ion- to adsorbate-limited regime, $\tilde{\tau} \cong 1 \rightarrow \tilde{\tau} \gg 1$, becomes very obvious: when too intense an ion flux is used, the deposition will turn out to be a milling experiment, since the sputter rate will not level off with ion flux but will continue to increase linearly until it reaches a point where more material is sputtered than deposited.

Typical adsorbate dissociation rates σf_0 achievable with focused electron or ion beams are in the range of 10^3–10^8 s^{-1}, based on typical cross-sections and beam intensity [17]. In comparison, typical molecule supply rates from a gas injection system to the substrate are in the range of $1 - 10^4$ s^{-1}, i.e., $sJ/N_0 = 1 - 10^4$ monolayers per second. Typical desorption rates at room temperature are situated mostly within the range $1/\tau = 10^3$ monolayers per second, although smaller desorption rates of $\leq 1s^{-1}$ were also reported occasionally. For the above ranges of parameters, the range of the irradiative depletion is found to vary approximately within $\tilde{\tau} = 1.01 - 10^{5...8}$.

The FEB deposition rate profiles—or shapes of FEB deposits at a given time—with varying irradiative depletion are shown in Figure 6.7. We remind the reader that the calculated rate profiles in the following figures apply when the deposits or holes have aspect ratios smaller than approximately one (see Section 1.2.2). It can be seen that with increasing irradiative depletion the deposits become broader than the beam distribution and change their shape from the initial incident or emitted distribution to a round/flat top shape. We have chosen a Gaussian as being the typical incident electron or ion beam distribution and an exponential distribution roughly describing the emitted secondary electron distribution from a plane substrate for a zero diameter impinging beam; see the discussion in Section 1.2.

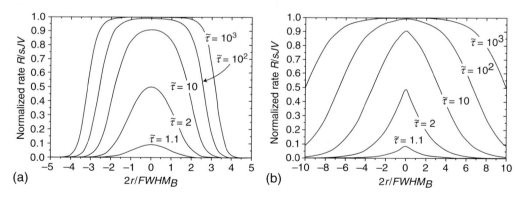

FIGURE 6.7: Calculated radial chemical deposition rates (no surface diffusion). Differing adsorbate depletions at the center of irradiation were simulated by changing the cross section while keeping all other parameters in Equation 8 constant. An inverted, but otherwise similar shape would be obtained for gas assisted electron etching and the chemical component of gas assisted ion milling. (a) For an incident Gaussian electron beam with constant full width at half maximum $FWHM_B$ and constant peak value f_0 (see Equation 3). Note that the full width at half maximum of $R(r)$ is increasing with increasing depletion. The distribution of $R(r)$ changes from a Gaussian shape for $\tilde{\tau} \approx 1$ to a round shape for $\tilde{\tau} \approx 10$ and finally to a flat top shape for $\tilde{\tau} \geq 10$. (b) For an exponential secondary electron flux $f(r) = f_0 \cdot \exp(-r/2a)$ with constant $FWHM_B$ and f_0. Note that the distinct peak distribution changes to a flat top shape with increasing adsorbate depletion. Compared to the Gaussian distribution, the FWHM of $R(r)$ increases more rapidly with increasing depletion.

The maximum chemical rate is given by the refreshment of adsorbates via gas transport from the gas injection system $R_{ch} = sJV$. This is the maximum deposition (etch) rate achievable when irradiative depletion prevails in the center of the incident (or emitted) electron distribution. In other words, the rate is adsorbate-limited inside the high intensity center. At the edges with much lower electron flux, the electron limited regime is operative and the chemical rate keeps the shape of the electron distribution. An immediate consequence when the adsorbate-limited regime prevails at the center is that the characteristic deposit or etch hole widths become *larger* than the electron distribution itself. Only if small irradiative depletions in the center, $\tilde{\tau} \approx 1$, prevail, then will the shape of the deposit or etch hole be similar to the electron distribution inducing the deposition (etch) reaction. The resulting scaling law is discussed in Section 2.1.2.

To obtain the shapes of gas-assisted ion beam deposits or etch holes, the physical sputter rate must be superimposed on the chemical rate according to Equation 1b. An example is shown in Figure 6.8, where a sputter rate of $R_S = 12 \times sJV$ was added to the chemical rates of Figure 6.7. In the case of stationary FIB deposition, the central part is milled due to too intense a focused ion beam, in agreement with experiments. "Too intense" means that at the center the chemical deposition rate regime is adsorbate-limited and overrun by the sputter (mill) rate. A deposit ring is formed only around the periphery of the milled central hole, where the ion limited regime prevails.

In the case of stationary FIB etching, an etch rate enhancement can be observed when $\sigma N_0/Y_S > 1$. From an experimental point of view this means that a suitable adsorbate was found, which dissociates under exposure to yield volatile reaction products with the substrate. The theoretical chemical etch rate enhancements (also noted as gas assisted enhancement GAE)

$$GAE = \sigma N_0/Y_S \qquad (12)$$

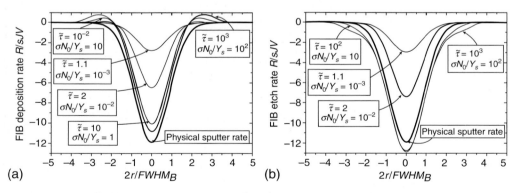

(a) (b)

FIGURE 6.8: Calculated FIB deposition (etch) rates (shapes) according to Equations 1b and 2 (no surface diffusion) for an incident Gaussian ion beam with constant full width at half maximum $FWHM_B$ and constant peak value f_0 (see Equation 3). Differing adsorbate depletions $\tilde{\tau}$ at the center of irradiation were simulated by changing the cross section while keeping all other parameters fixed. (a) For deposition with a stationary FIB, generally a hole is obtained in the center surrounded by a ring-shaped deposit. (b) An etch rate enhancement can be observed with respect to the physical sputter rate when $\sigma N_0/Y_S > 1$. However, when the irradiative depletion of adsorbates is large, the effective gas enhancement is small for a stationary beam. For $\sigma N_0/Y_S < 1$, the sputter rate is reduced due to adsorbate coverage.

given in Figure 6.8b are, however, not achievable when a large irradiative adsorbate depletion $\tilde{\tau} \gg 1$ exists at the beam center. A small widening at the hole entrance is observed when ion-limited etching takes place. Etch rate "enhancements" smaller than 1 in Figure 6.8b signify a bad choice of adsorbates in the sense that they inefficiently reacted with the substrate to form volatile products under ion exposure and that their coverage proportionally prevents physical sputtering.

In addition to the changes in shape that occur when different process regimes operate at different radial distances from the beam center, compositional changes in a deposit can also be obtained: when the number of electrons (ions) per adsorbates in the center is higher than at the peripheral regions, the dissociation of the adsorbate from its ligands can be more complete (and the deposit can become richer in metal when using an organometallic molecule) or, inversely, the co-dissociation of not-yet-desorbed ligands can lead to an increase in deposited ligand elements (and the deposit would become less rich in metal). However, a quantitative description of these intermediate processes is beyond the scope of this chapter.

2.1.2 Scaling law: Resolution versus irradiative depletion

The lateral resolution of gas-assisted focused electron (ion) beam processing can be defined as the non-dimensionally ratio of the full width at half maximum ($FWHM$) between the deposit (etch hole) and the incident beam:

$$\tilde{\varphi} = FWHM_D / FWHM_B, \tag{13}$$

see Figure 6.9. As discussed above, the $FWHM_B$ would also characterize the distributions of emitted secondary electrons or excited surface atoms generated by the incident electron (ion) distribution $f(r)$. $FWHM_D$ and $FWHM_B$ are given by the full widths at half maximum of $R(r) = sJ\tau_{eff}(r)V\sigma f(r)$ and $f(r)$. For Gaussian beams $f(r) = f_0 \exp(-r^2/2a^2)$ follows that $FWHM_B = 2a\sqrt{2\ln 2}$ and $FWHM_D = 2a\sqrt{2\ln(1+\tilde{\tau})}$. Then, the scaling law of lateral resolution as function of irradiative depletion $\tilde{\varphi}(\tilde{\tau})$ is conveniently expressed as [17]:

$$\tilde{\varphi} = \left\{ \log_2(1 + \tilde{\tau}) \right\}^{1/2}. \tag{14}$$

The $FWHM_D$ values of the deposit shapes illustrated in Figure 6.7a follow exactly this scaling law. For the exponential peak function $f(r) = f_0 \exp(-r/2a)$, roughly describing

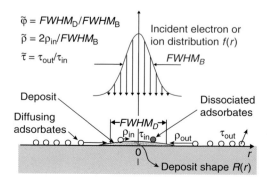

$\tilde{\varphi} = FWHM_D/FWHM_B$
$\tilde{\rho} = 2\rho_{in}/FWHM_B$
$\tilde{\tau} = \tau_{out}/\tau_{in}$

Incident electron or ion distribution $f(r)$

$FWHM_B$

Deposit

Diffusing adsorbates

Dissociated adsorbates

$FWHM_D$

ρ_{in} τ_{in} ρ_{out} τ_{out}

0   r

Deposit shape $R(r)$

FIGURE 6.9: Schematics of the full width at half maximum ($FWHM$) of incident beam and deposit. Additionally, dimensionless parameters, the effective residence time τ and the diffusion paths ρ inside and outside the irradiated area are indicated. Etch holes can be described with the same formalism.

the emitted SE distribution for a zero diameter impinging beam, the scaling law becomes $\tilde{\varphi} = \left\{ \log_2(1 + \tilde{\tau}) \right\}$ and is followed by the deposit shapes illustrated in Figure 6.7b. The point we wish to make here is that adsorbate depletion in the center of a beam distribution will directly result in a loss of beam size resolution. The above scaling law holds for deposits and etch pits having an aspect ratio smaller than approximately one. With higher aspect ratios the electron fluxes change on the curved surfaces. The same arguments apply to gas-assisted FIB deposition and etching; here the primary ion flux and the excited surface atom distribution change on the curved surfaces. In addition, redeposition of non-volatile sputtered material has to be taken into account in high-aspect ratio pit structures.

2.1.3 σ and τ determination from steady state exposures

Since deposition or etch rates, as well as the topography of deposits and etch holes, can be measured straightforwardly, the unknown parameters entering into Equations 1–2 can be extracted by comparison to shapes calculated by continuum models. We recall that there are eight parameters in Equations 1–2: the impinging precursor molecule flux J, the sticking probability s, the volume V of the deposited molecule (for etching this volume is defined by the etched material), the residence time τ, the integral cross-section σ, the surface diffusion coefficient D, the incident electron (ion) flux distribution $f(r)$, and the (maximum) monolayer density N_0 of adsorbates. Of these eight parameters, several can be determined independently, although experimental setups or calculations are sometimes complex: (a) $f(r)$ and the $FWHM_B$ can be determined with the knife edge method or a Fourier transform; see, for instance, reference [18]; (b) J can be calculated from MC simulations [19] (freeware: www.gissimulator.empa.ch); (c) V can be determined using a SEM-integrated cantilever based mass sensor [20]; (d) the (maximum) monolayer density N_0 can be estimated from the inverse of the molecule size. Four unknown parameters are left, requiring an equivalent number of four independent experiments for their determination. Surface diffusion can be "eliminated" by choosing large beam sizes (large compared to the diffusion path of adsorbates of around 100 nm to 1 μm). Often literature assumes the sticking probability s to be 100%, and one is left with the unknowns σ and τ. However, there is no rigid justification of the $s = 1$ simplification. The diffusion coefficient can be fitted into focused beam experiments once s, σ, and τ are known. For FIB, the physical sputter rate must also be known but can easily be determined experimentally. Obtaining a complete parameter set is obviously simple in theory. In practice, however, it is laborious to conduct such a series of eight well-defined experiments, and this is probably the reason why at present a complete data set extracted according to the above guidelines has not yet been established. Nevertheless, a few "incomplete" data sets were determined and will be presented in the following.

Varying the electron or ion flux f in an experiment allows testing the fundamental assumption of a second order kinetics relating R proportionally to the product Nf. Rearranging Equation 8 as:

$$\frac{f}{R} = \frac{sJ/N_0 + 1/\tau}{VsJ\sigma} + \frac{1}{VsJ}f \tag{15}$$

should result in a straight line in a Cartesian plot of f/R versus f (see figure 6.10a). The slope is sJV, and with known precursor flux J and known specific volume of the fixed decomposed adsorbate, the sticking probability s can be determined. The intersection point at $f = 0$ is $(sJ/N_0 + 1/\tau)/(VsJ\sigma) = (\sigma N_{out})^{-1}$. When the adsorbate density N_{out} outside the irradiated area

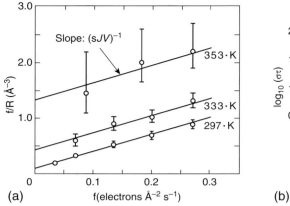
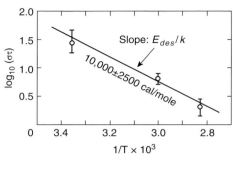

(a)

(b)

FIGURE 6.10: (a) f/R versus f plot for electron beam induced deposition using methylphenylpoly-siloxane (pump oil DC 704) and a 225 eV beam. A linear relation is obtained.

(b) Product of the electron impact cross-section with the residence time of the oil adsorbate on the deposit versus reciprocal absolute temperature. The slope gives the desorption enthalpy of 0.43 eV. Modified from Christy [1].

can be measured, for instance with a cantilever-based mass sensor [20], σ can be determined. Another experimental approach is to cool the substrate [2], which increases the residence time such that $1/\tau \ll sJ/N_0$ so that $N_{out} \to N_0$. Taking now N_0 as the inverse of the molecule area, FEB cross-sections can be determined. Cross-sections determined include those for the $Ru_3(CO)_{12}$ molecule ($\sigma = 2 \cdot 10^{-3}$ nm^2, 50 keV) [2] and benzene molecule ($\sigma = 0.35$ nm^2, 200 eV) [21]. As a comparison, FIB cross-sections include $W(CO)_6$ ($\sigma = 10$ nm^2, 42 keV Ga$^+$) [4] and Me_2-Au(hfac) ($\sigma = 52$ nm^2, 40 keV Ga$^+$) [22].

When $1/\tau \gg sJ/N_0$, the intersection point at $f = 0$ becomes $1/(VsJ\sigma\tau)$ and it is not possible to separate the unknowns σ and τ. Increasing the sample temperature will leave the cross-section unaltered but decrease the residence time according to $\tau = \tau_0 \cdot exp(E_{des}/kT)$, where τ_0 is an unknown characteristic time for the adsorbates used in gas-assisted FEB and FIB processing. However, desorption energies E_{des} of adsorbates (or adsorbate fragments) *on the irradiated deposit* can be determined from an Arrhenius fit to data points at different temperatures, see Figure 6.10b. The range of such determined desorption energies is in the 0.1 eV range, indicating weak van der Waals bonding, which allows adsorbates to desorb already at room temperature without having to break a chemical bond (which would sometimes need more than 500°C). The reversibility of surface adsorption is a prerequisite for local gas-assisted FEB (FIB) processing with near beam size resolution.

2.1.4 Time dependent solutions for pulsed irradiation

The term *pulsed irradiation* is well-known in laser processing where the focused photon beam is switched on and off with a given pulse frequency. A similar situation exists in gas-assisted FEB (FIB) processing. However, the focused beam is moved to an adjacent spot after having exposed a certain spot (with the dimension of the beam size) for a given dwell time. During the exposure dwell time the adsorbates are dissociated (depleted), and while the beam is moved around the depleted spot is refreshed by intact adsorbates. The time that passes until the spot is exposed again is known as the refresh time (synonyms include re-visit or loop time). When repeating the dwell/refresh cycles the beam becomes periodically pulsed on a given spot. Typical experimental values found for dwell times are microseconds, while refresh times can be milliseconds.

In the case of negligible surface diffusion the adsorption rate equation 2 can be solved analytically and gives the general time dependent solution:

$$N(t) = Ce^{-kt} + B. \tag{16a}$$

There are two sets of constants k, B, and C. One set holds for beam exposure (= dwell), i.e., while the adsorbates are depleted during the dwell time t_d, and the other set holds for adsorbate replenishment while the beam is off or blanked during a refresh time t_r. The two sets will be distinguished by the subscripts d and r, respectively. When exposing a fully refreshed area for a long time the conditions are $N_d(t = 0) = N_{out}$ and $N_d(t = \infty) = N_{in}$. For the subsequent long refresh cycle the conditions become: $N_r(t = 0) = N_{in}$ and $N_r(t = \infty) = N_{out}$. Solving Equation 16a for these conditions gives:

$$k = \begin{cases} k_d = 1/\tau_{in} = & sJ/N_0 + 1/\tau + \sigma f \quad dwell \\ k_r = 1/\tau_{out} = & sJ/N_0 + 1/\tau \qquad refresh \end{cases} \tag{16b}$$

and

$$B = \begin{cases} B_d = & sJ/k_d = N_{in} \quad dwell \\ B_r = & sJ/k_r = N_{out} \quad refresh \end{cases}. \tag{16c}$$

The constants C become $C_d = -C_r = B_r - B_d$. The solutions of Equation 16a for a *pulsed* beam must take into account that the adsorbate density will assume other values than N_{in} and N_{out} according to the dwell and refresh times applied. The corresponding periodic pulse conditions are thus $N_d(0) = N_r(t_r)$ and $N_r(0) = N_d(t_d)$. While k and B do not change, a new set of C values for pulsed beams is obtained:

$$C = \begin{cases} C_d = & (B_r - B_d) \cdot (e^{-k_r t_r} - 1) \cdot (e^{-k_r t_r} e^{-k_d t_d} - 1)^{-1} \quad on \\ C_r = & (B_d - B_r) \cdot (e^{-k_d t_d} - 1) \cdot (e^{-k_r t_r} e^{-k_d t_d} - 1)^{-1} \quad off \end{cases}. \tag{16d}$$

Note, Equations 16 hold true for any radial distribution of impinging beams, and the parameters N, B, k, and C can be considered as functions of r.

Figure 6.11 shows the normalized adsorbate coverage $N(t)/N_{out}$ for a continuous exposure in comparison to a pulsed exposure (one on/off beam cycle). Obviously, the adsorbate depletion can be effectively reduced when applying small dwell times in the order $t_d \approx k_d^{-1} = \tau_{in}$, followed by long replenishment (refresh) cycles, typically with $t_r = \tau_{out} \approx 1$–$10$ ms.

The chemically driven deposition or etch rate during the exposure interval t_d is given by:

$$R_{ch} = \frac{V\sigma f}{t_d} \int_0^{t_d} N(t)dt = V\sigma f \left\{ \frac{(B_r - B_d)}{k_d t_d} \frac{[1 - \exp(-k_r t_r)] \cdot [1 - \exp(-k_d t_d)]}{1 - \exp(-k_r t_r)\exp(-k_d t_d)} + B_d \right\}. \tag{17}$$

in units of unit distance per unit time. Similar expressions with differing notations can be found in [23; 24; 4; 25; 26; 27; 28]. Equation 17 also holds true when the parameters N, B, and C become functions of r.

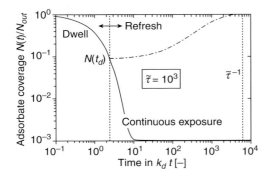

FIGURE 6.11: The normalized adsorbate density $N(t)/N_{out}$ for one on/off cycle and a continuous exposure. Typical dwell times are in the microsecond range whereas the refresh time range is around several milliseconds. Here we have chosen $t_d = 10^{-4} \cdot t_r$ and an irradiative depletion $\tilde{\tau} = 1,000$. A radial dependency is not considered here. Modified from Utke [7].

Figure 6.12(a-b) shows the adsorbate densities as calculated with Equation 16 for a Gaussian beam distribution having a $FWHM_B = 10$ nm and for an irradiative depletion $\tilde{\tau} = 1000$. It can be seen that for dwell times $t_d > 4\,\mu$s the adsorbate density becomes steady state, reaches the irradiative depletion $\tilde{\tau} = 1,000$ at the center $r = 0$ and does not change anymore visibly on the linear and logarithmic scale. Furthermore, for pulses with $t_d < 4\,\mu$s it can be seen that the effective irradiative depletion is reduced by orders of magnitude and comes very close to the limit of $\tilde{\tau} = 1$ for pulses with $t_d = 4$ ns.

Turning to the chemical rate calculations of Figure 6.12(c-d) one notices that the steady state shape $(R/sJV = 1)$ is obtained for pulses with dwell times $t_d \geq 40$ ms, while the adsorbate coverage is obtained already at around 4 μs dwell time. This is because the rate is integrated over the entire dwell time in Equation 17; in the beginning the rate values are very large and Gaussian in shape (since the refresh time was chosen large enough for full adsorbate replenishment by gas phase transport). Mathematically, $R \sim N_{out}f(r)$ for $t < k_d^{-1}$ and becomes $R \sim N_{out}f(r)/\tilde{\tau}(r)$ for $t > k_d^{-1}$. The deposition (etch) rate ratio between the electron-limited regime (very short pulse) and the adsorbate-limited regime (very long pulse) is R(electron (ion) limited)$/R$(adsorbate limited) $= \tilde{\tau}$, or, in different notation $R(t_d \to 0)/R(t_d \to \infty) = \tilde{\tau}$. In the example of Figure 6.12 the electron- (ion-) limited chemical deposition (etch) rate is 1,000 times higher than the adsorbate-limited rate.

The absolute deposit height (etch hole depth) will scale according to: $H = R(t_d) \times t_d \times$ (*Number of pulses*). For a given total time t_{tot} of the experiment, the number of pulses will be $n = t_{tot}/(t_r + t_d)$ and $H = R(t_d) \times t_{tot} \times t_d/(t_r + t_d)$.

For gas-assisted ion deposition and etching with pulsed beams the physical sputter rate must be superimposed:

$$R_{FIB} = Vf \left\{ \frac{\sigma}{t_d} \int_0^{t_d} N(t)dt \pm Y_S \left(1 - \frac{1}{t_d N_0} \int_0^{t_d} N(t)dt \right) \right\}. \tag{18}$$

As before, R, f, and N can be functions of the distance r. Figure 6.13 shows an example calculation for Equation 18. Basically, pulsing the FIB with decreasing dwell time during gas-assisted *deposition* allows changing from milled pits with shallow ring deposits to compact deposits having a round to Gaussian shape. Interestingly, the entrance hole diameter (at ordinate = 0) is self-limiting and can be tuned from zero to a maximum value given by the irradiative depletion and the sputter rate. This is due to the fact that deposition and milling are competing mechanisms; at a given distance r, both balance and the net rate is

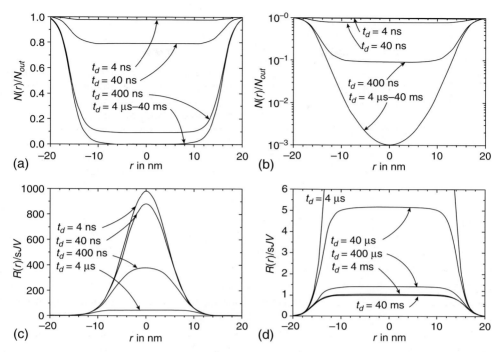

FIGURE 6.12: Normalized adsorbate density and chemical deposition (etch) rate for a pulsed electron beam according to Equations 16 and 17. The dwell time t_d was varied from 4ns to 40 ms for a constant refresh time $t_r = 4$ ms. The irradiative depletion was chosen to be $\tilde{\tau} = 1,000$. Other parameters were $\sigma = 0.06$ nm^2, flux $f_0 = 10^8$ nm^{-2}s^{-1}, $\tau = 1$ ms, $sJ/n_0 = 5000$ monolayers/s. The rate constants are thus $k_d = 6 \times 10^6$s^{-1} and $k_r = 6 \times 10^3$s^{-1}. (a) and (b) The normalized adsorbate density $N(r, t_d)/N_{out}$ at a given dwell time t_d. The Cartesian and logarithmic plots both show that for dwell times $t_d > 10k_d^{-1} = 1.7\,\mu$s the steady state adsorbate coverage is reached. The logarithmic plot shows furthermore, that $N_{in}\tilde{\tau} = N_{out}$ for steady state at the center. (c) and (d) The normalized chemical deposition (etch) rate during exposure is very close to the Gaussian beam profile for $t_d \leq 40$ ns $\approx (0.25)k_d^{-1}$ and becomes gradually flatter with increasing dwell time due to irradiative depletion. For $t_d = 40$ ms the normalized rate becomes 1 at its maximum and has its largest full width at half maximum. Note that at $r = 0$, $R(t_d = 4\ ns)/R(t_d = 40\ ms) \approx \tilde{\tau}$.

zero. Experimentally, such a self-limitation of hole diameters was found recently for tetraethyl-orthosilicate, TEOS, and for $(CH_3)_3Pt(CpCH_3)$ assisted FIB deposition on carbon membranes [29]. Interestingly, the competing processes of deposition and milling during gas assisted FIB deposition have an analog in gas-assisted FEB processing when "etch" and "deposition" adsorbates are simultaneously present on the surface; see here Chapter 7 by Lobo and Toth in this book, where similar radial features are derived from a multi-species continuum model for combined FEB deposition and etching.

The etch hole shapes in Figure 6.13 show a transition from Gaussian to round to Gaussian again with increasing dwell time. At low dwell times the chemical etching is ion-limited and will reproduce the incident Gaussian beam shape, increasing the dwell time will exhaust etching due to adsorbate depletion in the center part, and for very long dwell times in the microsecond range adsorbate depletion occurs in the whole beam. The chemical etch rate is then reduced by $\tilde{\tau}$, being 1,000 in Figure 6.13(c,d), and the physical sputter rate determines the etch profile.

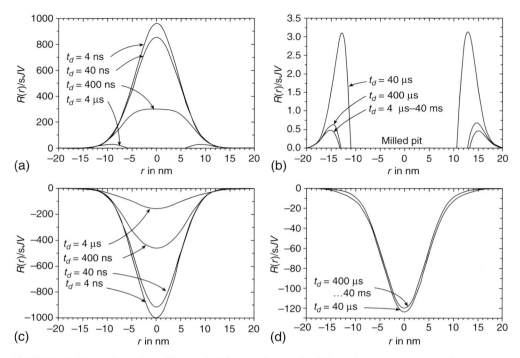

FIGURE 6.13: Normalized deposition and etch rates for a pulsed Gaussian ion beam according to eqn. 18. The dwell time t_d was varied from 4ns to 40 ms for a constant refresh time $t_r = 4$ ms. The irradiative depletion was chosen to be $\bar{\tau} = 1,000$ and the enhancement factor $\sigma N_0/Y_S = 10$. Other parameters were as for Figure 6.12.(a) and (b) The normalized FIB *deposition* rate shows that Gaussian shaped deposits can be obtained when low dwell times are applied for pulses. With increasing dwell time flat top deposits are obtained and finally ring deposits surrounding a milled hole in the beam center (negative values signify milling into the substrate and are not shown). (c) and (d) The normalized FIB *etch* rate shows an enhancement of 8.4 for a 4 ns pulse with respect to the physical sputter rate set to $R(r = 0)/sJV = 120$. For pulses with dwell times $t_d > 40\,\mu s$ no visible enhancement occurs since the adsorbates are depleted. The etch hole shape changes from Gaussian to rounded to Gaussian again with increasing dwell time.

The decrease of the peak height of a deposit and the decrease of the depth of an etch pit with increasing irradiative depletion becomes very clear in a plot of $R(r = 0)$ versus dwell time and refresh time in Figure 6.14.

Decreasing the pixel dwell time prevents adsorbate depletion from proceeding to its steady-state value. The deposition or etch rate (per pulse) therefore increases and will saturate for dwell times smaller than the effective residence time inside the irradiated beam spot $t_d \ll \tau_{in}$. Increasing the refresh time results in adsorbate replenishment before the next irradiation cycle begins. The deposition and etch rate increase (in absolute value) and finally saturate for refresh times larger than the effective residence time outside the irradiated area $t_r \gg \tau_{out}$. Obviously, for a dwell time $t_d < \tau_{in}$ and a refresh time $t_r > \tau_{out}$ the maximum rate $R = N_{out}V\sigma f$ is achieved due to negligible depletion.

Plots of rates versus dwell time and rates versus refresh time can be found in the literature with differing units. The rate in units of distance per unit time can be expressed in units of deposited (etched) volume per unit charge or as the number of deposited (etched) atoms

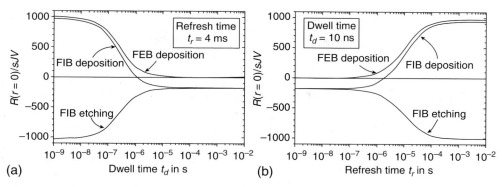

FIGURE 6.14: Normalized deposition (etch) rate for a pulsed Gaussian electron beam (Equation 17) and an ion beam (Equation 18) at the beam center $r = 0$. The irradiative depletion was chosen to be $\tilde{\tau} = 1000$ for both electron and ion beam, and for the ion beam the enhancement factor $\sigma N_0/Y_S = 10$ was set. Other parameters were as in Figure 6.12; specifically, $\tau_{in} = 1.7 \times 10^{-7}$s and $\tau_{out} = 1.7 \times 10^{-4}$s. (a) Rate versus dwell time plot; the dwell time t_d was varied from 1ns to 10 ms for a constant refresh time $t_r = 4$ ms. Note that for dwell times $t_d > \tau_{in}$ the absolute rates start to reduce considerably to their steady state minimum. The FEB deposition rate falls to $R/sJV = 1$ and the FIB deposition and etch rates become equal to the physical FIB sputter rate (chosen to be $R_S = -180\ sJV$) at large dwell times due to irradiative adsorbate depletion. The maximum absolute rates are defined by the refresh time. (b) Rate versus refresh time plot; the refresh time t_r was varied from 1ns to 10 ms for a constant dwell time $t_d = 10$ ns. Note that for refresh times $t_r > \tau_{out}$ the rates start to saturate to their maximum value (defined by the dwell time). Note that FEB etch rates can be obtained by mirroring the FEB deposition rates at the $R = 0$ line.

per incident electron (ion); the latter one being the yield Y. The conversions are $R[\text{m/s}] = R\,[\text{m}^3/\text{C}]/(e_0 f)$, e_0 being the elementary charge, and $Y = R[\text{m/s}]/(Vf)$.

2.1.5 Scaling law: Resolution versus dwell time

In Figure 6.15, the normalized lateral resolution $\tilde{\varphi}$ (see Equation 13) was plotted versus the dwell time using $FWHM$ values of $R(r)$ and the Gaussian incident electron beam $f(r)$ from Figure 6.12. $\tilde{\varphi}$ takes values from 1 to a maximum value of 3.16 found from the steady state discussion and Equation 14. It can be seen that the $FWHM_D$ of the deposit (etch pit) becomes *larger* than the $FWHM_B$ of the incident Gaussian electron beam for dwell times larger than $t_d \approx \tau_{in}$, since the center of irradiation becomes increasingly adsorbate depleted. The steady state FWHM is obtained for pulses with $t_d \approx 10^4 \tau_{in}$.

For small dwell times $t \leq k_d^{-1} \ln(\tilde{\tau})$, an approximate analytic scaling law for FEB deposition (etching) can be given [7]:

$$\tilde{\varphi} = \left\{ \log_2 \left(1 + \exp\left(k_d t_d \right) \right) \right\}^m, \tag{19}$$

with $m = \frac{1}{2}$ for the Gaussian distribution and $m = 1$ for the exponential distribution. Explicitly, when the lateral deposit or etch pit size should not exceed 10% of the $FWHM_B$ of a Gaussian incident beam, the condition $t_d \leq 0.27 \cdot \tau_{in}$ must be fulfilled.

For FIB deposition, the deposit shape changes from a wide ring to a Gaussian peak shape having a FWHM very close to the FWHM of the incident Gaussian ion beam for the smallest dwell time; see Figure 6.13(a,b). Also, the size of the hole, taken at the rate $R = 0$, in the central

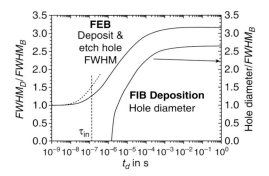

FIGURE 6.15: Plot of the FWHM of a FEB deposit from Figure 6.13 and of the hole (pit) diameter from a FIB deposit from Figure 6.14 normalized to the FWHM of the incident Gaussian beam. The dotted line is the approximation of eqn. 19. Note that hole (pit) diameters smaller than the beam diameter can be obtained in a small dwell time window.

irradiated part changes with dwell time. This hole size would be expected for a FIB deposition experiment on a membrane. The hole diameters from Figure 6.13(a,b) were plotted versus the dwell time in Figure 6.15. Evidently there is no hole etched as long as deposition occurs at the center. When adsorbates become depleted, at around $t_d \approx 10\tau_{in}$ in our example calculation, a small pit occurs and becomes rapidly larger when the dwell time is increased. Finally, the size of the hole becomes independent of the dwell time and takes a self-limited constant value, depending on irradiative depletion and sputter yield. In the above example of Figure 6.13(a,b) a self-limiting hole diameter of $2.6 \cdot FWHM_B$ was obtained.

2.1.6 Determination of σ and τ from raster scan exposures

During the refresh time the beam must not necessarily be switched off but can irradiate other spots instead. This can be done in a raster-like manner, so that large areas on the order of microns squared can be scanned with a nanometer-sized beam. Often exposure of rectangles is achieved via a serpentine scan as shown in Figure 6.16a. Also line by line or spiral scans are possible.

The beam is moved in increments along Δx and Δy on the surface, exposing each pixel with a dwell time t_d. After finishing one raster scan, the beam repeats the pattern after a refresh time t_r, sometimes also called loop or re-visit time. Clearly, the minimum refresh time that can be achieved in such a raster scan approximately equals the number of pixel exposures in the scan window times the dwell time for each pixel. Pixels will be exposed for a longer time than the dwell time t_d in the case of beam overlap in the x- or y-directions, see Figure 6.16. For zero overlap and flat top distribution $f(r)$ the deposition or etch rate of the raster scan is given by Equations 17 or 18. Raster scan exposures can be performed over micron-sized areas and yield easily measurable deposit (etch) topographies. For any new adsorbate in an experiment, the two plots of Figure 6.14 are very important to know: the variation of the deposition (etch) rate with residence time (for a given large dwell time $(t_d > \tau_{in})$) and with dwell time (for a given large refresh time $(t_r > \tau_{out})$). They give information on the process regime within which the gas-assisted FEB and FIB process is performed when choosing the exposure parameters, and thus about the lateral resolution to be expected. They also aid in the interpretation of compositions obtained for deposits.

Obviously, from $R(t_d)$ and $R(t_r)$ data sets the exponential exponents k_r and k_d can be fitted, hence the effective residence times τ_{in} and τ_{out}, see Equation 16b. If $\sigma f \gg sJ/n_0 + 1/\tau$, the cross section can be determined $\sigma \approx 1/(\tau_{in}f)$ and if $sJ/n_0 \ll 1/\tau$ the residence time can be estimated, $\tau \approx \tau_{out}$. Another condition is given by the minimum and maximum values of R

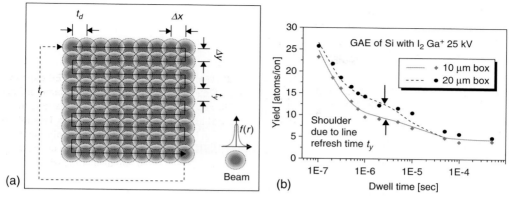

FIGURE 6.16: (a) Digital serpentine raster scan scheme with beam overlap. The (Gaussian) beam dwells for a time t_d at (x, y) and is then moved by Δx and Δy to the next exposure spot. The scan is repeated after a refresh time t_r. Reprinted with permission from [7]. Copyright 2008, American Institute of Physics. (b) Etch yield versus dwell time for I_2 assisted FIB etching (Ga-ions, 25 keV) of silicon. The refresh time was $t_r = 20$ ms, the step size $\Delta x = \Delta y = 240$ nm, and the (fitted) $FWHM_B = 920$ nm (defocused). The shoulder in the slope comes from the line refresh time. Since for the 10 μm box the line refresh time is smaller, the shoulder effect is less pronounced. Modified from Edinger and Kraus [27].

that scale with $V\sigma f$. The cross section could be determined when the atomic volume V of the deposited compound (or etched atoms) is known. For deposits composed of metal nanocrystals embedded in a matrix, this value must be determined via density measurements [20]. Often in the literature the deposit density is approximated with pure metal or compound values, which can introduce easily an uncertainty factor of 2 to 10 for the published yields or cross-sections on deposition, depending on composition. For etching the volume is generally known. The bulk of published measurements for FEB and FIB show that refresh times on the order of milliseconds are needed to achieve full replenishment of the exposed regions, thus implying that the residence time of adsorbates on the (irradiated) deposit is also in the *millisecond* range.

A numerical solution taking into account beam overlaps and Gaussian distributed beams [27; 5] was able to reproduce the shoulder effect seen in Figure 6.16b: this effect is related to the serpentine raster scan, which delivers portions of the ion dose to a given pixel at different times. While in the fast horizontal scan direction the beam passes over neighboring pixels in a continuous way, successive exposures of vertically adjacent pixels are separated approximately by the time it takes the beam to complete one line. During this line refresh time, t_y in Figure 3.16a, the pixel is replenished with precursor adsorbates. For Cl_2-assisted etching of Si with 50 keV Ga^+-ions, a cross-section of $\sigma = 14$ nm^2 was found, corresponding to a maximum chemical etch yield of 45 atoms/ion [27]. The same paper gives for I_2-assisted etching of Si with 25 keV Ga^+-ions the values $\sigma = 53$ nm^2 and a maximum chemical yield of 40 atoms/ion. For $W(CO)_6$ assisted FEB deposition the dependence of the cross-section on primary electron energy was reported in [30].

For gas-assisted FIB etching, the gas enhancement is often reported instead of cross-sections. The gas enhancement is obtained with respect to FIB milling (without gas assistance), see Equation 12. It represents a characteristic value for the substrate-adsorbate-ion system. Gas-assisted enhancement factors of FIB sputter rates are comprehensively summarized in a review by Utke et al. [7]. Of note is that large differences in gas enhancements for the same system can be found in different literature sources. This is likely due to the fact that adsorbate replenishment in the irradiated area was not achieved with the pulse conditions applied. In this case, the enhancement factors relate to the efficiency of the adsorbate-limited regime rather than representing the ultimate enhancement which could be obtained in the ion-limited regime.

2.2 DEPOSITION AND ETCH RATES INCLUDING ADSORBATE SURFACE DIFFUSION

In this section we introduce *diffusive replenishment* – the process of adsorbate replenishment by surface diffusion. We will show that this contribution depends on beam size and irradiative depletion and has important consequences for the process rate and the shapes of deposits and etch holes. We will confine the discussion in this section to gas assisted FEB processing. Historically, most of the theory was developed for FEB; however, analogies to gas-assisted FIB processing are straightforward.

2.2.1 Experimental facts from literature

Evidence that surface diffusion of adsorbates contributes to focused electron (ion) beam deposition came from several experiments where the primary beam size was varied. Already in 1967, Zhdanov [31] showed that reducing the beam diameter increased the deposition rate of hydrocarbons by one order of magnitude. Müller showed in 1971 [32] that the hydrocarbon deposition rate varied over *four* orders of magnitude and was inversely proportional to the square of the beam diameter. In 1978, Reimer and Wächter [33] showed ring deposits obtained from an approximately flat top distribution electron beam exposure (see Figure 6.17).

2.2.2 Steady state rates for uniform electron (ion) flux

Taking the diffusion term in Equation 2 into account and solving for $dN/dt = 0$ and rotational symmetry, an analytical solution was obtained for a cylindrical flat top distribution $f(r) = f_0$ for $|r| \leq FWHM_B/2$ by Müller back in 1971 [32]. The adsorbate density becomes (in our notation):

$$N(r) = N_{out} \left\{ \tilde{\tau}^{-1} + C_{diff} I_0 \left(r/\rho_{in} \right) \right\}, \tag{20a}$$

where I_0 is the modified Bessel function with $I_0(0) = 1$. Obviously, Equation 20a is composed of the solution for zero diffusion $N(r) = N_{out} \tilde{\tau}^{-1}$ known from Equation 7 and a term accounting for the contribution of surface diffusion. The surface diffusion contribution factor C_{diff} varies between 0 and 1 and is given in our notation as:

$$C_{diff} = \left(1 - \tilde{\tau}^{-1}\right) \frac{K_1\left(\tilde{\rho}_{out}^{-1}\right)}{I_0\left(\tilde{\rho}_{in}^{-1}\right) K_1\left(\tilde{\rho}_{out}^{-1}\right) + \tilde{\tau} \cdot I_1\left(\tilde{\rho}_{in}^{-1}\right) K_0\left(\tilde{\rho}_{out}^{-1}\right)}. \tag{20b}$$

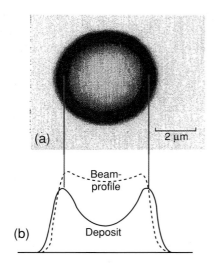

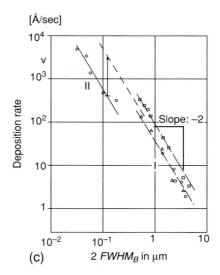

FIGURE 6.17: (a) TEM micrograph of a ring obtained from hydrocarbons by irradiating a circular area with approximately uniform current density of 0.1 A/cm². (b) Radial recording of mass thickness (solid line) and the current density (dashed line). Intact adsorbates from outside diffuse to the depleted area (where adsorbates are dissociated by the electron beam). A rim is observed when the diffusion path of the adsorbates is smaller than the beam FWHM *and* when the gas supply into the depleted area is small. Modified from Reimer and Wächter [33]. (c) Four orders of magnitude change in hydrocarbon deposition rate observed when decreasing the electron beam size. The quadratic increase of the rate with decreasing beam size is due to adsorbate surface diffusion. Modified from Müller [32].

K_n and I_n are modified Bessel functions. The normalized diffusion paths $\tilde{\rho}$ at the center of the irradiated region and the outside of the irradiated regions are defined as (see also Figure 6.9):

$$\tilde{\rho}_{in} = 2\rho_{in}/FWHM_B = 2\sqrt{D\tau_{in}}/FWHM_B \tag{21a}$$

$$\tilde{\rho}_{out} = 2\rho_{out}/FWHM_B = 2\sqrt{D\tau_{out}}/FWHM_B, \tag{21b}$$

where $FWHM_B$ is the full width at half maximum of $f(r)$, D the surface diffusion coefficient, and τ_{in} and τ_{out} the effective residence times as defined in Equation 9. Consequently, for a given irradiative depletion and a given diffusion path outside the irradiated region, the surface diffusion path at the center is given by:

$$\tilde{\rho}_{in} = \tilde{\rho}_{out} \cdot \tilde{\tau}^{-1/2}. \tag{21c}$$

Equation 21a states that the surface diffusion path inside the irradiated region scales inversely with the electron (ion) flux, $\rho_{in} \sim (\sigma f_0)^{-1/2}$. The dependence of the surface diffusion contribution factor on irradiative depletion and the normalized surface diffusion path, $C_{diff}\left(\tilde{\tau}, \tilde{\rho}_{out}\right)$, is shown in Figure 6.18. The maximum of the surface diffusion contribution increases with the mobility of the adsorbates (longer diffusion path).

The surface diffusion contribution tends to zero with increasing irradiative depletion since the adsorbates are dissociated and fixed at an increasingly high rate so that $\tilde{\rho}_{in} \ll 1$ according

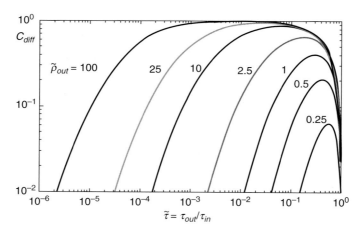

FIGURE 6.18: Plot of the contribution factor of surface diffusion C_{diff} versus irradiative depletion according to Equation 20b. The normalized surface diffusion path outside the irradiated area is varied as parameter. Note that inside the irradiated area the surface diffusion contribution changes according to the actual radial depletion.

to Equation 21c. In other words, the diffusion path inside the irradiated region becomes smaller than the beam size—the adsorbates are dissociated before reaching the center of the beam. The contribution of surface diffusion also tends to zero at very low irradiative depletion since there is no concentration gradient between the irradiated area and the surrounding. The diffusion contribution to the adsorbate density at the center of the irradiated region equals $N_{out} \cdot C_{diff}$ since at $r = 0$ the term $I_0\left(r/\rho_{in}\right) = 1$. The above relations are illustrated graphically in Figure 6.19.

2.2.3 Deposition and etch regimes in gas assisted FEB/FIB processing

As illustrated in Figure 6.19, adsorbate surface diffusion adds beam size dependent variations to the established adsorbate-limited regime and to the electron- (ion-) limited regime. The adsorbate-limited regime is characterised by an adsorbate density $N(r) < N_{out}$. At the center of the irradiated area, the adsorbate depletion becomes $N_{out}/N_{in} = \tilde{\tau}$ when adsorbate diffusion is negligible as for broad (approximately micron-sized) beams. Adsorbate replenishment occurs predominantly via gas phase transport, see Figure 6.19a. When the beam is focused to smaller size, adsorbates can diffuse into the irradiated centre and the irradiative depletion becomes smaller, $N_{out}/N_{in} = \tilde{\tau}/(1 + C_{diff}\tilde{\tau})$ due to the contribution of surface diffusion $N_{out}C_{diff}$, see Figure 6.19b. The deposition or etch rate increases accordingly.

The electron- and ion-limited regime is characterized by a fast replenishment rate compared to the adsorbate dissociation rate. When irradiative depletion $\tilde{\tau}$ is large, the diffusive replenishment can fully compensate if the beam size is chosen appropriately small (the appropriate size can be estimated from Equation 20b), see Figure 6.19c. For low intensity beams it can also happen that replenishment by gas transport is sufficient, meaning that the irradiative depletion remains very small, $\tilde{\tau} \approx 1$. Surface diffusion in this case becomes very

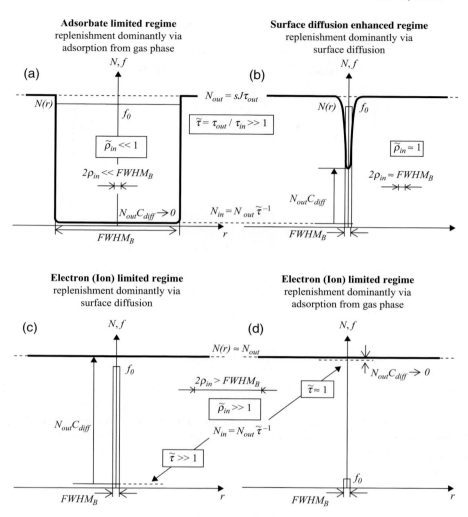

FIGURE 6.19: Graphical illustration of the adsorbate density $N(r)$ (bold line) of Equation 20 in the three different process regimes for deposition and etching with focused electron and ion beams. Note that the surface diffusion enhancement of the adsorbate density is beam size dependent. A large adsorbate density in the irradiated area and a large electron flux f_0 assure a high process rate. Note that the relations still hold when the flat top distribution of the electrons (ions) is replaced by a Gaussian; only the analytical expression for the diffusion distribution $N_{out}C_{diff}$ must be replaced with a refined numerical value.

small due to the very low concentration gradient of adsorbates inside and outside the irradiated area, see Figure 6.19d. In this case no beam size dependence exists.

The size dependent effects of diffusive replenishment on resolution and process rate are sometimes referred to as a third process regime: the (surface) diffusion-enhanced regime.

2.2.4 Scaling law: Deposition (etch) rate vs. diffusive replenishment

The diffusive replenishment deserves a special consideration because it is beam size dependent. The deposition or etch rate increases from $R_{ch} = (N_{out}/\tilde{\tau})V\sigma f_0$ for $\tilde{\rho}_{in} \ll 1$ to $R = N_{out}V\sigma f_0$

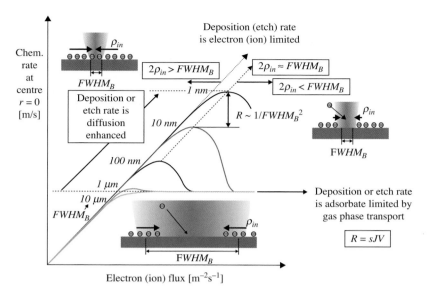

FIGURE 6.20: Generic plot of chemical rate versus electron (ion) flux according to Equation 20 at the center of the incident beam. The contribution of surface diffusion was varied by changing the beam size. Except for zero surface diffusion, all rates have a peak at around $\rho_{in} \approx FWHM_B/2$ exceeding the adsorbate limited process rate since surface diffusion of intact adsorbates replenishes the dissociated adsorbates. The decay for further increasing electron (ion) flux is due to the fact that the diffusion path scales with $\rho_{in} \sim (\sigma f_0)^{-1/2}$, consequently adsorbates will be dissociated at the rim—before they reach the center—and form a ring deposit; see Figure 6.17. The center will be replenished again by gas phase transport (at much lower rate). The related process regimes are indicated. For the sake of clarity, the physical sputter rate was omitted.

for $\tilde{\rho}_{in} \gg 1$. Obviously, when the irradiative depletion is very high and the surface diffusion contribution low, a low process rate results. If the adsorbate depleted zone could be refreshed via surface diffusion by making the beam size smaller, a much higher process rate can be achieved for this spot (see Figure 6.20). The scaling law of the chemical deposition (etch) rate as function of the full width at half maximum of the beam $FWHM_B$ can be derived from Equation 20 [32] and was experimentally verfied (see Figure 6.17b):

$$R(FWHM_B) \propto 1/FWHM_B^2. \tag{22}$$

Although this inverse square relation was derived for a flat top beam distribution, it can be shown to also hold for a Gaussian incident beam profile. The maximum diffusion enhancement in chemical deposition (etch) rate becomes $R(\tilde{\rho} \to \infty)/R(\tilde{\rho} = 0) = \tilde{\tau}$ at $r = 0$. Figure 6.20 illustrates the different process regimes that can become operative with increasing electron (ion) flux, i.e., with increasing irradiative replenishment, and for varying beam sizes.

2.2.5 Steady state rates for Gaussian electron (ion) flux

Replacing the uniform flat top distribution for the incident electron (ion) beam by a more realistic Gaussian distribution, an analytical solution cannot be derived, but numerical results can be obtained for the chemical rates of deposition (etching). Interestingly, indented (doughnut) shapes can be deposited or etched when surface diffusion is important. The indented shapes are due to surface diffusing precursor adsorbates fixed at the periphery of the impinging beam. Several deposit (etch hole) shapes can be observed: flat top shapes signify the adsorbate-limited regime; indented and round shapes are typical for the surface diffusion-enhanced regime; and Gaussian peaked shapes for the electron/ion-limited regime (see Figure 6.21a).

The physical sputtering by the FIB would manifest in Figure 6.21a as an indent at the center of the deposit with a "feature size" of approximately $[(1 - N/N_0)Y_S]/(\sigma N/N_0)$, see eqn. 1c. Hence, in the adsorbate limited regime the milling indent is largest since $N = N_{in}$ and for the ion limited regime it is smallest since $N = N_{out}$. Basically this means that the indentations in Figure 6.21a will be large for small diffusion paths $\tilde{\rho}_{in}$ and become smaller for larger $\tilde{\rho}_{in}$. For gas assisted FEB and FIB etching, similar arguments apply. Since for FIB etching the chemical and physical rate add to each other, the observation of ring-like etch pits is unlikely for heavy ions like gallium, but likely for light ions, like hydrogen or helium.

For completeness we mention that analytical continuum solutions have been also calculated for a high aspect ratio, freestanding rod deposited with a slowly one-directionally moving electron beam [34; 35] and for a uniform "line" exposure (beam is not sweeping) [36]. The latter exposure can be treated as a one-dimensional case of Equation 2.

2.2.6 Scaling law: Resolution vs. diffusive replenishment

From Figure 6.21a it is obvious that for a given irradiative depletion the deposit (etch pit) resolution improves with increasing diffusive replenishment. For $\tilde{\rho}_{in} \gg 1$, $R_{ch}(r) = sJ\tau_{out}V\sigma f(r)$ since any adsorbate depletion is entirely compensated by surface diffusion. In other words, the electron (ion)-limited regime is established and the deposit (etch pit) shape corresponds to the electron or ion beam distribution $f(r)$. A universal graph relating the dimensionless resolution to irradiative depletion and diffusive replenishment $\tilde{\varphi}(\tilde{\tau}, \tilde{\rho}_{in})$ is shown in Figure 6.21b. Note that diffusive replenishment can be achieved experimentally in two different ways: either by reducing the beam size $FWHM_B$ (using the focus of the beam) or by increasing the diffusion path ρ_{in} (keeping the same beam size but changing beam current or surface diffusion of the adsorbate).

The scaling law of resolution versus diffusive replenishment $\tilde{\varphi}(\tilde{\rho}_{in})$ which can be derived from Figure 6.21 b is [17]:

$$\tilde{\varphi} \cong \left\{\log_2\left(2 + 1/\tilde{\rho}_{in}^2\right)\right\}^{1/2} . \tag{23}$$

Equation 23 holds for a Gaussion distribution $f(r)$. For the exponential peak function $f(r) = f_0 \exp(-r/2a)$ introduced in Section 1.2.2, the relation $\tilde{\varphi} \cong \log_2\left(2 + 1/\tilde{\rho}_{in}^2\right)$ can be found.

The above scaling law of lateral resolution versus diffusive replenishment may vary when physical sputtering is involved as in gas-assisted FIB processing. Depending on the ratio of physical versus chemical rates these variations can become substantial. In Equation 23 the physical sputter rate is assumed to be zero.

2.2.7 Estimation of surface diffusion from continuous spot exposures

Any dot or etch pit experiment can be easily placed along the $\tilde{\varphi}$-axis of the universal graph in Figure 6.21b, provided the beam size $FWHM_B$ is known. According to Figure 6.21b, the

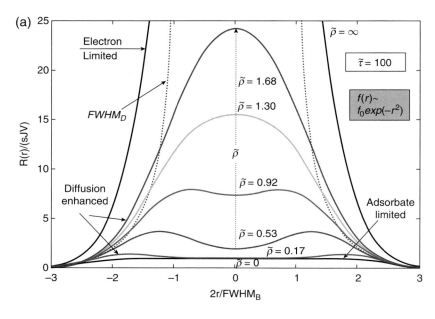

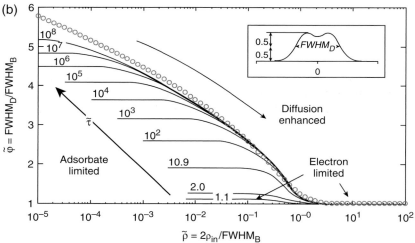

FIGURE 6.21: (a) Normalized steady-state chemical deposition (etch) rate representing the deposit (etch pit) shape at an irradiative depletion $\tilde{\tau} = 100$ and varying diffusion paths $\tilde{\rho}_{in}$. There is a shape transition from flat top: $\tilde{\rho}_{in} = 0$, indented: $\tilde{\rho}_{in} = 0.17$, round: $\tilde{\rho}_{in} = 1.3$, to Gaussian: $\tilde{\rho}_{in} = \infty$. The dotted line represents the deposit's (etch pit's) full width at half maximum ($FWHM_D$). Note the increase in deposition (etch) rate with increasing diffusive replenishment $\tilde{\rho}_{in}$. Reprinted with permission from [7]. Copyright 2008, American Institute of Physics. (b) Normalized deposit size $\tilde{\varphi}$ vs. normalized diffusion path $\tilde{\rho}_{in}$ for varying irradiative depletion $\tilde{\tau}$ (indicated). At $\tilde{\rho}_{in} = 2$ the diffusion path at the beam center equals the beam's full width at half maximum, $\rho_{in} = FWHM_B$. Circles represent the scaling law in Equation 23. The inset shows the $FWHM_D$ definition of indented deposits (etch pits). Reprinted with permission from [17]. Copyright 2007, American Institute of Physics. Note that for gas assisted FIB physical sputtering may substantially vary both graphs according to the ratio of sputter versus chemical rate. Here $R_S = 0$.

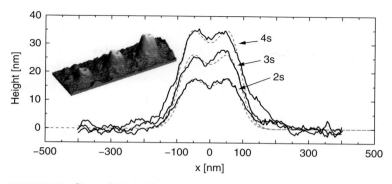

FIGURE 6.22: Shape fitting with continuum model. AFM image and line scans of FEB deposits from Cu(hfac)$_2$ precursor. Exposure times are indicated. Fits (dashed lines) were obtained with $\sigma = 0.6$ nm^2 (= adsorbate size), $\tau = 10^{-3}$ s (typical value), and $D = 4 \times 10^{-7}$ cm^{2}s^{-1}. Other deposition parameters: Gaussian beam with $FWHM = 110$ nm (5 keV) and $f_0 = 9 \times 10^4$ nm^{-2}s^{-1}. Precursor supply with $sJ/n_0 = 10$ ML/s. Reprinted with permission from [17]. Copyright 2007, American Institute of Physics.

same $FWHM$-ratio $\tilde{\varphi}$ can be achieved for several sets of irradiative depletion and diffusive replenishment $(\tilde{\tau}, \tilde{\rho}_{in})$ or, in other words, with several sets of the cross-section σ and surface diffusion coefficient D. For an exact calculation, we need the remaining parameters entering into $\tilde{\tau}$ and $\tilde{\rho}_{in}$ to be known: the molecule supply rate sJ/N_0 from simulations or measurements (see [19]), the maximum electron (ion) flux f_0 at the beam center, and the residence time τ from raster scans. Furthermore, the unique determination of the (σ, D)-set would require another experiment with a change in beam size or electron flux. The actual $\tilde{\tau}$-line in Figure 6.21b of a single continuous spot experiment can be roughly estimated by relying on a typical value of $\tau = 1$ ms and on the geometrical cross section as an upper bound for the electron impact dissociation cross-section σ. For ion-impact dissociation cross sections the geometrical cross-section can be regarded very roughly as a *lower* bound; see [7]. From these two assumptions, a rough idea of the surface diffusion coefficient can be obtained from Figure 6.21b. When topographic information of the indented shapes as shown in Figure 6.22 is available, further constraints to fit the parameters arise.

There are few published values of surface diffusion coefficients; see [7] for an overview. With the rediscovery of the importance of surface diffusion in the fabrication of small nanostructures, more data is definitely needed.

2.2.8 Time-dependent solutions for pulsed irradiation

As discussed in the related Section 2.1.4, short beam pulses result in less adsorbate depletion in the irradiated region. Taking into account surface diffusion, an additional increase in adsorbate density is obtained, equivalent to an increased effective residence time. A rough analytic estimate of the residence time at the beam center due to diffusion derived from the steady state solution of the flat top beam distribution is [7]:

$$\tau_{eff} = \tau_{in} + \tau_{diff}, \qquad \tau_{diff} \approx \tau_{out} \cdot C_{diff} \tag{24}$$

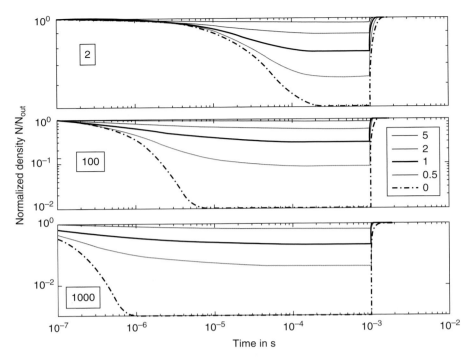

FIGURE 6.23: Time-dependent normalized adsorbate density at the beam center $r = 0$ for one on-off electron (ion) beam pulse obtained from solving Equation 2 for a Gaussian beam distribution. The dwell time period was set to $t_d = 1$ms to reach varying steady state adsorbate depletions $\tilde{\tau} = 2, 100, 1000$ (from top to bottom). The legend gives the values for $\tilde{\rho}_{in}$, see Equation 21a; $\tilde{\rho}_{in} = 0$ represents the analytical solution for zero surface diffusion from Equation 16 (dashed line). The actual diffusion coefficient scales with $\sqrt{\tilde{\tau}}$ for a given $\tilde{\rho}_{in}$. Note the varying ordinate axis range given by the irradiative depletion $1/\tilde{\tau}$. The time constants (effective residence times, see Equation 9) were chosen to be $\tau_{out} = 90\,\mu$s and $\tau_{in} = 45\,\mu$s, 900 ns, and 90 ns. They determine, in case of zero surface diffusion, the time scales for refreshment and irradiative depletion. Increasing surface diffusion has two effects: adsorbate depletion decreases considerably in value and time, and adsorbate refreshment proceeds more rapidly. Note that the steady state adsorbate coverage during irradiation is obtained more rapidly with increasing irradiative depletion.

where C_{diff} was introduced in eqn. 20b and τ_{out} in Equation 9b. The higher adsorbate coverage, i.e., lower depletion at the beam center, can be seen in the exact time dependent solutions of Equation 2 for a Gaussian beam distribution shown in Figure 6.23.

Another feature of Figure 6.23 is that steady state adsorbate coverage obtained by surface diffusion needs more time to be established during the exposure (dwell) cycle. This is attributed to the fact that surface diffusion is driven by the concentration gradient, which will be generated by irradiation with time. Refreshment during the non-exposure cycle is, however, achieved fastest with surface diffusion as compared to pure gas phase replenishment.

The deposit (pit) shapes resulting for pulsed exposures considering surface diffusion will obviously depend on the amount of material transport that can be achieved via surface diffusion. The shapes for continuous exposures were discussed in Section 2.2.5 and showed a pronounced dependence on the irradiative depletion $\tilde{\tau}$ and diffusive replenishment $\tilde{\rho}$ see Figure 6.21.

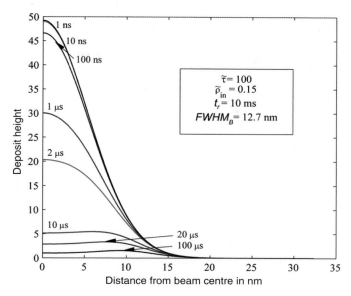

FIGURE 6.24: Deposit shape variation with varying dwell times. The irradiative depletion and diffusive replenishment were chosen to be $\tilde{\tau} = 100$ and $\tilde{\rho} = 0.15$ in order to obtain an indented shape for long dwell time exposures with a Gaussian beam of $FWHM_B = 12.7$ nm. The refresh time $t_r = 10$ ms during the exposure pulses was kept constant. The dwell time was varied from $t_d = 1$ ns to 100 μs. The total exposure time was kept constant at 100 μs. Note that small dwell times improve the lateral resolution of the deposit.

When now changing from a continuous exposure to a pulsed exposure, all the deposit (pit) shapes in Figure 6.21 will evolve to a Gaussian shape with decreasing dwell time (and a refresh time of about 10 ms), or, more generally to the shape of the distribution governing the chemical dissociation reaction. This is shown in Figure 6.24, where an indented shape deposit obtained for "continuous" exposure evolves through a round shape to a Gaussian shape with decreasing dwell time.

Except for one paper of Haraichi and Komuro [37], there is no report on implementing surface diffusion into simulations of raster scan experiments. This is probably due to the large computational efforts needed to solve the appropriate differential equations for each raster pixel.

2.3 CONDITIONS FOR THE ELECTRON OR ION LIMITED REGIME

In the previous sections we showed that this process regime, also referred to as the reaction limited regime, gives the highest resolution as well as the highest deposition or etch rate of the gas-assisted FEB/FIB process. This is why we here derive the conditions necessary for this regime. We start with a compilation of maximum and minimum values of FIB and FEB deposition cross-sections found in the literature, together with typical beam sizes and charged particle fluxes of field emitter SEMs and Ga-ion FIB columns shown in Table 6.1.

Table 6.1 Calculated minimum diffusion coefficients D and minimum exposure times t_d for the electron or ion limited regime. Typical ranges of incident peak flux f_0 and beam size $FWHM_B$ are taken for a focused electron beam (5 keV, field emission filament) and an ion beam (30 kV, Ga). Reprinted with permission from [7]. Copyright 2008, American Institute of Physics.

	f_0 (1/nm²s)	$FWHM_B$ (nm)	σ (nm²)	σf_0 (1/s)	D (cm²/s)	t_d (s)
FEB	$8 \cdot 10^6$	2.5	$2 \cdot 10^{-4}$	$1.6 \cdot 10^3$	$1 \cdot 10^{-10}$	$1.6 \cdot 10^{-4}$
	$5 \cdot 10^7$	100	0.2	$1 \cdot 10^7$	$1 \cdot 10^{-3}$	$2.6 \cdot 10^{-8}$
FIB	$2 \cdot 10^5$	7	10	$2 \cdot 10^6$	$9.8 \cdot 10^{-7}$	$1.3 \cdot 10^{-7}$
	$5 \cdot 10^6$	100	50	$2.5 \cdot 10^8$	$2.5 \cdot 10^{-2}$	$1.0 \cdot 10^{-9}$

According to Table 6.1, dissociation rates range between $\sigma f_0 = 10^3 - 10^8$ s^{-1}. In comparison, typical molecule supply rates from a gas injection system onto the substrate are in the range of $sJ/N_0 = 1 - 10^4$ s^{-1}. Typical desorption rates at room temperature are situated mostly within the kHz range, $1/\tau = 10^3$ s^{-1}. From the above parameter range, the irradiative depletion is found to vary approximately within $\tilde{\tau} = 1.01 - 10^8$.

For most typical cases, the adsorbate dissociation frequency σf_0 determines the effective residence time inside the irradiated area. Since $\tau_{in} = (k_d)^{-1} \approx (\sigma f_0)^{-1}$, we can readily estimate ranges of diffusion coefficients and exposure times needed for establishing the electron/ion-limited regime.

Compensation of depletion by surface diffusion requires an adsorbate diffusion path at least comparable to the beam size, $\rho_{in} \geq FWHM_B$. With $\rho_{in} = (D\tau_{in})^{1/2} \approx (D/\sigma f_0)^{1/2}$ we obtain the condition for the diffusion coefficient $D \geq \sigma f_0 \cdot FWHM_B^2$ to satisfy the electron or ion limited regime; for the corresponding values see Table 6.1. Vice versa, one can also start from a known surface diffusion coefficient for a given adsorbate and estimate the beam $FWHM_B$ needed to enter into the diffusion-enhanced regime and finally into the electron (ion) limited regime.

When the condition $\rho_{in} \geq FWHM_B$ cannot be met in a continuous exposure experiment, pulsed beams can be employed to operate in the electron (ion) limited regime. The exposure (dwell) time range can be estimated from the proportionality $\tau_{in} \sim \tau_{out} \cdot exp(-k_d t) \approx \tau_{out} \cdot exp(-\sigma f_0 t)$. Defining the criterion that the deposit or etch size should not exceed the beam size by 10%, i.e., $\tilde{\varphi} \leq 1.1$, translates into a low irradiative depletion $\tilde{\tau} = \tau_{out}/\tau_{in} \leq 1.3$ (from Equation 19) to be respected. Thus for the exposure time t_d the condition $t_d \leq \tau_{in} \cdot \ln(\tilde{\tau}) = (\sigma f_0)^{-1} \cdot 0.26$ is obtained to satisfy the electron or ion limited regime; for values, see Table 6.1. If larger size tolerances are accepted, the dwell times can be somewhat longer. It is interesting to calculate the number of primary electrons (ions) for the lowest exposure time of Table 6.1. Consider a beam current of $I_P = 1000$ pA, within a pulse $t_d = 1$ ns the number of charged particles incident on the surface becomes $I_P \cdot t_d/e_0 \approx 2$. This means that one reaches the limits of "shot noise" where statistical fluctuations of the number of charged particles during the exposure occur.

3 CONCLUSIONS AND OUTLOOK

The continuum model for a one-adsorbate species discussed in this chapter served to introduce important concepts that govern the deposition (etch) rate and resolution of gas-assisted FEB (FIB) processing. The first conceptual term is *irradiative depletion*, being the ratio of dissociated adsorbates and intact adsorbates continuously supplied from the gas phase. The second conceptual term is *diffusive replenishment*, being the ratio of the adsorbate diffusion path inside the irradiated area and the focused electron (ion) beam size(FWHM). Continuum models allow the derivation of analytical scaling laws for lateral resolution and rates thus providing insight into the gas assisted FEB (FIB) processes. Especially the electron- (ion-) limited, adsorbate-limited, and surface-diffusion regimes can be easily achieved, and the role of the related engineering parameters (beam current, beam size, molecule adsorption, exposure time, etc.) illustrated.

All the scaling laws derived above predict that the best lateral resolution is obtained in the electron- (ion-) limited regime, where there are always enough intact adsorbates available for dissociation. We would like to point out that the adsorbate-limited regime also can give high resolution. Indeed, when only one adsorbate is present in the irradiated region, the resolution will forcedly be "molecular"; and if the dissociated adsorbate can assume a catalytic activity, even small three-dimensional structures can be fabricated at molecular scale just by molecule supply. With respect to modeling it would be very desirable to compare the continuum approach with MC simulations to prove its applicability on curved surfaces (high aspect ratio structures), or to identify potential deviations caused by the energy dependence of dissociation cross-sections and by emitted electron fluxes and excited surface atom distributions. On the other hand, continuum modeling can provide reliable parameters of adsorbate kinetics, like the residence time, sticking probability, and surface diffusion coefficient, as well as total electron impact cross-sections to Monte Carlo models. In a recent paper from Fowlkes and Rack [38] a time-dependent continuum model accounting for an evolving nanopillar morphology (curved surface) under pulsed irradiation was established. The numerical solution allowed the surface diffusion coefficient, residence time, and sticking probability to be estimated from FEB experiments with $W(CO)_6$. Interestingly, for steady-state conditions a convergence of the Fowlkes-Rack model with the scaling laws of Utke et al. [17] (see Equations 14 and 23) was found, which suggests that the theoretical framework, or at least the grand principles, presented in this chapter also apply for curved apexes of pillars. Great progress in the fundamental understanding as well as in the process control of gas assisted FEB and FIB processing can be expected in future using continuum models which involve further experimental details, such as the raster scan strategy or background gases [39].

REFERENCES

[1] Christy, R. W. Formation of thin polymer films by electron bombardment. *Journal of Applied Physics* 31.9 (1960): 1680–1683.

[2] Scheuer, V., H. Koops, et al. (1986). Electron beam decomposition of carbonyls on silicon. *Microelectronic Engineering* 5.1–4 (1986): 423–430.

[3] Li, W., and D. C. Joy. Study of temperature influence on electron beam induced deposition. *Journal of Vacuum Science & Technology* A 24,3 (2006): 431–436.

[4] Petzold, H. C., and P. J. Heard. Ion-induced deposition for x-ray mask repair: Rate optimization using a time-dependent model. *Journal of Vacuum Science & Technology* B 9.5 (1991): 2664–2669.

[5] Edinger, K., and T. Kraus. Modeling of focused ion beam induced chemistry and comparison with experimental data. *Microelectronic Engineering* 57–58 (2001): 263–268.

[6] Haller, I., and P. White. Polymerization of butadiene gas on surfaces under low energy electron bombardment. *Journal of Physical Chemistry* 67.9 (1963): 1784–1788.

[7] Utke, I., P. Hoffmann, et al. Review: Gas-assisted focused electron beam and ion beam processing and fabrication. *Journal of Vacuum Science & Technology* B 26.4 (2008): 1197–1276.

[8] Xu, Z., K. Gamo, et al. Ion-beam assisted etching of SiO2 and Si3N4. *Journal of Vacuum Science & Technology* B 6.3 (1988): 1039–1042.

[9] Dubner, A. D., A. Wagner, et al. The role of the ion-solid interaction in ion-beam-induced deposition of gold. *Journal of Applied Physics* 70.2 (1991): 665–673.

[10] Ro, J. S., C. V. Thompson, et al. Mechanism of ion-beam-induced deposition of gold. *Journal of Vacuum Science & Technology* B 12.1 (1994): 73–77.

[11] Ibach, H. *Physics of surfaces and interfaces.* Berlin, Springer Verlag, 2006.

[12] Dubner, A. D., and A. Wagner. The role of gas-adsorption in ion-beam-induced deposition of gold. *Journal of Applied Physics* 66.2 (1989): 870–874.

[13] Mezhenny, S., I. Lyubinetsky, et al. Electron stimulated decomposition of adsorbed hexafluoroacetylacetonate Cu(I) vinyltrimethylsilane, Cu(I)(hfac)(vtms). *Journal of Applied Physics* 85.6 (1999): 3368–3373.

[14] Lassiter, M. G., and P. D. Rack. Nanoscale electron beam induced etching: A continuum model that correlates the etch profile to the experimental parameters. *Nanotechnology* 19.45 (2008): 455306.

[15] Humphreys, C. J., T. J. Bullough, et al. Electron-beam nano-etching in oxides, fluorides, metals and semiconductors. *Scanning Microscopy* (1990): 185–192.

[16] Ennos, A. E. The sources of electron-induced contamination in kinetic vacuum systems. *British Journal of Applied Physics* 5 (1954): 27–31.

[17] Utke, I., V. Friedli, et al. Resolution in focused electron and ion beam induced processing. *Journal of Vacuum Science & Technology* B 25.6 (2007): 2219–2223.

[18] Babin, S., M. Gaevski, et al. Technique to automatically measure electron-beam diameter and astigmatism: BEAMETR. *Journal of Vacuum Science & Technology* B 24.6 (2006): 2956–2959.

[19] Friedli, V., and I. Utke. Optimized molecule supply from nozzle-based gas injection systems for focused electron- and ion-beam induced deposition and etching: Simulation and experiment. *Journal of Physics D: Applied Physics* 42.12 (2009): 125305 (11pp).

[20] Friedli, V., C. Santschi, et al. Mass sensor for in situ monitoring of focused ion and electron beam induced processes. *Applied Physics Letters* 90.5 (2007): 053106 (4pp).

[21] Kunze, D., O. Peters, et al. Polymerisation adsorbierter Kohlenwasserstoffe bei Elektronenbeschuss. *Zeitschrift für angewandte Physik* 22.2 (1967): 69–75.

[22] Blauner, P. G., J. S. Ro, et al. Focused Ion-Beam Fabrication of Sub-Micron Gold Structures. *Journal of Vacuum Science & Technology* B 7.4 (1989): 609–617.

[23] Fritzsche, C. R. Time constants and reaction cross sections characterizing film deposition by beam-induced polymerization. *Journal of Applied Physics* 60.6 (1986): 2182–2184.

[24] Rudenauer, F. G., W. Steiger, et al. Localized ion-beam induced deposition of Al-containing layers. *Journal of Vacuum Science & Technology* B 6.5 (1988): 1542–1547.

[25] Harriott, L. R. Digital scan model for focused ion-beam-induced gas etching. *Journal of Vacuum Science & Technology* B 11.6 (1993): 2012–2015.

[26] Lipp, S., L. Frey, et al. Tetramethoxysilane as a precursor for focused ion beam and electron beam assisted insulator (SiOx) deposition. *Journal of Vacuum Science & Technology* B 14.6 (1996): 3920–3923.

[27] Edinger, K., and T. Kraus. Modeling of focused ion beam induced surface chemistry. *Journal of Vacuum Science & Technology* B 18.6 (2000): 3190–3193.

[28] Santschi, C., M. Jenke, et al. Interdigitated 50 nm Ti electrode arrays fabricated using XeF2 enhanced focused ion beam etching. *Nanotechnology* 17.11 (2006): 2722–2729.

[29] Chen, P., M. Y. Wu, et al. Fast single-step fabrication of nanopores. *Nanotechnology* 20.1 (2009): 015302 (6pp).

[30] Hoyle, P. C., J. R. A. Cleaver, et al. Ultralow-energy focused electron beam induced deposition. *Applied Physics Letters* 64.11 (1994): 1448–1450.

[31] Zhdanov, G. S., and V. N. Vertsner. On the mechanism of formation of hydrocarbon contaminants on electron-bombarded objects. *Dokladi Akademii Nauk SSSR* 176.5 (1967): 1040–1043.

[32] Muller, K. H. Elektronen-Mikroschreiber mit geschwindigkeitsgesteuerter Strahlführung. 1. *Optik* 33.3 (1971): 296–311.

[33] Reimer, L., and M. Wachter. Contribution to contamination problem in transmission electron-microscopy. *Ultramicroscopy* 3.2 (1978): 169–174.

[34] Aristov, V. V., N. A. Kislov, et al. Electron induced formation of carbon-containing nanometric structures in a transmission raster electron microscope. Nature of long-range growth. *Izvestiya Akademii Nauk SSSR Seriya Fizicheskaya* 55.8 (1991): 1523–1529.

[35] Kislov, N. A. Direct STEM fabrication and characterization of self-supporting carbon structures for nanoelectronics. *Scanning* 15.4 (1993): 212–218.

[36] Hirsch, P., M. Kassens, et al. Contamination in a scanning electron-microscope and the influence of specimen cooling. *Scanning* 16.2 (1994): 101–110.

[37] Haraichi, S., and M. Komuro. Broad-pulsed Ga ion-beam-assisted etching of Si with Cl2. *Japanese Journal of Applied Physics Part 1 – Regular Papers Short Notes & Review Papers* 32.12B (1993): 6168–6172.

[38] Fowlkes, J. D., and P. D. Rack. Fundamental Electron-Precursor-Solid Interactions Derived from Time-Dependent Electron-Beam-Induced Deposition Simulations and Experiments. *ACS Nano 4.3* (2010): 1619–1629.

[39] Bernau, L., M. Gabureac, R. Erni, and I. Utke. Tunable Nanosynthesis of Composite Materials by Electron-Impact Reaction. *Angew. Chem. Int. Ed.* 49 (2010): 8880–8884.

C H A P T E R   7

CONTINUUM MODELING OF ELECTRON BEAM INDUCED PROCESSES

Charlene J. Lobo and Milos Toth

1 INTRODUCTION

Gas-mediated focused electron beam induced etching (FEBIE) and deposition (FEBID), collectively referred to here as FEBIED, permit nanoscale modification of surface material via chemical reactions involving electron-dissociated precursor molecules. Electrons crossing the solid-vacuum interface usually possess a wide range of energies, are capable of breaking most bonds in typical precursor adsorbates and dissociation products, and can therefore generate a wide range of mobile, chemically active species. Adsorption, desorption, diffusion, and dissociation of these species all contribute to the development of nanostructures fabricated by FEBIED processes. Furthermore, these nanostructures are often electron-sensitive, and thus their structure evolves during deposition. The wide range of processes behind FEBIED yields very complex behavior that is yet to be modeled realistically fifty years after Christy first proposed a simple analytical model of deposition induced by a broad (defocused) electron beam [1]. A complete description of FEBIED requires a realistic model of electron-gas and electron-solid interactions, the spatial and energy distributions of secondary and backscattered electrons, electron interactions with adsorbates, and the behavior of adsorbates and dissociation products at the solid-vacuum interface.

Models published to date have not attempted to realistically simulate the nanostructure of materials grown by FEBIED. Instead, the models have focused on simulation of process rates and the geometries and sizes of deposited or etched features. Two types of modeling approaches are in use. The so-called continuum models of FEBIED [1–7] are comprised of differential equations for the rates of change of concentrations of all surface-adsorbed species thought to be involved in the deposition or etching process. The rate equations are functions of time and space and require specification of the molecular properties of each adsorbate, and the electron flux profile(s) at the solid-vacuum interface. Simple continuum models of FEBID and FEBIE can be solved analytically, yielding expressions that can be converted into laws that govern different process regimes. More complex processes, such as simultaneous FEBIED, require numerical solution of the rate equations. The numerical approach can cope with all levels of complexity encountered in FEBIED but is limited by knowledge of the applicable reaction pathways, properties of the molecular species considered by a given model, and difficulties associated with modeling the time-evolution of electron energy and flux distributions during FEBIED. Monte Carlo modeling is an alternate approach, in which electron-solid interactions and electron interactions with precursor adsorbates are simulated explicitly [8–15]. The most recent hybrid models also account for adsorbate transport at the solid-vacuum interface [16; 17]. In contrast to the continuum approach, the Monte Carlo method automatically accounts for the time-evolution of the electron flux profile at the solid-vacuum interface. However, the extent of this advantage is limited by complications associated with modeling low energy electron transport in solids (see section 2.4). In addition, the Monte Carlo approach with adsorbate transport is extremely calculation-intensive and limited to the simulation of small surface areas and relatively short FEBIED process times (impeding realistic modeling of surface diffusion and interactions of high energy electrons with bulk solids).

All existing models of FEBIED are limited by the problems highlighted in this chapter and incomplete knowledge of the applicable reaction pathways and properties of the molecular species considered by a given model. Each modeling approach has, however, provided some valuable insights into FEBIED. This chapter will focus on continuum models that are solved numerically, and use Monte Carlo simulations of electron-solid interactions only to generate some model input parameters.

2 ELECTRON-SOLID INTERACTIONS

Here we summarize aspects of electron-solid interactions that determine the electron-related input parameters required by continuum FEBIED models. Specifically, we focus on the flux and energy distributions at a solid-vacuum interface as these affect local deposition, etch, and adsorbate depletion rates. Electron flux profiles must be specified explicitly in spatially resolved models of FEBIED. Electron energy distributions play a role through the energy dependencies of cross-sections for electron-adsorbate interactions.

Most of the present discussion is illustrated by Monte Carlo data, which provide a good compromise between accuracy and convenience and are in a form appropriate for numerical FEBIED modeling. We limit the discussion to the case of stationary electron beams. Pulsed beams can be modeled by time-dependent electron flux profiles, and scanned beams require the use of Cartesian coordinates.

Discussions of the angular distribution of electrons emitted from a solid can be found in [18–20] and references therein, and are omitted here as they are neglected by existing continuum models of FEBIED. Furthermore, the discussion is limited to the case of a flat, conductive, bulk sample and a normal incidence stationary electron beam with a landing energy accessible by scanning electron microscopy (SEM). Time-evolution of the electron energy and flux distributions is discussed briefly in Section 2.5. Detailed discussions of charging in SEM can be found in references [21–28]. Transmission electron microscopy (TEM) is not discussed, but can be analyzed by the approach presented here by considering thin, electron transparent substrates and relativistic cross-sections for electron-solid interactions.

For completeness, we note that electron-solid interactions can cause changes in the nanostructure, density, and volume of materials grown by FEBID [29]. Such effects can alter the density, shapes, and sizes of the fabricated nanostructures, but have not been incorporated in existing FEBID models.

2.1 PRIMARY ELECTRONS

The primary electron flux profile at the final crossover of a focused SEM beam can typically be approximated by a Gaussian function [18]:

$$f(r) \propto Exp \left(\frac{-r}{r_0} \right)^2, \tag{2.1}$$

where r is radial distance from the beam axis and r_0 is the beam radius. The flux above and below the crossover can often be described by a top-hat (sometimes also referred to as a "flat-top" or "uniform") function (see, for example, the discussions in references [30, 31]):

$$f(r) \propto 1 / \left(Exp \left(\beta \left(\frac{r}{2r_0} - 1 \right) \right) + 1 \right), \tag{2.2}$$

where β determines the abruptness of the top-hat. The flux profile must be scaled to yield (upon integration over the solid-vacuum interface) the primary electron current I_P.

On most modern electron microscopes, FEBIED can be performed using a Gaussian or a top-hat electron beam over a wide range of beam currents simply by focusing or defocusing the beam in the plane of the substrate. Examples of deposits grown by each type of electron beam are shown in Figure 7.1.

Focused beams are needed to optimize process resolution. Defocused beams yield lower process resolution, but offer several advantages to fundamental studies: the two dimensional electron flux profile is very easy to measure, it can be changed in a well-controlled, quantitative manner during the course of systematic experiments, and it is possible to perform electron beam induced etching or deposition (EBIE or EBID) over a wide range of length scales and electron fluxes using a stationary electron beam.[a]

Knowledge of the electron flux profile $f(r)$ is a prerequisite to spatially resolved FEBIED modeling. Another parameter that must be known is the primary electron energy at the solid-vacuum interface (the so-called "landing energy," E_L, which is given by the accelerating voltage and the substrate surface potential). At the beam energies typically used for FEBIED, primary electron energy spread is negligible and E_L is a well-known, single-valued parameter [18].

2.2 ELECTRON-SOLID SCATTERING

Electron trajectories inside a solid and the shape and size of the electron-solid interaction volume are determined by elastic scattering with nuclei and inelastic scattering in the solid.

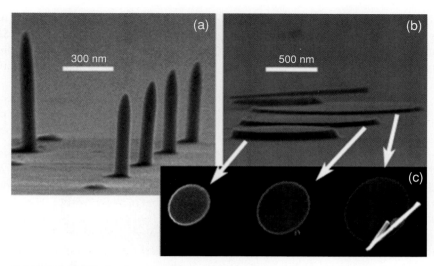

FIGURE 7.1: EBID structures fabricated using (a) focused Gaussian and (b, c) defocused top-hat electron beams. Images (a) and (b) were taken using a sample tilt of $\sim 85°$, and (c) is a top-down image of the deposits shown in (b).

Elastic scattering is characterized by negligible energy losses and high scatter angles, giving rise to straggle, which affects the shape and size of the interaction volume. Conversely, inelastic scattering is characterized by small scatter angles and causes energy transfer to the solid. These characteristics are exploited by the so-called "continuous slowing down approximation" Monte Carlo models [32, 33] which simulate elastic scattering explicitly and assume that electrons travel in straight lines and lose energy continuously between elastic scattering events. These simplifications enable the use of cross-sections and stopping powers that are functions only of the electron energy and the elemental composition and density of the solid. The models are efficient, require simple input parameters, and yield accurate electron penetration ranges, energy loss profiles, and backscattered electron energy and flux distributions. The models cannot simulate secondary electron yields or energy distributions directly, but can be used to approximate secondary electron flux profiles (see Section 2.4). The data that will be presented here was generated using the Monte Carlo program CASINO [33–35], which employs the continuous slowing down approximation.

The shape and size of the electron-solid interaction volume affect the secondary and backscattered electron flux profiles at the solid-vacuum interface. At electron beam landing energies in the range of 1 to 30 keV, the interaction volume in a bulk solid is approximately hemispherical, and the electron penetration range can be approximated by the Kanaya-Okayama equation [36]:

$$R_{KO} = \frac{27.6 A E_L^{5/3}}{\rho Z^{8/9}}, \tag{2.3}$$

where A is atomic weight (g/mol), E_L is landing energy (keV), ρ is density (g/cm^3), and Z is atomic number. Equation 2.3 provides a useful approximation to the scaling laws that govern the dependence of penetration range and interaction volume on the electron beam landing energy and basic properties of the solid. More accurate representations can be obtained by

Monte Carlo simulations. For example, Figure 7.2 shows the depth distribution of primary electron ranges in Au simulated for landing energies of 1, 5, 10, 20, and 30 keV, and in C at 1 and 30 keV.[b] The inset shows a projection of the trajectories of 200 primary electrons in Au simulated at 10 keV.

2.3 BACKSCATTERED ELECTRONS

By convention, backscattered electrons [18–20] are those emitted with energies greater than 50 eV, and are characterized by their energy distribution, shown in Figure 7.3 for Au, Mo, Cr, Al and C at a beam energy of 1 keV, and Au at 30 keV. A backscattered electron (BSE) spectrum contains a broad maximum at an energy E/E_L that increases with Z. The high energy tail extends up to a narrow elastically scattered peak at E_L (not resolved in the spectra shown in Figure 7.3), and the low energy tail extends down to and overlaps with the SE spectrum.

Backscattered electrons are emitted from approximately the top half of the interaction volume. This is illustrated in Figure 7.4 by the depth distribution of backscattered electron ranges, $\partial\eta/\partial(z/R_{99})$, in Au and C calculated for landing energies of 1 and 30 keV (R_{99} is the primary electron penetration range, defined here as the depth at which the kinetic energy of 99% of the primary electrons has fallen below 50 eV). For comparison, Figure 7.4 also shows the corresponding distributions of primary electron ranges, $\partial(1-\eta)/\partial(z/R_{99})$. The plots show that the backscattered electron penetration range is approximately equal to $0.5R_{99}$ for both low and high atomic numbers and primary electron landing energies.

We now turn to the radial distribution of backscattered electrons at the solid-vacuum interface, shown in Figure 7.5 for the case of Au and an electron beam landing energy

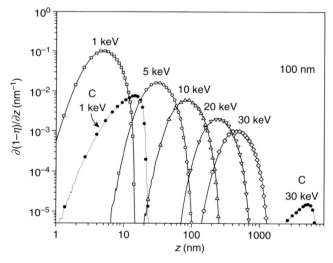

FIGURE 7.2: Depth distribution of primary electron ranges in Au calculated for landing energies of 1, 5, 10, 20, and 30 keV (open symbols), and in C at 1 and 30 keV. Inset: Trajectories of 200 primary electrons in Au simulated at 10 keV (black and grey colors show electrons that were backscattered into vacuum and those that stopped in the sample, respectively).

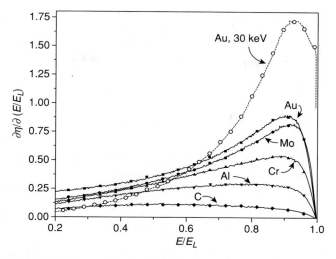

FIGURE 7.3: Energy spectra of backscattered electrons calculated for Au, Mo, Cr, Al, and C at a beam energy of 1 keV, and for Au at 30 keV.

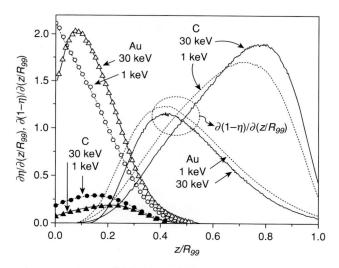

FIGURE 7.4: Depth distribution of backscattered electron ranges, $\partial\eta/\partial(z/R_{99})$, in Au and C calculated for landing energies of 1 and 30 keV. Also shown are the corresponding primary electron ranges, $\partial(1-\eta)/\partial(z/R_{99})$.

of 10 keV. The radial profile is expressed as the electron flux $f(r)$ corresponding to a backscattered electron current (ηI_p) of 100 pA. The flux profile has units of electrons (or charge) per unit area per unit time (as required by most continuum models of FEBIED), and can be scaled to yield any backscattered electron current.

Figure 7.5 illustrates that BSE flux profiles do not fall off to zero at a well-defined value of r. They must therefore extend to large values of r where the Monte Carlo data is noisy.[c]

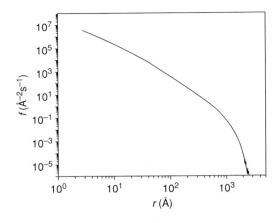

FIGURE 7.5: Backscattered electron flux profile, $f(r)$, calculated for Au irradiated by a Gaussian 10 keV electron beam with a full width at half maximum of 10 nm (backscattered electron current = 100 pA).

Hence, a flux profile calculated by Monte Carlo simulation is typically fitted by an analytic function that approaches zero at large values of r, and the analytic function is then used as an FEBIED model input parameter.

2.4 SECONDARY ELECTRONS

Secondary electrons [18–20] are, by convention, those emitted with energies below 50 eV. SE generation is usually described as a three-step process: (1) electron excitation along primary and backscattered electron trajectories, (2) transport of the excited electrons to the solid-vacuum interface, and (3) escape into vacuum. Excitation is relatively easy to simulate because, to a first approximation, the excitation rate is proportional to the primary (or backscattered) electron Bethe stopping power. More precisely, the excitation rate at each point in the solid is proportional to the power density (eV/Å^3/s) deposited by primary and backscattered electrons into the solid, and can be calculated directly by Monte Carlo methods. However, modeling of the transport and escape of low energy electrons is problematic for two reasons. First, the applicable scattering cross-sections are not known for most materials of interest. Second, the relevant surface barriers for electron escape into vacuum are also not well-known because they are strong functions of surface chemistry, such as the presence of thin oxide and carbon-rich films, and contaminant adsorbates encountered in typical FEBIED process chambers. These problems are compounded further by the complex structure of the nanocomposites typically grown by FEBID. Consequently, absolute SE yields and energy distributions used in FEBIED models are at best approximate, and are among the input parameters that limit the accuracy of existing models.

The secondary electron flux profile is, however, relatively easy to model . The probability (p) that an electron excited at depth z will diffuse to the surface with enough energy to overcome the surface barrier can be approximated by the empirical attenuation function $p = Be^{-z/s}$, where B is the electron escape probability and s is the mean SE escape depth. In most common materials, s is on the order of 1 to 10 nm [28; 37–40]. Most electrons excited at depths greater than s cannot reach the interface with enough energy to overcome the surface barrier, and the emission rate at the radial coordinate r can be assumed proportional to the corresponding power density deposited by primary and backscattered electrons into the top few nanometers of the solid. Hence, the secondary electron flux profile can be approximated by the sum of (1) the primary electron flux profile (typically a Gaussian or a top-hat function such as those discussed in Section 2.1) scaled by δ_I/δ_{II} and (2) the backscattered electron flux profile (see, for example, Figure 7.5) scaled by δ_{II}/δ_I; where δ_I and δ_{II} are the emission yields of secondary electrons excited by primary and backscattered electrons [18–20], respectively. The net SE flux profile must then be scaled such that its integral over the solid-vacuum interface equals the total SE current δI_p.

2.5 INPUT PARAMETERS TO CONTINUUM FEBIED MODELS

To summarize, in the case of a focused electron beam, the primary electron flux profile can be approximated by a Gaussian function, the BSE flux profile can be calculated using Monte Carlo simulations of electron-solid interactions, and the SE flux profile can be approximated by the primary and backscattered electron flux profiles scaled by δ_I/δ_{II} and δ_{II}/δ_I, respectively. The flux profiles must be scaled to yield (upon integration over the solid-vacuum interface) the primary, backscattered and secondary electron currents I_p, ηI_p and δI_p. The electron beam landing energy is used to select cross-sections for primary electron-adsorbate interactions, and overlap between the cross-sections and electron emission spectra are used to approximate single-valued cross-sections for BSE-adsorbate and SE-adsorbate interactions. BSE yields and spectra can be obtained from Monte Carlo simulations, while SE yields and spectra typically require approximations based on (theoretical or experimental) data from representative materials.

In the case of a defocused beam, the primary electron flux profile can often be represented by a top-hat function, and the corresponding BSE and SE parameters can be obtained as above by using the top-hat function as the primary beam profile in a Monte Carlo simulation of electron-solid interactions.

We note that the above approach neglects spatial dependencies of the electron emission spectra. However, this has a negligible effect on the accuracy of existing FEBIED models.

Another shortcoming of the above approach to specifying electron flux distributions is that it does not account for changes that occur during FEBIED. Deposition and etching by a focused electron beam locally modify the topography of the solid-vacuum interface, thereby altering the backscattered and secondary electron flux profiles. These changes can, in principle, be accounted for by the use of time-dependent flux profiles in continuum models of FEBIED. Changes in flux with surface topography are, however, a complex function of the primary beam landing energy and the composition of the solid, and have only been incorporated in one hybrid Monte Carlo/continuum FEBIED model [17]. In practice, complications caused by the time-evolution of electron flux profiles can be minimized by the use of defocused top-hat beams since they yield structures with a solid-vacuum interface that is flat over the majority of the deposit (see Figures 7.1 and 7.9).

3 ELECTRON-GAS AND ELECTRON-ADSORBATE INTERACTIONS

3.1 ELECTRON-GAS INTERACTIONS

Electrons can interact with both gas-phase and surface-adsorbed precursor molecules. Gas scattering can broaden the primary electron beam, and lead to the generation of chemically active dissociation products. The latter can cause etching or deposition of the substrate, but the processes are far less localized and occur at rates that are low compared to FEBIED arising from electron-adsorbate interactions. Hence, electron-gas interactions are neglected by existing continuum models of FEBIED. However, gas scattering can alter the primary electron flux profile at the solid-vacuum interface when the effective gas density along the beam path in the gas is high. Ideally, the beam gas path length should be much shorter than the primary electron mean free path in the gas. In practice, this condition is not satisfied when using very high precursor flow rates and/or low electron beam energies.

The gas pressure distribution along the beam path is a function of the gas delivery method. Precursors are typically delivered to the process chamber using one of two general approaches. A capillary with an inner diameter of 100–500 μm can be placed within $\sim$ 100 μm of the substrate surface to deliver the gas into a high vacuum SEM chamber [5; 41–44]. In this approach, a typical base pressure prior to gas delivery is approximately 10^{-4} Pa. A region of elevated pressure is created in a volume localized to the beam impact point at the substrate surface, and the chamber background pressure increases to approximately 10^{-3} Pa during gas delivery. The pressure at the beam impact point is not measured directly, but can be calculated from knowledge of the system geometry, gas flow rate into the chamber, and pumping speed of the vacuum pump [42–44]. An alternate approach entails the use of environmental scanning electron microscopy (ESEM) [45–47] where the entire process chamber is filled with a gas, and the high vacuum electron column is isolated from the chamber by pressure limiting apertures and a differential pumping system. ESEM can accommodate gas pressures up to the kPa regime, and has been used for a number of FEBIED studies [2; 29; 48–55]. The pressure at the beam impact point on the substrate surface can be made equal to the background chamber pressure and can therefore be measured directly. An additional advantage of ESEM is electrostatic stabilization of materials that charge under an electron beam, enabling high resolution processing and imaging of charging materials [55; 56].

3.2 ELECTRON-ADSORBATE INTERACTIONS

Continuum models of FEBIED require cross-sections for electron-adsorbate interactions. However, complete cross-section sets are not available over the required energy range (i.e., up to the primary electron landing energy) for most precursors used in FEBIED. Early polymer deposition studies reported values for dissociation cross-sections, σ_d, (or the product of the dissociation cross-section and adsorption time, $\sigma_d \tau$) as a function of temperature, but these leave a large degree of uncertainty about the value of σ_d [1]. Koops et al. [57] estimated the effective cross-section for FEBID of CpPtMe$_3$ using a 20 keV electron beam at 10^{-16} cm^{-2}. Hoyle et al. [58] estimated σ_d for W(CO)$_6$ at a number of energies between 0.06 and 20 keV, and reported a maximum of approximately 3.5×10^{-16} cm^{-2} between 0.1 and 1 keV. The fields of plasma physics and chemistry contain an extensive body of literature on cross-sections for electron interactions with a wide range of molecules in the gas phase. For example, Alman et al. [59] summarized experimental and theoretical cross-sections for 16 hydrocarbons and a range of electron dissociation pathways (the validity of these cross-sections at high energies has, however, been disputed in the FEBID literature [12]). Other molecules characterized in plasma processing and related literature include Cl$_2$, F$_2$, CF$_4$, SF$_6$, C$_2$F$_4$, NF$_3$, CH$_4$, SiH$_4$, Si$_2$H$_6$, CHF$_3$, Si(OCH$_2$CH$_3$)$_4$, H$_2$, and BCl$_3$, [60–82]. These data can be applied to FEBIED with the assumption that cross-sections for adsorbates are approximately the same as those for gas molecules. However, many of the molecules characterized in the plasma literature are

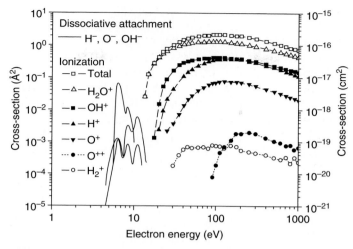

FIGURE 7.6: Cross-sections for the generation of several ions through electron interactions with gas-phase H_2O [98].

not appropriate for FEBIED, and most cross-sections are not available over a wide enough energy range. The field of surface chemistry contains detailed information on the mechanisms behind electron induced dissociation of adsorbates and FEBIED process rates (see, for example, References [83–97]). However, quantitative cross-section data are sparse and often limited to very narrow energy ranges (0 to approximately 20 eV), and the studies are performed under conditions that are not directly representative of the FEBIED processes discussed in this book (e.g., ultra-high vacuum process chambers, ultra-clean oxide-free substrates, low precursor pressures, or precursor films condensed to a cooled substrate).

Figure 7.6 shows several dissociation and ionization cross-sections for gas phase H_2O [98]. H_2O is a rare example of a precursor (used for FEBIE of carbon) for which the cross-sections are known for numerous dissociation pathways over a wide electron energy range. The graph illustrates a general trend in FEBIED: the low energy ($E < 50$ eV) secondary electrons can initiate FEBIED via dissociative attachment while the more energetic backscattered ($E > 50$ eV) and primary electrons contribute primarily through impact ionization.

4 ADSORBATE-SURFACE INTERACTIONS

Regardless of the experimental setup, precursor adsorbates arrive into the electron-irradiated area both from vacuum and by surface diffusion, with the respective arrival rates being determined by the partial pressure of the gas (multiplied by the sticking coefficient) and the surface diffusion rate, respectively. Both adsorption and diffusion have exponential dependencies on temperature, and thus FEBIED process rates are strong functions of substrate temperature. Hence, successful FEBIED modeling requires good estimates for precursor sticking coefficients, diffusion coefficients, and desorption and diffusion energies. Sticking, adsorption, desorption, and diffusion are all complex processes, and we will not attempt to discuss them comprehensively. Excellent reviews of the theoretical aspects and experimental studies of desorption [99] and diffusion [100; 101] can be found elsewhere. In the following, we will focus on processes and systems of most relevance to experimental FEBIED studies, and will outline the adsorption and diffusion behavior of hydrocarbons, metals, and common molecular etch precursors on metal and semiconductor surfaces.

4.1 STICKING

When an atom or molecule with a certain kinetic energy approaches a surface, it can either be scattered or it can be trapped. In the case of a molecule, such trapping can occur along with dissociation (termed *dissociative adsorption* or *dissociative chemisorption*[d]). The *sticking probability* or *sticking coefficient* measures the probability of adsorption, independent of the molecule flux. Sticking coefficients must be studied in a gas flow regime where the flow is determined by interactions of gas molecules with the surface. This is only the case when the characteristic length scale of the volume containing the gas is small compared to the molecular mean free path, a requirement which is generally only met at very low pressures. Measurements of sticking coefficients therefore utilize ultra-high vacuum techniques such as molecular beam scattering with angular and time-of-flight detection, Auger, LEED and helium atom reflectivity. In FEBIED, as in chemical vapour deposition (CVD), the sticking coefficient can affect the overall efficiency of the deposition or etching process. However, because neither FEBIED nor CVD are UHV techniques, determination of sticking coefficient data under appropriate conditions is often challenging.

For a small number of well-studied systems, such as hydrocarbons on metals (of interest in catalysis), halides (of interest in plasma processing) and water, sticking coefficient data is readily available in the literature. At low surface temperatures ($T_s < 200$ K) and low incident molecular translational energies, hydrocarbons physisorb even on reactive metal surfaces such as Pt and Ru. In this case, the sticking coefficient decreases with increasing temperature, as in the case of physisorption of alkanes on Au, a behavior that has been attributed to incomplete accommodation of the molecule by the surface at elevated temperatures [102]. At higher surface temperatures or translational energies, hydrocarbons can chemisorb on transition metal surfaces, a process that may be mediated by a physisorbed precursor. In this case, the energy barrier to direct dissociative chemisorption is overcome by the increased surface temperature or translational energy. Thus, the sticking coefficient can increase with increasing temperature, as in the dissociative chemisorption of methane and ethane on Pt. [103].

Water, a common etch precursor for carbon, sticks efficiently to metal surfaces in a non-activated process, and its sticking coefficient can therefore normally be assumed to be unity [104]. Oxygen chemisorption on catalytic metals such as Pt and Pd proceeds through a physisorbed precursor state and has been found to be coverage-dependent [105; 106]. Sticking coefficient data for halide etch precursors such as Cl_2, Br_2 and I_2 can be obtained fairly readily [107–109].

Sticking coefficients of the more complex organometallic precursors used in many CVD and FEBIED processes are much harder to measure, and there are only a few studies that provide quantitative data. In CVD and atomic layer deposition (ALD), a quantity termed the "reactive sticking coefficient," which is the total probability that a molecule will react after impingement on the surface, can be measured from the film thickness [110] or by laser-induced thermal desorption experiments [111; 112]. Yuuki et al. determined reactive sticking probabilities for SiH_3 and SiH_2 precursors used in the growth of amorphous silicon by using the sticking probabilities as an adjustable parameter in Monte Carlo simulations of cross-sectional profiles of films deposited in trenches [113]. This *trench-coverage analysis* procedure has recently been adapted in order to measure the sticking coefficients of ALD precursors [114]. CVD and ALD precursors are desired to have low reactive sticking coefficients, experiencing many adsorption and re-emission events before reacting, thus leading to homogenous, conformal growth on all exposed surfaces [115]. In FEBIED processes, in contrast, precursors are required to have very high sticking coefficients, in addition to long desorption times, both to ensure high process rates and high resolution. Thus, good CVD and ALD precursors are rarely good FEBID or FEBIE precursors, particularly since FEBIED precursors have an additional requirement of a relatively high vapor pressure near room temperature.

4.2 ADSORPTION AND DESORPTION

Desorption times and energies are normally measured under ultra-high-vacuum conditions using a variety of techniques, including temperature-programmed desorption, X-ray photo-electron spectroscopy, laser-induced thermal desorption, and helium atom reflectivity. The desorption time $\tau = \tau_0 e^{E_{des}/kT}$ (where E_{des} is the desorption energy and τ_0 is the desorption prefactor) is a measure of the mean adsorbate residence time on the surface before returning to vacuum. In FEBIED processes, reversible physisorption (where the adsorbate has a weak, van der Waals interaction with the surface) is generally the desired mode of adsorption. Chemisorption, in which the adsorbate forms a strong chemical bond with the surface, is often undesirable because it is irreversible and can result in delocalized (non-electron-induced) deposition or etching. However, in some cases, dissociative chemisorption is useful. This is the case in the beam-induced etching of SiO_2, Si_3N_4 or SiC with XeF_2 [116]. In these reactions, XeF_2 dissociates to produce adsorbed fluorine, which then reacts with the surface under the electron or ion beam to produce volatile products (predominantly SiF_4, SiF_2, SiF, and O_2 in the case of an SiO_2 surface [117]).

Due to the high level of interest in organic surface chemistry in the fields of catalysis, pharmaceutics, and tribology, to name but a few, extensive data are available on the physisorption of hydrocarbons on metals [118, 119]. Typical desorption energies of short-chain hydrocarbons on metals are of the order of 0.05–0.08 eV (5–8 kJ/mol) per methylene unit, with a "zero-chain-length intercept" of (16–30 kJ/mol) [119]. The desorption prefactor $\tau_0 = \frac{1}{\nu_0}$ is generally assumed to be constant for small molecules at $\tau_0 = 1 \times 10^{-13}$ s, and has been found to decrease with increasing chain length for longer-chain molecules [120; 121].

Adsorption of hydrocarbons on more complex surfaces—semiconductors, as well as metal oxides, nitrides, phosphides and carbides—has been less studied than on metals. However, there is sufficient data available in the literature to permit estimation of desorption times for select systems. For example, tribological studies of perfluoropolyalkylether (PFPE) lubricants used in magnetic storage media have provided data on the desorption energies of short- and long-chain oligomers and polymers on graphite substrates [122]. The desorption energies of organometallic deposit precursors such as trimethylgallium, Cu(hfac)$_2$, and Fe(CO)$_5$ on semiconductor and metal surfaces can be obtained from the literature on MOCVD growth [123–126].

Many molecular etch precursors, such as O_2 [127], Cl_2, F_2, and XeF_2 [128; 129], undergo dissociative adsorption or atom abstraction on transition metal and semiconductor surfaces, producing dissociation products that etch the surface spontaneously. However, there are some surfaces on which these precursors adsorb molecularly (such as in the adsorption of O and O_2 on Si using gaseous O_2 or N_2O), or dissociate into species that do not spontaneously etch

Table 7.1 Desorption energies and prefactors for some commonly studied etch precursor/surface combinations.

System	E_{des} (eV)	τ_0 (s)	References
H_2O/Ni, H_2O/Pt sub-ML	0.55	1×10^{-16}	[104, 136]
O_2/Au sub-ML	0.11	1×10^{-11}	[152]
N_2O/Pt sub-ML	0.24	1×10^{-13}	[153]
WF_6	0.32	1×10^{-13}	[154]
NH_3/Au sub-ML	0.59	1.22×10^{-3}	[155]

the surface, making them suitable for FEBIE processing. Table 7.1 lists some examples along with references for adsorption data in the sub-monolayer (ML) and multilayer regimes.

4.3 DIFFUSION

In general, surface diffusivity displays Arrhenius behavior, with the diffusion coefficient D given by $D = D_0 e^{-E_{diff}/kT}$. Accounting for the effect of surface diffusion of deposit and etch precursor molecules (and both together, in the case of simultaneous FEBIED) is necessary to accurately model process rates and nanostructure morphology, particularly if their evolution with time is of interest. While there are many established methods for measuring surface diffusion, most can be applied only to a small set of molecules. Seebauer and Allen have compiled a very useful set of experimental data on surface diffusion for a variety of atomic and molecular adsorbates, and have uncovered correlations that can be used as a basis for estimating the values of parameters such as activation energies and pre-exponential factors where experimental data do not exist [101]. In addition, J. V. Barth has recently reviewed both the theoretical and experimental literature on surface diffusion of non-metals on metal surfaces [100].

In the case of both metallic and non-metallic adsorbates on metal surfaces, References [100; 101] reveal that the diffusion energies form a fairly tight distribution. Using the distribution of corrugation factors $\Omega = \dfrac{E_{diff}}{E_{des}}$ as a measure of the dispersion in diffusion energies, Seebauer and Allen find an average of $\Omega = 0.23$ for non-metallic adsorbates on metal surfaces and $\Omega = 0.13$ for metallic adsorbates. They also report that the diffusion coefficients of both metallic and non-metallic adsorbates on metals display a Gaussian distribution with a centroid at $D_0 = 1 \times 10^{-3}$ cm^2/s. This figure was found to be fairly independent of the details of the diffusion mechanism and the presence or absence of surface steps. Diffusion data for semiconductor surfaces exhibits a much wider variation, but Ω appears to correlate with the number of directional bonds (M) an adsorbed atom has to the surface ($\Omega = 0.6/M$) [101].

The diffusion energies of short chain alkanes (up to chain length $n = 6$, where n is the number of carbon atoms) have been observed to increase linearly with chain length [130]. Diffusion data on longer-chain molecules (oligomers and polymers) on solid surfaces is somewhat harder to come by [131], and theoretical studies have produced conflicting results. Zeiri and Cohen [132] reported that the diffusion energy of alkane chains with length up to $n = 20$ was proportional to the square root of chain length, while the preexponential factor remained constant. In contrast, Raut and Fitchthorn found that the diffusion energy scaled linearly with chain length up to $n = 8$ and plateaued beyond $n = 10$ [133].

5 SINGLE SPECIES MODELS OF ELECTRON BEAM DEPOSITION AND ETCHING

In the following, we start by describing a simple delocalized model of FEBID, and gradually increase the model complexity by incorporating surface site occupation, a spatially resolved electron flux profile, and surface diffusion. Although we refer only to FEBID, the models described here apply equally well to FEBIE.

5.1 MODEL DESCRIPTION

An early model of electron beam induced deposition was reported by R. W. Christy to account for the deposition rate of a polymer film formed by irradiation of silicone oil by a broad electron source [1]. Christy assumed a first-order rate process, leading to the following equation for the deposition rate $\dfrac{dN_D}{dt}$:

$$\frac{dN_D}{dt} = \sigma_d f N_d \ (\text{molecules}/\text{Å}^2/s), \qquad (5.1a)$$

where N_d and N_D are the number of deposition precursor adsorbates and deposit molecules per unit area (molecules/Å^2), f is electron flux (electrons/Å^2/s), and σ_d (Å^2) is an electron energy-dependent cross-section for the electron-induced cross-linking process responsible for film growth. The rate of change of the concentration of precursor adsorbates is thus given by:

$$\frac{dN_d}{dt} = J_d - \frac{N_d}{\tau_d} - \frac{dN_D}{dt}$$

$$= J_d - \frac{N_d}{\tau_d} - \sigma_d f N_d \ (\text{molecules}/\text{Å}^2/s), \qquad (5.1b)$$

where J_d is the molecular flux of precursor molecules arriving at the sample surface (directly proportional to the precursor pressure P_d), and τ_d is the desorption time (mean residence time). This equation does not limit the number of adsorption sites but rather assumes that the molecular flux at the surface is determined solely by the gas pressure. It also assumes that the sticking coefficient is unity and the electron flux is constant over the substrate.

Incorporation of surface site occupation requires the assumption of a particular model of gas adsorption. In general, the Langmuir model is used when there is only one surface species. It assumes that there is no interaction between surface species, and also limits the concentration of adsorbed species to a single monolayer (i.e., the fractional surface coverage θ is limited to 1). The maximum concentration of adsorbates is thus $N_0 = \frac{1}{A_d}$, where A_d is the molecular surface area, and the fractional surface coverage θ is given by $\frac{N_d}{N_0} = N_d A_d$. The concentration of surface sites that are available for adsorption is thus $(1 - \theta) = \left(1 - N_d A_d\right)$.

Incorporation of Langmuir adsorption and sticking into Equation 5.1b gives:

$$\frac{dN_d}{dt} = s_d J_d (1 - N_d A_d) - \frac{N_d}{\tau_d} - \sigma_d f N_d,$$

where s_d is the sticking coefficient. In the absence of any electron flux, this equation has the solution:

$$N_d(t) = \frac{s_d J_d}{s_d J_d A_d + \dfrac{1}{\tau_d}} \left(1 - e^{-(s_d J_d A_d + \frac{1}{\tau_d})t} \right), \tag{5.2a}$$

which gives:

$$N_0 = \frac{s_d J_d}{s_d J_d A_d + \dfrac{1}{\tau_d}} \tag{5.2b}$$

as the equilibrium concentration of precursor molecules at the surface prior to electron irradiation (obtained by putting $t = \infty$ in equation 5.2a). We term this concentration the *initial coverage*.

Incorporation of a spatially resolved electron flux profile $f(r)$ requires further modification of Equations 5.1 to account for surface diffusion of molecular adsorbates. Assuming radial symmetry, this results in the following rate equations for deposition caused by electron induced dissociation of a single species [4]:

$$\frac{\partial N_d(r, t)}{\partial t} = s_d J_d (1 - N_d(r, t) A_d) - \frac{N_d(r, t)}{\tau_d}$$

$$+ D_d \left[\frac{\partial^2 N_d(r, t)}{\partial r^2} + \frac{1}{r} \frac{\partial N_d(r, t)}{\partial r} \right] - \frac{dN_D(r, t)}{dt} \tag{5.3a}$$

$$\frac{\partial N_D(r, t)}{\partial t} = \sigma_d f(r) N_d(r, t), \tag{5.3b}$$

where $\dfrac{\partial N_d(r, t)}{\partial t}$ and $\dfrac{\partial N_D(r, t)}{\partial t}$ are the rates of change of the adsorbate concentration and deposit coverage, respectively, r is the radial distance from the electron beam axis, and $f(r)$ is the electron flux. The four terms comprising Equation 5.3a account for adsorption of the precursor molecules, desorption, surface diffusion, and electron induced dissociation of the adsorbates. Equation 5.3b describes the precursor dissociation rate that gives rise to deposition. Here, σ_d is a single-valued adsorbate dissociation cross-section, evaluated over the entire energy spectrum of electrons crossing the solid-vacuum interface and summed over all possible dissociation pathways.

Because the diffusion length of the deposit precursor molecules can be much larger than the beam diameter, r must span many orders of magnitude, from $r = 0$ to $r \to \infty$. It is therefore often useful to change the variable r to a logarithmic radial parameter $l = \ln \dfrac{r}{L}$, where r is the radial distance from the beam axis and L is the beam diameter:

$$\frac{\partial N_d(l, t)}{\partial t} = s_d J_d [1 - A_d N_d(l, t)] - \frac{N_d(l, t)}{\tau_d} - \frac{\partial N_D(l, t)}{\partial t} + \frac{D_d}{e^{2l} L^2} \left[\frac{\partial^2 N_d(l, t)}{\partial l^2} \right] \tag{5.4a}$$

$$\frac{\partial N_D(l, t)}{\partial t} = \sigma_d f(l) N_d(l, t). \tag{5.4b}$$

Given a particular electron flux profile $f(l)$, beam diameter L, and beam current I_p, the system of Equations 5.4 can be solved numerically to yield the spatial and temporal evolution of the deposition precursor coverage $N_d(l, t)$, deposit coverage $N_D(l, t)$, growth rate $\dfrac{\partial N_D(l, t)}{dt} s^{-1}$, and deposit height h:

$$h(l, t) = V_D N_D(l, t) = V_D \int_0^t \frac{\partial N_D(l, t)}{\partial t} dt, \qquad (5.5)$$

where V_D is the volume of a deposited molecule. Equations 5.4 and 5.5 can be applied to FEBIE by replacing the molecular parameters relating to the deposition precursor by those of the etch precursor, and V_D by the volume removed by the molecule(s) evolved from the substrate. Solving these equations would then give the profile of the pit etched into the substrate.

The concentration of deposition precursor adsorbates in the region irradiated by the electron beam displays a characteristic "reverse-S" dependence on log time illustrated by Figure 7.7 (calculated at $r = 0$ using the electron flux profile shown in Figure 7.8). At very short times after the electron beam is "turned on" ($<10^{-8}$ s in Figure 7.7), the precursor coverage is equal to the initial value given by the competing rates of adsorption and desorption in the absence of electrons. Note that at the pressures shown in Figure 7.7, the initial coverage is saturated at a single monolayer (below 1 Pa, the initial coverage will fall below 1 ML), and that the pressures are characteristic of ESEM rather than the capillary gas injection method discussed in Section 3.1. Beyond 10^{-8} s, the precursor concentration decreases exponentially with time as an increasing number of precursor molecules are dissociated under the electron beam, until it reaches a steady state at which the adsorbate arrival and removal rates are equal. As shown in the figure, the steady state adsorbate concentration increases with pressure.

The growth rate profile $\dfrac{\partial N_D(l, t)}{\partial t}$ (i.e., the deposit growth rate as a function of radius) is directly proportional to the adsorbate concentration profile (Equation 5.4b), and generally displays a strong dependence on time, as shown in Figure 7.9. At very short times (e.g., 1.65 ns in this example), the growth rate profile is proportional to the electron flux profile.

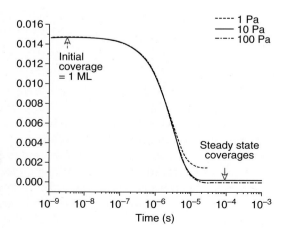

FIGURE 7.7: Adsorbate concentration at r=0 as a function of time for precursor pressures of 1 Pa, 10 Pa and 100 Pa ($I_P = 10$ nA, $s_d = 0.04$, $D_d = 1.42 \times 10^8$ Å^2/s, $\tau_d = 3.5$ s, $\sigma_d = 1$ Å^2, T = 290 K, top-hat electron flux profile shown in Figure 7.8).

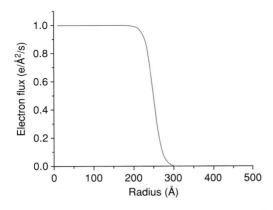

FIGURE 7.8: Top-hat electron flux profile (Equation 2.2 with $\beta = 25$ and $r_0 = L/2 = 250$Å) used to generate the curves shown in Figure 7.7.

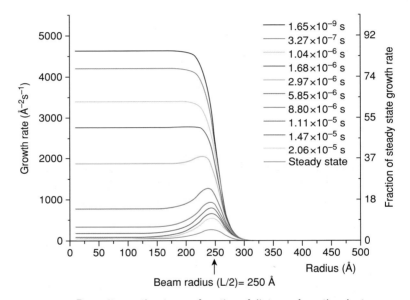

FIGURE 7.9: Deposit growth rate as a function of distance from the electron beam axis, plotted at a number of deposition times (the model input parameters are the same as in Figure 7.7). On the right-hand axis, the growth rates are given as a fraction of the steady state growth rate at $r = 0$.

However, at longer times, the adsorbates become depleted under the electron beam as the rate of removal (by desorption and electron-induced dissociation) exceeds the rate of replenishment (by adsorption and surface diffusion). Thus, the growth rate begins to decrease. At subsequent times, the adsorbate concentration—and thus the growth rate—is steadily reduced. At long times ($\geq 2.97 \times 10^{-6}$ s in this example), the effect of adsorbate diffusion into the irradiated region becomes apparent. These inward diffusing molecules are dissociated by electrons, resulting in a local increase in the growth rate near the beam radius ($r_0 = L/2 = 250$Å). In this example, a steady state growth rate of approximately 1% of the initial growth rate is reached in about 1 ms.

5.2 EXPERIMENTAL REGIMES

Removing the diffusion term from Equation 5.4 enables an analytical solution, which gives the steady state concentration of deposition precursor molecules as [6]:

$$(N_d)_{ss} = \frac{s_d J_d}{(s_d J_d A_d + \dfrac{1}{\tau_d} + \sigma_d f)} \text{ molecules/Å}^2 . \tag{5.6}$$

The steady state etch or deposition rate is then given by:

$$\left(\frac{dN_D}{dt}\right)_{ss} = \sigma_d f N_d = \frac{\sigma_d f s_d J_d}{(s_d J_d A_d + \dfrac{1}{\tau_d} + \sigma_d f)} \text{ molecules/Å}^2/s . \tag{5.7}$$

The FEBID rate is governed by the electron flux when the electron-induced dissociation rate is small with respect to the adsorbate arrival rate (i.e., the process is in the reaction-rate limited or electron-limited regime). In this electron-limited regime, the $\sigma_d f$ term is negligible compared to the other terms in the denominator of Equation 5.6, and the steady state etch or deposition rate is given by:

$$\left(\frac{dN_D}{dt}\right)_{ss} \approx \frac{\sigma_d f s_d J_d}{(s_d J_d A_d + \dfrac{1}{\tau_d})} \approx \sigma_d f N_0 . \tag{5.8}$$

Thus, at low electron flux, the etch or deposition rate increases linearly with electron flux. Conversely, in the limit of very high electron flux (i.e., when there are sufficient electrons to dissociate most adsorbates), the rate is limited by the concentration of adsorbed precursor molecules (the so-called precursor-limited or mass transport-limited regime). In this regime, the final term of the denominator of Equation 5.6 dominates and the steady state etch or deposition rate becomes:

$$\left(\frac{dN_D}{dt}\right)_{ss} \approx s_d J_d . \tag{5.9}$$

Thus, as the electron flux is steadily increased, the FEBID rate increases until it saturates at a point given by the adsorbate arrival rate. This shift from the electron flux–limited regime to precursor-limited regime has been observed in both deposition and etching [4; 6; 134].

Utke et al. [4] used the model of Equation 5.3 to determine scaling laws for the resolution of the FEBID process. As this model has been extensively discussed in a prior chapter, we will keep our discussion brief. Neglecting diffusion, the steady state solution to Equation 5.3 results in a volumetric deposit growth rate $R(r) = s_d J_d \tau_{eff}(r) V_D \sigma_d f(r)$, where $\tau_{eff}(r)$ is the effective residence time of the deposition precursor, given by $\tau_{eff}(r) = \left(\dfrac{s J_d}{N_0} + \dfrac{1}{\tau_d} + \sigma_d f(r)\right)^{-1}$.

Assuming a Gaussian beam, the authors defined three scaling laws relating the resolution of the focused electron or ion beam process to the molecular parameters (residence time and diffusion length). First, the achievable resolution can be related to the depletion level by:

$$\tilde{\varphi} = \left(\log_2(1 + \tilde{\tau})\right)^{1/2} , \tag{5.10}$$

where $\tilde{\tau} = \dfrac{\tau_{out}}{\tau_{in}}$ is the residence time ratio inside and outside the irradiated area.

This relation shows that the resolution is maximized ($\tilde{\varphi}$ is minimized) when there is zero depletion and $\tilde{\tau} = 1$ (i.e., in the electron-limited regime), and is degraded as the processing conditions transition into the precursor-limited regime. Replenishment of depleted molecules is possible by surface diffusion as well as adsorption, as quantified by the ratio of the molecule diffusion path inside the irradiated area with respect to the beam size $\tilde{\rho} = \dfrac{2\rho_{in}}{FWHM_B} = \dfrac{2\sqrt{D_d \tau_{in}}}{FWHM_B}$. When there is no diffusive replenishment ($\tilde{\rho} = 0$), the resolution is as defined by Equation 5.9 in the precursor-limited regime. With increasing diffusive replenishment, both deposition rate and resolution increase, until at $\tilde{\rho} \to \infty$, any depletion is entirely compensated for by surface diffusion, and the deposit shape corresponds to the electron flux profile. This situation marks the establishment of the electron-limited regime. The effect of diffusive replenishment on resolution can be expressed in a second scaling law:

$$\tilde{\varphi} = \left(\log_2(2 + \tilde{\rho}^{-2})\right)^{1/2}. \tag{5.11}$$

The minimum achievable deposit or etch resolution is given by the smaller value of Equations 5.9 and 5.10. Under most focused electron or ion beam processing conditions, the precursor-limited regime applies and Equation 5.9 governs the achievable resolution.

Recently, Rykaczewski et al. have developed a FEBID model for the case where the primary means of molecule arrival under the beam is via surface diffusion [17]. This applies to FEBID in some circumstances, such as contaminant film growth in ultra-high vacuum driven by surface-adsorbed residual hydrocarbons. The model comprises a continuum mass transport component, and a Monte Carlo algorithm for electron transport and scattering. The time-intensive Monte Carlo calculations are performed only when the deposit shape changes significantly, rather than at every time step. The model neglects adsorption and desorption of gas phase precursor molecules, and assumes an initial precursor molecule concentration equivalent to monolayer surface coverage. The diffusion coefficient was used as an adjustable parameter to investigate different growth regimes (mass transport limited, reaction rate limited, and mixed). As previously noted [4], the fastest growth rates and highest spatial resolutions were found to occur in the reaction-limited regime.

6 MULTISPECIES MODELS OF ELECTRON BEAM INDUCED DEPOSITION AND ETCHING

6.1 MULTIPLE PRECURSORS

The present authors have made a first attempt at developing a continuum model of simultaneous FEBIE and FEBID. Such multi-species modeling is relevant to FEBIE in the presence of contaminants such as hydrocarbons that give rise to unintended competing FEBID [2], processes in which FEBID and FEBIE precursors are mixed intentionally to generate specific process dependencies on electron flux [3], and FEBIE-mediated contamination reduction in electron microscopy [7]. The modeling approach is also applicable to any FEBIED processes that involve competing reactions, as will be discussed further in Section 7.

Full descriptions of two versions of the simultaneous FEBIED model can be found in References [2; 3]. The model extends Equations 5.3 to two adsorbate species (deposition and etch precursors), and can be extended to an arbitrary number of species (including reaction intermediates generated and consumed in multi-step reaction processes). Assuming Langmuir adsorption behavior in which the deposition and etch precursors share surface sites, the surface coverage is given by $\theta = N_d A_d + N_e A_e$, where N_e and N_d are the concentrations of etch and deposition adsorbates, respectively, and A_e and A_d are their molecular surface areas. Limiting the surface coverage θ to one monolayer gives $N_d A_d + N_e A_e \leq 1$. Assuming radial symmetry, we have the following equations for the rates of change of concentration of precursor adsorbates $\left(\dfrac{\partial N_e}{\partial t} \text{ and } \dfrac{\partial N_d}{\partial t} \right)$ and deposited molecules $\left(\dfrac{\partial N_D}{\partial t} \right)$ at the sample surface:

$$\frac{\partial N_e(r, t)}{\partial t} = s_e J_e [1 - (N_d A_d + N_e A_e)] - \frac{N_e(r, t)}{\tau_e} - \sigma_e f(r) N_e(r, t)$$

$$+ D_e \left[\frac{\partial^2 N_e(r, t)}{\partial r^2} + \frac{1}{r} \frac{\partial N_e(r, t)}{\partial r} \right] \tag{6.1a}$$

$$\frac{\partial N_d(r, t)}{\partial t} = s_d J_d [1 - (N_d A_d + N_e A_e)] - \frac{N_d(r, t)}{\tau_d} - \sigma_d f(r) N_d(r, t)$$

$$- [\sigma_e f(r) N_e(r, t)] \sigma_r N_d(r, t) + D_d \left[\frac{\partial^2 N_d(r, t)}{\partial r^2} + \frac{1}{r} \frac{\partial N_d(r, t)}{\partial r} \right] \tag{6.1b}$$

$$\frac{\partial N_D(r, t)}{\partial t} = \sigma_d f(r) N_d(r, t) - [\sigma_e f(r) N_e(r, t)] [1 - \sigma_{rd} N_d(r, t)] \sigma_{rD} N_D(r, t). \tag{6.1c}$$

The above model assumes that both the deposition precursor adsorbates and the deposit can be volatilized through FEBIE. The first term in Equations 6.1a and 6.1b accounts for adsorption of the etch and deposition precursor molecules, respectively; the second term accounts for desorption and the final term accounts for surface diffusion of the adsorbates where all parameters are as defined previously and the subscripts "e" and "d" relate to the etch and deposition processes, respectively. The terms $\sigma_e f(r) N_e(r, t)$ and $\sigma_d f(r) N_d(r, t)$ are the etch and deposition precursor dissociation rates, $[\sigma_e f(r) N_e(r, t)] \sigma_r N_d(r, t)$ (in Equation 6.1b) is the volatilization rate of deposition precursor adsorbates by FEBIE, and the last term of Equation 6.1c is the corresponding volatilization rate of the deposit. The reaction cross-sections σ_{rd} and σ_{rD} account for the effectiveness of collisions between the reactive etch precursor dissociation products (hereafter referred to as the etchant species) and the deposition precursor (σ_{rd}) or pinned deposit molecules (σ_{rD}) in leading to volatilization. Finally, in Equation 6.1c, the term $(1 - \sigma_{rd} N_d)$ discounts those etchant species that react with adsorbed deposition precursor molecules, and $N_D(r, t)$ is capped at $1/A_d \text{Å}^{-2}$ (making the assumption that the surface area of the pinned deposit molecule is equal to that of the deposition precursor molecule), since only molecules in the top monolayer of the deposit are available for volatilization by the etchant species.

As described previously, we change the variable r to a logarithmic radial parameter $l = \ln \left(\dfrac{r}{L} \right)$, discretize the equations using the Crank-Nicholson algorithm, and solve them numerically. The equations are solved for a time step Δt, and for two half-steps $\Delta t/2$. The coverages obtained are compared at the end of the calculation, and if they do not agree for all values of radius within a given fractional error (typically 10^{-3} as smaller errors yield negligible improvements in accuracy at the cost of increased computing time), the time step is halved and the calculation is run again. Conversely, if the coverages do agree, the time step is increased periodically by a given factor (e.g., an order of magnitude) to enable rapid approach to a steady state.

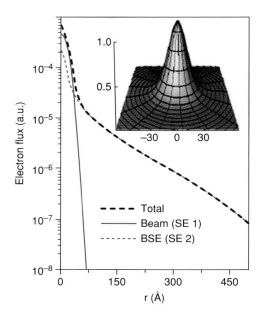

FIGURE 7.10: Electron flux profile used to obtain the simulation results shown in Fig. 6.2. Inset: surface plot of the normalized electron flux profile (linear scale). Reprinted with permission from Reference [3], Copyright (2008), Institute of Physics Publishing Ltd.

To illustrate the utility of this simultaneous FEBIED model, we simulate growth of a deposit under a focused electron beam of energy 5 keV and radius of 4 nm [3]. The electron flux was constructed using the approach described in Section 2, yielding the profile shown in Figure 7.10.

Gold was used as the substrate, nonane (C_9H_{20}) as the (carbon) deposition precursor molecule, and H_2O as the precursor for FEBIE of carbon. Both nonane and water adsorb molecularly on gold surfaces, and both have sufficient room temperature vapor pressures for FEBIED. In addition, this experimental system is one of the few for which most molecular parameters are available in the literature [101; 102; 104; 130; 133; 135–138] (linear extrapolations for surface areas, diffusion energies and diffusion prefactors were used where appropriate, and experimental data for similar transition metal surfaces were used when values for Au were unavailable). The values of the cross-sections for dissociation of the etch and deposit precursors, $\sigma_e = 10^{-2}Å^2$ and $\sigma_d = 1Å^2$, were estimated from the dissociation cross-sections in references [98] and [59; 139; 140], respectively. The reaction cross-section (σ_{rd} or σ_{rD}) is defined as the product of the reaction probability (assumed to be unity) and the surface area of the deposition precursor or deposit molecule that is available to the incoming etchant species. For simplicity, both σ_{rd} and σ_{rD} were set to the area of a single deposition precursor molecule, A_d.

Figure 7.11 shows the steady state growth rate plotted as a function of r at a number of beam currents I_p, calculated using deposit and etch precursor partial pressures of 10^{-2} Pa and 100 Pa, respectively. The total electron flux profile is also shown (dashed line). At low currents ($I_p < 0.4$ nA), FEBIE is negligible, so the FEBIED growth rate profiles are similar to that of the electron flux and their magnitude scales linearly with beam current, as in conventional FEBID. However, as I_p is increased from 0.4 to 1 nA, depletion of deposition precursor molecules under the beam causes the FEBIE rate to exceed that of FEBID, giving rise to a ring-like growth profile with an inner radius R (shown in Figure 7.11a). The three-dimensional shapes of full and ring-like deposition profiles are illustrated more clearly by the surface plots shown in Figure 7.11b.

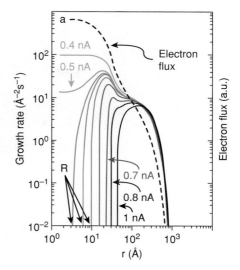

FIGURE 7.11: (a) Growth rate plotted as a function of radius (r) from the beam axis, calculated using beam currents of 0.4, 0.5, 0.52, 0.53, 0.55, 0.6, 0.7, 0.8 and 1 nA. Also shown is the electron flux profile. (b) Surface plots of the growth rate profiles at 0.4, 0.5 and 0.6 nA (linear scale). Reprinted with permission from Reference [3], Copyright (2008), Institute of Physics Publishing Ltd.

The transition from FEBIE (at $r < R$) to FEBID (at $r > R$) is caused by the change in electron flux with increasing r [3]. As will be discussed in more detail in Section 7, the transition occurs because the FEBID process is more efficient than FEBIE (*i.e.*, $\sigma_d \gg \sigma_e$) and the FEBID molecule arrival rate is much lower than that of the FEBIE precursor (in this case, because $P_d \ll P_e$).

At present, the main deficiencies of this model are that (1) it does not account for the effects of growth on SE and BSE emission and on the surface area available for mass transport of adsorbates, and (2) that it employs single-valued (energy-independent) cross-sections. The first assumption means that the calculated growth rates are only valid in the early stages of FEBIED (when the effects of changes in the geometry of the solid-vacuum interface are negligible). We note, however, that the key aspects of FEBIED behavior discussed here are the result of the presence of the sharp peak in the electron flux profile (centered on $r = 0$, Figure 7.10) and that the presence of this peak is not eliminated by a growing deposit.

6.2 REACTION INTERMEDIATES AND COEXISTING SPECIES

In some electron beam induced reactions, one or more reaction intermediates or coexisting species may exist. This situation can arise due to the existence of multiple adsorption states or multiple electron-induced fragmentation pathways. In both cases, the presence of multiple species leads to more complex process rate dependencies on the key experimental parameters, and complicates the development of reaction models. Soon after Christy's study of electron-induced polymerization of silicone oil, Haller and White studied the electron-beam induced polymerization of butadiene [141] and found that the reaction rate was proportional to the square root of the beam current density (and thus to the electron flux), rather than displaying a first-order relationship. The authors proposed that the polymer film grew by interaction of the adsorbed butadiene with active sites created by the electron beam, rather than by direct electron-induced cross-linking of the adsorbed molecules, as reported by Christy. Thus the rate-determining step is the formation of an adsorbed butadiene species that is active in the polymerization process. This "active" species has a finite but long lifetime compared to the adsorption time of butadiene. The square root dependence on current density was found to

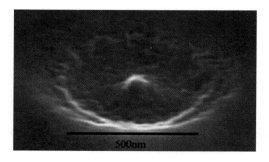

500nm

FIGURE 7.12: "Moat" effect observed in EBIE of SiO$_2$ under 2×10^{-2} Pa of XeF$_2$ and using a focused electron beam of energy 5 keV, current $I_p = 300$ pA and with a spot irradiation time of 2 minutes Reprinted with permission from Reference [6], Copyright (2008), Institute of Physics Publishing Ltd.

be consistent with an ionic mechanism involving a series of charge transfer steps between the metal substrate, polymer, and active species. Many years later, Fritzsche [142] developed a series of experiments to determine the deposition efficiency of a polymer film under both continuous and pulsed irradiation, and extracted cross-sections for electron-induced activation of the polymerization process and lifetimes of the active centers by fitting various analytical models to the experimental data.

Reaction intermediates also exist for much simpler molecules, for example in the case of electron or ion beam induced etching of SiO$_2$ using XeF$_2$ as the etch precursor. Ion or electron beam assisted XeF$_2$ etching of SiO$_2$ produces SiF$_4$ and O$_2$ as the major volatile products [117]. SiF$_2$ and SiF are volatile reaction intermediates that are observed as minor products. The gas-assisted etching reaction is believed to proceed through chemisorption of XeF$_2$ to produce surface-adsorbed SiF$_2$, which then reacts further with the XeF$_2$ to produce SiF$_4$.

In their recent experiments on electron beam induced XeF$_2$ etching of SiO$_2$, Lassiter et al. noted that the etch rate was not fastest at the beam center, as would be expected from the single-species etching model discussed in Section 6.1, but rather was close to zero (Figure 7.12) [6]. Having noted that this effect cannot be explained by incorporating the effect of surface diffusion, which would result in a ring-like etch profile (as discussed above), the authors introduced a finite residence time for the product of the etching reaction, rather than assuming that it immediately leaves the surface. This model therefore accounts for the formation of an adsorbed surface fluoride such as SiF$_2$, as experimentally observed. In addition, it is possible for the etch product to be dissociated by electrons during its residence time on the surface, resulting in redeposition. Neglecting diffusion, the equations describing the etch process then become:

$$\frac{\partial N_e(r, t)}{\partial t} = s_e J_e (1 - N_e(r, t) A_e - N_E(r, t) A_E) - \frac{N_e(r, t)}{\tau_e} - \sigma_e f(r) N_e(r, t) \qquad (6.2a)$$

$$\frac{\partial N_E(r, t)}{\partial t} = x \sigma_e f(r) N_e(r, t) - \frac{N_E(r, t)}{\tau_E} - \sigma_E f(r) N_E(r, t), \qquad (6.2b)$$

where the symbols are as defined above and e and E represent the etch precursor molecule and etch product molecule, respectively, and x is a stoichiometric factor relating the concentration of the etch product molecule to the concentration of the undissociated etch precursor molecule. In the electron limited regime, the etch rate is identical to that given by the single species model (Equation 5.7). However, in the mass transport limited regime, where $\sigma_e f \gg s_e J_e$ and both τ_E and σ_E are finite, the etch rate is inversely proportional to the electron flux, thus accounting for the decreased rather than increased etch rate observed at the beam center (where the electron flux is highest).

7 SIMULTANEOUS FEBIED: ELECTRON FLUX-CONTROLLED SWITCHING BETWEEN ETCHING AND DEPOSITION

Here we discuss a number of processes in which FEBID and FEBIE occur concurrently. Simultaneous FEBIED is encountered during the deposition of high purity metals using an organometallic precursor and an oxygen-containing background gas [49, 52]. It also occurs during most FEBIE processes due to the unintentional deposition of carbonaceous films via electron activated cross-linking of hydrocarbon contaminants present on the substrate surface. We will focus on two main applications of the model discussed in Section 6.1: determining imaging conditions that will lead to minimal contamination and surface modification in ESEM, and understanding and controlling the morphology and resolution of electron beam fabricated deposits.

7.1 REDUCING CONTAMINATION

Figure 7.13(a-c) shows images of a 100 nm Cr film (on a bulk quartz substrate) that was etched under 60 Pa of XeF_2 using electron beam diameters of $\sim$4, 50 and 200 nm [2]. An etch pit is clearly visible in Figure 7.13(a),[e] and the etch rate decreases with increasing beam diameter. At 200 nm (Figure 7.13(c)) etching is replaced with deposition of a carbonaceous film formed due to FEBID of background contaminants (Figure 7.13(d) shows an etch pit fabricated using an optimized FEBIE process discussed in Reference [2]).

The transition from etching to deposition has been attributed to the change in electron flux caused by defocusing the beam, and more specifically by differences in the efficiencies of the etch and deposition processes, and by saturation of the deposition rate with electron flux [2]. Based on this interpretation, the deposition process is more efficient than the etch process, but the deposition rate is limited by the hydrocarbon (contaminant) adsorbate arrival rate into

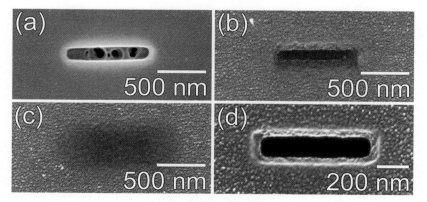

FIGURE 7.13: Cr-on-quartz photomask that was etched using XeF_2 precursor and electron beam diameters of $\sim$4, 50, 200 and 4 nm (a-d, respectively). The electron beam landing energy and etch time were 10 keV and 12 min in (a-c), and 20 keV and 20 min in (d). Reprinted with permission from Reference [2], Copyright (2007), American Institute of Physics.

the region irradiated by the electron beam. Conversely, the etch process is less efficient, but does not saturate at elevated electron flux, due to the high arrival rate of XeF_2 molecules at the sample surface. Below, the proposed behavior is illustrated using the model of simultaneous electron beam induced etching and deposition discussed in Section 6.1.

The model of Equations 6.1(a-c) was simplified by eliminating the effects of precursor molecule surface diffusion and of the beam revisit period (refresh time) [143–145] by considering the case of an electron-irradiated semi-infinite surface [2]. It was assumed that both etch and deposit precursors arrive from vacuum, with arrival rates determined by their respective partial pressures. Furthermore, due to the difficulty in obtaining adsorption data for XeF_2 on chrome, the etch precursor species was changed to H_2O; that is, the model describes FEBID of adsorbed hydrocarbons and competitive FEBIE of carbon-containing species by water [2]. The deposition precursor was taken to be hexane (C_6H_{14}), an intermediate-length hydrocarbon. Each H_2O molecule dissociated by an electron is assumed to produce one volatile and one reactive product. The latter can volatilize either adsorbed C_6H_{14} or a crosslinked deposit molecule. It was further assumed that both the etch and deposition precursor molecules are capable of forming multilayers (i.e., the model of Langmuir adsorption was removed from equations 6.1). The sticking coefficients s_e and s_d were set to unity, with $\tau_d = 2.57 \times 10^{-4}$ s and $\tau_e = 1.61 \times 10^{-7}$ s at 295 K respectively [104; 120]. The dissociation cross-sections were taken to be $\sigma_e = 10^{-2}$Å^2 and $\sigma_d = 1$Å^2 [59; 98] and both reaction cross-sections σ_{rd} and σ_{rD} were set to the area of a single hexane molecule (56.8 Å^2).

Figure 7.14 shows the steady state carbon film growth rate, dN_D/dt, calculated using a C_6H_{14} partial pressure of 10^{-3} Pa and H_2O partial pressures of 0 and 100 Pa, as a function of electron flux. In the absence of H_2O, the hexane pinning rate increases with electron flux until it reaches

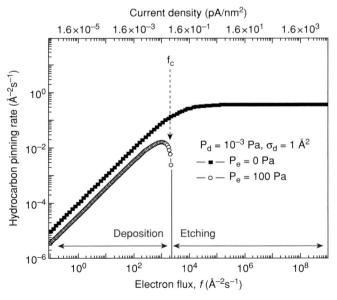

FIGURE 7.14: Steady state hydrocarbon pinning rate (i.e., EBID growth rate) calculated as a function of electron flux at H_2O partial pressures of 0 and 100 Pa. Reprinted with permission from Reference [2], Copyright (2007), American Institute of Physics.

saturation at f_s. When H_2O is introduced into the system, the pinning rate decreases at all values of f due to FEBIE of hexane. At low values of f, the pinning rate increases with f up to a critical point f_c, beyond which the volatilization rate exceeds the pinning rate, and the latter rapidly drops to zero, demonstrating the electron-flux induced transition from deposition to etching. The value of f_c is a function of all parameters that affect the deposition and etch rates.

The switching between FEBID and FEBIE occurs in all cases where the efficiency of one process is much higher and the adsorbate arrival rate (which is, in this case, governed by the partial pressure) is much lower. The abruptness of the switching transition depends on the relative magnitudes of the etch and deposition precursor partial pressures and dissociation cross-sections, as has been discussed in detail in Reference [2]. This study helps to explain the FEBIE behavior seen in Figure 7.13(a-c), and also explains why contamination buildup rates in ESEM, where H_2O is commonly used as the environmental gas, are low relative to high vacuum SEM, as will be discussed further in the following section.

The above analysis demonstrates how simplifying assumptions in a model of FEBIED can be used to identify the individual mechanisms behind behavior such as electron flux controlled switching between FEBIE and FEBID. A very simple model was used to demonstrate a switching mechanism that does not require surface site competition (as Langmuir adsorption was removed from the equations), surface diffusion, beam induced heating, or dynamic depletion effects caused by a scanned electron beam [2].

7.2 IMPROVING RESOLUTION

FEBIE and FEBID spatial resolution is limited by the electron flux profile at the substrate surface, which is determined by the electron beam diameter (which can be in the sub-1nm range), and by the spatial distribution of secondary and backscattered electrons emitted from the substrate, both of which extend beyond the electron beam. In FEBID, SE and BSE emission from deposit sidewalls gives rise to a lateral growth component and deposit broadening that occurs during growth [9; 10; 16]. FEBID on bulk substrates is typically performed using electron beam landing energies in the range of 1 to 30 keV. Under these conditions, the diameters of FEBID-fabricated structures such as pillars and wires are typically in the 10^1–10^2 nm range [55; 134; 146–148]. Greatest resolution is achieved by minimizing the electron dose, yielding dots and wires with diameters in the range of 4–30 nm [147; 149; 150]. In some studies, broad, FEBID-fabricated structures have been slimmed and sculpted by gas-mediated post-FEBID FEBIE [55], or a high energy (200 keV) electron beam (in a high vacuum environment) [151], to produce sub-5 nm and sub-1 nm feature dimensions, respectively.

FEBID and FEBIE resolution is also limited by adsorbate depletion, as was discussed in Section 6.1. Depletion occurs at high electron flux, when the molecule removal rate caused by electron induced processes is similar to the molecule arrival rate. In FEBID (rather than FEBIED), it gives rise to sub-linear dependencies of growth rates on electron flux [2; 16; 134], and leads to lateral broadening of the structures grown, as has been demonstrated by measurements [134] and simulations [4; 16] of the diameters of pillars grown by stationary electron beams.

In simultaneous FEBIED, however, adsorbate depletion can be used to beneficial effect, increasing rather than decreasing the resolution of electron beam induced deposition processes. This is the case when the electron flux induced transition from FEBIE to FEBID (which was discussed in Section 7.1) occurs *within* the electron flux profile produced by a stationary electron beam. Specifically, the electron flux has an intense peak at the beam axis (see,

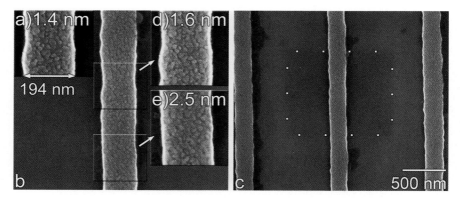

FIGURE 7.15: ESEM images of a Cr-on-quartz photolithographic mask acquired as a function of decreasing magnification: (a) high magnification image of a Cr absorber line; (b, c) the same region imaged at progressively reduced magnification. The insets (d) and (e) are digitally enlarged regions of image (b). Region (d) corresponds to the area imaged in (a). In (c), a dark rectangle (indicated by a dotted frame) corresponds to the area of image (b). Spatial resolution within the Cr grain structure seen in (a, d, e) is shown on the images. Reprinted with permission from Reference [7], Copyright (2009), American Institute of Physics.

for example, Figure 7.10) and the transition from FEBIE to FEBID can occur at some distance r that can be smaller than the electron beam radius, controlled by the electron beam current, as shown in Figure 7.11. This behavior has been studied in detail by the simultaneous FEBIED model of Equations 6.1 (a-c) which accounts for a spatially resolved electron flux profile, Langmuir adsorption, and surface diffusion of precursor adsorbates [3].

The switching between etching and deposition encountered in simultaneous FEBIED has also been used to explain changes in the resolution of ESEM images acquired as a function of magnification [7]. Environmental SEM imaging is typically performed in a low vacuum H_2O environment where H_2O mediated FEBIE of carbon competes with the growth of carbonaceous films caused by electron beam induced cross-linking of hydrocarbon contaminants. Carbonaceous films obscure surface features and reduce image resolution, an effect that can be mitigated or reversed by H_2O mediated FEBIE. To demonstrate, Figure 7.15 shows ESEM images of a fixed region of a Cr-on-quartz photolithographic mask acquired at successively reduced magnification (i.e., successively reduced electron flux). Image (a) shows a nominally 200 nm wide Cr absorber line. Images (b) and (c) were subsequently acquired at progressively reduced magnifications. Image (c) contains a dark rectangle corresponding to the area imaged in (b), characteristic of carbonaceous film growth caused by electron irradiation. Conversely, acquisition of image (a) did not produce an analogous dark rectangle in image (b), suggesting a reduced film growth rate at the elevated magnification used to acquire image (a). Figures 7.15 (d) and (e) show two digitally magnified areas of image (b). Image (d) corresponds to region (a), which had been imaged previously at elevated magnification. Image (e) is a neighboring region, which had not been imaged previously. Resolution within the Cr grain structure seen in Figures 7.15 (a), (d), and (e) was measured to be 1.4, 1.6, and 2.5 nm, respectively [7], illustrating that the resolution of (a) is similar to (d), and the resolution of both is significantly higher than that of image (e).

A model of simultaneous FEBIED was used to explain these observations. However, this time, the time-dependencies of solutions to Equations 6.1 calculated as a function of electron

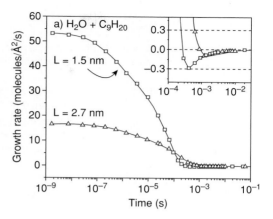

FIGURE 7.16: Simulated growth rates of carbonaceous deposits at the beam axis of a Gaussian electron beam with beam diameters (L) of 1.5 and 2.7 nm. The inset shows a close-up of the transitions from net material buildup (positive growth rates) to etching of the deposited contaminant (sub-zero growth rates). Reprinted with permission from Reference [7], Copyright (2009), American Institute of Physics.

beam diameter were used to analyze the experimental results (i.e., the images shown in Figure 7.15) obtained as a function of magnification using a scanned electron beam. A detailed discussion of this analysis can be found in Reference [7]. Key features of the model results are illustrated in Figure 7.16, which shows the growth rates of carbonaceous deposits as a function of time calculated using beam diameters of 1.5 and 2.7 nm. The two beam diameters translate to two electron fluxes which correspond to the effective electron flux per image pixel at the magnifications used to acquire the images shown in figure 7.15 (a) and (b), respectively. Both curves display a transition with irradiation time from net material buildup (positive growth rates) to etching of the deposited contaminant (sub-zero growth rates, seen clearly in the inset of Figure 7.16). The onset of the deposition-to-etching transition (i.e., the point where the growth rate equals zero) occurs within 250 μs at 1.5 nm, but is increased to ~1 ms at 2.7 nm. This difference in the switching time was used to interpret the transition from etching to deposition and the corresponding reduction in image resolution with decreasing magnification observed in Figure 7.15 [7]. We note that this modeling approach can also be applied to the XeF_2-mediated Cr etch experiments shown in Figure 7.13.

8 CONCLUSION

We described a continuum approach for modeling electron beam induced etching and deposition with an emphasis on the limitations of existing models. The continuum methodology is based on differential equations for the rates of change of molecular species at the substrate-vacuum interface. This approach can, in principle, account for all levels of complexity encountered in FEBIED. It is, however, limited by a lack of basic knowledge of applicable reaction pathways and the physical and chemical properties of the molecular species encountered in FEBIED. Other limitations arise from difficulties in calculating the time evolution of electron energy and flux distributions at the substrate-vacuum interface. Despite these shortcomings, this modeling approach has been used to explain complex behavior observed in electron beam induced etching, simultaneous etching and deposition, and contamination buildup and reduction in environmental SEM.

NOTES

a. Scanned electron beams are not always desirable for fundamental FEBIED studies because pixel, line, and frame dwell and revisit (refresh) times dramatically expand the FEBIED parameter space.

b. The depth distributions are expressed as $\partial(1 - \eta)/\partial z$ where η is the backscattering coefficient and z is depth below the solid-vacuum interface. The area under each curve equals the total fraction of incident electrons $(1 - \eta)$ that were not backscattered from the solid.

c. Most continuum models of FEBIED account for surface diffusion of adsorbates, which can affect adsorbate concentrations many microns away from the electron beam axis. Hence, premature termination of the BSE flux profile generates errors in solutions to the differential equations that constitute a continuum FEBIED model.

d. Chemisorbtion is typically undesirable as it usually compromises the localization of an etch or a deposition process.

e. The pitting in Figure 7.1(a) was caused by heterogeneous etching of the quartz underlayer.

REFERENCES

[1] Christy, R. W. Formation of thin polymer films by electron bombardment. *Journal of Applied Physics* 31 (1960): 1680–1683.

[2] Toth, M., C. J. Lobo, G. Hartigan, and W. R. Knowles. Electron flux controlled switching between electron beam induced etching and deposition. *Journal of Applied Physics* 101 (2007): 054309.

[3] Lobo, C. J., M. Toth, R. Wagner, B. L. Thiel, and M. Lysaght. High resolution radially symmetric nanostructures from simultaneous electron beam induced etching and deposition. *Nanotechnology* 19 (2008): 025303.

[4] Utke, I., V. Friedli, M. Purrucker, and J. Michler. Resolution in focused electron- and ion-beam induced processing. *Journal of Vacuum Science & Technology B* 25 (2007): 2219–2223.

[5] Utke, I., P. Hoffmann, and J. Melngailis. Gas-assisted focused electron beam and ion beam processing and fabrication. *Journal of Vacuum Science & Technology B* 26 (2008): 1197–1276.

[6] Lassiter, M. G., and P. D. Rack. Nanoscale electron beam induced etching: a continuum model that correlates the etch profile to the experimental parameters. *Nanotechnology* 19 (2008): 455306

[7] Toth, M., C. J. Lobo, M. J. Lysaght, A. E. Vlada'r, and M. T. Postek. Contamination-free imaging by electron induced carbon volatilization in environmental scanning electron microscopy. *Journal of Applied Physics* 106 (2009): 034306.

[8] Silvis-Cividjian, N., C. W. Hagen, and P. Kruit.Spatial resolution limits in electron-beam-induced deposition. *Journal Of Applied Physics* 98 (2005): 084905.

[9] Silvis-Cividjian, N., C. W. Hagen, P. Kruit, M. A. J. Van der Stam, and H. B. Groen. Direct fabrication of nanowires in an electron microscope. *Appl. Phys. Lett.* 82 (2003): 3514–3516.

[10] Silvis-Cividjian, N., C. W. Hagen, L. H. A. Leunissen, and P. Kruit. The role of secondary electrons in electron-beam-induced-deposition spatial resolution. *Microelectron. Eng.* 61–62 (2002): 693–699.

[11] Mitsuishi, K., Z. Q. Liu, M. Shimojo, M. Han, and K. Furuya. Dynamic profile calculation of deposition resolution by high-energy electrons in electron-beam-induced deposition. *Ultramicroscopy* 103 (2005): 17–22.

[12] J. D. Fowlkes, S. J. Randolph, and P. D. Rack. Growth and simulation of high-aspect ratio nanopillars by primary and secondary electron-induced deposition. *J. Vac. Sci. Technol. B* 23 (2005): 2825–2832.

[13] Liu, Z. Q., K. Mitsuishi, and K. Furuya. Modeling the process of electron-beam-induced deposition by dynamic Monte Carlo simulation. *Jpn. J. Appl. Phys. Part 1 – Regul. Pap. Brief Commun. Rev. Pap.* 44 (2005): 5659–5663.

[14] Liu, Z. Q., K. Mitsuishi, and K. Furuya. A dynamic Monte Carlo study of the in situ growth of a substance deposited using electron-beam-induced deposition. *Nanotechnology* 17 (2006): 3832–3837.

[15] D. A. Smith, J. D. Fowlkes, and P. D. Rack. Understanding the kinetics and nanoscale morphology of electron-beam-induced deposition via a three-dimensional Monte Carlo simulation: The effects of the precursor molecule and the deposited material. *Small* 4 (2008): 1382–1389.

[16] D. A. Smith, J. D. Fowlkes, and P. D. Rack. A nanoscale three-dimensional Monte Carlo simulation of electron-beam-induced deposition with gas dynamics. *Nanotechnology* 18 (2007): 1–14.

[17] Rykaczewski, K., W. B. White, and A. G. Fedorov. Analysis of electron beam induced deposition (EBID) of residual hydrocarbons in electron microscopy. *J. Appl. Phys.* 101 (2007): 054307.

[18] Reimer, L. *Scanning electron microscopy: Physics of image formation and microanalysis.* (Berlin: Springer, 1998).

[19] Goldstein, J. I., D. E. Newbury, P. Echlin, D. C. Joy, A. D. Roming, Jr., C. E. Lyman, C. Fiori, and E. Lifshin. *Scanning electron microscopy and X-ray microanalysis.* New York: Plenum Press, 1992.

[20] Seiler, H. Secondary-electron emission in the scanning electron-microscope. *Journal of Applied Physics* 54 (1983): R1–R18.

[21] Cazaux, J. Mechanisms of charging in electron spectroscopy. *J. Electron Spectrosc. Relat. Phenom.* 105 (1999): 155–185.

[22] Cazaux, J. Some considerations on the secondary electron emission, delta, from e(-) irradiated insulators. *Journal of Applied Physics* 85 (1999): 1137–1147.

[23] Cazaux, J. About the secondary electron yield and the sign of charging of electron irradiated insulators. *Eur. Phys. J.-Appl. Phys* 15 (2001): 167–172.

[24] Cazaux, J. Charging in scanning electron microscopy "from inside and outside." *Scanning* 26 (2004): 181–203.

[25] Cazaux, J. Scenario for time evolution of insulator charging under various focused electron irradiations. *Journal of Applied Physics* 95 (2004): 731–742.

[26] Fitting, H. J., E. Schreiber, and I. A. Glavatskikh. Monte Carlo modeling of electron scattering in nonconductive specimens. *Microsc. microanal.* 10 (2004): 764–770.

[27] Glavatskikh, I. A., V. S. Kortov, and H. J. Fitting. Self-consistent electrical charging of insulating layers and metal-insulator-semiconductor structures. *Journal of Applied Physics* 89 (2001): 440–448.

[28] Meyza, X., D. Goeuriot, C. Guerret-Piecourt, D. Treheux, and H. J. Fitting. Secondary electron emission and self-consistent charge transport and storage in bulk insulators: Application to alumina. *Journal Of Applied Physics* 94 (2003): 5384–5392.

[29] Li, J. T., M. Toth, V. Tileli, K. A. Dunn, C. J. Lobo, and B. L. Thiel. Evolution of the nanostructure of deposits grown by electron beam induced deposition. *Appl. Phys. Lett.* 93 (2008): 023130.

[30] Rempfer, G. R., and M. S. Mauck. A closer look at the effect of lens aberrations and object size on the intensity distribution and resolution in electron optics. *Journal of Applied Physics* 63 (1988): 2187–2199.

[31] Tuggle, D. W., L. W. Swanson, and M. A. Gesley. Current-density distribution in a chromatically limited electron-microprobe. *J. Vac. Sci. Technol. B* 4 (1986): 131–134.

[32] Joy, D. C. *Monte Carlo modeling for electron microscopy and microanalysis.* New York: Oxford University Press, 1995.

[33] Hovington, P., D. Drouin, and R. Gauvin. CASINO: A new Monte Carlo code in C language for electron beam interaction.1. Description of the program. *Scanning* 19 (1997): 1–14.

[34] Drouin, D., P. Hovington, and R. Gauvin. CASINO: A new Monte Carlo code in C language for electron beam interactions. 2. Tabulated values of the Mott cross section. *Scanning* 19 (1997): 20–28.

[35] Hovington, P., D. Drouin, R. Gauvin, D. C. Joy, and N. Evans. CASINO: A new Monte Carlo code in C language for electron beam interactions. 3. Stopping power at low energies. *Scanning* 19 (1997): 29–35.

[36] Kanaya, K., and S. Okayama. Penetration and energy-loss theory of electrons in solid targets. *J. Phys. D-Appl. Phys.* 5 (1972): 43–58.

[37] Cazaux, J. Correlation between the x-ray induced and the electron-induced electron emission yields of insulators. *Journal of Applied Physics* 89 (2001): 8265–8272.

[38] Cazaux, J. e-Induced secondary electron emission yield of insulators and charging effects. *Nuclear Instruments & Methods in Physics Research Section B-Beam Interactions with Materials and Atoms* 244 (2006): 307–322.

[39] Kanaya, K., S. Ono, and F. Ishigaki. Secondary-electron emission from insulators. *Journal of Physics D-Applied Physics* 11 (1978): 2425–2437.

[40] Schreiber, E., and H. J. Fitting. Monte Carlo simulation of secondary electron emission from the insulator SiO_2. *J. Electron Spectrosc. Relat. Phenom.* 124 (2002): 25–37.

[41] Randolph, S. J., J. D. Fowlkes, and P. D. Rack. Focused, nanoscale electron-beam-induced deposition and etching. *Crit. Rev. Solid State Mat. Sci.* 31 (2006): 55–89.

[42] Friedli, V., C. Santschi, J. Michler, P. Hoffmann, and I. Utke. Mass sensor for in situ monitoring of focused ion and electron beam induced processes. *Appl. Phys. Lett.* 90 (2007): 053106.

[43] Friedli, V., and I. Utke. Optimized molecule supply from nozzle-based gas injection systems for focused electron- and ion-beam induced deposition and etching: simulation and experiment. *Journal of Physics D-Applied Physics* 42 (2009): 125305.

[44] Utke, I., V. Friedli, S. Amorosi, J. Michler, and P. Hoffmann. Measurement and simulation of impinging precursor molecule distribution in focused particle beam deposition/etch systems. *Microelectron. Eng.* 83 (2006): 1499–1502.

[45] Danilatos, G. D. Foundations of environmental scanning electron microscopy. *Advances in Electronics and Electron Physics* 71 (1988): 109–250.

[46] Thiel, B. L., and M. Toth. Secondary electron contrast in low-vacuum/environmental scanning electron microscopy of dielectrics. *Journal of Applied Physics* 97 (2005): 051101.

[47] Stokes, D. J. *Principles and practice of variable pressure/environmental scanning electron microscopy (VP-ESEM).* New York: John Wiley and Sons, 2009.

[48] Adachi, H., H. Nakane, and M. Katamoto. Field-emission characteristics of a carbon needle made by use of electron-beam-assisted decomposition of methane. *Appl. Surf. Sci.* 76 (1994): 11–15.

[49] Folch, A., J. Tejada, C. H. Peters, and M. S. Wrighton. Electron-Beam Deposition Of Gold Nanostructures In A Reactive Environment. *Applied Physics Letters* 66 (1995): 2080–2082.

[50] Folch, A., J. Servat, J. Esteve, J. Tejada, and M. Seco. High-vacuum versus "environmental" electron beam deposition. *J. Vac. Sci. Technol. B* 14 (1996): 2609–2614.

[51] Ochiai, Y., J. Fujita, and S. Matsui. Electron-beam-induced deposition of copper compound with low resistivity. *J. Vac. Sci. Technol. B* 14 (1996): 3887–3891.

[52] Molhave, K., D. N. Madsen, A. M. Rasmussen, A. Carlsson, C. C. Appel, M. Brorson, C. J. H. Jacobsen, and P. Boggild. Solid gold nanostructures fabricated by electron beam deposition. *Nano Lett.* 3 (2003): 1499–1503.

[53] Molhave, K., D. N. Madsen, S. Dohn, and P. Boggild. Constructing, connecting and soldering nanostructures by environmental electron beam deposition. *Nanotechnology* 15 (2004): 1047–1053.

[54] White, W. B., K. Rykaczewski, and A. G. Fedorov. What controls deposition rate in electron-beam chemical vapor deposition? *Phys. Rev. Lett.* 97 (2006): 086101.

[55] Toth, M., C. J. Lobo, W. R. Knowles, M. R. Phillips, M. T. Postek, and A. E. Vladar. Nanostructure fabrication by ultra-high-resolution environmental scanning electron microscopy. *Nano Lett.* 7 (2007): 525–530.

[56] Toth, M., W. R. Knowles, and B. L. Thiel. Secondary electron imaging of nonconductors with nanometer resolution. *Appl. Phys. Lett.* 88 (2006): 023105.

[57] Koops, H. W. P., A. Kaya, and M. Weber. Fabrication and characterization of platinum nanocrystalline material grown by electron-beam induced deposition. *J. Vac. Sci. Technol. B* 13 (1995): 2400–2403.

[58] Hoyle, P. C., J. R. A. Cleaver, and H. Ahmed. Ultralow-Energy Focused Electron-Beam-Induced Deposition. *Appl. Phys. Lett.* 64 (1994): 1448–1450.

[59] Alman, D. A., D. N. Ruzic, and J. N. Brooks. A hydrocarbon reaction model for low temperature hydrogen plasmas and an application to the Joint European Torus. *Physics of Plasmas* 7 (2000): 1421–1432.

[60] Christophorou, L. G., and J. K. Olthoff. Electron interactions with C_2F_6. *Journal of Physical And Chemical Reference Data* 27 (1998): 1–29.

[61] Christophorou, L. G., and J. K. Olthoff. Electron interactions with C_3F_8. *Journal of Physical And Chemical Reference Data* 27 (1998): 889–913.

[62] Christophorou, L. G., and J. K. Olthoff. Electron interactions with Cl_2. *Journal of Physical And Chemical Reference Data* 28 (1999): 131–169.

[63] Christophorou, L. G., and J. K. Olthoff. Electron interactions with plasma processing gases: An update for CF_4, CHF_3, C_2F_6, and C_3F_8. *Journal of Physical and Chemical Reference Data* 28 (1999): 967–982.

[64] Christophorou, L. G., and J. K. Olthoff. Electron interactions with CF_3I. *Journal of Physical and Chemical Reference Data* 29 (2000): 553–569.

[65] Christophorou, L. G., and J. K. Olthoff. Electron interactions with SF_6. *Journal of Physical and Chemical Reference Data* 29 (2000): 267–330.

[66] Christophorou, L. G., and J. K. Olthoff. Electron interactions with $c\text{-}C_4F_8$. *Journal of Physical and Chemical Reference Data* 30 (2001): 449–473.

[67] Christophorou, L. G., and J. K. Olthoff. Electron interactions with BCl_3. *Journal of Physical and Chemical Reference Data* 31 (2002): 971–988.

[68] Christophorou, L. G., J. K. Olthoff, and M. Rao. Electron interactions with CF_4. *Journal Of Physical And Chemical Reference Data* 25 (1996): 1341–1388.

[69] Christophorou, L. G., J. K. Olthoff, and M. Rao. Electron interactions with CHF_3. *Journal of Physical and Chemical Reference Data* 26 (1997): 1–15.

[70] Christophorou, L. G. J. K. Olthoff, and Y. C. Wang. Electron interactions with CCl_2F_2. *Journal Of Physical And Chemical Reference Data* 26 (1997): 1205–1237.

[71] Font, G. I., W. L. Morgan, and G. Mennenga. Cross-section set and chemistry model for the simulation of $c\text{-}C_4F_8$ plasma discharges. *Journal of Applied Physics* 91 (2002): 3530–3538.

[72] Huo, W. M., and Y. K. Kim. Electron collision cross-section data for plasma modeling. *IEEE Transactions on Plasma Science* 27 (1999): 1225–1240.

[73] Morgan, W. L. A critical-evaluation of low-energy electron-impact cross-sections for plasma processing modeling. 1. Cl_2, F_2, and HCl. *Plasma Chemistry and Plasma Processing* 12 (1992): 449–476.

[74] Morgan, W. L. A critical-evaluation of low-energy electron-impact cross-sections for plasma processing modeling. 2. CF_4, SiH_4, and CH_4. *Plasma Chemistry and Plasma Processing* 12 (1992): 477–493.

[75] Morgan, W. L. Electron collision data for plasma chemistry modeling. *Advances in Atomic, Molecular, and Optical Physics* 43 (2000): 79–110.

[76] Morgan, W. L., C. Winstead, and V. McKoy. Electron cross section set for CHF_3. *Journal of Applied Physics* 90 (2001): 2009–2016.

[77] Morgan, W. L., C. Winstead, and V. McKoy. Electron collision cross sections for tetraethoxysilane. *Journal of Applied Physics* 92 (2002): 1663–1667.

[78] Perrin, J., O. Leroy, and M. C. Bordage. Cross-sections, rate constants and transport coefficients in silane plasma chemistry. *Contributions to Plasma Physics* 36 (1996): 3–49.

[79] Rozum, I., P. Limao-Vieira, S. Eden, J. Tennyson, and N. J. Mason. Electron interaction cross sections for CF_3I, C_2F_4, and CF_x (x=1–3) radicals. *Journal of Physical and Chemical Reference Data* 35 (2006): 267–284.

[80] Tawara, H., Y. Itikawa, H. Nishimura, and M. Yoshino. Cross-sections and related data for electron collisions with hydrogen molecules and molecular-ions. *Journal of Physical and Chemical Reference Data* 19 (1990): 617–637.

[81] Torres, I., R. Martinez, and F. Castano. Electron-impact dissociative ionization of the CH_3F molecule. *Journal of Physics B-Atomic Molecular and Optical Physics* 35 (2002): 4113–4123.

[82] Yoshida, K., S. Goto, H. Tagashira, C. Winstead, B. V. McKoy, and W. L. Morgan. Electron transport properties and collision cross sections in C_2F_4. *Journal Of Applied Physics* 91 (2002): 2637–2647.

[83] Bozso, F., and P. Avouris. Electron-induced chemical vapor-deposition by reactions induced in adsorbed molecular layers. *Appl. Phys. Lett.* 53 (1988): 1095–1097.

[84] Bozso, F., and P. Avouris. Thermal and electron-beam-induced reaction of disilane on Si(100)-(2x1). *Physical Review B* 38 (1988): 3943–3947.

[85] Ekerdt, J. G., Y. M. Sun, A. Szabo, G. J. Szulczewski, and J. M. White. Role of surface chemistry in semiconductor thin film processing. *Chemical Reviews* 96 (1996): 1499–1517.

[86] Guise, O., H. Marbach, J. Levy, J. Ahner, and J. T. Yates. Electron-beam-induced deposition of carbon films on Si(100) using chemisorbed ethylene as a precursor molecule. *Surf. Sci.* 571 (2004): 128–138.

[87] Huels, M. A., L. Parenteau, and L. Sanche. Substrate dependence of electron-stimulated O- yields from dissociative electron-attachment to physisorbed O_2. *J. Chem. Phys.* 100 (1994): 3940–3956.

[88] Joyce, B. A., and J. H. Neave. Electron beam-adsorbate interactions on silicon surfaces. *Surf. Sci.* 34 (1973): 401–419.

[89] Klyachko, D., P. Rowntree, and L. Sanche. Oxidation of hydrogen-passivated silicon surfaces induced by dissociative electron attachment to physisorbed H_2O. *Surf. Sci.* 346 (1996): L49–L54.

[90] Kunz, R. R., and T. M. Mayer. Electron-beam induced surface nucleation and low-temperature decomposition of metal-carbonyls. *J. Vac. Sci. Technol. B* 6 (1988): 1557–1564.

[91] Lu, Q. B., and L. Sanche. Enhancements in dissociative electron attachment to CF_4, chlorofluorocarbons and hydrochlorofluorocarbons adsorbed on H_2O ice. *J. Chem. Phys.* 120 (2004): 2434–2438.

[92] Mezhenny, S., I. Lyubinetsky, W. J. Choyke, and J. T. Yates. Electron stimulated decomposition of adsorbed hexafluoroacetylacetonate Cu(I) vinyltrimethylsilane, Cu(I)(hfac)(vtms). *Journal of Applied Physics* 85 (1999): 3368–3373.

[93] Sanche, L. Dissociative attachment and surface-reactions induced by low-energy electrons. *J. Vac. Sci. Technol. B* 10 (1992): 196–200.

[94] Sanche, L., and L. Parenteau. Dissociative attachment in electron-stimulated desorption from condensed NO and N_2O. *J. Vac. Sci. Technol. A-Vac. Surf. Films* 4 (1986): 1240–1242.

[95] Sanche, L., and L. Parenteau. Ion-molecule surface-reactions induced by slow (5–20 eV) electrons. *Physical Review Letters* 59 (1987): 136–139.

[96] Shek, M. L., S. P. Withrow, and W. H. Weinberg. Electron-beam induced desorption and dissociation of CO chemisorbed on Ir(111). *Surf. Sci.* 72 (1978): 678–692.

[97] Xu, J. Z., W. J. Choyke, and J. T. Yates. Enhanced silicon oxide film growth on Si(100) using electron impact. *Journal Of Applied Physics* 82 (1997): 6289–6292.

[98] Itikawa, Y., and N. Mason. Cross sections for electron collisions with water molecules. *J. Phys. Chem. Ref. Data* 34 (2005): 1–22.

[99] Lombardo, S. J., and A. T. Bell. A review of theoretical-models of adsorption, diffusion, desorption, and reaction of gases on metal-surfaces. *Surface Science Reports* 13 (1991): 1–72.

[100] Barth, J. V. Transport of adsorbates at metal surfaces: from thermal migration to hot precursors. *Surface Science Reports* 50 (2000): 75–149.

[101] Seebauer, E. G., and C. E. Allen. Estimating surface diffusion coefficients. *Progress in Surface Science* 49 (1995): 265–330.

[102] Wetterer, S. M., D. J. Lavrich, T. Cummings, S. L. Bernasek, and G. Scoles. Energetics and Kinetics of the Physisorption of Hydrocarbons on Au(111). *Journal of Physical Chemistry B* 102 (1998): 9266–9275.

[103] DeWitt, K. M., L. Valadez, H. L. Abbott, K. W. Kolasinski, and I. Harrison. Effusive molecular beam study of C_2H_6 dissociation on Pt(111). *Journal of Physical Chemistry B* 110 (2006): 6714–6720.

[104] Kuch, W., W. Schnurnberger, M. Schulze, and K. Bolwin. Equilibrium determination of H_2O desorption kinetic parameters of H_2O/K/Ni(111). *Journal of Chemical Physics* 101 (1994): 1687–1692.

[105] Sjovall, P., and P. Uvdal. Adsorption of oxygen on Pd(111): Precursor kinetics and coverage-dependent sticking. *J. Vac. Sci. Technol. A-Vac. Surf. Films* 16 (1998): 943–947.

[106] Artsyukhovich, A. N. Low temperature sticking and desorption dynamics of oxygen on Pt(111). *Surface Science* 347 (1996): 303–318.

[107] Behringer, E. R., H. C. Flaum, D. J. D. Sullivan, D. P. Masson, E. J. Lanzendorf, and A. C. Kummel. Effect of incident translational energy and surface temperature on the sticking probability of F_2 and O_2 on Si(100) 2x1 and Si(111) 7x7. *Journal of Physical Chemistry* 99 (1995): 12863–12874.

[108] Kastanas, G. N., and B. E. Koel. Interaction of Cl_2 with the Au(111) surface in the temperature-range of 120-K To 1000-K. *Applied Surface Science* 64 (1993): 235–249.

[109] Flaum, H. C., D. J. D. Sullivan, and A. C. Kummel. Mechanisms of halogen chemisorption upon a semiconductor surface-I_2, Br_2, Cl_2 and C_6H_5Cl chemisorption upon the Si(100) (2x1) surface. *Journal of physical chemistry* 98 (1994): 1719–1731.

[110] Cohen, S. L., M. Liehr, and S. Kasi. Surface-analysis studies of copper chemical vapor-deposition from 1,5-cyclooctadiene-copper(I)-hexafluoroacetylacetonate. *J. Vac. Sci. Technol. A* 10 (1992): 863–868.

[111] Coon, P. A., M. L. Wise, and S. M. George. Reaction-kinetics of $GeCl_4$ on Si(111) 7x7. *Surface Science* 278 (1992): 383–396.

[112] Coon, P. A., M. L. Wise, and S. M. George. Adsorption kinetics for ethylsilane, diethylsilane, and diethylgermane on Si(111) 7x7. *J. Chem. Phys.* 98 (1993): 7485–7495.

[113] Yuuki, A., Y. Matsui, and K. Tachibana. A study on radical fluxes in silane plasma CVD from trench coverage analysis. *Jpn. J. Appl. Phys. Part 1 – Regul. Pap. Short Notes Rev. Pap.* 28 (1989): 212–218.

[114] Rose, M., and J. W. Bartha. Method to determine the sticking coefficient of precursor molecules in atomic layer deposition. *Appl. Surf. Sci.* 255 (2009): 6620–6623.

[115] Burke, A., G. Braeckelmann, D. Manger, E. Eisenbraun, A. E. Kaloyeros, J. P. McVittie, J. Han, D. Bang, J. F. Loan, and J. J. Sullivan. Profile simulation of conformality of chemical vapor deposited copper in subquarter-micron trench and via structures. *J. Appl. Phys.* 82 (1997): 4651–4660.

[116] Winters, H. F. Etch products from the reaction of XeF_2 with SiO_2, Si_3N_4, SiC, and Si in the presence of ion-bombardment. *J. Vac. Sci. Technol. B* 1 (1983): 927–931.

[117] Winters, H. F., and J. W. Coburn. Etching of silicon with XeF_2 vapor. *Appl. Phys. Lett.* 34 (1979): 70–73.

[118] Ma, Z., and F. Zaera. Organic chemistry on solid surfaces. *Surface Science Reports* 61 (2006): 229–281.

[119] Lei, R. Z., A. J. Gellman, and B. E. Koel. Desorption energies of linear and cyclic alkanes on surfaces: anomalous scaling with length. *Surf. Sci.* 554 (2004): 125–140.

[120] Fichthorn, K. A., and R. A. Miron. Thermal desorption of large molecules from solid surfaces. *Phys. Rev. Lett.* 89 (2002): 196103.

[121] Tait, S. L., Z. Dohnalek, C. T. Campbell, and B. D. Kay. n-alkanes on Pt(111) and on C(0001)/Pt(111): Chain length dependence of kinetic desorption parameters. *J. Chem. Phys.* 125 (2006): 234308.

[122] Gellman, A. J., K. R. Paserba, and N. Vaidyanathan. Desorption kinetics of polyether lubricants from surfaces. *Tribology Letters* 12 (2005): 111–115.

[123] Koleske, D. D., A. E. Wickenden, R. L. Henry, W. J. DeSisto, and R. J. Gorman. Growth model for GaN with comparison to structural, optical, and electrical properties. *J. Appl. Phys.* 84 (1998): 1998–2010.

[124] Kodas, T. T., and M. J. Hampden-Smith (eds.). *The chemistry of metal CVD.* New York: VCH Publishers, 1994.

[125] Foord, J. S., and R. B. Jackman. Studies of adsorption and electron-induced dissociation of $Fe(CO)_5$ on Si(100). *Surface Science* 171 (1986): 197–207.

[126] Farkas, J., M. J. Hampdensmith, and T. T. Kodas. FTIR studies of the adsorption/desorption behavior of copper chemical-vapor-deposition precursors on silica. 1. Bis(1, 1, 1, 5, 5,5-Hexafluoroacetylacetonato)Copper(II). *Journal of Physical Chemistry* 98 (1994): 6753–6762.

[127] Nolan, P. D., M. C. Wheeler, J. E. Davis, and C. B. Mullins. Mechanisms of initial dissociative chemisorption of oxygen on transition-metal surfaces. *Accounts Chem. Res.* 31 (1998): 798–804.

[128] Seel, M., and P. S. Bagus. Ab initio cluster study of the interaction of fluorine and chlorine with the Si(111) surface. *Phys. Rev. B* 28 (1983): 2023–2038.

[129] Hefty, R. C., J. R. Holt, M. R. Tate, D. B. Gosalvez, M. F. Bertino, and S. T. Ceyer. Dissociation of a Product of a Surface Reaction in the Gas Phase: XeF_2 Reaction with Si. 92 (2004): 188302.

[130] Brand, J. L., M. V. Arena, A. A. Deckert, and S. M. George. Surface-diffusion of normal-alkanes on Ru(001). *J. Chem. Phys.* 92 (1990): 5136–5143.

[131] Weckesser, J., J. V. Barth & K. Kern. Direct observation of surface diffusion of large organic molecules at metal surfaces: PVBA on Pd(110). *Journal of Chemical Physics* 110 (1999): 5351–5354.

[132] Zeiri, Y., and D. Cohen. A theoretical study of the surface diffusion of large molecules. I. n-alkanetype chains on W(100). *J. Chem. Phys.* 97 (1992): 1531–1541.

[133] Raut, J. S., and K. A. Fichthorn. Diffusion mechanisms of short-chain alkanes on metal substrates: Unique molecular features. *J. Chem. Phys.* 108 (1998): 1626–1635.

[134] Choi, Y. R., P. D. Rack, S. J. Randolph, D. A. Smith, and D. C. Joy. Pressure effect of growing with electron beam-induced deposition with tungsten hexafluoride and tetraethylorthosilicate precursor. *Scanning* 28 (2006): 311–318.

[135] Firment, L. E., and G. A. Somorjai. Surface structures of normal paraffins and cyclohexane monolayers and thin crystals grown on the (111) crystal face of platinum: A low energy electron diffraction study. *J. Chem. Phys.* 66 (1977): 2904–2913.

[136] Blaszczyszyn, R., A. Ciszewski, M. Blaszczyszynowa, R. Bryl, and S. Zuber. The interaction of water with surfaces of Pt and Ir field emitters. *Applied Surface Science* 67 (1993): 211–217.

[137] Mitsui, T., M. K. Rose, E. Fomin, D. F. Ogletree, and M. Salmeron. Water diffusion and clustering on Pd(111). *Science* 297 (2002): 1850–1852.

[138] Thiel, P. A., and T. E. Madey. The interaction of water with solid-surfaces – Fundamental-aspects. *Surface Science Reports* 7 (1987): 211–385.

[139] Tsai, Y. L., and B. E. Koel. Importance of hydrocarbon fragment diffusion in the formation of adsorbed alkyls' via EID of multilayers on Pt(111). *J. Phys. Chem. B* 101 (1997): 4781–4786.

[140] Abe, H., T. Shimizu, A. Ando, and H. Tokumoto. Electric transport and mechanical strength measurements of carbon nanotubes in scanning electron microscope. *Physica E* 24 (2004): 42–45.

[141] Haller, I., and P. White. Polymerization of butadiene gas on surfaces under low energy electron bombardment. *Journal of Physical Chemistry* 67 (1963): 1784–1788.

[142] Fritzsche, C. R. Deposition of thin-films by beam induced polymerization of divinyl benzene. *J. Appl. Phys.* 53 (1982): 9053–9057.

[143] Ding, W., D. A. Dikin, X. Chen, R. D. Piner, R. S. Ruoff, E. Zussman, X. Wang, and X. Li. Mechanics of hydrogenated amorphous carbon deposits from electron-beam-induced deposition of a paraffin precursor. *Journal Of Applied Physics* 98 (2005): 014905.

[144] Kohlmann-von Platen, K. T., L. M. Buchmann, H. C. Petzold, and W. H. Brunger. Electron-beam induced tungsten deposition – Growth-rate enhancement and applications in microelectronics. *Journal Of Vacuum Science & Technology B* 10 (1992): 2690–2694.

[145] Beaulieu, D., Y. Ding, Z. L. Wang, and W. J. Lackey. Influence of process variables on electron beam chemical vapor deposition of platinum. *Journal of Vacuum Science & Technology B* 23 (2005): 2151–2159.

[146] Barry, J. D., M. Ervin, J. Molstad, A. Wickenden, T. Brintlinger, P. Hoffmann, and J. Melngailis. Electron beam induced deposition of low resistivity platinum from $Pt(PF_3)_4$. *J. Vac. Sci. Technol. B* 24 (2006): 3165–3168.

[147] Hubner, U., R. Plontke, M. Blume, A. Reinhardt, and H. W. P. Koops. On-line nanolithography using electron beam-induced deposition technique. *Microelectron. Eng.* 57–58 (2001): 953–958.

[148] Rotkina, L., S. Oh, J. N. Eckstein, and S. V. Rotkin. Logarithmic behavior of the conductivity of electron-beam deposited granular Pt/C nanowires. *Phys. Rev. B* 72 (2005): 233407.

[149] Guise, O., J. Ahner, J. Yates, and J. Levy. Formation and thermal stability of sub-10-nm carbon templates on Si(100). *Appl. Phys. Lett.* 85 (2004): 2352–2354.

[150] Komuro, M., H. Hiroshima, and A. Takechi. Miniature tunnel junction by electron-beam-induced deposition. *Nanotechnology* 9 (1998): 104–107.

[151] Frabboni, S., G. C. Gazzadi, and A. Spessot. Transmission electron microscopy characterization and sculpting of sub-1 nm Si-O-C freestanding nanowires grown by electron beam induced deposition. *Appl. Phys. Lett.* 89 (2006): 113108.

[152] Gottfried, J. M., K. J. Schmidt, S. L. M. Schroeder, and K. Christmann. Spontaneous and electron-induced adsorption of oxygen on Au(110)-(1 x 2). *Surface Science* 511 (2002): 65–82.

[153] Avery, N. R. An EELS study of N_2O adsorption on Pt(111). *Surface Science* 131 (1983): 501–510.

[154] Jackman, R. B., and J. S. Foord. The interaction of WF_6 with Si(100) – Thermal and photon induced reactions. *Surface Science* 201 (1988): 47–58.

[155] Surplice, N. A., and W. Brearley. The adsorption of carbon monoxide ammonia, and wet air on gold. *Surface Science* 52 (1975): 62–74.

CHAPTER 8

MONTE CARLO METHOD IN FEBID PROCESS SIMULATIONS

Kazutaka Mitsuishi

1 INTRODUCTION

Electron beam induced deposition (FEBID) is a technique that produces nanostructures by decomposing precursor gases with a focused electron beam. Its development is directed at improving the spatial resolution, process speed, material properties, and functionalities. In each of these topics, detailed understanding of the process is required, which can be gained through FEBID simulations.

FEBID involves many physical phenomena such as scattering of electrons, generation of secondary electrons, as well as adsorption, diffusion, and dissociation of the precursor gas; all those properties are related in the actual deposition process. Even if all physical laws governing each of these processes are known or can be approximated by models, the properties of the deposit, such as the size or the shape, cannot be predicted because they are a result of complex interaction between those processes. Only computer simulation can suggest, for example, how widely the electron trajectories are spread, or what kind of energy distribution of electron will be. Through simulation we can study the relation between the various experimental parameters and the properties of the resultant deposit, and obtain feedback for optimization of those parameters.

Although the basic physical laws governing each process are known, making a "reasonably accurate" model is still difficult. This is because of yet incomplete knowledge of the experimental parameters required for these models. For example, the dissociation cross-sections (which defines the rate of molecular dissociation by electron beam) are yet unknown for many precursor gases used in FEBID. Nevertheless, the simulation has been, and will be, providing profound understanding of the FEBID process. In this chapter, the basics of Monte Carlo method used in simulating the FEBID process are given for the reader unfamiliar with the technique. The text focuses on the simulation techniques, assumptions, and the physical interpretation of the simulations results.

2 MONTE CARLO METHOD

Monte Carlo (MC) method is a technique to reproduce a physical situation using random numbers; in the FEBID simulations, it is used to describe electron trajectory, secondary electron generation, precursor dissociation, precursor diffusion, and so forth.

The simulation of FEBID process by MC methods is an active field. The dynamic process simulation was first performed by Silvis-Cividjian et al. [1–3]. The precursor molecules were assumed to cover the material surface at any instant. However, during the FEBID process, precursor gas is supplied mainly by the surface diffusion, which depends on various factors. Recent MC simulation revealed the importance of precursor gas dynamics in the process simulation [4–10]. In this chapter, we start with describing electron scattering, and then outline FEBID simulation using the trajectory calculation by MC methods, with and without including the precursor gas dynamics. We shall also mention the dissociation cross-section that defines how the electron beam decomposes the gas molecules.

3 ELECTRON TRAJECTORY SIMULATION BY THE MONTE CARLO METHOD

The study of the electron interaction with materials has a long history in relation with the image interpretations of transmission electron microscopy (TEM) and scanning electron microscopy (SEM). A solid can be viewed as an aggregate of atoms of an order of Avogadro's number. Because electrons strongly interact with materials, they are scattered many times from one atom to another. Although the actual number of atoms that interact with incident electron is much smaller than the Avogadro's number, it is still impossible to treat each scattering event quantum-mechanically, except for very limited cases. For example, injecting electrons along a low-index zone axis of crystalline sample generates diffraction into certain sets of angles. In this special case, even though many atoms are involved in the scattering process, only limited diffraction angles need to be considered, and it is possible to solve, either analytically or numerically, the corresponding Schrödinger equation. This kind of calculation is called "dynamical calculation," and the word *dynamical* in this context indicates the multiple electron scattering process [11].

Solving the Schrödinger equation is virtually impossible in a situation when many atoms, located at random positions, participate in the electron scattering. This is exactly the case of FEBID, where the deposited materials are usually amorphous. Furthermore, the structure of the material also changes dynamically as the FEBID process proceeds.

On the other hand, because of the random configuration of atoms, we do not have to worry about the coherent interference of electron as a wave. This allows treating the scattering of electron by an atom in the same way as scattering of a particle in classical mechanics. In this treatment, although each electron scattering event appears random, its angular distribution coincides with that expected from quantum mechanics. Once the scattering nature of electron is parameterized, each scattering event can be expressed by a random number assigned by a computer, and we shall see later how this can be done. This is how MC calculation is performed. Using this technique, we can learn what happens when an electron impinges a material, namely how it moves through the material and what its energy distribution is. This allows us to characterize the FEBID process.

Several MC codes for electron trajectory simulations are described in the literature. Reimer outlined his FORTRAN 77 code in 1986. [12]. D. C. Joy described MC simulation details and related issues in his book in 1995 [13]. More recent Monte Carlo codes, which use Mott scattering factor, are discussed by Hovington et al. and Drouin et al. [14–16], and their executable files are available on the web [17]. Very recently, Kieft and Bosch developed a code combining the Mott cross-section with phonon-scattering based cross-sections for the elastic scattering and adopting a dielectric function theory approach for inelastic scattering and generation of secondary electrons [18].

We should note that in some special cases when the wave nature of electron is important, the MC simulations could result in completely false conclusions. This is of no concern for our FEBID simulations, but care must be taken when FEBID is performed in high-voltage TEM or when using crystalline samples where electron diffraction is important.

3.1 SCATTERING CROSS-SECTION

Scattering cross-section is the central value in the Monte Carlo electron trajectory simulation. The differential cross-section defines scattering into a solid angle Ω by an atom with atomic number Z. The total cross-section is obtained by integrating the differential cross-section over all angles and it indicates the strength of the scattering. In the simplest case, the differential cross-section $d\sigma$ is expressed by the Rutherford scattering cross-section σ as [19],

$$\frac{d\sigma}{d\Omega} = \frac{e^4 Z^2}{4(4\pi\varepsilon_0)^2 E^2} \left(\frac{E + m_0 c^2}{E + 2m_0 c^2}\right)^2 \frac{1}{(1 - \cos\theta)^2}.$$

Here e, m_0 and E are the electron charge, mass and energy, respectively; ε_0 is vacuum permittivity and θ is scattering angle. The screening effect of the Coulomb potential can be considered by including a characteristic angle θ_0 of elastic scattering or its square α as [19; 20],

$$\frac{d\sigma}{d\Omega} = 5.21 \times 10^{-21} \cdot \frac{Z^2}{E^2} \frac{1}{(1 - \cos\theta + 2\alpha)^2}, \alpha = 0.034 Z^{2/3} / E \text{ (keV)}.$$

This Rutherford cross-section is a good approximation for low-Z elements and high electron energies. However, it does not work well when the electron energy is smaller than the K shell ionization energy of the scattering atoms [21; 22]. For those low energies Mott cross-section can be used [20], but it is rather complex and is difficult to implement into the Monte Carlo simulations. Various efforts [23] have been made to simplify the use of Mott scattering cross-section, for example by adjusting the screening parameters or parameterizing the ratio between the Rutherford and Mott cross-sections. Recent analytical approximations of the total Mott cross-section and the parameterized differential-cross-section are available from Gauvin and Drouin [24, 25].

For even lower kinetic energies of below ~100 eV, the Mott cross-section is no longer valid because the de Broglie wavelength of electron becomes large so that the picture of isolated scattering event on ions breaks down. For the accurate treatment of this energy range, see Kieft and Bosch [18].

3.2 MONTE CARLO TREATMENT OF ELECTRON SCATTERING

Monte Carlo simulations treat electron scattering as scattering of classical particles. The nature of scattering is characterized by the above-mentioned scattering cross-section, which defines two characteristic values of scattering. One is the mean free path λ, which corresponds to the average distance that an electron travels before scattering and is inversely proportional to the scattering cross-section σ_E:

$$\lambda = \frac{m_A}{N_A \rho \sigma_E} \text{ [cm]}.$$

Here N_A is Avogadro's number, ρ is the target density in g/cm^3 and m_A is the atomic weight of the target in g/mol. This relation can be understood intuitively as: the larger the scattering cross-section, the more frequently the scattering will occur.

The distance that an individual electron travels between two scattering events may vary for each step (Figure 8.1); i.e., it can be considered a random process. In the Monte Carlo simulation, this step length is decided by a random number that is generated by a program while keeping the step distance equal to the mean free path when it is averaged over many steps.

The number of atoms in a target of area A and thickness dx is $N_A \rho A dx / m_A$. Assuming each atom has a cross-section of σ_E, the total effective cross-section becomes $N_A \rho A \sigma_E dx / m_A$. If there are n incident electrons in the beam, then the number of electrons dn that interact in the thickness dx of the target is given by:

$$\frac{\text{(interacting electrons)}}{\text{(incident electrons)}} = \frac{\text{(cross section)}}{\text{(total area)}}.$$

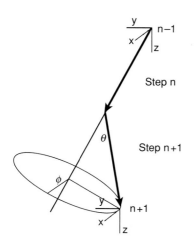

FIGURE 8.1: Definition of the coordinate system used in Monte Carlo simulation. Electron scattering is divided into steps whose lengths and corresponding scattering angles are decided by a scattering cross-section and a random number.

Therefore:

$$\frac{dn}{n} = -\frac{N_A \rho A \sigma_E dx}{m_A A} = -\frac{N_A \rho \sigma_E dx}{m_A}.$$

Integration gives:

$$n(x) = n_0 e^{-\frac{N_A \rho \sigma_E x}{m_A}} = n_0 e^{-\frac{x}{\lambda}},$$

where n_0 is the number of electrons entering the target. Therefore the mean free path is the sample thickness which reduced the number of incident electrons by $1/e = 0.368$. Using this value, we can estimate the probability of electron to travel a step length s.

The probability of electron to scatter between x and $x + dx$ is expressed as:

$$p(x)dx = \frac{1}{\lambda} e^{-\frac{x}{\lambda}} dx.$$

Integral of this probability function from $x = 0$ to infinity corresponds to a certain scattering, and indeed it gives 1. The expectation value of x can be calculated as:

$$\int_0^{\infty} x p(x) dx = \lambda,$$

meaning the average distance is λ. This is actually the original meaning of the mean-free-path that the electron scatters once while traveling the distance λ. The probability that an electron travels a certain distance s before scattering can be obtained by taking a ratio of integrals of $p(x)$ taken up to s and up to infinity as:

$$P(s) = \frac{\int_0^s p(x)dx}{\int_0^{\infty} p(x)dx} = 1 - \exp(-s/\lambda)$$

Here we introduce a random number RND that uniformly takes any value between 0 and 1. Equating $P(s)$ and RND expresses a step distance from a random number.

$$RND = 1 - \exp(-s/\lambda).$$

Thus, the step distance can be expressed as:

$$s = -\lambda \log(1 - RND).$$

Because either RND or $(1 - RND)$ is a random number, which uniformly takes any value between 0 and 1, we can write [13; 19]:

$$s = -\lambda \log(RND).$$

This formula relates the step distance and a random number.

The other characteristic value derived by the scattering cross-section is the angular distribution of scattering event. The probability that the electron scattered into an angle between Ω and $\Omega + d\Omega$ is the differential cross-section mentioned above, $\frac{d\sigma_E}{d\Omega}$.

The angle of electron scattering at a certain step can be expressed by a random number *RND* in the same manner as for the mean free path. Thus the equation that will be equated to a random number *RND* is:

$$P(s) = \frac{\int_0^s \frac{d\sigma_E}{d\Omega} d\Omega}{\int_0^\infty \frac{d\sigma_E}{d\Omega} d\Omega} = RND.$$

This equation can be solved analytically [13; 19; 26] for the Rutherford cross-section as:

$$\cos\theta = 1 - \frac{2\alpha RND}{1 + \alpha - RND}.$$

Solving the above equation for the Mott cross-section, however, requires a numerical integration and the scattering angle θ cannot be expressed as a simple function. Drouin [25] gives a parameterized form of the function as:

$$\cos(\theta_i^\beta) = 1 - \frac{2\alpha R^*}{1 + \alpha - R^*}$$

where,

$$\log_{10}(\alpha) = a + b\log_{10}(E) + c\log_{10}^2(E) + \frac{d}{e^{\log_{10}(E)}}$$

$$\beta^* = a + b\sqrt{E}\log(E) + \frac{c\log(E)}{E} + \frac{d}{E}$$

$$\beta = 1 \quad \text{if } \beta^* > 1$$
$$\beta^* = \beta^* \quad \text{if } \beta^* \leq 1$$

and the fitting parameters a, b, c, and d are given as a table in the paper [25].

As for the azimuthal angle, for each scattering event it is simply given by:

$$\varphi = 2\pi \times RND.$$

If the target consists of more than one atom species, the condition

$$RND > \sum_{i=1}^n \frac{F_i\sigma_i}{\sum_{j=1}^n F_j\sigma_j}$$

is evaluated at the beginning of each scattering event, where σ_i is the total scattering cross-section by an atom of element i, and F_i is the atomic fraction of element i [14]. This decides which atomic species scatters the electron. The corresponding mean free path is modified to:

$$\lambda = \frac{1 \times 10^{14}}{N_A} \frac{\sum_{i=1}^n \frac{C_iA_i}{\rho}}{\sum_{i=1}^n F_i\sigma_i} (cm),$$

where C_i is the weight fraction and A_i is the atomic weight of element i. This is a mean free path that is averaged by the ratio of each atomic species.

3.3 INELASTIC SCATTERING

Inelastic scattering consists of discrete scattering events where the scattering atoms experience transitions into excited states; for example, their core electrons may become free electrons. In Monte Carlo simulation, inelastic scattering is usually treated by an average energy loss, which the electron experiences upon traveling the step distance s. This average energy loss value is called *stopping power*; the approach is called *continuous slowing down approximation* and was introduced by Bethe back in 1930 [27] and modified by others [13; 19; 28].

The energy of an electron at i-th step can be expressed by the energy of the $(i-1)$ step as:

$$E_i = E_{i-1} + \frac{dE}{dS}s.$$

Emission of secondary electrons is also inelastic scattering, but it will be treated independently. Bethe approximation is used for the stopping power [13; 14; 19; 27; 28] as:

$$\frac{dE}{dS} = -\frac{7.85 \times 10^4 \cdot \rho}{E} \frac{CZ}{A} \ln\left(\frac{1.166E}{J}\right) \text{ [keV/cm]},$$

with the mean ionization potential:

$$J = 9.76Z + 58.5Z^{-0.19} \quad \text{or} \quad J = 11.5Z \text{ for } Z \le 12.$$

In the low-energy range, the formula fails because as the energy decreases, the number of inelastic events that the incident electron can excite decreases too. Joy and Luo (1989) [29] suggested an empirical approach that replaces the J value by J* calculated as:

$$J^* = \frac{J}{1 + k\dfrac{J}{E}}.$$

Here k is a function of Z [30]. More recently, stopping powers computed from experimental measurements have become available in a tabulated form [16]. They show significant deviation from the values of Joy and Luo at low energies [16].

3.4 THE SECONDARY ELECTRON MODELS

3.4.1 Streitwolf's source function model

Understanding the secondary electron generation and its distribution is crucial for modeling of the FEBID process. The generation and propagation of secondary electrons have been studied in relation to the spatial resolution of SEM images [13]. The details of the secondary electron emission mechanism are very complicated and are far beyond the scope of this chapter. However, there are two basic models that are often used in the Monte Carlo method. One is by Koshikawa and Shimizu [31] that employs the Streitwolf's source function as the mechanism of secondary electron generation. In this model, electron-electron interaction in the cascade process is taken into account by using the experimental mean free path values. The energy excitation function $S(E)$ is described using perturbation theory as:

$$S(E) = \frac{e^4 k_F^3}{3\pi E_p (E - E_F)^2},$$

where E_p is the primary electron energy, E_F is the Fermi energy, and k_F is the wave vector of the Fermi energy. $S(E)$ is the number of secondary electrons excited per unit energy into an energy interval between E and $E + dE$ per unit path length of the fast electron; its unit is $cm^{-1}\,eV^{-1}$. The energy of an excited secondary electron E can be obtained in the same manner using random numbers, by taking the ratio of two integrals of $S(E)$, one from $E_F + \Phi$ to E and another from $E_F + \Phi$ to E_p, where Φ is the work function. Equating it to a random number as:

$$RND = \frac{\int_{E_F+\phi}^{E} S(E)dE}{\int_{E_F+\phi}^{E_p} S(E)dE},$$

we obtain the following expression for E as a function of RND:

$$E = \frac{RND \cdot E_F - A(E_F + \phi)}{RND - A}, \quad \text{where} \quad A = \frac{(E_p - E_F)}{(E_p - E_F - \phi)}.$$

The position of an excitation is determined at random within a step of the primary electron trajectory. The angular distributions of secondary electrons are assumed to be spherically symmetric and are expressed using the classical binary collision model as:

$$E' = E \cos^2 \theta,$$

for one electron and for the other as:

$$E'' = E \sin^2 \theta,$$

where θ is the scattering angle.

3.4.2 Fast secondary model

Another model that is often used in MC simulations is the fast secondary model originally proposed by Murata [32]; it treats the production of secondary electron as a knock-on collision of a primary electron and a free electron. Within that model, the cross-section is described as:

$$\frac{d\sigma}{d\varepsilon} = \frac{\pi e^4}{E^2}\left(\frac{1}{\varepsilon^2} + \frac{1}{(1-\varepsilon)^2}\right).$$

Here $\varepsilon = \Delta E/E$ is the normalized energy transfer and e is electron charge. There are several choices for the cross-section [32], for example, quantum mechanical treatment of this interaction has been performed by Mott and by Moller. Gryzinsky gives the cross-section using a coulombic collision of two moving particles based on the classical binary-encounter theory. These models differ in the assumptions, but result in rather similar cross-sections, provided the energy of the secondary electron is sufficiently high so that it can be considered as a free electron. For convenience, we can restrict the value ε within a range of $0 < \varepsilon < 0.5$ defining the higher energy electron as the primary.

The energy of fast secondary electron is computed by solving the equation:

$$RND = \frac{\int_{\varepsilon_C}^{\varepsilon} \left(\frac{d\sigma}{d\varepsilon}\right) d\varepsilon}{\int_{\varepsilon_C}^{0.5} \left(\frac{d\sigma}{d\varepsilon}\right) d\varepsilon},$$

where ε_C is the cut-off energy. There is no large difference between the result for $\varepsilon_C = 0.01$ and 0.001 [32]. Therefore, usually ε_C is chosen as 0.01 [13]. Then:

$$\varepsilon = 1/(1000 - 998^* \, RND).$$

The deflection angles of the primary electron are expressed as:

$$\sin^2 \theta = \frac{2\varepsilon}{(2 + T - T\varepsilon)}, \quad \sin^2 \varphi = \frac{2(1 - \varepsilon)}{(2 + T\varepsilon)},$$

where T is the kinetic energy of the electron normalized to its rest mass (511 keV).

4 CELLULAR AUTOMATA METHOD FOR FEBID MODELING

Kunz et al. [33] discussed as early as 1987, using MC electron trajectory simulation, the importance of secondary electron distribution to the size of deposits obtainable by FEBID. Silvis-Cividjian et al. [2] reproduced their calculation for small beam diameters by considering only the secondary electron distribution. Figure 8.2 presents a MC simulation of the radial electron distribution on the surface of 10 nm thick Cu target irradiated by a zero-diameter 20 keV electron beam with zero diameter; $N(r)$ is the total secondary electron current, $I(r)$ is the radial surface density distribution and $N_{tot}(r)$ is the total number of secondary electrons emerged from the circle of radius r [2].

The distribution of secondary electrons is well focused at the beam center, and the diameter that includes 50% of secondary electrons is about 3 nm.

The gas decomposition process can be taken into account by considering the energy dependent function that provides the probability of decomposition of the gas by electron flux. The function is called dissociation cross-section $\sigma_{diss}(E)$. The amount of the deposit can be expressed as [14; 33; 34]:

$$R(x) = \int_0^{E_0} f(x, E)\sigma_{diss}(E)N dE,$$

where function $f(x, E)$ describes the energy and position dependent electron flux, N is the density of the precursor molecules. As mentioned above, it is difficult to obtain an appropriate functional form of the dissociation cross-section $\sigma_{diss}(E)$ for the gases which are typically used

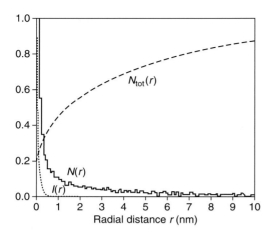

FIGURE 8.2: Monte Carlo simulations of secondary-electron emission, showing the radial spatial distribution of secondary electrons on the top surface of a 10-nm-thick Cu target irradiated by a zero-dimensional 20 keV electron beam. $N(r)$ is the total SE current, $I(r)$ is the radial surface density distribution, and $N_{tot}(r)$ is the total number of secondary electrons emerged from a circle of radius r. Reprinted with permission from Silvis-Cividjian et al., *J. Appl. Phys.* 98 (2005): 084905. Copyright (2005), American Institute of Physics.

in FEBID, such as $W(CO)_6$. Silvis-Cividjian et al., therefore, used hydrocarbon gases for their simulation [35]. This will be discussed more in detail further in this chapter.

Figure 8.3 shows their result of cross-sectional profiles of deposits generated by the primary electron beams of 2 and 0.2 nm diameter [2]. They used as the criteria of resolution the area on the target containing 50% of the dissociated molecules and found that the role of secondary electrons in the lateral resolution of the deposit is negligible when the primary beam diameter is 2 nm, but it becomes significant when the beam diameter is 0.2 nm. Note that the distribution of dissociated molecules is not identical to the distribution of the secondary electrons obtained in the previous figure; this demonstrates the importance of combining $f(x, E)$ and $\sigma_{diss}(E)$ in the simulation.

This simulation still contradicts the experimental results that a 2 nm diameter beam results in a deposit with a lateral size of 15 to 20 nm [36; 37]. The broadening was attributed to the secondary electrons emerged from the substrate. A new calculation technique was required that simulates the profile evolution during the deposition. A simulation program was developed that used two-dimensional cell-automata to predict the spatial evolution of the deposit. This way the simulation could account for the scattered electron distribution $f(x, E)$ as a result of electron scattering by the deposit.

This method could reproduce the experimental lateral growth of the deposited dots, namely a rapid diameter increase at an earlier stage (called intermediate stage) followed by a saturation (Figures 8.4 and 8.5) [2]. Therefore it was concluded that secondary electrons play an important role and impose a fundamental limit on the spatial resolution of FEBID.

This cell-automata treatment of FEBID process was later repeated by several authors [5; 6; 38–42]. Here we demonstrate the simulation details taking examples of Mitsuishi et al. [38].

In the simulation, the material is considered as an aggregation of cells. Each cell is treated as a cube having six {100} faces. A trade-off between the computational time and accuracy determines whether or not to add {110} and/or {111} faces.

Figure 8.6 presents an illustration of the program display window during the calculation. Each cube has a Boolean variable that indicates whether or not the cube has an exposed surface [38]. In each step of MC simulation, the program checks whether or not the electron trajectory crosses the exposed surface, and if it does, the product of electron flux and the dissociation cross-section is compared with a random number. If that random number exceeds the product,

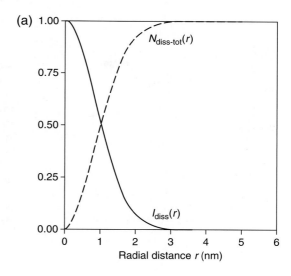

(a)

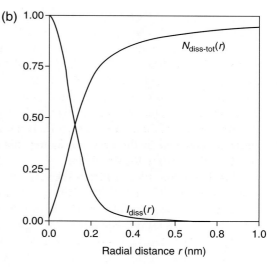

(b)

FIGURE 8.3: The normalized profile of a single dot fabricated using EBID by a 20-keV electron beam with (a) 2-nm diameter and (b) 0.2-nm diameter on a 10-nm-flat Cu target. Here $I_{diss}(r)$ is the normalized radial density distribution of the dissociated molecules on the surface and $N_{diss-tot}(r)$ is the normalized radial integral function, showing the number of dissociated molecules on a disk of radius r. The very beginning of the growth process is analyzed. Reprinted with permission from Silvis-Cividjian et al., *J. Appl. Phys.* 98 (2005): 084905. Copyright (2005), American Institute of Physics.

a new cell is added to the face. The Boolean information on whether or not the cell has an exposed surface is updated for the neighboring cells.

Using this procedure and regarding the substrate as an aggregation of cells, it is easy to start with an inhomogeneous sample shape that is hardly possible to treat with an analytical approach. For the example shown in Figure 8.7, FEBID processes are started with a small number of cells to see the effect of the substrate shape at different energies of the incident electrons. Using this simulation, Mitsuishi et al. and Liu et al. discussed the effect of the electron acceleration voltage, substrate, and lift-off angle upon lateral scanning during FEBID [38–41].

Some examples are shown in Figures 8.8 and 8.9. Here, the growth of a tungsten tip on a 10-nm-thick tungsten film is studied for two different acceleration voltages. Figure 8.8 presents the cross-sectional profiles of the tips, simulated with different numbers of impinging electrons. For 1–5 million of 200 keV electrons (panel a), which penetrate the substrate, the deposition mostly occurs at its rear side. However, for 50–250 thousand of 20 keV electrons (panel b),

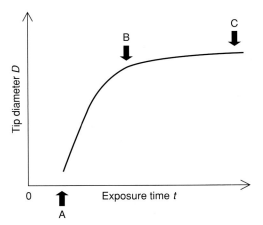

FIGURE 8.4: Experiments show this type of evolution of the FEEBID spot exposure response, $D(t)$. Three regimes can be distinguished: the nucleation stage (0-A), when no significant growth is observed; an intermediate regime (A-B), characterized by a fast growth of the diameter D; and the saturation regime (B-C), where the diameter attains a more or less constant value. Note that here the term *nucleation* is not associated with the thermodynamical concept of a critical nucleus. Here it is rather the minimum observable size determining the period 0-A. Reprinted with permission from Silvis-Cividjian et al., *J. Appl. Phys.* 98 (2005): 084905. Copyright (2005), American Institute of Physics.

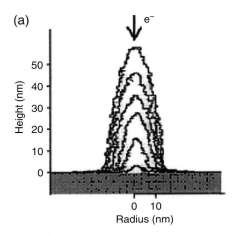

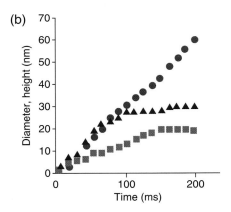

FIGURE 8.5: A sequence of simulated profiles for a single contamination dot grown by a zero-diameter 200-keV electron beam, at normal incidence on a 10-nm thick carbon foil; the time intervals corresponding to each profile are 18, 54, 90, 126, 162, 198 ms. (a) Sequence of cross-sectional profiles. (b) Time evolution of the dot geometry: the dot height (circles), the dot diameter measured at the base (triangles), and the dot diameter measured at half maximum of its height (squares). Saturation in the diameter is clearly observed. Reprinted with permission from Silvis-Cividjian et al., *J. Appl. Phys.* 98 (2005): 084905. Copyright (2005), American Institute of Physics.

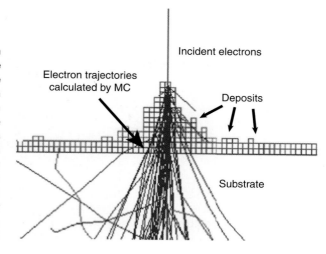

FIGURE 8.6: A schematic illustration of the calculation process of the dynamic profile simulator. Substrate and the deposit are considered as an aggregation of cells. Electron trajectories are calculated by Monte Carlo simulations, and a new cell is added when the electrons exit from the cells at a certain probability which is decided by the dissociation cross section function. Reprinted from Mitsuishi et al., *Ultramicroscopy* 103 (2005): 17–22. Copyright (2005), with permission from Elsevier.

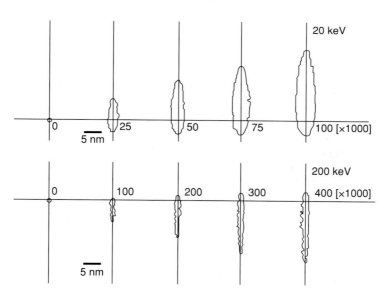

FIGURE 8.7: Cross-sections of deposits calculated by the dynamic profile simulation for two accelerating voltages. Zero-diameter beam falls on the substrate parallel to the page along the vertical lines. The number of injected electrons is indicated. Reprinted from Mitsuishi et al., *Ultramicroscopy* 103 (2005): 17–22. Copyright (2005), with permission from Elsevier.

the rear growth is small and diffuse whereas most deposition is occurs at the front surface. The shape of the tips shown in Figure 8.8 was characterized by such parameters as upside length, downside length, and diameters taken as full widths at half maximum (FWHM). These parameters are plotted for 200 and 20 keV electrons in Figures 8.9(a) and 8.9(b), respectively. The x-coordinate is the number of electrons bombarding the substrate, and the y-coordinate is the tip size (FWHM, upside length, and downside length) within the total range of 70 nm.

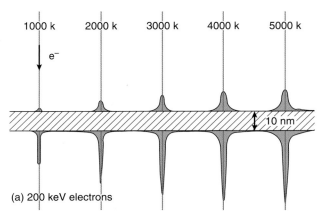

FIGURE 8.8: Cross-sectional profiles of tungsten tips produced by dynamic Monte Carlo simulation on a 10-nm-thick tungsten film. (a) Tips simulated with different numbers of 200 keV primary electrons show the saturation of the downside part on the bottom surface. (b) Tips simulated with different numbers of 20 keV primary electrons show the linear growth of upside part on the top surface Reprinted with permission from Liu et al., *Jpn. J. Appl. Phys.* 44 (2005): 5659–5663. Copyright (2005), Japan Society of Applied Phyics.

To distinguish the growth direction, a negative value is assigned for the downward growth. Figure 8.9(a) reveals that the upside length is small and it increases very slowly with the number of 200 keV electrons. However, the downside length increases abruptly at the initial stage and then saturates. Saturation is also observed for the FWHM diameters with the increase in the number of incident electrons. If a tip is grown on both sides of the film, the diameters are similar for the front (circles) and the rear parts (diamonds). Thus the rear part has a higher aspect ratio than the front part. For the 20 keV electrons in Figure 8.9(b), the downside length saturates quickly at a very small value compared with that in Figure 8.9(a). The upside FWHM diameter saturates after irradiation with 100 thousand electrons, and the upside length increases almost linearly with the number of electrons. According to these MC based dynamic growth simulations, the growth model of a deposit can be deduced as follows. If a substrate is semitransparent to electrons, deposits nucleate both at its top and bottom surfaces. After nucleation, a deposit preferentially grows downward with a high growth rate, and then the growth saturates. The upside part grows linearly with the number of electrons, and its length exceeds that of the downside part at a later stage. The lateral size of the deposit also saturates with the number of impinging electrons.

Silvis-Cividjian and Mitsuishi et al. did not consider gas dynamics and assumed that the surface is always covered with precursor gas molecules. In more recent studies, the growth regime was divided into two categories depending on molecule supply rates and molecule dissociation rates; the smallest sizes of FEB deposits also changed accordingly (see the chapter by Utke et al. in this book). In that sense, the simulations mentioned above correspond to the reaction-limited (electron-limited) regime and to the highest achievable resolution under the condition of no precursor molecule depletion.

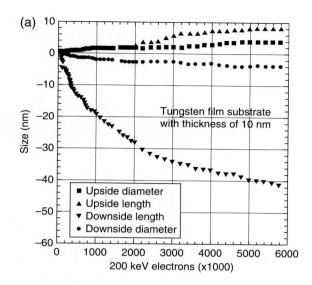

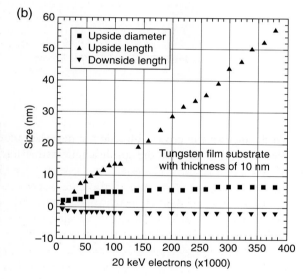

FIGURE 8.9: Plots of FWHM diameter, upside length, and downside length of simulated tungsten tips as functions of the numbers of (a) 200 keV and (b) 20 keV electrons. The diameter and downside length saturate, while the upside length increases linearly with the increasing number of impinging electrons Reprinted with permission from Liu et al., *Jpn. J. Appl. Phys.* 44 (2005): 5659–5663. Copyright (2005), Japan Society of Applied Physics.

5 DISSOCIATION CROSS-SECTION

Besides considering the gas dynamics, efforts are being made in improving the dissociation cross-section model, which is discussed in the next section. Compared to Chapter 4 by Moore et al. in this book, the energy dependencies used in FFEBID MC simulations were very simple but constituted a reasonable pragmatic approach. One should keep in mind that the energy-dependencies outlined below will not enable to simulate deposit compositions but rather to model deposition rates (deposit shapes).

The form of dissociation cross-section is given by the Bethe theory. The cross-section for n-state excitation can be written within the first Born approximation as [43]:

$$\sigma_n = \frac{4\pi a_0^2}{E/R} \frac{R}{E_n} f_n \ln\left[\frac{4c_n E}{R}\right],$$

where a_0 is Bohr radius, R is Rydberg energy, f_n is dipole oscillator strength and c_n are constants. The total cross-section can be obtained by a sum over all possible n as:

$$\sigma_{tot} = \sum_n \sigma_n = \frac{4\pi a_0^2}{E/R} M_{tot}^2 \ln\left[\frac{4c_{tot}E}{R}\right].$$

A similar formula is used for the ionization cross-section. [44–46]

Because the dissociation cross-section is unknown for the common precursor gases, such as $W(CO)_6$, Mitsuishi et al. used the asymptotic form as [38]:

$$\sigma_{tot} = \frac{A}{E} \ln\left[\frac{E}{B}\right],$$

where A and B are constants decided by the peak energy and threshold energy as 100 and 35.5 eV, respectively. Quantitative evaluation is therefore not possible, i.e., the volume of deposit per electron is arbitrary. This is generally the case when dissociation cross-sections are not precisely known in MC simulations.

Silvis-Cividjian et al. [1–3] used a general equation for the dissociation cross-section of hydrocarbon (C_xH_y) gases given by Alman et al. [35]. They adjusted the dissociation threshold energy to 3–4 eV using the experimental results obtained for C_2H_5 by scanning tunneling microscopy [47].

$$\sigma_{diss}(E) = \begin{cases} 0, & E \le E_{th} \\ \sigma_{max}\left(1 - \frac{(E_{max} - E)^2}{(E_{max} - E_{th})^2}\right), & E_{th} < E < E_{max} \\ \sigma_{max}e^{(E-E_{max})/\lambda}, & E \ge E_{max} \end{cases},$$

with

$$\sigma_{max} = (1.89\#C + 0.33\#H - 0.505) \times 10^{-2} \text{ nm}^2,$$

where . C and . H indicate the number of carbon and hydrogen atoms in the molecule.

Relying on the fact that the dissociation cross-section has similar functional form with the ionization cross-section [35; 43], Fowlkes et al. used the latter for the dissociation cross-section, with a small modification. First, the experimentally obtained ionization cross-section of WF_6 [48] is fitted as:

$$\sigma_{diss}(E) = \left(A_1\left(1 - \frac{1}{E}\right) + A_2\left(1 - \frac{1}{E}\right)^2 + A_3\left(\frac{\log E}{E}\right) + A_4 \log E\right)\frac{1}{E},$$

where A_1, A_2, A_3, and A_4 are constants. Then its behavior in the low-energy region is modified by normalizing the σ_{diss} with the ratio of dissociation and ionization cross-sections for C_2H_5

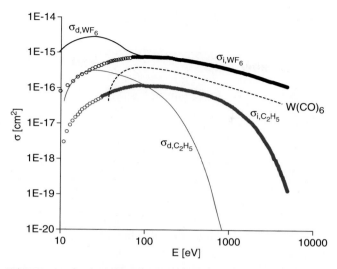

FIGURE 8.10: Approximate dissociation cross section for WF_6 gas ($\log \sigma$ vs $\log E$) estimated from the experimental determined ionization cross section for WF_6 and the estimated dissociation and ionization cross-sections for C_2H_5 based on extrapolations from experimental data on CH_x species. The dissociation cross-section for $W(CO)_6$ of Mitsuishi et al Ref. [38] is also added. Reprinted with permission from Fowlkes et al., *J. Vac. Sci. Technol.* B 23 (2005): 2825. Copyright (2005), American Vacuum Society.

obtained by Alman [35] (and used by Silvis-Cividjian) and equating it to the ionization cross-section above 100 eV, such that:

$$\sigma_{diss,WF_6} = \left[\frac{\sigma_{d,C_2H_5}}{\sigma_{i,C_2H_5}} \right] \cdot \sigma_{i,WF_6} \qquad \text{for} \quad \frac{\sigma_{d,C_2H_5}}{\sigma_{i,C_2H_5}} \geq 1,$$

$$\sigma_{diss,WF_6} = \sigma_{i,WF_6} \qquad \text{for} \quad \frac{\sigma_{d,C_2H_5}}{\sigma_{i,C_2H_5}} < 1 \quad \text{and for} \quad E > 100\,\text{eV}.$$

This complex procedure was performed because of the belief that Alman strongly underestimated the dissociation cross-section at high energies.

Figure 8.10 presents a comparison between those cross-sections [42]. Because σ_{diss} for WF_6 is set equivalent with the ionization cross-sections in the high-energy region, its value decreases with energy much slower than σ_{diss} of C_2H_5.

Using this dissociation cross-section, and the probability of electron-induced molecular dissociation per electron Q as:

$$Q = \frac{A \cdot S}{n_{PE}} \cdot \int^{E_f} n_x(E) \cdot \sigma(E)dE,$$

Fowlkes et al. [42] estimated the contributions of particular electron species x, which included primary electrons (PE), backscattered electrons (BSE), and secondary electrons produced from the incident beam (SE^I) and from the backscattered electrons (SE^{II}). Here θ is the percentage surface coverage of adsorbed precursor molecules, S is the density of the atomic surface sites, $n_x(E)$ is the electron energy distribution of species x calculated by MC simulations.

Table 8.1 Simulated probability of dissociation (Q) and the probability of dissociation normalized to the interaction area (Q_{area}) [Ref 42]

	Q	Q_{area} normalized
PE	0.183	1.00
SEI	0.155	0.85
SEII	0.100	1.09E-01
BSE	0.019	4.51E-05

The simulation assumes a uniform coverage of precursor molecules, i.e., the reaction-limited regime. The results are summarized in Table 8.1 [42]. It is interesting to note a large contribution of primary electrons, which is even larger than that of secondary electrons SE[I].

The column Q_{area} in Table 8.1 is the probability that a monolayer of atoms will be deposited normal to the substrate surface per electron species. The contribution of back-scattered electrons is small and becomes virtually negligible when normalized by the large BSE area.

Figure 8.11 plots the simulated contribution of primary and secondary electrons to the vertical growth rate of a tungsten nanopillar as a function of the incident electron energy [42]. The curves are normalized to the contribution of primary electrons to the vertical growth rate at 500 eV. The simulation is performed in the reaction-limited (electron-limited) regime. For all electron beam energies, the primary electron contribution is the largest, and the secondary electron contribution has a maximum at the energy of 5 keV.

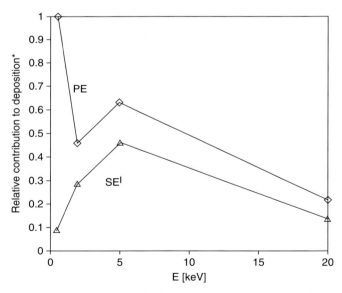

FIGURE 8.11: Relative contribution to vertical nanopillar growth for PE and SE[I] electron species. The simulation was carried out at E_0 = 500, 2000, 5000, and 20000 eV. The plot has been normalized to the contribution of PEs to the vertical growth rate at E_0 = 500 eV. Reprinted with permission from Fowlkes et al., *J. Vac. Sci. Technol.* B 23 (2005): 2825. Copyright (2005), American Vacuum Society.

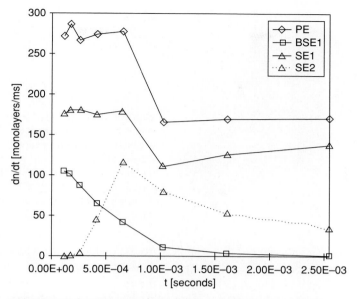

FIGURE 8.12: Growth rates in monolayers of tungsten per second separated based upon the electron species responsible for deposition. The primary and secondary electrons most significantly contribute to the vertical growth of the W nanopillar. Reprinted with permission from Fowlkes et al., *J. Vac. Sci. Technol.* B 23 (2005): 2825. Copyright (2005), American Vacuum Society.

Figure 8.12 is a plot of the nanopillar growth rate as a function of time for each contributing electron species [42]. The vertical nanopillar growth rate is highest for primary electrons. The contribution of secondary electrons is smaller for the vertical growth while it is larger for the volume of tungsten atoms, which aggregate and thereby increase the radius and the surface area of the nanopillar.

While the secondary electrons are responsible for the pillar shape in the carbon deposit simulated by Silvis et al., for tungsten deposition, primary electrons determine the deposition rate and the secondary electrons affect the volume. This difference is due to the dissociation cross-section used in the simulations, and it shows the importance of the dissociation cross-section model. Care must be taken when analyzing the simulation results, even for a qualitative understanding of which electrons cause the deposition. As shown below, it is now well recognized that the precursor gas dynamics drastically alter the shape of the deposit [4–7]. In that sense, the simulation considered so far excluded gas dynamics, and should be used for limited situations when the gas depletion could be neglected.

6 GAS DYNAMICS

6.1 CONTINUUM APPROACH

So far, no gas dynamics were considered and the precursor molecules were assumed to be covering the surface of a cell at all times. This corresponds to a situation where deposition is limited by the number of electrons and is called *reaction limited*[4].

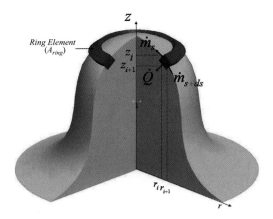

FIGURE 8.13: Schematic representation of a nanopillar with axial symmetry. The problem can be described by a single curvilinear coordinate with running along the surface of the deposit. Reprinted with permission from Rykaczewski et al., *J. Appl. Phys.* 101 (2007): 54307. Copyright (2007), American Institute of Physics.

In reality, the gas supply itself is a complex process including adsorption, diffusion, and reaction. The partial pressure of precursor gas should be small enough not to damage the electron source. Therefore, the gas diffusion from the area surrounding the electron beam plays an important role in the FEBID process, especially when the deposit shape (e.g., pillar) limits the diffusion path and the deposit shape changes during the deposition. When the diffusion is relatively slow the process is called *diffusion limited*.

The gas dynamics itself can be treated by a simple rate equation with the diffusion from the outside as sources, and adsorptions and reactions as sinks [4]. This rate equation can be combined with MC simulations as either continuum [7; 8] or cell models [5; 6].

6.2 CONTINUUM APPROACH OF FEBID MODELING

Pure continuum models are discussed Chapter 6 by Utke and Chapter 7 by Lobo and Toth in this book.

Rykaczewski et al. [8] developed a dynamic model of FEBID. It coupled surface mass transport, electron transport, scattering, and decomposition by a continuous representation of the deposit and the substrate. The surface transport equation is solved with a finite difference method, and an MC method is used for electron trajectory simulations.

The continuous approach has several advantages, including straightforward coupling with the surface mass transport equation and significant reduction of computing time compared to the above-mentioned cellular automata method.

Considering a cylindrically symmetric pillar growth (Figure 8.13), the problem can be described by a single curvilinear coordinate running along the surface of the deposit [8]. In the simulation, it is assumed that adsorption and desorption have reached an equilibrium and do not affect the surface concentration. It is also assumed that the shape of the pillar remains constant during the time step Δt.

Figure 8.14 shows the shape difference between the reaction-limited (electron-limited) and mixed (diffusion-enhanced) regimes [8]. When mass transport limits the deposition, it broadens the shape of the deposit. The highest spatial resolution is achieved for purely reaction-limited (electron-limited) regime. Generalized scaling laws of resolution depending on the FFEBID process regime could be derived using a pure continuum approach, see chapter by Utke in this book.

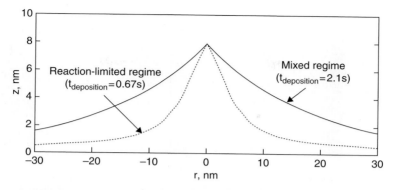

FIGURE 8.14: Comparison of deposit shapes simulated for the reaction-limited and mixed diffusion-reaction regime grown to the same height. The growth of the mixed regime deposit takes much longer (0.67 s) as compared to that (2.1 s) for the reaction-limited deposit. Reprinted with permission from Rykaczewski et al., *J. Appl. Phys.* 101 (2007): 54307. Copyright (2007), American Institute of Physics.

So far the coupled MC/continuum simulations were performed only for exposure times leading to low-aspect ratio deposits and did not yet prove their potential for simulation of high-aspect ratio shapes like pillars.

6.3 MONTE CARLO APPROACH FOR GAS DYNAMICS

Monte Carlo methods also allow the solution of the rate equations of gas dynamics. That was done by Smith et al. for a simulation of tungsten deposition from WF_6 [5] and SiO_2 deposition from a tetraethyl-orthosilicate (TEOS) precursor [6; 49]. This is advantageous when the substrate shape is non-planar with respect to the electron excitation volume, like high-aspect ratio pillars. Then continuum approaches are quantitatively limited.

In the simulation, diffusion is treated by random-walk motion and the number of gas jump per electron is determined by the surface diffusion coefficient, by the elapsed time for the gas to move, and by the cell separation distance. Eighteen possible <100> and <110> jump directions are considered while excluding eight <111> directions.

The number of gas molecules that impinge on the surface is:

$$N_{gas} = \Gamma_{gas} \cdot N_{site} \cdot \Delta x^2 \cdot t,$$

where N_{site} is the number of total gas surface sites within the simulated area, and the precursor gas flux is calculated based on the classical kinetic theory as:

$$\Gamma_{gas} = \frac{N_A P}{\sqrt{2\pi MRT}}.$$

Here the N_A is Avogadro's number, M is molecular weight of the gas, P is gas pressure, T is temperature, and R is the universal gas constant. As the deposit grows, the number of gas molecules adsorbed on the surface increases and the surface area increases. After determining the number of gas molecules that impinge on the surface, these gas molecules are randomly placed, and a molecule is adsorbed with a certain probability if the site is not occupied.

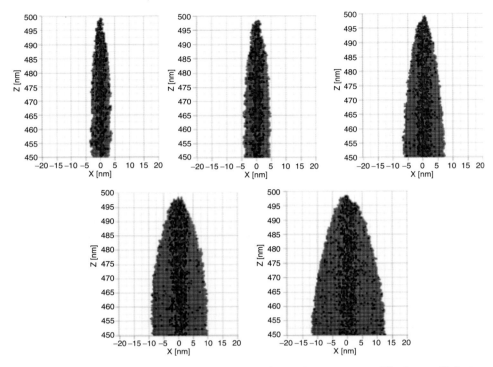

FIGURE 8.15: Cross-sections through the top 50 nm of the pillars at varying diffusion coefficients. Top row, left to right: 1.0×10^{-8}, 1.0×10^{-9}, and 1.0×10^{-10} cm^2 s^{-1}. Bottom row, left to right: 1.0×10^{-11}, and 0.0 cm^2 s^{-1}. The color represents the type of electron that induced its deposition: red = primary electron (PE); blue = backscattered electron (BSE); green = forward scattered electron (FSE); yellow = secondary electron type I (SEI); and cyan = secondary electron type II (SEII). Reprinted with permission from D. A. Smith et al., *Nanotechnology* 19 (2008): 415704. Copyright (2008), IOP Publishing Ltd.

Figure 8.15 presents an example of simulation results obtained by this technique, namely a cross-section of the top 50 nm of pillars formed using different diffusion coefficients, under diffusion-limited regime [49]. The SiO$_2$ molecules are colored by the type of electrons that induced its deposition: red = primary electrons (PE); blue = backscattered electrons (BSE); green = forward scattered electrons (FSE); yellow = secondary electrons type I (SE^I); and gray = secondary electrons type II (SE^{II}). The pillar broadens as the diffusion coefficient decreases, and the vertical growth slows down.

The number of deposits induced by FSE and SE^{II} is quite large for all diffusion coefficient values [5, 49]. Primary and SE^I contributions are strong when diffusion coefficient is large, but they decrease with decreasing diffusion coefficient. This can be explained by that PE and SE^I components contribute to deposition near the top of the pillars, and higher diffusion coefficient results in the higher coverage of the top area.

Figure 8.16 shows an example of how a diffusion coefficient affects the morphology of the deposit [49]. When the beam size is relatively large and the diffusion length is small, the precursor gas is consumed before it reaches the center of the irradiated area and the volcano (indented) shape is formed as observed in the experiment [50].

It is very interesting to compare these MC simulated shapes to similar shapes calculated from a pure continuum approach [4]. Performing both at the same lateral sub-10-nm scale

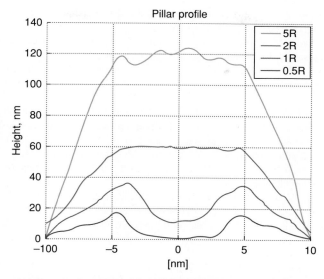

FIGURE 8.16: Comparison of resultant profiles for the boundary source simulations for the root mean square distances equal to 0.5r, 1.0r, 2.0r, and 5.0r runs after 750000 electrons, where r is the beam radius. Reprinted with permission from D. A. Smith et al., *Nanotechnology* 19 (2008): 415704. Copyright (2008), IOP Publishing Ltd.

would reveal when significant deviations appear due to secondary electron distribution effects, the energy dependence of the cross-sections, and the high-aspect ratio.

7 CONCLUSIONS

This chapter overviews Monte Carlo (MC) simulation techniques used in simulating the FEBID process, focusing the simulation procedures and validity of the assumptions. We saw that MC simulations are used in various aspects of FEBID processes and are considered as the only available method to treat the realistic growth conditions. Optimization of the shape of precursor supply nozzle using MC based gas flow simulation has been considered recently [51], which is a good example of expanding MC applications to FEBID.

Modeling difficulties still remain, not only quantitative, but also qualitative—mainly due to the lack of experimental data on the dissociation cross-section. This situation should be improved rapidly because of the expanding interests in the FEBID technique. Then the FEBID simulation could move on to more quantitative predictions, where further improvements of the techniques will be necessary.

REFERENCES

[1] Silvis-Cividjian, N., C. W. Hagen, P. Kruit, M. A. J. vd Stam, and H. B. Groen. Direct fabrication of nanowires in an electron microscope. *Appl. Phys. Lett.* 82 (2003): 3514–3516.

[2] Silvis-Cividjian, N., C. W. Hagen, and P. Kruit. Spatial resolution limits in electron-beam-induced deposition. *J. Appl. Phys.* 98 (2005): 084905.

[3] Hagen, C. W., W. F. van Drop, P. A. Crozier, and P. Kruit. Electronic pathways in nanostructure fabrication. *Surf. Sci.* 602 (2008): 3212–3219.

[4] Utke, I., V. Friedli, M. Purrucker, and J. Michler. Resolution in focused electron- and ion-beam induced processing. *J. Vac. Sci. Technol.* B25 (2007): 2219–2223.

[5] Smith, D. A., J. D. Fowlkes, and P. D. Rack. A nanoscale three-dimensional Monte Carlo simulation of electron-beam-induced deposition with gas dynamics. *Nanotechnology* 18 (2007): 265308.

[6] Smith, J. D., D. A. Fowlkes, and P. D. Rack. Understanding the kinetics and nanoscale morphology of electron-beam-induced deposition via a three dimensional monte carlo simulation: The effects of the precursor melocule and the deposited material. *Small* 4 (2008): 1382–1389.

[7] Lobo, C. J., M. Toth, R. Wagner, B. L. Thiel, and M. Lysaght. High-resolution radially symmetric nanostructures from simultaneous electron beam induced etching and deposition. *Nanotechnology* 19 (2008): 025303–025306.

[8] Rykaczewski, K., W. B. White, and A. Fedorov. Analysis of electron beam induced deposition (EBID) of residual hydrocarbons in electron microscopy. *J. Appl. Phys.* A101 (2007): 054307–054319.

[9] Rykaczewski, K., A. Marshall, W. B. White, and A. G. Fedorov. Dynamic growth of carbon nanopillars and microrings in electron beam induced dissociation of residual hydrocarbons. *Ultramicroscpy* 108 (2008): 989–992.

[10] Fedorov, A., K. Rykaczewski, and W. White. Transport issues in focused electron beam chemical vapor deposition. *Surface & Coatings Tech.* 201 (2007): 8808–8812.

[11] [11].J. C. H. Spence and J. M.Zuo, *Electron microdiffraction.* New York: Plenum Press, 1992.

[12] Reimer, L., and D. Stelter. FORTRAN 77 Monte-Carlo program for microcomputers using Mott cross-section. *Scanning* 8 (1986): 265–277.

[13] Joy, D. C. Monte Carlo modeling for electron microscopy and microanalysis. New York: Oxford University Press, 1995.

[14] Hovington, P., D. Drouin, and R. Gauvin. CASINO: A new ear of Monte Carlo code in C language for the electron beam interactions-Part I: Description of the program. *Scanning* 19 (1997): 1–14.

[15] Drouin, D., P. Hovington, and R. Gauvin. CASINO: A new ear of Monte Carlo code in C language for the electron beam interactions-Part II: Tabulated values of the Mott cross section. *Scanning* 19 (1997): 20–28.

[16] Hovington, P., D. Drouin, and R. Gauvin. CASINO: A new ear of Monte Carlo code in C language for the electron beam interactions-Part III: Stopping power. *Scanning* 19 (1997): 29–35.

[17] The web page of CASINO'i http://www.gel.usherbrooke.ca/casino/'j

[18] Kieft, E., and E. Bosch. Refinement of Monte Carlo simulations of electron-specimen interaction in low-voltage SEM. *J. Phys.* D 41 (2008): 215310.

[19] Reimer, L., and R. Snekel. Monte Carlo simulations in low voltage scanning electron microscopy. *Optik* 98 (1995): 85–94.

[20] Reimer, L., and B. Lödding. Calculation and tabulation of Mott cross-sections for large angle electron scattering. *Scanning* 6 (1984): 128–151.

[21] Kyser, D. F. Monte Carlo calculations for electron microscopy. *Microanalysis and lithography: Electron beam interaction with solid*, pp. 119–135 (SEM Inc, AMF O'Hare 1982).

[22] Böngeler, R., U. Golla, M. Kässens, L. Reimer, B. Schindler, R Senkel, and M. Spranck. Electron-specimen interactions in low-voltage scanning electron microscopy. *Scanning* 15 (1993): 1–18.

[23] Browning, R. Universal elastic scattering cross section for electrons in the range 1–100 keV. *Appl. Phys. Lett.* 58 (1991): 2845–2847.

[24] Gauvin, R., and D. Drouin. A formula to compute total elastic Mott cross-sections. *Scanning* 15 (1993): 140–150.

[25] Drouin, D., R. Gauvin, and D. C. Joy. Computation of polar angle of collisions from partial elastic Mott cross-sections. *Scanning* 16 (1994): 219–225.

[26] Newbury, D. E., and R. L. Myklebust. Monte Carlo electron trajectory simulation of beam spreading in thin foil targets. *Ultramicroscopy* 3 (1979): 391–395.

[27] Bethe, H. Zur Theorie des Durchganges schneller Korpuskularstrahlen durch Materie. *Ann. Phys.* 5 (1930): 325–400.

[28] Murata, K., T. Matukawa, and R. Shimizu. Monte Carlo calculations on electron scattering in a solid target. *Jpn. J. Appl. Phys.* 10 (1971): 678–686.

[29] Joy, D. C., and S. Luo. An empirical stopping power relationship for low-energy electrons. *Scanning* 11 (1989): 176–180.

[30] Gauvin, R., and G. L'Espérance. A Monte Carlo code to simulate the effect of fast seconday electron on kAB factors and spatial resolution in the TEM. *J. Microsc.* 168 (1992): 152–167.

[31] Koshikawa, T., and R. Shimizu. A Monte Carlo calculation of low-energy secondary electron emission from metals. J. *Phys.* D 7 (1974): 1303–1315.

[32] Murata, K., D. F. Kyse, and C. H. Ting. Monte Carlo simulation of fast secondary electron production in electron beam resists. *J. Appl. Phys.* 52 (1981): 4396–4405.

[33] Kunz, R. R., T. E. Allen, and T. M. Mayer. Selective are deposition of metals using low-energy electron beams. *J. Vac. Sci. Technol.* B5 (1987): 1427–1431.

[34] Koops, H. W. P., A. Kaya, and M. Weber. Fabrication and characterization of platinum nanocrystalline material grown by electron-beam induced deposition. *J. Vac. Sci. Technol* B 13 (1995): 2400–2403.

[35] Alman, D. A., D. N. Ruzic, and J. N. Brooks. A hydorocarbon reaction model for low temperature hydrogen plasmas and application to the Joint European Torus. *Phys. Plasma* 7 (2000): 1421–1432.

[36] Kohlman-von Platen, K. T., J. Chiebek, M. Weiss, K. Reimer, H. Oertel, and W. H. Bürger. Resolution limits in electron-beam induced tungsten deposition. *J. Vac. Sci. Technol.* B 11 (1993): 2219–2223.

[37] Hiroshima, H., and M. Komuro. High growth rate for slow scanning in electron-beam-induced deposition. *Jpn. J. Appl. Phys. Part I* 36 (1997): 7686–7690.

[38] Mitsuishi, K., Z. Q. Liu, M. Shimojo, M. Han, and K. Furuya. Dynamic profile calculation of deposition resolution by high-energy electrons in electron-beam induced deposition. *Ultramicroscopy* 103 (2005): 17–22.

[39] Liu, Z. Q., K. Mitsuishi, and K. Furuya. Modeling the process of electron-beam-induced deposition by dynamic Monte Carlo simulation. *Jpn. J. Apl. Phys.* 44 (2005): 5659–5663.

[40] Liu, Z. Q., K. Mitsuishi, and K. Furuya. A dynamic Monte Carlo study of the in situ growth of a substance deposited using electron-beam-induced deposition. *Nanotechnology* 17 (2006): 3832–3837.

[41] Liu, Z. Q., K. Mitsuishi, and K. Furuya. Dynamic Monte Carlo simulation on the electron-beam-induced deposition of Carbon, Silver and Tungsten supertips. *Microsc. Microanal.* 12 (2006): 549–552.

[42] Fowlkes, J. D., S. J. Randolph, and P. D. Rack. Growth and simulation of high-aspect ratio nanopillars by primary and seconaday electron-induced deposition. *J. Vac. Sci Technol.* B 23 (2005): 2825–2832.

[43] Winters, H. F., and M. Inokuti. Total dissociation cross section of CF4 and other fluroalkanes. *Phys. Rev.* A 25 (1982): 1420–1430.

[44] Inokuti, M. Inelastic collisions of fast charged particles with atoms and molecules—The Bethe theory revisited. *Rev. Mod. Phys.* 43 (1971): 297–347.

[45] Miller, W. F., and R. L. Platzman. On the theory of the inelastic scattering of electron by helium atoms. *Proc. Phys. Soc. London* A 70 (1957): 299–303.

[46] Schram, B. L., M. J. van der Wiel, F. J. de Heer, and H. R.Moustafs. Absolute gross ionization cross sections for electrons (0.6–12 keV) in hydrocarbons. *J. Chem. Phys.* 44 (1966): 49–54.

[47] Uesugi, K., and T. Yao. Nanometer-scale fabrication on graphite surfaces by scanning tunneling microscopy. *Ultramicroscopy* 42–44 (1992): 1443–1445.

[48] Kwitnewski, S., E. Pasinska-Denga, and C. Szmytkowski. Relationship between electron-scattering grand total and ionization total cross sections. *Radiat. Phys. Chem.* 68 (2003): 169–174.

[49] Smith, D. A., J. D. Fowlkes, and R. D. Rack., Simulating the effect of surface diffusion on electron beam induced deposition via a three-dimensional Monte Carlo simulation. *Nanotechnology* 19 (2008) 415704–11pp.

[50] Amman, M., J. W. Sleight, D. R. Lombardi, R. E. Welser, M. R. Deshpande, M. A. Reed, and L. J. Guido. Atomic force microscopy study of electron beam written contamination structures. *J. Vac. Sci. Technol.* B 14 (1996): 54–62.

[51] Friedli, V., and I Utke. Optimized molecule supply form nozzle-based gas injection systems for focused electron- and ion-beam induced deposition and etching: simulation and experiment. *J. Phys.* D. 42 (2009): 125035–125045.

PART II

APPLICATIONS

CHAPTER 9

FOCUSED ELECTRON BEAM INDUCED PROCESSING (FEBIP) FOR INDUSTRIAL APPLICATIONS

Thorsten Hofmann, Nicole Auth, and Klaus Edinger

1 INTRODUCTION

Scanning electron beam and ion beam microscopes can be easily turned into nanostructuring tools by adding a suitable gas delivery system and beam steering control. Such systems combine precise material removal and deposition capabilities with direct, high-resolution imaging for navigation, alignment, and real time process control. While this features makes them ideal research tools for prototyping and testing of nanostructures and devices, the fact that they rely on a relatively slow, serial direct write process limits their commercial application range. In terms of commercial scale production, focused ion and electron beam nanostructuring has been utilized for only a few mix-and-match type applications, such as high-resolution, high aspect ratio AFM tips by electron beam induced deposition or trimming of magnetic write heads for hard disk drives by focused ion beam sputtering.

An area where focused ion beam nanostructuring has been very successful is semiconductor manufacturing, where the FIB tool are used in mask repair and circuit editing, as well as for the preparation of cross-sections and TEM samples for device failure analysis.

The requirements for those applications are quite different. The preparation of cross-sections and TEM lamellas requires fast material removal without material selectivity between different layers within the sample under preparation. This is a typical application for focused ion beam sputtering, where the removal rate scales linearly with the beam current and shows relatively little material dependence. The repair of photolithographic masks, on the other hand, requires high-resolution nanostructuring with high material selectivity to avoid damage to underlying layers. This is an area where electron beam based nanostructuring has clear advantages over conventional FIB processing and in fact has replaced ion beams as the tool of record for repair of advanced photo masks.

Photolithographic masks consist of a patterned layer on a highly transparent Quartz plate. Typically the pattern represents a 4:1 enlargement of the structure projected onto the wafer and consists of either a completely opaque (so-called binary masks) or semitransparent material that shifts the phase of the transmitted light by 180 degrees (so-called attenuated phase shift masks). In general, these masks can have two types of defects. If absorber is missing, light can pass through and expose the resist (clear defect) or areas that should be transparent are covered with absorber material (opaque defect). In addition, phase defects are created if the phase shifting material has not the correct thickness. All these types of defects can be repaired by beam induced processing, either removing the absorber by etching or adding absorber or phase shifter by depositing material from a suitable precursor [1–5].

In order to keep pace with the ever shrinking feature size of integrated circuits, the technology employed for repair tools has changed from laser-based ablation, to gas-assisted focused ion beam processing and absorber removal by proximal probes, to focused electron beam etching and deposition. This evolution was not only driven by minimum feature sizes and thus higher resolution and increased placement requirement, but also by the need to maintain the optical properties of the mask, such as transmission and phase shift errors. Here, electron beam induced chemical processing has a clear advantage over focused ion beams, because of the absence of ion implantation into the underlying Quartz substrate (e.g., *gallium staining*) the transmission can be kept above 97% of the pre-repair value. In addition, because the etch process is solely chemically induced with no sputter contribution, the removal of absorber material can be precisely terminated at the Quartz interface, provided a suitable precursor chemistry for the material systems exist. A comprehensive overview of the current state of the art in electron beam based mask repair is given in the next section.

In the area of circuit editing, focused ion beam tools have been widely used for debugging and rewiring of prototype integrated circuit. This allows design modifications to be tested without going through a long and expensive mask making and fabrication cycle. For this application, the case for electron beam based processing is not so clear-cut. On one hand, the rewiring of a working device requires extremely high precision and the ability to create high resolution, high aspect ratio access holes combined with exact endpointing at deep lying interconnect metal lines. This are again areas where electron beam based tools can extend the limits of FIB based processing. However, beside standard materials such as silicon, dielectric and copper modern circuits contain numerous, often proprietary material stacks and thin layers for which individual etch chemistries need to be developed. As compared to a conventional FIB circuit-editing tool, where those layers can be simply sputtered away, the variety of materials makes the development of a processing work flow for a particular device generation rather tedious and increases the complexity of tool operation. These issues are addressed in more detail in Section 3.

2 FOCUSED ELECTRON BEAM BASED MASK REPAIR

2.1 TECHNOLOGY ENABLERS

2.1.1 E-beam system

As the performance of scanning electron microscopes evolved, new applications for e-beam induced surface processes emerged. As mentioned above, the main drawbacks of the predecessing focused ion beam technology if used for photo mask repair are the limited resolution and Gallium staining. To overcome this limitation, an electron beam instead of the ion beam can be used. Since the virtual source size of field emitter SEMs is notably smaller than that of a liquid metal ion source, one can achieve much smaller focus diameters. However, the diameter of the primary beam is not the only parameter that determines the ultimate resolution of the mask repair processes. The effective process diameter strongly depends on the spatial distribution of the energy loss (e.g., secondary particles) around the point of impact of the primary beam. To decrease the size of the area where most of the secondary electrons leave the sample and contribute to the process, it is necessary to decrease the penetration depth of the primary electrons—hence working with low primary energies.

On the other hand, typical SEMs deliver their optimum performance at higher acceleration voltages. Therefore the best effective resolution of these SEMs is usually at beam energies well above 5 kV. Due to the interdependence of process resolution and primary energy, this is too high to meet the specifications of mask repair tools that can cover current and future nodes and also leads to pronounced charging (see Section 2.1.4). This issue can be solved by using special electron beam columns that combine electric and magnetic lens elements like the Gemini® design by Carl Zeiss NTS (see Figure 9.1). SEMs based on this design are able to deliver a resolution of below 2 nm at a primary energy of 1 keV, which makes them the ideal choice as a base system for high end mask repair systems.

The ultimate process resolution is only one of the factors that have a strong influence on the mask repair application. The performance of the image acquisition is of almost equal significance. Starting with the repair pattern generation, which relies on images of the defects,

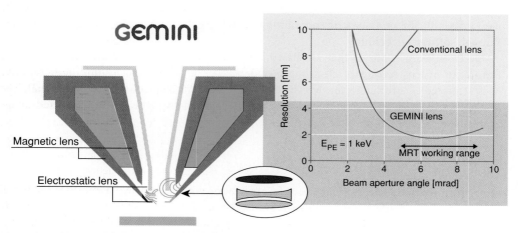

FIGURE 9.1: Comparison of optimum aperture and resolution for a conventional magnetic lens and the GEMINI electric-magnetic lens. MRT is mask repair tool.

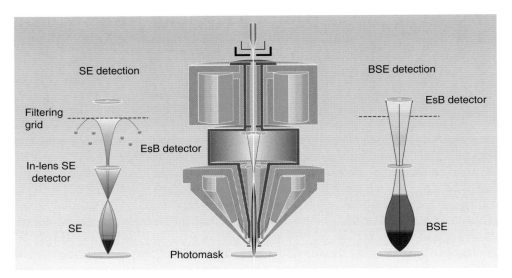

FIGURE 9.2: Gemini SEM column with in-lens SE and BSE detection, EsB is Energy selective Backscatter Detector.

it is important to be able to acquire artifact-free images having high contrast and low noise level. Combined with the requirement of small beam currents and limited total electron doses, this drives the need of having very efficient electron detectors. Since the above-mentioned low primary energies require small working distances the only viable choice is using in-lens detectors.

During mask repair it is necessary to have online information about the progress of the process. This information is generated by simultaneously acquiring images during the process. The related signal intensities are derived from the acquired images. Even though the secondary electron yield at typical beam conditions is much higher than the yield of the backscattered electrons, it turns out to be advantageous to use the backscattered electron signal for specific tasks due to its lack of topographical information. During etching the backscattered electron signal usually changes significantly at interfaces between different materials without being influenced strongly by surface roughness or other topographical effects. Therefore it can be used efficiently for an automatic endpointing at that specific interface by evaluating the signal through software algorithms.

Electron beam columns that are based on the Ultra-SEM series of the Carl Zeiss NTS feature an in-lens secondary electron detector as well as an intra-column energy selective back scattered electron detector, which can be operated simultaneously (see Figure 9.2). The availability of these two technologies enables the operator to make use of optimized image acquisition parameters for every task of the application.

2.1.2 System layout

Other than tools that are built for scientific applications in laboratories, systems for the use in semiconductor manufacturing facilities require a lot of additional bits and pieces surrounding the SEM. First, a lot of certification regulations apply, such as SEMI S2 or CSA. Additionally, the tool must be equipped with an enclosure that fulfills the demands of an industry-grade clean room environment. The same applies for subsystems of the tool, like particle-free automated handling of the masks or the ban of certain materials that are considered to be harmful for the processes used within the facilities.

Moreover, common SEM platforms cannot provide the mechanical stability that would allow the application to meet the specifications required by the current technology nodes. During mask repair processes, which take from a few up to 30 minutes, the beam needs to stay within the defined region without being disturbed by external influences. Typical specifications of the 32 nm node require placement accuracies of the final repairs in the order of 1.5 nm (3σ deviation). The susceptibility of electron beam columns, vacuum chambers, and sample changes stages to mechanical vibrations needs to be addressed by extremely rigid chambers and highly sophisticated active mechanical damping. Moreover, an efficient acoustical damping is needed to minimize the impact of the noisy surrounding of the tool in a production environment.

Another source of placement deviations are thermally induced drifts. Even using sophisticated drift compensation algorithms (see Section 2.1.4) thermal drift can result in misplacements in the order of a couple of nanometers. This can be addressed by using a thermally stabilized environment for the SEM system and the minimization of heat sources in the thermally controlled area. To avoid additional drifts due to temperature differences between the clean room environment and the thermal chamber of the tool, the set point must be adjustable.

These measures, together with the requirements regarding reliability and serviceability, result in the fact that the footprint of a current e-beam based mask repair tool can exceed 15 square meters.

2.1.3 Contamination control

To achieve full control over the conditions during the electron beam induced etching and deposition, it is mandatory to reduce residual contamination to an absolute minimum. In case of etching, contamination can hinder the process completely by generating depositions which dominate the etching process. The deposition processes cannot be controlled in a repeatable way if the composition of the used precursors is not precisely controlled. A couple of measures must be taken to achieve the required conditions.

During development all parts of the system that are exposed to the vacuum must be chosen carefully. Especially seals, glues, and lubricants must meet a couple of specifications regarding outgassing. Moreover, all materials have to be tested against the used precursors and cleaning methods to avoid chemical reactions that might also lead to outgassing, corrosion of parts, or the formation of particles.

All vacuum exposed parts must be cleaned meticulously after manufacturing by chemical and physical methods. Additionally, it is necessary to clean the vacuum chamber after the integration of all parts by heating and other physical and chemical methods.

The vacuum system itself must be chosen such that it works absolutely oil-free. Since the electron-induced processes are specifically sensitive to carbon-containing residual gases, the base pressure of the vacuum system is less important than the chemical composition of the remaining substances. Some precursors tend to stay in the chamber for a fairly long time and can lead to cross-contamination between the processes, specifically between deposition and etching processes. To avoid a negative impact on the repeatability of the repair, all precursor combinations have to be tested against such a behavior.

To ensure repeatable conditions every day of operation, a mask repair tool must also be equipped with automated chamber cleaning means. This prevents contamination formed by the operation of the tool from cumulating and thereby influencing the repair results.

2.1.4 Mitigation of charging

The main source of disturbances of the beam placement accuracy is charging of isolated or floating sample areas. The charging effect is due to the implantation of the primary electrons in insulating materials and unequal quantities of the incoming primary and the emitted secondary electrons.

The electric fields generated by charges in insulating species influence the trajectories of primary and secondary electrons as well as the yields of the secondary particles [6]. The influences on secondary particles only affect the image contrast, which most of the time can be compensated by appropriate scanning techniques and image manipulation algorithms. The influences on the primary beam, however, directly compromise the repair quality by introducing positional and focus drifts. Even worse than plain isolating samples are structures of conducting or semiconducting materials on insulating samples. Due to the electrically floating character of these structures and finite leakage currents to other structures or to the ground, potential complex and time-dependent charging effects occur, which lead to strong beam deflection.

A conventional approach to mitigate this phenomenon is the use of flood guns. This technique is widely used in FIB systems where low energy flood electrons compensate the positive charging induced by the impinging Ga ions and the emitted secondary electrons. To maintain the capability of acquiring online images during the processes, it is necessary to multiplex the primary and the flood beam because of the inevitable flooding of the secondary electron detector during charge compensation phase. Since electron impingement can lead to positive as well as negative surface charging, flooding electrons cannot be used with the same ease as with FIB tools. Interactions between the flooding electrons and the precursors during the repair might also contribute unwanted side effects.

Another common technique to reduce surface charging is the use of fairly high gas pressures at the sample surface that may neutralize surface charges (see Figure 9.3). Unfortunately, the gas pressures typically used for electron-induced processes are not high enough to exhibit the neutralization effect to a satisfactory extent, and applying an additional gas influences the process itself. Moreover, it can be shown that the scattering of the primary electrons at the gas molecules leads to a background dose at the sample surface which can exceed several percent of the total electron dose. This is irrelevant for most imaging applications but might lead to halos, ghost images or other unwanted effects during surface processes.

Therefore NaWoTec developed a proprietary charge-blocking device that does not make use of any active compensation means but shields the primary beam from the influence of charged surface areas. Combined with the operation of the primary beam around the so-called

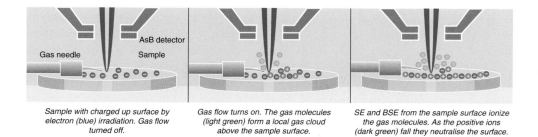

Sample with charged up surface by electron (blue) irradiation. Gas flow turned off.

Gas flow turns on. The gas molecules (light green) form a local gas cloud above the sample surface.

SE and BSE from the sample surface ionize the gas molecules. As the positive ions (dark green) fall they neutralise the surface.

FIGURE 9.3: Working principle of charge compensation by local application of gas using the example of the ULTRAPlus column by Carl Zeiss NTS.

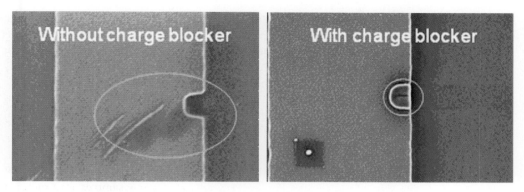

FIGURE 9.4: Repair attempts with and without the charge blocker technology.

neutral point[a] at a secondary electron yield around 1, it could be demonstrated that even heavily charging masks with isolated metal features can be repaired within specification.

Since all these measures still do not reduce charging effects to zero, one needs a drift compensation to address the remaining drifts, which typically are around a couple of nanometers per minute. This is done by regularly acquiring a reference image and comparing this reference image with original data of the pre-repair image.

In Figure 9.4 a comparison between two repair attempts is shown. On the left side a repair attempt on a heavily charging mask with isolated chromium structures is shown. The fine lines on the chromium indicate the attempt to deposit a dot for measurement purposes. Since this was not possible due to heavy drifting an end of the line was used instead as the reference for the repair. The dark grey area on the chromium illustrates the unsuccessful attempts to deposit into the line intrusion defect. The right picture shows the same type of defect on the same mask with the charge blocker technology applied. It can easily be seen that the drift compensation reference is a sharp dot and the placement of the repair deposition is excellent. Figure 9.5

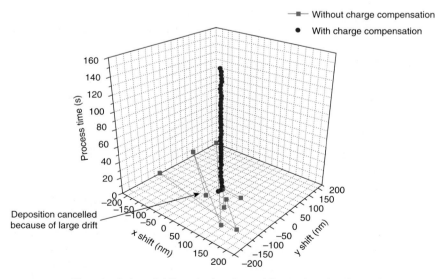

FIGURE 9.5: Charging induced drift on isolated metal line with and without charge blocker technology.

shows the quantitative drift data of both repairs. The x and y axes illustrate the drift along both coordinate axes of the mask in units of nanometers, while the z axis depicts the elapsed process time. The deposition process without the charge blocker technology had to be stopped after approximately 20 seconds because the drift compensation reference marker was completely lost. Drifts in the order of more than 100 nanometers occur almost instantly after starting the process. With the charge blocker technology on the other hand, the drift remains very low and can easily be corrected by the drift compensation algorithm. This results in an excellent placement accuracy, as can be seen on the right side of Figure 9.4.

2.1.5 Versatile beam scanning system

As mentioned before, one of the key demands on a mask repair tool is the repeatability of the repair results. To make sure that etch or deposition rates do not depend on the pattern size or shape, the scanning system must be capable of controlling the beam position extremely accurately. On the other hand, optimizations of the process yield or other parameters that are relevant for the application often require short dwell times. Together with increasing pattern fidelity and tighter repair specifications, these requirements put very high demands on the scanning system.

The following figures illustrate the connection between the performance of the scan generator and the pattern fidelity and repeatability. In Figure 9.6 typical scan generator signals for a simple serpentine scans are depicted. The scan signal of the fast scan axis ideally has a saw tooth type shape. The amplitude of the saw tooth determines the width of the scan pattern while the line frequency is given by the dwell time.

Typical scan systems of scanning electron microscopes do not require extremely fast scanning since they can, for instance, implement over-scan strategies and just display a sub-frame in the center of the scanning area where the scanning is linear and at constant speed. For electron beam processing, however, this is of course not possible, which means that the beam position has to be accurately controlled throughout the entire scan field, especially around the turning points. Figure 9.7 displays how typical sawtooth signals are altered by a signal chain of

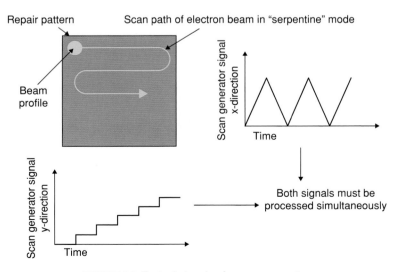

FIGURE 9.6: Typical signals of a scan generator.

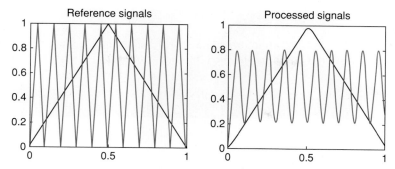

FIGURE 9.7: Scanning signals processed by a low bandwidth. The blue curve shows a slow signal, the red curve shows a fast signal.

limited bandwidth. The most prominent effect can be seen for the red, high-frequency signal. The reference signal has an ideal triangular shape, while the processed signal looks more like a distorted sine signal with severely reduced amplitude (so-called dynamical biasing effect). For slow signals one can find that even if the amplitude remains more or less unaltered, the signal shape at the turning point gets distorted significantly. This results in placement issues of the repair shape boundaries and changed effective process parameters, depending on the pattern shape.

The impact of these bandwidth effects on the dose distribution of real repairs is shown in Figure 9.9. The repair shape represents the lower part of an extension type defect on a "line and space" pattern (see Figure 9.8). The center image displays the dose distribution with realistic repair parameters if scanned by a standard SEM beam deflection system in serpentine mode. The result is an inhomogeneous pattern that shows significant edge roughness and distinct dose maxima in the vicinity of the turning points of the electron beam. Moreover, the dynamical biasing effect is clearly visible. Since the amount of the biasing depends on the shape and size of the repair pattern, it is very difficult to address by fully automatic algorithms which is a strong requirement for the mask repair application (see also Section 2.3). Scanning the same defect shape with sufficient bandwidth results in a dose distribution that is very homogeneous (see right image). The only still visible effect at the shape edges is due to the non-zero focal diameter of the Gaussian electron beam.

The same bandwidth considerations of course are valid for the beam blanker (which is used for switching the beam on and off during processing) as well. Typical electrostatic beam

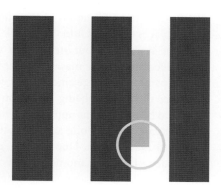

FIGURE 9.8: Extension defect (light blue) on a "line and space" pattern. Yellow circle indicates the region for the pattern analysis (see Figure 9.9).

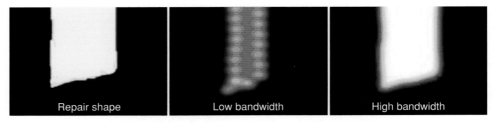

FIGURE 9.9: Electron dose distributions for real repair shapes depending on the deflection bandwidth.

blankers have rise times in the order of several hundreds of nanoseconds. Considering the usage of dwell times of some tens of nanoseconds, this implies that the beam is not at the required spot for several dwell times after each blanking cycle. Since most processes require refresh times that are not equal to the natural duration of a scanned frame, an additional blanking time is required after every loop.[b] Therefore special blanking strategies are required to achieve repeatable scanning performance even in the corners of the shape.

2.2 WORK FLOW AND AUTOMATION

Automation is a major requirement for semiconductor fab tools. In most fabs the mask repair application has to be executed by staff on the operator level. Therefore the application's front end must guide the operator very closely and should limit user interaction to an absolute minimum. This is the reason that most software graphical user interfaces on fab tools use wizard-style layouts and a large number of different sensors are distributed over the entire tool to make sure that the current status of the machine is well-known at any time. In the following section, the work flow of a mask repair tool is described.

2.2.1 Loading and navigation

In a typical mask shop, photo masks are handled within containers (so-called pods) where the mask is protected from contamination and particles. Almost every fab tool has an interface to this standardized pod to allow automated handling.

After being automatically extracted from the pod, the mask is typically handled by a robot system in an ultra-clean mini-environment. This robot typically is able to rotate and flip the mask since the orientation of the photomasks within the pod might vary between fabs. This robot places the mask into the airlock, which can be evacuated to allow the mask to be shuttled into the system without breaking its vacuum. The mask is then placed on an interferometer-controlled stage since mask repair requires very accurate navigation. The entire procedure is accompanied by a large amount of safety checks by optical and mechanical position sensors.

After the mask is being placed on the stage and the correct position is verified, the mask needs to be put into the appropriate distance to the objective lens of the electron beam column (which can also be done automatically by distance sensors) and aligned. Figure 9.10 shows a typical mask layout. Due to the accuracy demands, hence pixel sizes of the SEM image in the order of a few nanometers, the field of view of a mask repair tool is just a couple of micrometers.[c] To be able to navigate on a 6" mask within sub-micron accuracy, the mask is

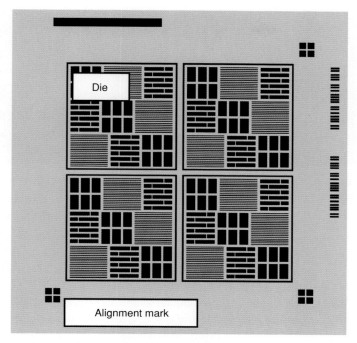

FIGURE 9.10: Schematic layout of a photo mask containing four dice and three alignment marks. (Note: a die (pl. dice) typically is a functional unit on a semiconductor wafer like a singe CPU or a memory chip. After processing, the dice are cut out of the wafer and put in the well-known IC packages.)

aligned by global alignment marks with well-known positions. This is done by consecutively centering the alignment points in the field of view and saving the so-called native stage coordinates of the mark positions together with the according user coordinates given by the mask layout. Every location on the mask is now described with the user coordinates relative to these alignment points.

To find the defects on a mask, the entire reticle gets checked by an optical inspection tool after the manufacturing process. The inspection tool generates a list of coordinates with potential defects. This so-called defect file is transferred into the mask repair software. The navigation to the defects is hence done just by clicking on the respective entry on the defect list.

2.2.2 Repair pattern generation

Following the alignment and the navigation, an automatic optimization of the electron-optical imaging parameters like focus and stigmation is performed and a high-resolution image is grabbed and displayed. Based on that image, the repair pattern has to be generated.

While scientific applications usually make use of geometrical shapes generated by drawing or programming tools, real defects are typically of an irregular shape. Simple free-hand drawing of course could be a solution, but in most cases the tight repeatability specifications in the order of a few nanometers prohibit this approach. Therefore an automated pattern generation is needed. This can be done by image comparison algorithms.

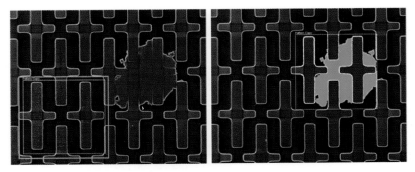

FIGURE 9.11: Semiautomatic generation of complex repair patterns by comparison of reference and defect structure.

Figure 9.11 illustrates the work flow of a semiautomatic pattern generation making use of in-die reference structures. The image on the left-hand side shows a fairly large defect spreading over several cells of a cross-type structure, while a defect-free area of the original mask structure is still available within the same field of view. As a first step, the operator selects the reference structure by drawing a rectangle, which should be somewhat larger than the defect. In a second step the selected area can be dragged to the defect area. After coarse alignment the copied reference structure is dropped on the defect. Software algorithms now automatically adjust the shape position by means of comparing the adjacent features with the reference structure. After that, the repair shape is extracted by comparing the aligned structures of the defect and the repair area. As a result, a repair pattern is generated and visualized in green.

In cases where the reference area cannot be found within the same field of view, other work flows may be used. If the defect is located on a unique structure on the mask or an even higher level of automation is needed, the reference structure can be extracted from the original CAD file that has been used to generate the mask with e-beam lithography. These CAD files, however, show ideal patterns which are substantially altered by the exposure and process steps to fabricate the mask. Therefore the pattern generation software needs to render the generated structures quite heavily to get appropriate reference patterns. To avoid human error, all these steps need to run automatically.

2.2.3 Automated repair work flow

A mask repair tool is typically able to address a couple of mask technologies; both etching and deposition (see Section 2.3). Together with supporting processes like passivation or polishing steps, this can add up to more than 10 processes with unique sets of parameters and gas mixtures. To make such a complex application work with operator personnel of semiconductor fabs and guarantee certain specifications, it is necessary to reduce user interaction to an absolute minimum. This requirement is solved by a guided tool setup procedure. The operator tells the tool which kind of mask technology is currently loaded and whether a clear or an opaque defect is currently being addressed.[d] Based on this information, all repair and supportive shapes are generated, the needed gas channels are prepared[e] and the process parameters are set. The chosen parameters might depend on the shape size or other repair-specific properties. Therefore some process parameters cannot be saved into static recipes but need to be calculated based on the meta-information of the repair shape. Since these parameters deviate marginally between different tools, the tuning of these recipes is part of the tool qualification process.

Having everything set up appropriately, the repair process itself can be started. This is done by simply clicking "start" in the mask repair GUI. To eliminate uncertainties due to unstable gas conditions, the entire gas handling of the mask repair tool is automated. So, hitting the start button does not instantly start the process, but initiates the gas activation procedure. The activation of the required gas channels comprises mostly opening some valves, setting flow rates, and checking pressures to avoid any unforeseen issues caused by tool failures or other incidents. After a certain waiting time which assures a stable gas flow, the process can be started.

The end of the process of course needs to be initiated automatically as well. Depending on the defect type and mask technology, the involved steps vary. Most deposition processes are controlled by electron dose and gas flux. In order to achieve repeatable deposition heights, it is required to maintain constant deposition rates over the lifetime of a precursor reservoir and specific requalification procedures after service or maintenance work.

In an ideal world, one would have highly selective etching processes for every mask material. In this case, endpointing (i.e., stopping the process at the required etch depth) for etching would also be used, based on the applied electron dose, which would be chosen in a way that every potential defect material is etched away entirely. Since selective processes that deliver sufficient safety margin with respect to the substrate damage specifications are typically not available for all types of material combinations, automated termination of the etching has to be supported by other methods. Etching processes can in most cases make use of sample signals achieved during processing. Most commonly used is the detection and evaluation of secondary or backscattered electron signals while etching. Since the structured layer typically consists of either metals or semiconducting compounds while the substrate is made of Quartz, a more or less distinct material contrast can be expected. If all boundary conditions are well controlled and repeatable, a quantitative analysis of the electron signals can be utilized as a trigger for automatic endpointing algorithms. The signal may not only be generated by integration of the signal over the entire shape, but also spatially resolved over the repair pattern. In this case the endpointing can also be executed in a spatially resolved manner. This allows, for instance, automated etching of defects that do not have constant height.

Some mask technologies employ masks where the bulk substrate material gets structured; for instance, imprint or alternating phase shift masks (see Sections 2.3.2 and 2.3.3). Since the patterned layer and the underlying substrate imperatively are made out of the same material, an electron signal based endpointing cannot solve the endpointing problem. In this case, the only solution is to achieve three-dimensional information about the defect (for instance by an AFM) and to generate a repair pattern that contains this volumetric information in form of a dose map. If the pattern generation software accounts for scanning artifacts of the 3D imaging and potential alignment issues between different imaging tools, this approach can be used to repair all kinds of three-dimensional defects on all available mask technologies.

2.3 REPAIR MODULES

Based on the above considerations, it has become obvious that a functional mask repair process involves more than just picking an appropriate precursor and a set of scanning parameters. For example, a commercial e-beam based mask repair tool by Carl Zeiss provides one so-called Application Module per mask type. This Application Module contains all relevant tool settings, guides the work flow and links all needed recipes each with more than 200 parameters controlling the e-beam process itself. In this section we will describe several relevant mask types and the corresponding repair processes. The main emphasis will be put

---

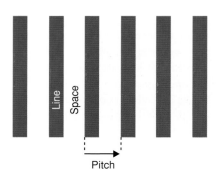

FIGURE 9.12: Schematic view of a dense line-and-space pattern with a line width ratio unequal to 1:1.

on masks for optical lithography at a wavelength of 193 nm, which is the workhorse for high-volume production, at least until the 22 nm node. Considering that the structures of this technology node are approximately ten times smaller than the lithography wavelength, one can imagine that quite a number of so-called resolution enhancement techniques (RET) need to be employed to push the limits of optical lithography further and further. One can make use of the quantitative knowledge of the diffraction effects by introducing non-printing assist features, optical proximity correction (OPC), or even more advanced RET measures summarized under keywords like "computational lithography" or "design for manufacturing" [7; 8].

To understand the real size relations on the mask and the wafer, it is helpful to know that an "x" nm node does not necessarily mean that all features on the wafer use line thicknesses of "x" nm. The nomenclature leads to different real feature sizes, depending on the application; i.e., flash memory uses different line widths at the, for instance, 22 nm node from those used by DRAM. For manufacturers of logic ICs like CPUs or graphics chips the situation is even more complex. At a rule of thumb we can nevertheless assume that the node number reflects the so-called half pitch on a dense line-and-space pattern. The half pitch describes the average value of the line and space widths (see Figure 9.12). Since optical lithography usually uses a reduced imaging of 4:1, one has to keep in mind that mask structures are typically four times larger than the according patterns on the chip.

2.3.1 Binary masks

Lithography with binary masks (Figure 9.13) makes use of the straightforward approach of projecting a binary (i.e., black and white without grey levels) pattern onto a work piece. The absorber usually is made of metallic chromium that has been treated at the surface to form an anti-reflective coating by converting the metallic chromium into, for instance, Cr-O-N compounds of a specific thickness.

At feature sizes considerably larger than the wavelength of the lithography system, it can be assumed that the patterns on the mask are reproduced identically on the work piece. As soon as the pattern size approaches the wavelength, one has to take diffraction effects into account. The main result of the diffraction effects is the reduction of the contrast of the imaged features. Since the resist process on the wafer is a binary process, i.e., the resist does not support grey levels, the reduced contrast can be leveraged to a certain extent. The process window of the lithographic process (the sensitivity of the process to parameter variations like source power, focus, etc.), however, is more and more reduced. Therefore, with feature sizes getting smaller, binary masks are less frequently used for the high-end layers.

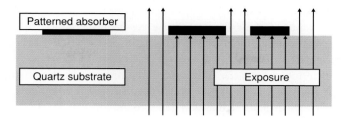

FIGURE 9.13: Schematic cross section of a binary mask.

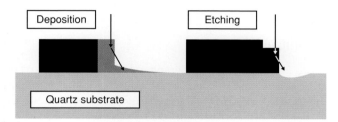

FIGURE 9.14: Generation of halos during deposition (left) and river-bedding during etching (right) due to secondary electrons leaving the absorber.

Since chromium masks are considerably cheaper than more advanced phase shifting masks (see below), they still play an important role at nodes (or layers, respectively) around 65 nm and above.

A mask repair tool has to provide two different processes for binary masks: deposition and etching. As mentioned above, defects with missing material are called clear defects and need to be repaired by deposition, while defects with excess material are called opaque defects and need to be repaired by etching.

Deposition on binary masks is pretty much straightforward. The deposits need to be opaque and should have a reasonable sidewall-angle. Taking a closer look, it turns out that it is also important that the repair deposits should not be much taller than the reference absorber, which typically has a thickness about 70 nm. At 193 nm UV irradiation, it is preferable to deposit metals with low carbon content in order to obtain completely opaque depositions. Moreover, significant carbon content can lead to durability issues during the aggressive wet cleaning used in the industry. One additional complication is caused by the so-called halo that is usually generated around depositions due to the presence of secondary electrons in the vicinity of the repair (see Figure 9.14). Depending on the deposition process, one has to introduce an additional repair step after the deposition, which selectively removes these halos from the quartz surface.

Etching of chromium masks is more challenging than deposition since the strongest arguments for chromium are its excellent adhesion on quartz, its durability, and chemical inertness, which make it difficult to etch with electron induced processes as well. Very aggressive substances like the above mentioned XeF_2 usually etch the quartz substrate much faster than the chromium layer on top of it. This so-called inverse selectivity requires extremely sophisticated endpointing strategies to meet the tight specifications regarding quartz damage. On non-flat defects or with etching processes that do not etch chromium in a perfectly flat manner, the situation is even worse. Additionally, the same effect that causes halos during deposition processes is now responsible for the so-called river-bedding, which is a strongly localized substrate damage just around the defect edges. Using inversely selective processes then requires elaborate scanning strategies to keep the damage in a tolerable regime.

The mask repair tool is not the only tool that suffers from the chemical durability of Chromium. This leads to issues, like a pronounced line edge roughness of the structures, which limit the performance of the masks. Therefore the industry is investigating the usage of other absorber materials for binary masks. One promising candidate is the so-called "Opaque MoSi on Glass–OMOG" technology, where the chromium has been substituted by an opaque Mo-Si alloy.

2.3.2 Phase shifting masks

As mentioned before, the achievable lithographic contrast with binary masks is limited as structures are getting smaller and smaller. To extend the usability of 193 nm lithography, one can make use of interference effects by using partially transparent absorbers. This leads to the so-called attenuated phase shift masks, Figure 9.15.

The most common attenuated PSM masks consist of an absorber made of Mo-Si compound with approximately 6% bulk transmission at a wavelength of 193 nm for a 70 nm absorber layer. Using such a low transmission ensures that larger absorber patches still exhibit enough absorption to avoid resist exposure. On the other hand, even 6% transmission leads to pronounced edge effects at the boundaries of the patterns. The thickness of the absorber is tuned in such a way that the light which passes the Mo-Si layer has a 180° phase shift, compared to the light passing through the clear areas. At the edges of patterns, wavelets of both phases destructively interfere with each other, leading to darker shape edges. At shapes smaller than a couple of hundred nanometers at mask level, the bulk transmission of the Mo-Si absorber does not show up any more and hence the contrast of the features is increased by the interference effect (see Figure 9.16). To enhance the contrast even further, PSM masks with less absorbing Mo-Si can be used.

Repairing phase shift masks has its own challenges. As MoSi is not as inert as chromium, it is easier to find a chemistry that etches the absorber layer sufficiently fast. On the other hand, the secondary electron yield of the Mo-Si compound (and its anti-reflective coating on top, which usually is an Mo-Si-O-N compound) is much closer to that of the SiO_2 substrate. This makes endpointing processes that make use of the secondary electron signal even more difficult than for chromium-based masks.

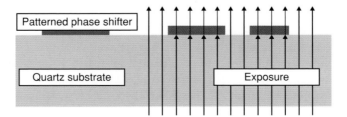

FIGURE 9.15: Schematic cross-section of an attenuated phase shift mask.

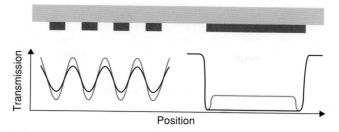

FIGURE 9.16: Transmission of MoSi (grey) and chromium (black) pattern depending on pattern size. Note the enhanced contrast on MoSi line and space patterns (left) due to the interferometric edge contrast visible on larger structures (right).

The repair of clear defects on attenuated PSM masks is completely different from repairing binary masks. For relaxed nodes of about 130 nm and up, Mo-Si masks have been repaired by the deposition of binary material and positively bias the repair shape in a way that the repairs print darker. This measure, of course, does not bring the contrast of the repaired features to the level of the reference Mo-Si features but still can be successful at large pitches. At narrow pitches this low-contrast repair technique impacts the process window of the wafer exposure so much that it is not tolerable any more. The only way to repair such defects is the deposition of material with optical properties similar to the reference absorber. This requires excellent control of the deposition process, since both absorption and phase shift of the repair have to be within tight specifications of a few percent. It has been proposed to repair such defects by etching into the quartz to meet the phase shift and then depositing an absorber material in the etched region to meet the transmission specifications. But it turns out that with this technique it is not possible to meet the required out-of-focus specifications that are needed to maintain an appropriate process window during the wafer printing process.

Another technology that makes use of interference effects is the so-called alternating phase shift mask (Figures 9.17 and 9.18), where again chromium is used as the absorber. The destructive interference is introduced by having two different quartz levels at alternating positions. The absorber is flanked by a normal space on one side and by an etched quartz substrate on the other side. The etch depth is controlled to account for 180° phase shift with respect to the un-etched quartz. This technique is capable of delivering very high contrast by just using the same materials as in standard binary masks; however, this requires certain design rules since the effect cannot be used at isolated lines.

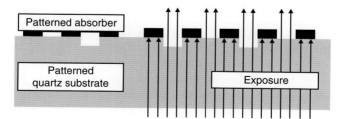

FIGURE 9.17: Schematic cross-section of an alternating phase shift mask.

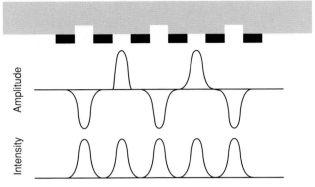

FIGURE 9.18: Imaging principle of an alternating phase shift mask.

FIGURE 9.19: Schematic 3D view of a quartz bump defect on an APSM mask. For a typical dense 45 nm structure, the width of a single space would be in the order of 180 nm.

While repairing absorber defects on alternating phase shift masks is very similar to binary masks, repairing at the quartz level requires the knowledge of the 3D shape of the defects. A common defect on APSM masks is a so-called quartz bump (see Figure 9.19). This describes an area on the lower quartz level where the quartz is not sufficiently etched. These bumps can be arbitrarily shaped. Additionally, it is not possible to use a classical endpointing due to missing material contrast during etching quartz on quartz. Therefore the electron dose used for the repair must be controlled in a spatially resolved manner, i.e. the dose distribution has to be matched to the 3D profile of the defect. This technique of course requires excellent control over and long-term stability of the etch rate. Repair of clear defects on quartz level needs ultrapure quartz deposition. So-called chromeless phase lithography masks (CPL) are quite similar to the alternating phase shift masks—at least regarding their impact on repair technology. This mask type uses interference contrast only and consists only of a patterned quartz mask without any conventional absorber.

To improve the pattern fidelity even further, the industry is looking into more and more sophisticated mask technologies. One candidate is a combination of the above-mentioned technologies. This so-called tri-tone mask uses high transmission MoSi phase shifter together with patterned quartz to generate complex phase distributions (Figure 9.20). The high-T MoSi has to be covered with a stronger absorber to achieve real opacity where necessary. Repairing these masks typically requires a combination of the processes used for the less complex mask types.

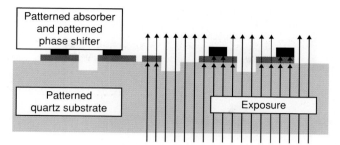

FIGURE 9.20: Schematic cross-section of a tri-tone mask with two quartz levels.

In contrast to the mask types described before, the tri-tone adds up one more complication. A potential absorber defect can affect the chromium or the MoSi absorber or even a combination of both. Moreover, quartz defects may occur as well. Besides the repair work itself, this also puts high demands on the automated pattern generation. The software has to distinguish between different mask tones, even though there might be no material contrast at all (e.g., for quartz-on-quartz defects).

2.3.3 Special processes

Classical 193 nm lithography has been extended to smaller and smaller nodes because no successor technology was available. Being already certain that any of the potential successors will not be used at 32 nm node, as of 2011 it is still unclear if the insertion point can be 22 or even 16 nm. This puts an enormous pressure on the existing 193 nm technology. Amazingly enough, 193 nm litho is still able to keep up with the demands while being pushed further and further.

The most promising candidate for 22 nm lithography is the combination of immersion, computational lithography, and, depending on the desired application of the integrated circuit, double patterning[f] as well. Considering that at 22 nm half pitch the resolution at wafer exceeds the wavelength by almost a factor of ten makes it pretty obvious that this can only achieved with sophisticated measures. In this section we will give a brief overview about the currently used methods to extend the existing lithography and potential successors with focus on the impact on mask repair.

The most straightforward approach to increase the resolution of optical imaging is to use higher numerical apertures. This is currently done with water immersion, which provides an NA of approx. 1.4. This technique affects lithography mainly on the wafer side and therefore its impact on mask repair besides feature shrinking is fairly low. Other litho techniques like source mask optimization—where, for instance, the illumination pattern of the mask is varied depending on the used structure—have comparably low impact on mask repair.

The next step when squeezing lithography to smaller feature sizes is the use of so-called resolution enhancement techniques (RET). The easiest form of RET is the usage of so-called optical proximity correction (OPC). This technology makes use of the quantitative knowledge of diffraction effects by optimizing the mask structure such that the printed result on the wafer is as close as possible to the desired pattern, see Figure 9.21.

The implementation of OPC is more or less the end of manual shape drawing for mask repair—at least for extended defects. A pattern copy algorithm is needed to ensure the correct

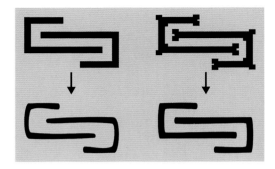

FIGURE 9.21: Schematic view of complex mask structures (top) and printing results (bottom) due to diffraction. Without OPC (left), line widths shrink and the resulting circuit may fail. With OPC (right), the printing results are close to the intended pattern. Black color represents opaque areas of the structure.

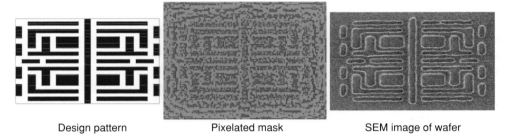

Design pattern   Pixelated mask   SEM image of wafer

FIGURE 9.22: Imaging with pixelated masks. The original design is converted into an extremely complex pixel pattern on a CPL mask. Red and green colors represent the two different quartz levels. The wafer image of the pixelated mask reproduces the original design. Images used with kind permission of Yan Borodovsky/Intel.

shape of the repair pattern (see Section 2.3.2). The main impact on the FEBIP processes is the increased pattern complexity, which requires excellent process control. The RET technology can be escalated to a level where the mask structure does not even give a hint to the human eye which kind of structure is actually printed with it. The term "computational lithography" is used for this technology. In 2008 Intel reported a very extensive manifestation of RET called pixelated phase mask (see Figure 9.22). This technology uses supposedly irregularly placed pixels on a CPL mask [8], which puts high demands on the minimum repair size. On top of the resolution challenges, the irregularity of the mask layout makes even conventional pattern copy impossible. This can be solved by using database pattern copy which analyzes the CAD data of the mask, extracts the repair region, processes the data with specific algorithms to simulate mask patterns due to the manufacturing process, and uses the processed data as pattern copy reference.

If pattern shrinking remains the number one tuning knob regarding IC performance and manufacturing cost per item it is inevitable to substitute 193 nm litho by a technology that delivers a substantial resolution jump. As of 2011 there are three potential technologies in sight that might be able to bridge this gap: EUV lithography, nano imprint lithography (NIL), and "direct-write". Direct write makes use of established technologies like electron-beam-lithography for chip manufacturing and does not require masks any more. Both NIL and direct-write face fundamental issues like defectivity and throughput and will most likely only be used within specific niches of semiconductor manufacturing. Although the EUV technology as well is extremely challenging in some aspects, it is the most promising candidate for high volume manufacturing (HVM) as of 2011.

EUV lithography still uses optical imaging, but with a significantly shorter wavelength of 13.5 nm in the "extreme ultraviolet" regime. For this radiation, neither transparent masks nor glass lenses can be used for imaging any more. Therefore all optical elements are reflective optics; the photo mask is used in reflection geometry as well. Due to the fact that reflective optical elements still suffer aberrations to a larger extend than conventional lenses, some of the benefits of the shorter wavelength are consumed already. This is even more the case since the insertion point of EUV keeps on slipping because the technology still has issues fulfilling the demands regarding throughput, stability, and performance. This means that when EUV will be introduced for HVM, the pattern sizes already approach the wavelength again; hence RET technologies come into place for EUV as well. This fact is even more amplified because 193 nm lithography already operates at numerical apertures in the order of 1.4 while realistically the first EUV steppers will work at numerical apertures not higher than 0.32.

Since metal layers do not exhibit usable reflectivity in the EUV regime, a multilayer acting as dielectric mirror is used for the optical elements as well as for the reticles. The multilayer consists of alternating layers of molybdenum and silicon; each layer just a couple of nanometers thick. Compared to phase shifting Mo-Si masks where the constituents form an alloy, the layers of the EUV mirror stay separated.

EUV reticles exist in several styles. In the most common setup, the multilayer is covered by a protective layer (mostly ruthenium), on top of which the structured absorber layer (TaN or TaBN) can be found (see Figure 9.23).

Defects occurring on EUV reticles can be separated into two different classes: multilayer defects and absorber defects. Repairing ML defects still is a very complex matter, and several approaches are being discussed. Absorber repair already has been demonstrated with adequate performance. We will focus here on the latter type of defects.

During repair of the absorber layer it is absolutely crucial to leave the underlying multilayer untouched. The mirror stack spontaneously reacts with a large variety of gases (for instance, water in normal air), leading to unrepairable damage of the reticle. Hence, reliable endpointing combined with decent process selectivity is a prerequisite for successful EUV absorber repair. To make matters worse, the absorber layer consists of several material alloys as well. The TaN layer, for instance, contains a significant amount of oxygen in the vicinity of the surface. Some of the substances of the absorber layer etch extremely fast with common etch gases like XeF_2. During etching one finds significant damage to the area surrounding the defect size, the amount of damage being dependent on the shape of the repair pattern, its size, and so on. To overcome this effect one has to use certain passivation techniques during repair similar to the technology used for via etching during circuit editing (see section 3).

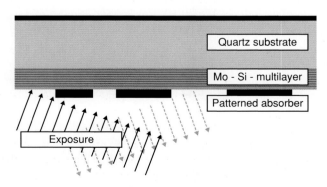

FIGURE 9.23: Schematic cross-section of a EUV mask.

NIL, on the other hand, uses mechanical pattern transfer. NIL masks operate as stamps; i.e., the structure has been established on the mask surface. The mask is then pressed on the wafer, which is coated by a special resist. The pattern on the mask is hereby transferred (imprinted) to the wafer resist. After this point the wafer processing can continue as usual. NIL masks are patterned quartz substrates; therefore defect repair for these mask types employs the same processes as for APSM or CPL masks. Side effects that can play a significant role in optical lithography do not apply for mechanical pattern transfer, which may relax specification at certain points somewhat. However, in contrast to the optical masks, imprint technology of course requires 1x masks, i.e., the feature sizes on the mask and on the wafer are equivalent. This leads to considerable increased requirements regarding minimum repair feature size.

2.3.4 Future challenges

Since a couple of decades, Moore's law has been the motor of the entire semiconductor industry. It was declared dead more than a dozen times; however, it is still alive. Many semiconductor vendors follow a two-year node cycle. This means that every four years feature sizes are decreased by a factor of two. In 2011 the major technology driving companies were in high volume production between 32 and 22 nm, depending on the application of the ICs. Following the two-year schedule, we will roughly see 16 nm in 2013 and 8 nm in 2017. However, the official technology roadmaps of the semiconductor industry already predict a slowdown in the upcoming years, not seeing 8 nm technology before 2020. Obviously there must be an end some day. However, it is still unclear when this end will be reached. It seems probable that 3D-stacking will open up new methods of increasing IC complexity after scaling has reached its limit.

Now, what is the impact on mask repair? Squeezing the lithography brought new mask materials and complex mask technologies. Scaling tightens the specifications for repeatability, endpointing, and minimum repair size. It is currently unclear if e-beam technology will reach a fundamental limit before other links of the manufacturing chain run out of steam. Repeatability and in parts even endpointing can be addressed by using more rigid platforms and sophisticated software. Minimum repair size, on the other hand, depends strongly on the available chemistry. Typically the required minimum repair size equals the node number, i.e., for the 45 nm node, repair capability for isolated 45 nm features on mask level is needed. With decreasing feature sizes it becomes more and more difficult to find reliable process parameters that can ensure reproducible results on such small features. Therefore a significant amount of fundamental research needs to be done to identify precursors that still work under such extreme process conditions. The above-mentioned successor technologies will generate new challenges like particle-free handling (EUV), or extreme durability demands (NIL), or other issues that are not even visible as of 2011.

3 CIRCUIT EDITING

As a result of Moore's law the increasing complexity of semiconductor fabrication is bringing up new challenges with every chip generation. One aspect is that with decreasing feature size, the efforts for testing and debugging grow. Once a prototype has been produced and tested, local modification is employed. During the debug phase, circuit logic errors or bottlenecks which cause timing faults can be localized and corrected by nano-machining. This approach enables short time-to-market and substantially reduces the costs for expensive mask sets.

In principle the aim of circuit editing is to cut and rebuilt electrical connections in an operational chip according to the requirements of design improvement. Another task is to prepare a chip for electrical probing by depositing probing pads and connecting them to buried structures. While FIB machining covers several orders of magnitude from hundreds of micrometers down to about one tenth of a micrometer, e-beam based circuit editing mainly aims at extending the processible range to below 100 nm. The need for nanometer scale features is triggered by the drastic increase of the integration density, which in turn leads to a step-up in the number of metal layers. Consequently backside editing was introduced. On one hand, this technique allows direct access to the active regions (transistor level), while on the other, there is little free space for connecting to metals below.

3.1 WORK FLOW

Unlike mask repair, in circuit editing the tool is operated by engineers, and the processes themselves are more complex. Therefore the operator has comprehensive access to the tool parameters and is required to make decisions as part of the editing work flow; for example, in many cases, endpointing is done manually.

Before a circuit edit is performed, several preparation steps are necessary. These can be subdivided in two groups. First, the edit has to be planned on the design level. Once a bug is found in testing, the layout is checked and a possible solution suggested. Subsequently a circuit edit is designed. Usually there are different possibilities for executing one and the same edit. However, as the dimensions get tighter, more edits are simply not possible with the established FIB technology.

Secondly the chip itself has to be prepared. With state-of–the-art flip chip packaging, the silicon chip is bonded to the carrier through metal bumps, so the connection to the metal lines is very short and the whole backside is covered by interconnects. Thus only the bulk silicon is freely accessible (see Figure 9.24). In a mechanical polishing step, the die is thinned down to less than 100 μm. Local thinning to below 10 μm silicon thickness can for example be achieved by laser chemical etching (LCE). Trenching is done on those regions that need to be prepared for modification or serve for global alignment. Now the chip is ready to be transferred into the micro-machining tool.

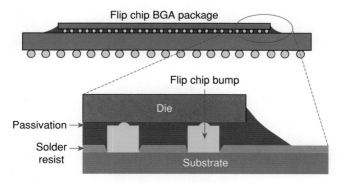

FIGURE 9.24: Flip Chip Ball Grid Array (BGA) package with solder bumps.

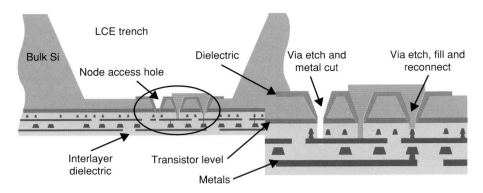

FIGURE 9.25: Possible circuits edit illustrating the major steps of the work flow.

Due to the high voltage of the particle beam, it is vital to ground the chip in order to prevent electrostatic discharge events, which would ruin the device. Therefore the whole package is placed on a suitable holder and loaded into the vacuum chamber of the FIB or SEM tool. An infrared microscope that is aligned to the field of view of the particle beam allows the operator to navigate to the global alignment markers in the corners of the die. Once the remaining silicon is etched, the chip structures can be aligned to the layout database.

With database driven navigation, the edit region can be addressed. Optical distortions during the lithography process lead to small deviations between layout and the actual chip. However, placement accuracy requirements are so high that at least one local alignment is performed within each trench to correct for these distortions. Remembering that the transistors are just 10 μm below the surface, each operation requires high precision.

In the following step the remaining silicon is removed in a small area. The bottom of the micro trench may measure less than a micron. As part of this procedure, transistors can be trimmed or eliminated. If rewiring of the circuitry is necessary, a dielectric is deposited as a separation layer. This can be done locally in the micro-machining tool or globally in a dedicated evaporation tool, which in this case requires an additional unload/load cycle including alignment.

Real nano-processing is done through the dielectric at the bottom of the micro trench. Via etching is either followed by metal cutting or the via is refilled with conducting material to establish a new connection. The lateral distance to the adjacent transistors can be as little as 20 nm while the vias themselves can be less than 100 nm wide. Details of these processes will be discussed in Section 3.3. A schematic illustration of an edit in cross-sectional view is shown in Figure 9.25.

3.2 DEMANDS ON TECHNOLOGY AND HARDWARE

E-beam based mask repair and circuit editing share basic requirements concerning the performance of the e-beam column as well as system stability, contamination, and charge control. Therefore the same platform as described in Section 2.2 has been used to build up an e-beam based circuit editing tool. It should be noted that the system is designated to operate in a laboratory environment. Consequently the requirements for particle control and acoustic isolation are relaxed compared to mask repair.

On the other hand, navigation accuracy is of particular importance for circuit editing. As the via dimensions aim at sizes below 100 nm, the placement accuracy has to comply. A reasonable upper limit for the allowed deviation is 20 nm within a field of view of several microns. In a real system there are two main contributions to the placement error. One of them is the tolerance of the laser interferometer controlled stage. Less obvious is the inaccuracy in box placement which is performed relative to a SEM micrograph. By scanning the sample surface the charge distribution is changed dynamically resulting in image distortions that are small but not negligible considering the sub 10 nm position. Though processing introduces far more charges than the imaging, these distortions can be compensated by a suitable drift correction algorithm, which allows an on-line check of the box placement relative to the initial position.

While mask repair deals with planar structures, where non-planar real defects are the exception, circuit editing requires high aspect ratios and 3D processing like pyramidal shaped trenches. High aspect ratios are a typical attribute of via etching, which will be discussed in Section 3.3. A simple approach, software-sided, to generate 3D structures is to expose several shapes at the same time so that their dosage is summarized in the overlapping regions. Clearly there are different possibilities, hardware-sided, regarding how to raster the beam across the surface in order to achieve the required dose distribution. Considering the influence of parameters like dwell time or frame refresh time, it is important to develop a clever scanning routine for this purpose.

Regarding the material properties of deposited structures, mask repair and circuit editing have differing requirements. Therefore the most suitable gas precursors have to be identified specifically for each application. For mask repair the UV optical quality is crucial, whereas circuit editing demands electrical insulators of low conductance and high breakdown voltage as well as conductors with less than 100 $\mu\Omega$cm resistivity. Both have been demonstrated previously with FIB and FEB [9]. Especially FEB has the potential to produce high purity materials as there is no implantation of parasitic species like Ga+. However, it is also more likely in FEB that insulating barriers are formed around conducting grains. This can lead to a significant degradation of the conductivity even if nearly pure metal is deposited. Therefore it is crucial to work in a low contamination environment.

3.3 PROCESSES

Micro trenching is the last step of Si removal. While XeF_2 can be employed for etching a wide range of materials, it is not suitable for Si etch due to spontaneous reaction. Therefore another precursor gas or some kind of gas mixture has to be found which inhibits spontaneous etch through sidewall passivation. This effect is well-known in plasma etching [10]. Using the 3D processing mode described in the previous section, FEB trenches of several μm depths have been produced. The breakthrough to the dielectric can be clearly distinguished, as shown in Figure 9.26.

Via etching is a basic process for any rewiring task. It is characterized by small lateral dimensions and a high aspect ratio. The via is typically drilled into interlayer dielectric, which is made up of a proprietary material stack containing, for example, high k dielectric and diffusion barrier layers. Sometimes sacrificial transistors, gate poly silicon or metal oxide, have to be intersected as well. So in general a multilayer of different materials has to be etched. As mentioned before, XeF_2 is a good candidate for removing a variety of materials. However, a major task is the development of convenient passivation strategies in order to attribute for the different etching yields.

(a)

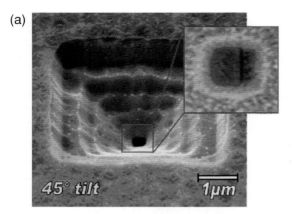

(b)

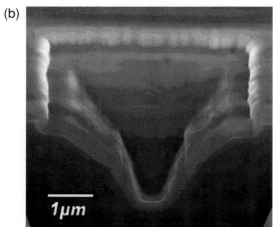

FIGURE 9.26: Left: Micro trench with sub-micron opening. In the inset the top-down view of the inner region is shown, revealing a clear signal change when hitting the dielectric. Right: Cross-section of a similar trench in bulk silicon covered with dielectric deposition.

Due to the absence of sputtering, e-beam etching can produce vertical sidewalls or even undercuts—a property of high potential, considering the tight dimensions and the aspect ratio requirements. Depending on gas flux, selectivity, and scanning parameters, the resulting shape can be tuned. A via etch with optimized parameters is shown in Figure 9.27.

Conducting material is deposited into the via to establish an electrical connection between a buried metal line and some structure on the surface for probing or to interconnect two signals. Though nano-machining is not expected to reach bulk metal conductivity, the specifications are rising. Copper has been introduced in microchip fabrication, and its superior conductivity opens the way for higher clock rates. Therefore an upper limit is set for the via resistance.

Highly conductive deposits can be achieved on the sample surface and in tight vias under conditions that allow sub-micron processing with FIB and FEB [9]. Nevertheless, via filling is a critical processing step. As dimensions are tightened, the gas diffusion into the via is hindered. Therefore longer refresh times have to be implemented. Furthermore, placement accuracy of a few nm is required. Misplacement, however, will lead to enhanced edge deposition. As a result, more material is deposited at the top edges of the via rather than inside. Thus the via is further tightened and it becomes more difficult to establish a continuous fill. The formation of voids within the via is a typical cause of high resistance connections and, moreover, the resistance variation in a series of similar vias will be increased dramatically. Therefore an automatic drift

(a)   (b)

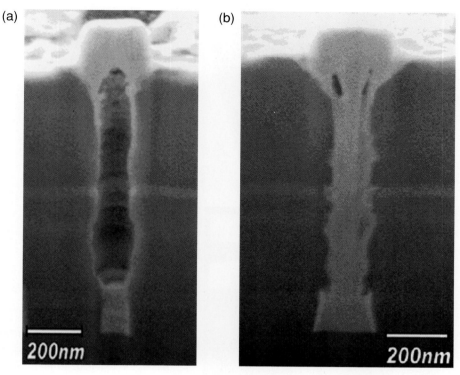

FIGURE 9.27: Left: FEB via etch with vertical sidewalls. Right: FEB via and metal refill.

correction routine is required that reduces placement errors to a minimum and eliminates human error in this highly critical process step.

The complementary process, metal cutting, completes the nano-processing tasks. It is interesting to look at FIB and FEB processes as separate approaches.

The electroplated copper in current generation microchips typically has a poly-crystalline structure. Due to channeling effects the different crystal orientations can be distinguished by the respective secondary electron yield in FIB. Moreover, the sputtering yield also depends on the crystal orientation and leads to an uneven removal of the copper, which is an issue for metal cutting due to the high integration density. A solution is the use of a special scanning strategy combined with WCO_6 during copper milling [11]. When introducing the gas precursor, a tungsten scattering layer is formed on the processed surface. Thereby the channeling is reduced to a large extent and a flat etching profile can be achieved. However, as the copper removal is done by physical sputtering, the process is not very selective. Therefore it involves a high risk to either overetch or end up with an incomplete cut in the case of very small structures. Lately it has been shown that working with H_2O can increase the selectivity dramatically for low ion energies [12]. So the endpoint detection is enhanced, but at the expense of a lower resolution.

Cutting of thin copper lines through small vias is the particular competence of e-beam processing. Material removal in FEB is chemical by nature. Though it is non-trivial to find a gas chemistry for this task, the resulting etch strategy has a good selectivity relative to the surrounding interlayer dielectric (see Figure 9.28). DC measurements have revealed a resistance of more than 50 GΩ. However, some residual non-conductive material can be found in the vicinity of the cuts.

(a)

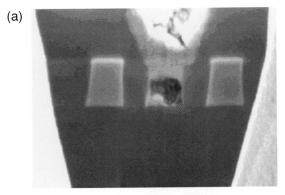

(b)

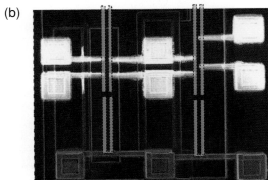

FIGURE 9.28: Top: cross-section of FEB metal cut. Bottom: FIB voltage contrast image of four FEB metal cuts. The position of the metal line is illustrated as an overlay.

A fast verification of the cut success can be obtained with voltage contrast imaging in an FIB tool as shown in Figure 9.28. Unconnected structures are charged positively by the Ga+ bombardment and secondary electron emission. Therefore the secondary electron yield from these areas is reduced relative to metal structures which are connected to ground potential.

Local dielectric deposition is employed to form an isolation layer between the active region around a micro trench and the via itself. A typical issue in SiO_2 deposition is the carbon content, which reduces the mechanical and electrical stability of the deposit. Working in an oxidizing atmosphere with low contaminant contribution reduces the carbon parasitic phase markedly (see Figure 9.29). Due to the absence of Ga+ ion implantation, e-beam deposits have reached superior values for the resistivity compared to FIB [9]. Consequently, a thinner layer can be employed for good insulation. This is of vital interest in order to keep the required via aspect ratios as small as possible.

3.4 CHALLENGES

Further reduction of the structure sizes will impose stringent specifications on the nano-machining capabilities in circuit editing. The established FIB technology, however, has already been pushed to the limits of the technically possible. E-beam processing is suited to extend the method into the sub-100 nm regime.

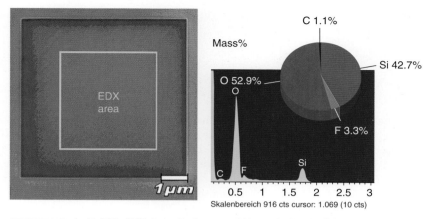

FIGURE 9.29: Left: SiOx FEB deposit of approx. 200 nm thickness. Right: EDX result of marked area demonstrating low carbon content.

The research and development time needed in order to prepare for a new chip generation is not only driven by miniaturization. Also the implementation of new materials poses a challenge, especially on the chemically enhanced processes. Therefore dual beam tools would be desirable.

4 CONCLUSIONS

Micro- and nanostructuring tools using FIBIP technology have been well established in labs as well as industrial environments for decades. Due to the limitations of this technology, especially regarding ultimate resolution and sample invasiveness (staining, etc.), FEBIP tools are catching up, particularly at high-end disciplines. In this chapter two examples for industrial applications of FEBIP have been given: photo mask repair and circuit editing. These two applications represent two extremes of industrial demands.

Photo mask repair is a productive application within semiconductor fabs with the well-known requirements like automation and clean room compatibility. These requirements lead to extra efforts regarding platform design, safety regulations, environmental control, production worthiness, easy operator interface, etc., which go far beyond a university lab tool used for FEBIP research. Circuit editing, on the other hand, does not imply such restrictive boundary conditions since this application is used in failure analysis labs, which do not use clean rooms or exhibit such high environmental noise.

However, circuit editing deals with a much higher scope of materials, which have to be removed and deposited. The different processes are handled very individually and have to be adapted to the particular edit. This results in a much more complex user interface and requires extraordinary versatility.

Both applications share high demands regarding accuracy and repeatability. The semiconductor industry's race after lower costs and higher performance that is described by Moore's law tightens specifications approximately by a factor of two every four years. Additional technology moves (like EUV for the mask business or high-k metal gate for circuit editing) sometimes generate extra efforts for the tool suppliers. As long as tightening specifications is the motor of this multi-hundred-billion dollar industry, it will remain challenging for the R&D departments around the world to keep pace.

NOTES

a. At the neutral point the incoming primary beam current and the emitted secondary beam current is equal. By neglecting higher order effects like the spatial distribution of the involved charged regions, one can assume that the equality of the currents leads to vanishing charging.
b. The natural duration of a scanned loop is given by the sum over all dwell times of the pixels of the processed pattern. If, for instance a shape consists of 10,000 pixels (e.g., a square pattern of 100 by 100) with each pixel having a dwell time of 10 nanoseconds, the natural duration of a loop would be 100,000 nanoseconds, i.e., 0.1 milliseconds. The required refresh time, however, can be in the order of several milliseconds.
c. Considering a pixel size of one nanometer, a field of view of 10 by 10 micrometers would already have 100 megapixel.
d. If a defect consists of missing absorber, it is called clear defect, hence the optical image is too bright. A defect with residual absorber in areas where it should have been removed during mask manufacturing is called an opaque defect, hence it prints too dark.
e. During the preparation all necessary channels are checked for pressure, temperature conditions, and clean state.
f. The keyword "double patterning" summarizes a variety of techniques that make use of repeated process steps to increase resolution instead of pushing the lithography. This can, for instance, be done by chemically generating so-called spacers on both sides of every line and subsequently removing the line itself. Since every line is used to produce two spacers, the pitch of the pattern can effectively be halved.

REFERENCES

[1] Ehrlich, C., U. Buttgereit, K. Boehm, T. Scheruebl, K. Edinger, and T. Bret. Integrated photomask defect printability check, mask repair, and repair validation procedure for phase-shifting masks for the 45-nm node and beyond. *Proc. SPIE* 6730 (2007): 67301Z.
[2] Ehrlich, C., U. Buttgereit, K. Boehm, T. Scheruebl, K. Edinger, and T. Bret. Phase-shifting photomask repair and repair validation procedure for transparent and opaque defects relevant for the 45 nm node and beyond. *Proc. SPIE* 6792 (2008): 67920H.
[3] Garetto, A., C. Baur, J. Oster, M. Waiblinger, and K. Edinger. Advanced process capabilities for electron beam based photomask repair in a production environment. *Proc. SPIE* 7122 (2008): 71221K.
[4] Garetto, A., J. Oster, M. Waiblinger, and K. Edinger. Challenging defect repair techniques for maximizing mask repair yield. *Proc. SPIE* 7488 (2009): 74880H.
[5] Waiblinger, M., K. Kornilov, T. Hofmann, and K. Edinger. E-beam technology and EUV photomask repair–A perfect match. *Proc. SPIE* 7823, 782304 (2010) .
[6] Cazaux, J. Some considerations on the electric field induced in insulators by electron bombardment. *J. Appl. Phys.* 59 (1986): 1418–1430.
[7] Webb, C. 45 nm Design for Manufacturing. *Intel Technology Journal* 12.2 (2008), http://www.intel.com/technology/itj/2008/v12i2/5-design/1-abstract.htm.
[8] Borodovsky, Y. Pixelated phase mask as novel lithography RET. *Proc. SPIE* 6924 (2008): 69240E–1–14.
[9] Utke, I., P. Hoffmann, and J. Melngailis. Gas-assisted focused electron beam and ion beam processing and fabrication. *J. Vac. Sci. Technol. B* 26 (2008): 1197–1275.

[10] Oehrlein, G. S. In *Handbook of plasma processing technology – Fundamentals, Etching, Deposition, and Surface Interactions*, eds. S. M. Rossnagel, J. J. Cuomo, and W. D. Westwood. Park Ridge, NJ: Noyes Publications, 1990, Chapter 8: Reactive Ion Etching, pp. 196ff.

[11] Casey, J. David, Jr., M. Phaneuf, C. Chandler, M. Megorden, K. E. Noll, R. Schuman, T. J. Gannon, A. Krechmer, D. Monforte, N. Antoniou, N. Bassom, J. Li, P. Carleson, and C. Huynh. Copper device editing: Strategy for focused ion beam milling of copper. *J. Vac. Sci. Technol. B* 20 (2002): 2682–2685.

[12] Rue, C., R. Shepherd, R. Hallstein, and R. Livengood. Low keV FIB applications for circuit edit. *ISTFA2007* (ASM International 2007): 312–318.

CHAPTER 10

FOCUSED ION BEAM AND DUALBEAM™ TECHNOLOGY APPLIED TO NANOPROTOTYPING

Oliver Wilhelmi and Hans Mulders

1 INTRODUCTION

Nanofabrication with the focused ion beam (FIB) in a DualBeam instrument, which integrates an FIB column and an SEM column in one single instrument, creates fascination—probably as its most prominent feature. Seeing nanopatterns emerge live on the screen of a scanning electron microscope never fails to fascinate scientists and engineers, and is a source of inspiration and dedication for students and operators who drive nanotechnology projects. The flexibility to machine almost any material at a nanometer scale and the ability to inspect the result in great detail and to immediately use the just-gathered findings to optimize the patterning process have made DualBeam instruments highly valuable tools for successful nanoprototyping across a wide range of diverse applications. This chapter summarizes the basics of patterning with ion and electron beams, shows some practical aspects in application examples, and outlines how the capabilities of DualBeam instruments can best be used to complement batch nanofabrication techniques.

The two most relevant techniques are focused ion beam (FIB) milling of materials, including the use of etch gases to selectively enhance the FIB milling process, and the beam induced deposition of materials, using either the FIB or the electron beam and a gaseous precursor as a source of the material to be deposited. Within the content of this chapter we are taking an empirical approach, showing how different process parameters affect the resulting pattern, and based on this, derive a set of best-known methods that will allow instrument operators to successfully create nanoprototype devices. Current boundaries and limitations of the techniques are discussed, as well as future trends for possible improvements.

The FIB columns in DualBeam instruments generate a focused beam of Ga^+-ions from a liquid metal ion source, the term FIB will therefore be used to describe a Ga^+-FIB.

2 BASIC PATTERNING TECHNIQUES FOR THE FIB MILLING OF NANOSTRUCTURES

The understanding of the basic patterning techniques for the FIB milling of nanostructures builds on a working knowledge of the characteristics of the FIB, the interaction of the FIB with the substrate material, and, if relevant, the interaction of the FIB with the gas molecules of a deposition precursor gas or a mill-enhancing etch gas.

For the FIB, a Gaussian beam profile as shown in Figure 10.1 can be assumed as a good approximation for the radial distribution of ions in the ion beam. The ion source, the ion optics, and the ion beam profile have been studied more extensively [1]; in practice, however, the exact beam shape will be affected by the true shape of the beam-defining apertures in the ion column, the ion column alignment, and the vacuum in the ion column. In daily routine operation, none of these parameters will exactly be known; hence, a Gaussian beam is a well suitable approximation for nanoprototyping.

For direct milling of the substrate material (material removal) with the FIB, the ions impinge typically with 30 keV energy on the substrate surface. The ions will penetrate the surface and transfer energy by elastic collisions to the nuclei in the substrate material. These elastic collisions of the ions with the nuclei of the substrate will set a collision cascade in motion [2; 3], which will also transfer momentum to surface atoms, which in turn will leave the substrate. This material removal is what is referred to as milling (the more common term for FIB applications) or sputtering in general. The surface atoms can leave the substrate as atoms (neutral) or ions (either positively or negatively charged), but the charge state of the leaving particle is not

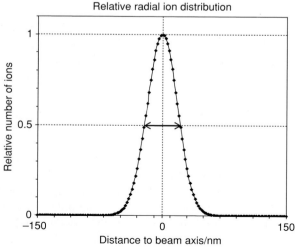

FIGURE 10.1: A Gaussian distribution as an approximation of the radial distributions of ions in the beam of an FEI Sidewinder™ FIB column for an accelerating voltage of 30 kV and a beam current of 93 pA. The arrow illustrates the full width at half maximum (FWHM) of 24 nm.

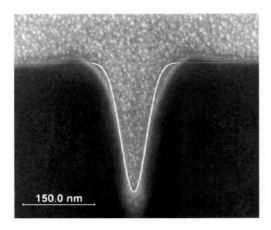

FIGURE 10.2: The profile of a single pixel line milled with a 30 kV, 93 pA FIB into an Si surface. The dark material is the Si-substrate, the brighter grainy material, is used to fill the trench during the FIB-cross-sectioning. A Gaussian profile is overlaid to validate the assumption of a Gaussian FIB by comparing it to the real trench profile.

relevant for the depth of the FIB milled pattern, as the depth will only depend on the total number of removed atoms.

The profile that an FIB generates in a substrate material is compared to the assumed Gaussian profile of the ions in the FIB in Figure 10.2. Apart from the profile of the FIB itself, there are primarily two effects of the ion-substrate interaction that will have an impact on the actual line profile milled into a substrate material:

(a)   the change in ion incidence angle as the line evolves;
(b)   the redeposition of sputtered material.

To (a): the material removal rate for FIB milling depends on the incidence angle of the ion beam. For a grazing incidence beam, the surface atoms can leave the substrate in almost the same direction as the incident ion beam, making direct momentum transfer efficient. The situation is different for an ion beam hitting on a surface under normal incidence, because a collision cascade is needed to transfers momentum to surface atoms, which will leave the surface in the almost opposite direction with respect to the incident ion beam, yielding a less efficient momentum transfer to surface atoms. The consequence is that the FIB mills material from a pattern sidewall much quicker than from a flat surface. The dependence of the sputter yield on the incidence angle of the ions is displayed in Figure 10.3.

The consequence for the profile of an FIB milled line is that, while the line is milled to its final depth, a trench with sloped sidewalls starts to form. Ions hitting the sloped sidewalls have a dynamically changing incidence angle, as the trench is getting deeper with respect to the beginning of the milling process. The angles as defined for the calculation in Figure 10.3 are shown in Figure 10.4 for different beam incidence positions on a trench profile.

The considerations above have only dealt with the first phase, the removal of substrate material by the FIB. In order to gain a better understanding of the FIB milling process, the question of where the removed material goes needs to be addressed, leading to the above-mentioned redeposition of sputtered material.

To (b): Inside the vacuum chamber the neutral atoms that are being removed from the substrate will move on a straight line unless deflected by a collision. The directions of the sputtered particles follow a cosine distribution [2]. The trajectories of secondary ions as sputtered particles will be influenced by any electrical fields inside the DualBeam instrument,

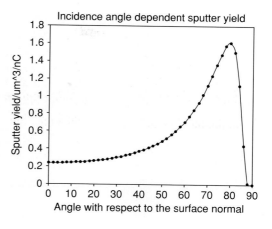

FIGURE 10.3: The sputter yield for Si (in $\mu m^3/nC$) is plotted against the angle between the incidence ion beam and the surface normal; calculated with the Yamamura formula [4]. 0° corresponds to normal incidence, angles close to 90° to grazing incidence. The sputter yield drop-off at angles close to 90° is caused by an increasing probability for backscattering of the primary ion beam at glancing angles.

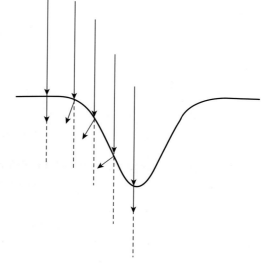

FIGURE 10.4: A virtual snapshot of the FIB milling of a line during the process. The arrows from the top indicate the directions of the incident ions; the dashed line the progression into the material. The short arrows indicate the normal vector to the substrate surface. The angle between incident beam and surface normal is largest where the forming line has the steepest sidewall. The sputter yield will be highest at these positions.

for instance electrical fields introduced by bombarding the substrate surface with the Ga^+-ions. When a line as in Figure 10.2 is milled into the substrate, the increasing depth of the trench will make it more difficult for ions to escape from the bottom of the trench. Instead, more and more sputtered particles that are ejected from one sidewall will redeposit on the opposing sidewall until this mechanism will stop the trench from getting deeper. The redeposition inside a deep FIB pattern will eventually pose a limit to the maximum achievable aspect ratio.

Redeposition is not directly evident in the image of the FIB milled line in Figure 10.2, but it does have a major influence on devising appropriate milling strategies for successful nanoprototyping, a topic to be discussed in Section 5.

In general, the most important parameter for the operator is the material removal rate, as this will allow the control of the depth of an FIB milled nanostructures. The material removal rate is best defined as the volume of the respective substrate material that is removed by a certain number of the singly charged ions, or more conveniently a certain charge. For FIB work

Table 10.1 Sputter rates in $\mu m^3/nC$ for a 30 kV Ga^+-FIB under normal incidence. The above values are indications that have been derived from different experiments [5]. Actual sputter rates can vary depending on lateral size and aspect ratio of a structure or the crystal orientation of the substrate material.

Material	Sputter rate	Material	Sputter rate	Material	Sputter rate
Al	0.29	Fe_2O_3	0.25	Si	0.24
Al_2O_3	0.08	GaAs	0.61	Si_3N_4	0.16
Au	1.50	InP	1.2	SiO_2	0.24
Bronze	0.20	MgO	0.15	Stainl. Steel	0.32
Brass	0.30	Mo	0.12	Ta	0.32
C	0.18	Ni	0.14	Ti	0.46
Cr	0.10	Pb	2.87	TiN	0.15
Cu	0.25	PMMA	0.4	TiO	0.15
Fe	0.29	Pt	0.23	W	0.12

the unit $\mu m^3/nC$ has proven itself as most practical. Table 10.1 lists material removal rates for some frequently used substrate materials.

The material removal rate is related to the surface binding energy of the surface atoms to the substrate. Since the surface binding energy is also a determining factor for the melting point of a material, a low melting point is usually a good indicator of a high material removal rate.

For the case of normal incidence on a homogeneous substrate material, the depth of a FIB milled pattern is:

$$Depth = n \times (I \times T) \times \frac{A}{\Delta x \times \Delta y} \times R \times \frac{1}{A}, \tag{1}$$

with n being the number of passes in which the pattern is repeated and I (FIB current) times T (dwell time) the charge applied to each individual pattern pixel. A (total pattern area) divided by $\Delta x \times \Delta y$ (pitch or center-to-center distance of adjacent pattern pixels in x and y) is the total number of pattern pixels in the pattern, R the material removal rate, and A again the total area of the pattern.

R is a constant for a given substrate material. The FIB current will usually be chosen such that the FIB diameter, usually defined as the full width at half maximum (FWHM), is small enough to produce the pattern with sufficiently well-defined side wall roughness, corner radius, and critical dimension control. The selection of a suitable FIB current is thus dependent on the characteristics of the FIB column on the instrument (Figure 10.5 shows the chart of the FIB diameters as a function of FIB current for a FEI Sidewinder™ column as an example).

The pitch parameters $\Delta x \times \Delta y$ are ideally chosen such that the beam is advanced with half of its diameter from one pattern pixel to the next pattern pixel, creating a 50% overlap of the beam diameters when the beam is placed on two adjacent pixels in order to achieve a good, but not unnecessarily dense filling of a pattern (Figure 10.6). Hence the pitch will be determined with the selection of an FIB current. The dimensions of the pattern, together with the substrate material, will thus confine the parameter space for Equation 1 and the operator will have to find a suitable combination of number of passes n and dwell time T.

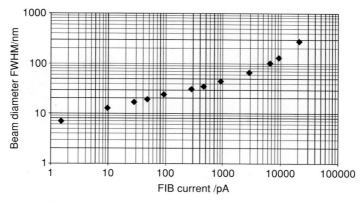

FIGURE 10.5: FIB diameter as function of FIB current for the FEI Sidewinder™ FIB column.

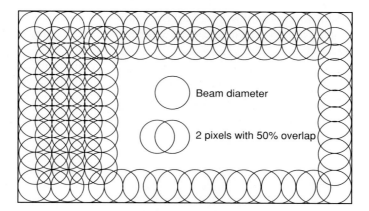

FIGURE 10.6: A rectangle pattern filled with circles representing the beam diameter at each position. For better clarity the beam diameter and two beam positions (pixels) with a 50% overlap of the beam diameters are drawn individually. The filling of the pattern is shown on the left-hand side and the beam diameters for all pixels along the rectangle outline show how the contour is reproduced.

Executing patterns in multiple passes is of particular importance for FIB milling and beam induced deposition processes, and the necessity of multiple pass patterning is probably the most prominent distinction from electron beam lithography, which appears closely related in terms of being a beam-writing technique with a focused particle beam. The effect of executing a pattern with different number of passes is illustrated in Figures 10.7a and 10.7b. They show a Y-junction of three individual trenches milled into Si in parallel with a short dwell time T and a large number of passes n (Figure 10.7a) and a long dwell time T and a single pass n = 1 (Figure 10.7b) with otherwise identical FIB current and pitch in X and Y. Both parameter sets yield a drastically different quality of the result.

(a) 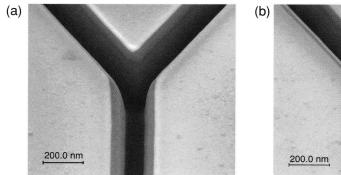   (b)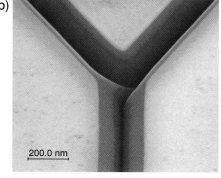

FIGURE 10.7: The Y-junction in image (a) was executed with a pixel dwell time of 1 μs and the entire pattern was repeated in 2131 passes. The same Y-junction was patterned with 1 pass and a dwell time of 2.131 ms in image (b).

The explanation for the different result lies in the effect of redeposition. When the beam is milling out one pattern pixel, the material removed at this position will redeposit in the vicinity, including the already milled parts of a nanopattern. Redeposition will be discussed in more detail in Section 5.

3 THE BEAM PATTERNING ENGINE AND ITS CHARACTERISTICS

A pattern design defines the position and shape of all individual pattern elements that are to be patterned with the DualBeam instrument. The patterning software will then calculate how the shapes have to be filled with discrete pattern pixels, taking into account the actual beam current which determines the respective beam diameter for setting the pitch, as discussed in Section 2. Each pattern pixel is assigned a position within the field of view. In order to steer the FIB or electron beam in a DualBeam instrument accordingly, the position of each pattern pixel is sent with its xy-coordinate to the digital patterning board. The digital patterning board is fitted with a buffer memory that can hold a certain amount of individual pattern pixels and with digital-to-analog converters (DAC) for x and y coordinates, which generate analog output voltages that are applied to the beam deflection circuitry of the respective column. The data chain is illustrated in Figure 10.8.

Modern instruments are fitted with patterning boards that have 16-bit DAC resolution. Hence the field of view is divided into $2^{16} = 65536$ pixels, providing a fine pixel grid, which ensures that the beam can physically be positioned with very high accuracy on its target position. In a 200-μm field of view, the beam can be positioned with an accuracy of 200 μm/65536 = 3 nm. Taking into considerations that the pitch is often selected as 50% of the beam diameter and the smallest beam diameter in Figure 10.5 of 7 nm, the patterning board resolution is sufficient to not restrict patterning the smallest possible features even in 200-μm writing fields.

Another figure of merit for patterning boards is the fastest achievable patterning speed, or the shortest possible pixel dwell time. This is of particular relevance for the beam chemistry

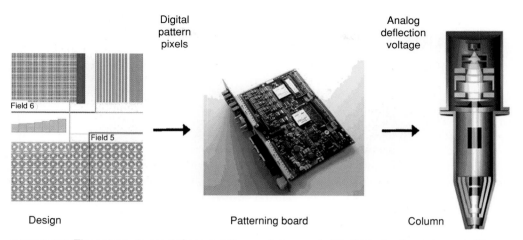

Digital pattern pixels   Analog deflection voltage

Design   Patterning board   Column

FIGURE 10.8: The pattern design defines positions and shapes of all individual pattern elements. The patterning software will calculate the individual pattern pixels and send them to the digital patterning board. The board then uses digital-to-analog converters (DAC) to generate analog voltages which are applied to the deflection circuits in the respective column.

processes in Section 6. If a gaseous precursor is introduced into the vacuum chamber of a DualBeam by a gas injection system, the maximum gas flow is limited by the vacuum level that needs to be maintained in order to operate FIB and electron beam. If the depletion of gas molecules by beam current of FIB or electron beam exceeds the rate at which precursor molecules can be fed to the substrate surface, the beam chemistry process becomes ineffective. The patterning speed therefore determines the maximum beam current that can be used in combination with a gas injection system, without exhausting the precursor molecules on the substrate surface. With a high patterning speed and a larger number of passes or repetitions of the pattern, the beam dwells on each position for a very short time, as long as there are still precursor molecules present. By the time the beam revisits a position in the following pass, the coverage of the surface with molecules is refreshed. Some of the well-established recipes that are routinely being used for beam chemistry applications use pixel dwell times of 100 ns, which makes a patterning speed of 10 MHz a minimum requirement for a suitable pattern engine.

It has been pointed out above (Figures 10.7a and 10.7b) that executing patterns in multiple passes is one of the essential features for DualBeam patterning processes. In practice, the number of passes can range up to well above 100,000, which raises the importance of data handling and time overheads. The buffer memory of the digital patterning board can hold a certain number of pattern pixels. Once these pixels are loaded into the patterning board, they can be looped in any number of passes without introducing any time overhead. If the number of pattern pixels exceeds the capacity of the buffer memory, the patterning software needs to stream data to the patterning board: a first set of pattern pixels is sent to the patterning board, one pass is being executed, the pattern pixels are cleared from the buffer memory, the second set of pattern pixels is sent, executed, cleared, and so on, until one single pass is complete. The entire procedure will have to be repeated until all passes are completed. This streaming of pattern pixels can delay the actual patterning process significantly, and for that reason the size of the memory buffer on the patterning board is an important characteristic of the beam-patterning engine.

4 DEFINITION OF NANOPROTOTYPING AND COMPARISON WITH CONVENTIONAL NANOFABRICATION

Conventional nanofabrication applies lithography processes (electron beam lithography or photolithography) or nanoimprint techniques for each design layer and requires an entire suite of specialized tools, for instance: a resist spin-coater, a hot plate, the lithography tool, resist development, deposition or etching equipment for pattern transfer, resist strippers, and inspection tools to monitor the process. Since nanodevices often consist of several pattern layers, the lithography tools have to ensure accurate alignment of the layer stack. The diversity of nanotechnology applications, the demand of ever smaller pattern dimensions, and the increasing availability of novel materials often make the established process recipes no longer applicable, and the entire nanofabrication process needs to be reinvented. In practice, the iterations to reinvent a process take up a considerable amount of operator and tool time, and cause project delays, especially in facilities where the equipment is shared among many research groups.

The nanoprototyping capabilities of DualBeam instruments hold the potential to cut short turn-around times in nanotechnology projects and enable early tests of prototype functionality. Beam induced deposition of different materials can be combined with FIB milling without the need of several aligned lithography steps; stacks of dissimilar materials can be structured in one single milling process; patterns can be added to existing structures on a substrate or existing patterns can be modified. The patterned substrates are immediately available for further processing or characterization.

Once a prototype has been tested successfully, a batch nanofabrication process can be qualified by integrating electron beam lithography with the SEM column of the DualBeam instrument as patterning step. The ability to do FIB nanoprototyping and do resist patterning for nanofabrication in one instrument facilitates the delivery of a proof of concept for a device much faster and allows the identification of a process for volume manufacturing at the same time. Figure 10.9 compares the process flow for FIB nanoprototyping with conventional nanofabrication.

5 ARTIFACTS, ISSUES AND SOLUTIONS FOR FIB MILLING

The direct material removal when milling a nanostructure with an FIB into a substrate differentiates FIB nanoprototyping from conventional nanofabrication techniques, as already discussed above. The significance of taking the specifics of the ion-substrate interaction into account when developing a patterning strategy is further detailed here and we describe techniques that can minimize frequently encountered patterning artifacts.

5.1 REDEPOSITION

FIB milling directly removes, or sputters material at the moment the ions impinge on the substrate. The sputtered substrate material is ejected from the position of the pattern pixel

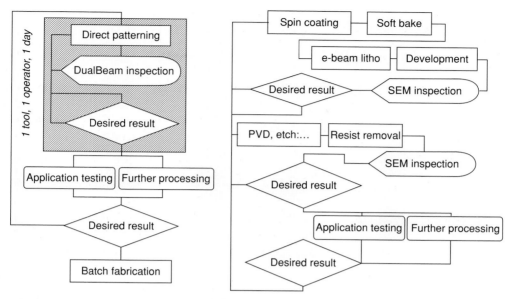

FIGURE 10.9: Left: a flow diagram of nanoprototyping with a DualBeam instrument. Right: flow diagram of a conventional batch nanofabrication process (a DualBeam instrument can replace the SEM inspection, adding site selective cross-sections, 3D analysis and TEM sample preparation).

where the FIB beam is dwelling and will redeposit on any surface of the sample that can be reached in a straight line from the ejection point. The deeper an FIB pattern is, the more sidewall surface area becomes susceptible to redeposition. The effect of redeposition in the patterning process is illustrated in Figure 10.10, showing the combined effects of the incidence-angle dependent sputter rate and redeposition of milled material on a trench pattern when milling a 500-nm trench pattern in a single pass from left to right.

When a pattern is being executed with multiple passes and a short pixel dwell time at each pass, the amount of redeposited material in each pass is very small and is being removed when milling the pattern in the subsequent pass. When the FIB milling process has finished, the pattern will only have a negligible amount of redeposition in a single pass. The dramatic difference between multiple-pass milling versus single-pass milling is prominently visible in Figure 10.11.

In addition, redeposition can often be seen at surfaces that do not connect in a straight line to the ejection points of the removed material, but on areas that have just been exposed to the ion beam. An example is shown in Figure 10.12. While the exact root cause of this effect is still under investigation, the observations do suggest that charging of a patterned area can bend the trajectories of secondary ions, leading to charge-driven redeposition in addition to the redeposition of the electrically neutral sputtered material.

5.2 DIFFERENTIAL MILLING

For poly-crystalline substrate materials the milling rate depends on the crystal orientation of the different grains due to the channeling of ions in a crystalline material [3]. Consequently, the pattern depth will not be uniform across a pattern. The difference in pattern depth in relation

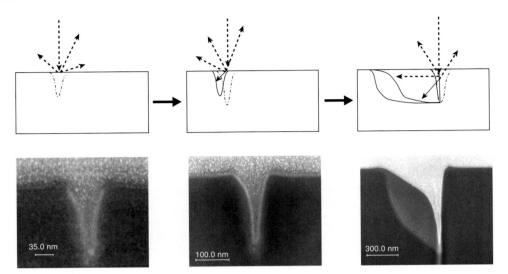

FIGURE 10.10: Left: the Gaussian FIB hits the sample surface and creates an almost Gaussian line profile. Center: as the pattern advances to the next position in the pattern, the FIB is dwelling on the sloped sidewall of the first pattern. Because the sputter rate is higher on the sloped sidewall, the pattern is deeper on the second line. Right: Toward the right hand-side of the wider pattern, the angle-dependent increase of the sputter rate saturates. The arrows indicate the directions in which sputtered material can leave the sample and the directions in which the material will deposit on the left-hand side slope of the pattern. In the cross-section of the trench the redeposited material is well visible as a brighter zone.

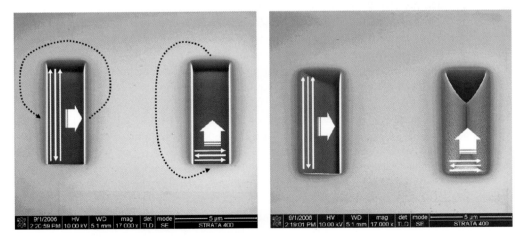

FIGURE 10.11: Two rectangular boxes of $2 \times 10 \ \mu m^2$ were milled into Si with a target depth of $1 \ \mu m$—one with a serpentine sweep parallel to the long side, one with a serpentine sweep parallel to the short side as indicated by the white arrows. The image on the left shows that a multiple pass FIB strategy yields good results independent of the sweep direction. The image on the right shows the results when exactly the same patterns are milled in a single pass. The redeposition of sputtered material inside the pattern is obvious.

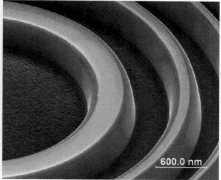

FIGURE 10.12: For the Fresnel zone plate on the left, all zones were milled in parallel with a multiple-pass strategy, resulting in a smooth pattern bottom. The Fresnel zone plate on the right was milled with a multiple-pass strategy, but this time one zone after the other, starting from the center. Especially the central zone has clearly a rougher bottom, an observation attributed to redeposition of material from the milling of the outer zones. The original Si surface in between the milled surfaces does not show any apparent difference.

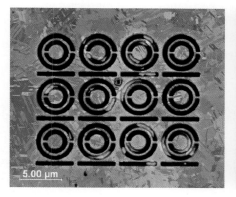

FIGURE 10.13: An array of split ring resonators milled into Cu. The backscattered electron images show the grain structure as orientation contrast. A standard FIB milling recipe creates an uneven depth due to different milling rates for different grain orientation; seen as remaining material inside the pattern on the left. The image on the right shows the same pattern milled with intermediate FIB induced deposition into the pattern. This planarization of the pattern gives better pattern depth uniformity.

to the grain orientation can be seen in Figure 10.13 (left image). The backscattered electron images show the orientation of the different grains with differing brightness, and the image on the left for a standard FIB milling process shows the residual material inside the pattern for certain grain orientations.

One potential approach to overcome pronounced artifacts due to differential milling in poly-crystalline bulk materials is the use of a beam-induced deposition process in order to planarize the pattern bottom at intermediate steps—the result is shown in the image at the right in Figure 10.13. The times for the milling and deposition intervals will vary for each new combination of pattern design and FIB patterning parameters and will need to

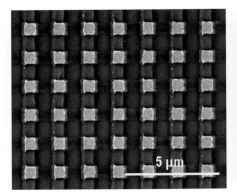 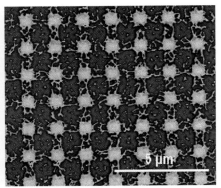

FIGURE 10.14: An array of gold squares on glass substrate, which was milled out of a continuous thin film. The image on the left shows a single-pass milling with 200 μs pixel dwell time. The image on the right shows a multi-pass milling with 1 μs pixel dwell time. The artifacts in the glass around the squares from the single-pass milling are tolerable, while the speckles of remaining gold in the image on the right will disturb the functionality of the square array.

be determined experimentally. High pattern aspect-ratios pose a practical limitation to the process applicability, as deep and narrow patterns create a shadowing effect for the precursor gas molecules being fed into deep structures for the intermediate deposition.

Thin films of polycrystalline material on an amorphous or single-crystalline substrate are best milled with a small number of passes and a long pixel dwell time. The obtainable definition of the pattern will in most cases outweigh the FIB milling artifacts in the substrate due to long dwell times (see the discussion of redeposition above). Figure 10.14 shows an array of squares that have been milled from a gold thin film on glass. When the thin film in the area around the gold squares was milled away with a single pass at a long pixel dwell time, the squares are well defined but the glass itself shows single-pass artifacts. The multiple-pass milling with a short pixel dwell time leaves speckles of gold on the glass between the squares, making the single-pass milling the better choice for many thin film applications.

5.3 SURFACE DAMAGE

The effect of impinging ions starting a collision cascade in the substrate material as the physical mechanism behind FIB milling also causes disorder in the FIB milled surfaces. The damage layers caused by FIB milling of various materials have mostly been studied in relation to the FIB preparation of TEM samples [6]. Amorphization or recrystallization occurs in the surface layers of crystalline substrate materials, as shown in the example of the damage layer on the sidewall of an FIB milled trench in InP in Figure 10.15. Should the change of the crystalline structure on the pattern surface alter the functionality of a nanoprototype device, surface cleaning techniques with a low-energy FIB can be adopted from TEM sample preparation and can minimize the thickness of the surface damage layer [7].

The implantation of Ga^+-ions into the surfaces of a nanopattern generally raises more concern with respect to the functionality of an FIB milled nanodevice. Significant Ga-concentrations in the sidewalls of photonic crystals can cause a device failure by increasing the absorption of light, although gas-enhanced FIB milling has been reported to largely reduce

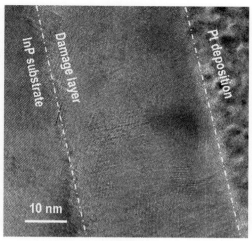

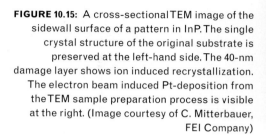

FIGURE 10.15: A cross-sectional TEM image of the sidewall surface of a pattern in InP. The single crystal structure of the original substrate is preserved at the left-hand side. The 40-nm damage layer shows ion induced recrystallization. The electron beam induced Pt-deposition from the TEM sample preparation process is visible at the right. (Image courtesy of C. Mitterbauer, FEI Company)

the Ga-implantation in InP substrates for photonic crystal applications [8; 9]. In the case of nanofluidic devices, Ga could be chemically incompatible with molecules or cells in the liquid. Conformal thin film coatings (for instance, by electron beam induced deposition in the DualBeam) can be applied to FIB milled fluidic devices in order to avoid direct contact of the liquid or the dispersed molecules with the Ga in the surfaces. Basic properties of a nanofluidic prototype, such as sealing of the nanofluidic channels, the connections to pumps and reservoirs, throughput and particle filtering, can be tested regardless of the Ga-content in the pattern sidewalls.

Low-energy FIB cleaning of pattern surfaces, gas-enhanced FIB milling processes, and additional coatings represent some of the available techniques for minimizing damage layers on the pattern surfaces. In order to allow balancing the implied increased complexity of the design of experiment with the benefit, the actual extent of Ga implantation in pattern surfaces needs to be taken into consideration. The chart in Figure 10.16 displays the total amount of Ga in weight % as a function of distance from the sidewall surface for an InP substrate and different ion energies.

The Ga^+ ions may not only be implanted into the substrate material, excess Ga can also form an unstable phase in conjunction with the substrate material, forming droplets on the pattern surfaces, for instance on GaAs substrates. These droplets can be avoided by gas-enhanced FIB milling, for which the introduction of an etch gas into the vacuum chamber allows the excess material on the pattern surfaces to react, forming a volatile component. Another example is shown in Figure 10.17, where a honeycomb pattern was milled into InP. The P in the substrate material is removed at a higher rate, leaving the excess In to form droplets. The formation of these droplets can also be avoided by the use of gas enhanced FIB milling.

6 ELECTRON AND ION BEAM INDUCED BEAM CHEMISTRY

Application examples of ion beam induced beam chemistry have been briefly introduced in the previous section as ion beam induced deposition to overcome differential milling artifacts and the use of gas-enhanced milling to maintain clean surfaces of FIB milled nanostructures.

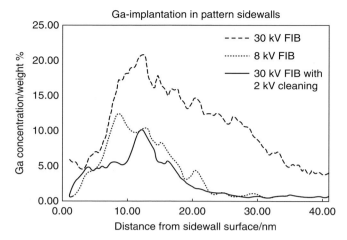

FIGURE 10.16: The Ga-concentration in a pattern sidewall (here InP) measured with an EDX line scan on the cross-sectional TEM image in Figure 10.15. The Ga-concentration is relatively low at the surface itself and peaks a few nm inside the sidewall. The extension of the Ga-implantation and the peak concentration are lower for a lower FIB energy (30 kV vs. 8 kV). The effect of a 2kV-cleaning step of a 30-kV pattern is shown as well.

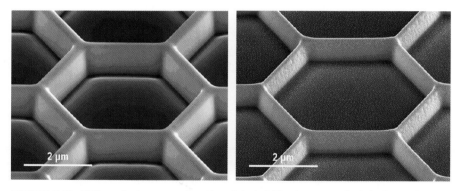

FIGURE 10.17: Milling of a honeycomb pattern into InP. The image on the left shows the result with XeF2 assisted FIB milling. The image on the right shows FIB milling without the assistance of an etch gas. The P mills quicker than the In, leaving excess In close to the pattern surface, which forms the bright droplets on sidewalls and pattern bottom. (Images courtesy of D. Wall of FEI Company)

Beam chemistry in general refers to processes related to the interaction between a (charged) particle beam and precursor molecules adsorbed onto the substrate. These precursor molecules will be modified by the local interaction between beam and sample, and depending on the precursor type, two situations can be distinguished:

(a)   a precursor molecule that ideally is split into volatile and non-volatile parts. The non-volatile part forms a deposit on the sample surface, while the volatile part is

Dicobalt(O) octacarbonyl

FIGURE 10.18: Example of a point deposition of Co created with a Co2(CO)8 precursor (right) on a silicon AFM tip, allowing use of the AFM as a magnetic force microscope (MFM). The small Co tip is typically created in 8 s.

pumped out of the vacuum chamber. An example of such a reaction is:

$$Co_2(CO)_8 + \text{electrons} \rightarrow 2Co + 8CO \uparrow + \text{electrons}$$

In this case the precursor molecule is $Co_2(CO)_8$, the volatile part is the CO group, a gas that is pumped out of the system. The non-volatile part is the cobalt metal, which forms a deposition under the beam. An example of an MFM tip, made this way as a point deposition, is shown in Figure 10.18.

(b) a precursor molecule that under reaction with the beam releases a component which is able to react chemically with the substrate and by this reaction attacks the substrate to form another volatile part that can be pumped away. An example is the known etching capability of water vapor on carbon-based materials under influence of the beam. In general, such a reaction could be described by:

$$C_nH_m + \text{electrons} + H_2O \rightarrow x\,CO_2 + y\,CH_4 + \text{electrons}$$

In this case the result is an etching phenomenon where sample material is removed from the substrate by the beam interaction and converted into volatile products.

It should be noted that the reactions mentioned above show the ideal process, i.e., there are only desired non-volatile and volatile parts. In reality, however, the processes are not ideal and as a consequence the decomposition is not complete. This results in a deposition that does not only contain the desired element, but also a remainder of the precursor molecule, which in many cases consists of C, O, and H atoms that are included in the deposition. As a result, the purity of the deposition is diminished and hence the property of the deposit deviates (in some cases substantially) from the known bulk property of the material of interest. An extensive overview article can be found in the literature [10].

Although the above discussed processes are described for electrons, the same principle applies for (Ga⁺) ions from a FIB where, in the case of (a) the deposition also contains ions from the primary Ga⁺ beam, and in the case of (b) the intrinsic ion milling can be amplified by the additional chemistry in such a way that the milling rate strongly increases, while at

Table 10.2 Global comparison between IBID and EBID characteristics.

	IBID	EBID
deposition yield	fast, typically around $0.1 - 0.5\,\mu m^3/nC$	slow, typically between 5×10^{-4} and $5 \times 10^{-3}\,\mu m^3/nC$
initial substrate milling	yes	No
deposition contaminated	yes with primary ions	No
Minimum deposition size	~ 15 nm	~ 1 nm
atomic purity (excluding Ga)	higher e.g., 34 at% Pt from MeCpPtMe3	lower e.g., 16 at% Pt from MeCpPtMe3

the same time the redeposition effect can be reduced. The overall milling performance can be improved for the right combinations of substrate material and precursor molecules, since the chemical amplification of FIB milling is of course material specific. For beam induced deposition processes, the substrate material is far less relevant and deposition is possible on almost any substrate, although local charge buildup on non-conductive substrates may complicate the clear formation of a very small structure.

The ion beam induced deposition is typically done at low beam currents because two processes occur simultaneously: deposition from the precursor gas and milling of the substrate. If the ion beam current is increased, the deposition will level off (gas limited regime), while at the same time the milling rate will continue to increase with the beam current. As a consequence the deposition rate will go back to zero or even to negative values when for a given beam current the milling rate is higher than the deposition rate. A brief comparison between ion and electron beam induced deposition is given in Table 10.2.

In practical applications the two techniques can be combined to make use of their respective advantages: for instance, electron beam induced deposition (EBID) can be applied first to avoid initial substrate milling and then ion beam induced deposition (IBID) is applied to speed up the layer growth. This technique is commonly practiced when preparing TEM samples with an intact top surface.

It should be noted that for most precursors the EBID yield remains constant up to around 0.5 nA (this means that the deposition rate scales linearly with the current), which indicates a beam current limited process. At higher currents, local supply of precursor molecules is too low, and the yield decreases. The yield of the EBID process is also strongly related to the energy of the electron beam: at lower electron energies the yield is higher. For Pt-deposition with MeCpPtMe$_3$ this maximizes at around 140 eV primary electron energy [11].

Beam chemistry in FIB and DualBeam instruments is a well-established technology. Its first applications were found in mask repair and circuit edit, two applications that have strongly been driven by the demands of the semiconductor industry. The primary requirements in this industry are the local and site specific removal of material—usually by the ion beam—and the local deposition of either a conductor or an isolator (circuit edit, see Figure 10.19) or an opaque structure as used in mask repair. In both applications an operator will directly control the FIB instrument, guided by a proven recipe. The cost and time savings that can be achieved with these applications by shortening (re)design cycles of integrated circuits have made circuit edit and mask repair common applications in the semiconductor industry at an early stage in the development of beam chemistry. Consequently, much of today's commercial beam chemistry has a focus on semiconductor applications mostly for silicon based processes. An overview

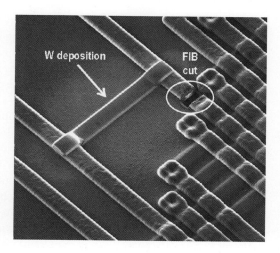

FIGURE 10.19: Example of a local circuit modification. One track has been cut with the FIB and reconnected to another track by W deposition.

Table 10.3 Commercially available beam chemistries and their main applications and characteristics.

Name	Precursor	Characteristics for FIB use	Applications
Pt deposition	MeCpPtMe$_3$	$\rho = 1000 - 2000\,\mu\text{Ohm.cm}$	Contacts, TEM lamella (protection/contrast)
W deposition	W(CO)$_6$	$\rho = 200\,\mu\text{Ohm.cm}$	Contacts
C deposition	naphthalene	$\rho = 4500\,\mu\text{Ohm.cm}$	Large area contact, TEM lamella (protection/contrast)
Insulator deposition	TEOS + H$_2$O	$\rho = 10^{+12} - 10^{+14}\,\mu\text{Ohm.cm}$	Insulator
Delineation etch	—	layer selectivity	increased layer visibility by small edge formation
Insulator enhanced etch	XeF$_2$	layer selectivity	increased removal yield for SiO$_2$ (∼7x)
Enhanced etch	I$_2$	layer selectivity	increased removal yield for metals, Si

of today's beam chemistry processes can be found in Table 10.3, together with their main application.

6.1 BEAM CHEMISTRY AND NANOTECHNOLOGY

Some of the characteristics listed in Table 10.3 do not only depend on the actual substrate and selected precursor, but also on the operating conditions such as beam energy and current, the patterning strategy, applied geometry, and even on less obvious parameters such as vacuum conditions. In literature there is a wide variety of results and a general concern is the poor repeatability, which only implies that not all parameters of beam chemistry processes have been fully recognized and hence not recorded and reported in publications. An example is the residual water vapor in the chamber during the beam-induced deposition of material. For the application of beam chemistry and the acceptance of the technology, the repeatability of the processes and the obtainable functionality at the nanoscale will be crucial issues; consistency of results and a sound understanding of the whole process are necessary to bring beam chemistry to a next level.

For nanotechnology applications, the emphasis lies more on electron beam induced deposition (EBID) than on ion beam induced deposition (IBID). The main reasons for the preference for EBID, despite the overall slower deposition rates, are the better beam definition (to allow the creation of smaller structures), the absence of (Ga^+) ions in the deposited material, and the absence of invasive sample milling at the onset of the deposition process. For example, direct contacting of nanotubes by IBID is practically impossible because the FIB mills the original nanotube in a fraction of a second and hence destroys the structure of interest. In addition, nanotechnology applications require a wider variety of materials and different material characteristics than semiconductor related applications. The research of domain wall movement in spintronics can be studied by directly making prototypes of nanoscale magnetic material with beam induced deposition of materials such as Fe, Co, or Ni. In other fields of interest, such as plasmonics and photonic crystal applications, the need for creating small Au structures both by FIB milling and EBID is very important.

With more nanoprototyping applications for DualBeam instruments emerging, the request for a wider range of materials is evident. Sometimes, in a function such as a contact point, it can be critical to avoid Schottky barriers in the contact; sometimes the need for local electrodes to lock certain molecules in position is important, and in this way the chemical activity of the deposited material is important. Therefore, new processes and new beam chemistries are being developed, for example for the deposition of Au, Co, Fe, Pd, and Ag. In addition to this, the current existing processes can and will be improved, by either switching a precursor (for example using $Pt(PF_3)_4$ instead of $MeCpPtMe_3$ for Pt-EBID) or by additional (post) anneal processes [12] and better control of the water content during deposition.

The water content is naturally present in any SEM or DualBeam chamber and only slowly diminishes after pump down, while it is known to play an important, yet unquantified role in the deposition process. The water content may form a positive contribution to the process (e.g., higher purity) or a negative contribution (e.g., slower deposition, oxidation). Therefore the amount of water vapor present during the deposition process has to be under control (repeatable water vapor pressure). As an example of how this parameter changes over time, the decrease of the partial water vapor pressure as a function of time is given in Figure 10.20. A tighter control of the residual water vapor pressure can be achieved by longer pump-down times, the additional use of a cold trap for capturing the water vapor, and to some extent by a sample load-lock.

6.2 SELECTION CRITERIA FOR PRECURSOR GASES FOR EBID AND IBI

When a new precursor material has to be defined, the following characteristics are important to allow the realization of successful applications in nanotechnology.

(a)   The element of interest must be present in the precursor molecule. Preferably the precursor molecule should have the lowest possible number of carbon atoms.

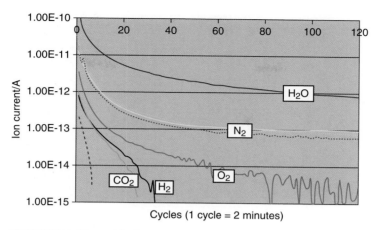

FIGURE 10.20: The residual gas composition as measured by the RGA ion current (y axis), after vacuum Ok signal as a function of time (x-axis). As can be observed it takes around 4 hours, before the water vapor content has reduced two orders of magnitude.

(b) If a non-carbon based precursor is available, this certainly must be considered. For example, Pt can be deposited using $Pt(PF_3)_4$ (tetrakis Platinum) which is a carbon-free precursor. Although the Pt-deposits from this precursor do not contain any carbon, it has other disadvantages, such as the high toxicity and the release of PF_3 and F during the deposition process, which may not be compatible with substrate material or components inside the vacuum chamber. During tests on a DualBeam, it was found that the EBID process improved, but even with the lowest FIB current the IBID process did not create any deposition at all. This is likely due to the fact that the release of F during the deposition enhances the FIB milling process so much that it exceeds the overall deposition process, even at the lowest currents.

(c) In most cases the precursor is solid or liquid at standard temperature and pressure, allowing the use of its vapor as the gaseous supply to the sample (with the exception of etch gases such as F, Cl, and Br). Precursors are artificially produced molecules that tend to decay at higher temperature, which implies an upper limit to the operating temperature range. The precursor of interest should have a high enough vapor pressure within the operating temperature range in order to ensure a sufficient delivery of precursor molecules for the target deposition rate. Liquid precursors usually have high vapor pressures and require the reduction of the gas flow to the sample by a flow screening aperture in the needle of the gas injection system.

Before using a precursor, the air has to be pumped out of the gas injection system to a partial pressure below 0.1% of the vapor pressure of the precursor itself. During this pump-out, some of the precursor material is lost and the overall design of the gas injection system has to balance the requirements of a sufficient gas flow of precursor molecules with a short pump-out time without shortening the reservoir service life.

In practice, the vapor pressure of a precursor is often in the range of 0.05 to 50 mbar and can be adjusted over the temperature as long as the stability of the precursor molecules

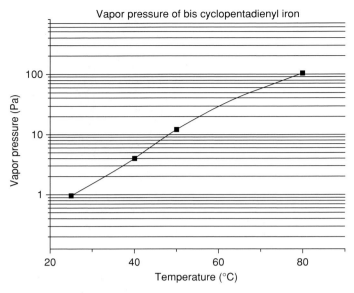

FIGURE 10.21: Vapor pressure curve for ferroscene, generated using data from MSDS data sheet (40°C) and from [Monte, M. J. S., et al., *J. Chem. Eng. Data* 51 (2006): 757–766]. By controlling the temperature, the vapor pressure can be adjusted and set to an optimum between gas flow and chemical stability.

is assured. The material safety data sheets (MSDS) of the precursor gases rarely provide vapor pressure curves or at least one value for the respective precursor. An example of a useful and known curve is given in Figure 10.21 for bis(cyclopentadienly) iron or ferroscene.

(a) The adhesion of the precursor molecule to the sample surface has to fulfill certain conditions. For deposition of a material from a precursor, the precursor molecules have to interact with secondary electrons from the substrate surface, which are locally released by the incident beam. In order to enable an effective deposition the precursor molecules need to adsorb onto the surface at least for a time (the so-called residence time) that is equivalent to the time constant for the release of secondary electrons from the surface. The molecule residence time can be increased by lowering the substrate temperature. On the other hand, if cooled too much, condensation will occur, i.e., the precursor molecules build up a continuous multilayer film, even in the absence of an electron or ion beam. This can be observed when opening a regular GIS with the sample at very low, cryogenic temperature, e.g., −140°C (Figure 10.22). Since the adhesion kinetc is hard to predict when selecting a new precursor, experiments will have to determine whether a certain molecule is suitable or not.

(b) The precursor should not be hazardous to the instrument, operators or service engineers. By nature, most precursors are toxic, ranging from low toxicity (for example naphthalene) to very high toxicity for $Ni(CO)_4$. The potential hazards are strongly dictated by the possible release and the characteristics of the ligands such as CO, which is dangerous to humans, even in low concentration. The hazards are more

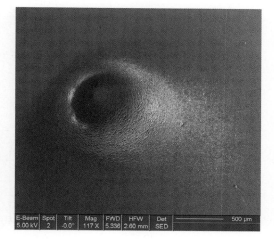

FIGURE 10.22: Capturing of a standard Pt precursor (MeCPPtMe3) on a very cold surface (−140°C). The molecules "freeze" onto the substrate and the thin film growth rate is very large—typically >200 nm/s. The output area of the GIS is clearly visible as the precursor layer almost grows into the needle.

a practical obstruction then a scientific problem, but safe work practice needs to be observed in several phases of the work with a precursor:

- Loading of the precursor into the gas injection system—this should minimally be done in a fume hood but preferably in a glove box with a protective nitrogen or argon atmosphere.
- During operation the exhaust of the microscope pre-vacuum pump should be secured and brought to outside of the building to dilute the already low amounts of material (typically 5–20 mg/hour).

(c) Loading the precursor into the gas-injection system inside a glove box under a protective atmosphere will also protect the precursor from reactive agents such as water vapor or even the oxygen in air. For example, the precursor $Co_2(CO)_8$ for beam induced Co deposition is very sensitive to reaction with water vapor with a possibility of spontaneous combustion. A secure environment must be available for the handling of this precursor, or the precursor must be supplied with a protective hexane addition (a so-called stabiliser) with the benefit of safer handling, but at the expense of high carbon content in the deposit during the start-up phase. An alternative precursor could be $Co(CO)_3NO$, which does not have these complications during loading, but will result in a less pure deposit.

(d) The precursor gas should not be harmful to the instrument. This refers to chemical reactions between precursor and instrument parts. For example, an etching agent such as I_2 will spontaneously attack bronze and copper parts inside the vacuum system. Instruments that are designed for beam chemistry work take the possible precursor gases into account in the bill of material for parts and components. However, retrofitting a DualBeam instrument with additional detectors or new precursor gases will require a check of the compatibility of instrument and gas chemistry.

(e) The precursor should not be excessively expensive in its daily use. This refers to the actual amount of "used hours" per gram and the cost per gram and includes such factors as the chemical inherent stability and not only the pump-out speed. Typical precursor prices can range from 10 Euro/gram to 3000 Euro/gram.

(f) The precursor should preferably be one chemical and not a mix of two independent chemicals, even if these chemicals are stable with respect to each other. A mix of

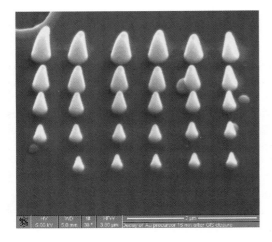

FIGURE 10.23: Single pixel depositions of Au-C (precursor is $(CH_3)_2$-Au(acac)) with a dwell time of 30 seconds each—starting from top left—after closing the valve of the gas injection system. After 15 minutes the Au deposits are a lot smaller, but still present. The precursor molecules come from the inside of the gas injection system needle and/or the vacuum chamber walls.

two precursors is more complicated to handle (two reservoirs, two delivery systems, and twice the number of control parameters). In the case of a deposition of an alloy, however, the simultaneous use of two precursors could be the most logical approach.

(g)   The precursor should not leave any traces in the DualBeam instrument after use. Since DualBeam instruments are not exclusively used for EBID nanoprototyping, the microscopy and FIB performance in high-resolution imaging and analysis must not be compromised. Today's commercially available precursors can all be completely pumped out of the vacuum chamber when the gas injection system is closed. The pump-out time will depend on the respective precursor gases and can range from around 10 minutes for the W deposition precursor $W(CO)_6$ to up until 40 minutes for the SiO_2 deposition precursor TEOS. An illustration of the post delivery of a precursor after closure of the valve in the gas injection system is given in Figure 10.23 where single pixel depositions have been made with a Me_2Auacac precursor for Au-EBID. After 15 minutes the deposited dots are a lot smaller, but still present.

6.3 INSTRUMENT REQUIREMENTS AND CONFIGURATION

There are certain additional requirements on the tool and best work practices when working with precursor gas delivery and EBID experiments:

a)   Penning vacuum gauges are known to contaminate over time, an effect that can be accelerated when working with precursor gases. A contaminated Penning gauge will cause the measurement of the pressure inside the vacuum chamber to be too low. Since the reproducibility and consistency of EBID processes will also depend on the background pressure during deposition, a reliable measurement of the actual pressure should be recorded when doing EBID experiments. Extra maintenance and cleaning of the Penning gauge might be required for extensive EBID work. Regular recordings of the vacuum level after overnight pumping will show a trend that is indicative for the condition of the Penning gauge.

b) The geometry of the gas delivery system has a major impact on the deposition process: the angles under which the needles of the gas injection systems are mounted, and the distance of the needles to the position on the sample where the deposition process is carried out, determine the density and uniformity of the precursor molecules on the substrate surface. In a DualBeam, the gas injection systems are set up such that the coincidence point of FIB and electron beam (which is also the tilt-eucentric point of the sample stage) is the geometric centre of the needle alignment. The needles are in a retracted position during stand-by and are only inserted for the delivery of the precursor molecules. In the inserted position they are close above the substrate surface, typically at a distance of 100–200 μm, in order to locally concentrate the flux of precursor molecules to the area of the deposition. Software interlocks warrant a safe operation of a DualBeam instrument with one or several gas injection systems. Supplying more than one precursor gas through the same needle relaxes geometrical restrictions, but creates a risk of cross-contamination.

c) For many EBID experiments it is strongly advised to have *in situ* analytical capabilities—most common is energy dispersive X-ray spectroscopy (EDS)—as this helps to immediately characterize the chemical composition of the deposits. An EBID deposit can easily be made with an area of 2×2 μm^2 and a height of 400–600 nm, a size that is sufficient for an accurate EDS-analysis. Low electron energies for the EDS-analysis help to minimize the influence of X-rays from the substrate. Dedicated multi-layer programs for EDS-analysis such as STRATGEM provide more accurate quantitative results for thin films. Determining the carbon content in a high purity EBID deposit requires special attention, because the W, Pt, and Au N-lines, which overlap in the spectrum with the carbon peak, need to be taken into account in the deconvolution of the X-ray spectrum. In general, the high over-correction of the low energy C Kα-line makes an absolute quantification of the carbon content in an EBID deposit difficult. (Wavelength dispersive X-ray spectroscopy (WDS) is not recommended here, because the relatively long exposure of the deposit to high beam currents can be regarded as a post treatment that can potentially change the composition of the deposit.) Another useful analytical capability is a manipulator for *in situ* measurement of the electrical conductivity of the deposited material, especially since many deposits change their properties when exposed to air (oxygen) during venting of the vacuum chamber. The comparison of *in situ* measurements versus *ex situ* measurements outside the DualBeam instrument will show the oxidation of a deposit in air. One way to prevent oxidation of a deposit is to deposit a thin SiO$_2$ layer over it in a two-step process before venting the vacuum chamber. It has been shown that this prevents oxidation of the small Pt grains in Pt-EBID and maintains the specific resistivity of the deposited material over time [13; 14].

6.4 FUTURE DEVELOPMENTS FOR EBID AND IBID

On the way of making EBID and IBID even more powerful enablers for nanotechnology, some of the problems of the current state of the technology must be solved. Higher purity of the beam induced deposits is one of the primary goals. Being able to rely on the material properties close to those of pure bulk materials to build nanoprototypes in a DualBeam instrument will expand the number of functional devices that can be fabricated at the nanoscale. Achieving purities of 100% will not be necessary in most cases, as demonstrated with the

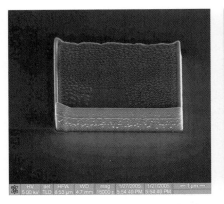 

FIGURE 10.24: IBID of Pd from the Pdhfac precursor with two different beam currents and uniform exposure. The IBID with 90 pA in the image on the left shows that the thickness of a deposition reduces from edges to the center, and there is no deposition in the pattern center, which is attributed to precursor depletion. The IBID with 32 pA in the image on the right shows a uniform height of the deposition across the full pattern. The composition of each deposition is uniform and includes large amounts of carbon as the main contaminant.

successful realization of nanoscale magnetism [15] or the use of Au-EBID as a catalyst to grow III/V-semiconductor nanotubes after reducing the carbon content in the Au-deposits in an annealing step. The thermal annealing of EBID deposits as a post-process treatment has proven to increase the purity of the deposited materials significantly [16] and is considered an important path into the future. However, the removal of carbon from the deposited material in the annealing process can cause shrinkage and shape changes of the nanostructures and substrate materials such as polymers are incompatible with the typical annealing temperatures in excess of 250°C.

The range of materials available for EBID and IBID will definitely grow and will include more magnetic materials, low Schottky-barrier contact materials, optically relevant materials, (bio)chemically active materials and probably materials with better mechanical properties (increased stiffness). A better understanding of the basic processes by dedicated research in this field will help to improve the concepts for precursor gas delivery and to further optimize patterning strategies. Current research focuses on higher purity [16], the smallest possible feature size [17], and the basic understanding of the growth mechanisms and modelling of shape control [18]. The need for a better understanding of the growth mechanism is illustrated in figure 10.24 with the example of IBID with a low-flux Pd precursor. A relatively small FIB current difference causes a very obvious difference between an evenly deposited structure and a partially milled structure, giving Pd-IBID a very narrow process latitude.

Very important at this point is that present and future work needs to focus more on consistency of the many experiments that are done worldwide. The reproducibility of results is a key requirement, before the technology can be applied to its full extent. As a technology for nanoprototyping, EBID and IBID will have to come out of the "trial and error" mode. Parameters such as the background vacuum pressure will need to be part of the design of experiment for EBID processes and will need to be included in the control parameter set for nanoprototyping with EBID and IBID. The deposition of iron in an ultra-high vacuum (UHV) environment has been reported to produces high purity and consistent results [19], indicating the importance of background vacuum pressure and residual water vapor.

FIGURE 10.25: The SEM image in the center shows a top-down view on four Au-IBID squares deposited with different FIB currents onto an Si substrate. The Au deposit creates the bright contrast. Two AFM measurements of the bottom left and top right squares (next to the square they were taken from) show that the squares have been milled into the Si substrate with a distinct wall along the edge. (AFM measurements courtesy of L. Belova, KTH, Stockholm, Sweden)

Apart from the material properties of a deposit, the actual control of the minimum feature size and the true 3D shape of a deposited structure need to be under control as well. In many deposition experiments the shape of a deposited nanostructure is being judged based on the SEM image from the DualBeam instrument. While SEM images directly provide valuable information, the interpretation of contrasts can be misleading. The analysis of the 3D topography of a deposit with an atomic force microscope (AFM) can complement the characterisation of EBID and IBID results with accurate height measurements of very thin deposits and surface roughness information. Figure 10.25 compares an SEM image with two corresponding AFM measurements.

Further understanding of the basics of the growth process will definitely help to get a better control of the 3D shape of a nanostructure and will help implementing additional patterning strategies dedicated to EBID and IBID in nanoprototyping applications.

Combining the capabilities of EBID and IBID with other deposition techniques can be a welcome addition to the scientists' toolbox. An example of this is atomic layer deposition (ALD), a thin film technique producing very pure films of for instance Pt. The ALD process is a two-cycle process of a precursor and an activator in a self-limiting reaction. A material layer is built by ALD in atomic mono-layers, and the control of the total thickness is defined by the number of ALD cycles. Since the ALD process will in principle apply to the entire substrate surface, EBID of thin Pt-nanostructures can be employed for creating fine seed structures for an area selective ALD process, i.e., the ALD grows Pt only on those areas that contain a Pt seed layer. The successful combination of the excellent lateral resolution of EBID with the high purity and excellent thickness control of the ALD process has been demonstrated experimentally [20].

Another approach toward creating 3D-nanostructures with EBID is the use of a cooling stage in order to largely increase the adhesion of the precursor molecules to the substrate which allows condensing a thick precursor coating on the substrate. The variation of the electron energy during electron beam patterning controls how deep the electrons penetrate into the condensed precursor coating and opens the opportunity to control down to which depth the electrons will interact with the precursor molecules. For EBID experiments on a

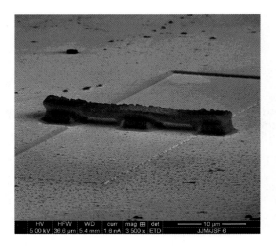

FIGURE 10.26: A suspended micro-bridge created by exposing two 1 μm thick condensed precursor layers with 10-kV electrons one after the other on a cold substrate and then removing the unexposed precursor molecules by warming up the substrate to room temperature.

cold substrate, the precursor flow from the gas injection setting is reduced with respect to the room temperature settings, permitting a more accurate thickness control of the precursor layer. With the standard precursor for Pt-EBID (MeCpPtMe$_3$) a growth rate of 17 nm per second was achieved for the precursor layer at a substrate temperature of $-100°C$. The precursor layer was deposited up to a thickness that an electron beam with the highest electron energy available in DualBeam instrument (30 keV) still can fully penetrate. After the patterning step and electron beam exposure of the condensed precursor layer, it is possible to use the process repetitively to build a 3D-nanostrucutre out of multiple condensed precursor layers. After all exposure steps the substrate is warmed up to room temperature. The precursor molecules that have been exposed to the electron beam form the Pt-EBID nanostructures that will remain on the substrate, while all unexposed precursor molecules will desorb from the substrate surface. Building a 3D-nanostructure in several consecutive patterning steps requires patterning software that has full automated control over the electron beam and FIB, the gas injection system and provides overlay functionality for accurate pattern placement of each individual pattern layer. A first example is the suspended micro-bridge over a milled groove in Figure 10.26. The EBID process was carried out with two exposures of two subsequently condensed precursor layers with fully automated process control. The total process time was 430 seconds. Expanding this technique to more layers has the potential to prototype more complex 3D-nanostructures in a fast and direct way.

7 PROCESS AUTOMATION FOR ADVANCED DUALBEAM NANOPROTOTYPING

Although individual nanoprototype devices can be fabricated in a very short time in a DualBeam instrument, the process time for applications that aim at the repeated fabrication of a nanostructure at many sites on a wafer or for applications that require complex nanostructures with a large number of pattern elements, can limit the feasibility of DualBeam nanoprototyping. The speed of the different patterning processes–FIB milling, IBID, EBID– needs to be optimized by the instrument operator through selecting beam currents and process

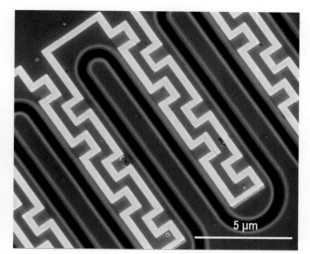

FIGURE 10.27: A FIB milled nanofluidic channel (dark meander) interdigitated with an IBID Pt-line (bright). The FIB milling was done with 300 pA, the IBID with 50 pA. An automated alignment was used to correct the beam shift resulting from the change of FIB currents and the insertion of the gas injection system for the Pt-deposition.

conditions that represent the best compromise of a reasonable process time and the desired control of critical dimensions, sidewall roughness and sidewall verticality. Once an optimized parameter set has been defined, the patterning process can be automated, allowing unattended FIB and electron beam patterning over many hours.

With the long and unattended operation of DualBeam systems, the overall instrument stability (beam drift, stage drift and sample drift) becomes critical. Therefore drift correction functions have become integral part of automated nanoprototyping systems. The drift correction function will acquire an image of an alignment mark outside the actual nanodevice area in regular time intervals and will match the acquired image with a reference image. Any mismatch will be fed back to the FIB or electron beam column as a beam shift correction. Having this alignment functionality available has also led to an increased use of mix-and-match techniques, in which a substrate is pre-patterned with a conventional nanofabrication process and DualBeam nanoprototyping is applied only on selected areas of the existing pattern. The capability to register the position of alignment marks allows positioning the FIB patterns accurately on a target position without exposing any other part of the substrate to unwanted ions. Furthermore, the pattern alignment is crucial for automated pattern execution, when the instrument is programmed to automatically change beam currents or insert and retract gas injection systems, since any of these changes can introduce beam position offsets which need to be corrected. Figure 10.27 shows an area of a nanofluidic device with a FIB milled channel, interdigitated with a Pt-IBID line. Automated pattern alignment ensures that the two process steps are nested accurately.

Patterns that extend beyond a single writing field rely on stitching techniques for continuity of the pattern across the writing field borders. The direct patterning with the FIB offers in this respect the unique capability to incorporate alignment marks into the pattern that can be read back after a stage move. The actual position of the pattern in the previous writing field can thus be passed on in a nano-relay to the following writing field. Accurate stitching can be achieved this way, while maintaining the flexibility of the sample stage with five axes of freedom. A stitched photonic crystal pattern is shown in Figure 10.28, where the alignment marks indicate where the photonic crystal has been stitched together.

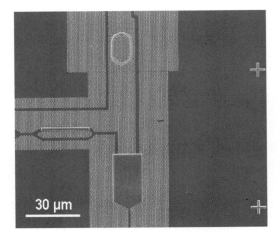

FIGURE 10.28: A photonic crystal pattern milled in three different writing fields at different stage positions. The crosses at the right are two of the alignment marks used for an alignment based stitching. The stitching error at the position of the lower alignment mark is not visible at this magnification; a fine horizontal line at the position of the upper alignment mark shows a very small residual stitching error.

8 OUTLOOK

DualBeam nanoprototyping is a unique patterning technique as it includes pattern depth for milling or pattern height for deposition processes as a control parameter for entire pattern layers or individual pattern entities and makes thus an entry into the world of 3D nanodevices. Using the sample stage of a DualBeam instrument with five degrees of freedom also opens opportunities to use the FIB at different angles to create 3D nanostructures. The flexibility to assign a different depth or height to various parts of a design, either in discrete steps or as a gradual change combined with different substrate orientations for different patterning steps does overstretch the present design tools and DualBeam nanoprototyping intelligence will need to be built into future design tools. The foundation for 3D-nanoprototyping however has been laid, with a sound understanding of the FIB-substrate interactions and a more thorough investigation of the principles of beam chemistry; and with pattern engines that have the speed and precision to execute dedicated DualBeam patterning strategies and provide nm-accuracy to position patterns on a target position.

Venturing into the third pattern dimension for nanoprototyping remains a goal for scientists across all application areas of nanotechnology and the rapidly increasing use of FIB nanoprototyping has made DualBeam technology the spearhead of the push toward 3D-nanoprototyping.

REFERENCES

[1] Orloff, J., L. Swanson, M. Utlaut. Ion optics for LMIS. In *High resolution focused ion beams: FIB and applications.* New York: Kluwer Academic/Plenum Publishers, 2003.

[2] Giannuzzi, L. A., B. I. Prenitzer, B. W. Kempshall. Ion-solid interactions. In *Introduction to focused ion beams*, eds. L. A. Giannuzzi and F. A. Steve. New York: Springer, 2005.

[3] Volkert, C. A., and A. M. Minor. Focused ion beam microscopy and micromachining. *MRS Bulletin* 32 (2007): 389.

[4] Adams, D. P., and M. J. Vasile, Accurate focused ion beam sculpting of silicon using a variable pixel dwell time approach. *J. Vac. Sci. Tech.* B 24.2 (2006): 836.

[5] FEI DualBeam User Manual and J. J. L. Mulders, D. A. M. De Winter, W. J. H. C. P. Duinkerken. Measurement and calculations of FIB milling yield of bulk materials. *Microelectronic Engineering* 84 (2007): 1540.

[6] Mayer, J., L. A. Giannuzzi, T. Kamino, and J. Michael. TEM sample preparation and FIB-induced damage. *MRS Bulletin* 32 (2007): 400.

[7] Wilhelmi, O., S. Reyntjens, C. Mitterbauer, L. Roussel, D. J. Stokes, and D. Hubert. Rapid prototyping of nanostructured materials with a focused ion beam. *Jpn. J. Appl. Phys.* 47.6 (2008): 5010.

[8] Callegari, V., P. M. Nellen, J. Kaufmann, P. Strasser, F. Robin, and U. Sennhauser. Focused ion beam iodine-enhanced etching of high aspect ratio holes in InP photonic crystals. *J. Vac. Sci. Technol.* B 25.6 (2007): 2175.

[9] Schrauwen, J., D. van Thourhout, and R. Baets. Iodine enhanced focus-ion-beam etching of silicon for photonic applications. *J. Appl. Phys.* 102 (2007): 103104.

[10] Utke, I., P. Hoffmann, and J. Melngailis. Gas-assisted focused electron beam and ion beam processing and fabrication. *J. Vac. Sci. Technol.* B 26 (Jul/Aug 2008): 1197.

[11] Botman, A., D. A. M. de Winter, and J. J. L. Mulders, Focused electron beam induced deposition at very low landing energies. *Second International Workshop on Electron Beam Induced Processing,* Thun, Switzerland, 2008.

[12] Botman, A., J. J. L. Mulders, R. Weemaes, and S. Mentink. Purification of platinum and gold structures after electron-beam-induced deposition. *Nanotechnology* 17 (2006): 3779.

[13] Botman, A., M. Hesselberth, and J. J. L. Mulders, Investigation of morphological changes in platinum-containing nanostructures created by electron-beam-induced deposition. *J. Vac. Sci Technol.* B 26.6 (2008): 2464–2467.

[14] Porrati, F., R. Sachser, and M. Huth. The transient electrical conductivity of W-based electron-beam-induced deposits during growth, irradiation and exposure to air. *Nanotechnology* 20 (2009): 195301.

[15] Fernandez-Pacheco, A., J. M. de Teresa, R. Cordoba, and M. R. Ibarra. Magnetotransport properties of high-quality cobalt nanowires grown by focused-electron-beam-induced deposition. *J. Phys. D: Appl. Phys.* 42 (2009): 055005.

[16] Botman, A., J. J. L. Mulders, and C. W. Hagen. Creating pure nanostructures from electron beam induced deposition using purification techniques: A technology perspective. *Nanotechnology* 20 (2009) 372001.

[17] van Dorp, W. F. Sub 10 nm focused electron beam induced deposition. Thesis, TU Delft, 2008.

[18] Smith, D., J. D. Fowlkes, and P. D. Rack, Simulating the effects of surface diffusion on electron beam induced deposition via a three-dimensional Monte Carlo simulation. *Nanotechnology* 19 (2008): 415704.

[19] Lukasczyk, T., M. Schirmer, H. P. Steinrueck, and H. Marbach. Electron-beam-induced deposition in ultrahigh vacuum: lithographic fabrication of clean iron nanostructures. *Small* 4.6 (2008): 841–846.

[20] Mackus, A. J. M., J. J. L. Mulders, M. C. M. van de Sanden, and W. M. M. Kessels. Local deposition of high-purity Pt nanostructures by combining electron beam induced deposition and atomic layer deposition. *J. Appl. Phys.* 107 (2010): 116102.

CHAPTER 11

REVIEW OF FIB TOMOGRAPHY

Lorenz Holzer and Marco Cantoni

1 INTRODUCTION

Over the last few years, FIB tomography has rapidly evolved from a specialized method that was initially used only in a few specialized laboratories, to a common microscopy technique which is nowadays applied in numerous disciplines of materials and life sciences. This successful evolution of FIB-tomography was accompanied by significant improvements of imaging resolution, machine stability, and introduction of user-friendly automation procedures. Due to the combination with a variety of detection modes (low kV BSE, EDX, EBSD, SIMS), FIB tomography has become a versatile method that is now an integral part of the standard equipment of commercially available FIB/SEM machines.

In comparison to other 3D-microscopy techniques, FIB tomography is currently occupying a niche with respect to its resolution and to the volume of the material that can be analyzed (see Figure 11.1; also compare discussion in Mobus and Inkson, 2007). The typical resolutions (i.e., voxel sizes) that can be reached are in the range of tens of nm down to 5 nm. The resolution of FIB tomography is thus clearly better than that provided by X-ray tomography or by the other serial sectioning techniques. Yet even higher resolutions can be achieved with electron tomography by TEM and with atom probe tomography. However, these high resolution 3D-techniques suffer from the drawback of a much smaller sample volume that can be analyzed.

A general problem in quantitative 3D microscopy is representativity. This problem is caused by the limited volume of the analyzed materials. Therefore, resolution and image window size (i.e., volume) must be adjusted according to the feature sizes of interest. For a representative 3D analysis from samples with a wide particle size distribution, a large volume (e.g., cube edge length of 50 μm) and a high resolution (e.g., 10 nm) are required at the

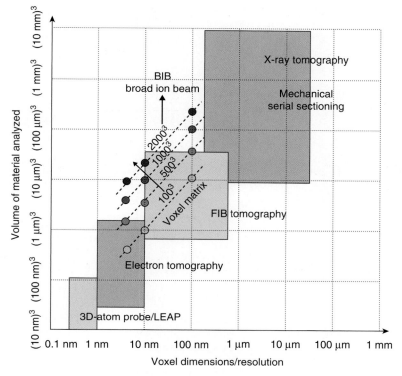

FIGURE 11.1: Comparison of FIB-tomography with other 3D-microscopy methods (modified after Uchic et al., 2007). As illustrated, FIB-tomography is evolving toward larger volumes at higher resolutions, which require a larger voxel matrix.

same time. However, these are contradicting requirements, since the voxel matrix of all 3D techniques is limited by detector capacities and reasonable acquisition time (typically 1000^3 voxels).

Figure 11.2 represents an illustration of the range of cube sizes that can typically be produced with FIB tomography at different resolutions. The particle size distribution of the cement powder under investigation was too wide for a single representative analysis with FIB tomography. Therefore, the powder was separated into five different grain size fractions with average particle diameters (D_{50}) ranging from 0.6 to 14.2 μm. The resolutions were then optimized between 12 to 116 nm for representative analyses, which resulted in cube sizes with edge lengths between 5 and 65 μm. Each cube contains between 1,000 to 4,000 particles, which is the basis for statistical size and 3D shape analysis (Holzer et al., 2006c). It is important to note that the number of images in these analyses was limited to 200–300 images.

In the meantime, the number of images that can be acquired by FIB serial sectioning has significantly increased. Superstacks with 1,000 to 2,000 images (each image with 2048*1536 pixels) can be produced nowadays at a rate of 40 to 60 slices per hour (Cantoni, 2009). Thus, FIB tomography has now reached almost the same resolution as electron tomography (i.e., 5 nm), but with the advantage that it can be applied to much larger sample volumes. Hence, with the current state of FIB tomography, large cubes with a matrix of up to $2,000^3$ voxels can be produced. When imaging with a voxel resolution of 20 nm, this matrix will correspond to a

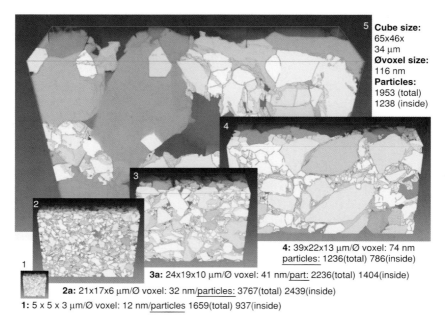

FIGURE 11.2: 3D-analysis of five grain size fractions from ordinary portland cement, adapted from Holzer et al. (2006c). For representative analysis, resolution and volume have to be adapted according to the average grain sizes (D_{50}), which are ranging from 0.6 to 14.2 μm.

cube with edge length of 40 μm. Such a volume would be sufficiently large for the representative analysis of samples with a relatively wide size distribution, which could include all five particle size fractions from Figure 11.2.

Modern FIB tomography thus enables the acquisition of microstructural (BSE), chemical (EDX), and crystallographic (EBSD) 3D-information at resolutions of 5 nm–×100 nm, but for sample volumes with an edge length of x*10 μm. (Note that resolutions are lower for EDX and EBSD). These unique capabilities for 3D imaging with FIB are currently used in a variety of applications, such as:

- Subsurface analysis for diagnostics and failure analysis of semiconductors;
- Investigations of deformation patterns in nanolayered materials;
- Study of the permeability in porous building materials or in oil reservoir rocks;
- Mapping of crystallographic grain orientation in alloys;
- Quantification of triple phase boundaries and percolation in fuel cell electrodes;
- 3D-connectivity of neurons and synapses in brain tissue.

In summary, FIB tomography is still a relatively young microscopy technique. But due to the broad applicability it is spreading rapidly. In this chapter we intend to give a short introduction to the FIB-tomography technique and an overview of the current applications. In the next section, basic aspects of the FIB serial sectioning technique are discussed. These aspects should be taken into account for successful 3D imaging with FIB. In the subsequent section, the succession of methodological innovation steps that led to the current state of FIB tomography is summarized chronologically. Since these innovations opened new possibilities, the present review also includes an overview of the emerging application fields of materials and life sciences. Finally, in the last section, we briefly discuss the current limitations of FIB tomography and try to formulate the corresponding needs for further developments.

2 FIB SERIAL SECTIONING PROCEDURE

Modern FIB/SEM machines are equipped with both, ion and electron optical columns, which make them perfect tools for serial sectioning at high resolution (see Figure 11.3).

Serial sectioning is an alternating process of sectioning and imaging. Thereby the ion beam (y-direction) is used for sequential erosion of thin layers in the 10 nm range. The sample is placed at the eucentric height, so that the imaging plane (x-y-directions) can be scanned with the electron beam under an angle of 52° without changing sample position. During the acquisition of the image stack, the imaging plane is moving step by step in z-direction due to the sequential ion-milling. The entire serial sectioning procedure includes three phases that are described below: (1) cube preparation and optimization of parameters, (2) serial sectioning, and (3) data processing.

2.1 CUBE PREPARATION AND OPTIMIZATION OF SERIAL SECTIONING PARAMETERS

Before starting the FIB serial sectioning experiment, several imaging and sectioning parameters that are dependent on each other have to be optimized. As discussed in the introduction to

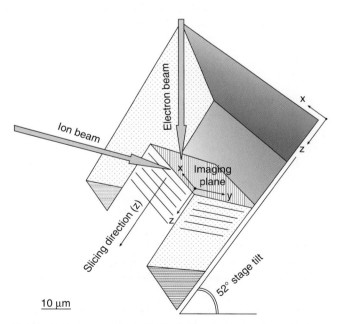

FIGURE 11.3: Illustration of the sample cube geometry, optimized for FIB serial sectioning with a dual beam FIB/SEM machine (modified after Holzer et al., 2004).

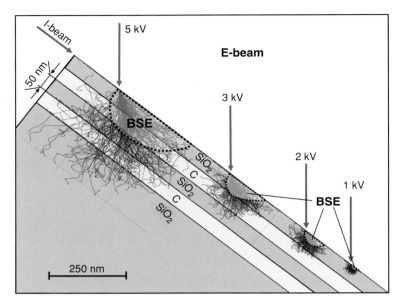

FIGURE 11.4: Electron trajectories in a virtual multilayered material, based on a simulation with Casino-software (http://www.gel.usherbrooke.ca/casino/What.html). Note that the size of the interaction volume is strongly depending on the accelerating voltage and that its shape is asymmetric due to the oblique incident angle in FIB/SEM.

this chapter, resolution and volume of analysis have to be chosen according to the size of the microstructural features in the sample. For the definition of an optimum magnification, the relationship between depth penetration of the electron beam and excitation volume of BSE and X-rays also must be taken into account. This relationship is illustrated in Figure 11.4 for a multilayer carbon-SiO_2 structure. At 5 kV the depth penetration of the electrons is 300 nm, and thus the electron beam protrudes the entire layered structure. In addition, the oblique angle of the incident electron beam also leads to an asymmetrical shape of the interaction volume. The excitation volume for backscattered electrons (indicated by the semitransparent white overlay) is much smaller than the total interaction volume.

Depending on the accelerating voltage, the size of the BSE interaction volume can put drastic limitations to the effective resolution of FIB tomography. In our example (Figure 11.4), the depth of the BSE excitation volume at 5 kV is still larger than 100 nm, which is too large for the investigation of samples with 50 nm layers. In contrast, at 1 kV the BSE excitation volume and the corresponding spatial resolution are in the range of a few nm. For high resolution FIB tomography the ability of low kV imaging thus becomes crucial. The critical aspect for imaging at such low acceleration voltages is then whether the sample and the BSE detector can provide sufficient material contrast. In this context, the latest generation of sensitive BSE detectors, which are optimized for low kV imaging, open new possibilities for 3D imaging at resolutions in the 10 nm range (see, for example, De Winter et al., 2009). Of course, the relationship between depth penetration and accelerating voltage also has to be taken into account for the optimization of the serial sectioning step size. For example, serial sectioning at $\geq$ 5 kV (SEM) with a step size in the 10 nm range will lead to a strong oversampling without improving the information density of the acquired data volume.

After defining the optimum parameters for serial sectioning, a cube of appropriate size has to be prepared by using the FIB milling capabilities at high beam currents (e.g., 10 nA). About 10 μm wide trenches are milled around the cube in order to enable access of the electron beam to the imaging plane and to reduce shadowing effects in the lower part of the images. The typical cube geometry is shown in Figure 11.3. For serial sectioning in combination with EDX- or with EBSD-detection, this cube preparation is not sufficient and more rigorous preparations techniques such as block lift-out (Schaffer and Wagner, 2008) or milling of a cantilever-shape (Uchic et al., 2006) have been proposed. For the last step of cube preparation, a Pt-layer of approximately 1 μm thickness is produced with *in situ* I-beam deposition in order to protect the cube surface during serial sectioning. In addition, for non-conductive samples a thick metal-coating also should be sputtered (*ex situ*) in order to suppress eventual charging effects. After cube preparation, the sample is ready for serial sectioning.

2.2 SERIAL SECTIONING

The aim of serial sectioning is to produce a regular stack of images, which can directly be transformed into a voxel-based data volume. For this purpose, the thickness of eroded layers should have similar dimensions to the pixel resolution in the imaging plane (i.e., approximately 10 nm). Therefore, the stepwise erosion should be repeated with high precision at a constant z-step size. Because the acquisition of hundreds of images lasts for 20 hours or even longer, drift can become significant. Without correction, the drift in z-direction causes distortions in the reconstructed 3D-microstructure. In contrast, drift components in x- and y-directions are automatically compensated during the off-line data processing with image alignment.

For high resolution FIB tomography an automated serial sectioning with integrated drift correction was introduced (Holzer et al., 2004). However, drift compensation by pattern recognition is affected by continuous degradation of the reference marks during imaging of the x-z-surface with ion-induced secondary electrons (ISE). Hence, the precision of the automated drift correction is decreasing continuously during the sectioning procedure. Fortunately, modern FIB/SEM systems have become much more stable, and usually after a period of stabilization and thermal equilibration drift becomes very small or even negligible. Due to the improved stability in modern FIBs, serial sectioning can be performed without drift correction. Remaining z-drift components can be measured for image sequences of 10, 50, or 100 images and then be compensated linearly during 3D reconstruction. When FIB serial sectioning is combined with EDX, image acquisition takes more time and drift compensation becomes more important. For this purpose an alternative drift correction procedure using electron images at two different magnifications ("overview" and "detail") was introduced (Schaffer et al., 2007).

In addition to the drift compensation, further correction procedures are needed because of the geometrical peculiarities in dual beam systems: dynamic focus correction is required because of the oblique imaging angle (52°), which also leads to a distortion of x-y-dimensions that have to be corrected accordingly. During the serial sectioning, the imaging plane is shifted in z-direction and therefore focus tracking is required in order to correct for the increasing working distances. When shifting the working distance and imaging under oblique angle, the region of interest is shifted out of the field of view. This has to be compensated with automated region tracking. In modern dual beam FIB/SEM systems, all these phenomena are compensated with the automated sectioning procedure. In this way, image stacks of high quality can be acquired.

2.3 DATA PROCESSING

Extraction of scientifically valuable data from the 3D image volume is a challenging task which is worth paying much attention. Unfortunately, the processing of each material type and each detection mode requires its specific treatment and hence no standard procedures for quantification can be proposed. Nevertheless, the basic procedures of stack processing can be described as follows:

- 3D-reconstruction by stack alignment and image registration;
- Correction of image defects (e.g., noise reduction with gauss-filters);
- Segmentation and recognition of individual objects for subsequent statistical analysis;
- Visualization of voxel-based grey-scale data or surface visualization of segmented features by triangulation;
- Quantitative analysis and statistical measurement of features: e.g., particle/pore size distributions, surface area, triple phase boundary lengths, feature counting, etc.

A more detailed description of image analysis principles is given elsewhere (Russ, 1999). Selected examples for quantitative analysis of granular textures and porous networks based on 3D data from FIB are presented in section 3.3.

3 CHRONOLOGICAL EVOLUTION OF FIB TOMOGRAPHY

Today, commercial FIB/SEM machines provide automated and user-friendly solutions for 3D characterization by serial sectioning. This advanced stage of FIB-tomography has evolved over a period of nearly three decades. Initially the FIB machines were not optimized for automated serial sectioning but rather for user-operated cross-sectioning and TEM-lamella preparation. The development of advanced serial sectioning capabilities is based on numerous methodological innovation steps, which typically were initiated in pilot projects of pioneering research labs. In a second step, usually the methodological improvements were automated by means of scripting routines. Finally, when these routines proved to operate successfully for important scientific and technological applications, then they were incorporated into the standard equipment of commercially available FIB machines. The improvements of the FIB serial sectioning also opened new possibilities for 3D microstructure characterization and hence new fields of applications emerged. This evolution, which is illustrated in Figure 11.5, is discussed in the following sections.

3.1 PIONEERING STAGES OF FIB TOMOGRAPHY: SUBSURFACE ANALYSIS WITH 3D-SIMS

Initially, 3D-analysis using a focused ion beam was performed mainly in combination with secondary ion mass spectrometry (SIMS) for subsurface chemical analysis and for diagnostics of artifacts in IC-production. In these fields, 3D analysis with in-depth profiling by SIMS was introduced already in the early 1980s (Patkin and Morrison, 1982) and improved later (e.g., by Hutter and Grasserbauer, 1992). At that time the lateral resolution was typically above the μm-scale. The corresponding depth-resolution was ill defined because of differential sputtering,

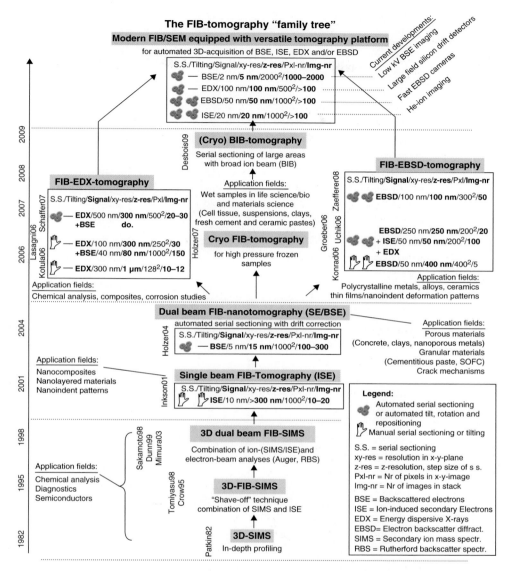

FIGURE 11.5: Chronological evolution of FIB-tomography and the corresponding application fields.

which occurs during in-depth profiling with the ion beam orientation being perpendicular to the surface (see Figure 11.6, left).

During the 1990s, FIB columns with improved spatial resolution and with higher current densities became available. These improvements enabled the acquisition of sequential SIMS-mappings at sub-μm spatial resolution. Such FIB-SIMS systems were used, e.g., for the localization of contaminating nanoparticles within multilayer IC structures (Sotah et al., 1991) and for analyses of semiconductor materials (Crow et al., 1995). For the correction of the artifacts from depth profiling the so-called "shave-off method" was then introduced (Tomiyasu et al., 1998). In the shave-off technique the sample surface is tilted parallel to the ion beam and the uneven surface layer can be polished by "in-plane" erosion (see Figure 11.6).

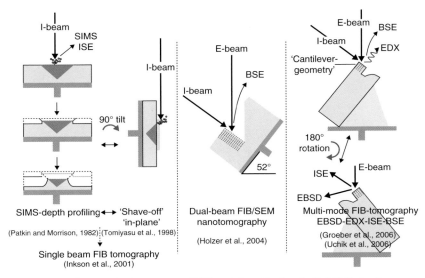

FIGURE 11.6: Geometrical concepts of FIB serial sectioning. Left: SIMS-depth profiling and shave-off for single beam FIB. Middle) FIB-nanotomography with dual beam FIB (no tilting and repositioning necessary). Right: Multimode FIB tomography for 3D-EBSD and optional combination with ISE, EDX and BSE using cantilever geometry and sample rotation.

The combined procedure of sequential depth profiling and shave-off erosion with single beam FIB-SIMS systems included tilting and repositioning between the alternating steps. Because this procedure is very time consuming, usually small stacks with only 5 to 20 maps were acquired. In addition, mechanical instabilities related to the stage tilting and sample repositioning still imposed limitations to the z-depth resolution, which was typically several hundreds of nm.

The first prototypes of dual beam FIB-SEM systems were then introduced, which enabled efficient 3D acquisition of combined chemical and structural information by using quadrupol SIMS together with secondary electron imaging (Sakamoto et al., 1998) and/or with auger spectroscopy (Dunn and Hull, 1999). Nevertheless, since these systems were mainly dedicated to the detection of secondary ions (SIMS) or ion-induced secondary electrons (ISE), the serial sectioning procedure still required time-consuming tilting and sample repositioning between the alternating "shave-off" and imaging steps, which were both performed with the ion beam at different stage positions.

In the retrospective, 3D-SIMS can be considered as a pioneering sub-discipline of FIB tomography. Numerous technological innovations have been introduced in the field of FIB-SIMS five to ten years earlier than in the more "conventional" field of FIB/SEM-tomography. For example, high resolution SIMS tomography with 20 nm isometric voxels was reported already before the year 2000 (Dunn and Hull, 1999). Alternatively, serial sectioning with a large mapping size in the order of $\times 100\ \mu m^2$ was achieved by combining FIB-SIMS with chemically assisted ion beam etching (Tanaka et al., 2003). Last but not least, 3D-SIMS was often also combined with various other beam techniques such as Auger spectroscopy (Sakamoto et al., 1998) and Rutherford backscattering (Mimura et al., 2003). These combinations provided interesting complementary 3D information for combined microstructure and chemical 3D-characterization. Nevertheless, the early papers about 3D-FIB-SIMS were usually focusing on specific methodological and analytical details, whereas in-depth investigations of the microstructure-property relationships were missing. For a broader application of the FIB tomography technique in materials science, further improvements of the imaging capabilities and automation of the serial sectioning were required.

3.2 SERIAL SECTIONING WITH SINGLE BEAM FIB: INITIATION OF FIB TOMOGRAPHY *SENSU STRICTO*

The term *FIB tomography* was first used in conjunction with serial sectioning that was performed manually with a single beam FIB (Inkson et al., 2001b). By taking a series of 15 sequential images that were separated by irregular step sizes between 260 and 660 nm, the subsurface nanoindentation deformation of Cu-Al multilayers could be visualized. Thereby the plastic thickening of the Al-layer and its delamination from the Al_2O_3-substrate could be analyzed by so-called displacement maps that were extracted from the corresponding 3D-reconstructions. This single beam FIB-tomography technique was subsequently also applied for 3D grain shape analysis in Fe-Al nanocomposites (Inkson et al., 2001a), for 3D reconstruction of MOSFET gate morphology (Inkson et al., 2003) and for subsurface damage analysis in Al–Al/SiC nanocomposites (Wu et al., 2003).

A major drawback of the single beam FIB tomography was the limitation in z-resolution, related to the manual stage tilting. In fact, special 3D reconstruction techniques were required for the z-correlation of specific objects that could be identified in the single 2D images (x-y-plane). A new technique for shape based interpolation (SBI) was then introduced, which enabled accurate grain shape reconstruction of γ'-phase precipitates in Ni-based superalloys (Kubis et al., 2004). However, 3D analysis of small objects in the sub-μm range was still limited by the relatively large z-spacing in the 100 nm range. Thus, for 3D analysis of nanostructured materials, it was necessary to develop a serial sectioning procedure with repetitive z-step size in the lower nm-range.

3.3 AUTOMATED DUAL BEAM FIB NANOTOMOGRAPHY

With the introduction of dual beam FIB/SEM systems new possibilities for automated serial sectioning arose. In these FIB/SEM systems no tilting is necessary for serial sectioning because the dual beam geometry is optimized for synchronous operation of ion and electron beams under an angle of 52° (see Figure 11.3 and Figure 11.6, middle). Automated serial sectioning without tilting and repositioning enabled the acquisition of larger stacks with hundreds of images at higher precision and in shorter time. By using the machine-related scripting language, a fully automated "FIB nanotomography" procedure was developed for stack acquisition by overnight processing with a repeatable z-step size in the 10 nm range (Holzer et al., 2004). The automated FIB nanotomography procedure included drift correction based on pattern recognition. This drift correction was important because the early dual beam FIB/SEM systems were not yet optimized for long time stability and therefore various sources of electromagnetic and mechanical drift became apparent when sectioning at z-step sizes below 50 nm.

An alternative version of drift-corrected FIB nanotomography was based on shape analysis of reference marks (so-called z-spacer trenches), which were applied before serial sectioning by using the precise FIB milling capabilities (Bansal et al., 2006). Based on the profile shape of the spacer trenches in the serial images, uneven step sizes could be identified, and this information was then used for drift-corrected 3D reconstruction. However, since this procedure requires time-consuming FIB milling of reference marks, only a limited number of very thin samples ($\times 100$ nm total thickness) were analyzed in this way, e.g., for the study of θ'-precipitates in Al-alloy.

In contrast, the "normal" FIB nanotomography technique was continuously improved over the last few years and is now commercially available, so that it is used in a growing number of investigations (see e.g. Jeanvoine et al., 2006 or McGrouther and Munroe, 2007). The next few subsections of this chapter contain a short summary of FIB tomography applications for the 3D-investigation of cementitious building materials, ceramics, solid oxide fuel cells, and biological tissue.

3.3.1 3D-analysis of porous materials

FIB nanotomography is often used for the quantitative characterization of porous and granular microstructures. In this context it is important to note, that the geometrical concepts for the quantification of porous and granular microstructures are fundamentally different. Granular microstructures are composed of discrete objects, whereas pore structures usually form a continuous network. These contradicting geometrical concepts were mathematically introduced together with a description of algorithms for the quantification of continuous and discrete size distributions (Münch and Holzer, 2008), which were then used in combination with FIB nanotomography for the study of granular and porous microstructures.

In porous materials, permeability and corresponding transport properties are controlled by topological features such as connectivity, (de-)percolation, and dimensions of the pore necks. Quantification of these features can only be performed on the basis of suitable 3D-images. Thereby, FIB nanotomography opens new possibilities for analysis of pore structures with dimensions on the sub-μm scale, which was demonstrated for ceramic $BaTiO_3$ (Holzer et al., 2004), for hydrated cement (Holzer et al., 2006a; Holzer et al., 2006b; Münch et al., 2006b; Münch and Holzer, 2008; Trtik et al., 2008), for clay-rich materials and porous rocks (Desbois et al., 2008; Desbois et al., 2009; Holzer et al., 2010) and for crack growth in metals (Holzapfel et al., 2007).

An example of a quantitative pore structure analysis based on FIB tomography is illustrated in Figure 11.7 for hydrated cement paste (Holzer et al., 2006a). Thereby, the results from 3D analysis and from mercury intrusion porosimetry give significantly different pore size distributions (PSD). The PSD curve of mercury porosimetry has a discontinuous shape, whereas the PSD curves based on 3D analysis are nearly linear (Figure 11.7, bottom left). The same pattern with systematically different results from 3D-FIB and from mercury porosimetry was observed in other porosity investigations (compare Münch and Holzer, 2008). This different behavior is explained by the ink-bottle effect in mercury porosimetry, which leads to an overestimation of the smaller pores with sizes close to the so-called breakthrough diameter. Hence, more reliable size distributions are obtained from 3D microscopy, which is not affected by the ink-bottle effect. In addition, not only the bulk porosity can be characterized with tomography. By careful analysis of the 3D-images, the total porosity in cement paste (Figure 11.7, top left) can be subdivided into two different types of pores. The so-called "internal" porosity is hosted inside the hydrated cement grains (Figure 11.7, middle left, dark gray regions), whereas the "matrix" porosity is hosted in the interstitial region between the hydrated cement grains. The connectivity between the two types of pores can be characterized based on the analysis of the skeletonised 3D-pore network (top right). Thereby, the connections

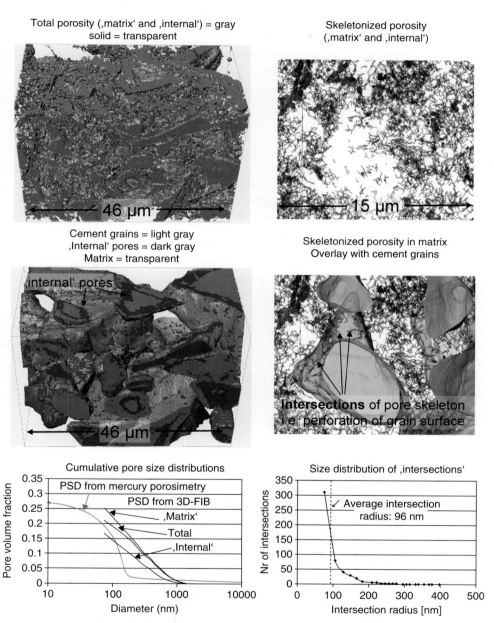

FIGURE 11.7: 3D-analysis of porosity in hydrated cement paste (modified after Holzer et al., 2006a). 3D images allow the distinction of "matrix" and "internal" pore types. The connectivity between the two pore types is quantified by statistical analysis of the intersections where the pore network is perforating the particle surfaces.

between the "internal" and the "matrix" porosity are represented by the intersections of the pore skeleton through the particle surfaces (middle right). A statistical description of the intersections (i.e., size distributions and number of intersections per surface area) can be extracted from the skeletonized pore structure (Figure 11.7, bottom right). Since this information is crucial for the understanding of transport and associated degradation processes, this kind of information should be incorporated in future computational models on concrete and cement permeability and degradation.

3.3.2 3D-analysis of granular materials

In granular materials, not only the particle size distributions but also the particle shape and parameters related to the particle packing (i.e., coordination numbers, particle-particle interfaces) are relevant for the understanding of mechanical properties (strength, elasticity) and flow behavior (rheology). These parameters are specific aspects of the 3D microstructure. Before such complex topological parameters can be extracted from tomography data, the individual particles have to be identified reliably. Unfortunately, recognition of discrete objects (i.e., individual particles) in the images from FIB tomography represents a major challenge, because BSE contrast usually does not reveal the grain boundaries in polycrystalline materials. Therefore, specific computational methods have been introduced for object recognition in densely packed granular materials (based on "splitting"), together with stereological procedures for the correction of boundary truncation effects (Münch et al., 2006a). An example of the object recognition by particle splitting is illustrated in Figure 11.8 (see also Figure 11.2).

FIB nanotomography in combination with accurate object recognition was then used for quantitative studies of particle size distributions and surface area fractions in densely packed cement powder (Figure 11.9). Also more complex parameters such as particle shape and particle-particle contacts were elaborated with FIB nanotomography (Holzer et al., 2006c; Holzer et al., 2007; Zingg et al., 2008; Holzer and Münch, 2009). From the particle shape

FIGURE 11.8: Granular texture of densely packed cement powder (modified after Holzer and Münch, 2009). Left: 3D-reconstruction from original FIB/SEM gray scale images. Note the poor contrast between the cement particles. Right: Object recognition by splitting. Individual particles are labelled with different colours (Münch et al., 2006a).

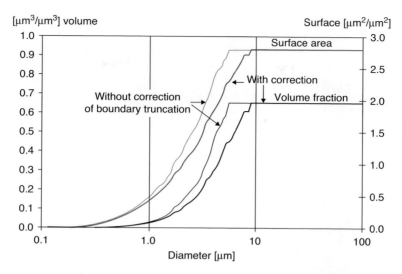

FIGURE 11.9: Quantification of particle size distributions related to cumulative volume fraction (black) and cumulative surface area (gray) of the cement sample shown in Figure 11.8 (modified after Holzer and Münch, 2009).

analysis, unique description of inertia axes for different particle size classes could be obtained. This information was then used as input for simulations of granular flow in cementitious suspensions by means of dissipative particle dynamics (Martys, 2005). Furthermore, in metallurgical applications FIB nanotomography was also used to classify complex graphite morphologies in steel (Velichko et al., 2008).

3.3.3 3D analysis of electrodes in solid oxide fuel cells (SOFC)

Solid oxide fuel cells (SOFC) efficiently convert chemical energy into electrical power. The performance of the fuel cells is strongly related to the microstructure of the porous electrodes. Thereby, 3D connectivity of the pore network is controlling transport of fuel and air, whereas ionic and electronic currents are conducted by the percolating metal- and ceramic-phases (McLachlan et al., 1990). The electrochemical charge transfer reactions take place at the triple phase boundaries. Also, catalytic activity of pore-solid interfaces and associated surface area are important. Thus, the electrode performances are strongly dependent on 3D microstructural parameters, which can only be determined based on high-resolution tomography. In this context, the recent improvements of FIB tomography enable the collection of unique topological information that gives new insight into the complex microstructure-property relationships of SOFC electrodes. Consequently, numerous research institutes have started to use FIB nanotomography for fuel cell investigations (Boukamp, 2006; Wilson et al., 2006; Gostovic et al., 2007; Izzo et al., 2008; Wilson et al., 2009a; Wilson et al., 2009b; Smith et al., 2009; Gostovic et al., 2011; Holzer et al., 2011).

An example of a 3D-microstructure analysis from a composite porous SOFC anode is illustrated in Figure 11.10. After segmentation, the three principal components (Ni, CGO, porosity) can be visualized and quantified separately. The skeletonization shows the potential transport paths whereby Ni (for electron conduction) is much coarser than CGO (Ce-Gd-Oxide, ionic conduction) and the pores (transport of fuel). The challenge for future 3D

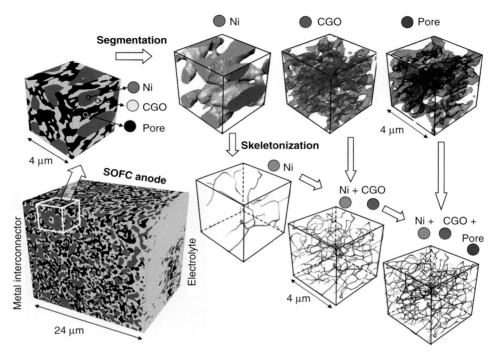

FIGURE 11.10: 3D-microstructure of a composite anode in a solid oxide fuel cell (SOFC) (adapted from Holzer et al., 2011). The segmented Nickel, Ce-Gd-Oxide (CGO), and pore phases are shown for a small subvolume (4 μm edge length). The skeletonised structures illustrate the potential transport pathways for electrons (Ni), oxygen-ions (CGO), and fuel (pore). The skeleton overlay (bottom right) shows the dense intercalation of the three phases and transport pathways.

investigations of fuel cells is to elaborate the way in which the different transport and electrochemical processes are related to the various microstructural parameters and how this knowledge can be used for the improvement of the reaction kinetics and the overall cell performance.

3.3.4 FIB tomography in life science applications

For many years, the exploration of cells and tissues with electron microscopy has been limited to imaging a sample's topography using SEM, and thin section imaging with serial sectioning TEM. Tilt series electron tomography provides high-resolution 3D imaging over a limited volume, but now FIB-SEM tomography has the potential to explore cells and tissue on a much larger scale. The studies so far have mainly been limited to the use of traditional resin embedding approaches in which material has been chemically fixed with aldehydes, heavy metal stained, dehydrated, and resin embedded. These sample blocks are therefore robust and easily manipulated in the FIB/SEM. Imaging the stained cellular material relies therefore on differences in the densities of stained and unstained structures. Backscattered images have so far shown contrast that is comparable to the one from thin sections in the TEM, though at a lesser resolution. However, these are still sufficient for observing all membrane bound structures within different types of cells.

Heymann et al. (2006) used ion beam milling to explore both animal and plant cells. More recently, the same group has shown how this technique is improving rapidly with the ability of seeing the 3D arrangement of cell organelles (Heymann et al., 2009). This higher resolution capability has been used to analyze the inner architecture of chromosomes (Schroeder-Reiter et al., 2009) as well as the connectivity between neurons in the brain (Knott et al., 2008). This latter study has brought the FIB/SEM microscopes to the attention of neuroscientists as synaptic connections are too small to be seen with optical approaches. Until recently the only way of trying to make sense of the highly complex interconnectivity between neurons in the brain was with TEM. Denk and Horstmann (2004) demonstrated how backscattered imaging of a block face can be used to image neural tissue, and today FIB/SEM is drawing increasing interest as a tool for exploring the brain as well as other tissues.

Most of these studies were carried out on conventionally prepared material. Only Heymann et al. (2006) has so far used frozen hydrated samples. In this study, cryo-fixed yeast cells were analyzed by generating images of the membranous structures. This was achieved by carefully allowing the surface of the milled block face to sublimate slightly, and the relief of this face being seen with the imaging beam. Again, this method provided clear images of internal membrane structures. However, achieving tomographic image series will require some further developments.

As described earlier in this chapter, "close to surface" information is required for high-resolution serial sectioning. Thereby the interaction volume of the electron beam in the sample is defined by the beam energy and the sample density. An energy-selective backscattered detector (EsB) on the ZEISS NVision40 FIB can be used at low accelerating voltages (below 2kV) and tuned so that backscattered electrons that have lost only a few hundred eVs are detected. In this way the escape depth of the backscattered electrons is below 5nm. This allows for details such as cell membranes to be seen, with a contrast and resolution that are normally only seen with serial sectioning TEM (ssTEM). We have used these capabilities for imaging of mammalian tissue samples as well as cultured cells grown on various artificial substrates. These were prepared using aldehyde fixation and then post fixed and stained with osmium tetroxide and uranyl acetate. The samples were then embedded in Durcupan resin, and once cured, blocks were trimmed with glass knives using an ultramicrotome. The following examples show the clear potential of FIB nanotomography in the "bio-nano" field:

Example A/Bio-compatibility (Figure 11.11) shows a cell grown on ceramic coated steel. 600 slices were milled through resin embedded cell/ceramic layer/steel with 20 nm slice thickness at 10 nm image pixel size. The volume covers $20 \times 15 \times 12$ µm with a voxel size of $10 \times 10 \times 20$ nm. The reconstruction permits the 3D morphological analysis of the cell and its organelles as well as the visualization of the ceramic surface and grain boundaries in the steel substrate below. The inspection of the interface between the cell and the ceramics reveals that in this particular case the cell is only attached at a few spots.

For Example B/Brain tissue of rat hippocampus (Figure 11.12), 1,600 slices of 5nm thickness were cut and imaged in 40 hours by automated acquisition. The image pixel size is 5 nm ($2,048 \times 1,536$ pixels per image). The volume of this stack covers $10 \times 8 \times 8$ µm with a voxel size of $5 \times 5 \times 5$ nm. Cell membranes, vesicles and mitochondria with their internal membranes are perfectly resolved. This allows for the identification of different types of neurons and the synaptic contacts that exist between them. Within this volume there are more than 800 synapses present. Future research activities will focus on the integration of the FIB-3D-information from brain tissue into computational models from neuroinformatics.

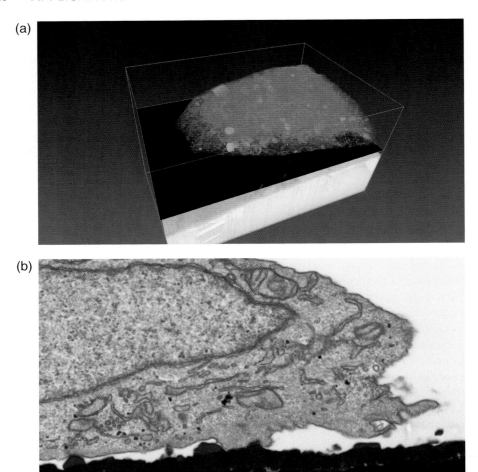

(a)

(b)

EHT = 2.00 kV	Aperture size = 60.00 μm	Signal A = ESB	Noise reduction = Line avg	Date: 23 Jun 2008 Time: 13:48:28
Mag = 50.00 K X	Image pixel size = 2.7 nm	FIB imaging = SEM	Scan speed = 7 N = 6	FIB fil current = 14.00 mA
Width = 5.600 μm ⊢——⊣ 300 nm	WD = 5.1 mm	FIB lock mags = No	Cycle time = 2.0 mins	System vacuum = 1.80e-006 mbar

FIGURE 11.11: Bio-compatibility study of cell-substrate interface (a): 3D-reconstructed cell on ceramic coated steel. (b): detailed view of a single SEM image of the FIB-stack. Contrast inverted for better visualization of membrane structures.

3.4 CRYO FIB TOMOGRAPHY AND COMBINATIONS WITH BROAD ION BEAM (BIB)

For the investigation of wet samples, cryo FIB tomography was introduced (Holzer et al., 2007). Thereby, cryo-techniques from life-sciences such as high-pressure freezing, cryo-transfer, and cryo-SEM were combined with FIB serial sectioning. In the first cryo FIB tomography studies, this methodology was used for the quantification of microstructural changes in complex cementitious suspensions. The microstructural changes, including particle-number-densities, surface area, and size distributions, could be correlated with reactions of the early cement hydration (precipitation, agglomeration, dispersion by surfactants) and with the corresponding

(a)

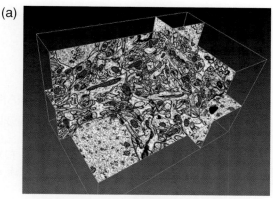

(b)

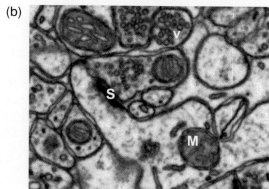

FIGURE 11.12: 3D-investigation of brain tissue (a): orthogonal slices representing a volume of $10 \times 10 \times 8$ μm. (b): detailed view of a 3×2.5 μm area inside the FIB-stack which illustrates a synapse (S), vesicles (v) and mitochondria (M).

rheological properties of fresh cement pastes (Holzer et al., 2007; Zingg et al., 2008; Holzer and Münch, 2009).

As mentioned in the introduction, a larger field of view is necessary for representative analysis of heterogeneous samples. This is of particular interest in geological applications of natural rocks (e.g., oil/gas reservoir rocks), which typically have wide pore size distributions that are spanning from nm- to mm-scales. For such investigations the capabilities of cryo-serial sectioning was extended by a combination of SEM with broad ion beam (BIB), which enables 2D-imaging and serial sectioning of larger sample areas (100 nm^2 to mm^2) due to much higher sputter rates (Desbois et al., 2009). Thin layers of $\times 100$ nm thickness can be removed by BIB erosion under a shallow angle. Thereby, deeper levels of the sample are protected with a massive metal blend. The geological applications of cryo-FIB and cryo-BIB include pore studies in gas/oil reservoirs, brine-filled grain boundaries in rock salt (Desbois et al., 2008), engineered bentonite barriers, and porous clay host rocks of radioactive waste repositories (Desbois et al., 2009; Holzer et al., 2010).

The above review of FIB tomography and the corresponding applications in materials, life, geo- and environmental sciences is by far not complete. Further FIB-3D studies were performed for example on the topic of crack formation (Elfallagh and Inkson, 2008; Elfallagh and Inkson, 2009). The large number of applications documents the unique imaging capabilities of the FIB tomography method. However, one of the major restrictions of FIB tomography is given by the relatively low BSE-contrast in polycrystalline ceramics and alloys (see discussion in Cao et al., 2009). Consequently there is a need to combine FIB serial sectioning with alternative detection modes that deliver additional information on the internal structure such as spatial distribution of chemical elements (EDX) and crystal orientations (EBSD). The combinations of FIB tomography with these detection modes are discussed in the subsequent sections.

3.5 FIB-EDX-TOMOGRAPHY: TOWARD 3D SPECTRAL IMAGING (TSI)

By combining FIB-serial sectioning with EDX-detection, information about the 3D elemental distribution at μm- and sub-μm scales can be acquired. However, for the FIB-EDX tomography some peculiarities that are inherent to the EDX-method have to be taken into account: (a) EDX cannot be performed at low kV, which results in a relatively large excitation volume (see Figure 11.4). Consequently the corresponding spatial resolution is in the range of $\times 100$ nm up to >1 μm. (b) Acquisition time with SiLi-detectors is much longer (typically 0.5 up to >1 hour per 2D-mapping). The recent introduction of silicon drift detectors with wide collecting angle enables higher count rates and faster acquisition. (c) Significant charging generally occurs in non-conductive materials due to high accelerating voltage and slow scanning rates. (d) Significant drift components are observed due to long milling time (large z-step-size), the slow EDX-acquisition and the mentioned charging effects. (e) Artificial x-ray counts are produced from scattered electrons which interact with the surrounding walls of the sample.

Due to these problems, automation of FIB-EDX-tomography is more challenging than "normal" FIB-SEM-tomography. Therefore, serial sectioning with EDX was initially performed only manually. In this way the 3D structure of CuAl-alloys and corrosion products in layered CuNiAu samples were investigated (Kotula et al., 2006) by acquisition of 10 to 12 sequential EDX maps. The individual imaging planes were separated by 1 μm in z-direction, whereas the pixel resolutions in x-z-directions were 300 to 400nm. Already with the limited manual acquisition, a relatively large amount of chemical data (full spectral information for each voxel in a matrix of $128 \times 128 \times 10$) was generated. For efficient processing and accurate identification of the principal components, computational methods based on multivariate statistical analysis (MSA) were introduced (Kotula et al., 2003; Kotula et al., 2006).

Subsequently an automated serial sectioning procedure was presented which enabled synchronized acquisition of sequential EDX-mappings in combination with SE/BSE-tomography (Schaffer et al., 2007). It turned out that drift correction represents the most challenging problem for 3D-EDX-automation. Due to high ion beam currents which are used for efficient erosion of relatively thick material layers (z-step size: $\times 100$nm to 1 μm) drift correction based on reference pattern recognition with ISE imaging was not possible because the reference pattern would be eroded during the analysis. Therefore a post-acquisition drift correction method, which is based on image registration of BSE "overview" and "detail" images, was developed (Schaffer et al., 2007). The automated FIB-EDX-tomography was then successfully applied for the analysis of the 3D microstructures in CaMgTiO$_x$ and in FeAl-alloy (see Figure 11.13). For further improvement of the EDX data quality, the small cube of analysis was transferred to a TEM-holder by using the block lift-out technique (Schaffer and Wagner, 2008). In this way artificial X-ray signal from the surrounding sample walls and shadowing effects can be suppressed.

Combined FIB-SEM-EDX-tomography was then also used for the characterization of eutectic microstructures in AlSi-alloys (Lasagni et al., 2006; Lasagni et al., 2007; Lasagni et al., 2008a). Due to the combination of BSE images (z-resolution 50 nm) with EDX mappings (z-resolution 500 nm, but with better contrast), complementary information could be acquired, which improved segmentation and recognition of minor components that were otherwise not

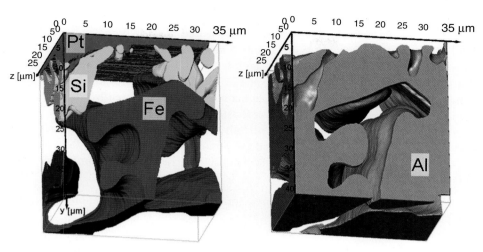

FIGURE 11.13: 3D-resonstruction of FeAl-alloy based on FIB-EDX tomography (adapted from Schaffer et al., 2009).

traceable (Lasagni et al., 2008a). The combination of BSE- and EDX-tomography thus provides important microstructural information at the sub-μm level which cannot otherwise be achieved easily (Engstler et al., 2008). At this stage, the missing piece for complete 3D microstructural characterization of polycrystalline materials was grain orientation contrast.

3.6 3D ELECTRON BACKSCATTER DIFFRACTION (EBSD) BASED ON FIB SERIAL SECTIONING

Electron backscatter diffraction (EBSD) enables the acquisition of crystallographic orientation maps at a maximum resolution of 50 nm. In polycrystalline (but monomineralic) materials, neighboring grains with the same composition cannot be distinguished with BSE- or with EDX-analysis. For such materials the orientation contrast of EBSD opens unique possibilities for the characterization of granular textures. EBSD is particularly well suited for the study of deformation mechanisms such as recrystallization in thermo-mechanical experiments. Since these processes usually result in anisotropic microstructures, 3D microscopy is necessary for a thorough texture analysis. EBSD was therefore also combined with serial sectioning techniques, first with mechanical polishing (Rollett et al., 2007) and later also with FIB-serial sectioning (Konrad et al., 2006; Rowenhorst et al., 2006).

In contrast to the other FIB tomography procedures, EBSD is performed under forward scattering conditions, whereby the angle between surface normal and electron beam directions (incident and scattered beams) is 70°. In order to ensure accessibility of the FIB prepared surface under 70°, a so-called "cantilever beam geometry" was proposed (Groeber et al., 2006; Uchic et al., 2006). As an alternative approach, the block lift-out technique (Schaffer and Wagner, 2008) and other preparation methods (Lasagni et al., 2008b) were developed. During the serial sectioning procedure with the cantilever geometry, the sample is rotated 180° (see Figure 11.6, right) in order to change between the "sectioning position" (in plane with ion beam) and the "EBSD position" (70° between electron beam and normal to FIB prepared surface).

Initially, FIB-EBSD serial sectioning was performed manually for the study of the microstructural evolution in warm-rolled Fe_3Al-based alloy (Konrad et al., 2006). Small stacks of about 5 sections were acquired with a z-spacing of 400nm and the acquisition time was comparatively slow (3 hours per image). It was shown that secondary particles, which are added for mechanical strengthening at elevated temperatures, lead to the formation of strong orientation gradients at the particle-matrix interface. Apparently, the areas with high orientation gradients act as nucleation sites for strain induced recrystallization.

Scripting for serial EBSD acquisition with fully automated sample rotation and repositioning were then introduced, which also allowed the optional combination with ISE-, BSE- and/or EDX-imaging (Groeber et al., 2006; Uchic et al., 2006). The FIB-EBSD tomography was then used in numerous studies such as:

- Quantification of granular textures (grains size, grain orientation, grain surface/perimeter) in Ni-base alloys with γ'-precipitates (Uchic et al., 2006).
- Grain rotation below a conical nanoindentation was described for a (111)Cu single crystal (Zaafarani et al., 2006). Thereby, FIB-EBSD tomography enables insight into the plastic/elastoplastic deformation mechanisms at the crystal lattice level. (Note: in previous investigations with FIB-SEM tomography the deformation patterns could only be investigated at larger scales, e.g., change of layer thickness under nanoindents (Inkson et al., 2001b; Soldera et al., 2007).
- Further FIB-EBSD investigations include microstructural analyses of perlites in carbon steel and fatigue cracks in steel and alloy (Zaefferer et al., 2008; Xu et al., 2007; Motoyashiki et al., 2007).
- 3D characterization of granular textures in thin films with various compositions such as diamond, Ni-Co and Ni-Pt (Liu et al., 2008; Bastos et al., 2008; Jeanvoine et al., 2008; West and Thomson, 2009).
- Last but not least, also polycrystalline ceramic materials were analyzed with 3D-FIB-EBSD (Dillon and Rohrer, 2009).

More detailed descriptions of this specific methodology are presented in recent reviews about FIB-EBSD-tomography (Zaefferer and Wright, 2009) and the corresponding computational processing of 3D-EBSD data (Groeber et al., 2009).

4 CONCLUSIONS: CURRENT STATUS AND NEEDS FOR FURTHER DEVELOPMENTS

FIB tomography is nowadays a versatile method that enables automated acquisition of large image stacks at resolutions below 10 nm. In addition, the ability to perform FIB serial sectioning with different detection modes enables the synchronous acquisition of microstructural (BSE, ISE), chemical (EDX) and crystallographic (EBSD) information from the same sample. These unique 3D-imaging capabilities open new possibilities for 3D microstructural investigations in materials, geo-, environmental, and life sciences. Nevertheless, the current state of FIB serial sectioning suffers from a few, but distinct limitations, which impose the necessity for further improvements:

- Resolution vs. volume: As discussed in the introduction (see also Figure 11.1), the representative analysis of heterogeneous microstructures infers contradicting requirements of a large volume and high magnifications at the same time. FIB tomography is mainly restricted in the size of the volume due to limitations of ion-milling capabilities. Improvements of the focused ion beam columns should therefore allow faster sputter rates by maintaining the high precision. In this context it is also important to note that alternative techniques for sequential erosion of large areas have been introduced recently, which include broad ion beam (BIB) (Desbois et al., 2009) and mechanical systems (e.g., Micromiller- and RoboMet.3D-systems described by Alkemper and Vorhees, 2001; Spowart et al., 2003). In the near future, intelligent combinations of FIB tomography with BIB or with mechanical polishing techniques may open new possibilities for the analysis of larger volumes at high resolutions.

- Acquisition time and detector sensitivity: FIB serial sectioning is currently strongly limited by relatively long acquisition times. Improved detector sensitivities are therefore required, which enable faster stack acquisition. Recent developments in this field include energy sensitive BSE-detectors for low kV imaging and Silicon-Drift Detectors with wide solid angle for EDX-acquisition with higher acquisition rates. Further improvements of the acquisition rate could enable new "high throughput" applications, such as routine testing of nano-materials and failure analysis.

- Charging and drift: For non-conductive samples, charging of the FIB-milled surface is still a serious problem. Especially for FIB-EDX tomography (which requires high kV and high beam currents) charging of insulating materials induces image distortions and drift. Therefore, in-situ charge neutralization and improved drift-compensation procedures are required.

- 3D-data processing and intelligent automation: Image acquisition and visualization (i.e., "slice and view") represent only half of the task that has to be solved in order to establish meaningful microstructure-property relationships. Numerous standard software packages for the "off-line" data processing (3D-reconstruction and–visualization) are available. However, suitable software for reliable quantification of complex topological parameters is largely missing (e.g. characterization of skeletonized pore networks or statistical analysis of particle shape and counting of particle-particle contacts). In addition, there are a number of data processing problems that are specific for FIB tomography, especially with regard to analytical FIB-EDX-EBSD tomography. For example, in EDX stacks which were collected at z-step sizes that are smaller than the excitation volume, depth information can be used for spectral deconvolution and improvement of the depth resolution. These new data processing procedures are not yet established.

Furthermore, new possibilities for intelligent automation of multimode FIB serial sectioning can be developed. For example, online "feature recognition" can be used during data acquisition at a coarse and fast serial sectioning. Upon recognition of a specific feature (e.g., failure), the online information can be used to switch to a different serial sectioning mode, e.g., higher magnification or different analytical information (EDX, SIMS, EBSD). In summary, there is much space for further improvements in the fields of intelligent automation of the serial sectioning procedure and for the subsequent quantitative analysis of 3D-microstructures.

REFERENCES

Alkemper, J., and Vorhees, P. W., 2001. Quantitative serial sectioning analysis. *Journal of Microscopy* 201: 388–394.

Bansal, R. K., Kubis, A., Hull, R., and Fitz-Gerald, J. M., 2006. High-resolution three-dimensional reconstruction: A combined scanning electron microscope and focused ion-beam approach. *Journal of Vacuum Science and Technology* B 24: 554–561.

Bastos, A., Zaefferer, S., and Raabe, D., 2008. Three dimensional EBSD study on the relationship between triple junctions and columnar grains in electrodeposited Co-Ni films. *Journal of Microscopy* 230: 487–498.

Boukamp, B., 2006. Anodes sliced with ions. *Nature Materials* 5: 517–518.

Cantoni, M., 2009. FIB nanotomography: 3D Superstacks, isometric voxels. In S. Abolhassani et al. (eds.), *Interdisciplinary Symposium on 3D Microscopy*. SSOM/PSI, Interlaken, Switzerland, pp. 151.

Cao, S., Tirry, W., Van Den Broek, W., and Schryvers, D., 2009. Optimization of a FIB/SEM slice-and-view study of the 3D distribution of Ni4Ti3 precipitates in Ni-Ti. *Journal of Microscopy* 233: 61–68.

Crow, G. A., Christman, L., and Utlaut, L., 1995. A focused ion beam secondary ion mass spectroscopy system. *Journal of Vacuum Science and Technology* B 13(6): 2607–2612.

De Winter, D. A. M., et al., 2009. Tomography of insulating biological and geological materials using focused ion beam (FIB) sectioning and low-kV BSE imaging. *Journal of Microscopy* 233: 372–383.

Denk, W., and Horstmann, H., 2004. Serial block-face scanning electron microscopy to reconstruct three-dimensional tissue nanostructure. *PLoS Biology* 2: 11.

Desbois, G., Urai, J. L., Burkhardt, C., Drury, M. R., Hayles, M., and Humbel, B., 2008. Cryogenic vitrification and 3D serial sectioning using high resolution cryo-FIB SEM technology for brine-filled grain boundaries in halite: first results. *Geofluids* 8: 60–72.

Desbois, G., Urai, J. L. and Kukla, P. A., 2009. Morphology of the pore space in claystones – evidence from BIB/FIB ion beam sectioning and cryo-SEM observations. *eEarth* 4: 15–22.

Dillon, S. J., and Rohrer, G. S., 2009. Characterization of the grain-boundary character and energy distributions of yttria using automated serial sectioning and EBSD in the FIB. *Journal of the American Ceramic Society* 92: 1580–1585.

Dunn, D. N., and Hull, R., 1999. Reconstruction of three-dimensional chemistry and geometry using focused ion beam microscopy. *Applied Physics Letters* 75: 3414–3416.

Elfallagh, F., and Inkson, B. J., 2008. Evolution of residual stress and crack morphologies during 3D FIB tomographic analysis of alumina. *Journal of Microscopy* 230: 240–251.

Elfallagh, F., and Inkson, B. J., 2009. 3D analysis of crack morphologies in silicate glass using FIB tomography. *Journal of the European Ceramic Society* 29: 47–52.

Engstler, M., Velichko, A., Klöpper, F., and Mücklich, F., 2008. Automated FIB-EDS-nanotomography and its application to Sr-modified Al-Si foundry alloys. In J. Hirsch, B. Strotzki, and G. Gottstein (eds.), *Aluminium alloys: Their physical and mechanical properties*. Hoboken, NJ: Wiley VCH, pp. 795–800.

Gostovic, D., Smith, J. R., Kundinger, D. P., Jones, K. S., and Wachsman, E. D., 2007. Three-dimensional reconstruction of porous LSCF cathodes. *Electrochemical and Solid State Letters* 10: B214–B217.

Gostovic, D., Vito, N. J., O'Hara, K. A., Jones, K. S., and Wachsman, E. D., 2011. Microstructure and connectivity quantification of complex composite solid oxide fuel cell electrode three-dimensional networks. *Journal of the American Ceramic Societ*: 94: 620-627.

Groeber, M. A., Haley, B. K., Uchic, M. D., Dimiduk, D. M., and Ghosh, S., 2006. 3D reconstruction and characterization of polycrystalline microstructures using a FIB-SEM system. *Materials Characterization* 57: 259–273.

Groeber, M. A., Rowenhorst, D. J., and Uchic, M. D., 2009. Collection, processing and analysis of three-dimensional EBSD data sets. In A. J. Schwartz, M. Kumar, B. L. Adams, and D. P. Field (eds.), *Electron backscatter diffraction in materials science*. New York: Springer, pp. 123–138.

Heymann, J., et al., 2006. Site-specific 3D imaging of cells and tissues with a dual beam microscope. *Journal of Structural Biology* 155: 63–73.

Heymann, J. A. W., et al., 2009. 3D imaging of mammalian cells with ion-abrasion scanning electron microscopy. *Journal of Structural Biology* 166: 1–7.

Holzapfel, C., Schäf, W., Marx, M., Vehoff, H., and Mücklich, F., 2007. Interaction of cracks with precipitates and grain boundaries: Understanding crack growth mechanisms through focused ion beam tomography. *Scripta Materialia* 56: 697–700.

Holzer, L., et al., 2007. Cryo-FIB-nanotomography for quantitative analysis of particle structures in cement suspensions. *Journal of Microscopy* 227: 216–228.

Holzer, L., Gasser, P., and Münch, B., 2006a. Quantification of capillary pores and hadley grains in cement paste using FIB-nanotomography. In M.S. Konsta-Gdoutos (ed.), *Measuring, monitoring and modeling concrete properties.* Dordrecht: Springer, pp. 509–516.

Holzer, L., Indutnyi, F., Gasser, P., Münch, B., and Wegmann, M., 2004. 3D analysis of porous BaTiO$_3$ ceramics using FIB nanotomography. *Journal of Microscopy* 216(1): 84–95.

Holzer, L., Iwanschitz, B., Hocker, Th., Prestat, M., Wiedenmann, D., Vogt, U., Holtappels, P., Sfeir, J., Mai, A., and Graule, Th., 20011. Microstructure degradation of cermet anodes for solid oxide fuel cells: Quantification of nickel grain growth in dry and in humid atmospheres. Journal of Power Sources 196: 1279–1294.

Holzer, L., Muench, B., Leemann, A., and Gasser, P., 2006b. Quantification of capillary porosity in cement paste using high resolution 3D-microscopy: Potential and limitations of FIB-nanotomography. In J. Marchand, B. Bissonnette, R. Gagne, M. Jolin, and F. Paradis (eds.), *Advances in concrete.* Quebec: RILEM, pp. 247 (electronic media).

Holzer, L., Muench, B., Rizzi, M., Wepf, R., and Marschall, P., 2010. 3D-microstructure analysis of hydrated bentonite with cryo-stabilized pore water. *Applied Clay Science*: 45: 330–342.

Holzer, L., and Münch, B., 2009. Towards reproducible three-dimensional microstructure analysis of granular materials and complex suspensions. *Microscopy and Microanalysis* 15: 130–146.

Holzer, L., Münch, B., Wegmann, M., Flatt, R., and Gasser, P., 2006c. FIB-nanotomography of particulate systems – Part I: particle shape and topology of interfaces. *Journal of the American Ceramic Society* 89: 2577–2585.

Hutter, H., and Grasserbauer, M., 1992. Three dimensional ultra trace analysis of materials. *Mikrochimica Acta* 107: 137–148.

Inkson, B. J., Mulvihill, M., and Möbus, G., 2001a. 3D determination of grain shape in a FeAl-based nanocomposite by 3D FIB tomography. *Scripta Materialia* 45: 753–758.

Inkson, B. J., Olsen, S., Norris, D. J., O'Neill, A. G., and Möbus, G., 2003. 3D determination of a MOSFET gate morphology by FIB tomography, *Microsc. Semicond. Mater. Conf. Inst. Phys. Conf.*, Cambridge, pp. 611–616.

Inkson, B. J., Steer, T., Möbus, G., and Wagner, T., 2001b. Subsurface nanoindentation deformation of Cu-Al multilayers mapped in 3D by focused ion beam microscopy. *Journal of Microscopy* 201: 256–269.

Izzo, J. R., et al., 2008. Nondestructive reconstruction and analysis of SOFC anodes using x-ray computed tomography at sub-50 nm resolution. *Journal of the Electrochemical Society* 155: B504–508.

Jeanvoine, N., Holzapfel, C., Soldera, F. and Mücklich, F., 2006. 3D investigations of plasma erosion craters using FIB/SEM Dual Beam techniques. *Practical Metallography* 43: 470–482.

Jeanvoine, N., Holzapfel, C., Soldera, F. and Mücklich, F., 2008. Microstructure characterization of electrical discharge craters using FIB/SEM dual beam techniques. *Advanced Engineering Materials* 10: 973–977.

Knott, G., Marchman, H., Wall, D., and Lich, B., 2008. Serial section scanning electron microscopy of adult brain tissue using focused ion beam milling. *Journal of Neuroscience* 28: 2959–2964.

Konrad, J., Zaefferer, S., and Raabe, D., 2006. Investigation of orientation gradients around a hard Laves particle in a warm-rolled Fe3Al-based alloy using a 3D EBSD-FIB technique. *Acta Materialia* 54: 1369–1380.

Kotula, P. G., Keenan, M. R., and Michael, J. R., 2003. Automated analysis of SEM X-ray spectral images. *Microscopy and Microanalysis* 9: 1–17.

Kotula, P. G., Keenan, M. R., and Michael, J. R., 2006. Tomographic spectral imaging with multivariate statistical analysis: Comprehensive 3d microanalysis. *Microscopy and Microanalysis* 12: 36–48.

Kubis, A. J., Shiflet, G. J., Dunn, D. N., and Hull, R., 2004. Focused ion-beam tomography. *Metallurgical and Materials Transactions* A, 35A: 1935–1943.

Lasagni, F., Lasagni, A., Engstler, M., Degischer, H. P., and Mücklich, F., 2008a. Nano-characterization of cast structures by FIB-tomography. *Advanced Engineering Materials* 10: 62–66.

Lasagni, F., Lasagni, A., Holzapfel, C., and Mücklich, F., 2008b. Sample preparation of Al-alloys for 3D FIB nano-characterization. *Practical Metallography* 45: 246–250.

Lasagni, F., Lasagni, A., Holzapfel, C., Mücklich, F. and Degischer, H. P., 2006. Three-dimensional characterization of unmodified and Sr-modified Al-Si eutectics by FIB and FIB EDX tomography. *Advanced Engineering Materials* 8: 719–723.

Lasagni, F., et al., 2007. Three dimensional characterization of "as-cast" and solution-treated AlSi12(Sr) alloys by high-resolution FIB tomography. *Acta Materialia* 55: 3875–3882.

Liu, T., Raabe, D., and Zaefferer, S., 2008. A 3D tomographic EBSD analysis of a CVD diamond thin film. *Sci. Technol. Adv. Mater.* 9: 035013.

Martys, N. S., 2005. Study of a dissipative particle dynamics based approach for modelling suspensions. *Journal of Rheology* 49: 401–424.

McGrouther, D., and Munroe, P. R., 2007. Imaging and analysis of 3-D structure using a dual beam FIB. *Microscopy Research and Technique* 70: 186–194.

McLachlan, D. S., Blaszkiewicz, M., and Newnham, R., 1990. Electrical resistivity of composites. *Journal of the American Ceramic Society* 73: 2187–2203.

Mimura, R., Takayama, H., and Takai, M., 2003. Nanometer resolution three-dimentional microprobe RBS analyses by using a 200 kV focused ion beam system. *Nuclear Instruments and Methods in Physics Research*, B210: 104–107.

Mobus, G., and Inkson, B. J., 2007. Nanoscale tomography in materials science. *Materials Today* 10: 18–25.

Motoyashiki, Y., Bruckner-Foit, A., and Sugeta, A., 2007. Investigation of small crack behaviour under cyclic loading in a dual phase steel with an FIB tomography technique. *Fatigue & Fracture of Egineering Materials & Structures* 30: 556–564.

Münch, B., Gasser, P., Flatt, R., and Holzer, L., 2006a. FIB-nanotomography of particulate systems – Part II: Particle recognition and effect of boundary truncation. *Journal of the American Ceramic Society* 89: 2586–2595.

Münch, B., Gasser, P., and Holzer, L., 2006b. Simulation of drying shrinkage based on high-resolution 3D pore structure analysis of cement paste. In J. Marchand (ed.), *International Symposium on Advances in Concrete Through Science and Engineering.* RILEM, Quebec, CRIB.

Münch, B., and Holzer, L., 2008. Contradicting geometrical concepts in pore size analysis attained with electron microscopy and mercury intrusion. *Journal of the American Ceramic Society* 91: 4059–4067.

Patkin, A. J., and Morrison, G. H., 1982. Secondary ion mass spectrometric image depth profiling for three-dimensional elemental analysis. *Analytical Chemistry* 54: 2–5.

Rollett, A. D., Lee, S. B., Campman, R., and Rohrer, G. S., 2007. Three-dimensional characterization of microstructure by electron backscatter diffraction. *Annu. Rev. Mater. Res.* 37: 627–658.

Rowenhorst, D. J., Gupta, A., Feng, C. R. and Spanos, G., 2006. 3D crystallographic and morphological analysis of coarse martensite: Combining EBSD and serial sectioning. *Scripta Materialia* 55: 11–16.

Russ, J. C., 1999. *The image processing handbook.* Boca Raton, FL: CRC Press.

Sakamoto, T., Cheng, Z., Takahashi, M., Owari, M., and Nihei, Y., 1998. Development of an Ion and Electron Dual Focused Beam Apparatus for three Dimensional Microanalysis. *Japanese Journal of Applied Physics* 37: 2051–2056.

Schaffer, M., and Wagner, J., 2008. Block lift-out preparation for 3D experiments in a dual beam focused ion beam microscope. *Microchimica Acta* 161: 421–425.

Schaffer, M., Wagner, J. and Grogger, W., 2009. 3D-energy-dispersive X-ray spectrometry in a dual-beam FIB (3D-FIB EDXS). In S. Abolhassani et al. (eds.), *Interdisciplinary Symposium on 3D Microscopy,* SSOM/PSI, Interlaken, Switzerland, pp. 43–44.

Schaffer, M., Wagner, J., Schaffer, B., Schmied, M., and Mulders, H., 2007. Automated three-dimensional X-ray analysis using a dual-beam FIB. *Ultramicroscopy,* 107: 587–597.

Schroeder-Reiter, E., Pérez-Willard, F., Zeile, U., and Wanner, G., 2009. Focused ion beam (FIB) combined with high resolution scanning electron microscopy: A promising tool for 3D analysis of chromosome architecture. *Journal of Structural Biology* 165: 97–106.

Smith, J. R., et al., 2009. Evaluation of the relationship between cathode microstructure and electro-chemical behavior for SOFCs. *Solid State Ionics* 180: 90–98.

Soldera, F., Gaillard, Y., Anglada Gomila, M., and Mücklich, F., 2007. FIB-tomography of nanoindentation cracks in zirconia polycrystals. *Microscopy and Microanalysis* 13: 1510–1511.

Sotah, H., Owari, M., and Nihei, Y., 1991. Three-dimensional analysis of a microstructure by submicron secondary ion mass spectrometry. *Journal of Vacuum Science and Technology* B9: 2638–2640.

Spowart, J. E., Mullins, H. M., and Puchala, B. T., 2003. Collecting and analyzing microstructures in three dimensions: A fully automated approach. *JOM* 55: 35–37.

Tanaka, Y., et al., 2003. Development of a chemically assisted micro-beam etching system for three-dimensional microanalysis. *Applied Surface Science* 203: 205–208.

Tomiyasu, B., Fukuju, I., Komatsubara, H., Owari, M., and Nihei, Y., 1998. High spatial resolution 3D analysis of materials using gallium focused ion beam secondary ion mass spectrometry (FIB SIMS). *Nuclear Instruments and Methods in Physics Research* B 136–138: 1028–1033.

Trtik, P., Dual, J., Muench, B., and Holzer, L., 2008. Limitation in obtainable surface roughness of hardened cement paste: "virtual" topographic experiment based on focussed ion beam nanotomography datasets. *Journal of Microscopy* 232: 200–206.

Uchic, M. D., Groeber, M. A., Dimiduk, D. M., and Simmons, J. P., 2006. 3D microstructural characterization of nickel superalloys via serial-sectioning using a dual beam FIB-SEM. *Scripta Materialia* 55: 23–28.

Uchic, M. D., Holzer, L., Inkson, B. J., Principe, E. L., and Munroe, P., 2007. Three-dimensional microstructural characterization using focused ion beam tomography. *MRS Bulletin* 32: 408–416.

Velichko, A., Holzapfel, C., Siefers, A., Schladitz, K., and Mücklich, F., 2008. Unambiguous classification of complex microstructures by their three-dimensional parameters applied to graphite in cast iron. *Acta Materialia* 56: 1981–1990.

West, G. D., and Thomson, R. C., 2009. Combined EBSD/EDS tomography in a dual-beam FIB/FEG-SEM. *Journal of Microscopy* 233: 442–450.

Wilson, J. R., et al., 2009a. Quantitative three-dimensional microstructure of a solid oxide fuel cell cathode. *Electrochemistry Communications* 11: 1052–1056.

Wilson, J. R., et al., 2009b. Three-dimensional analysis of solid oxide fuel cell Ni-YSZ anode interconnectivity. *Microscopy and Microanalysis* 15: 71–77.

Wilson, J. R., et al., 2006. Three-dimensional reconstruction of a solid-oxide fuel-cell anode. *Nature Materials* 5: 541–544.

Wu, H. Z., Roberts, S. G., Möbus, G., and Inkson, B. J., 2003. Subsurface damage analysis by TEM and 3D FIB crack mapping in alumina and alumina/5vol%SiC nanocomposites. *Acta Materialia* 51: 149–163.

Xu, W., Ferry, M., Mateescu, N., Cairney, J. M., and Humphreys, F. J., 2007. Techniques for generating 3-D EBSD microstructures by FIB tomography. *Materials Characterization* 58: 961–967.

Zaafarani, N., Raabe, D., Singh, R. N., Roters, F., and Zaefferer, S., 2006. Three-dimensional investigation of the texture and microstructure below a nanoindent in a Cu single crystal using 3D EBSD and crystal plasicity finite element simulations. *Acta Materialia* 54: 1863–1876.

Zaefferer, S., and Wright, S. I., 2009. Three-dimensional orientation microscopy by serial sectioning and EBSD-based orientation mapping in a FIB-SEM. In A. J. Schwartz, M. Kumar, B. L. Adams, and D. P. Filed (eds.), *Electron backscatter diffraction in materials science*. New York: Springer, pp. 109–122.

Zaefferer, S., Wright, S. I., and Raabe, D., 2008. Three-dimensional orientation microscopy in a focused ion beam-scanning electron microscope: A new dimension of microstructure characterization. *Metallurgical and Materials Transactions* A, 39A: 374–389.

Zingg, A. et al., 2008. The microstructure of dispersed and non-dispersed fresh cement pastes – New insight by cryo-microscopy. *Cement and Concrete Research* 38: 522–529.

CHAPTER 12

IN SITU MONITORING OF GAS-ASSISTED FOCUSED ION BEAM AND FOCUSED ELECTRON BEAM INDUCED PROCESSING

Ivo Utke, Martin G. Jenke, Vinzenz Friedli, and Johann Michler

1 INTRODUCTION

Gas-assisted focused electron beam (FEB) and focused ion beam (FIB) processing methods are well established techniques for local deposition and etching of materials that rely on the decomposition of surface adsorbed precursor molecules by irradiation.

These high-resolution nanostructuring techniques have various applications in nano-science and industry comprising attach-and-release procedures in nanomanipulation, fabrication of sensors (magnetic, optical, and thermal) for scanning probe microscopy, fabrication of electrical contacts and devices, and photomask repair for the lithography of next generation electronic chips below the 32 nm node. Readers interested in a detailed listing of FEB and FIB applications as well as the underlying fundamentals are referred to a comprehensive review of Utke et al. [1].

Control of the process with respect to endpoint detection is fairly well established; however, a complete physical and chemical understanding of the gas assisted FEB and FIB processes is hampered by the lack of suitable means to monitor and to access the numerous interrelated process parameters (deposition and etch rate, yield, molecule flux, and adsorption/desorption).

This chapter will discuss *in situ* monitoring methods for gas-assisted focused electron beam and focused ion beam deposition and etching using various signals and related setups. It will not cover *in situ observation* of the process using dual or cross beam machines with both a focused electron and ion beam. Generally, such an approach will produce unwanted co-deposits from the observing beam, which destroys or modifies the fabricated structures.

2 MONITORING OF SAMPLE CURRENT AND SECONDARY ELECTRON SIGNAL

The easiest way to realize an *in situ* process control is to record the sample current signal and the secondary electron signal. Sometimes the term *stage current* is used in literature as a synonym for the sample current measurement.

2.1 PRINCIPLE

The sample current I_S is collected through the sample stage (see Figure 12.1) and determined via the general current balance:

$$I_S = I_P - I_{SE} - I_{BSE}, \qquad (1)$$

where I_P is the current of the incident primary electron (PE) beam or primary ion beam (PI), I_{SE} is the current of the emitted secondary electrons (SE), and I_{BSE} is the current of the backscattered electrons (BSE). The latter current is zero for ion beams so that $I_S = I_P - I_{SE}$ since the majority of sputtered atoms leaves the sample as neutrals. Many modern microscopes allow to monitor the I_S signal in their standard software. Alternatively, the I_S signal can be measured simply by means of an external pico-amperemeter, which is connected to the microscope stage. The beam current I_P is accessible using a faraday cup. Since all SEs and BSEs are trapped in a Faraday cup the measured sample current in the cup is equal to the beam current I_P (see Figure 12.1a). The I_{SE} and I_{BSE} signals are accessible through the respective detectors; however, the detector signals are electronically amplified and need absolute calibration. Since the amount of the SEs and BSEs change with the sample's topography and material, the deposition or etch process can be monitored. The yields of SEs and BSEs are well documented for planar surfaces [2–4], where two types of secondary electrons are distinguished. Secondary electrons of type 1, SE1, are generated by incident primary electrons; secondary electrons of type 2, SE2, are generated by backscattered electrons. On non-planar surfaces, the forward scattered electrons (FSE—not to be confused with fast secondary electrons) generate additional SEs and BSEs, their yields strongly and differently depend on the geometry and material at the place they are generated. Therefore, any three-dimensionally (3D) deposited or etched structure has its unique sample current versus time graph, i.e., a fingerprint due to its specific shape. In the following we will present examples of these fingerprints for typical structures. In a first run these fingerprints need to be calibrated via post-process observation and analysis with their actual shape and composition. Ultimately, they can serve as benchmarks for models of SE, FSE, and BSE yields in 3D structures and give important input into models of gas-assisted FEB and FIB processing; see Chapter 6 by Utke, Chapter 7 by Lobo and Toth, and Chapter 8 by Mitsuishi in this book. An analytical approach to arrive at parameterized quantification of the above yields is given in [5].

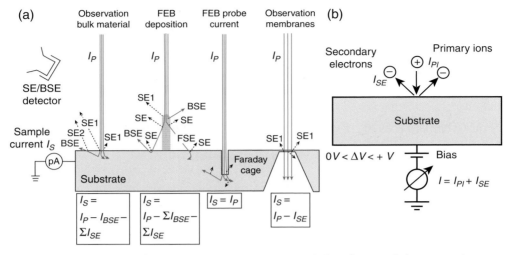

FIGURE 12.1: (a) Principle of sample current measurements and classification of electrons and current balances during gas-assisted focused electron beam structuring. SE = secondary electrons, BSE = backscattered electrons, FSE = forward scattered electrons, I_P = beam current (of primary electrons). Recording the stage current signal and the secondary electron signal is an easy and efficient *in situ* process control. (b) Principle of sample current measurements for FIB irradiation; I_{PI} is the current of primary ions (positively charged). Varying the substrate bias potential allows the recapture of the low-energy SEs and measurement of the SE energy spectrum (both for FIB and FEB irradiation).

Biasing a substrate with a variable potential of a few volts recaptures secondary electrons leaving the surface with low energy (see Figure 12.1b). Such substrate bias effects were investigated for gas-assisted FEB deposition and etching by Choi et al. [6] and for Ga$^+$-FIB by Chen et al. [7], who recorded SE spectra in dependence of the angle of incidence and material.

2.2 LAYERED STRUCTURES

The deposition of layers or films using gas-assisted focused electron beam induced deposition (FEBID) is usually limited to small areas, typically in the range of a few hundred μm^2. During the deposition the electron beam is multiply raster-scanned over the deposition area. Between each raster scan there is generally a (programmed) "refresh" period of a few milliseconds, which allows new intact molecules to re-cover the irradiated area. Bret et al. [8] used time-resolved sample current measurements to monitor the deposition of a hydrogenated amorphous carbon film from an acrylic acid molecule on an Si and Au substrate. The deposit composition was determined as $C_9H_2O_1$ (sum formula) [9] and the constant deposition rate was 67 nm/min on both substrates, with the conditions given in Figure 12.2a. A relatively large fraction of the beam current leaves the flat sample as backscattered electrons; yields at 10 keV are ~16% on Si and ~46% on Au, respectively [10]. Due to the low backscattered electron yield of a carbon film (~4%) an increasing fraction of primary electrons is absorbed in the growing deposit film and contributes dominantly to the sample current increase as the C film grows. This evolution is illustrated in Monte-Carlo simulations of the electron trajectories (see Figure 12.2b).

The primary electrons penetrate ~260 nm in Au, but down to ~1.3 μm in the C film. Consequently, an 800 nm thick carbonaceous film "absorbs" all substrate generated BSEs, and

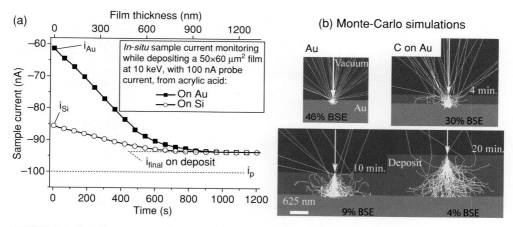

FIGURE 12.2: Sample current evolution for FEBID of large-area carbon films (100nA, 10 keV): (a) *In situ* sample current monitoring. The bare gold (Au) and silicon (Si) substrates have different sample currents. With increasing carbon (C) film thickness the substrate signal vanishes. (b) Corresponding Monte-Carlo simulations of 10 keV electronic trajectories and computed backscattered electron (BSE) yields for several pure C film thicknesses. Reprinted with permission from [8]. Copyright (2006), Elsevier.

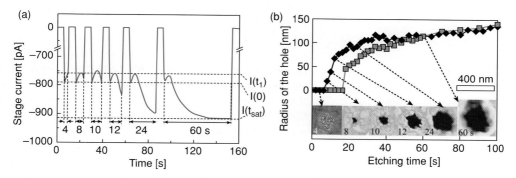

FIGURE 12.3: (a) Time resolved sample current measurement for H_2O-assisted etching of a carbon membrane; six successive FEB exposures (stationary, 5 keV, 924 pA) with varying dwell time. (b) Radius of etch holes as a function of FEB exposure time using H_2O (diamonds) and XeF_2 (squares) as precursor molecule. Inset: SEM images of nanoholes with corresponding exposure time in seconds. Reprinted with permission from [11]. Copyright (2008), American Institute of Physics.

the film behaves as bulk material with respect to electron signals. The change of secondary electron emission is confined within the first few nanometers of the deposited film due to their small escape depth.

This *in situ* monitoring principle can be used for "endpoint detection" in gas-assisted FEB/FIB etching. Once a buried layer in the structure is reached by etching, the signals will change due to the differing SE and BSE emission of materials and the process can be stopped. However, this technique is limited to holes having an aspect ratio that still permits a significant fraction of the secondary electrons to leave the surface.

Figure 12.3 shows an example for a sample current monitoring during water-assisted FEB etching of a carbon membrane, which enabled sub-beam-diameter resolution [11]. The etched

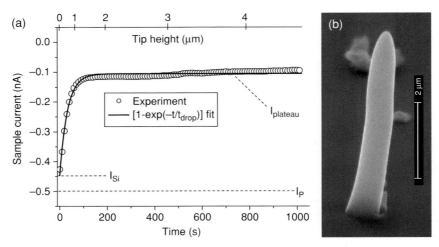

FIGURE 12.4: (a) *In situ* sample current evolution during deposition of the pillar in (b). Deviations from the plateau current indicate beam or stage drifts during pillar deposition as witnessed by post-deposition SEM observation. (b) SEM image of the final pillar. The exposure parameters were: 25keV, 500pA, molecule [IrCl(PF$_3$)$_2$]$_2$, substrate Si, pillar composition IrClP. Modified reprint with permission from [8]. Copyright (2006), Elsevier.

nanoholes had a size of 20–40% of the FWHM[1] of the incident beam. The interesting rise of the stage current curve during the first 4 seconds results from the roughening of the membrane surface.

2.3 PILLAR STRUCTURES

Vertical pillars or tips can be grown by keeping the FEB beam at a fixed position. The deposited pillars mostly have diameters narrower than the Bethe electron range, due to the fact that the vertical deposition rate is larger than the lateral deposition rate. As shown in Figure 12.1, there are now more types of electrons involved in the current balance which give rise to the typical current versus time curve shown in Figure 12.4a.

During pillar deposition, the collected sample current decreases exponentially in absolute value and reaches a constant value as soon as the pillar cone is completed. This prevents the access to further information on the continuing vertical cylinder growth . Still, deviations from the plateau current value indicate drifts (see Figure 12.4b), vibrations, material change, or other instabilities of the growth process. The monitoring thus allows a quantitative *in situ* control and an optimization of the initial stages of FEB induced pillar deposition. The sample current during the pillar growth can be well described by the equation [8]:

$$I_S(t) = I_{Plateau} + \left(I_{Substrate} - I_{Plateau}\right) \exp\left(-t/t_{drop}\right), \tag{2}$$

[1] Full Width at Half Maximum.

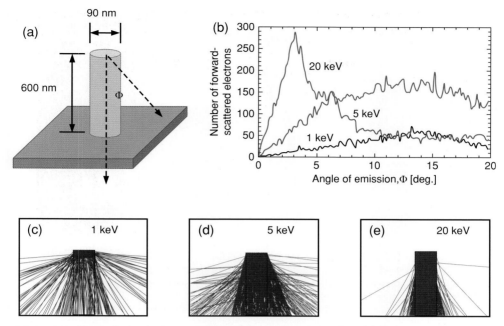

FIGURE 12.5: Monte Carlo electron-solid simulations in tungsten pillars (a) Simulation geometry. (b) Angular distribution of forward scattered electrons at 1, 5 and 20 keV beam energy. (c)-(e) Illustration of 200 electron trajectories. Reprinted with permission from [12]. Copyright (2007), IOP Publishing Ltd.

where $I_{Substrate}$ is the sample current on the substrate, $I_{Plateau}$ the sample current as soon as the pillar cone is formed, t is the time, and t_{drop} is a characteristic time during which cone formation is accomplished by 63% ($=1/e$).

The plateau value can be derived from the theory of electron scattering in the cone part of the pillar. Apart from being *back* scattered inside the cone, a considerable part of the primary electrons are *forward* scattered by the pillar with an angle distribution centred around the main forward scattering angle Φ, see Figure 12.5. The main forward scattering angle Φ obviously changes and gets larger as the pillar cone grows in volume. It depends generally on the primary electron energy E (the higher E the lower Φ), the cylinder diameter of the pillars (the thicker the larger Φ), and on the pillar material (as heavier the material the larger the cone angle for a given primary electron energy) [13]. However, once the pillar cone is formed and the vertical cylinder growth proceeds, the main forward scattering angle Φ remains constant and the sample current plateaus. Obviously, the main forward scattering angle will also largely determine the cone opening angle and the cone height. The pillar apex curvature will, however, be determined by the incident primary electron beam profile and the availability of adsorbates for electron (ion) induced dissociation, see Chapter 6 by Utke and Chapter 8 by Mitsuishi in this volume. Rack et al. [12] calculated the forward scattering angle distribution by Monte Carlo simulations, see Figure 12.5. An analytical derivation was given by Bret et al. [5].

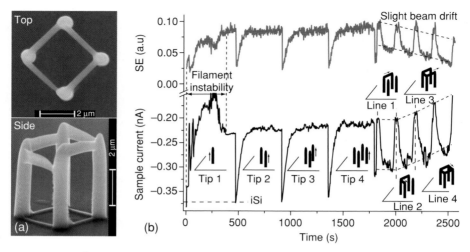

FIGURE 12.6: "Stool base" deposition from (hfac) CuVTMS, at 25 keV, 500 pA, on Si: (a) Ex-situ SEM views. (b) In-situ monitoring of SE signal and sample current. The deposition sequence is indicated. Reprinted with permission from T. Bret [5].

2.4 THREE DIMENSIONAL (3D) STRUCTURES

Complex 3D structures can be deposited by scanning the beam in a controlled fashion during deposition [8; 14; 15]. Figure 12.6 shows a "stool base" structure, which consists of four vertical pillars and four horizontal rods. The corresponding stage and secondary electron current signals are presented in the same figure. The 3D deposition progress can thus be monitored and controlled. Below the horizontal rods, "secondary" deposits are created because the primary electrons transmit through the rod material as it is being deposited. These co-deposits are hardly avoidable during 3D deposition of freestanding horizontal rod deposits [8] at high electron energies (5-30 keV) while Ga-FIB 3D deposition encounters considerably less problems with co-deposits due to the small penetration range of heavy ions in matter.

3 MASS SENSING

3.1 PRINCIPLE

The typical mass of deposited or removed material by an FEB or an FIB ranges from picograms (10^{-12} g) to zeptograms (10^{-21} g). This range of mass resolution is reached by cantilever-based vibrating sensor structures [16]. However, special care must be taken to stabilize the temperature of the setup in the millikelvin range in order to get a suitable signal to noise ratio [17]. The principle is given in Figure 12.7a: It consists of a cantilever with piezoresistive readout being actuated by a piezoelectric actuator. The frequency response to mass changes must be calibrated due to the unknown cantilever force constant. Once this calibration step is performed, the mass measurements inside the microscope chamber are quantitative.

(a)   (b)

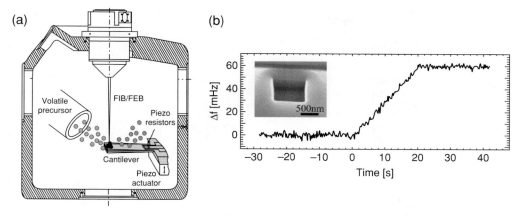

FIGURE 12.7: (a) Schematic diagram of the SEM integrated cantilever mass sensor for FIB/FEB induced process monitoring with local precursor supply from a micro-tube gas injection system. The mass added to or removed from the cantilever is detected as a negative or positive resonance frequency shift, respectively. (b) Evolution of the cantilever resonance frequency during FIB (30 kV, 100 pA) sputtering of a $1 \times 1 \ \mu m^2$ pit. Inset: SEM tilt view (45°) of the sputtered pit. Reprinted with permission from [16]. Copyright (2007), American Institute of Physics.

3.2 FIB MILLING AND FIB ETCHING

FIB milling and FIB etching removes material of the cantilever and thus the frequency change becomes positive. Figure 12.7b shows the continuously recorded resonance frequency shift obtained during FIB milling of a $1 \times 1 \ \mu m^2$ pit using a focused Ga^+ ion beam (30 kV, 50 pA) at the free end of the cantilever. Due to the fact that the dimensions of the milling can be easily measured by FIB or FEB microscopy, this method was used to perform a mass response calibration by Friedli et al. [16]. The mass response at the cantilever end can be calculated by:

$$\left(\frac{\Delta f}{\Delta m}\right)_{end} = \left(\frac{\Delta f}{(m_{Ga} - V \cdot \rho_{Si})}\right), \tag{3}$$

where Δf is the frequency shift, m_{Ga} the mass of the implanted gallium, V the volume of the removed material, and ρ_{Si} is the silicon density. The best mass resolution obtained in [16] with piezoresistive cantilevers was 1 femtogram corresponding to a silicon half sphere with a radius of 60 nm.

3.3 FIB/FEB DEPOSITION

Gas-assisted FIB/FEB deposition adds mass on the cantilever. Figures 12.8b shows the mass increase of the cantilever by writing a $1 \times 1 \ \mu m^2$ square and using a $(CH_3)_3PtCpCH_3$ precursor. The mass deposition rate obtained from Figure 12.8b is 23 fg/s in average. It can be seen that the deposited mass increases linearly with the deposition time during irradiation.

In literature, often the volume deposition yield Y_V in units of volume per charge is reported, since it is easily accessible from SEM observation. In contrast, mass sensing during FEB and FIB deposition permits calculating the *mass* deposition yield (Y_m) in units of mass per

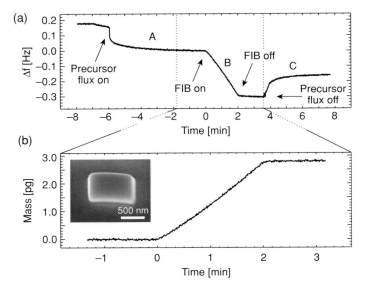

FIGURE 12.8: (a) FIB induced deposition experiment using the $(CH_3)_3PtCpCH_3$ precursor: (A) Mass loading due to adsorption and (C) mass loss due to desorption. (B) FIB exposure of a $1 \times 1\ \mu m^2$ rectangle. Mass loading due to FIB induced deposition. (b) Evolution of the FIB (30 kV, 10 pA) deposited mass corresponding to part (B) in (a). Inset: SEM tilt view (52°) of the FIB deposit. Reprinted with permission from [16]. Copyright (2007), American Institute of Physics.

incident charge. From this value correct *atom yields* can be derived. Taking Figure 12.8a as an example, the yield of deposited platinum atoms per incident electron or ion, Y_{Pt}, can be calculated from the mass yield by using:

$$Y_{Pt} = Y_m \cdot e_o \cdot \frac{x_{Pt}}{m_{Pt}} = \frac{m_{Deposit}}{I_P \Delta t} \cdot e_o \cdot \frac{x_{Pt}}{m_{Pt}}, \tag{4}$$

where x_{Pt} is the atomic fraction of Pt content in the deposit (accessible via EDX), m_{Pt} is the Pt atom mass, Δt is the irradiation time, and $I_P \Delta t$ is the irradiation dose.

3.4 ADSORPTION/DESORPTION KINETICS

The mass load/loss curves A and C from Figure 12.8a taken without irradiation are related to the mass of reversibly adsorbed molecules and of possible remaining chemisorbed molecules on the cantilever surface. From the mass load curve A the adsorbate *surface residence time* τ can be determined. This requires an assumption about the specific adsorption isotherm, see Chapter 6 by Utke in this volume. Taking the non-dissociative Langmuir isotherm, the adsorbate coverage is given as:

$$\theta(x, y) = \frac{sJ(x, y)\tau}{sJ(x, y)\tau + N_0} \approx sJ(x, y)\tau/N_0, \tag{5}$$

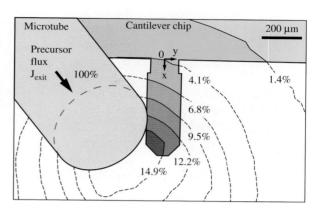

FIGURE 12.9: Monte Carlo simulated precursor iso-flux lines of an impinging molecule flux distribution on a cantilever chip (for clarity the contours are extended as dashed lines over the whole plane). Top-view of the experimental arrangement. The gas injection tube is 52° inclined to the chip plane and its exit surface center is at a distance of 600 μm to the cantilever chip plane. Isoflux percentages are indicated with respect to $J_{tot} = 3.3 \times 10^{18}$ cm^{-2}s^{-1} = 100%. Reprinted with permission from [16]. Copyright (2007), American Institute of Physics.

where N_0 is the adsorbate surface density of a complete monolayer, s the sticking probability, and τ the surface residence time. The right hand term approximation holds for low incident molecule flux $sJ\tau \ll N_0$. The incident molecule flux distribution $J(x, y)$ on the cantilever surface can be calculated via a Monte Carlo based algorithm described in [18]. An example for a distribution is shown in Figure 12.9.

The mass distribution of the adsorbed molecules is $m(x, y) = m_0\theta(x, y)$, with m_0 as the mass of a complete adsorbate monolayer on the cantilever surface, $m_0 = N_0 \cdot A \cdot M/N_A$; A is the cantilever surface, M the molecular weight of the adsorbate, and N_A the Avogadro constant. The related frequency shift Δf of the cantilever is obtained by integrating the product of mass distribution and cantilever mass response, $R_p(x, y)$, over the surface A of the cantilever:

$$\Delta f = \frac{1}{A} \int_0^A m(x, y)R_P dA \cong \frac{m_0 s\tau}{N_0 A} \int_0^A J(x, y)\frac{\Delta f}{\Delta m}(x, y)dA, \qquad (6)$$

Equation 6 can be simplified for symmetric molecule flux distributions where the relation $R_p(x, y) = \Delta f/\Delta m(x, y) \approx (\Delta f/\Delta m)_{end}(x/l)$ can be used. $(\Delta f/\Delta m)_{end}$ can be measured, see Equation 3, so that $s\tau$ (sticking probability times surface residence time) can be determined from Equation 6 once the integral is solved numerically. A value of $s\tau = 29 \times 10^{-6}$ s was determined for the $(CH_3)_3PtCpCH_3$ adsorbate [16]. Setting $s = 1$ (100%) results in $\tau = 29$ μs which represents a minimum value of the adsorbates' mean surface residence time before their spontaneous thermal desorption.

The above values were obtained on the cantilever material. To measure adsorption on the deposit surface there should be a few monolayers of the precursor molecule deposited on the cantilever prior to the measurement. In order to determine the type of the adsorption isotherm (see chapter 6 by Utke) it is necessary to repeat such mass measurements at varying temperature. From the comparison of the mass load curve A and the desorption curve C in Figure 12.8a follows that a small difference in mass persists. This might be attributed to chemisorption of a few adsorbates with the substrate (cantilever); such adsorbates established a chemical bond with the surface and will not desorb spontaneously at room temperature.

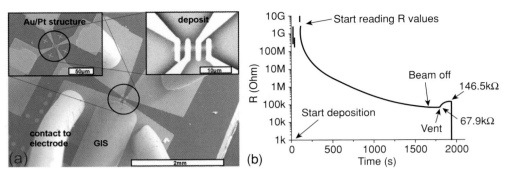

FIGURE 12.10: (a) 4-point *in situ* electrical measurement setup. Au/Pt electrodes are predefined on an SiO$_2$/Si substrate (main figure) and provide access to a four-finger structure (top-left inset). A straight line is deposited across the fingers (top-right inset) while its resistance is continuously monitored. (b) Resistance evolution during deposition and subsequent venting with air.

4 *IN SITU* ELECTRICAL RESISTANCE MEASUREMENTS

Time-resolved resistance measurements during gas-assisted FEB deposition can be performed using pre-deposited electrodes on an insulating substrate. They will be "connected" by scanning the electron or ion beam back and forth between them, which will deposit a wire. The resistance of the wire depends on the deposited material by electron (ion) impact adsorbate dissociation and the deposited thickness. Of important note is that also charge carriers generated during primary electron (/ion) irradiation contribute non-negligibly to the electrical conductance of the deposit being measured *in situ* [5, 19]. *In situ* resistance measurements during FEB/FIB deposition were reported for the molecules W(CO)$_6$ [20–23], (CH$_3$)$_3$PtCpCH$_3$ [24–27], Pd(F$_6$acac)$_2$ [23] and acrylic acid [1]. *In situ* measurements of FIB deposited wires from (CH$_3$)$_3$PtCpCH$_3$ show the formation of a Ga-phase [22] due to Joule heating, while FEB deposited *freestanding* wires show electro-migration of platinum [27]. We will confine this section to a brief description of the method and refer readers to Chapter 22 by Gazzadi and Frabboni, Chapter 23 by De Teresa et al., and Chapter 24 by Huth in this volume, who present an in-depth analysis of the results obtained.

4.1 REAL-TIME FOUR POINT ELECTRICAL RESISTANCE MEASUREMENTS

Real-time four point electrical resistance measurements allow studying the "electrical thickness" of the deposit during FEB depositing. Figure 12.10 shows an example of a four point electrical resistance real-time measurement while FEB depositing with a Co$_2$(CO)$_8$ precursor.

For constant length, width, and homogeneous material the resistance R is inversely proportional to the deposited wire thickness:

$$R(t) = \rho(t) \cdot l/[w \cdot h(t)], \tag{7}$$

where ρ denotes the deposit resistivity, and l, w, and h the deposit length, width, and thickness, respectively. In order to extract the material resistivity from the *in situ* resistance measurements, the growing thickness of the deposit has to be known during deposition, either by assuming a linear dependence with the deposition time or by calibrating the thickness-time correlation by investigation of deposits produced with different doses.

According to Equation 7, for a constant deposition rate an inverse relationship between the resistance and the deposition time is obtained. Deviations from this trend point to changes affecting the charge carrier generation and kinetics, for example, by the formation of intermediate species. Also, post-irradiation and air exposure phenomena can be studied and related to eventual oxidation processes. Two point measurements can be uniquely attributed to the deposit only if the contact resistance between electrodes and deposit is low compared to the deposit resistance. Alternatively, four point probe measurements would cancel out contact resistance effects.

5 REFLECTOMETRY

5.1 PRINCIPLE

Real-time *in situ* reflectometry during FEB deposition requires that a laser beam is allowed to enter into the microscope chamber. The monochromatic laser is then focused on the growing transparent or semi-transparent deposit inside the SEM chamber, and the reflected optical signal is collected by a photodiode. The intensity of the reflected laser beam depends on the optical thickness of the deposit and its optical absorption (Figure 12.11a). The related equations were derived in [28]. Neglecting light scattering, the total intensity resulting from the addition of the directly reflected and refracted fractions of light, can be written as follows:

$$I_{tot} = I_{R1} + I_{R2} + 2 \left(I_{R1} I_{R2}\right)^{1/2} \cos \Delta\Psi, \tag{8}$$

with the reflected intensities $I_{R1} = R_{dep} I_0$, $I_{R2} = T^2 R_S I_0 \exp(-4 d_{dep}/l_{abs})$, and the optical penetration depth $l_{abs} = 1/4\pi k_{dep}$. I_0 is the incident intensity, R_{dep} and R_S are the reflectivity coefficients of the deposit (R_{dep} is constant, i.e., the deposit is considered homogeneous) and the substrate, respectively. T is the transmission coefficient of the vacuum-deposit interface, d_{dep} is the deposit thickness, n_{dep} and k_{dep} are the real part and the imaginary part of the complex-refractive index $\tilde{n}_{dep} = n_{dep} + i k_{dep}$ of the deposit material at the wavelength λ. The phase shift $\Delta\psi$ of the refracted light exiting from the deposit (with Θ as the incident angle of light) is given by:

$$\Delta\Psi = 2 \left(\frac{2\pi \cdot n_{dep} \cdot d_{dep}}{\lambda \cdot \cos\Theta} \right), \tag{9}$$

This intensity signal is periodic with deposition time (deposit thickness) and results from constructive and destructive interferences. The theoretical deposit thicknesses $d_{dep}^{\pm}$ for the constructive and destructive interferences as function of the deposit's refractive index n_{dep} can be calculated by:

$$d_{dep}^{\pm} = \frac{m \cdot \lambda}{4 n_{dep}} \sqrt{\left(1 - \left(\frac{1}{n_{dep}}\right)^2 \sin^2\Theta\right)}, \tag{10}$$

where m is an integer. Odd and even values of m lead to destructive and constructive interferences, respectively. Using the reflectometry signal as a function of time and calibrating

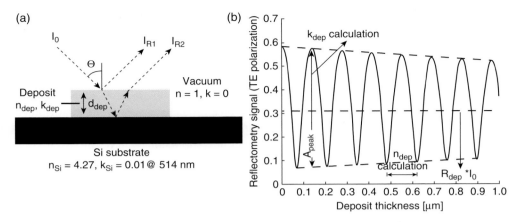

FIGURE 12.11: (a) Principle of reflectometry with three media. (b) Theoretical reflectometry curve expected for a Si_3N_4 ($\bar{n}_{Si3N4} = 2.05 + i0.01$ at the 514 nm wavelength) film growing on a silicon substrate using Equation 8. This curve represents the variation of a transverse electric (TE) polarized wave. Dashed lines represent the envelopes and the limit value reached for large thickness, $R_{dep}I_0$. Reprinted with permission from [28]. Copyright (2006), American Institute of Physics.

the deposit thickness as a function of time by using an AFM, one can plot the reflectometry signal as a function of the deposit thickness d_{dep}, from which values of $d_{dep}^{\pm}$ and corresponding values of m are directly collected (Figure 12.11b). If we use a setup with $\Theta = 45°$ this leads to:

$$\frac{1}{n_{dep}^2} = 1 - \sqrt{1 - \frac{32d_{dep}^2}{m^2\lambda^2}},\tag{11}$$

n_{dep} can be obtained from this equation by inserting measured ($d_{dep}^{\pm}$, m) couples and solving it for each couple. Deposition of absorbing material leads to an exponential decay of the peak amplitude A_{peak} in the reflectometry signal as a function of time (Figure 12.11b). The absorption coefficient k_{dep} of the deposited material can be calculated from two successive A_{peak} values and their corresponding deposit thicknesses:

$$k_{dep} = \frac{\lambda \ln\left(A_{peak1}/A_{peak2}\right)}{8\pi\left(d_{peak2}/d_{peak1}\right)},\tag{12}$$

Light scattering effects were not included but would only add a signal decrease to the reflectometry background component and not perturb the modulated signal.

A derivation taking into account multiple refractions by using the Fresnel algorithm was outlined in [29].

5.2 *IN SITU* REFLECTOMETRY MEASUREMENT OF SIO$_2$ DEPOSITS

FEBID of materials transparent in the UV and the visible spectra is challenging since the decomposition of carbon-containing precursor adsorbates usually leads to co-deposition of

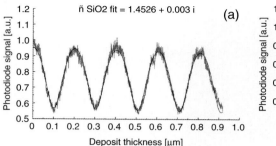
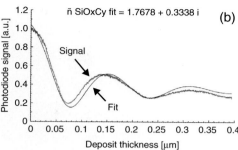

FIGURE 12.12: Experimental reflectometry signals ($\lambda = 514$ nm) acquired during gas-assisted FEBID with tetramethyl orthosilicate (TMOS). (a) FEB deposited SiO_2 using additional reactive oxygen $[O_2]/[TMOS] = 0.6$. (b) FEB deposited $SiO_xC_yH_\alpha$ using TMOS alone. The fit was performed using the Fresnel theory framework. Reprinted with permission from A. Perentes [29].

carbon with a content of typically 18–60 at.% [30; 31]. These carbon impurities lead to a measurable absorption and process parameters can be changed to counteract co-deposition. Figure 12.12 shows the *in situ* reflectometry results for a transparent SiO_2 FEB-deposit fabricated from tetramethyl orthosilicate (TMOS: $Si(OCH_3)_4$) and an additional oxygen flux. In comparison, TMOS deposited material without additional oxygen shows a high optical absorption (large imaginary part of the refractive index) due to carbon co-deposition.

Measurements at 514 nm laser wavelength of FEB deposits with other molecules without reactive gas addition gave the following refractive indices: $1.51 + i0.055$ (HCOOH) and $2.19 + i0.013$ ($Ti(NO_3)_4$) [28]. The smallest absorption was obtained with oxygen addition, see Figure 12.12a and from the carbon-free molecule ($Ti(NO_3)_4$).

6 *IN SITU* ANNULAR DARK FIELD SIGNAL

6.1 PRINCIPLE

In situ annular dark field (ADF) measurements during gas assisted FEB induced deposition (FEBID) can be performed in an environmental scanning transmission electron microscope (STEM) [32] (see Figure 12.13). The tolerated gas pressure can be locally up to 8 Torr and the beam is often operated at 200 kV with a nominal beam spot size of about 0.3 nm. The acquired ADF signal depends linearly on the deposited mass in case of a constant deposit composition. The smallest sampling time of the ADF signal monitoring is presently about 40 ms.

6.2 DYNAMIC FEB DEPOSITION CONTROL WITH ADF

A precise control on FEBID at nanoscale is not straightforward because several nonlinear effects can occur. Among others, they comprise proximity effects [33] on the curved surfaces of the growing deposit which continuously change the incident angle of primary electrons and result in varying growth rates [34]. The ADF intensity signal was used by van Dorp et al. [34] as a feedback in the deposition process to compensate "online" for these nonlinear deposition effects (see Figure 12.14).

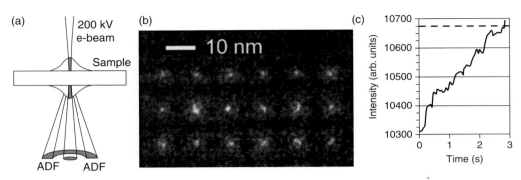

FIGURE 12.13: (a) Schematic drawing of the area from which the ADF signal originates. Note that the growing dot volume is only partially probed by the electron beam. (b) Dot array written with ADF control. (c) ADF signal curve of an individual dot from the array. Reprinted with permission from [32]. Copyright (2007), American Institute of Physics.

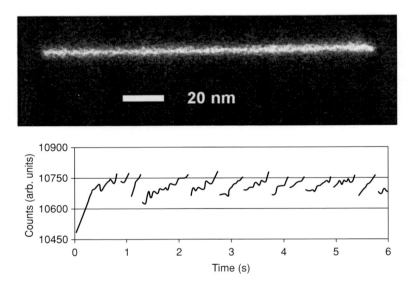

FIGURE 12.14: A line (400 points over 160 nm length) was deposited while the deposition process was controlled *in situ* by using the ADF signal as feedback. At an ADF threshold of 10750 counts the dwell time was reduced for subsequent exposure points. The graph below shows the first seconds of the recorded ADF signal. Reprinted with permission from [32]. Copyright (2007), American Institute of Physics.

A 160 nm long line was deposited in a single scan. One can observe that only the initial exposure of the line pattern was on the flat substrate and that any following scan exposure deposited on top of already deposited material (increase of ADF signal). Once the threshold for the ADF intensity was passed, the dwell time of the subsequent line pixel was reduced and hence less material deposited. Van Dorp et al. [32] suggested that the ultimate mass resolution that can be achieved with this method is a single molecule. This would be the highest resolution of all the *in situ* techniques employed today. A similar threshold feedback should be feasible with the sample current signal discussed in Section 1; with the advantage that it can be employed in scanning electron microscopes (SEMs). From Figures 12.3 and 12.4 a high sensitivity to initial deposition (etch) stages can be expected.

7 ENERGY DISPERSIVE X-RAY SPECTROSCOPY (EDX)

In EDX an electron beam is directed onto the deposited nanostructure which excites the characteristic X-ray emission of the elements (K, L, M lines). The information depth of the X-rays depends on the electron penetration depth and the energy of the X-rays excited. A parameterized equation for the information depth R_X was given by Kanaya and Okayama [48]:

$$R_x = 2.76 \cdot 10^{-4} \frac{M[g/mol](E_{PE} - E_X)^{5/3}[eV]}{Z^{8/9}\rho[g/cm^3]} \quad [nm] \ , \tag{13}$$

where E_{PE} is the energy of the incident primary electrons, E_X the energy of the X-ray photon, and Z the atom number. Typical values for 3 keV incident electrons are in the submicron range, whereas X-rays from 10 keV incident electrons carry information from a micron-sized depth. These values also represent the spatial and depth resolution of EDX.

EDX is often used as *ex situ* method to determine the chemical composition of the deposited structure. Rack et al. [35] used an energy dispersive X-ray spectrometer (EDX) attached to a Hitachi S-3500N environmental scanning electron microscope (ESEM) to provide qualitative *in situ* compositional measurements during electron beam induced deposition of tungsten. In order to quantify the spectra, however, an (*ex situ*) calibration step is necessary to take into account the varying thickness of the deposit for X-ray generation.

8 SEM INTEGRATED MECHANICAL MEASUREMENTS

Although being "SEM-integrated", the following measurement methods are not strictly *in situ* as described in the previous sections, because they are performed after the fabrication of the structures. They allow performing precise mechanical measurements on FEB or FIB deposits, as well as other nanostructures, like nanowires or nanotubes. The measurements setups involve advanced nanomanipulation setups comprising x, y, z stages, vibrational components, force sensors, and the corresponding electronics. A more detailed overview on such systems and their applications in scanning electron/ion microscopes can be found in [36; 37].

8.1 BENDING TESTS

SEM integrated bending experiments for the determination of the spring constant and the Young's modulus of micro- and nanostructures rely on the physical interactions between the micro- and nanostructure and a micromachined reference AFM cantilever used as the force sensor. Figure 12.15 shows an example were the stiffness of a gas-assisted focused electron

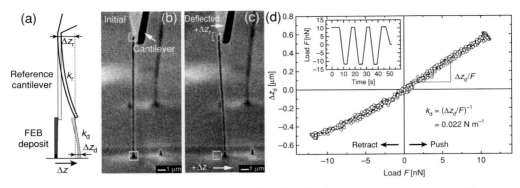

FIGURE 12.15: SEM integrated bending test on an FEB pillar-like deposit using a reference cantilever as the force sensor. (a) shows a schematic drawing of the setup (not to scale). (b) and (c) show SEM images of the bending test together with the positions (solid boxes) used for determining Δz and Δz_r. (d) presents the resulting graph of deflection vs. applied force . Reprinted with permission from [38]. Copyright (2009), IOP Publishing Ltd..

beam deposited pillar is measured by deflecting it via a reference AFM cantilever with a known spring constant. The applied force at the contact point of cantilever (index r) and deposit (index d) is:

$$F = \Delta z_r \cdot k_r = \Delta z_d \cdot k_d, \tag{14}$$

with k as force constant and z as deflection. Knowing that a stage displacement of Δz results in a displacement of the cantilever and the deposited beam with the relation:

$$\Delta z = \Delta z_r + \Delta z_d, \tag{15}$$

allows us to plot a graph of the applied force vs. the deflection of the deposited beam and to calculate the spring constant k_d from the pushing part of Figure 12.15d.

The Young's modulus, E, of a single clamped cylindrical pillar consisting of an elastically isotropic material and with a uniform diameter, d, and length, l, can be calculated according to the Euler-Bernoulli beam model for bending by:

$$E = \frac{64l^3}{3\pi d^4} k_d, \tag{16}$$

Mechanical properties of pillar deposits from phenanthrene, $W(CO)_6$, and $Cu(hfa)_2$ were measured from different groups and summarized in [36]. The typical range of the Young's modulus is in the order of 10 GPa for FEB deposition to up to a few 100 GPa for FIB deposition [39], whereas dependencies on electron or ion dose and energy were found. Special tensile stage setups can be used as well to determine the tensile strengths and the Young's modulus [40].

8.2 VIBRATION/RESONANCE TESTS

Similar to the static bending tests, it is possible to perform dynamic resonance tests on micro- and nanostructures. These tests allow one to determine the resonance frequency and

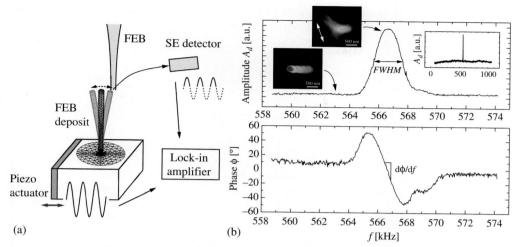

FIGURE 12.16: (a) Measurement principle in a piezo-driven resonance experiment based on the stationary beam technique. Locking the sample deflection signal from the secondary electron (SE) detector to the excitation signal reveals the deflection amplitude $A_d(f)$ and the phase characteristics $\Phi(f)$ at resonance. (b) Spectrum of the fundamental resonance of the pillar. Insets: top-view SEM images show the slightly tilted pillar excited at off-resonance and at resonance. The inset on the right shows a full range spectrum up to 1 MHz of the amplitude response which locates the peak at $f_0 = 566.6$ kHz. The phase slope at resonance corresponds to a quality factor of 274. Reprinted with permission from [38]. Copyright (2009), IOP Publishing Ltd.

the quality factor of the micro- and nanostructures. Figure 12.16 shows an example for a resonance test performed on a gas-assisted focused electron beam deposited pillar. The pillar is mechanically excited at its fundamental resonance mode by using a piezo actuator, as shown in Figure 12.16a.

If the stationary electron beam irradiates the maximum amplitude position, a peak in the integrated SE signal is detected at resonance while sweeping the excitation frequency. This is due to the increasing time the vibrating sample dwells inside the beam [41]. Phase locking of the time-resolved SE and excitation signal allows extracting the response of amplitude and phase at resonance from the noisy SE signal. The position of the peak maximum and the zero phase corresponds to the resonance frequency f_0, which in turn is proportional to $\sqrt{E/\rho_d}$. Knowing the elastic modulus or spring constant from the static test described in Section 8.1, the deposit density, ρ_d, of a (perfectly) cylindrical pillar can be calculated by [36]:

$$\rho_d = 0.133 \frac{k_d}{l \cdot d^2 \cdot f_0^2}, \tag{17}$$

where k_d is the measured spring constant, l is the length of the pillar, d is the average diameter of the pillar and f_0 is the measured resonance frequency. Equation 18 must be replaced by finite element calculations when the pillar is non-homogeneous in diameter (shape) and material [36].

The quality factor is correctly obtained from the phase slope at resonance:

$$Q = \frac{f_0}{2 \left| d\Phi/df \right|_{f_0}},\qquad(18)$$

8.3 NANOINDENTATION

Nanoindentation combined with a scanning electron microscope (SEM) allows to visually observe the interaction between the indenter and the specimen and to determine the elastic modulus and hardness [42–46]. An example for such a measurement is shown in Figure 12.17. The main advantages of SEM integrated indentation techniques are as follows: a continuous observation of the surface deformation during the application of an external load on the specimen surface by an indenter, the correction of the reduced elastic modulus and hardness evaluated by the Oliver and Pharr method [47] (by accounting for pile-up and sink-in phenomena), and submicron positioning capability allowing indentation of micron and submicron features. However, some physical phenomena that might take place underneath the surface, such as phase transformation, because of high pressure applied by the indenter, or nucleation and propagation of dislocations, are hardly visible by SEM observation.

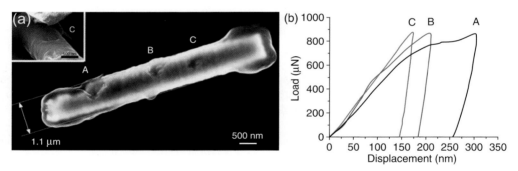

FIGURE 12.17: (a) High-resolution SEM image of three (A to C) indentations in an electroplated cobalt wire with 1.1 μm diameter. The imprints left by the three indentations are performed with the same maximum load but at different distance between the axes of the indenter and wire. (b) The load-displacement curve of the three indents. The elastic modulus, 29 GPa, of the nanowire was determined by a Hertzian analysis (effective radius curvature of 150 nm) of the shown curves. This is much lower than the elastic modulus, 78 GPa, of a flat cobalt surface and it highlights that materials can have at nanoscale very different properties from the ones they have at the macroscale. Reprinted with permission from [42]. Copyright (2009), Wiley.

REFERENCES

[1] Utke, I., P. Hoffmann, and J. Melngailis. Gas-assisted focused electron beam and ion beam processing and fabrication. *Journal of Vacuum Science & Technology* B 26.4 (2008): 1197–1276.
[2] Reimer, L., and P. W. Hawkes. *Scanning electron microscopy: Physics of image formation and microanalysis.* Berlin; Heidelberg; New York: Springer-Verlag, 1998, p. 527.

[3] Goldstein, J., et al. *Scanning electron microscopy and X-ray microanalysis*. Berlin; Heidelberg; New York: Springer-Verlag, 2003, p. 689.

[4] Lin, Y. H., and D. C. Joy. A new examination of secondary electron yield data. *Surface and Interface Analysis*, 37.11 (2005): 895–900.

[5] Bret, T. *Physico-chemical study of the focused electron beam induced deposition process*. Faculté Sciences et Techniques de l'Ingénieur. EPFL (École Polytechnique Fédérale de Lausanne): Lausanne, 2005, p. 220.

[6] Choi, Y. R., et al. Effect of electron beam-induced deposition and etching under bias. *Scanning* 29.4 (2007): 171–176.

[7] Chen, P., H. W. M. Salemink, and P. F. A. Alkemade. Roles of secondary electrons and sputtered atoms in ion-beam-induced deposition. *Journal of Vacuum Science & Technology* B 27.6 (2009): 2718–2721.

[8] Bret, T., et al. Electron range effects in focused electron beam induced deposition of 3D nanostructures. *Microelectronic Engineering* 83.4–9 (2006): 1482–1486.

[9] Bret, T., et al. Characterization of focused electron beam induced carbon deposits from organic precursors. *Microelectronic Engineering* 78–79 (2005): 300–306.

[10] Niedrig, H. Electron backscattering from thin-films. *Journal of Applied Physics* 53.4 (1982): R15–R49.

[11] Miyazoe, H., et al. Controlled focused electron beam-induced etching for the fabrication of sub-beam-size nanoholes. *Applied Physics Letters* 92.4 (2008): 043124.

[12] Rack, P. D. In situ probing of the growth and morphology in electron-beam-induced deposited nanostructures. *Nanotechnology* 18.46 (2007): 465602.

[13] Bret, T., et al. In situ control of the focused-electron-beam-induced deposition process. *Applied Physics Letters* 83.19 (2003): 4005–4007.

[14] Bret, T., et al. Periodic structure formation by focused electron-beam-induced deposition. *Journal of Vacuum Science & Technology* B 22.5 (2004): 2504–2510.

[15] Matsui, S., and R. Kometani, Three-dimensional nanostructure fabrication by focused-ion-beam chemical vapor deposition and its applications. *IEICE Transactions on Electronics* E90c.1 (2007): 25–35.

[16] Friedli, V., et al. Mass sensor for in situ monitoring of focused ion and electron beam induced processes. *Applied Physics Letters* 90.5 (2007): 053106.

[17] Friedli, V. *Focused electron- and ion-beam induced processes: In situ monitoring, analysis and modeling*. Faculté Sciences et Techniques de l'Ingénieur. EPFL (École Polytechnique Fédérale de Lausanne): Lausanne, 2008, p. 157.

[18] Friedli, V., and I. Utke. Optimized molecule supply from nozzle-based gas injection systems for focused electron- and ion-beam induced deposition and etching: Simulation and experiment. *Journal of Physics D-Applied Physics* 42.12 (2009): 125305.

[19] Rykaczewski, K., et al. The effect of the geometry and material properties of a carbon joint produced by electron beam induced deposition on the electrical resistance of a multiwalled carbon nanotube-to-metal contact interface. *Nanotechnology* 21.3 (2010): 035202.

[20] Hoyle, P. C., J. R. A. Cleaver, and H. Ahmed. Electron beam induced deposition from W(CO)(6) at 2 to 20 keV and its applications. *Journal of Vacuum Science & Technology* B 14.2 (1996): 662–673.

[21] Porrati, F., R. Sachser, and M. Huth. The transient electrical conductivity of W-based electron-beam-induced deposits during growth, irradiation and exposure to air. *Nanotechnology* 20.19 (2009): 195301.

[22] Reguer, A., et al. Structural and electrical studies of conductive nanowires prepared by focused ion beam induced deposition. *Journal of Vacuum Science & Technology* B 26.1 (2008): 175–180.

[23] Spoddig, D., et al. Transport properties and growth parameters of PdC and WC nanowires prepared in a dual-beam microscope. *Nanotechnology* 18.49 (2007): 12.

[24] Botman, A., M. Hesselberth, and J. J. L. Mulders. Investigation of morphological changes in platinum-containing nanostructures created by electron-beam-induced deposition. *Journal of Vacuum Science & Technology* B 26.6 (2008): 2464–2467.

[25] De Teresa, J. M., et al. Origin of the difference in the resistivity of as-grown focused-ion- and focused-electron-beam-induced Pt nanodeposits. *Journal of Nanomaterials* 2009 (2009): 936863.

[26] Fernandez-Pacheco, A., et al. Metal-insulator transition in Pt-C nanowires grown by focused-ion-beam-induced deposition. *Physical Review* B 79.17 (2009): 174204.

[27] Gazzadi, G. C., and S. Frabboni. Structural evolution and graphitization of metallorganic-Pt suspended nanowires under high-current-density electrical test. *Applied Physics Letters* 94.17 (2009): 173112.

[28] Perentes, A., et al. Real-time reflectometry-controlled focused-electron-beam-induced deposition of transparent materials. *Journal of Vacuum Science & Technology* B 24.2 (2006): 587–591.

[29] Perentes, A. *Oxygen assisted focused electron beam induced deposition of silicon dioxide*. Faculté Sciences et Techniques de l'Ingénieur. EPFL (École Polytechnique Fédérale de Lausanne): Lausanne, 2007, 226.

[30] Perentes, A., et al. Focused electron beam induced deposition of a periodic transparent nano-optic pattern. *Microelectronic Engineering* 73–74 (2004): 412–416.

[31] Lipp, S., et al. A comparison of focused ion beam and electron beam induced deposition processes. *Microelectronics and Reliability* 36.11–12 (1996): 1779–1782.

[32] Van Dorp, W. F., et al. In situ monitoring and control of material growth for high resolution electron beam induced deposition. *Journal of Vacuum Science & Technology* B 25.6 (2007): 2210.

[33] Hiroshima, H., and M. Komuro. Fabrication of conductive wires by electron-beam-induced deposition. *Nanotechnology* 9.2 (1998): 108–112.

[34] van Dorp, W. F., et al. Solutions to a proximity effect in high resolution electron beam induced deposition. *Journal of Vacuum Science & Technology* B 25.5 (2007): 1603–1608.

[35] Rack, P. D., et al. Nanoscale electron-beam-stimulated processing. *Applied Physics Letters* 82.14 (2003): 2326–2328.

[36] Friedli, V., et al. AFM sensors in scanning electron and ion microscopes: tools for nanomechanics, nanoanalytics, and nanofabrication. In *Applied Scanning Probe Methods VIII*, B. Bushan, H. Fuchs, and M. Tomitori, eds. Berlin: Springer, 2008, pp. 247–287.

[37] Fatikow, S. *Automated nanohandling by microrobots*. Springer Series in Advanced Manufacturing. Berlin: Springer-Verlag, p. 346.

[38] Friedli, V., et al. Dose and energy dependence of mechanical properties of focused electron-beam-induced pillar deposits from Cu(C5HF6O2)(2). *Nanotechnology* 20.38 (2009): 385304.

[39] Fujita, J., et al. Growth of three-dimensional nano-structures using FIB-CVD and its mechanical properties. *Nuclear Instruments and Methods in Physics Research Section B: Beam Interactions with Materials and Atoms* 206 (2003): 472–477.

[40] Utke, I., et al. Tensile strengths of metal-containing joints fabricated by focused electron beam induced deposition. *Advanced Engineering Materials* 8.3 (2006): 155–157.

[41] Fujita, J., et al. Observation and characteristics of mechanical vibration in three-dimensional nanostructures and pillars grown by focused ion beam chemical vapor deposition. *Journal of Vacuum Science & Technology* B 19.6 (2001): 2834–2837.

[42] Ghisleni, R., et al. In situ SEM indentation experiments: Instruments, methodology, and applications. *Microscopy Research and Technique* 72.3 (2009): 242–249.

[43] Rabe, R., et al. Observation of fracture and plastic deformation during indentation and scratching inside the scanning electron microscope. *Thin Solid Films* 469–470 (2004): 206–213.

[44] Moser, B., et al. Strength and fracture of Si micropillars: A new scanning electron microscopy-based micro-compression test. *Journal of Materials Research* 22.4 (2007): 1004–1011.

[45] Moser, B., J. F. Loffler, and J. Michler. Discrete deformation in amorphous metals: An in situ SEM indentation study. *Philosophical Magazine* 86.33–35 (2006): 5715–5728.

[46] Moser, B., et al. Observation of instabilities during plastic deformation by in-situ SEM indentation experiments. *Advanced Engineering Materials* 7.5 (2005): 388–392.

[47] Oliver, W. C., and G. M. Pharr. Measurement of hardness and elastic modulus by instrumented indentation: Advances in understanding and refinements to methodology. *Journal of Materials Research* 19.1 (2004): 3–20.

[48] K. Kanaya and S. Okayama, Penetration and energy-loss theory of electrons in solid targets. *Journal of Physics D: Applied Physics* 5 (1972) 43–58.

C H A P T E R   1 3

CLUSTER BEAM DEPOSITION OF METAL, INSULATOR, AND SEMICONDUCTOR NANOPARTICLES

Adam M. Zachary, Igor L. Bolotin, and Luke Hanley

1 INTRODUCTION

Nanostructures have immense potential due to their unique electronic, optical, and chemical properties that arise from quantum confinement and high surface area. This chapter describes cluster beam deposition (CBD) and related methods for the preparation of clusters in the gas phase and deposition either onto a solid surface or into a matrix of a distinct material [1–5]. The resultant nanoparticles are of enormous interest for a wide range of fundamental studies and applications.

Much effort has focused on the wet chemical or colloidal synthesis of nanoparticles [6; 7]. Colloidal or liquid phase methods of synthesizing nanoparticles typically use capping agents to limit their growth. However, tuning surface chemistry of these colloidal nanoparticles often requires separate steps that can alter the nanoparticle size or other properties. Casting of colloidal nanoparticles into films can also lead to their unintended agglomeration. Finally, preparation of films from colloidal nanoparticles can expose them to atmospheric oxidation. CBD can overcome all these shortcomings.

Many properties of nanoparticles are controlled by their surface chemistry and/or matrix environment [7]. Gas phase deposition of nanoparticles allows direct control of this environment since cluster formation occurs before any interaction with matrix or substrate [1–4]. CBD either does not utilize a matrix or the matrix is introduced independently of the cluster formation and deposition process. The presence or absence of a matrix can be used to modulate the cluster-cluster and/or cluster-matrix interaction in a fashion that optimizes desirable properties. Careful control of matrix-to-cluster ratio can also eliminate nanoparticle agglomeration in the deposited film.

CBD of nanoparticles typically occurs in vacuum, mitigating reaction of atmospheric oxygen with nanoparticle surfaces that can otherwise lead to unintended property changes. For example, films of surface oxidized PbS nanoparticles behave as photodetectors while their unoxidized counterparts behave as photovoltaics [8]. CBD is a vacuum-based method that is readily integrated into traditional lithographic strategies such as electron and ion beam methods. CBD can also be used for combinatorial studies of device performance or optimization of other physical properties.

The ability to select cluster size and/or kinetic energy, especially for ionic clusters, is a further advantage of CBD. Control of kinetic energy affects the interaction of an impacting cluster with a target substrate, varying this interaction from a soft landing to sputtering event [9]. Nanostructured films formed with clusters with 0.1 versus 10 eV/atom kinetic energies display very different morphologies and compactness: the former can lead to films with low surface adhesion and rough surfaces, while the latter can show high surface adhesion and relatively smooth surfaces [2; 10]. Mass-selection of a specific cluster size also allows deposition of nanostructured films with variable properties. For example, metallic clusters deposited on supports can display size-dependent catalytic behavior [11–13].

The following sections discuss in exemplary fashion both CBD and related sources used to produce gaseous clusters for deposition as nanoparticles. The effect of selecting cluster mass and kinetic energy on the nanostructured film is examined in the presence or absence of matrices. Different sources of matrices for film formation are highlighted. Finally, several applications of clusters produced by CBD are discussed, including surface smoothing, surface analysis, catalysis, sensors, electronic circuitry, and optoelectronic devices. Nanostructures can be formed by a variety of strategies employing focused electron and atomic ion beams, but these are discussed elsewhere in this book.

2 SOURCES OF GASEOUS CLUSTERS

2.1 VAN DER WAALS CLUSTER SOURCES

Van der Waals cluster sources are somewhat outside the purview of this chapter, but are mentioned here because they are the prototypes for the most common varieties of CBD sources [14]. Figure 13.1 shows a typical gaseous van der Waals cluster ion beam apparatus which produces a 30 keV cluster ion beam and possesses mass-selection capabilities [15]. Figure 13.1 displays the first pumping stage housing the liquid nitrogen cooled cluster formation region, source gas input, nozzle, and skimmer; the second pumping region containing ion optical components including the ionizer and mass selector with a Faraday cup for current measurement; and the main chamber containing the accelerator and deflector lenses to guide the cluster ions onto a solid target. The cluster source initially produces neutral gaseous clusters via supersonic expansion of gaseous atoms or molecules from a high pressure environment through a room temperature nozzle into vacuum [15]. The beam of neutral clusters is

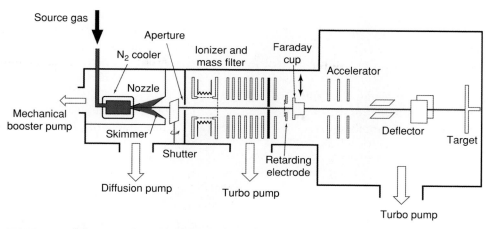

FIGURE 13.1: Schematic of typical gaseous van der Waals cluster ion beam source with 30 keV cluster ion beam, mass filter, Faraday cup, and ion optics for acceleration, focusing, and deflection. Reprinted from *Materials Science and Engineering: R: Reports* 34, I. Yamada, J. Matsuo, N. Toyoda, and A. Kirkpatrick, 231–295. Copyright 2001, with permission from Elsevier.

then ionized by electron bombardment and undergoes high voltage acceleration to impact a target [15].

Many sources of metal, insulator, and semiconductor clusters share several aspects of this van der Waals cluster source: condensation of gaseous atoms into clusters, ionization, electrostatic guidance of cluster ions toward deposition, and several stages of differential pumping. The following descriptions of cluster sources focus on the sources themselves, largely ignoring issues of differential pumping and post-source ion optics.

2.2 VAPORIZATION/ELECTRON IMPACT IONIZATION SOURCES

Simple electron impact ionization sources combined with a thermal vaporization capability can be used to produce C_{60}^+, C_{60}^{2+}, and C_{60}^{3+} from solid C_{60}. C_{60} is loaded into a cylinder, which is heated by electron impact, leading to the vaporization of gaseous C_{60} into the ionization region [16]. Figure 13.2 shows a schematic of a C_{60} source, which consists of the C_{60} reservoir encased in a mesh anode and includes a filament for both evaporation and electron generation for ionization. This source produces up to ~1 nA of filtered 2–18 keV C_{60} ions after passage of the beam through a Wien filter and associated ion optics. A commercial 10–120 keV C_{60} ion source has been developed with an oven for evaporation and a separate filament for ionization whose ion beam can be focused to an ~200 nm diameter spot onto a target [17; 18].

Electron impact ionization sources can also be used to produce molecular ions from volatile precursors such as $C_4H_4S^+$ from thiophene, $B_{10}H_{14}^+$ from decaborane, and SF_5^+ from sulfur hexafluoride. The extent of dissociation typically depends upon the thermal stability of the precursor and the stability of its radical cation, as well as source conditions such as temperature, electron impact energy, and electron density [19]. However, many clusters cannot be produced by these sources because most metal, insulator, and semiconductor clusters are not thermally stable and cannot be evaporated as intact gaseous species. There is also a trade-off in focus versus dissociation in electron impact sources. Low pressure electron impact

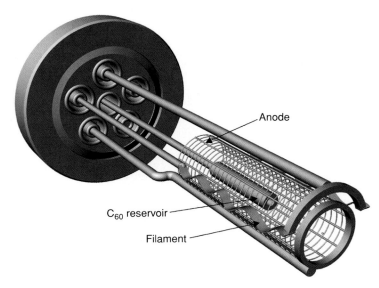

FIGURE 13.2: Schematic of a C_{60} ion source consisting of a C_{60} reservoir encased in anode cage and a filament for electron impact. Illustrated by Gerald L. Gasper, University of Illinois at Chicago.

sources produce more intact parent ions, but their beams are relatively unfocused. Magnetically confined duoplasmatrons induce more dissociation but produce focused beams.

2.3 VAPORIZATION/AGGREGATION SOURCES

Vaporization/aggregation sources involve two steps to cluster formation: volatilization of atoms followed by gas phase condensation into clusters. Haberland and colleagues developed the most popular version of this cluster formation strategy in the magnetron sputtering/aggregation source, then used it to produce thin films via a continuous beam of metal clusters ions and neutrals [1]. This source constituted a major advance when it was first reported because its cluster flux was orders of magnitude larger than other sources available at that time, vastly improving the possibility for depositing clusters for practical applications.

2.3.1 Volatilization

Magnetron sputtering is used for volatilization of a solid target in the typical magnetron sputtering/aggregation source, depicted schematically in Figure 13.3, which was adapted from a commercially available source (Nanogen-50, Mantis Deposition Ltd., Oxfordshire, UK). Figure 13.3 shows the gas inlets for He and Ar, a linear translator for adjusting magnetron path length, the cooling system for the magnetron and condensation chamber, magnetron assembly, nozzle, skimmer, and the expansion zone. Atoms are magnetron sputtered from a solid target into a flow of argon and/or helium as a carrier gas [1; 3; 4]. Conducting targets can be sputtered with DC magnetrons, but insulating targets require radio frequency magnetrons.

Other volatilization methods include thermal evaporation, laser vaporization, and plasma discharges and are used in a wide variety of sources that are otherwise quite similar to the

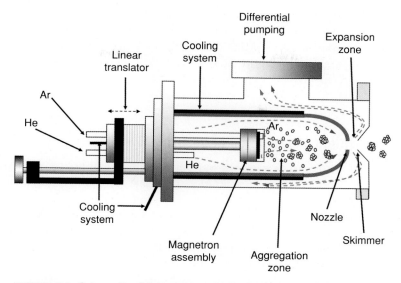

FIGURE 13.3: Schematic of the magnetron sputtering/aggregation cluster beam deposition (CBD) source similar to that developed by Haberland and coworkers [1]. Components include gas inlets for He and Ar carrier gas, magnetron assembly, linear translator, cooling system for magnetron and condensation chamber, nozzle, skimmer, and differential pumping.

Haberland source. Sublimation of a powder within a tube furnace at ∼600°C –700°C at atmospheric pressure has been used to form atoms which are condensed in a nitrogen gas flow to yield clusters [20; 21]. A similar process with addition of oxygen is used for industrial scale synthesis of metal oxide nanoparticles (i.e., see nanophase.com). Laser vaporization sources were originally developed by Smalley and colleagues [2; 4; 22–24], preceding the magnetron sputtering/aggregation source developed by Haberland [1]. These sources focus a pulsed laser beam onto a target rod that is confined to a very small volume through which a pulsed valve releases a burst of carrier gas, as depicted in Figure 13.4 adapted from

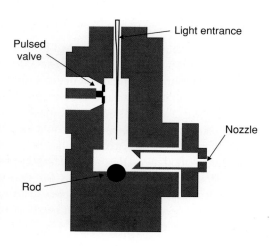

FIGURE 13.4: Schematic of a laser vaporization cluster source containing a pulsed valve, target rod and exit nozzle. Reprinted with permission from P. Milani and W. A. de Heer, *Review of Scientific Instruments* 61, "Improved pulsed laser vaporization source for production of intense beams of neutral and ionized clusters," 1835–1838. Copyright 1990, American Institute of Physics.

Reference [22]. Laser ablation forms a high density atomic vapor of the target material, which is carried by an inert gas to form clusters that are thermalized in the near-sonic flow as they escape the expansion nozzle. High repetition rate pulsed lasers are required to obtain high cluster fluxes from laser vaporization sources. Plasma discharge sources operate in a similar fashion, except that the target is volatilized by a plasma driven by a high voltage pulse [4; 25].

2.3.2 Condensation and reaction of gaseous atoms

Atoms and other small particles formed by sputtering, evaporation, or other volatilization methods enter the condensation chamber and are cooled via collisions with the carrier gas to initiate cluster nucleation and growth. Condensation theory explains cluster formation to a metastable phase based on the free energy of the system [2]. The system reaches a supersaturated state from the decrease in temperature and an equilibrium vapor pressure via collisions with the carrier gas. Three body collisions between gaseous atoms play a role whereby excess energy from dimer formation is transferred to a third atom, allowing further condensation [1]. Additional collisions eventually allow cluster-cluster nucleation to form larger species. Helium or argon carrier gases both cool and transport the clusters to the skimmer. The central portion of the gas mixture undergoes expansion due to a drop in pressure when traveling from the condensation chamber into a second, differentially pumped chamber. Van der Waals cluster sources operate in the supersonic regime [15], but magnetron sputtering/aggregation cluster sources are most often effusive due to their more open nozzle configuration and lower differential pumping [4]. The cluster size range and beam intensity can be controlled at least in part by optimizing the length of the condensation chamber, gas flow, nozzle and/or aperture diameter, and discharge power from the magnetron.

One additional step can be exploited during cluster formation: reaction with an additional feed gas during the volatilization/condensation steps to form oxide, sulfide, nitride, or other compound cluster species. For example, hydrogen sulfide gas can be leaked into the plasma through the Ar-sputtering gas outlet (see Figure 13.3). H_2S will then react with metallic atoms and/or clusters to form metal sulfide clusters [5; 26].

2.4 DIRECT LASER ABLATION

Direct laser ablation is distinguished from laser ablation combined with condensation in that the former sometimes produces clusters directly when a laser ablates a target material. The event directly injects a plume of neutrals and ions into a vacuum chamber for deposition onto a substrate, although variants have achieved small cluster ion formation by laser ablation of a target within a low pressure radio frequency ion trap [27].

Laser ablation of nanoparticle-containing targets can also produce cluster beams [28; 29]. However, this method requires that both sources of stable nanoparticles be produced by another method for casting onto the target by suspension in a solvent, and also the nanoparticles survive the laser ablation process for gaseous transfer to a second target. The relatively infrequent use of this strategy implies that most nanoparticles either do not survive the laser ablation process, or their films can be prepared directly from suspensions in solvents, thereby bypassing the laser ablation step.

2.5 THERMOSPRAY AND ELECTROSPRAY SOURCES

Several groups have reported the use of electrospray and related sources to deposit nanoparticles at atmospheric pressure. A modified electrospray organometallic chemical vapor deposition technique was used to deposit CdSe nanoparticles into a ZnS matrix [30; 31]. Nanoparticles were first synthesized colloidally and dispersed in an acetonitrile:pyridine solution. Next, the solution was injected via a capillary into a chemical vapor deposition reactor. A high voltage was applied to the capillary to charge the injected nanoparticle solution and generate a highly dispersed spray via electrospray. Precursors for the ZnSe matrix were combined with the nanoparticles in the mixing zone of the reactor inlet. It has been demonstrated how the overlayer prevented cluster aggregation and, through surface passivation, prevented a reduction in luminescence efficiency [30].

Semiconductor nanoparticles were produced using a similar method described above, but without the chemical vapor deposition reactor, to produce unsupported and uncoated nanoparticles [32]. This method involved thermospray nebulization of monodispersed droplets of semiconductor salt solutions which form solid nanoparticles upon solvent evaporation. The precursor droplets of the semiconductor salt moved through the stream, were isolated, and eventually, through evaporation, spontaneously became saturated. It was postulated that each semiconductor nanoparticle resulted from a single, isolated droplet of the precursor from which the solvent had completely evaporated.

2.6 LIQUID METAL ION SOURCES

The production of Ga^+ and other metal ions by field emission from liquid metal ion sources (LMIS) is described elsewhere in this book. However, LMIS has also been used to produce <100 nm diameter beams of Au_{2-3}^+, Bi_{2-7}^+, and a few other small metal clusters [33]. LMIS can be defocused further to produce yet higher mass clusters such as Au_{400}^{4+} [34].

3 SELECTION OF NARROW SIZE DISTRIBUTIONS OF CLUSTERS

Size often dictates the properties of nanoparticles, therefore the ability to choose the cluster size prior to deposition is invaluable when fabricating devices or for applications such as catalysis. Many of the sputtered particles in a CBD source are ionized in the plasma before they undergo condensation, producing a beam with a large fraction of singly charged clusters [1; 35]. Size-selection of ionized clusters employs a ion deflecting lens system with an applied potential and a mass filter, analogous to the configuration pictured in Figure 13.1. Time-of-flight, quadrupole mass filters, and Wien velocity filters as well as magnetic sectors have all been used for mass-selection of cluster ions [3; 12; 35–38]. Wien filters have often been used with cluster ion sources despite their low mass resolution because they have much higher transmission compared to other continuous wave devices such as the quadrupole mass filter. One example of cluster size-selection is demonstrated by coupling a CBD source with a time-of-flight (TOF) mass selector [35]. Figure 13.5 shows the mass spectra given when the pulsed gate was (a) open allowing passage of a wide range of Cu_n^+ sizes and (b) closed to allow passage of only a single cluster size, Cu_{60}^+.

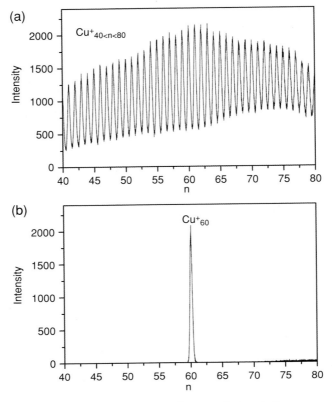

FIGURE 13.5: (a) Mass spectrum of copper clusters in the range of 40 to 80 atoms per cluster. (b) Mass spectrum after selection and production of a pure beam of copper clusters containing 60 atoms. Reprinted with permission from O. Kamalou, J. Rangama, J-M. Ramillon, P. Guinement, and B. A. Huber, "Production of pulsed, mass-selected beams of metal and semiconductor clusters," *Review of Scientific Instruments* 79, 063301. Copyright 2008, American Institute of Physics.

Finally, mass-selection of higher charged cluster ions allows access to higher kinetic energy cluster-surface impacts for a given acceleration voltage. For example, mass-selection of C_{60}^{2+} and C_{60}^{3+} has been used to double and triple the kinetic energy relative to the target of C_{60} ions formed in electron impact sources [16–18].

Size control of neutral nanoparticle beams has also been demonstrated by using aerodynamic lenses of various diameters and capacity [4; 39; 40]. Aerodynamic lens systems use axisymmetric flow contractions in the condensation region to focus the cluster beam before reaching nozzle expansion. Multiple lenses allow the neutral cluster beam to be focused, provided the particle sizes are less than a critical value.

4 ION KINETIC ENERGY EFFECTS

Figure 13.6 illustrates some of the events that occur when a cluster of low charge-to-mass ratio impacts a target surface, including dry etching, surface cleaning and smoothing, shallow

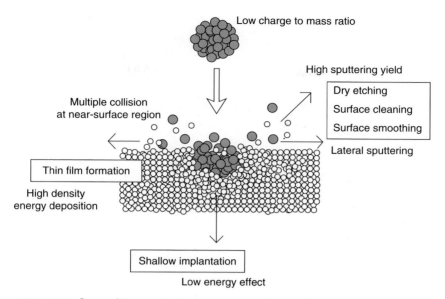

FIGURE 13.6: Some of the events that occur when a cluster of low charge-to-mass ratio impacts a target surface include dry etching, surface cleaning and smoothing, shallow implantation, and thin film formation. Reprinted from *Materials Science and Engineering: R: Reports* 34, I. Yamada, J. Matsuo, N. Toyoda, and A. Kirkpatrick, 231–295. Copyright 2001, with permission from Elsevier.

implantation, and thin film formation [15]. Cluster impact at the surface can induce a variety of phenomenon including high density energy deposition, sputtering of surface atoms, smoothing of surface morphology, and shallow implantation of cluster atoms.

These phenomena depend on the incident kinetic energy of the cluster and its amplitude with respect to binding energies of surface atoms or molecules and between cluster atoms. Controlling cluster kinetic energy allows for deposition of smooth or nanostructured thin films with a wide range of physical and chemical properties, but it is difficult to predict the results of a given cluster-surface collision, which depend upon the specifics of binding energies, surface chemistry, and surface structure.

Table 13.1 summarizes how incident ion energy can control thin film growth using intact clusters by dictating surface roughness, film quality, and other properties [9]. 0.1–1 eV/atom kinetic energy clusters tend to land nearly intact in a random stacking orientation [2]. In the absence of a matrix material, there is often high surface diffusion of clusters on the surface. Small clusters typically have higher surface mobility than larger clusters, so the former tend to coalesce to larger structures upon surface landing, as explained by mean ripening theory [41]. Substrates such as silicon and highly oriented pyrolytic graphite have been found to contribute favorably to cluster diffusion due to their relatively low binding energies [42–44].

Clusters impacting the target surface at 0.1 to 10 eV/atom kinetic energies can initiate thin film growth, but the morphology of the resultant film also depends upon the rate of surface atom diffusion [2] and deposition [9]. Nanostructured films have been produced by CBD with high crystallinity without post-deposition thermal annealing [2]. Deposition of intact clusters is preferred at 0.1 eV/atom kinetic energy, but increasing the cluster kinetic energy to 10 eV/atom leads to more implantation and sputtering. Cluster impact imparts

Table 13.1 Kinetic energy dependence of surface phenomena
resulting from cluster-surface collisions

Cluster Energy	Surface Phenomenon
0.1–1 eV/atom	High surface diffusion
	Poor surface adhesion
	Very porous films
	No cluster-surface atom intermixing
1–10 eV/atom	Better surface adhesion
	Less porous films
	Cluster-surface atom intermixing
> 10 eV/atom	Surface pinning and/or shallow implantation
	Surface sputtering and/or smoothing
	Strong surface adhesion
	Dense films
	Strong cluster-surface atom intermixing

energy closer to the top of a surface compared with atoms impacting at similar energy per atom [9; 14; 15]. Multiple secondary collisions occur between the constituent atoms of both partners in a cluster-surface collision and the results are energy dependent. Sputtering of surface atoms into various angles can facilitate surface smoothing when cluster kinetic energy exceeds 10 eV/atom. Shallow surface implantation of the cluster can also occur by dislodging of the top layer of surface atoms.

Molecular dynamics (MD) calculations are often employed to describe the results of cluster-surface collisions [9]. For example, MD simulations of a single Mo_{1043} cluster impacting a Mo(001) surface with kinetic energies of 0.1, 1, and 10 eV/atom exhibited a soft landing, strong flattening against the surface, and cratering comparable to a meteor impact, respectively [10]. Clusters with 0.1 eV/atom kinetic energy landed intact on the surface, stacking up to form large cavities within the film. Furthermore, these 0.1 eV/atom clusters did not intermix with surface atoms and displayed poor surface adhesion. Clusters with 1 eV/atom kinetic energy showed a more ordered film growth and a denser film compared to lower energy clusters. Cluster-surface atom mixing was predicted for 1 eV/atom energy clusters, which lead to better adhesion and small cavities in the films. Finally, 10 eV/atom clusters formed dense, less porous, strongly adhered, and more ordered films with thorough intermixing between cluster and surface atoms. Furthermore, MD showed that a single cluster impacting at 10 eV/atom never ejected surface atoms, but did induce surface melting and was completely dissociated upon impact.

For thin film formation at increased energies, the deposited cluster ions can diffuse via layer-by-layer growth if it is thermodynamically favored [2]. Smoother films are formed due to sputtering and a complete dissociation of the cluster upon impact [2]. MD simulations showed an ordered growth in some cases when kinetic energy/atom approached cluster-binding energies [2].

MD simulations are quite powerful, but can exclude important interactions that occur experimentally in cluster-surface collisions, especially the near-thermal chemical reactions that occur at the edges of the collision cascade initiated by the cluster impact. Therefore, the influence of ion kinetic energy on film morphology is demonstrated below by comparison of experimental data for low versus high kinetic energy cluster-surface collisions.

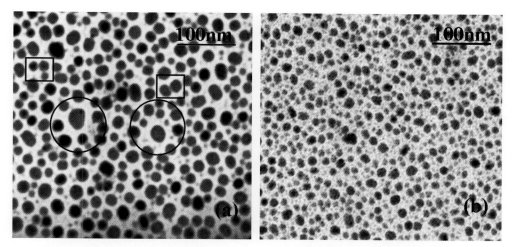

FIGURE 13.7: TEM images of nanoparticle assemblies from Sn cluster beam deposition at normal incident angles. (a) Neutral beam of Sn clusters with a few eV kinetic energy formed nanoparticles with evidence of Ostwald ripening (circles) and annealing (rectangles). (b) 10 keV kinetic energy Sn cluster ion beam formed nanoparticles without agglomeration or annealing. Reprinted from *European Physical Journal D—Atoms, Molecules, Clusters & Optical Physics* 34 (2005): 251–254, "Morphological studies of nanostructures from directed cluster beam deposition," J. B. Chen, J. F. Zhou, A. Häfele, C. R. Yin, W. Kronmüller, M. Han, and H. Haberland, Figure 2. Copyright 2005, with kind permission of Springer Science and Business Media.

One example is the morphology of Sn clusters deposited across a range of kinetic energies onto amorphous carbon films and silicon single crystal surfaces at normal and glancing incidence angles [43]. Figure 13.7 shows TEM images of cluster assemblies from CBD for two different kinetic energies [43]. In Figure 13.7(a), films deposited from neutral Sn clusters with a few eV kinetic energy displayed smooth surfaces with spherical, polydisperse nanoparticles. Evidence for Ostwald ripening was apparent in the growth of large clusters (marked by the circles in Figure 13.7(a)) from smaller clusters by monomeric exchange that resulted from the latter's higher free energy [41; 45]. However, the films deposited from 10 keV Sn cluster ions and shown in Figure 13.7(b) displayed nanoparticles that were pinned to the surface and remained irregularly shaped due to the absence of cluster diffusion. Further experiments performed at glancing incident angles supported these results.

Surface pinning was also observed with $Co_{50\pm5}$ clusters impacting at relatively high kinetic energies on highly oriented pyrolytic graphite (HOPG), although Ostwald ripening was not observed [24; 44]. Figure 13.8 shows STM images of cobalt films on HOPG from incident $Co_{50\pm5}$ clusters before and after annealing. The STM images in Figures 13.8(a), 13.8(c), and 13.8(e) represent the Co films after $Co_{50\pm5}$ cluster impact at energies of 97, 30, and 9 eV/atom, respectively, and images in Figures 13.8(b), 13.8(d), and 13.8(f) correspond to the films after the initial STM study when samples were annealed at ambient atmosphere to smooth the sample surface. The 97 eV/atom impact formed ∼2 nm craters upon landing while the 30 eV/atom impact transformed the initial defect craters to small bumps with a ∼3 nm diameter. In the latter case, the impact energy was high enough to dissociate or implant the clusters but too low to form a crater. Finally, the lowest energy per atom of 9 eV/atom formed surface bumps with ∼1 nm diameter and 0.3 nm height due to intermixing between the atoms of the cobalt cluster atoms and those displaced from the surface upon impact.

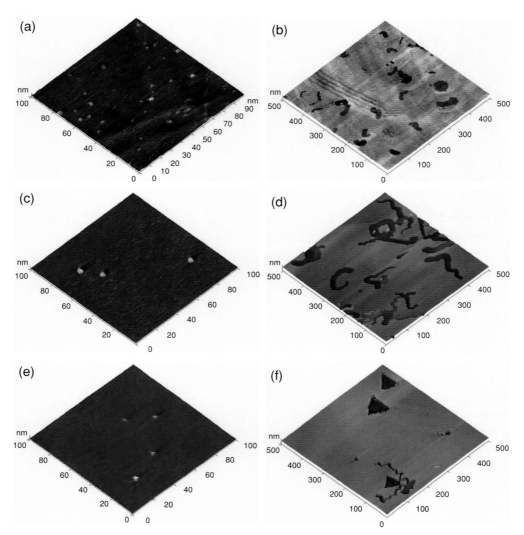

FIGURE 13.8: STM images (a), (c), and (e) of substrates after Co cluster impact at kinetic energies of 97, 30, and 9 eV/atom, respectively. Images (b), (d), and (f) represent the samples produced by kinetic energies of 97, 30, and 9 eV/atom, respectively, after annealing to 600°C for 3 minutes. Reprinted from *The European Physical Journal D - Atomic, Molecular, Optical and Plasma Physics* 52 (2009) 107–110, "Pinning of size-selected Co clusters on highly oriented pyrolytic graphite" S. Vučković, J. Samela, K. Nordlund, and V. N. Popok, Figure 2. Copyright 2009, with kind permission of Springer Science and Business Media.

5 SOURCES OF MATRICES

The environment into which a cluster is deposited has a critical effect upon the nature of the nanoparticles formed and the resultant film properties. Clusters deposited directly onto a substrate can undergo agglomeration or surface pinning, albeit with a dependence upon kinetic energy [2; 10; 24; 43; 44]. In the presence of a matrix, the nanoparticles can avoid agglomeration and surface binding, often achieving a spherical shape [5; 26]. For example, TEM images of

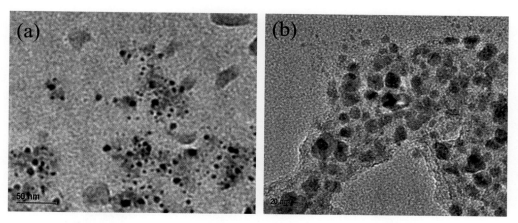

FIGURE 13.9: TEM images of mostly neutral Pb clusters deposited with a few eV kinetic energy in the (a) presence and (b) absence of titanyl phthalocyanine organic matrix.

CBD produced Pb nanoparticles deposited into an organic oligomer matrix in Figure 13.9(a) showed distinct spherical shapes with no agglomeration, in contrast to highly agglomerated Pb nanoparticles without the matrix shown in Figure 13.9(b) [5].

Matrix can be evaporated directly onto the substrate before, after, or simultaneously with CBD. Evaporation of matrix compounds can be achieved with a resistively heated alumina crucible, which thermally evaporates the desired material [46]. Energetic deposition of matrices from hyperthermal atomic ions can also be used to deposit thin films [9]. Target sputtering, plasma polymerization, and laser ablation have all been used for film formation and are well established methods that have been reviewed previously [9]. Ion-assisted deposition with broad beam ion sources [19] provides an additional method of preparing a matrix from evaporated neutrals [9].

Surface polymerization by ion-assisted deposition (SPIAD) of organic films is one variant on ion-assisted deposition that can be used for deposition of organic matrices. Figure 13.10 shows a general representation of the SPIAD method in which a molecular feed gas undergoes electron impact to form 5–200 eV ions which impinge on a substrate simultaneously with thermally evaporated organic molecules. Evaporated molecules are polymerized on the substrate surface by the ion beam to form conducting polymer films [47; 48]. Ion energy and ion-to-neutral ratio control polymerization and film morphology [49; 50]. Polymerization and fragmentation of the incident ions and/or neutral species were also found by combined experiments and MD simulations to be critical steps in the SPIAD process [51].

FIGURE 13.10: Schematic representation of surface polymerization by ion-assisted deposition (SPIAD). The feed gas undergoing electron impact is shown to form 5–200 eV ions, which impinge on a substrate simultaneous with thermally evaporated organic molecules.

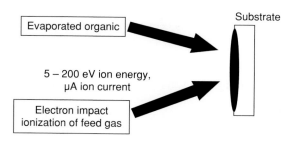

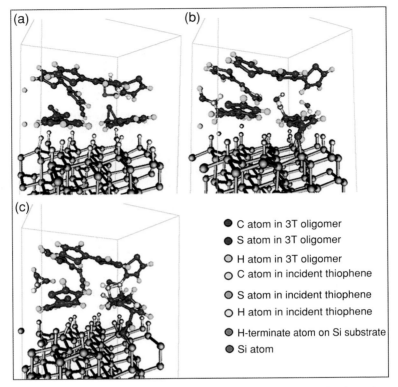

FIGURE 13.11: Snap shots from density functional theory-molecular dynamics (DFT-MD) simulations of thiophene ion ($C_4H_4S^+$) deposited at (a) 100 eV, (b) 200 eV, and (c) 250 eV onto an α-terthiophene (3T) film on a Si substrate. Reprinted with permission from *Journal of Physical Chemistry* C 111, W-D. Hsu, S. Tepavcevic, L. Hanley, and S. Sinnott, "Mechanistic studies of surface polymerization by ion-assisted deposition," 4199–4208. Copyright 2007, American Chemical Society.

Figure 13.11 shows the results of density functional theory (DFT)-MD simulations of SPIAD of thiophene ions deposited at various kinetic energies interacting with α-terthiophene films and Si substrates. Snapshots of the thiophene ion at (a) 100, (b) 200, and (c) 250 eV deposition caused the thiophene ring to dissociate into several fragments and lead to several polymerization events [51]. Other organic films formed via SPIAD have exhibited promising optical characteristics relevant to optoelectronic devices [52; 53].

6 SELECTED APPLICATIONS

6.1 SURFACE SMOOTHING BY ENERGETIC CLUSTER IMPACT

Sputtering with van der Waals cluster ion beams has been shown to display high sputter yields and improved surface smoothing, compared to sputtering with atomic ion beams at similar absolute kinetic energy or energy/atom [15]. Clusters allow simultaneous influx of atoms at the same impact location, producing multiple collisions between the incident and surface atoms,

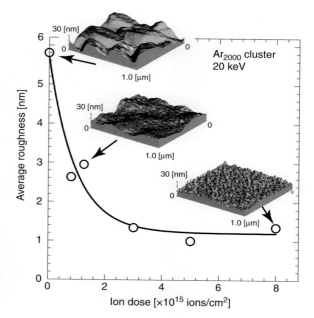

FIGURE 13.12: The bombardment of Cu surfaces with 20 keV Ar cluster ion beams at normal incidence angle showing an average roughness dependent on Ar cluster ion fluence. Reprinted from *Materials Science and Engineering: R: Reports* 34, I. Yamada, J. Matsuo, N. Toyoda, and A. Kirkpatrick, 231–295. Copyright 2001, with permission from Elsevier.

leading to lateral sputtering and redeposition of surface atoms. Van der Waals cluster sputtering at keV kinetic energies and subsequent surface atom redeposition and diffusion can be used to produce extremely smooth surfaces [14; 15; 54]. MD simulations confirmed the surface smoothing process occurred via sputtered atoms migrating to or redepositing into preformed valleys within the film and removing surface protrusions to form a smooth solid surface [54]. Furthermore, there is a dependence on cluster ion fluence in these smoothing events [15; 54]. Figure 13.12 illustrates the bombardment of Cu surfaces with 20 keV Ar cluster ion beams at normal incidence angles and its average roughness dependence on Ar ion fluences [15]. AFM images taken at each ion dose are also shown with 2,000 atoms/cluster and demonstrate that the average roughness decreased proportionally from ~6 to 1.3 nm as the Ar cluster ion fluence increased.

Smoothing events can also be achieved with metal cluster impacts, although the additional possibility exists of atoms from the cluster binding to the surface. Figure 13.13 represents the roughness evolution by AFM of an initially rough Cu film onto which an additional Cu overlayer was deposited from 10 keV Cu clusters with a mean size of 2000 atoms/cluster [55]. Figure 13.13(a) shows the initially rough Cu film (average thickness was ~200 nm) prior to cluster deposition and also shows the roughness after deposition of (b) 15 nm, (c) 22 nm, and 72 nm thick films from Cu clusters. These energetic Cu clusters smoothed the initially rough Cu film surfaces to give roughness values as low as ~0.8 nm, depending on the deposited Cu film thickness [55].

6.2 MASS SPECTROMETRY AND SURFACE ANALYSIS USING KEV CLUSTER ION PROJECTILES

The detection of secondary ions in mass spectrometry (SIMS) that result from atomic ion impact has well-established analytical utility [56]. SIMS with 2–200 keV cluster ion projectiles

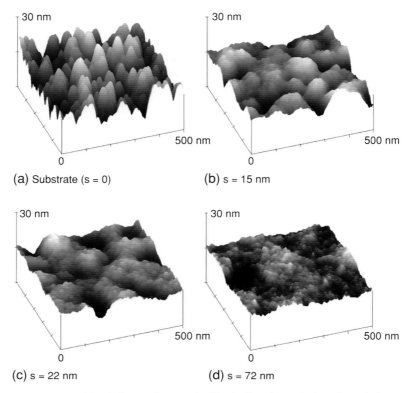

FIGURE 13.13: Atomic force micrographs illustrating the evolution of rough Cu films used as the substrate and made smooth via Cu CBD. (a) The initial film without modification, (b) after 15 nm, (c) 22 nm, and (d) 72 nm deposition of 10 keV Cu clusters. Reprinted with permission from O. Rattunde, M. Moseler, A. Häfele, J. Kraft, D. Rieser, and H. Haberland, "Surface smoothing by energetic cluster impact." *Journal of Applied Physics* 90 (2001): 3226–3231. Copyright 2001, American Institute of Physics.

such as C_{60}, Au_{2-4}, SF_5, and Bi_{2-7} ions displays several advantages compared with SIMS with atomic ion projectiles, as discussed in several reviews [33; 57; 58]. Improvements in focused sources of C_{60}, Au_3, and Bi_{2-7} ions [17; 18; 33] have been particularly useful for these experiments since the primary beam diameter defines the spatial resolution in SIMS instruments that operate in microprobe mode. The three major advantages of cluster SIMS are higher secondary ions yields, higher sputtering efficiencies, and improved depth profiling.

The enhancement of secondary ion yields resulting from sputtering with cluster projectile ions is especially pronounced for the higher mass ions that are often most useful in SIMS analysis of organic, polymeric, and biological materials [17; 33; 57; 59]. For example, up to 10^3 enhancements in yields of the highest mass secondary ions from various polymer targets have been observed for 10 keV C_{60}^+ projectiles, compared with 10 keV Ga^+ projectiles [17; 33]. However, less impressive secondary ion yield enhancements are often observed in cluster SIMS of inorganic materials or small molecules absorbed on metal surfaces [57; 59].

Secondary ion yields combine both the sputtering of species from a solid surface and the ionization of those secondary species, but ionization is notoriously complicated and remains

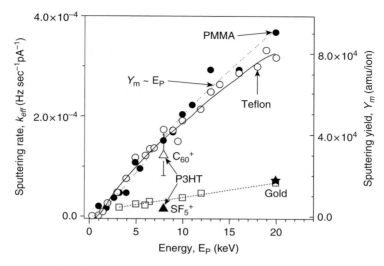

FIGURE 13.14: Sputtering yields and rates for polycrystalline gold, poly(methyl methacrylate) or PMMA, and Teflon as a function of C_{60} ion kinetic energy (Ep). Empty and solid triangles correspond to sputtering yields for poly(3-hexylthiophene) or P3HT by 8 keV C_{60} or SF_5 ions, respectively. Star corresponds to literature value of polycrystalline gold sputtering by 20 keV C_{60} ions [61]. Reprinted with permission from *The Journal of Physical Chemistry* C 111, I. L. Bolotin, S. H. Tetzler, and L. Hanley, "XPS and QCM studies of hydrocarbon and fluorocarbon polymer films bombarded by 1–20 keV C_{60} Ions," 9953–9960. Copyright 2007, American Chemical Society.

resistant to predictive models [57]. The total sputtering yield corresponds to the total removal of material from a surface (both neutral and charged species) resulting from projectile impacts and can be measured accurately by a quartz crystal microbalance [60; 61] or estimated by MD [62] or Monte Carlo simulations [57]. Figure 13.14 shows the sputtering yields for polycrystalline gold and several polymers as a function of C_{60} projectile ion kinetic energy [16].

Teflon and poly(methyl methacrylate) showed similar sputtering yields, which increased essentially linearly with C_{60} ion kinetic energy. A similar linear increase was seen for gold, but the overall sputtering yield of the polymers was roughly five times higher at similar projectile kinetic energy. The sputtering yield of poly(3-hexylthiophene) by 8 keV SF_5^+ was much lower than by 8 keV C_{60}^+, but the latter was similar to that of the other two polymers [63]. Prior work found that 3 keV SF_5^+ was approximately twice as efficient at sputtering poly(methyl methacrylate) than 3 keV Ar^+ [60]. MD simulations have explained the enhancement in sputtering yields for C_{60} versus Ga projectile ions as deriving mostly from collective excitation, the enhanced reflection of C vs. Ga atoms, and other effects [62; 64; 65].

The higher sputtering yields of cluster projectiles combines with the confinement of damage to the uppermost region of a surface to improve the potential for depth profiling with incident clusters compared with atomic projectiles [33]. Another way of phrasing this effect is that cluster projectiles have the ability to remove damaged layers of the substrate at comparable rates as the damage is produced (in at least some materials) [16]. This is an especially significant improvement for the analysis of many organic, polymeric, and biological materials that cannot be depth profiled at all by atomic projectile SIMS. Finally, the combination of focused cluster ion sources with their ability to depth profile have allowed cluster SIMS to be applied to the three dimensional imaging of biological material with depth and spatial resolution approaching 20 and 200 nm, respectively [18; 66; 67].

6.3 FUNDAMENTAL STUDIES OF CATALYSIS

Catalytic materials composed of supported clusters have been developed for use in electric fuel cells, monopropellant thrusters, and gas generators [11; 13]. For example, iridium clusters on an aluminum oxide support make up the standard commercial catalyst (Shell 405) used for gas generators because of their high thermal stability and other factors [12]. The size dependence of catalytic properties of metallic clusters on metal oxide supports is well established, in part by experiments utilizing size-selected deposited clusters performed by Anderson, Heiz, and their respective colleagues [11; 68]. Clusters formed via CBD and mass-selection techniques have displayed the versatility needed to vary metal loading and metal cluster size to investigate their effect upon catalytic activity while displaying strong cluster adhesion to the support.

CO oxidation is one system studied on model catalysts prepared by deposition of size-selected Au_n (n = 1–7) clusters on TiO_2 and Al_2O_3 supports [11; 12]. CO oxidation activity at room temperature was found to be cluster size dependent and related to the Au_n cluster's ability to bind O_2 [12]. Figure 13.15 illustrates the cluster size dependence on CO oxidation activity by displaying the activity at vacancy sites on the TiO_2. The activity scale is in units of $C^{16}O^{18}O$ molecules per Langmuir of incident $C^{16}O$ per deposited Au atom (where gas exposure of 10^{-6} torr sec^{-1} = 1 Langmuir). Data are shown for clean TiO_2 support and Au_n clusters (n = 1–7) deposited onto the TiO_2 support. For the clean TiO_2 support, some CO oxidation was attributed to reactive oxygen, while for deposited Au^+, almost no CO oxidation activity was observed and O_2 interaction was suppressed. However, the catalytic activity increased significantly for Au_6^+ and dramatically for Au_7^+. The Au_2^+ cluster, when bound at vacancies, poisoned CO oxidation activity. Au_3^+ had substantial CO oxidation activity due to high CO_2 levels. The Au_4^+ and Au_5^+ clusters showed lower activity than Au_3^+ but still higher than activity solely from support vacancies. Additionally, support vacancies increased as the cluster size increased.

In addition to having an effect on CO oxidation, size-selected clusters have shown the capacity to control hydrazine decomposition activity. Mass-selected iridium clusters, Ir_n (n = 1, 3, 5, 7, 10 and 15), were deposited onto a planar Al_2O_3 support and catalytic activity was examined at various reaction temperatures [12; 13]. Catalytic activity was confirmed in

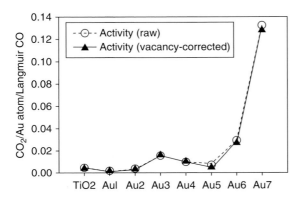

FIGURE 13.15: CO oxidation activity dependence on Au_n cluster size, both raw and corrected data for activity at vacancy sites on the TiO_2. Reprinted with permission from S. Lee, C. Fan, T. Wu, and S. L. Anderson, "Cluster size effects on CO oxidation activity, adsorbate affinity, and temporal behavior of model Au_n/TiO_2 catalysts," *Journal of Chemical Physics* 123 (2005): 124710. Copyright 2005, American Institute of Physics.

all Ir_n clusters, but the extent of decomposition activity was distinct with larger Ir_n clusters being more reactive and displaying a significant increase in activity from Ir_5 to Ir_7. Small Ir_n (n = 1, 3, and 5) clusters demonstrated activity only at temperatures < 200 K, in comparison to the Ir_n (n = 7, 10, or 15), which displayed hydrazine decomposition activity regardless of temperature.

6.4 MAGNETIC THIN FILMS

The ability to manipulate clusters in the gas phase before reaching the substrate allows for fabrication of various types of electronic media based upon the magnetic properties of nanoparticles simultaneously co-deposited with atomic matrices [69; 70]. Even clusters deposited without atomic matrices display novel magnetic properties that are sensitive to cluster size [69]. The deposition of magnetic clusters into a matrix often changes their properties based either on the particle-matrix interaction or in films with high densities of nanoparticles, on the particle-particle interaction. Exploitation of these effects allows CBD to tune the magnetic properties of nanostructured films. The co-deposition technique produces granular matrices that have higher magnetization than homogeneous matrices. Moreover, energetically deposited clusters have also shown stronger magnetization than lower energy (0.1 eV/atom) deposited clusters even though both types were ferromagnetic [71]. Annealing induced by temperature spikes from energetic cluster deposition lead to film recrystallization and higher magnetization [71].

Magnetic films of CBD fabricated nanoparticles have adaptable properties and through variations in nanoparticle size and fractional volume within the film, high performance materials can be produced [70]. For example, magnetic recording tape has been fabricated using a magnetron sputtering/aggregation source by the process shown in Figure 13.16 [72]. Figure 13.16(a) shows the cluster source producing nanoparticles, a heater for recrystallization, and two sputtering guns focused on the target surface. Figure 13.16(b) displays the complete magnetic media fabrication process with a substrate, the magnetic nanoparticles on top of the adhesion layer and the final overcoat layer. CoPt clusters formed by CBD were subsequently heated for in-flight crystallization because gas phase particles are usually polycrystalline or in a magnetically soft-phase [72]. These crystallized clusters were then co-deposited with polymer materials and thereby stabilized on the substrate as spherical CoPt nanoparticles with narrow size distributions that compared favorably to those produced by other state-of-the art manufacturing techniques. Magnetron sputtering/aggregation sources have also been used to produce ferromagnetic chromium oxide films that displayed magnetic hysteresis loops similar to those for bulk CrO_2 [73].

6.5 SENSORS AND ELECTRONIC CIRCUITRY COMPONENTS

Metal-oxide based gas sensors have attracted attention for their potential use in detection of toxic emissions and chemical warfare agents in environmental air quality monitoring [74; 75]. Sensors with nanostructured materials as active layers have been studied with the goal of overcoming the poor chemical selectivity common to metal-oxide gas sensors, such as their frequent inability to distinguish oxidizing and reducing species [76]. The micro-hotplate is a popular micromachined platform for gas sensors that uses metal oxide materials.

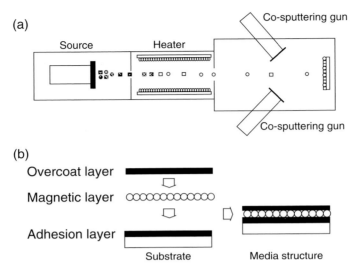

FIGURE 13.16: (a) Schematic representation of the cluster source producing nanoparticles for deposition of magnetic films, a heater for nanoparticle recrystallization, and two sputtering guns focused on the target surface. (b) The magnetic media fabrication process displayed with the substrate, the magnetic nanoparticles on top of the adhesion layer and the final overcoat layer. Reprinted with permission from J.-M. Qiu, Y.-H. Xu, J. H. Judy, and J.-P. Wang, "Nanocluster deposition for high density magnetic recording tape media," *Journal of Applied Physics* 97 (2005): 10P704. Copyright 2005, American Institute of Physics.

First developed by Semancik and coworkers, the micro-hotplate consists of multiple layers of materials of micron thicknesses, which extend upward from a silicon wafer [77].

Nanostructured films of metal oxides such SnO_2, TiO_2, and MoO_3 produced via CBD addressed the problem of poor chemical selectivity in sensors and also improved their sensitivity by reduction in oxide grain size [4]. For example, nanostructured WO_3 was studied as the sensing material due to its deposition rate and electrical properties [76]. Figure 13.17 shows the components involved in fabricating the micro-hotplate: (a) the mask, (b) the auto-aligning concept: mask to micro-hotplate coupling, and (c) deposition of the nanomaterial prepared by CBD. It was found that the device performed well in terms of linearity and sensitivity in detecting ethanol and NO_2.

Nanostructured TiO_2 was also examined for use in micro gas sensors for volatile organic compound detection [78]. The nanostructured TiO_2 was a mixture of anatase and rutile nanoparticles, where the rutile nanoparticles behaved as growth seeds in controlling the rutile/anatase ratio. Individual sensing arrays of nanostructured TiO_2 were then produced and characterized to build a database whose purpose was to correlate chemical selectivity to rutile-to-anatase ratio [78].

Micro gas sensors have been produced using a stencil masking method similar to the scheme presented in Figure 13.17 [79]. Two classes of silicon structures, micro-hotplates and micro-bridges, were first fabricated using a micromachining method. Next, arrays of nanostructured metal oxides, WO_3 and Fe_2O_3, were deposited on these silicon microstructures

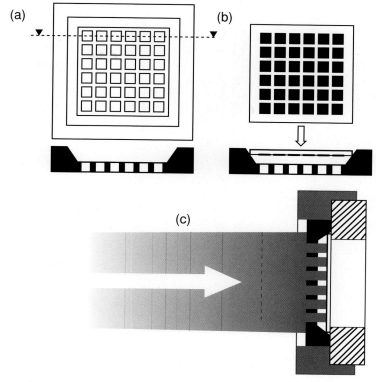

FIGURE 13.17: Micro-hotplate assembly for chemical sensing shown with (a) mask, (b) auto-aligning concept with mask to micro-hotplate coupling, and (c) deposition of the nanomaterial by super CBD. Reprinted from E. Barborini, S. Vinati, M. Leccardi, P. Repetto, G. Bertolini, O. Rorato, L. Lorenzelli, M. Decarli, V. Guarnieri, C. Ducati, and P. Milani, "Batch fabrication of metal oxide sensors on micro-hotplates," *Journal of Micromechanics and Microengineering* 18 (2008): 055015. Copyright 2008 with permission from IOP Publishing.

using a silicon hard mask and a metallic shadow mask for the two sensing layers. Nanostructured Fe_2O_3 on the micro-bridge is shown in Figure 13.18 with (a) the heater and electrode for contact to the sensing layer and (b) a closer magnification of the micro-bridge. The films were annealed and analyzed, yielding ~90% working microstructures. Cluster-assembled films performed favorably under mechanical stresses compared to the overly rigid oxide films usually employed as sensors [79].

A method for producing micron-wide, cluster-assembled wires has been developed on SiN passivated Si substrates [80]. Clusters were deposited by a CBD source through aperture-slots in SiN membranes which are aligned over planar contacts on SiN passivated substrates. Figure 13.19 displays a PMMA coated SiN membrane after cluster beam deposition. Figure 13.19(a) illustrates the types of stencils used in SiN membranes, and Figure 13.19(b) shows the Bi clusters along with PMMA defects and the location where the stencil was located during deposition. Electrical measurement of the final device displayed contact-to-contact current indicative of completion of a conducting cluster-assembled wire.

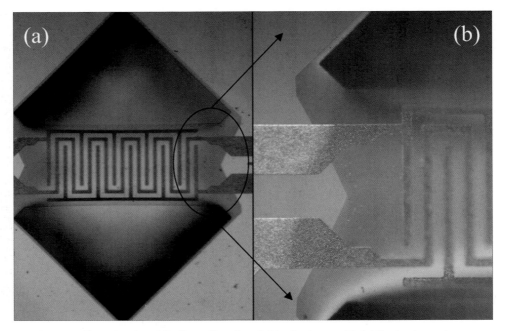

FIGURE 13.18: Nanostructured Fe_2O_3 on the micro-bridge is shown with (a) the heater and electrode for contact to the sensing layer and (b) closer magnification of the bridge. Reprinted from *Microelectronic Engineering* 86, M. Decarli, L. Lorenzelli, V. Guarnieri, E. Barborini, S. Vinati, C. Ducati, and P. Milani, "Integration of a technique for the deposition of nanostructured films with MEMS-based microfabrication technologies: Application to micro gas sensors," 1247–1249. Copyright 2009 with permission from Elsevier.

6.6 OPTOELECTRONICS AND PHOTOVOLTAICS

Nanoparticles embedded in films are candidates for use in many optoelectronic devices, such as photodetectors and photovoltaics, due to their unique linear optical properties [6–8]. The linear optical properties of many types of nanoparticles can be varied with their size, taking advantage of quantum size effects [6; 7].

Nonlinear optical properties of nanoparticles are also of interest for various devices such as optical switches [81]. The nonlinear optical absorption of films of PbS nanoparticles embedded in titanyl phthalocyanine (TiOPc) have been recorded by Z-scan measurements with 5 ns long laser pulses at 532 nm [81]. PbS clusters were fabricated in a reactive magnetron sputtering/aggregation source with a Pb target and a H_2S/Ar carrier gas mixture [5; 26]. These mostly neutral PbS clusters were deposited simultaneously with thermally evaporated TiOPc to form a film containing PbS nanoparticles. Figure 13.20 displays the nonlinear absorption at 532 nm of this film of PbS nanoparticles in the TiOPc matrix, after coating with a protective overlayer of spin-coated polystyrene to reduce laser ablation during Z-scan optical measurement [81]. The normalized transmittance is plotted as a function of incident energy (bottom axis) and fluence (top axis). As laser fluence increased, the transmission through the sample decreased in both samples, but only in the presence of PbS nanoparticles did the system shift to a form of nonlinear activity known as reverse saturable absorption, displaying a lower threshold for the onset of nonlinear optical absorption [81].

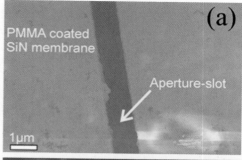

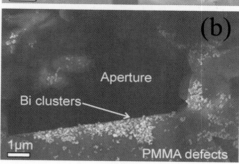

FIGURE 13.19: PMMA coated SiN membrane after cluster beam deposition. (a) Shows the aperture slot and (b) deposited Bi nanoparticles along with PMMA defects. Reprinted from *Microelectronic Engineering* 83, J. G. Partridge, D. M. A. Mackenzie, R. Reichel, and S. A. Brown, "Electrically conducting Bi cluster-assembled wires formed using SiN nanostencils," 1460–1463. Copyright 2006, with permission from Elsevier.

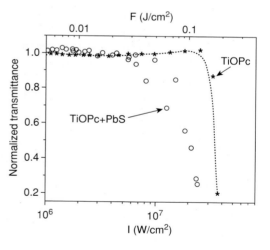

FIGURE 13.20: Nonlinear optical absorption of titanyl phthalocyanine (TiOPc) films with and without PbS nanoparticles (prepared by CBD), recorded by the Z-scan technique [81]. Improved nonlinear optical absorption at 532 nm was observed in the TiOPc films containing PbS nanoparticles.

Surface properties of nanoparticles influence both their linear [82] and nonlinear optical properties [81]. Direct control of nanoparticle surface chemistry has been demonstrated for PbS nanoparticles in the aforementioned magnetron sputtering/aggregation source by adjusting the flow ratio of H_2S reactive gas to Ar sputtering gas [5; 26]. Figure 13.21 shows (a) the hypothesized PbS nanoparticle structure and the surface analysis of the PbS nanoparticles in thermally evaporated TiOPc films using soft X-ray photoelectron spectroscopy (soft-XPS). Figure 13.21(b) and 13.21(c) display the soft-XPS of the $S2p_{1/2}$ and $S2p_{3/2}$ core level components for PbS nanoparticles prepared with H_2S:Ar flow ratios of 0.5:1 and 2:1, respectively (points). The fitted spectral components (lines in Figure 13.21) were assigned as S_{core} from sulfide (S^{-2})

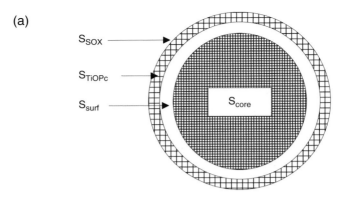

(a)

S_{SOX}
S_{TiOPc}
S_{surf}
S_{core}

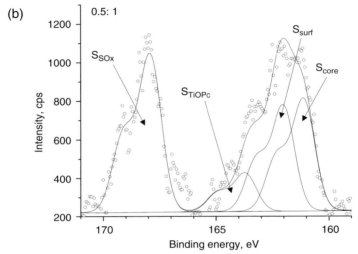

(b) 0.5: 1

S_{SOx}
S_{TiOPc}
S_{surf}
S_{core}

Intensity, cps

Binding energy, eV

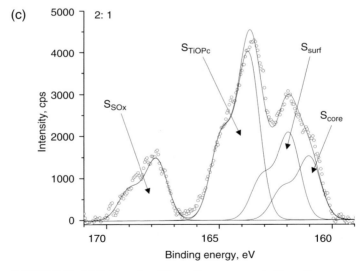

(c) 2: 1

S_{SOx}
S_{TiOPc}
S_{surf}
S_{core}

Intensity, cps

Binding energy, eV

FIGURE 13.21: (a) Schematic of the sulfur components S_{core}, S_{surf}, S_{TiOPc} and S_{ox} of PbS nanoparticles in TiOPc. S2p core levels from soft X-ray photoelectron spectroscopy of PbS nanoparticles in TiOPc prepared with H_2S:Ar ratios of (b) 0.5:1 and (c) 2:1 in the CBD source, demonstrating the source's ability to tune the surface chemistry of nanoparticles.

bound to Pb^{2+} in the nanoparticle core; S_{surf} from sulfur at the nanoparticle surface; S_{TiOPc} from sulfur bound to oxygen at the nanoparticle surface; and S_{SOx} for oxidized surface sulfur of the form SO_3 and/or SO_4 [26]. The large variation in the S_{TiOPc} component indicates that the nanoparticle surface chemistry was controlled by the reactive to sputtering gas flow ratio: S_{TiOPc} increased with the H_2S:Ar ratio, varying the nanoparticle surface from Pb-rich to S-rich.

The ability of CBD to control nanoparticle surface chemistry shows particular promise for depositing photovoltaic and photodetector films. Work with colloidal nanoparticles has shown that film surface chemistry, such as ligand capping or the extent of oxidation, can vary the behavior of those films from photovoltaic to photodetector as well as affect their efficiency in each mode [8; 83]. The minimization of oxidation in most CBD methods is particularly attractive in this regard, as is its ability to produce films of other nanoparticles of interest for photovoltaic applications. The unique optical properties of nanoparticles combine with the ability of CBD to produce intense nanoparticle beams, to control film thickness, and to adapt existing lithography manufacturing processes, showing promise for producing novel and efficient photovoltaic devices.

7 CONCLUSIONS

Cluster beam deposition has shown the capability to produce various types of metal, insulator, and semiconductor nanoparticles. The unique physical, chemical, and electronic properties of deposited nanoparticles make them attractive materials for a variety of applications. CBD has demonstrated that it can play an important role in producing novel nanoparticle-based films that can be competitive with other methods for producing such nanostructured materials.

REFERENCES

[1] Haberland, H., M. Karrais, M. Mall, Y. Thurner. Thin films from energetic cluster impact: A feasibility study. *J. Vac. Sci. Technol.* A 10 (1992): 3266.

[2] Milani, P., and S. Iannotta. *Cluster beam synthesis of nanostructured materials.* Berlin: Springer, 1999.

[3] Pratontep, S., S. J. Carroll, C. Xirouchaki, M. Struen, and R. E. Palmer. Size-selected cluster beam source based on radio frequency magnetron plasma sputtering and gas condensation. *Rev. Sci. Instrum.* 76 (2005): 045103.

[4] Wegner, K., P. Piseri, H.V. Tafreshi, and P. Milani. Cluster beam deposition: A tool for nanoscale science and technology. *J. Phys. D: Appl. Phys.* 39 (2006): R439.

[5] Asunskis, D. J., I. L. Bolotin, A. T. Wroble, A. M. Zachary, and L. Hanley. Lead sulfide nanocrystal-polymer composites for optoelectronic applications. *Macromol. Symp.* 268 (2008): 33.

[6] Burda, C., X. Chen, R. Narayanan, and M. A. El-Sayed. Chemistry and properties of nanocrystals of different shapes. *Chem. Rev.* 105 (2005): 1025.

[7] Rogach, A. L., A. Eychmüller, S. G. Hickey, and S. V. Kershaw. Infrared-emitting colloidal nanocrystals: Synthesis, assembly, spectroscopy, and applications. *Small* 3 (2007): 536.

[8] Sargent, E. H. Solar cells, photodetectors, and optical sources from infrared colloidal quantum dots. *Adv. Mater.* 20 (2008): 3958.

[9] Hanley, L., and S. B. Sinnott. The growth and modification of materials via ion-surface processing. *Surf. Sci.* 500 (2002): 500 and references therein.

[10] Haberland, H., Z. Insepov, and M. Moseler. Molecular-dynamics simulation of thin-film growth by energetic cluster impact. *Phys. Rev.* B 51 (1995): 11061.

[11] Aizawa, M., S. Lee, and S. L. Anderson. Deposition dynamics and chemical properties of size-selected Ir clusters on TiO$_2$. *Surf. Sci.* 542 (2003): 253.

[12] Lee, S., C. Fan, T. Wu, and S. L. Anderson. Cluster size effects on CO oxidation activity, adsorbate affinity, and temporal behavior of model Aun/TiO2 catalysts. *J. Chem. Phys.* 123 (2005): 124710.

[13] Fan, C., T. Wu, W. E. Kaden, and S. L. Anderson. Cluster size effects on hydrazine decomposition on Ir$_n$/Al$_2$O$_3$/NiAl(110). *Surf. Sci.* 600 (2006): 461.

[14] Yamada, I., and N. Toyoda. Current research and development topics on gas cluster ion-beam processes. *J. Vac. Sci. Technol. A* 23 (2005): 1090.

[15] Yamada, I., J. Matsuo, N. Toyoda, and A. Kirkpatrick. Materials processing by gas cluster ion beams. *Mater. Sci. Engin. Rep.* 34 (2001): 231.

[16] Bolotin, I. L., S. H. Tetzler, and L. Hanley. XPS and QCM studies of hydrocarbon and fluorocarbon polymer films bombarded by 1–20 keV C$_{60}$ ions. *J. Phys. Chem. C* 111 (2007): 9953.

[17] Weibel, D., S. Wong, N. Lockyer, P. Blenkinsopp, R. Hill, and J. C. Vickerman. A C$_{60}$ primary ion beam system for time of flight secondary ion mass spectrometry: Its development and secondary ion yield characteristics. *Anal. Chem.* 75 (2003): 1754.

[18] Fletcher, J. S., S. Rabbani, A. Henderson, P. Blenkinsopp, S. P. Thompson, N. P. Lockyer, and J. C. Vickerman. A new dynamic in mass spectral imaging of single biological cells. *Anal. Chem.* 80 (2008): 9058.

[19] Kaufman, H. R., and R. S. Robinson. *Operation of broad-beam sources*. Alexandria, VA: Commonwealth Scientific Corp, 1987.

[20] Kruis, F. E., N. Kornelius, H. Fissan, B. Rellinghaus, and E. F. Wassermann. Preparation of size-classified PbS nanoparticles in the gas phase. *Appl. Phys. Lett.* 73 (1998): 547.

[21] Nanda, K. K., F.E. Kruis, H. Fissan, and M. Acet. Band-gap tuning of PbS nanoparticles by in-flight sintering of size classified aerosols. *J. Appl. Phys.* 91 (2002): 2315.

[22] Milani, P., and W. A. deHeer. Improved pulsed laser vaporization source for production of intense beams of neutral and ionized clusters. *Rev. Sci. Instrum.* 61 (1990): 1835.

[23] Parent, D. C., and S. L. Anderson. Chemistry of metal and semimetal cluster ions. *Chem. Rev.* 92 (1992): 1541.

[24] Vuèkoviæ, S., M. Svanqvist, and V. N. Popok. Laser ablation source for formation and deposition of size-selected metal clusters. *Rev. Sci. Instrum.* 79 (2008): 073303.

[25] Barborini, E., P. Piseri, and P. Milani. A pulsed microplasma source of high intensity supersonic carbon cluster beams. *J. Phys. D: Appl. Phys.* 32 (1999): L105.

[26] Zachary, A. M., I. L. Bolotin, D. J. Asunskis, A. T. Wroble, and L. Hanley. Cluster beam deposition of lead sulfide nanocrystals into organic matrices. *ACS Appl. Mater. Interf.* 1 (2009): 1770.

[27] Hanley, L., and S. L. Anderson. Production and collision induced dissociation of small boron cluster ions. *J. Phys. Chem.* 91 (1987): 5161.

[28] Lill, T., H.-G. Busmann, F. Lacher, and I. V. Hertel. The influence of the surface nature on scattering, fragmentation and deposition processes in C$_{60}^+$ collisions with solid surfaces. *Chem. Phys.* 193 (1995): 199.

[29] Ganeev, R. A., L. B. Elouga Bom, and T. Ozaki. Deposition of nanoparticles during laser ablation of nanoparticle-containing targets. *Appl. Phys. B* 96 (2009): 491.

[30] Danek, M., K. F. Jensen, C. B. Murray, and M. G. Bawendi. Preparation of II-VI quantum dot composites by electrospray organometallic chemical vapor deposition. *J. Cryst. Growth* 145 (1994): 714.

[31] Heine, J. R., J. Rodriguez-Viejo, M. G. Bawendi, and K. F. Jensen. Synthesis of CdSe quantum dot-ZnS matrix thin films via electrospray organometallic chemical vapor deposition. *J. Cryst. Growth* 195 (1998): 564.

[32] Amirav, L., and E. Lifshitz. Thermospray: A method for producing high quality semiconductor nanocrystals. *J. Phys. Chem. C* 112 (2008): 13105.

[33] Wucher, A. Molecular secondary ion formation under cluster bombardment: A fundamental review. *Appl. Surf. Sci.* 252 (2006): 6482.

[34] Tempez, A., J. A. Schultz, S. Della-Negra, J. Depauw, D. Jacquet, A. Novikov, Y. Lebeyec, M. Pautrat, M. Caroff, M. Ugarov, H. Bensaoula, M. Gonin, K. Fuhrer, and A. Woods. Orthogonal time-of-flight

secondary ion mass spectrometric analysis of peptides using large gold clusters as primary ions. *Rap. Comm. Mass Spectrom.* 18 (2004): 371.

[35] Kamalou, O., J. Rangama, J.-M. Ramillon, P. Guinement, and B. A. Huber. Production of pulsed, mass-selected beams of metal and semiconductor clusters. *Rev. Sci. Instrum.* 79 (2008): 063301.

[36] Hanley, L., S. A. Ruatta, and S. L. Anderson. Collision induced dissociation of aluminum cluster ions: Fragmentation patterns, bond energies, and structures for $Al_2^+ - Al_7^+$. *J. Chem. Phys.* 87 (1987): 260.

[37] Goldby, I. M., B. von Issendorff, L. Kuipers, and R. E. Palmer. Gas condensation source for production and deposition of size-selected metal clusters. *Rev. Sci. Instrum.* 68 (1997): 3327.

[38] Banerjee, A. N., and B. Das. Size controlled deposition of Cu and Si nano-clusters by an ultra-high vacuum sputtering gas aggregation technique. *Appl. Phys.* A 90 (2008): 299.

[39] Liu, P., P. J. Ziemann, D. B. Kittelson, and P. H. McMurry. Generating particle beams of controlled dimensions and divergence: I. Theory of particle motion in aerodynamic lenses and nozzle expansions. *Aero. Sci. Technol.* 22 (1995): 293.

[40] Liu, P., P. J. Ziemann, D. B. Kittelson, and P. H. McMurry. Generating particle beams of controlled dimensions and divergence: II. Experimental evaluation of particle motion in aerodynamic lenses and nozzle expansions. *Aero. Sci. Technol.* 22 (1995): 314.

[41] Carlow, G. R. Ostwald ripening on surfaces when mass conservation is violated: Spatial cluster patterns. *Physica* A 239 (1997): 65.

[42] Palmer, R. E., S. Pratontep, and H.-G. Boyen. Nanostructured surfaces from size-selected clusters. *Nat. Mater.* 2 (2003): 443.

[43] Chen, J. B., J. F. Zhou, A. Häfele, C. R. Yin, W. Kronmüller, M. Han, and H. Haberland. Morphological studies of nanostructures from directed cluster beam deposition. *Eur. Phys. J.* D 34 (2005): 251.

[44] Vuèkoviæ, S., J. Samela, K. Nordlund, and V. N. Popok. Pinning of size-selected Co clusters on highly oriented pyrolytic graphite. *Eur. Phys. J.* D 52 (2009): 107.

[45] Zinke-Allmang, M., L. C. Feldman, and M. H. Grabow. Clustering on surfaces. *Surf. Sci. Rep.* 16 (1992): 377.

[46] Yates, J. T., Jr. *Experimental innovations in surface science: A guide to practical laboratory methods and instruments.* New York: Springer-Verlag, 1998, p. 198.

[47] Choi, Y., S. Tepavcevic, Z. Xu, and L. Hanley. Optical and chemical properties of polythiophene films produced via surface polymerization by ion assisted deposition. *Chem. Mater.* 16 (2004): 1924.

[48] Choi, Y., A. Zachary, S. Tepavcevic, C. Wu, and L. Hanley. Polyatomicity and kinetic energy effects on surface polymerization by ion-assisted deposition. *Inter. J. Mass Spectrom.* 241 (2005): 139.

[49] Tepavcevic, S., Y. Choi, and L. Hanley. Growth of novel polythiophene and polyphenyl films via surface polymerization by ion-assisted deposition. *Langmuir* 20 (2004): 8754.

[50] Tepavcevic, S., A. M. Zachary, A. T. Wroble, Y. Choi, and L. Hanley. Morphology of polythiophene and polyphenyl films produced via surface polymerization by ion-assisted deposition. *J. Phys. Chem.* A 110 (2006): 1618.

[51] Hsu, W.-D., S. Tepavcevic, L. Hanley, and S. B. Sinnott. Mechanistic studies of surface polymerization by ion-assisted deposition. *J. Phys. Chem.* C 111 (2007): 4199.

[52] Wroble, A. T., J. Wildeman, D. J. Asunskis, and L. Hanley. Growth of phenylene vinylene thin films via surface polymerization by ion-assisted deposition. *Thin Solid Films* 516 (2008): 7386.

[53] Zachary, A. M., Y. Choi, M. Drabik, I. L. Bolotin, H. Biederman, and L. Hanley. Dimerization of titanyl phthalocyanine in thin films prepared by surface polymerization by ion-assisted deposition. *J. Vac. Sci. Technol.* A 26 (2008): 212.

[54] Chen, H., S. W. Liu, X. M. Wang, M. N. Iliev, C. L. Chen, X. K. Yu, J. R. Liu, K. Ma, and W. K. Chu. Smoothing of ZnO films by gas cluster ion beam. *Nucl. Instr. Meth. Phys. Res.* B 241 (2005): 630.

[55] Rattunde, O., M. Moseler, A. Häfele, J. Kraft, D. Rieser, and H. Haberland. Surface smoothing by energetic cluster impact. *J. Appl. Phys.* 90 (2001): 3226.

[56] A. Benninghoven, F. G. Rudenauer, H. W. Werner. *Secondary ion mass spectrometry: Basic concepts, instrumental aspects, applications, and trends.* New York: John Wiley & Sons, 1987.

[57] Hanley, L., O. Kornienko, E. T. Ada, E. Fuoco, and J. L. Trevor. Surface mass spectrometry of molecular species. *J. Mass Spectrom.* 34 (1999): 705 and references therein.

[58] Winograd, N. The magic of cluster SIMS. *Anal. Chem.* 77 (2005): 143A.

[59] de Mondt, R., L. Van Vaeck, A. Heile, H. F. Arlinghaus, N. Nieuwjaer, A. Delacorte, P. Bertrand, J. Lenaerts, and F. Vangaever. Ion yield improvement for static secondary ion mass spectrometry by use of polyatomic primary ions. *Rap. Comm. Mass Spectrom.* 22 (2008): 1481.

[60] Fuoco, E., G. Gillen, M. B. J. Wijesundara, W. E. Wallace, and L. Hanley. Surface analysis studies of yield enhancements in secondary ion mass spectrometry by polyatomic projectiles. *J. Phys. Chem. B* 105 (2001): 3950.

[61] Wucher, A., S. Sun, C. Szakal, and N. Winograd. Molecular depth profiling of histamine in ice using a buckminsterfullerene probe. *Anal. Chem.* 76 (2004): 7234.

[62] Postawa, Z., B. Czerwinski, N. Winograd, and B. J. Garrison. Microscopic insights into the sputtering of thin organic films on Ag{111} induced by C_{60} and Ga bombardment. *J. Phys. Chem. B* 109 (2005): 11973.

[63] Tetzler, S. Characterization of poly-3(hexylthiophene) films bombarded by keV C_{60} and SF_5 ions. M.Sc. thesis, University of Illinois at Chicago, 2007.

[64] Ryan, K. E., and B. J. Garrison. Energy deposition control during cluster bombardment: A molecular dynamics view. *Anal. Chem.* 80 (2008): 5302.

[65] Garrison, B. J., Z. Postawa, K. E. Ryan, J. C. Vickerman, R. P. Webb, and N. Winograd. Internal energy of molecules ejected due to energetic C_{60} bombardment. *Anal. Chem.* 81 (2009): 2260.

[66] McDonnell, L. A., and R. M. A. Heeren. Imaging mass spectrometry. *Mass Spectrom. Rev.* 26 (2007): 606.

[67] Walker, A. V. Why is SIMS underused in chemical and biological analysis? Challenges and opportunities. *Anal. Chem.* 80 (2008): 8865.

[68] Heiz, U., and E. L. Bullock. Fundamental aspects of catalysis on supported metal clusters. *J. Mater. Chem.* 14 (2004): 564.

[69] Bansmann, J., S. H. Baker, S. H. Baker, C. Binns, J. A. Blackman, J. P. Bucher, J. Dorantes-Dávila, V. Dupuis, L. Favre, D. Kechrakos, A. Kleibert, K. H. Meiwes-Broer, G. M. Pastor, A. Perez, O. Toulemonde, K. N. Trohidou, J. Tuaillon, and Y. Xie. Magnetic and structural properties of isolated and assembled clusters. *Surf. Sci. Rep.* 56 (2005): 189.

[70] Binns, C., K.N. Trohidou, J. Bansmann, S. H. Baker, J. A. Blackman, J.-P. Bucher, D. Kechrakos, A. Kleibert, S. Louch, K.-H. Meiwes-Broer, G. M. Pastor, A. Perez, and Y. Xie. The behaviour of nanostructured magnetic materials produced by depositing gas-phase nanoparticles. *J. Phys. D: Appl. Phys.* 38 (2005): R357.

[71] Zhao, S. F., M. L. Yao, J. G. Wan, Y. W. Mu, J. F. Zhou, and G. H. Wang. The microstructure and magnetic behavior of Co nanostructured film prepared by energetic cluster beam deposition. *Eur. Phys. J. D.* 52 (2009): 163.

[72] Qiu, J.-M., Y.-H. Xu, J. H. Judy, J.-P. Wang. Nanocluster deposition for high density magnetic recording tape media. *J. Appl. Phys.* 97 (2005): 10P704.

[73] Chen, J.-H., S.-C. Lo, C.-G. Chao, and T.-F. Liu. A study on fabrication, morphological and optical properties of lead sulfide nanocrystals. *J. Nanosci. Nanotech.* 8 (2008): 967.

[74] Kohl, D. Function and applications of gas sensors. *J. Phys. D: Appl. Phys.* 34 (2001): R125.

[75] Yamazoe, N. Towards innovation of gas sensor technology. *Sensors Actuat. B* 108 (2005): 2.

[76] Barborini, E., S. Vinati, M. Leccardi, P. Repetto, G. Bertolini, O. Rorato, L. Lorenzelli, M. Decarli, V. Guarnieri, C. Ducati, and P. Milani. Batch fabrication of metal oxide sensors on micro-hotplates. *J. Micromech. Microeng.* 18 (2008): 055015.

[77] Benkstein, K. D., C. J. Martinez, L. Guofeng, D. C. Meier, C. B. Montgomery, and S. Semancik. Intergration of nanostructured materials with MEMS microhotplate platforms to enhance chemical sensor performance. *J. Nanopart. Res.* 8 (2006): 809.

[78] Mazza, T., E. Barborini, I. N. Kholmanov, P. Piseri, G. Bongiorno, S. Vinati, P. Milani, C. Ducati, D. Cattaneo, A. L. Bassi, C. E. Bottani, A. M. Taurino, and P. Siciliano. Libraries of cluster-assembled titania films for chemical sensing. *Appl. Phys. Lett.* 87 (2005): 103108.

[79] Decarli, M., L. Lorenzelli, V. Guarnieri, E. Barborini, S. Vinati, C. Ducati, and P. Milani. Integration of a technique for the deposition of nanostructured films with MEMS-based microfabrication technologies: Application to micro gas sensors. *Microelec. Engin.* 86 (2009): 1247–1249.

[80] Partridge, J. G., D. M. A. Mackenzie, R. Reichel, and S. A. Brown. Electrically conducting Bi cluster-assembled wires formed using SiN nanostencils. *Microelec. Engin.* 83 (2006): 1460.

[81] Asunskis, D. J., I. L. Bolotin, and L. Hanley. Nonlinear optical properties of PbS nanocrystals grown in polymer solutions. *J. Phys. Chem.* C 112 (2008): 9555.

[82] Guyot-Sionnest, P., B. Wehrenberg, and D. Yu. Intraband relaxation in CdSe nanocrystals and the strong influence of surface ligands. *J. Chem. Phys.* 123 (2005): 074709.

[83] Nozik, A. J. Multiple exciton generation in semiconductor quantum dots. *Chem. Phys. Lett.* 457 (2008): 3.

CHAPTER 14

ELECTRON- AND ION-ASSISTED METAL DEPOSITION FOR THE FABRICATION OF NANODEVICES BASED ON INDIVIDUAL NANOWIRES

Francisco Hernández-Ramírez, Román Jiménez-Díaz, Juan Daniel Prades, and Albert Romano-Rodríguez

1 INTRODUCTION

The advent of first integrated circuits (ICs) in the second half of the last century paved the way to the later miniaturization of electronic systems, which has become the pillar of present high-tech industries. To date, cell phones, laptops, and many other electronic systems have permeated our modern life, reaching a market size of billions of euros [1; 2]. In parallel, the beginning of the nanoscience and nanotechnology era, which is defined by a unit of length, the nanometer (1 nm $= 10^{-9}$ m), and encompasses several disciplines of traditional science and engineering, aims to attain materials and devices with new physical, chemical, and electrical properties derived from the phenomena at this small scale [3–5].

The fundamental properties of one-dimensional (1D) materials have been intensively studied for the last decade, leading to design of innovative and reliable devices for manifold applications, such as gas sensors, solar cells, batteries, lasers, etc. [6–8]. Nevertheless, the transfer of all this knowledge to an industrial level compatible with IC technologies at reasonable manufacturing cost entails well-controlled assembly and contact procedures of 1D nanomaterials (i.e., nanowires and nanotubes) [9].

This chapter deals with the issue of obtaining reliable electrical contacts to individual nanomaterials, exploring the use of focused ion beam (FIB)-based lithography techniques to accomplish such an ambitious goal. In spite of the fact that these nanofabrication protocols are only suitable for research prototyping, since they are not scalable, and therefore do not fulfill the requirements to take a leap in industry, they have played an important role in the preliminary electrical characterization of different metallic and semiconductor nanowires and, as a consequence, in demonstrating the performance of some advanced proof-of-concept devices based on them [9–11]. Here, a nanofabrication process that combines focused electron and focused ion beam induced deposition (from now on, FEBID and FIBID, respectively [12]) of platinum (Pt) merits special attention. The advantages and disadvantages of using the above-mentioned methodology to carry out the fabrication and subsequent characterization of nanowires-based devices are discussed; and the most significant scientific and technical results achieved so far are presented, emphasizing the applicability of nanowires for gas detection, and energy harvesting, as well as optoelectronics.

2 FOCUSED ION BEAM AS A NANOFABRICATION TOOL FOR DEVICES BASED ON INDIVIDUAL NANOWIRES

Novel electronic and optical properties of nanomaterials are strongly influenced by their dimensions, making them a natural bridge between single molecules and bulk materials [10; 13]. In this context, experimental studies on individual nanowires are an excellent tool to elucidate the advantages and limitations of individual nanostructures as building blocks of complex circuit architectures [9].

FIB nanolithography has been traditionally used to electrically connect nanometer-sized materials [14–22]. Nevertheless, damage and sample amorphization induced by ion bombardment (typically, Ga^+ ions accelerated to 30 kV), necessary to decompose the organometallic precursor for fabricating the electrical contacts [23], has restrained their use [10]. As a result, fabrication alternatives were preferred to accomplish electrical measurements on nanomaterials free of this undesired effect, such as e-beam, shadow-mask, and photolithography techniques [24].

In the last decade, the development of the "dual-beam systems," also known as "cross-beam systems" (conventional FIB which integrates an electron column), has boosted a renewed interest in FIB nanolithography; since they give the chance to the *in situ* capture of electron images and to perform simultaneous dissociation of organometallic compound precursors with secondary electrons (SE) generated by the incident electron beam, giving rise to well-controlled FEBID contacts [10]. Given the fact that the interaction between electrons and the sample is less destructive than the alternative based only on the use of ions [10], performing FEBID-based contacts on the nanostructure for the electrical contacts and the later fabrication of the rest of the contacts with ions (FIBID process) was demonstrated to be an extremely useful methodology to avoid undesired modifications of nanowires [10; 25; 26] (see Figure 14.1).

However, and in spite of this intrinsic advantage, the combination of FEBID and FIBID remains sparsely exploited, because of some problems and weaknesses, which will be discussed in detail below. Thus, most of the devices based on individual nanowires fabricated with

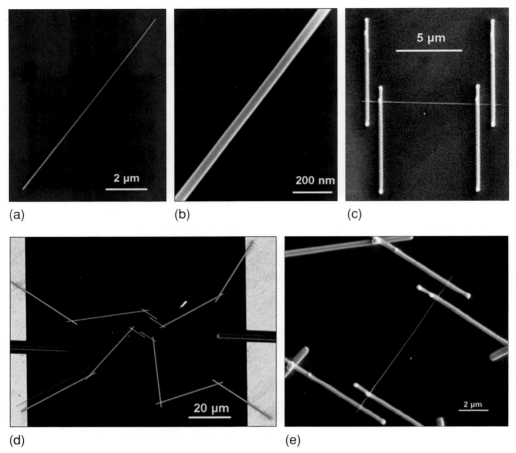

FIGURE 14.1: SnO_2 nanowire contacted using FEBID and FIBID nanolithography techniques. Dimensions: L = 11 μm (length) and D = 55 ± 5 nm (diameter). (a) & (b) Detail of the nanowire in a high-resolution SEM image revealing the uniformity of the diameter. (c) Image of the four FEBID platinum depositions fabricated in the proximity of the nanowire to minimize damage and contamination produced by ion bombardment. (d) General view of the final device. Au/Ti/Ni pre-patterned microelectrodes can be seen at both the left and right side of the image. (e) High resolution image of the final device. Reprinted with permission from Reference [10] F. Hernández-Ramírez, et al., *Nanotechnol.* 16 (2006): 5577. Copyright IOP, Institute of Physics (2006).

FIB-related nanolithography techniques are still obtained following the standard FIBID process [14–22], since ohmic and low resistance contacts are usually formed. The origin of this favorable behavior is related to the amorphization and localized doping of the nanowires-contact area by impinging Ga^+ ions, as was demonstrated by Fischer et al. [27].

Though these devices have enabled us to acquire a good comprehension of the fundamental physical principles of nanowires, there are some nanomaterials whose electrical response needs to be monitored without external modifications (i.e., defects originated by ions); otherwise the estimation of their key parameters is strongly biased. On this matter, the case of metal oxides is paradigmatic, since their response toward gases and photons is directly related to the state of the surface, which is strongly influenced by the ion beam [28]. For this reason, the

combination of FEBID and FIBID nanolithography is considered the best alternative to work
with this sort of material.

3 FEBID AND FIBID NANOLITHOGRAPHY, A METHOD OF CONTACTING INDIVIDUAL NANOWIRES

FEBID and FIBID involve a specific chemical vapor deposition process that is assisted
by electron and ion beams, respectively [12]. Roughly speaking, precursor gas molecules
(containing the material to be deposited) flow from an injector toward the substrate and become
adsorbed to it (see Figure 14.2). Then, the precursor gas molecules are decomposed by the
electron or ion beams, and the material is locally deposited, due to different physical/chemical
processes [12].

The typical precursor used to obtain Pt nanoelectrodes is trimetyl-methylcyclopentadienyl-
platinum $[(CH_3)_3CH_3C_5H_4Pt]$ [10; 11]. Both FEBID and FIBID can be described as carbon
matrixes (content of carbon higher than 60%) doped by platinum nanoparticles of size smaller
than 5 nm. It is noteworthy that the total content of platinum can be even lower than 30%,
depending on the deposition conditions. This particular morphology is behind the resistivity
values of FEBID and FIBID, which can be several orders of magnitude higher than bulk
platinum [25] (see Figure 14.3).

The main difference between FEBID and FIBID lies in the presence of gallium in the latter
(content higher than 5%) introduced by the ion beam during the nanofabrication process,
which reduces the electrical resistivity between 100 and 1,000 times. Moreover, the better
efficiency of the precursor decomposition induced by impinging ions and secondary electrons
generated during the process translates into a higher deposition rate of FIBID [25].

For the reasons described thus far, the combination of FEBID and FIBID to electrically
contact nanowires is slower than following the standard FIBID process; final devices exhibit

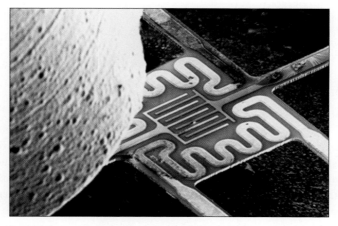

FIGURE 14.2: SEM image of a suspended microhotplate onto
which the FIB nanofabrication process is done. The microneedle
used as injector of the organometallic precursor inside the FIB
chamber is shown at the left part of the image.

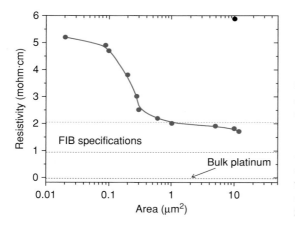

FIGURE 14.3: Electrical resistivity of Pt-FIBID as function of the cross-section area; experimental values demonstrate that Pt-FIBID is not as good conductor as bulk platinum. It is noteworthy that FIB depositions exhibit increasing resistivity with smaller areas, reaching values worse than those determined by FIB specifications. Here, data provided by FEI Company at Strata DB 235 (Product Datasheet) are shown.

higher contact resistances and typically non-ohmic responses; and it is necessary to take a great deal of care to prevent imaging the nanowires with the ion beam during the nanofabrication process in order to avoid damaging the sample (see Figure 14.1) [9].

In spite of the above-mentioned issues, the combination of FEBID and FIBID has been successfully used to study many nanomaterials in the past. Gopal et al. took the first step with individual carbon nanotubes (CNTs) [29]; and later, Valizadeh et al. [26] demonstrated that the electrical resistivity of a single metallic nanowire can be easily estimated using devices obtained by means of this technique. Nevertheless, the real application field has concentrated on the characterization of semiconductor nanowires and nanoparticles, mainly metal oxides [9]. Hernández-Ramírez et al. showed that functional devices based on SnO_2, ZnO, NiO, WO_3, and In_2O_3 are obtained in a well-controlled and reproducible way following this nanofabrication strategy [9; 10; 11; 24] (see Figure 14.4); and recent studies revealed that attaining prototypes with performances as good as the monitored with nanodevices fabricated with more conventional technologies (i.e., e-beam lithography) is feasible [30; 31]. Thus, it can be asserted that the combination of FEBID and FIBID to contact nanomaterials has already become a mature nanofabrication lithography.

4 ELECTRICAL CHARACTERIZATION OF NANODEVICES FABRICATED WITH FEBID AND FIBID NANOLITHOGRAPHY

The use of nanomaterials as building blocks of new circuit architectures requires both advanced nanofabrication techniques to assemble them in complex systems, and a wide comprehension and control of their electrical properties. This second condition can be only fulfilled from the accurate characterization of simple prototypes based on these nanomaterials. As a result, the electrical parameters of individual nanowires contacted with the combination of FEBID and FIBID nanolithography need to be accurately evaluated and modeled in order to gain a good comprehension of the electrical phenomena at the nanoscale.

Performing electrical measurements on these tiny devices is not a straightforward process, since a large number of technological issues arise. Although some of them are common to

(a)

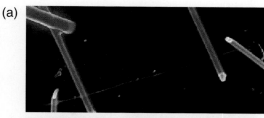

(b)

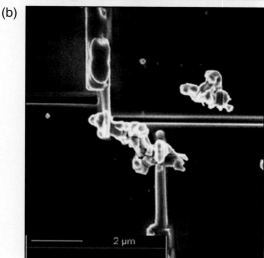

FIGURE 14.4: Functional devices based on different metal oxides nanomaterials have been fabricated so far, following the combination of FEBID and FIBID nanolithography. The use of this technique is not limited to nanowires and nanotubes (a), but can be applied also to agglomerates of nanoparticles (b). Here, a SnO_2 nanowire (a), and an agglomerate of WO_3 nanoparticles are shown. Reprinted with permission from Reference [28] F. Hernández-Ramírez, et al., *Sensors and Actuators B: Chemical*, 118 (2006) 198–203. Copyright Elsevier (2006).

their micro counterparts, the reduced dimensions of nanowires bring about new ones [9]. For this reason, it becomes crucial to gather enough information about them to develop new and better devices, but first and foremost to prevent damaging the nanowires during their life span [32].

In general, metallic and polymeric nanowires exhibit low contact resistances and excellent ohmic responses when they are electrically contacted with FEBID and FIBID nanolithography [26] (see Figure 14.5). Thus, their characterization does not entail significant problems and simple two-probing DC measurements are usually enough to estimate the electrical resistivity. On the contrary, the case of semiconductor nanowires is far more complicated, since the fabrication of electrical nanocontacts with high stability, low contact resistance (R_C), and ohmic behavior remains as an open issue [32]. Rectifying contacts are usually formed between the semiconductor nanowire and typical metal contacts [32], whereby it becomes essential that their contribution to the overall response is correctly modeled to avoid wrong estimations of their key electrical parameters due to contact resistance underestimations [10].

For instance, rectifying I-V curves are usually found in two-probing DC measurements when Pt-SnO_2 contacts are present at one of these devices (see Figure 14.6). This is simply explained by the formation of Schottky barrier of height $\Phi_B = 0.75 \pm 0.10$ eV in the absence of interface states at the metal-semiconductor junction, owing to the differences between the work function of Pt ($\Phi_B = 5.65$ eV) and the electron affinity of SnO_2 ($\chi_{SnO2} = 4.9 \pm 0.1$ eV) [32]. Thus, SnO_2 and many other semiconductor nanowires measured in two-probe DC measurements can be generally described as a back-to-back Schottky circuit (see Figure 14.7).

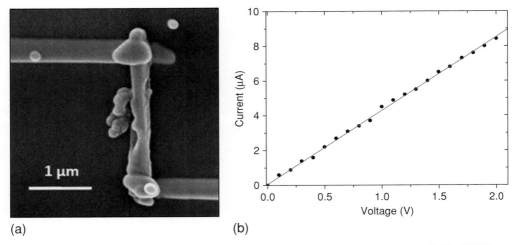

(a)   (b)

FIGURE 14.5: (a) (Inset) SEM image detail of an iron-doped polymer nanowire contacted with FEBID and FIBID nanolithography. (b) Metal and metal-doped nanowires exhibit perfect ohmic and low contact resistance responses in two-probing measurements.

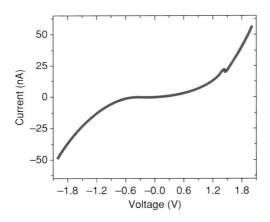

FIGURE 14.6: Room temperature two-probing I-V curve of the SnO_2 nanowire shown in Figure 14.1. A symmetric rectifying response is clearly observed.

According to this assumption, the applied bias V distributes as:

$$V = V_C + V_{SD} + V_{NW} + V_{SI}, \tag{1}$$

where V_C is the voltage along the parasitic elements of the systems, such as cables, electrodes; V_{SD} is the voltage drop in the direct biased metal-nanowire junction; V_{NW} is the voltage drop along the nanowire, and V_{SI} is the voltage drop in the reverse biased junction.

Electrical conduction through this type of device is determined by reverse-biased junctions, which is the major contribution to the total contact resistance. On the other hand, forward-biased metal-nanowire interfaces usually exhibit good electrical conduction and low contact resistance [32]. As a result, for two-probe DC measurements, the device resistance R can

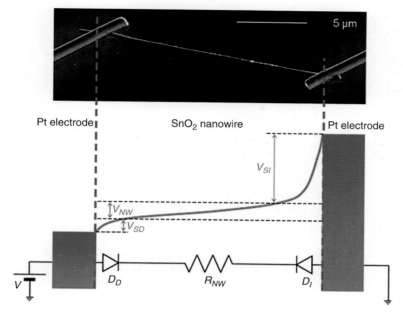

FIGURE 14.7: Band diagram of the equivalent back-to-back Schottky circuit, which corresponds to a typical semiconductor nanowire measured in two-probing dc measurements. Voltage drops at each element of the circuit are indicated. Here, an SnO_2 nanowire is shown as example.

be described as the sum of the reverse-biased contact resistance R_{SI} and the nanowire resistance R_{NW}:

$$R = R_{SI} + R_{NW}. \tag{2}$$

In spite of the typically non-ohmic characteristics of semiconductor nanodevices, their overall resistance R can be locally estimated as follows:

$$R = \left(\frac{\partial I}{\partial V}\right)_{V_0}^{-1}, \tag{3}$$

where I is the probing current, V the applied bias, and V_0 the voltage at a fixed point (see Figure 14.8).

Moreover, voltage drop at the reverse-biased junction V_{SI} can be determined with the help of the expression:

$$V_{SI} = V\left(1 - \frac{R_{NW}}{R}\right) = V\left(1 - \frac{R_{NW}}{R_{SI} + R_{NW}}\right). \tag{4}$$

Therefore, the modeling of Schottky junctions at metal-nanowire interfaces demands the estimation of R_{NW}, otherwise erroneous conclusions are found, especially if R_{NW} and R_{SI} are comparable. For this reason, four-probe DC measurements become extremely useful to carry out the electrical characterization of nanodevices due to the fact that the contribution of

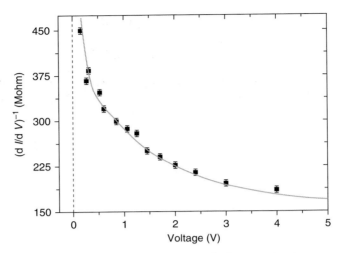

FIGURE 14.8: Two-probe resistance of the SnO_2 nanowire shown in Figure 14.1 as a function of the applied voltage at $T = 298$ K. Reduction of R occurs when electrons overcome the reverse-biased metal-semiconductor interface more easily because of the decreased barrier height caused by increasing voltage.

contact resistance is minimized. Nevertheless, they are far more complex than standard two-probe DC measurements, because an extra pair of nanocontacts needs to be fabricated onto a single nanowire [10].

The combination of two- and four-probing DC measurements in a unique nanowire paves the way to determine the I–V_{SI} characteristics (Equation 4) and thus to model the conduction mechanisms through the metal-semiconductor interface [32]. In a rough approximation, the current flow in Schottky junctions is originated by the transport of charge carriers from the semiconductor to the metal by means of different mechanisms, such as thermionic emission over the barrier, tunneling assisted by interface states located at the metal-semiconductor interface, and pure tunneling through the barrier [33].

At room temperature, I–V_{SI} curves of FIB-fabricated nanodevices fit the theoretical response predicted by a combination of thermionic emission and tunneling assisted by interface states. According to this model, the barrier height Φ_{BE} diminishes, and then the current I rises with increasing bias as:

$$I = AA^{**}T^2 \exp\left(\frac{-q\phi_{BE}}{k_B T}\right), \tag{5}$$

where

$$\phi_{BE} = \phi_{B0} - \sqrt{\frac{qE}{4\pi\varepsilon_S}} - \frac{q}{\varepsilon_S}\sqrt{\frac{N_S d}{4\pi}} \tag{6}$$

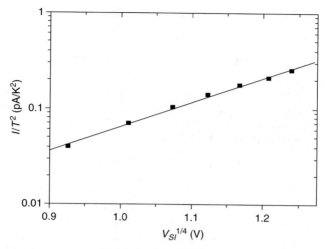

FIGURE 14.9: $\log(I)$ vs $(V_{SI})^{1/4}$ plot of the nanowire shown in Figure 14.1 at $T = 298$ K. The linear behavior demonstrates that experimental data fit with the proposed conduction model. Nanowire free carrier concentration ($n_d \approx 10^{18}$ cm^{-3}) was estimated from the slope. Reprinted with permission from Reference [32] F. Hernandez-Ramirez, et al., *Phys. Rev.* B 76 (2007): 085429. Copyright American Physical Society (2007).

and

$$E = \sqrt{\frac{2qn_d}{\varepsilon_S}\left(V_{SI} + \phi_{bi} - \frac{k_B T}{q}\right)}. \tag{7}$$

Here, A is the contact area, A^{**} is the effective Richardson constant, Φ_{BE} is the effective barrier height, Φ_{B0} is the ideal barrier height in the absence of image force, E is the maximum electric field at the junction, ε_S and n_d are the dielectric constant and free electron concentration of SnO$_2$, Φ_{bi} is the built-in potential, N_S is the interface states concentration, and d is the depth of these defects.

According to Equations 5, 6, and 7, the relationship given by the expression $\ln(I)$ vs $(V_{SI})^{1/4}$ should be linear at the temperature range in which the proposed conduction mechanism is applicable.

Figure 14.9 shows the validity of this model for an SnO$_2$ nanowire; similar results were found for ZnO nanowires as well [34]. It should be pointed out that the height of Φ_{BE} can be roughly estimated by plotting the dependence of R_{SI} at a fixed voltage V_{SI} as a function of the effective temperature T. These two magnitudes are linked by the following activation energy law

$$R_{SI} \propto \exp\left(\frac{q\phi_{BE}}{k_B T}\right) \tag{8}$$

In the particular case of clean Pt-SnO$_2$ junctions, effective barrier heights (Φ_{BE}) below 0.4 V are found [32], which is significantly lower than the expected theoretical value

($\Phi_{B0} = 0.75 \pm 0.10$ eV). This suggests that tunneling assisted by interface states is not a negligible term in the conduction of reverse-biased junctions. In this scenario, the necessity of considering both tunneling and thermoionic mechanisms is attributed to the existence of defects in the contact areas induced during the nanofabrication process. Furthermore, the presence of intrinsic defects in the nanowire originated at the synthesis stage is also plausible. This last result demonstrates the importance of optimizing the nanofabrication technique outlined in the previous section of this chapter, so that uncontrolled modifications of the nanowire are minimized as much as possible.

If the concentration of interface states at the metal-semiconductor junction rises, tunneling current becomes more important, and as a result it superimposes on the thermoionic emission current. Consequently, effective Schottky barrier (Φ_{BE}) tends to disappear, leading to linear I–V curves at room temperature. In some cases, these nanodevices even exhibit non-symmetric I–V curves due to the different modulation of the barrier heights at both extremes of the nanowire because of the discrepancy between the concentration of interface states at each nanocontact [32]. For this reason, well-controlled synthesis and nanofabrication methodologies are required in order to circumvent this complicated scenario [9].

Finally, it should be pointed out that the electrical characterization of individual semiconductor nanowires, combining in the same device the above-mentioned analysis of the rectifying contacts and data obtained with four-probe DC measurements, allows a reliable estimation of their free electron concentration (n_d) and electron mobility (μ), whose experimental values conform to those obtained with more complex approaches [35], such as field effect transistor (FET) experiments. This demonstrates that nanowires electrically contacted with the combination of FIBID and FEBID make up an excellent experimental platform to study the intrinsic electrical phenomena at the nanoscale.

5 DEVICE FABRICATION WITH NANOWIRES CONTACTED WITH FEBID AND FIBID NANOLITHOGRAPHY

Our interest in nanomaterials lies in the excellent and sometimes unique properties derived from their size and well-defined structure, which can be exploited to fabricate better devices than using standard macro and micro technologies [9]. To date, most of the prototypes based on nanomaterials obtained with FEBID and FIBID nanolithography represent gas sensors and photodetectors, since a large family of metal oxide nanowires with excellent and well-controlled properties have been synthesized so far [9; 10]. It is well-known that metal oxides provide an excellent platform to evaluate the electrical response of semiconductor materials toward different external stimuli, such as gas species and photons [11; 31].

In particular, SnO_2 has been traditionally used to design and fabricate conductometric gas sensors, whose detection principle is the variation of the electrical resistance after exposure to gases at temperatures higher than 100°C [24; 36]. Our search of alternatives which contribute to the miniaturization process of present technologies is increasing interest in the development of nanowire-based gas detectors; Hernández-Ramírez et al. demonstrated that reproducible and long-term stable gas sensors could be fabricated with individual nanowires electrically contacted onto micro-hotplates [11] (see Figure 14.10).

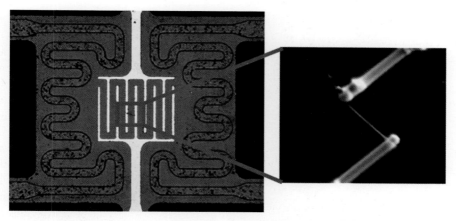

FIGURE 14.10: (Left) Optical image of a micro-hotplate with interdigitated platinum electrodes in the middle. After the FIB nanolithography, a SnO$_2$ nanowire was electrically contacted (right).

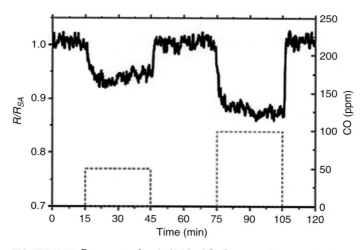

FIGURE 14.11: Response of an individual SnO$_2$ nanowire contacted onto the microhotplate shown in Figure 14.10 toward different carbon monoxide concentrations (50 and 100 ppm). Here, the dissipated power at the heater was selected to set the effective temperature at $T = 120°$C. The dashed lines indicate the CO pulses. Reprinted with permission from Ref. [11] F. Hernandez-Ramirez, et al., *Nanotechnology* 18 (2007): 495501. Copyright IOP (2007).

The use of the integrated heater allowed a fast and precise modulation of the effective temperature, and as a consequence, setting to values appropriate for gas sensing applications (see Figure 14.11). This was one of the first steps toward the development of portable microsystems with integrated nanowires in them. Later, Prades et al. showed that gas responses, nearly identical to those obtained with an external heater, were found if only self-heating effect

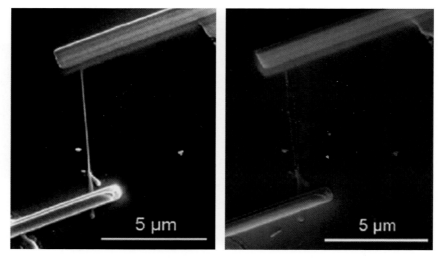

FIGURE 14.12: SnO$_2$ nanowire before (left) and after (right) the electrical characterization. The nanowire is completely destroyed due to self-heating effects. Reprinted with permission from Reference [31] J. D. Prades, et al., *J. Phys. Chem.* C, 112 (2008): 14639. Copyright ACS (2008).

was used to increase the temperature of the device in a controlled way [30]. Dissipated power in nanowires (radii below 50 nm) often causes important self-heating effects, and as a result, aging and failure of the devices (see Figure 14.12, [32]).

This result was considered a major breakthrough in the development of gas sensors based on individual nanowires, since the feasibility of taking advantage of self-heating opened the door to develop ultralow power consumption integrated devices. For instance, SnO$_2$ nanowires operated under optimal conditions for nitrogen dioxide (NO$_2$) sensing required less than 20 μW to both bias and heat them, which is significantly lower than 140 mW required using external microheaters [30]. According to this, commercial energy harvesting modules would provide enough power to operate self-heated gas nanosensors, overcoming the need of external energy sources, such as batteries, which today restrains the use of metal oxide gas sensors in portable and autonomous applications [9].

The second main field of application of electrically contacted metal oxide nanowires is the development of photodetectors. These materials are gaining an increasing interest because of their potential applications in optoelectronics [9]. Among the large family of metal oxides, SnO$_2$ and especially ZnO have been widely studied.

The operating principle of these devices is strongly influenced by their geometry as well as the chemisorbed species at their surface, which play a central role in regulating the final photosensitivity. In a rough approximation, after exposure to ultraviolet light, photogenerated holes migrate to the metal oxide surface and discharge the adsorbed molecules through surface electron-hole recombination. At the same time, photogenerated electrons significantly increase the conductivity of the nanowires, which is easily monitored using the electrical characterization described before (see Figure 14.13). Here, it should be highlighted that Prades et al. modeled the response of FIB-fabricated photodetectors based on individual ZnO nanowires, and identified the best design strategies to optimize their response towards impinging photons [31; 34].

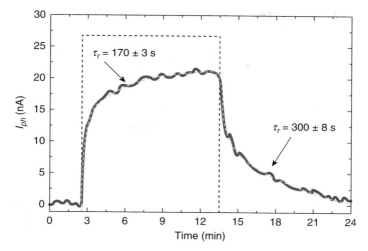

FIGURE 14.13: Dynamic behavior of the photoresponse I_{ph} measured with an individual ZnO nanowire when a UV pulse is applied (dashed line). Here, a photon flux of $\Phi = 3.3 \times 10^{18}$ ph m^{-2}s^{-1}, and wavelength $\lambda = 340 \pm 10$ nm was used. The response fits with theoretically deduced exponential laws, whose characteristic time constants are clearly indicated. Reprinted with permission from Reference [31] J. D. Prades, et al., *J. Phys. Chem.* C, 112 (2008): 14639. Copyright ACS (2008).

6 CONCLUSIONS

The development of low-cost complex circuit architectures based on the use of nanomaterials as building blocks, particularly semiconductor nanowires and nanotubes, entails the development of advanced nanofabrication methodologies as well as precise characterization techniques. At present, most of the nanofabrication routes are only suitable for research purposes since a limited number of prototypes can be fabricated at each time. For this reason, the use of nanowires remains restrained at the academic level, a long way off from industry. Nevertheless, and in spite of this preliminary stage of development, current prototypes represent a giant leap for the study of nanomaterials' properties.

Among available nanofabrication techniques, the use of FIB lithography is considered one of the best alternatives for production of devices based on individual nanomaterials because of the high precision it provides. Besides, the combination of FIBID and FEBID, which is technically viable in dual-beam systems, reduces damage and amorphization of samples by impinging ions, typical of the standard FIBID process. For this reason, it is preferred to contact nanomaterials whose electrical and physical properties are highly dependent on the state of their surfaces.

To date, FIBID and FEBID lithography has been extensively used to study metal oxide nanowires, demonstrating that resulting devices make up an excellent platform to analyze the phenomena at the nanoscale. Moreover, it has been found that most of these devices exhibit excellent functionalities as gas sensors and photodetectors, which could be used in the mid-term future in manifold safety, security, and optoelectronics applications. Here, it should be highlighted that the main advantage that nanowire technology provides over its standard counterparts is the significant reduction of power consumption. For this reason, its integration in commercial technologies would mean a step forward in our race to increase the efficiency of our everyday systems and gadgets.

Finally, it should be pointed out that FIBID and FEBID nanolithography protocols evaluated so far are only suitable for research prototyping, requiring the development of advanced and affordable techniques (i.e., self assembly approaches such as dielectrophoresis and ink-jet printing) to solve the problem of scalability of nanowire-based devices and thus to edit complex circuit architectures. Nevertheless, FIB lithography has played a key role in our present comprehension of the phenomena that determine the physical and electrical laws at this tiny level.

ACKNOWLEDGMENTS

First of all, we would like to thank our families. Without their full support we would not have been able to accomplish this work.

The collaboration with colleagues who have actively participated in the execution of the results presented in this work is appreciated. A considerable part of this scientific research has been possible thanks to collaborations with present and past colleagues at the Department of Electronics of Universitat de Barcelona and with researchers formerly at the INM of the Leibnitz-Institute for New Materials in Saarbrücken (Germany).

Different funding agencies have supported parts of this work: the Spanish Ministries of Education and Science and of Science and Innovation (projects CROMINA (TEC2004–06854-C03–01), ISIS (TEC2007–67962-C04–04), FPU fellowships (F.H.-R., J.D.P. and R.J.-D.) and mobility grants (A.R-R.) and European Union (projects NANOS4, NMP-001528, and Transnational Access to Major Research Infrastructures, HPRI-CT-2002–00192).

Timely discussions with T. Westmalle and W. Hoegaarden have been especially valuable.

REFERENCES

[1] Zheng, C. *Micro-Nanofabrication: technologies and applications.* Beijing: Higher Education Press, distributed by Springer, 2005.
[2] *International Technology Roadmap for Semiconductors.* 2007 edition (http://public.itrs.net).
[3] Di Ventra, M., S. Evoy, and J. R. Heflin. *Introduction to nanoscale science and technology.* New York: Springer, 2004.
[4] Jortner, J., and C. N. R. Rao. Special topic issue on the theme of nanostructured advanced materials. *Pure Appl. Chem.* 74 (2002): 1489–1783.
[5] Schmidt, G. *Nanoparticles: From theory to applications.* New York: Springer, 2004.
[6] Law, M., J. Goldberger, and P. Yang. Semiconductor nanotubes and nanowires. *Annu. Rev. Mater. Res.* 34 (2004): 83.
[7] Kolmakov, A., and M. Moskovits. Chemical sensing and catalysis by one-dimensional metal-oxide nanostructures. *Annu. Rev. Mater. Res.* 34 (2004): 151.
[8] Comini, E., C. Baratto, G. Faglia, M. Ferroni, A. Vomiero, and G. Sberveglieri. Quasi-one dimensional metal oxide semiconductors: preparation, characterization and application as chemical sensors. *Prog. Mater. Sci.* 54 (2009): 1.
[9] Hernandez-Ramirez, F., J. D. Prades, R. Jimenez-Diaz, T. Fischer, A. Romano-Rodriguez, S. Mathur, and J. R. Morante. On the role of individual metal oxide nanowires in the scaling down of chemical sensors. *Phys. Chem. Chem. Phys.* 11 (2009): 7105.

[10] Hernandez-Ramirez, F., J. Rodriguez, O. Casals, A. Tarancón, A. Romano-Rodriguez, J. R. Morante, S. Barth, S. Mathur, T. Y. Choi, D. Poulikakos, V. Callegari, and P. M. Nellen. Fabrication and electrical characterization of circuits based on individual tin oxide nanowires. *Nanotechnology* 17 (2006): 5577.

[11] Hernandez-Ramirez, F., J. D. Prades, A. Tarancón, S. Barth, O. Casals, R. Jimenez-Diaz, E. Pellicer, J. Rodriguez, M. A. Juli, A. Romano-Rodriguez, J. R. Morante, S. Mathur, A. Helwig, J. Spannhake, and G. Mueller, "Portable microsensors based on individual SnO_2 nanowires," *Nanotechnology* 18 (2007): 495501.

[12] De Teresa, J. M., R. Córdoba, A. Fernández-Pacheco, O. Montero, P. Strichovanec, and M. R. Ibarra. Origin of the difference in the resistivity of as-grown focused-ion-beam and focused-electron-beam-induced Pt nanodeposits. *J. Nanomaterials* (2009): 936863.

[13] Jortner, J., and C. N. R. Rao. Nanostructured advanced materials: Perspectives and directions. *Pure Appl. Chem.* 74 (2002): 1489.

[14] Ebbesen, T. W., H. J. Lezec, H. Hiura, J. W. Bennett, H. Gaemi, and H. Tio. Electrical conductivity of individual carbon nanotubes. *Nature* 382 (1996): 54.

[15] Cronin, S. B., Y.-M. Lin, O. Rabin, M. R. Black, J. Y. Ying, M. S. Dresselhaus, P. L. Gai, J.-P. Minet, and J.-P. Issi. Making electrical contacts to nanowires with a thick oxide coating. *Nanotechnology* 13 (2002): 653.

[16] Vajtai, R. Building carbon nanotubes and their smart architectures. *Smart Mater. Struct.* 11 (2002): 691.

[17] Ziroff, J., G. Angello, J. Rullan, and K.Dovidenko. Localized fabrication of metal contacts and electrical measurementson multiwalled carbon nanotubes. *Mater. Res. Soc. Symp. Proc.* 772 (2002): M.8.8.1.

[18] Xiao, K. High-mobility thin-film transistors based on aligned carbon nanotubes. *Appl. Phys. Lett.* 83 (2003): 150.

[19] De Marzi, G., D. Iacopino, A. J. Quinn, and G. Redmond. Probing intrinsic transport properties of single metal nanowires: Direct-write contact formation using a focused ion beam. *J. Appl. Phys.* 96 (2004): 3458.

[20] Nam, C. Y., D. Tham, and J. E. Fischer. Disorder effects in focused-ion-beam-deposited Pt contacts on GaN nanowire. *Nano Lett.* 5 (2005): 2029.

[21] *Strata DB235.* Product data sheet, FEI Company.

[22] Rotkina, L., J.-F. Lin, and J. P. Bird. Nonlinear current-voltage characteristics of Pt nanowire transistors fabricated by electron-beam deposition. *Appl. Phys. Lett.* 83 (2003): 4426.

[23] Romano-Rodriguez, A., and F. Hernandez-Ramirez. Dual-beam focused ion beam (FIB): A prototyping tool for micro and nanofabrication. *Microelectronic Engineering* 84 (2007): 789.

[24] Hernandez-Ramirez, F., J. D. Prades, R. Jimenez-Diaz, O. Casals, A. Cirera, A. Romano-Rodriguez, J. R. Morante, S. Barth, and S. Mathur. Fabrication of electrical contacts on individual metal oxide nanowires and novel device architectures. In *Nanotechnology: Nanofabrication, Patterning and Self Assembly.* New York: Nova Science Publisher, 2009.

[25] Vilà, A., F. Hernandez-Ramirez, J. Rodriguez, O. Casals, A. Romano-Rodriguez, J. R. Morante, and M. Abid. Fabrication of metallics contacts to nanometer-sized materials using a Focused-Ion-Beam (FIB). *Materials Science and Engineering C* 26 (2006): 1063.

[26] Valizadeh, S., M. Abid, F. Hernandez-Ramirez, A. Romano-Rodriguez, K. Hjort, and J. A. Schweitz. Template synthesis and forming electrical contacts to single Au nanowires by focused ion beam techniques. *Nanotechnology* 17 (2006): 1134.

[27] Tham, D., C.-Y. Nam, and J. E. Fischer. Microstructure and Composition of Focused-Ion-Beam-Deposited Pt Contactsto GaN Nanowires. *Adv. Mater.* 18 (2006): 290.

[28] Hernandez-Ramirez, F., J. Rodriguez, O. Casals, E. Russinyol, A. Vilà, A. Romano-Rodriguez, J. R. Morante, and M. Abid. Characterization of metal-oxide nanosensors fabricated with Focused Ion Beam (FIB). *Sensors & Actuators B: Chemical* 118 (2006): 198.

[29] Gopal, V., V. R. Radmilovic, C. Daraio, S. Jin, P. Yang, and E. Stach. Rapid prototyping of site-specific nanocontacts by electron and ion beam assisted direct-write nanolithography. *Nano Lett.* 4 (2004): 2059.

[30] Prades, J. D., R. Jimenez-Diaz, F. Hernandez-Ramirez, S. Barth, A. Cirera, A. Romano-Rodriguez, S. Mathur, and J. R. Morante. Ultralow power consumption gas sensors based on self-heated individual nanowires. *Appl. Phys. Lett.* 93 (2008): 123110.

[31] Prades, J. D., R. Jimenez-Diaz, F. Hernandez-Ramirez, L. Fernandez-Romero, T. Andreu, A. Cirera, A. Romano-Rodriguez, A. Cornet, J. R. Morante, S. Barth, and S. Mathur. Towards a systematic understanding of photodetectors based on individual nanowires. *J. Phys. Chem. C* 112 (2008): 14639.

[32] Hernandez-Ramirez, F., A. Tarancón, O. Casals, E. Pellicer, J. Rodriguez, A. Romano-Rodriguez, J. R. Morante, S. Barth, and S. Mathur. Electrical properties of individual tin oxide nanowires contacted to platinum electrodes. *Phys. Rev. B* 76 (2007): 085429.

[33] Sharma, B. L. *Metal semiconductors Schottky barrier junctions and their applications.* New York: Plenum, 1984.

[34] Prades, J. D., F. Hernandez-Ramirez, R. Jimenez-Diaz, M. Manzanares, T. Andreu, A. Cirera, A. Romano-Rodriguez, and J. R. Morante. The effects of electron-hole separation on the photoconductivity of individual metal oxide nanowires. *Nanotechnology* 19 (2008): 465501.

[35] For a detailed description of this methodology, see Reference 32.

[36] Hernandez-Ramirez, F., A. Tarancón, O. Casals, J. Arbiol, A. Romano-Rodriguez, and J. R. Morante. High response and stability in CO and humidity measures using a single SnO_2 nanowire. *Sensors and Actuators B: Chemical* 121 (2007): 3.

CHAPTER 15

FOCUSED ION BEAM FABRICATION OF CARBON NANOTUBE AND ZnO NANODEVICES

Guangyu Chai, Oleg Lupan, and Lee Chow

1 INTRODUCTION

In the last 15 years, the focused ion beam (FIB) instrument has gained widespread recognition as an important tool for nanoscience and nanotechnology research. Modern dual beam FIB instruments typically combine the focused ion beam column with a scanning electron microscope column and gas delivery nozzles, making it one of the most versatile tools in a material research laboratory for (a) material removal (sputtering), (b) material deposition, and (c) imaging. This chapter is mostly concerned with using the single beam FIB instrument for fabrication of carbon nanotube and ZnO nanodevices and nanosensors. It will start with a brief introduction of the basic properties of carbon nanotube and ZnO nanostructured materials, and then the challenges and difficulties of FIB fabrication will be outlined and the techniques used to overcome these challenges will be discussed. In the following section, the procedures to fabricate carbon nanotube (CNT) nanodevices using the FIB instrument without damaging the CNT will be described. A new configuration of CNT that was used to fabricate CNT devices in an FIB instrument is introduced. In Section 3, the fabrication of ZnO nanostructure-based devices using FIB is described. In Section 4, results are summarized.

Carbon nanotubes have been probably the most studied material of the last two decades. They consist of graphene layers wrapped around themselves to form a tubular structure. This tubular structure can be either open-ended or close-ended. Multiwall carbon nanotubes

(MWCNT) consist of several to several dozen layers of graphene sheets with the spacing between each layer to be 0.34 nm [1]. This can be compared with the spacing of graphene sheets of graphite, which is about 0.335 nm [2]. The single wall carbon nanotube (SWCNT) consists of only one single layer of graphene with a diameter around 1.0–1.5 nm, and a typical aspect ratio of $10 \sim 1000$, while MWCNT can have a diameter of a few nanometers to ~ 100 nm.

The intense interests in CNT are due to the fact that CNT, particularly SWCNT, is a fantastic material that possesses many excellent material properties. The SWCNT is the hardest material known to man. It has a Young's modulus of 1.25 TPa [3]. Carbon nanotube is also one of the best thermal conductors at room temperature. The thermal conductivity of a MWCNT has been measured [4] recently to be at least 3,000 W/mK, which is comparable to that of diamond. The thermal conductivity of individual SWCNTs has been measured in a similar range [5]; however, one theoretical calculation [6] indicated that it could be around 6,600 W/mK. The electrical transport in CNT is also very interesting. Conducting SWCNT behaves almost like a perfect 1-D conductor showing single electron charging, resonance tunneling, and proximity induced superconductivity at low temperature [7]. For example, a single SWCNT with a low resistance contact can carry a current density as high as 10^9 A/cm^2 [8]. This is three orders of magnitude higher than the critical current density of most superconductors.

On the other hand, ZnO has been used commercially for more than 100 years [9], including applications in medicine, paints, adhesives, foods, and batteries [10–12]. However, current research on ZnO as a wide band gap semiconductor is focused on new applications in nano- and microelectronics, sensors, biomedicine, and optoelectronic devices [11; 13–14]. ZnO is a unique material that exhibits a combination of optical, piezoelectric, semiconducting, and magnetic properties [15–17] that can be employed for different types of applications. ZnO has a high exciton binding energy of 60 meV [18], a room-temperature direct bandgap of $\Delta E_g = 3.37$ eV [19], and it is transparent in the visible region [15; 20]. ZnO is chemically more stable and capable of operation at higher temperatures than Si and Ge [21]. It possesses high optical gain (300 cm^{-1}) [19], a high melting point (2248 K), high mechanical and thermal stability [22], and it is radiation resilient [23]. It has been shown that the room-temperature bombardment by electrons [24], protons [25], and heavy ions [26] caused much less damages in ZnO than in other semiconductors. Recently, it has been demonstrated that ZnO nanorods/nanowires show enhanced radiation hardness versus bulk ZnO layers [27; 28]. This is a very important point here, since in the ion beam processing of nanostructured ZnO material, accidental exposure of ZnO nanostructures to the Ga ion beam is inevitable.

2 FOCUSED ION BEAM FABRICATION OF CARBON NANOTUBE NANODEVICES

In the following, we will confine this review mostly to experimental works applying FIB technique to making CNT nanodevices. Using the ability of FIB to pattern nanostructures was one of the early applications of FIB in CNT research. Several groups [29–31] have demonstrated patterned growth of carbon nanotubes. Javoenapibal et al. [31] were able to use FIB patterning to grow suspended SWCNT across a trench with both ends clamped down. The suspended SWCNTs were used in a nanoelectro-mechanical system (NEMS) to measure the Young's modulus of SWCNT.

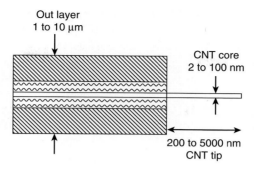

FIGURE 15.1: A schematic diagram of a fiber coated carbon nanotube (F-CNT).

Since the major effect of exposure to ion beam on carbon nanotube is the creation of defects and damages, a group in Japan [32; 33] took advantage of these particular characteristics and used the localized defects created in carbon nanotubes to form quantum dot devices. Kurokawa et al. [32] in 2004 developed an FIB-based technique to fabricate a MWCNT quantum dot device. The FIB was used both to deposit W contacts on the MWCNT and also to create two trenches with a separation of 200 nm. This isolated middle portion of CNT acted as a quantum dot. In 2007 Maehashi et al. [33] used FIB to create two tunnel barriers on an SWCNT with a separation of 50 nm. They demonstrated clearly a source-drain current oscillation at room temperature, which was a result of Coulomb blockade effect.

In the rest of this section, we will discuss in detail several of the carbon nanotube devices fabricated using FIB technique without damaging the carbon nanotube. As mentioned earlier, the carbon nanotubes are easily damaged by the Ga ion beam. These damages will modify/deteriorate the electrical and mechanical properties of the CNTs that are essential for the CNT devices. This problem is solved by adding a protective graphitic outer layer on each individual CNTs, such that they can "survive" the ion beam bombardment during fabrication using FIB. A chemical vapor deposition technique was developed to grow fiber-coated individual carbon nanotubes (F-CNT) [34]. Figure 15.1 shows schematically the F-CNT cross-section. The nanotube was enclosed by a layer of amorphous carbon and an outer layer of graphitic material. The geometry of these F-CNTs makes them rather easy to handle and at the same time the graphitic layer protects the CNT core from damages during FIB processing.

2.1 INDIVIDUAL CNT ELECTRON FIELD EMITTER

To fabricate CNT electron field emitter, we used an *in situ* micro manipulator to pick up the fiber-protected CNT. In Figure 15.2, the procedure to prepare the F-CNT field emitter is shown.

We start out with a 200 μm diameter tungsten wire and etched the tip down to a radius of less than 0.1 μm. In order to have an extra degree of freedom to manipulate the CNT samples, an auxiliary or "intermediate" fiber is first picked up with the micromanipulator tip (Figure 15.2(a)). The pick-up procedure is as follows: first etching a slot on the tip of the micro-manipulator and then moving it to an auxiliary carbon fiber such that the fiber is rested in the slot. The Ga ion beam energy is fixed at 30 keV, and only the beam current is changed for different operations. For etching the slot, a beam current of 1 nA is used. Then platinum is deposited at the fiber-tip contact with a beam current of 50 pA to create a rigid bond between the tip and the micromanipulator. Then the fiber is cut off from the substrate

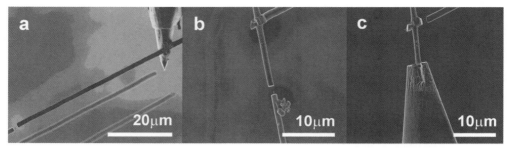

FIGURE 15.2: (a) Secondary electron image of the micromanipulator picking up an auxiliary carbon fiber. (b) Secondary electron image of the auxiliary fiber picking up F-CNT. (c) Secondary electron image of welding the F-CNT onto a tungsten tip. Reprinted from *Carbon* 43 (2005): 2083. Copyright 2005, with permission from Elsevier.

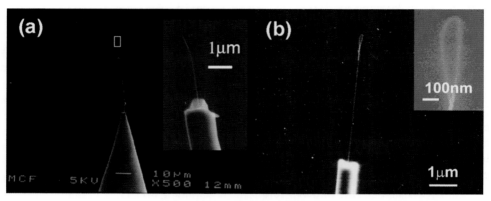

FIGURE 15.3: (a) SEM image of a CNT tip field emitter. (b) SEM image of a CNT loop field emitter. Reprinted from *Carbon* 45 (2007): 281. Copyright 2007, with permission from Elsevier.

with a beam current of 500 pA (Figure 15.2(a)). Next, we used this auxiliary fiber to pick up a CNT-containing carbon fiber (Figure 15.2(b)). We then move the CNT-containing carbon fiber into the W slot prepared earlier. After depositing platinum to weld the fiber to the tungsten tip, the auxiliary fiber is cut off from the CNT-containing fiber (Figure 15.2(c)). The applied beam current in this step is 500 pA. Details of the pick-up procedure can be found in [35–37]. Figure 15.3(a) shows the scanning electron image (SEM) of the fabricated CNT field emitter at the tip of the etched W wire. The insert in Figure 15.3(a) shows the nanotube at the carbon fiber tip [35].

During the FIB-assisted CNT tip pick-up process, the CNT tip can also be curved by the exposure to the Ga ion beam. Individual CNTs can be severely damaged with a beam current of 500 pA in a few seconds. When the beam current is reduced to several pA, by adjusting the exposure time, CNT loops with different radii can be produced, as shown in Figure 15.3(b). The insert in Figure 15.3(b) shows a CNT loop with a radius of 40 nm.

The I–V characteristic curve of the field emission from both the CNT tip field emitter and CNT loop field emitter agreed with the standard Fowler-Nordheim relationship. Compared with those of the CNT tip emitter, the field emission properties of the CNT loop emitter, such as field enhancement factor and the turn-on voltage, are improved. It is believed that the FIB-induced defects contribute to the improvement of the field emission properties [35–36].

FIGURE 15.4: (a) Secondary electron image of a fiber-coated carbon nanotube stabilized with Pt deposition on a Si substrate. (b) Trenches are cut with the ion beam along a selected tube section to isolate the tube section for lift-out. (c) The isolated tube section is picked up with a micromanipulator and reattached to a TEM grid with Pt deposition. Reprinted from *Appl. Phys. Lett.* 91 (2007): 103101. G. Chai H. Heinrich, L. Chow, and T. Schenkel, Electron Transport through Single Carbon Nanotubes. Copyright 2007, with permission from AIP.

2.2 CNT NANOAPERTURE/NANOPORE

Another area of research using an FIB instrument is fabrication of carbon nanotube nano-apertures or nanopores. To fabricate a CNT nano-aperture or nanopore, an F-CNT is first identified in an FIB system, and then standard lift-out techniques are applied to isolate an encapsulated tube section. Figure 15.4 shows the three stages of this lift-out process, where a selected F-CNT section is coated with platinum and trenches are cut with a focused ion beam to release the sample and transfer it onto a TEM copper grid for the electron transport experiments [38] inside a scanning transmission electron microscope (STEM), or integrate into a sealed aqueous chamber for nanopore test.

Figure 15.5(a) shows a low magnification TEM image of the F-CNT on a silicon substrate where the F-CNT is mostly surrounded by the platinum deposit. The gap between the F-CNT

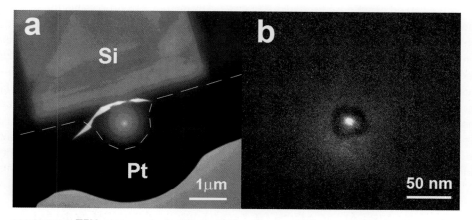

FIGURE 15.5: TEM micrograph of a 700 nm CNT collimator. (a) Low magnification image of the F-CNT section with Pt coating and the Si substrate. (b) High magnification image of the aligned CNT core. Reprinted from *Appl. Phys. Lett.* 91 (2007): 103101. G. Chai H. Heinrich, L. Chow, and T. Schenkel, Electron Transport through Single Carbon Nanotubes. Copyright 2007, with permission from AIP.

and silicon substrate results from a shadowing effect during the Pt deposition onto the F-CNT segment. The high magnification image in Figure 15.5(b) shows electron transport through a single and well aligned F-CNT with a tube length of 0.72 μm [38]. The recorded diffraction pattern of the transmitted electrons with zero energy loss suggested a mechanism of electron channeling along the carbon nanotube hollow core or between the atomic layers surrounding the hollow core of the MWCNT.

Here it is demonstrated that focused ion beam technique can be used to fabricate CNT nanodevices. The FIB fabrication process is aided by the unique fiber-protected carbon nanotube configuration. Two different types of CNT-based nanodevices, namely CNT field emission tips and individual CNT nanocollimator devices were demonstrated.

3 FIB FABRICATION OF ZnO NANODEVICES

Besides our own works on ZnO-based devices fabricated using FIB [37; 39–43], there has been only one recent work on using FIB to fabricate ZnO nanodevices. In 2009, He et al. [44] reported the transport properties of a single ZnO nanowire-based device manufactured by FIB. They used Pt deposition to fabricate the electric contacts and investigate UV response of the ZnO nanowires. They found a high photoconductive gain in the ZnO nanowire, even after exposure to FIB ion beam during nanodevice fabrication.

Here, we will describe our works on the FIB processing of ZnO nanodevices using different ZnO nanostructures, including single ZnO micro/nanorods [37; 39], ZnO branched nanorods [40], ZnO crosses [41], and ZnO tetrapods [42; 43]. Unlike fiber-protected carbon nanotubes described above, these ZnO nanostructures were naked, without any protection. During FIB processing, these ZnO nanostructures were constantly exposed to the ion beam. Fortunately, there are two important factors that make FIB processing of ZnO nanodevices possible: (1) the ability for ZnO to resist radiation damages as was demonstrated earlier [23–28]; (2) it has been shown that Ga ions are good n-type dopants for ZnO due to their similar size and low electro-negativity that allows ions to incorporate into Zn sites [45]. A recent work [46] has demonstrated that FIB Ga ion beam only started to affect the electric properties of ZnO when the Ga ion beam dose exceeded 10^{15} cm^{-2}. During our FIB processing, the exposure to the Ga ion beam is mainly due to low magnification broad ion beam imaging and accidental exposures to the tail of the Gaussian beam profile when the focused ion beam is near the ZnO nanostructures. For example, with an imaging beam current of 50 pA, 1000$\times$ magnification and an imaging area of 300 μm by 300 μm, the ion dose is 3×10^{11} cm^{-2}s^{-1}. We estimated that our ZnO nanostructures were never exposed to Ga ion beam more than 10^{14} cm^{-2}; therefore the effect of FIB processing on our ZnO nanostructures has been minimal. This makes the ZnO-based nanodevices presented below possible. In performed experiments, the typical process time was less than 20 minutes. In our latest experiment, we were able to reduce the process time to less than 9 minutes. This improves the sensitivity of the fabricated ZnO nanodevices.

3.1 FIB FABRICATION OF ZnO NANOSENSORS BY *IN SITU* LIFT-OUT TECHNIQUE

In recent reports [36; 38–43], we presented the fabrication of a single ZnO nanorod-based sensor using an FIB *in situ* lift-out technique. Typical secondary electron images of as-grown

FIGURE 15.6: Secondary electron images of the aqueous-solution-synthesized ZnO: (a) branched nanorods on an initial glass substrate; (b) single and branched nanorods transferred to a Si/SiO$_2$ substrate; insets are different regions of substrate showing the rods distributed over the surface; and (c) attempt to pick up an individual nanorod from an agglomeration of ZnO nanorods. Reprinted from *Microelectronics Journal*, 38 (2007): 1211. Copyright 2007, with permission from Elsevier.

ZnO nanostructures are shown in Figure 15.6(a). In order to simplify the sensor fabrication, it is usually advantageous to lower the ZnO nanorods density on a substrate. This was accomplished through a direct contact process by touching the as-grown substrate with a clean oxidized silicon wafer. By doing so, we were able to transfer a small amount of ZnO nanorods to the new clean substrate, as shown in Figure 15.6(b). These transferred ZnO nanostructures are loosely attached to the new substrate surface and are ready for pick-up.

It is important to emphasize the difficulty of picking up a single nanorod from an agglomeration as shown in Figure 15.6(c). The difficulty of picking up nano-objects from a non-conducting substrate (like glass) is mainly due to the charging effect, that hampers (or even makes impossible) the process of imaging and picking up the individual nanorods. In modern FIB instruments, an electron flood gun is available as an option for surface charge neutralization.

The *in situ* lift-out technique to fabricate a nanosensor from a single ZnO nanorod or a tripod (or a cross) is described. The Si wafer was used as an intermediate substrate for ZnO nanorods/tripods transferring and distribution in order to avoid charging problems. Low density of nanorods is required in order to select and attach them more easily to the micromanipulator tip. Samples on the stage can be rotated, tilted perpendicular to the ion beam, enabling easy arrangement of single ZnO nanorods on the nanosensor template.

In Figure 15.7, detailed images of FIB *in situ* lift-out procedures are shown. In the first step, the micro-manipulator tip was lowered and positioned at one end of an intermediate nanorod as shown in Figure 15.7(a). The tip was then moved until touching the nanorod. Then the nanorod was welded to the FIB needle end, as shown in Figure 15.7(b), with deposition of a 0.5 μm thick Pt layer.

The next step was to scan the Si/SiO$_2$ substrate for selection of a well-positioned ZnO nanorod. Once the desired nanorod is identified, it is recommended to push the nanorod in order to make sure that it is not firmly attached to the substrate and is strong enough to be transferred. After alignment of the intermediate nanorod see Figure 15.7(c) with one end of the selected nanorod, they were welded together again with Pt deposition (following this step, the needle and specimen were raised away from the substrate). Figure 15.7(d) shows a square hole fabricated on the glass substrate by FIB milling. The insert shows a nanosensor substrate template (glass substrate with Al contact electrodes). Using a micromanipulator, the nanorod

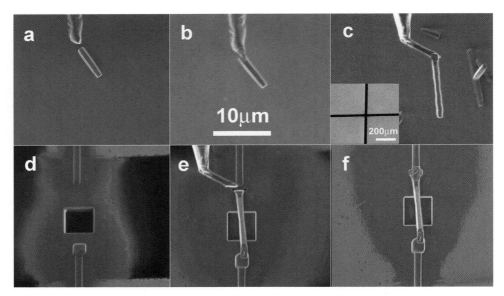

FIGURE 15.7: Secondary electron images showing steps of the *in situ* lift-out fabrication procedure in the FIB system. (a) an intermediate ZnO nanorod on Si substrate, next to the FIB needle, (b) an intermediate ZnO nanorod—picked up by the needle; (c) a single ZnO nanorod selected for sensor fabrication. Inset shows nanosensor substrate template (glass with Al contacts as contact electrodes); (d) a square hole cut on the glass; (e) the ZnO nanorod placed over the hole; (f) single nanorod welded to both electrode/external connections as final nanosensor. Reprinted from *Microelectronic Engineering*, 85 (2008): 2220. Copyright 2008, with permission from Elsevier.

is carefully positioned over the square hole Figure 15.7(e). This way, the nanorod sensor can sense gases in all directions. In the last step, the nanorod was fixed to one of the pre-deposited external contacts. The intermediate nanorod is cut and the micro-manipulator tip raised away from the substrate. Figure 15.7(f) shows a fabricated single nanorod-based nanosensor. The typical time taken to perform this *in situ* lift-out FIB nanofabrication is about 20 minutes. The process yield using this method is higher than 90%.

The procedure to fabricate nanosensor by using a single branched nanorod is very similar to the above method. However, branches between nanorods that are not strong enough can make the further fabrication of such devices impossible. The micro-manipulator tip has to be lowered and its tip positioned at the close end of the branched nanorod, as shown in Figure 15.8(a). Then the needle is lowered and welded to the end of the FIB needle, as shown in Figure 15.8(b), using Pt deposition of 0.5 μm thickness. Following this step (Figure 15.8(b)), the needle and specimen were moved away from the substrate to the pre-deposited Al external electrodes (see inset in Figure 15.8(b)).

3.2 FIB FABRICATION OF UV PHOTODETECTOR

A novel single ZnO crossed nanorod-based UV photodetector fabricated by this *in situ* lift-out technique in the FIB instrument was demonstrated by Chai et al. in 2008 [41]. Figure 15.9(a) shows a single nanorod cross attached to an intermediate ZnO nanorod on the tungsten needle tip. In order to simplify the procedure, pre-patterned external electrodes

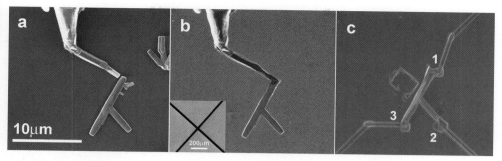

FIGURE 15.8: Secondary electron micrographs showing the steps of the *in situ* lift-out fabrication for: (a) Branched ZnO nanorod transferred to Si/SiO$_2$ substrate and positioned FIB needle with intermediate single ZnO nanorod connection. (b) Branched ZnO nanorod pick-up by FIB needle, insert shows substrate template (glass substrate with Al contacts as external electrodes); (c) branched ZnO nanorod placed on the template and welded to three electrodes. Reprinted from *Microelectronics Journal* 38 (2007) 1211. Copyright (2007), with permission from Elsevier.

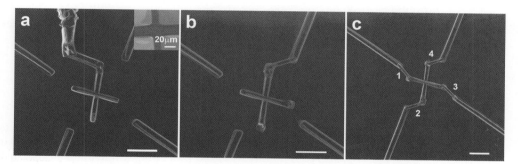

FIGURE 15.9: Secondary electron images showing steps of the *in situ* lift-out fabrication procedure in the FIB system. (a) the tungsten needle with an intermediate nanorod is attached to a single cross ZnO nanorod; inset is the view of a substrate template (glass with Cr/Au contacts as contact electrodes); (b) the cross ZnO nanorod is placed on the substrate and connected to the external electrodes; (c) the single cross ZnO nanorod after welding to the four electrodes. The scale bar is 3 μm. Reprinted from *Sensors and Actuators A: Physical* 150 (2009): 184. Copyright 2009, with permission from Elsevier.

were first deposited. In Figure 15.9(b), a single crossed ZnO nanorod is shown as it is placed on the substrate and attached to one of external electrodes. In Figure 15.9(c), all four contacts on each end of the ZnO nanocross are connected to Pt strips deposited using the ion induced chemical vapor deposition method.

Using this technique, photodetectors of different shapes have been prepared [39; 42; 43], including one based on a single tetrapod [42]. The *in situ* lift-out procedure in an FIB chamber was almost the same as described above. The main difficulties that have to be overcome have been outlined in [42]: a single ZnO tetrapod has to be transferable to another substrate; the tilt angles in the FIB chamber must be the same for initial substrate and templates substrate; and FIB needle with tetrapod must be positioned very accurately with nm precision in order to avoid break of the tetrapod junction.

In summary, in order to facilitate FIB processing of nanostructured materials, the following conditions must be met: (1) the nanostructured material is resistant to irradiation by an FIB ion beam; (2) the density of nanostructures must be controlled at a low level; (3) the substrate must be conductive to avoid charging effects; (4) the use of intermediate nanorods facilitates further handling of nanorods used in devices.

4 CONCLUSIONS

In this chapter, several types of CNT- and ZnO-based nanodevices fabricated using focused ion beam instrument are described. We also point out the difficulties in handling nanoscale objects inside an FIB process chamber and the techniques used to overcome these obstacles. Due to the effects of ion beam exposure on the nanostructured materials, such as carbon nanotubes or ZnO nanorods, it is necessary to develop strategies to minimize their exposure to the Ga ion beam. Here, it has been demonstrated that fiber-protected carbon nanotubes and radiation-resistant ZnO nanorods are excellent nanomaterials for focused ion beam processing due to their radiation resilience. In contrast to single beam FIB instruments, the dual beam FIB/SEM instruments can carry out the imaging function with SEM. This will greatly reduce the sample's exposure to ion beam and will make it possible to process a wider class of nanomaterials using FIB/SEM for future nanodevices.

REFERENCES

[1] Iijima, S., T. Ichibashi, and Y. Ando. Pentagons, heptagons and negative curvature in graphite microtube growth. *Nature* 356 (1992): 776–778.

[2] Trickey, S. B., F. Muller-Plathe, G. H. F. Diercksen, and J. C. Boettger. Interplanar binding and lattice-relaxation in a graphite dilayer. *Phy. Rev. B* 45 (1992): 4460–4468.

[3] Krishnan, A., E. Dujardin, T. W. Eddesen, P. N. Yianilos, and M. M. J. Treacy. Young's modulus of single-walled nanotubes. *Phys. Rev. B* 58 (1998): 14013–14019.

[4] Kim, P., L. Shi, A. Majumdar, and P. L. McEuen. Thermal transport measurements of individual multiwalled nanotubes. *Phys. Rev. Lett.* 87 (2001): 215502.

[5] Pop, E., D. Mann, Q. Wang, K. Goodson, and H. Dai. Thermal conductance of an individual single-wall carbon nanotube above room temperature. *Nano Lett.* 6 (2006): 96.

[6] Berber, S., Y. K. Kwon, and D. Thomanek. Unusually high thermal conductivity of carbon nannotubes. *Phy. Rev. Lett.* 84 (2000): 4613–4616.

[7] Tsuneta, T., L. Lechner, and P. J. Hakonen. Gate-controlled superconductivity in a diffusive multiwalled carbon nanotube. *Phy. Rev. Lett.* 98 (2007): 087002-1–087002-4.

[8] Yao, Z., C. L. Kane, and C. Dekker. High-field electrical transport in single-wall carbon nanotubes. *Phy. Rev. Lett.* 84 (2000): 2941–2944.

[9] Brown, H. E. *Zinc oxide rediscovered.* New York: The New Jersey Zinc Company, 1957.

[10] Brown, H. E. *Zinc oxide properties and applications.* New York: Pergamon, 1976.

[11] Padmavathy, N., and R. Vijayaraghavan. Enhanced bioactivity of ZnO nanoparticles: An antimicrobial study. *Sci. Technol. Adv. Mater.* 9 (2007): 035004.

[12] Li, Q., S. Chen, and W. Jiang. Durability of nano ZnO antibacterial cotton fabric to sweat. *J. Appl. Polymer Sci.,* 103 (2006): 412–416.

[13] Steiner. Todd D. *Semiconductor nanostructures for optoelectronic applications.* Boston: Artech House Publishers, 2004.

[14] Huang, M. H., S. Mao, H. Feick, H. Yan, Y. Wu, H. Kind, E. Weber, R. Russo, and P. Yang. Room-Temperature Ultraviolet Nanowire Nanolasers. *Science* 292 (2001): 1897–1899.

[15] Özgür, Ü., Ya. I. Alivov, C. Liu, A. Teke, M. A. Reshchikov, S. Dogan, V. Avrutin, S. J. Cho, and H. Morkoç. A comprehensive review of ZnO materials and devices. *J. Appl. Phys.* 98 (2005): 041301.

[16] Wang, Z. L. Zinc oxide nanostructures: growth, properties and applications. *J. Phys.: Cond. Matter* 16 (2004): R829–R858.

[17] Bai, X. D., P. X. Gao, Z. L. Wang, and E. G. Wang. Dual-mode mechanical resonance of individual ZnO nanobelts. *Appl. Phys. Lett.* 82 (2003): 4806–4808.

[18] Thomas, D. G. The exciton spectrum of zinc oxide. *J. Phys. Chem. Sol.* 15 (1960): 86.

[19] Chen, Y., D. M. Bagnall, H. J. Koh, K. T. Park, K. Hiraga, Z. Zhu, and T. Yao. Plasma assisted molecular beam epitaxy of ZnO on c-plane sapphire: Growth and characterization. *J. Appl. Phys.* 84 (1998): 3912–3918.

[20] Jagadish, C., and S. J. Pearton. ZnO- Zinc oxide bulk, thin films and nanostructures: Processing, properties, and applications. In N. H. Nickel, E. Terukov, eds. *Zinc oxide: A material for micro- and optoelectronic applications* (NATO Science Series II: Mathematics, Physics and Chemistry). Netherlands: Springer, 2005.

[21] Cui, Y., Z. H. Zhong, D. L. Wang, W. U. Wang, and C. M. Lieber. High performance silicon nanowire field effect transistors. *Nano Lett.* 3 (2003): 149–152.

[22] Wang, R. C., C. P. Liu, J. L. Huang, S.-J. Chen, and Y.-K. Tseng. ZnO nanopencils: Efficient field emitters. *Appl. Phys. Lett.* 87 (2005): 013110.

[23] Peale, R. E., E. S. Flitsiyan, C. Swartz, O. Lupan, L. Chernyak, L. Chow, W. G. Vernetson, and Z. Dashevsky Neutron transmutation doping and radiation hardness for solution-grown bulk and nano-structured ZnO In *Performance and reliability of semiconductor devices*, M. Mastro, J. LaRoche, F. Ren, J.-I. Chyi, J. Kim (eds.), Materials Research Society Symposium Proceedings 1108 (2009): 55–60.

[24] Look, D. C., D. C. Reynolds, J. W. Hemsky, R. L. Jones, and J. R. Sizelove. Production and annealing of electron irradiation damage in ZnO. *Appl. Phys. Lett.* 75 (1999): 811–813.

[25] Auret, F. D., S. A. Goodman, M. Hayes, M. J. Legodi, H. A. van Laarhoven, and D. C. Look. Electrical characterization of 1.8 MeV proton-bombarded ZnO. *Appl. Phys. Lett.* 79 (2001): 3074–3076.

[26] Kucheyev, S. O., C. Jagadish, J. S. Williams, P. N. K. Deenapanray, M. Yano, K. Koike, S. Sasa, M. Inoue, and K. Ogata. Implant isolation of ZnO. *J. Appl. Phys.* 93 (2003): 2972–2976.

[27] Burlacu, A., V. V. Ursaki, V. A. Skuratov, D. Lincot, T. Pauporté, H. Elbelghiti, E. Rusu, and I. M. Tiginyanu. The impact of morphology upon the radiation hardness of ZnO layer. *Nanotechnology* 19 (2008): 215714.

[28] Burlacu, A., V. V. Ursaki, D. Lincot, V. A. Skuratov, T. Pauporte, E. Rusu, and I. M. Tiginyanu. Enhanced radiation hardness of ZnO nanorods versus bulk layers. *Physica Status Solidi-rapid research letters* 2 (2008): 68–70.

[29] Chen, Y., H. Chen, J. Yu, J. S. Williams, and V. Craig. Focused ion beam milling as a universal template technique for patterned growth of carbon nanotubes. *Appl. Pphys. Lett.* 90 (2007): 093126.

[30] Peng, H. B., T. G. Ristroph, G. M. Schumann, G. M. King, J. Yoon, V. Narayanamuru, and J. A. Golovchenko. Patterned growth of single-walled carbon nanotube arrays from a vapor-deposited Fe catalyst. *Appl. Phys. Lett.* 83 (2003): 4238–4240.

[31] Javoenapibal, P., Y. Jung, S. Evoy, and D. E. Luzzi. Electromechanical properties of individual single-walled carbon nanotubes grown on focused-ion-beam patterned substrates. *Ultramicroscopy* 109 (2009): 167–171.

[32] Kurokawa, Y., Y. Ohno, S. Kishimoto, T. Okazaki, H. Shinohara, and T. Mizutani. Fabrication technique for carbon nanotube single-electron transistors using focused ion beam. *Jpn. J. Appl. Phys.* 43 (2004): 5669–5670.

[33] Maehashi, K., H. Ozaki, Y. Ohno, K. Inoue, K. Matsumoto, S. Seki, and S. Tagawa. Formation of single quantum dot in single-walled carbon nanotube channel using focused-ion-beam technique. *Appl. Phys. Lett.* 90 (2007): 023103.

[34] Kleckley, S., G. Y. Chai, D. Zhou, R. Vanfleet, and L. Chow. Fabrication of multilayered nanotube probe tips. *Carbon* 41 (2003): 833–836.

[35] Chai, G., and L. Chow. Focused ion beam assisted fabrication of multiwall carbon nanotube field emitters. *Carbon* 43 (2005): 2083–2087.

[36] Chai, G., and L. Chow. Electron emission from the side wall of individual multiwall carbon nanotube. *Carbon* 45 (2007): 281–284.

[37] Lupan, O., L. Chow, and G. Chai. Novel hydrogen gas sensor based on single ZnO nanorod. *Microelectr. Eng.* 85 (2008): 2220–2225.

[38] Chai, G., H. Heinrich, L. Chow, and T. Schenkel. Electron transport through single carbon nanotubes. *Appl. Phys. Lett.* 91 (2007): 103101.

[39] Lupan, O., L. Chow, G. Chai, L. Chernyak, O. Lopatiuk-Tirpak, and H. Heinrich. Focused-ion-beam fabrication of ZnO nanorod-based UV photodetector using the in-situ lift-out technique. *Phys. Stat. Sol. (a)* 205 (2008): 2673–2678.

[40] Lupan, O., G. Chai, and L. Chow. Fabrication of ZnO nanorod-based hygrogen gas nanosensor. *Microelectr. J.* 38 (2007): 1211–1216.

[41] Chai, G., O. Lupan, L. Chow, and H. Heinrich. Crossed zinc oxide nanorods for ultraviolet radiation detection. *Sens. Actuator. A: Phys.* 150 (2009): 184–187.

[42] Lupan, O., L. Chow, and G. Chai. A single ZnO tetrapod based sensor. *Sens. Actuator. B: Chem.* 141 (2009): 511.

[43] Chow, L., O Lupan, and G. Chai. FIB fabrication of ZnO nanotetrapod and cross-sensor. *Phys. Stat. Sol. (b)* 247 (2010): 1628–1632.

[44] He, J. H., H. Chen, P. C. Y. Chen, and K. T. Tsai. Electrical and optoelectronic characterization of a ZnO nanowire contacted by focused-ion-beam-deposited Pt. *Nanotechnology* 20 (2009): 135701.

[45] S. Kohiki, M. Nishitani, and T. Wada. Enhanced electrical-conductivity of zinc-oxide thin-films by ion-implantation of gallium, aluminum, and boron atoms. *J. Appl. Phys.* 75 (1994): 2069–2072.

[46] Weisenberg, D., M. Durrschuabel, G. Gerthsen, F. Perez-Willard, A. Reiser, G. M. Prinz, M. Feneberg, K. Thonke, and R. Sauer, Conductivity of single ZnO nanorods after Ga ions implantion. *Appl. Phys. Lett.* 91 (2007): 132110.

C H A P T E R 1 6

FOCUSED ION AND ELECTRON BEAM INDUCED DEPOSITION OF MAGNETIC NANOSTRUCTURES

Mihai S. Gabureac, Laurent Bernau, Ivo Utke, Amalio Fernández-Pacheco, and Jose María De Teresa

1 INTRODUCTION

Focused electron beam and focused ion beam (FEB/FIB) induced deposits using organometallic precursors usually exhibit various levels of carbon and oxygen contamination and relatively low metallic concentrations [1], which affects the electrical transport through the deposits and limits their use in electronic devices [2]. However, for the cobalt carbonyl ($Co_2(CO)_8$) precursor, high atomic Co concentrations have been shown to be accessible when depositing in the molecule limited regime, either by increasing the dwell time per pixel [3] or the beam current [4]. This opens the way for use of focused electron beam and ion beam induced deposition (FEBID/FIBID) fabricated, cobalt-containing nanostructures in magnetic sensing devices [5; 6].

For most applications it is highly desirable to produce FEB/FIB metal and dielectric deposits with the highest purity, for use as nanocomponents in building or repairing electrical circuits or optical masks. In contrast, for use in magnetic sensing applications based on the Hall effect, lower purity deposits could show enhanced field sensitivities close to the metal-insulator transition, due to the Giant Hall Effect (GHE) [7; 8]. For example, for Co deposits with small grain sizes, the optimum composition for magnetic sensing was found to be around 65 at.% Co [6].

In the following section of this chapter, we analyze the characteristics of Co_xC_{1-x} FEBID/FIBID Hall nanosensors with Co ratios of $0.5 \leq x \leq 0.8$ and sizes in the 200 nm range. In Section 3, we discuss the magnetic properties of high purity Co nanowires, and in Section 4, we discuss Fe deposits.

2 HALL SENSORS WITH HIGH SPATIAL RESOLUTION

Super-paramagnetic beads, used as markers in medical diagnostic protocols and biological research [9], or the magnetic grains used for magnetic recording [10], are small objects in the micrometer or sub-micrometer size range that can be assimilated to dipolar field sources producing a spatially inhomogeneous magnetic field:

$$\vec{B}(x, y, z) = \frac{\mu_0}{4\pi} \frac{3\vec{r}(\vec{\mu} \cdot \vec{r}) - \vec{\mu}r^2}{r^5}, \tag{1}$$

where $\mu_0 = 4\pi \times 10^{-7}$ H/m is the permeability of free space, μ is the magnetic moment of the bead and r is the distance between the center of the bead and center of the Hall sensor. As can be seen in Figures 16.1 and 16.2, the field to be detected on the sensor surface rapidly vanishes on the scale of the bead diameter. Detection of such highly localized magnetic fields requires sensors with similarly sized sensing areas. Sub-micrometer or nanometer sized magnetic sensors also offer a high spatial resolution which is essential in many magnetic imaging or field sensing applications [11–13].

2.1 INTRODUCTION TO PARAMAGNETIC BEAD DETECTION

Micrometer sized paramagnetic beads (e.g., Dynabeads from Dynal Biotech or Micromer beads from Micromod), selectively attached to specific molecules (proteins, DNA), are

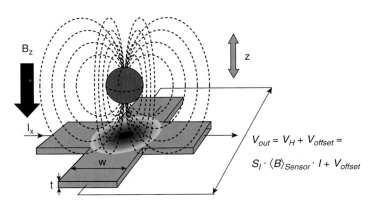

FIGURE 16.1: Schematics of a Hall sensor used for bead detection: the field on the sensor is highly non-uniform and depends on the distance between the bead and the sensor surface. For a given height, the average magnetic field on the sensor plane, $\langle B \rangle_{Sensor}$, depends on the Hall cross area.

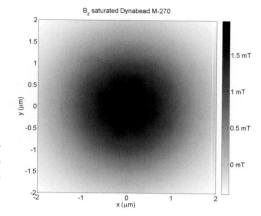

FIGURE 16.2: Simulation of the perpendicular component of the magnetic field at the sensor plane from a vertically saturated Dynabead M-270. The bead is 2.8 μm in diameter and suspended at 1 μm above the sensor.

commonly used as labels in many biological applications, for the manipulation or detection at the molecular level [9]. Using a magnetic sensor at a given position below the magnetic bead as a transducer from the biological to an electrical signal [13] offers an alternative to classical optical detection [14], for on-chip integration low cost array applications, and opens the way for using smaller bead sizes in single-molecule handling.

Typical bead-detection setups include free-flowing beads in microfluidic channels for counting-type experiments [15] or molecules tethered to an Au pad using specific biochemistry for studies of enzymatic action [16]. Handling of the microbeads is provided using magnetic field gradients (magnetic tweezers [17]) or viscous forces in the case of microfluidic platforms. In all setups, an external magnetic field is required to polarize the bead (Figure 16.1), which can be either perpendicular or parallel to the sensor surface.

The voltage output V_H of the magnetic sensor will then depend on the bead position, with a magnitude proportional to the sensor current sensitivity S_I: $V_H = S_I B_{avg}(x, y, z)I$, where I is the measurement current employed, and B_{avg} is the magnetic signal of the bead averaged over the sensor active area [11]. Therefore, when measuring inhomogeneous localized fields (i.e., for typical bead detection setups), the performance of the magnetic sensor is given by the value of the minimum change in the magnetic flux that can be detected $\Phi_{min} = B_{min} \times A$, where A is the sensor area and B_{min} is the magnetic field resolution $B_{min} = \delta B_z = \delta V / S_I \cdot I$ estimated from uniform field measurements of the S_I and the voltage noise δV on the Hall probes. For frequencies in the white noise regime, the voltage noise is fundamentally limited by the thermal noise: $\delta V_{th} = \sqrt{4k_B R_{Hall} T}$ [6], where k_B is the Boltzmann constant, T is the temperature, and R_{Hall} is the resistance of the Hall voltage line.

The maximum Hall signal is obtained when the bead lays on the surface $V_{Hmax} = V_H(0, 0, R)$, where R is the bead radius. However, in applications, it is limited by the height of the passivation layer above the sensor (z_0) and the position of the bead within this channel (x_0, y_0): $V_{Hmax} = V_H(x_0, y_0, z_0 + R)$. Because the dipolar magnetic field of the paramagnetic bead falls off sharply with the z distance as $1/z^3$, the beads can be detected only up to a maximum height z_{max} for which $V_{Hmin}(x_0, y_0, z_{max}) = \delta V_{th}$. The corresponding B_{max} and B_{min} define the detection range of the magnetic sensor, while its vertical resolution is encoded in the field resolution $B_{min} = \delta B_z(x_0, y_0, \delta z)$.

However, considering the external field necessary to saturate the paramagnetic beads, the working field range has to be even larger than B_{max}. In this respect, Hall sensors, although

Table 16.1 Room temperature characteristics of different types of Hall sensors classified by their magnetic flux resolution Φ_{min}. Although the Co-C sensors have a rather large field resolution B_{min} compared to the semiconductor sensors, their flux resolution is the best because of their small sensing area. All sensors other than the Co-C were produced with standard techniques (UHV deposition, lithography, wet etching, or FIB milling). Symbols were explained in text.

Reference	[19]	[20]	[21]	[12]	[22]	[23]	[24]	[6]
Type	InAs	InGaAs	n-Si	FePt	Bi	InSb	InAsSb	Co-C
Width (nm)	1000	2000	2400	160	50	500	1000	100
S_I (Ω/T)	616	700	175	325	4	–	2750	0.15
B_{min} ($nT/Hz^{1/2}$)	4300	400	200	166000^1	80000	720	51	900
Φ_{min}/Φ_0	2×10^{-3}	8×10^{-4}	5.5×10^{-4}	1.6×10^{-4}	10^{-4}	9×10^{-5}	2.5×10^{-5}	5×10^{-6}

[1] B_{min} is estimated from the published resistivity and the geometrical dimensions of the FePt sensor.

having lower sensitivities than magnetoresistive sensors, are better suited for bead detection experiments, because they have linear sensitivities up to very large fields (B_{sat}).

The key characteristics for several, recently published micro- and nano-Hall sensors are given in Table 16.1 above. For the measurement of uniform magnetic fields, the minimum detectable field change B_{min} is giving the sensing resolution. However, for bead detection and tracking applications, the determinant parameter is the minimum change in the magnetic flux $\delta\Phi$, as the magnetic field of the microbead is highly inhomogeneous. The references given in Table 16.1 are sorted according to Φ_{min}/Φ_0 (see also [13; 24]), where $\Phi_0 = h/2e = 2.07 \times 10^{-15}\,Tm^2$ is the quantum flux.

Granular Hall sensors have a better Hall sensitivity compared to polycrystalline Co sensors [25] due to enhanced scattering. They, are scalable below the 100 nm range (in contrast to semiconductor based sensors), and exhibit a high field range compatible with bead detection, which is making them a suitable candidate for a magnetic biosensor.

2.2 HALL EFFECT ENHANCEMENT IN GRANULAR MATERIALS

The classical Hall effect stems from the Lorentz force deflecting the electron trajectories, which generates a voltage proportional to the local magnetic field (B) and the current density (j). This is expressed using the (ordinary) Hall coefficient $R_{OH} = E_y/j_x B_z = (ne)^{-1}$, where E_y is the electric field perpendicular both to the current density j_x and the transverse *external* magnetic field $B_z = \mu_0 H$, n is the density of charge carriers and e is the electron elementary charge.

In a ferromagnetic material, however, because the magnetic flux density is given both by the external field and the spontaneous magnetization, $B = \mu_0[H + (1 - D)M]$, where D is the demagnetizing factor accounting for geometrical effects (for a thin film, $D = 1$ when the field is perpendicular to the plane, and $D = 0$ when the field is in plane), there are two contributions to the total Hall effect; one is the so-called ordinary Hall effect (OHE), due to the Lorentz forces generated by the external applied field, and the other is the so-called extraordinary Hall effect (EHE), being proportional to the spontaneous magnetization and originating from spin-dependent scattering of charge carriers [26; 27]. The Hall resistivity is

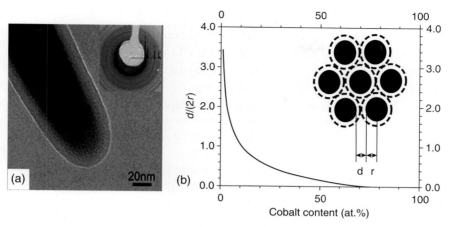

FIGURE 16.3: (a) TEM bright field image of rod apex deposited at 100 pA, reprinted with permission from [37], Copyright (2004) Elsevier; inset: indexed diffraction of cubic Co indicating nanocrystals of 1–2 nm in size in carbonaceous matrix; (b) dependence of the average distance d of nanocrystals normalized to their size $2r$ as a function of overall Co-content using equation 4; the percolation threshold ($d/2r = 0$) is at 75 at.% Co; inset: closed packed model for Co-C.

therefore given by [28]:

$$\rho_{\mathrm{H}} = \rho_{\mathrm{OH}} + \rho_{\mathrm{EHE}} = \mu_0(R_{\mathrm{OH}}H + R_{\mathrm{EH}}M_z) = \mu_0(R_{\mathrm{OH}} + \chi R_{\mathrm{EH}})H \tag{2}$$

in units $\Omega \cdot$m, where χ is the magnetic susceptibility and R_{EH} is the EHE coefficient in units of $\Omega \cdot$ m/T.

In a metal having a high charge density, ρ_{OH} is small ($R_{\mathrm{OH}} \to 0$ for $n \to \infty$), so that below saturation, its contribution can be neglected with respect to the spontaneous part, ρ_{EHE}. The Hall resistivity is found to vary at saturation with the longitudinal resistivity ρ as [29]:

$$\rho_{EHE\text{-}sat} = \gamma \rho_0 \rho_S + \gamma \rho_S^2, \tag{3}$$

where γ is a model coefficient and where the longitudinal resistivity $\rho = \rho_0 + \rho_S$ in units of $\Omega \cdot$m, has been decomposed using Matthiessen's rule in a magnetic part ρ_S that stems from spin-orbit (S-O) scattering and a spin-independent scattering contribution ρ_0 [29].

The electrical transport in granular metal-insulator films depends on the metal atomic concentration, average grain size, and spacing. Below the percolation limit, electrons travel by hopping from grain to grain [30]; in other words, the transport is strongly influenced by the inter-grain scattering. The longitudinal resistivity depends both on the nanoparticle size $2r$ (see Figure 16.3), because the surface scattering is proportional to the interfacial area per unit volume, and the average nanoparticle spacing d: $\rho_0 \sim d/2r$. Therefore, the EHE is expected to be very sensitive both to the nanoparticle mean size ($2r$) and to the metal atomic concentration (through the average nanoparticle spacing d).

The ratio of nanoparticle spacing and nanoparticle size $d/2r$ can be estimated for Co-C using a closed packed model and assuming a mono-disperse size distribution of the nanoparticles, as shown in Figure 16.3(b):

$$\frac{d}{2r} = \sqrt[3]{\frac{1}{P}\left(\frac{(1-x)\,\rho_{np}M_m}{x\rho_m M_{np}} + 1\right)} - 1, \tag{4}$$

where x is the metallic atomic concentration in the carbonaceous matrix, M denotes the molar masses, ρ denotes the densities, and $P \approx 1.35$ for close packing geometry. The subscripts m and np denote the insulating matrix and the metallic nanoparticles, respectively.

The EHE contribution of deposits with the same atomic concentration (x) will then vary only because of the spin depending contribution ρ_S, which should increase for small nanoparticle sizes $2r$ because the number of S-O scattering events is increased. Close to the metal-insulator transition (x_C) [7], the Hall signal will be enhanced mainly because of the increase in the spin independent contribution $\rho_0 \sim d$ [31]. This behavior, observed for the first time in co-sputtered granular ferromagnetic metal-insulator thin films (Ni-SiO$_2$), is known as the giant Hall effect (GHE) [32].

Above the percolation threshold, the EHE can be also enhanced if the film thickness is smaller than the mean free path of the charge carriers, because of the extra contribution to ρ_0 given by the surface scattering. Although this effect can be neglected in granular films, below the percolation threshold, a similar enhancement should be observed in this case when the sensor width becomes comparable to the mean free path.

2.3 GRANULAR CO-C HALL NANOSENSORS

TEM characterization of Co-FEBID tips [33; 34] showed that the deposited material is granular, consisting of Co-nanocrystals, a few nm in size, embedded in an amorphous carbonaceous matrix: Co$_x$(CO)$_{1-x}$. This nano-composite structure is similar to that of co-sputtered heterogeneous granular magnetic films embedded in an insulating matrix. For sizes below 10 nm, the Co nanoparticles are single domains [35] forming films that are superparamagnetic at room temperature [36]. These films show an important EHE [8]. In this section, we present an example of a magnetic sensing device realized using FEBID/FIBID of cobalt-carbon material and discuss the characteristics of the sensor in terms of composition and nano-structure.

2.3.1 FEB fabrication

We have used standard UV and e-beam lithography to predefine Pt-Au electrodes, having thicknesses below 100 nm and a gap l between electrodes below 1.5 μm on Si wafers with an oxide layer of 200 nm (see Figures 16.4 and 16.5).

Granular Co-C nano-Hall sensors were deposited between the Pt-Au electrodes by FEBID of Argon-stabilized Co$_2$(CO)$_8$ (from VWR International, Dietikon, Switzerland). The active area of the sensor is given by the crossover of the current and voltage line, making sizes between 200 nm and 500 nm readily accessible. The Hall sensors had thicknesses between a few tens up to a few hundreds of nanometers. For sensors with thicknesses smaller than the Pt/Au electrode, we have defined FEBID pads with a thickness larger than the central active area, in order to avoid coverage problems at the electrode edges, see Figure 16.6, while for thicker or FIBID sensors we have used a uniform thickness (Figure 16.7).

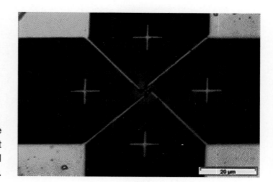

FIGURE 16.4: Optical image of Hall sensor device with large Au bonding pads (corners), alignment marks (crosses) and a FEB deposited Co-C Hall sensor in the center.

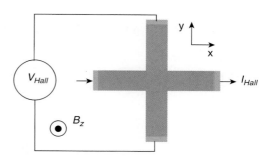

FIGURE 16.5: Scheme for Hall measurements: the external magnetic field B_z is applied perpendicular to the surface and is uniform across the active area.

EMPA_Thun 3.0kV 14.7mm x13.0k SE(L)   4.00um

FIGURE 16.6: SEM image of Hall sensor with sub 100 nm square sensing area and thicker CoC contact pads on top of Au/Cr predefined electrodes. The motif was repeated 100 times. The contact pads were written at a higher dose in order to improve the interface conductance (ohmic contacts).

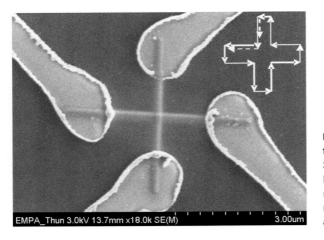

EMPA_Thun 3.0kV 13.7mm x18.0k SE(M)   3.00um

FIGURE 16.7: Hall sensors with uniform thickness. Inset: writing strategy. Starting with the outer contour, the beam is iteratively scanned with narrowing cross widths. This also minimizes the carbon contamination and the asymmetry in the two lines.

2.3.2 Specifications of FEB deposited material

The composition of the FEBID nanostructures depends on several deposition parameters and can be tuned by varying the deposition parameters, such as the precursor flow, beam current, residual chamber pressure, dwell time (i.e., scan speed) and refreshment rate [3]. As can be seen in Figure 16.8, the EDX analysis showed a large variation in the Co-C ratio of these deposits depending on the dwell time, which was confirmed by *in situ* resistance measurements.

However, a similar variation can also be artificially obtained depending on the base vacuum pressure, because of the co-deposition of C from hydrocarbons present as residual gases in the SEM chamber. This contamination arises in our SEM chamber for pressures above 5×10^{-5} mbar, corresponding to approximately 20 monolayer/s of hydrocarbons impinging on the substrate. For comparison, the $Co_2(CO)_8$ molecule flux delivered to the substrate is of 570 monolayers/s at the point of irradiation. Therefore a good control of all the deposition parameters is needed in order to eliminate background pressure artifacts.

The oxygen contamination is difficult to assess from *ex situ* EDX measurements, because of the surface oxidation when exposing the deposit to air. Electrical measurements have shown an important degradation of the conductance during exposure to air for the high

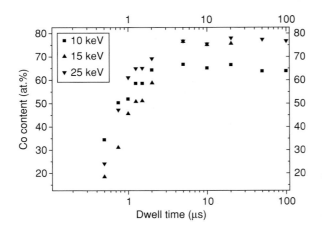

FIGURE 16.8: EDX analysis of Co content in FEBID deposits obtained with different dwell times for 1nA beam current at different acceleration voltages.

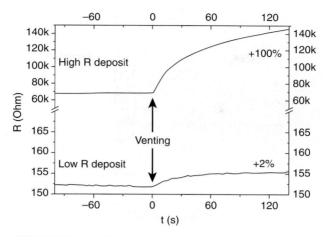

FIGURE 16.9: 4-point resistance measurements during air venting ($t = 0$s) for FEBID nanowires showing large or small changes depending on the value of the "as grown" resistance.

resistance deposits, while the deposits with low resistance had changes of only a few percent, as can be seen in Figure 16.9. In the following, we based our analysis on the assumption that the deposit is made of pure Co nanoparticles embedded in a carbon matrix (Co_xC_{1-x}), and we ignored any size related anisotropy effects when estimating the mean magnetic moment of the nanoparticles ($< \mu > \sim r^3$).

In order to differentiate between deposits with similar Co atomic concentrations, EDX analysis is not sufficient, as it gives no insight on their structure (mean size r and relative distance d of the nanoparticle). This information can be obtained from Hall measurements, which provide a local characterization method giving access to the average nanoparticle size r, and from longitudinal resistivity measurements (as $\rho_0 \sim d/2r$). Alternatively, TEM analysis can provide similar structural information, but it requires the preparation of TEM lamellas, which is both time consuming and an irreversible characterization method.

2.4 HALL CHARACTERIZATION OF FEB-FIB DEPOSITS

Although it is possible to obtain deposits with similar Co ratios (x) by using different sets of deposition parameters giving the same $d/2r$ ratio (see Equations 4 and 5 and Figure 16.10), their magnetic characteristics (sensitivity S_I and saturation field B_{sat}; see Figure 16.11) can be very different, as shown below in Figures 16.12 and 16.13.

Magnetic characterization of our nanodeposits by standard measurements (VSM, MOKE, SQUID) is not possible, given the minimal magnetic material volume deposited. Furthermore, magnetic properties of the deposits depend intrinsically on the chosen geometry (via dwell time, refresh rate, shadowing effects). A local characterization of the magnetic properties of the nano-sized FEBID deposits is therefore required and has been provided by decoding the information embedded in the Hall voltage, by using the Langevin model for superparamagnetic nanoparticles as in Figure 16.10.

The Langevin analysis provided a plausible explanation of the relatively large spread of values we found in the magnetic characteristics of different sensors at similar Co atomic

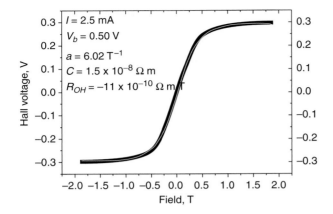

FIGURE 16.10: Langevin fit using equation 5 for a FEBID sensor, with the parameter a giving access to the mean magnetic moment μ, and to the ordinary Hall constant R_{OH} (see Equations 2 and 5); the values for the fit parameters C and R_{OH} were evaluated for a thickness t of 100 nm. R_{OH} was found negative as it has to compensate for the positive slope of the Langevin curve at saturation, but its value is probably overestimated because of the slight disaccord at 2T.

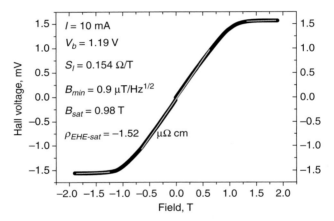

FIGURE 16.11: Field variation of Hall voltage for FIBID sensors; the fit around zero field (EHE) gives access to the current sensitivity S_I and, combined with the linear fit at saturation (OH-region with a field > 1.25T), to the saturation field B_{sat}. Note that R_{OH} cannot be estimated from this fit because of the positive Langevin contribution. B_{min} is calculated for a uniform magnetic field and I is the maximum longitudinal current before any thermal drift is noticeable on the bias voltage V_b. The Hall resistivity $\rho_{EHE-sat} = V_H(B_{sat})t/I$ was estimated for a thickness $t = 100$ nm.

concentration x, because of the spreading found in the mean magnetic moment $<\mu>$ (the parameter a in Figure 16.10). In the same time, using linear fits at zero field and saturation, as in Figure 16.11, we were able to calculate the saturation field B_{sat} which was found to scale inversely with $<\mu>$ ($B_{sat} \sim 1/<\mu>$, see [6]), therefore confirming the validity of our approach.

The Hall voltage was measured in a homogeneous magnetic field perpendicular to the sensor at room temperature (see Figure 16.4). If the magnetic interactions between nanoparticles are negligible (below the percolation threshold), it is possible to use the Langevin

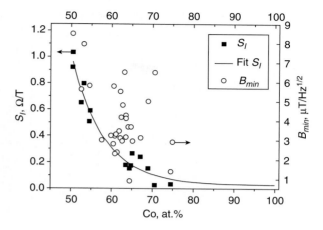

FIGURE 16.12: Magnetic current sensitivity S_I and field resolution B_{min} for sensors with different growth parameters (hence the results spreading for B_{min}) as a function of the EDX atomic Co concentration; S_I increases exponentially close to the metal-insulator threshold x_c (granular Hall Effect enhancement).

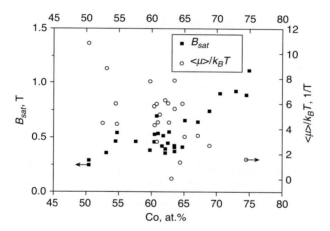

FIGURE 16.13: Working field range B_{sat} and normalized mean magnetic moment as a function of the EDX atomic Co concentration for sensors with different growth parameters. The spreading in B_{sat} can be explained by the difference in the size ($\mu \sim r^3$), smaller nanoparticles having higher saturation fields.

model $M = M_S L(\mu B/k_B T)$, in order to fit the $V_H(B)$ curves using Equation 2:

$$V_H = \rho_H Il/(wt) = \mu_0(R_{OH}H + R_{EH}M_z)Il/(wt) = [R_{OH}B_z + CL(aB_z)]Il/(wt) \qquad (5)$$

with $a = <\mu>/k_B T$, where $<\mu>$ is the mean magnetic moment of the nanoparticles and l, w, and t are the length, width, and thickness, respectively, of the current line, $C = \mu_0 \cdot R_s \cdot M_s$ and $B_z = \mu_0 H$ is the external magnetic field applied (as in Figure 16.5). Assuming that there are no variations in the surface anisotropy with the nanoparticle size, this also provides an insight on the mean nanoparticle size $2r$ ($\mu \sim r^3$) [36], which can then be correlated with the values of the saturation field B_{sat} from Figure 16.11, or with TEM measurements.

Because of the linear region around zero fields, and the hysteresis free magnetization reversal, these deposits can be used as magnetic sensors, for bead detection, provided they have a good field sensitivity S_I. Due to the internal structure of the Co-C deposits, favoring the EHE, sensitivities up to 100 times larger than polycrystalline Co sensors [25] can be achieved, while retaining similarly high saturation fields B_{sat}.

Because of the intricate dependence of S_I I_{max}, and δV_{th} on the longitudinal resistivity, the best field resolution B_{min} in Figure 16.12 was obtained for an optimum atomic Co concentration around 65%. For most of the sensors, B_{min} was in the range $1 - 10\,\mu T/Hz^{0.5}$, with active areas of $200 \times 200\,nm^2$, giving a resolution for the magnetic flux $\delta\Phi$ in the range $2 \times 10^{-5}\Phi_0$, limited by the thermal noise above 1 kHz. The typical Hall resistivity ($\rho_{EHE\text{-}sat} = V_H(B_{sat})t/I$) of the Co-C deposits (Figures 16.10, 16.11, and 16.13) was in the range $0.5 - 10\,\mu\Omega\cdot cm$, and the longitudinal resistivity (ρ) was in the range $10^3 - 10^4\,\mu\Omega\cdot cm$, depending on the Co concentration ($d/2r$ ratio).

3 FEBID OF CO NANOWIRES

The creation of magnetic nanoelements is nowadays a major research topic owing to their wide range of applications in magnetic storage and sensing [38; 39]. The study of magnetization reversal processes in nanoscale structures has been very intense for the last decade [40–43] because the performance of devices based on these nanoelements depends on their magnetic switching properties. In particular, the domain-wall manipulation in magnetic nanowires (NWs) is of special relevance, as it can be addressed by the application of external magnetic fields as well as by injecting large spin-polarized currents [44–46]. Generally, the creation of magnetic NWs is performed by electron-beam lithography or other top-down techniques using resists and lift-off or etching processes [47]. The use of gas precursors and direct writing of magnetic NWs by means of focused electron beams is of great interest due to high flexibility and resolution of the technique. Recently, a few publications have demonstrated that it is possible to grow high-purity Co NWs by FEBID (>95%) [4] and some of their functional properties have been described [5; 48]. Hereafter, we summarize the main properties of such NWs.

For the growth, an automatized gas-injection system (GIS) inside a FEI Nova 200 Nanolab© equipment, with dicobalt octacarbonyl [$Co_2(CO)_8$] as the precursor material, has been used. The GIS tip is positioned about 150 μm away from the region of interest in the z direction and about 50 μm away in the x/y direction. The GIS and the substrate are at room temperature during the process. The chamber base pressure is about 10^{-6} mbar or lower before the growth, and it increases up to $2 - 5 \times 10^{-6}$ mbar during the growth. The structures have been grown with beam energies in the range 5–30 kV, and beam currents in the range 0.13–9.5 nA. Typical dimensions for the deposits are length = 9 μm × width = 500 nm × thickness = 200 nm, with deposition time of the order of a few minutes when using high currents and of the order of one hour when using low currents.

We have performed *in situ* energy-dispersive X-ray (EDX) microanalysis of several Co nanodeposits in the range of voltages and currents specified above in order to study any correlation between the growth conditions and the nanodeposit composition. Within the range of parameters used, it can be concluded that the beam current is the main parameter governing the nanodeposit content. At all beam energies, for beam currents above ~2 nA, the Co content in atomic percentage is around 90% or more, whereas for lower currents the Co content is typically around 80% or less. The rest of constituents of the nanodeposits are C and O. As an

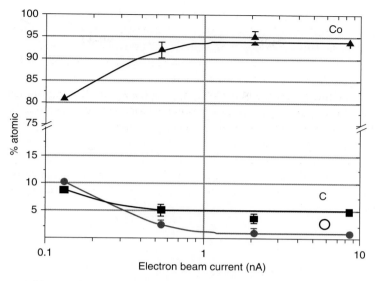

FIGURE 16.14: Atomic percentage of Co (triangles), C (squares), and O (dots) as a function of the electron-beam current in nanodeposits grown with the FEBID technique at beam energy of 10 kV using $Co_2(CO)_8$ as the precursor gas.

example, in Figure 16.14 the atomic chemical composition of the deposits is represented as a function of the beam current at 10 kV beam energy. The Co content increases monotonously from about 80% for a beam current of 0.13 nA to about 95% for a beam current above 2 nA.

A previous study by Utke et al. also found an increase in the Co content as a function of the beam current, but it never exceeded 80% [34]. In that case, currents above 80 nA were necessary for this high concentration, which limits the lateral resolution of the deposits. In that work, if current about 1 nA is used, the Co content does not go beyond 60%, in contrast to the findings of the present study. It has been argued that beam-induced heating is the main effect responsible for this behavior as the precursor gas molecule, $Co_2(CO)_8$, can be thermally decomposed at relatively low temperature [34, 49]. As a consequence, it could be expected that the use of high current would favor the local heating, enhancing the precursor molecule dissociation and the high Co content. It can be pointed out that in the present case, the electron beam source is a Schottky field-emission gun, which will likely produce smaller beam spots on the substrate than in the case of thermionic-tungsten-filament sources, as is the case in Reference [34]. Thus, the electron beam current will be focused on a smaller area, producing larger local heating effects. A recent study of the influence of the local substrate temperature on the Co content of this type of deposits has been performed [49]. For that, a commercial heater has been used in order to control the surface temperature. When performing the Co growth with low beam currents on that heated surface, the Co content can be increased up to values as high as those obtained using high beam currents. These experiments confirm the physical origin of the enhanced cobalt content in the deposits when using high beam currents and open the possibility of growing highly pure Co nanodeposits with a higher resolution.

In Chapter 23 by de Teresa et al. in this book, it is shown that the temperature dependence of resistivity in Co NWs with high Co content is the typical one of metallic systems, and the low-temperature resistivity, $\rho(0) \approx 27\ \mu\Omega\cdot cm$, can be described quantitatively using the bulk

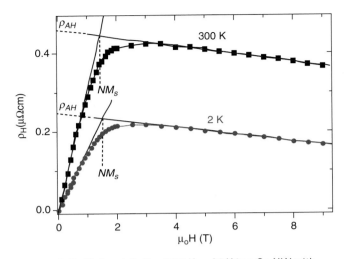

FIGURE 16.15: Hall resistivity at 300 K and 2 K in a Co NW with dimensions length $= 9\,\mu$m, width $= 500$ nm, thickness $= 200$ nm. The magnetic character of the wire is manifested by the presence of the anomalous Hall effect. The ordinary part is determined by the slope at high H, whereas the anomalous constant is calculated by extrapolating this slope at zero field. From the intersection of the linear regions of the ordinary and the anomalous Hall effect the value of NM_S ($N =$ demagnetizing factor, $M_S =$ saturation magnetization) is obtained, which allows the determination of M_S.

parameters of Co. Hereafter, we will focus on the description of the magnetotransport properties of these NWs.

In Figure 16.15 we show the Hall effect resistivity ($\rho_H = V_H t/I$) as a function of field for the maximum and minimum temperature measured for a Co NW grown under 2.1 nA current. ρ_H increases at low fields until reaching magnetic saturation. For higher fields, only the ordinary part variation (ρ_{OH}) is observed, with opposite sign to the anomalous one (ρ_{EHE}). This behavior has also been observed in magnetron sputtered polycrystalline cobalt films [25]. The ordinary part is obtained by the standard method of fitting the slope at high fields, whereas the anomalous one is determined by extrapolating to H=0 the fit of ρ_{OH} (see Figure 16.15). From the intersection of the linear regions of the ordinary and the anomalous Hall effect [50] (1.42 T at 300 K) and taking into account the demagnetizing factor for the created geometry, we find a saturation magnetization $M_s = 1329 \pm 20$ emu/cm^3, corresponding to 97% of the bulk Co value. From the ordinary part at 2 K, the density of electrons is calculated to be $n_e = 7 \times 10^{28}$m^{-3}. Using this value, together with the experimental value of the resistivity at low temperature, $\rho(0) \approx 27\,\mu\Omega$cm, one can estimate the mean-free-path as in reference [25]: $l(0) = hk_F/[n_e e^2 \rho(0)] \approx 4$ nm. This value is of the order of the grain size determined by high-resolution-transmission-electron microscopy-images [51], suggesting that the scattering is mainly produced at grain boundaries.

In Figure 16.16(a), magnetoresistance (MR(%) $= 100 \times [R(H)-R(H=0)]/R(H=0)$) measurements at 300 K are shown with three different geometries: H perpendicular to the thin film plane (perpendicular geometry: PG), H in plane and parallel to the current I (longitudinal geometry: LG), H in plane and perpendicular to I (transversal geometry: TG). The measured MR in the

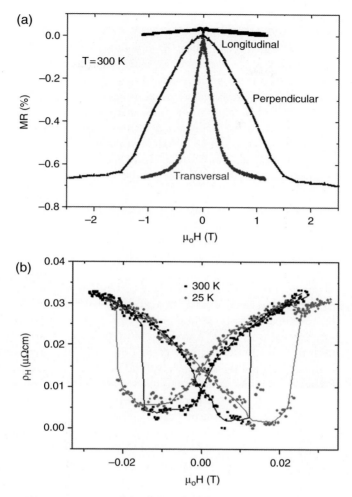

FIGURE 16.16: (a) Magnetoresistance (MR) measurements at 300 K in longitudinal, transversal and perpendicular geometries of the same Co NW of Figure 16.15. The observed MR effects are explained by the AMR of cobalt. (b) Planar Hall effect (PHE) (which has the same origin as the AMR) measurement performed in a Co NW. The magnetic field is applied in the substrate plane, forming 45° with the current. This figure is a zoom-in of the measurement at low magnetic fields in order to single out the abrupt switches occurring at the corresponding coercive fields. The lines are guides to the eye.

three geometries is in perfect agreement with previous results in Co NWs fabricated by an electron beam lithography process [43]. The magnetization tends to align along the NW axis due to the strong shape anisotropy (length about 18 times the width and 45 times the thickness). The differences in the saturation fields for TG and PG are due to the width-thickness ratio (>1). The slight decrease of the MR at high magnetic fields is probably caused by small misalignments of the wire direction with H. The different MR observed between the LG and any of the others is caused by the anisotropic magnetoresistance, AMR $= 100(\rho_{//} - \rho_{\perp})/\rho$. The value at 300 K is positive, around 0.65–0.7%, as previously found in Co NWs [42] and thin films [52].

The AMR was further studied as a function of temperature by means of the Planar Hall effect (PHE), in a 45° configuration between I and H, where the transversal voltage is maximum [53]. In Figure 16.16(b) the transversal resistivity ρ_{xy} in this configuration is shown for several temperatures. From ρ_{xy} and under saturation, the AMR ratios can be calculated as AMR(%) $= 200\rho_{xy}/\rho$. The AMR slightly increases when lowering temperature, from 0.65% at 300 K up to 0.8% at 25 K, in good agreement with EBL-fabricated Co NWs [42]. Abrupt switches occur around 150 Oe, the switching field increasing when temperature diminishes.

For applications, it is preferable to study in detail the magnetization reversal in narrow and thin Co NWs. Local magnetometry is today possible with several techniques. In particular, we have used the magneto-optical Kerr effect (MOKE) of individual Co NWs using focused laser beams of size about 5 µm as described in reference [38]. It has been possible to grow narrow and thin Co NWs in single-pixel mode and in raster scan mode. As an example, in Figure 16.17(a) NW with the following dimensions are shown: width of 150 nm, thickness of 40 nm, and length of 4 µm. Obviously, NWs with larger dimensions are easier to grow, and we have explored the magnetization reversal of NWs with thickness of 50 and 200 nm and width in the range 150 nm to 2 µm. The MOKE signal in all the studied NWs is good enough to determine the coercive fields in a magnetization reversal process. The MOKE signal in the smallest NW studied is displayed in Figure 16.17(b), showing an abrupt change when the magnetization reversal takes place. A systematic study of the coercive field as a function of the NW width at fixed thickness has been carried out [48]. The coercivity decreases with the width and increases with the thickness. This trend has been previously observed in wires of different materials [38; 41–43], and can be understood in magnetostatic terms, since the demagnetizing field depends in a first order approximation on the thickness/width ratio. This indicates that magnetocrystalline anisotropy plays a secondary role in these Co NWs due to their polycrystalline nature, and thus coercivity can be explained in a first approximation by the shape anisotropy. Micromagnetic simulations [48] together with electron-energy-loss spectroscopy results [54] indicate that the halo of the Co NW is likely non-ferromagnetic due to the lower Co content, all the ferromagnetism being concentrated on the central part of the NW.

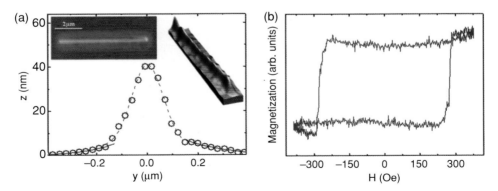

FIGURE 16.17: (a) The cross-sectional AFM profile of a Co NW with aspect ratio $= 26$ and the corresponding fits to a Gaussian dependence (the central part) and an exponential dependence (the edge part, also called halo). The insets show the 2D-SEM and 3D-AFM images of this NW. (b) MOKE hysteresis loop of the same NW showing an abrupt coercive field slightly below 300 Oe.

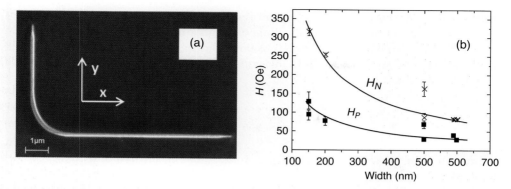

FIGURE 16.18: (a) SEM image of one of the grown L-shaped Co NWs (width = 200 nm) for the study of the domain-wall conduit behavior. (b) Dependence of the domain-wall nucleation and propagation fields in the L-shaped Co NWs as a function of the wire width.

Most interestingly, the domain walls in these NWs show good conduit behavior, which is required for many of the proposed applications of magnetic NWs [55]. L-shaped NWs such as that shown in Figure 16.18(a) are suitable to study propagation and nucleation fields of domain walls. By using appropriate field routines, it is possible to investigate the field required to move a domain wall initially at the corner of the L-shaped NW (propagation field H_p) as well as the field required to nucleate a domain wall (nucleation field H_N) [5]. For several applications, we are interested in the propagation of domain walls without the creation of new ones, the so-called good conduit behavior, which implies lower propagation than nucleation fields. As shown in Figure 16.18(b), these Co NWs have lower propagation fields than nucleation fields for widths below 600 nm. This is very promising for manipulation of domain walls, even though the absolute field values obtained are higher than in Permalloy [55], which is a softer magnetic material. MFM studies on these NWs [56] indicate that domain walls produce an intense MFM contrast, resulting in an exciting playground for basic studies of magnetization reversal, types of domain walls, domain-wall pinning and depinning, etc.

In summary, it is currently possible to grow high-purity Co NWs with the FEBID technique. Their dimensions can be small enough to perform basic nanomagnetism studies. In particular, the good domain-wall conduit behavior seems very promising for applications involving the manipulation of domain walls.

4 FEBID OF FE-C-O NANOSTRUCTURES

The partial oxygen pressure in the chamber during FEBID can lead to the partial oxidation of the deposits affecting their magnetic properties as shown in [57]. In contrast, oxygen free Fe or Fe-C deposits have been reported when using FEBID in UHV conditions [58–60], but their magnetic properties were not characterized. Therefore, besides the magneto-transport characterization techniques reported above in the case of Co-C and Co deposits, we know of only one other example of magnetic characterized *nanostructures* deposited with this technique. This has been done by Takeguchi and colleagues [61–63], who have deposited freestanding nanorods with widths around 10 nm of Fe and FePt embedded in the carbon matrix.

In order to characterize the magnetic properties they have used the electron holography method inside a TEM. Unlike the image formed in an SEM, this technique provides full information about both the amplitude and the phase of electrons, using the interference between a direct electron beam and the beam passing through the nanorod. Any shift in the reconstructed phase image at the nanorod position indicates the presence of an electromagnetic field (Aharonov-Bohm effect), which allows the calculation of the residual magnetic flux density B_r in the nanorod:

$$\Delta\varphi = \pi \frac{\phi}{\phi_0} = 2\pi \frac{e}{h} B_r S. \tag{6}$$

Before the measurement, the freestanding nanorods were magnetized in the magnetic field of objective lens of the transmission electron microscope, which was then switched off.

The deposited material was in amorphous phase, containing iron, carbon, and oxygen elements in the volume and iron oxide nanocrystals near their surfaces. The measured phase shift $\Delta\phi$ of 7.3 rad was calculated to correspond to a remanent magnetic field $B_r = 0.61$ T (see Figure 16.19). After that, a post-deposition heat treatment at about 600°C in UHV (10^{-7} Pa) resulted in the transformation into a crystalline alpha-iron phase, while their shapes were maintained. The calculated remanent magnetic field was smaller, $B_r = 0.45$ T, which could be due to a partial oxidation of the Fe crystals in the surface, or to the change of their size/shape.

In Figure 16.20, the residual magnetic flux density B_r is shown to depend linearly with the Fe atomic concentration, which has been estimated using EDX measurements from the

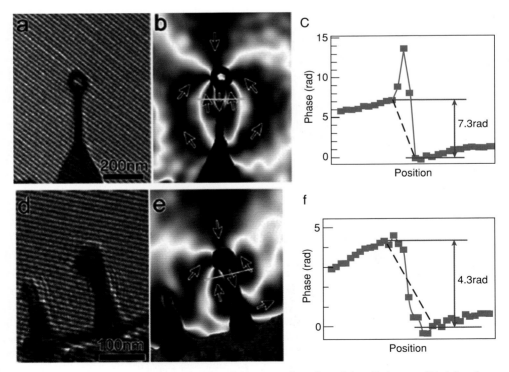

FIGURE 16.19: Electron holograms (a), (d) and corresponding phase intensity images (b), (e) and profiles of the phase distribution along the yellow line crossing the nanorod (c), (f), as grown (top) and after annealing (bottom). Reprinted with permission from [61], copyright (2005), IOP.

intensity of the Fe-L line, on samples without thermal treatment. In [63], the FePt alloys were first annealed *in situ* at 600°C, before any magnetic or structural characterization was performed. The EDX analyses indicated that there were iron, platinum and carbon elements inside the nanorod, while high resolution TEM (Figure 16.21) showed that the annealed nanorod was composed of crystalline $L1_0$-FePt nanoparticles encapsulated in a thin layer of carbon-containing sheath. The residual magnetic flux density B_r for these deposits was calculated to be 1.53T, revealing a hard magnetic nature.

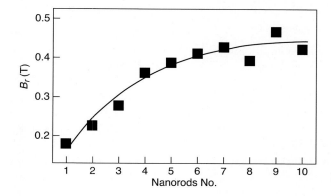

FIGURE 16.20: Increasing remanent magnetic flux density B_r for as grown deposits with increasing iron concentration. Reprinted with permission [62], copyright (2006), Springer.

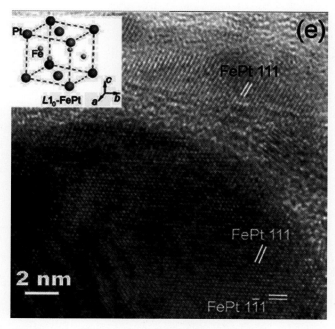

FIGURE 16.21: High-resolution TEM image of a FePt nanorod after annealing—viewing along FePt [10–1] orientation—and the inset is an atomic structure model of $L1_0$-FePt. Reprinted with permission from [63], copyright (2005), American Institute of Physics.

5 CONCLUSIONS

Magnetism brings a new dimension to the FEB-FIB induced deposits of nanostructures, as magnetic functional nanostructures have been recently grown. For instance, controlling the deposition parameters allows access to granular nanomaterials, which can be tuned for the detection of low magnetic fields in Hall sensor devices. Such nano-Hall sensors are very attractive for application in biosensing, where high spatial resolution is required. On the other hand, high-purity Co and Fe deposits grown by FEB and FIB show characteristics close to pure UHV deposited metals, opening the way for their use in magnetoelectronic devices. In particular, the manipulation of domain walls in Co nanowires has been identified as a promising topic with applications in magnetic logic, sensing and storage.

ACKNOWLEDGMENTS

EMPA acknowledges support by EC funding through the 6th Framework NEST Pathfinder (Synthetic Biology) program for the BIONANO-SWITCH Project Grant No 043288. J.M.D.T. and A.F.-P. acknowledge the collaboration on this topic with R. Córdoba, M. R. Ibarra, D. Petit, L. O'Brien, H. T. Zeng, E. R. Lewis, D. E. Read, and R. P. Cowburn. This work was supported by the Spanish Ministry of Science (through projects MAT2008–03636-E and MAT2008–06567-C02, including FEDER funding), and the Aragon Regional Government.

REFERENCES

[1] Utke, I., P. Hoffmann, and J. Melngailis. Gas-assisted focused electron beam and ion beam processing and fabrication. *J. Vac. Sci. Technol. B* 26 (2008): 1197–1276.

[2] Botman, A., J. J. L. Mulders, and C. W. Hagen. Creating pure nanostructures from electron-beam-induced deposition using purification techniques: a technology perspective. *Nanotechnology* 20 (2009): 372001.

[3] Bernau, L., M. S. Gabureac, I. Utke, J. Michler, and P. Hoffmann. Tunable Nanosynthesis of Composite Materials by Electron-Impact Reaction. *Angew. Chem. Int. Ed.* 49 (2010): 8880–8884.

[4] Fernandez-Pacheco, A., J. M. De Teresa, R. Cordoba, and M. R. Ibarra. Magnetotransport properties of high-quality cobalt nanowires grown by focused-electron-beam-induced deposition. *J. Phys. D: Appl. Phys.* 42 (2009): 055005.

[5] Fernandez-Pacheco, A., et al. Domain wall conduit behavior in cobalt nanowires grown by focused electron beam induced deposition. *Appl. Phys. Lett.* 94 (2009): 192509.

[6] Gabureac, M. S., L. Bernau, I. Utke, and G. Boero. Granular Co-C nano-Hall sensors by focused-beam-induced deposition. *Nanotechnology* 21 (2010): 115503.

[7] Pakhomov, A. B., X. Yan, and B. Zhao. Giant Hall effect in percolating ferromagnetic granular metal-insulator films. *Appl. Phys. Lett.* 67 (1995): 3497–3499.

[8] Boero, G., et al. Submicrometer Hall devices fabricated by focused electron-beam-induced deposition. *Appl. Phys. Lett.* 86 (2005): 042503.

[9] Togawa, K., H. Sanbonsugi, A. Sandhu, M. Abe, H. Narimatsu, K. Nishio, and H. Handa. Detection of magnetically labeled DNA using pseudomorphic AlGaAs/InGaAs/GaAs heterostructure micro-Hall biosensors. *J. Appl. Phys.* 99 (2006): 08P103.

[10] Freitas, P. P., R. Ferreira, S. Cardoso, and F. Cardoso. Magnetoresistive sensors. *J. Phys.: Condens. Matter* 19 (2007): 165221.

[11] Dolabdjian, C., A. Qasimi, D. Bloyet, and V. Mosser. Spatial resolution of SQUID magnetometers and comparison with low noise room temperature magnetic sensors. *Physica C: Superconductivity* 368 (2002): 80–84.

[12] Matveev, V. N., V. I. Levashov, V. T. Volkov, O. V. Kononenko, A. V. Chernyh, M. A. Knjazev, and V. A. Tulin. Fabrication and use of a nanoscale Hall probe for measurements of the magnetic field induced by MFM tips. *Nanotechnology* 19 (2008): 475502.

[13] Novoselov, K. S., S. V. Morozov, S. V. Dubonos, M. Missous, A. O. Volkov, D. A. Christian, and A. K. Geim. Submicron probes for Hall magnetometry over the extended temperature range from helium to room temperature. *J. Appl. Phys.* 93 (2003): 10053–10057.

[14] Lionnet, T., S. Joubaud, R. Lavery, D. Bensimon, and V. Croquette. Wringing out DNA. *Phys. Rev. Lett.* 96 (2006): 178102.

[15] Gijs, M. A. M. *Magnetic particle handling in lab-on-a-chip microsystems*, in *Conference of the NATO Advanced Study Institute on Magnetic Nanostructures for Micro-Electromechanical Systems and Spintronic Applications*, B. Azzerboni, et al., eds. Springer: Catona, Italy: Springer, 2006, pp. 153–165.

[16] Seidel, R., and C. Dekker. Single-molecule studies of nucleic acid motors. *Current Opinion in Structural Biology* 17 (2007): 80–86.

[17] Neuman, K. C., T. Lionnet, and J. F. Allemand. Single-molecule micromanipulation techniques. *Annual Review of Materials Research* 37 (2007): 33–67.

[18] Candini, A., G. C. Gazzadi, A. Di Bona, M. Affronte, D. Ercolani, G. Biasiol, and L. Sorba. Hall nano-probes fabricated by focused ion beam. *Nanotechnology* 17 (2006): 2105–2109.

[19] Mihajlovic, G., P. Xiong, S. von Molnar, K. Ohtani, H. Ohno, M. Field, and G. J. Sullivan. Detection of single magnetic bead for biological applications using an InAs quantum-well micro-Hall sensor. *Appl. Phys. Lett.* 87 (2005): 112502.

[20] Pross, A., A. I. Crisan, S. J. Bending, V. Mosser, and M. Konczykowski. Second-generation quantum-well sensors for room-temperature scanning Hall probe microscopy. *J. Appl. Phys.* 97 (2005): 096105.

[21] Besse, P. A., G. Boero, M. Demierre, V. Pott, and R. Popovic. Detection of a single magnetic microbead using a miniaturized silicon Hall sensor. *Appl. Phys. Lett.* 80 (2002): 4199–4201.

[22] Sandhu, A., K. Kurosawa, M. Dede, and A. Oral. 50 nm Hall sensors for room temperature scanning Hall probe microscopy. *Jpn. J. Appl. Phys. Part 1 - Regul. Pap. Short Notes Rev. Pap.* 43 (2004): 777–778.

[23] Sandhu, A., H. Sanbonsugi, I. Shibasaki, M. Abe, and H. Handa. High sensitivity InSb ultra-thin film micro-hall sensors for bioscreening applications. *Jpn. J. Appl. Phys.* 43 (2004): L868–L870.

[24] Bando, M., T. Ohashi, M. Dede, R. Akram, A. Oral, S. Y. Park, I. Shibasaki, H. Handa, and A. Sandhu. High sensitivity and multifunctional micro-Hall sensors fabricated using InAlSb/InAsSb/InAlSb heterostructures. *J. Appl. Phys.* 105 (2009): 07E909.

[25] Kotzler, J., and W. Gil. Anomalous Hall resistivity of cobalt films: Evidence for the intrinsic spin-orbit effect. *Phys. Rev. B* 72 (2005): 060412(R).

[26] Berger, L., and G. Bergmann. The Hall effect of ferromagnets. In *The Hall effect and its applications (Proceedings of the Commemorative Symposium; Baltimore, Md., November 13, 1979)*, C. L. Chien and C. R. Westgate, eds. New York and London: Plenum Press, 1980, pp. 55–76.

[27] Fert, A., and D. Lottis. Magnetotransport phenomena. In *Concise encyclopedia of magnetic & superconducting materials*, J. Evetts, R. W. Cahn, and M. B. Bever, eds. New York: Pergamon Press, 1992, pp. 287–296.

[28] Gerber, A., et al. Extraordinary Hall effect in magnetic films. *J. Magn. Magn. Mater.* 242–245 (2002): 90–97.

[29] Gerber, A., A. Milner, A. Finkler, M. Karpovski, L. Goldsmith, J. Tuaillon-Combes, O. Boisron, P. Mélinon, and A. Perez. Correlation between the extraordinary Hall effect and resistivity. *Phys. Rev. B* 69 (2004): 224403.

[30] Porrati, F., R. Sachser, and M. Huth. The transient electrical conductivity of W-based electron-beam-induced deposits during growth, irradiation and exposure to air. *Nanotechnology* 20 (2009): 195301.

[31] Brouers, F., A. Granovsky, A. Sarychev, and A. Kalitsov. The influence of boundary scattering on transport phenomena in ferromagnetic metal—dielectric nanocomposites. *Physica A: Statistical and Theoretical Physics* 241 (1997): 284–288.

[32] Socolovsky, L. M., C. L. P. Oliveira, J. C. Denardin, M. Knobel, and I. L. Torriani. Nanostructure of granular Co-SiO2 thin films modified by thermal treatment and its relationship with the giant Hall effect. *Phys. Rev. B* 72 (2005): 184423.

[33] Lau, Y. M., P. C. Chee, J. T. L. Thong, and V. Ng. Properties and applications of cobalt-based Material produced by electron-beam-induced deposition. *J. Vac. Sci. Technol. A*, 20 (2002): 1295–1302.

[34] Utke, I., J. Michler, P. Gasser, C. Santschi, D. Laub, M. Cantoni, P.th A. Buffat, C. Jiao, and P. Hoffmann. Cross section investigations of compositions and sub-structures of tips obtained by focused electron beam induced deposition. *Adv. Eng. Mater.* 7 (2005): 323–331.

[35] Park, I. W., M. Yoon, Y. M. Kim, Y. Kim, H. Yoon, H. J. Song, V. Volkov, A. Avilov, and Y. J. Park. Magnetic properties and microstructure of cobalt nanoparticles in a polymer film. *Solid State Communications* 126 (2003): 385–389.

[36] Knobel, M., W. C. Nunes, L. M. Socolovsky, E. De Biasi, J. M. Vargas, and J. C. Denardin. Superparamagnetism and other magnetic features in granular materials: A review on ideal and real systems. *J. Nanosci. Nanotechno.* 8 (2008): 2836–2857.

[37] Utke, I., T. Bret, D. Laub, P. Buffat, L. Scandella, and P. Hoffmann. Thermal effects during focused electron beam induced deposition of nanocomposite magnetic-cobalt-containing tips. *Microelectronic Engineering* 73–74 (2004): 553–558.

[38] Cowburn, R. P. Property variation with shape in magnetic nanoelements. *J. Phys. D-Appl. Phys.* 33 (2000): R1–R16.

[39] Chappert, C., A. Fert, and F. N. Van Dau. The emergence of spin electronics in data storage. *Nat. Mater.* 6 (2007): 813–823.

[40] Wernsdorfer, W., E. B. Orozco, K. Hasselbach, A. Benoit, B. Barbara, N. Demoncy, A. Loiseau, H. Pascard, and D. Mailly. Experimental evidence of the Neel-Brown model of magnetization reversal. *Phys. Rev. Lett.* 78 (1997): 1791–1794.

[41] Uhlig, W. C., and J. Shi. Systematic study of the magnetization reversal in patterned Co and NiFe nanolines. *Appl. Phys. Lett.* 84 (2004): 759–761.

[42] Leven, B., and G. Dumpich. Resistance behavior and magnetization reversal analysis of individual Co nanowires. *Phys. Rev. B* 71 (2005): 064411.

[43] Brands, M., R. Wieser, C. Hassel, D. Hinzke, and G. Dumpich. Reversal processes and domain wall pinning in polycrystalline Co-nanowires. *Phys. Rev. B* 74 (2006): 174411.

[44] Cowburn, R., and D. Petit. Spintronics: Turbulence ahead. *Nat. Mater.* 4 (2005): 721–722.

[45] Parkin, S. S. P., M. Hayashi, and L. Thomas. Magnetic domain-wall racetrack memory. *Science* 320 (2008): 190–194.

[46] Beach, G. S. D., M. Tsoi, and J. L. Erskine. Current-induced domain wall motion. *J. Magn. Magn. Mater.* 320 (2008): 1272–1281.

[47] Martin, J. I., J. Nogues, K. Liu, J. L. Vicent, and I. K. Schuller. Ordered magnetic nanostructures: fabrication and properties. *J. Magn. Magn. Mater.* 256 (2003): 449–501.

[48] Fernandez-Pacheco, A., et al. Magnetization reversal in individual cobalt micro- and nanowires grown by focused-electron-beam-induced-deposition. *Nanotechnology* 20 (2009): 475704.

[49] Córdoba, R., J. Sesé, J. M. De Teresa, and M. R. Ibarra. In-situ purification of cobalt nanostructures grown by focused-electron-beam-induced deposition at low beam current. *Microelectronic Engineering* 87 (2010): 1550.

[50] Rubinstein, M., F. J. Rachford, W. W. Fuller, and G. A. Prinz. Electrical transport-properties of thin epitaxially grown iron films. *Phys. Rev. B* 37 (1988): 8689–8700.

[51] Fernández-Pacheco, A. Studies of nanoconstrictions, nanowires and Fe3O4 thin films. *Springer Thesis*, 2011.

[52] Gil, W., D. Gorlitz, M. Horisberger, and J. Kotzler. Magnetoresistance anisotropy of polycrystalline cobalt films: Geometrical-size and domain effects. *Phys. Rev. B* 72 (2005): 134401.

[53] Campbell, I. A., and A. Fert. Transport properties of ferromagnets. In *Ferromagnetic materials*, E. P. Wohlfarth, ed. Amsterdam: North-Holland, 1982, pp. 747–804.

[54] Córdoba R. et al., submitted to Nanoscale Research Letters.

[55] Allwood, D. A., G. Xiong, C. C. Faulkner, D. Atkinson, D. Petit, and R. P. Cowburn. Magnetic domain-wall logic. *Science* 309 (2005): 1688–1692.

[56] M. Jaafar, L. Serrano-Ramón, O. Iglesias-Freire, A. Fernández-Pacheco, M. Ricardo Ibarra, J. M de Teresa and A. Asenjo, Hysteresis loops of individual Co nanostripes measured by magnetic force microscopy. Nanoscale Research Letters 6 (2011): 407.

[57] Lavrijsen R., R. Cordoba, F. J. Schoenaker, T. H. Ellis, B. Barcones, J. T. Kohlhepp, H. J. M. Swagten, B. Koopmans, J. M. De Teresa, C. Magen, M. R. Ibarra, P. Trompenaars, J. J. L. Mulders. Fe:O:C grown by focused-electron-beam-induced deposition: magnetic and electric properties. *Nanotechnology* 22.2 (2011): 025302.

[58] Tanaka M., M. Shimojoa, M. Takeguchia, K. Mitsuishia, K. Furuya. Formation of iron nano-dot arrays by electron beam-induced deposition using an ultrahigh vacuum transmission electron microscope. *Journal of Crystal Growth* 275 (2005): e2361–e2366.

[59] Lukasczyk T., M. Schirmer, H.-P. Steinrück, and H. Marbach. Electron-Beam-Induced Deposition in Ultrahigh Vacuum: Lithographic Fabrication of Clean Iron Nanostructures. *Small*, 4 (2008): 841–846.

[60] Walz M.-M., M. Schirmer, F. Vollnhals, T. Lukasczyk, H.-P. Steinrück, and H. Marbach "Electrons as 'Invisible Ink': Fabrication of Nanostructures by Local Electron Beam Induced Activation of SiOx", *Angewandte Chemie International Edition*, 49 (2010): 4669–4673.

[61] Takeguchi, M., M. Shimojo, and K. Furuya. Fabrication of magnetic nanostructures using electron beam induced chemical vapour deposition. *Nanotechnology* 16 (2005): 1321–1325.

[62] Takeguchi, M., M. Shimojo, K. Mitsuishi, M. Tanaka, R. Che, and K. Furuya. Fabrication of nanostructures with different iron concentration by electron beam induced deposition with a mixture gas of iron carbonyl and ferrocene, and their magnetic properties. *Journal of Materials Science* 41 (2006): 4532–4536.

[63] Che, R. C., M. Takeguchi, M. Shimojo, W. Zhang, and K. Furuya. Fabrication and electron holography characterization of FePt alloy nanorods. *Appl. Phys. Lett.* 87 (2005): 223109.

C H A P T E R   1 7

METAL FILMS AND NANOWIRES DEPOSITED BY FIB AND FEB FOR NANOFABRICATION AND NANOCONTACTING

Alfredo R. Vaz and Stanislav A. Moshkalev

1 INTRODUCTION

Development of high-resolution microscopy techniques during the last decades made possible the discovery and systematic studies of new nanostructured materials like nanoparticles, nanotubes, nanowires, nanosheets, and others. Currently, in the process of integration of nanomaterials in micro and nanodevices, the development of versatile and reliable techniques for manipulation, assembling, functionalization, mechanical and electrical joining (contacting) is one of the main challenges in nanotechnologies. In this context, the formation of reliable electrical contacts between nanostructures and other device components using new-generation microscopy tools is an important issue that must be addressed.

Carbon nanotubes (CNTs) and metal and metal oxide nanowires (NWs) are considered as possible building blocks in future nanoelectronics for applications in field-effect transistors (FETs), interconnects, sensors and actuators, etc. [1]. There are two alternative methods of integration of nano-objects in electronic devices: controlled placement (using previously fabricated, selected, purified, modified nanostructures) and controlled synthesis directly on a desired location. Controlled synthesis of CNTs or NWs (with predetermined position, growth direction, and properties) continues to be a very challenging task. For instance, for growth of high quality carbon nanotubes using a chemical vapor deposition (CVD) process directly on a silicon chip, quite high temperatures (up to 600°C–800°C) are required, practically incompatible with a conventional Si planar technology. Lower temperatures are usually required for NW synthesis. Alignment and controlled placement of previously grown and processed 1D nanostructures can be done using a number of techniques such as deposition from solutions using di-electrophoresis (DEP) with AC electric fields, alignment in gas or liquid flows, surface acoustic waves [2–4], optical tweezers [5; 6], mechanical or magnetic (when applicable) manipulation [7; 8], and others. Joining and formation of contacts in nanoscale devices can be achieved using several techniques: atomic diffusion, soldering, welding, and contact fabrication using FIB and FEB [9–15]. Each method has advantages and specific areas of application. In contrast to "cold" processing using beams, the first three methods usually require relatively high temperatures, which can be provided by conventional thermal heating, laser and Joule heating (by electric current), or by ultrasonic processing. Among these methods, Joule heating provides better localization of energy in the contact area to be improved, and thus seems to be the most promising. It should be noted also that the use of nanoparticles in soldering pastes or inks [11] can allow for significant reduction of the process temperature. Higher resolution (smaller contacting areas) is another advantage of focused particle beam based methods; however, the sequential processing by beams implies a low process throughput.

The DEP process has been successfully applied to fabricate individual devices based on nanotubes or other 1D and 2D polarizable objects (like NWs, thin graphite, or graphene sheets) [15; 16], and can be also scaled up for massive parallel deposition of these objects in multiple sites in a chip [3]. It should be noted, however, that when liquid solutions are used for dispersion of nanotubes, surfactants are usually employed to avoid their agglomeration due to van der Waals forces. The presence of surfactants, surface oxide layers, and other impurities impedes the formation of high-quality ohmic contacts between electrodes and nanotubes directly in a DEP process. Then, additional steps like electroless or electrochemical metal film deposition to cover the NTs and/or thermal annealing in vacuum are needed to improve the quality of contacts and get low electrical resistances [3; 4]. Note that in some cases, when the heating of nanotubes by Joule effect is significant, the quality of electrical and thermal contacts can be also important for the entire device performance [17].

For the fabrication of electrical contacts in microdevices, optical and electron beam lithography techniques, followed by thin metal film deposition and lift-off, are routinely used. In order to meet the requirements of fast prototyping with nanoscale resolution, methods of metal deposition by focused ion and electron beams were successfully introduced during the last decade. In particular, a dual beam FIB/SEM technology enabling maskless processing by ion beams and real time scanning electron imaging of samples was proved to be very efficient and fast compared with other current micro/nanofabrication methods [18; 19]. "Direct writing" by FIB is extensively used for fabrication of nanostructures and making interconnects for their electrical and mechanical characterization [18]. Two different kinds of strategies are utilized for the electrical characterization of nanostructures using FIB or FEB: (1) contacts are fabricated over the existing object (in this case, possible damage of probed structures by energetic particles, especially ions, should be considered); or (2) the pads and/or contact structure are created first, followed by the controlled placement of the object over the contacts.

It is important to note that a number of secondary effects like amorphization of surface layers by energetic ions, material redeposition, and "halo" effect (also called extended metal deposition or overspray [19]) observed around the main deposition site can seriously affect the performance of nanocontacts or even make the adequate characterization of nanostructures impossible.

The choice of metal for nanocontacting is determined by a number of factors like chemical and physical interactions between the metal and tested material, low contact resistance, chemical and thermal stability of the interface region, and the effect of possible impurities. Thus, in principle, different metals should be used to get the best quality contacts for different nanomaterials. Of course, in practice the choice is often restricted by established technologies and is a result of a trade-off between various technical requirements.

Concerning contacting with carbon nanostructured materials, Matsuda et al. [20] considered the resistance of metal electrode-carbon nanotube contacts for five metals, using first-principles quantum mechanical methods. Only side contacts were considered. They found that Ti leads to the lowest contact resistance, followed by Pd, Pt, Cu, and Au. This correlates well with the predicted cohesive strength of the electrode-carbon interface, and basically confirms existing experimental trends [21]. However, the high reactivity of Ti electrodes in contact with a nanotube (graphene layer) can distort the nanotube structure and also may lead to the electrode material oxidation. Pt-C layers deposited by FEB (and FIB, when tubes are deposited over pre-fabricated contacts, as FIBID can damage nanotubes) thus should be considered as an appropriate material to contact nanotubes. Note that better results can be expected for Pd (or Pd-C) and especially for Ti (or Ti-C) layers, both not available commercially for FIBID or FEBID at present. Tungsten (W) also can form stable carbides and its deposition by FIB/FEB therefore is interesting for CNT contacting; however, most of the results already published are obtained with Pt deposition.

It should be emphasized that whereas FIBID is not used to fabricate nanocontacts *over* nanotubes due to its damaging character, this process can be successfully employed for contacting nanowires like GaN. Tham et al. [22] reported unusually low resistance and reduced Schottky barrier heights for contacts fabricated by Pt FIBID to n-type GaN NWs. This was attributed to formation of a thin (2–3 nm) amorphous layer just beneath the contact in GaN, and the possible creation of interface states that could pin the Fermi level within the nanowire.

2 FIB AND FEB INDUCED METAL DEPOSITION

The fundamentals of the metal deposition by FIBID and FEBID are considered in detail in other chapters of this book, in particular in Chapter 23 by de Teresa et al. For purposes of nanocontacting, films based on Pt and W are currently employed, and the mostly used gas precursors for deposition are carbon containing $(CH_3)_3Pt(CpCH_3)$ or $(CH_3)_3CH_3C_5H_4Pt$ and $W(CO)_6$, respectively. There are also recent reports of deposition of Pt using carbon-free gases like $Pt(PF_3)_4$ [23], with very promising characteristics but not yet widely available.

2.1 PT FIBID

There are a number of reports on the structural and electrical properties of Pt thin films obtained using FIBID and carbon-containing gas sources [24–28]. Deposits contain not only Pt but considerable amounts of C, Ga, and O. Due to the complexity of the process, results depend on many factors (such as beam current and energy, local gas pressure, dwell and refresh time), some of them being sometimes difficult to control and compare (vacuum conditions, surface properties, gas impurities, beam quality/profile), so they can vary significantly from one laboratory to another. Film resistivities determined in different experiments [4; 24–29] show relatively wide variation, being basically within the range $(1 - 5) \times 10^{-3}$ Ω cm for relatively thick films ($\geq$50–100 nm). These values are about two orders of magnitude higher compared with that for bulk Pt, 1.06×10^{-5} Ω cm. Atomic composition of films was analyzed using an EDX method, and deposits were shown to contain roughly from 20 to 50 at.% Pt, with small fractions of oxygen (from 0 to 5 at.%) and Ga (5–15 at.%), while the major fraction is carbon, varying from 45–55 to 70 at.%. In FIBID experiments, the energy of Ga ions is usually fixed at 30 keV, beam current at low level (10–30 pA), dwell time of 0.2–1 μs, room substrate temperature. For higher currents and longer dwell times, milling rather than deposition occurs [30].

Comparative analysis of results obtained by different groups has been done recently by Fernandez-Pacheco et al. [28]. Electrical and structural film properties have been investigated, and detailed analysis of atomic composition in the films by EDX has shown that it is thickness dependent, and changes from Pt:C:Ga=18:68:14 at.% to Pt:C:Ga=35:40:25 at.% for 25–50 nm and 250 nm film thickness, respectively. That means also that the interfacial area is carbon rich (it also can contain some amount of Si and O from the underlying oxidized Si substrate), and as the film becomes thicker, the Pt content increases together with Ga, which is implanted in the film. This composition gradient can be explained by the changing conditions of a beam interaction with a substrate during the process (indeed, the substrate is SiO_2 in the beginning, and changes to a Pt-C-Ga film after a $\sim$50 nm thick film is deposited), and also by possibly changing conditions of heat dissipation.

Structural analysis by TEM performed by Langford et al. [25] has shown the presence of 6–8 nm size Pt grains (nanocrystals) surrounded by amorphous carbon matrix. Smaller grain sizes, 2–4 nm, were reported by de Teresa et al. [27; 29]. XPS measurements have shown that a main part of carbon deposit ($\sim$55%) [27] has sp^2 hybridization (graphitic form), with smaller sp^3 contribution.

Figure 17.1 shows cross-sections of a Pt-C layer deposited by FIBID over Si. Before cross-sectioning by FIB, a FEBID Pt protective layer and Au layer were deposited over the FIBID layer. A similar procedure was employed by de Teresa et al. [27]. Important features can be observed in the images: (1) metal grains (brighter spots) in the superior part of the FIBID Pt-C film (Figure 17.1(b)) tend to agglomerate forming chains (partly, this agglomeration could occur during final milling by Ga ions); (2) an interfacial layer is formed between the Si substrate and the deposited film (Figure 17.1(b)), where smaller grains can be seen, without visible agglomeration; the thickness of the interfacial layer can be estimated roughly as 15–25 nm for 30 keV ion energy; (3) amorphized Si surface layer with a thickness of $\sim$40–50 nm for 30 keV (Figure 17.1(b)) and of 25–30 nm for 10 keV (Figure 17.1(d)) ion energy, respectively; (4) lateral deposition produces a metal film with gradually decreasing thickness ("halo"), spreading up to 300–500 nm from the deposited pattern edge (Figures 17.1(c,d)).

In accordance with compositional changes, electrical properties of films also change strongly with film thickness, demonstrating two different types of behavior. Basically, as the thickness increased, an *in situ* film resistivity was found first to decrease abruptly, from very high values to $\sim(2 - 3) \times 10^{-3}$ Ω cm and then to change very slowly, reaching $(0.7 - 1.0) \times 10^{-3}$ Ω cm for 100–200 nm thick films [4; 27]. The critical value of thickness t_{crit} when the nanowire resistivity starts to change slowly (truncation of the resistivity vs. thickness curve), was reported to be $\sim$50 nm [28; 29]. A smaller value, $t_{crit} \sim 15 - 20$ nm, was found in experiments by Vaz et al. [4] (note that even smaller values were reported in studies of a "halo" effect by Notargiakomo et al. [26]). In part, the difference could be attributed to the fact that measurements were performed under different conditions: films with lateral dimensions of 150 μm $\times$ 150 μm with relatively high ion current of 0.5 nA [4], and 500 nm wide NWs at

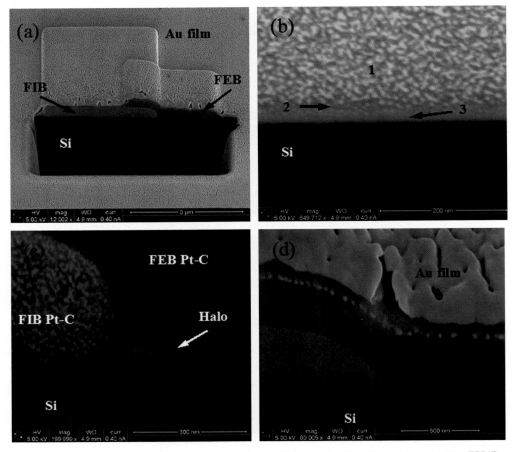

FIGURE 17.1: SEM images of cross-sections of Pt deposits made by FIB milling. (a) Protective FEBID deposit is made to cover partly the 300 nm thick FIBID layer. Before cross-sectioning, the area was also covered by a thin Au layer, to avoid the charging effect. (b) Cross-section fragment showing the fine structure of FIBID Pt deposit in the vicinity of a Si substrate: 1- main FIBID layer, 2–intermediate layer, 3–amorphized Si surface layer, (c,d,): Profiles of the FIBID Pt deposit. Ga ion energy of 30 keV (a-c) and 10 keV (d).

low current of $\sim$10 pA [27]. It should be noted also that the film composition, as determined by EDX, changes only slightly for thickness between 25 and 50 nm [27].

Further, in experiments with 30-μm long Pt linear resistors with cross-sections varying from 1×1 to $0.05 \times 0.05\,\mu m^2$, Vaz et al. [4] found that the measured resistance followed well the linear fitting ($R = \rho L/A$, where ρ, L, and A are the resistivity, length, and cross-section area, and $A = wt$, w and t are the resistor width and thickness) using the ρ values as determined from the four points thin film measurements practically for all cross-sections except for 50×50 nm, when the resistance started to grow much faster (Figure 17.2). On the other hand, in an earlier work, Smith et al. [24] observed slow changes of resistivity only for film thickness exceeding 200 nm (the minimum film thickness used was 100 nm). The reasons of these discrepancies are not clear at the moment, but it is likely that even small variations in film composition in the interfacial layer could be responsible for significant changes in the

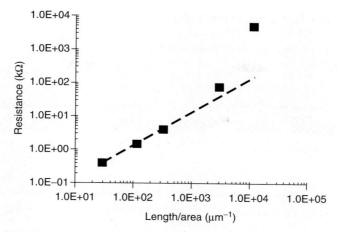

FIGURE 17.2: Resistance of 30 μm long FIBID Pt resistors vs. L/A ratio, dashed line – linear fitting [4]. Reprinted with permission from Springer Science+Business Media: *J. Materials Science* 43 (2008): 3429–3434, A. R. Vaz et al.

film resistance and critical film thickness. To sum up, for reliable contacting of nanostructures using FIBID, Pt-C nanowires with thickness at least $t = 50$ nm (and width $w = 2 \div 3t$) should be used.

Annealing of the 200 nm thick Pt-C films deposited by FIBID (500°C, one hour, forming gas atmosphere) did not result in significant improvement in a film resistance [25], indicating the absence of structural and compositional changes in thick films.

2.2 FEBID OF PLATINUM

For platinum films deposited by FEB using the same gas precursor $(CH_3)_3Pt(CpCH_3)$, much higher resistivities and wider variation of results were reported: $10^{-1} - 1$ Ω cm [25], $10^{-2} - 0.6$ Ω cm [4], $1 - 10$ Ω cm [28]. Typical deposition conditions are: 5 keV, 1 nA, 1 μs dwell time [25]. It should be noted that in spite of different mechanisms responsible for the precursor molecules dissociation and Pt deposition using ion and electron beams, certain similarities can be found between high-resistance (low thickness) FIB and FEB deposits [27; 28]. In both cases, amorphous carbon is the main component in film composition, and the average size of Pt nanocrystals is about the same (a few nanometers), but implanted Ga is only present for FIBID.

For 200 nm thick FEBID Pt-C films, Langford et al. [25] found higher carbon (65–75 at.%) and lower Pt content (15–25 at.%) and slightly smaller Pt grains sizes (3–5 nm) compared with FIBID.

A structural difference between FIB and FEB deposited Pt layers can be clearly seen in Figure 17.3, where a high-resolution SEM image of a sample prepared by sequential Pt deposition of FIBID, FEBID, and Au metal layers (the metal layer was deposited to reduce charging), followed by FIB cross-sectioning, is presented. Smaller grain sizes, no agglomeration of grains, and apparently higher amorphous carbon content (darker image background) are characteristics of a FEBID Pt layer, as compared with a FIBID.

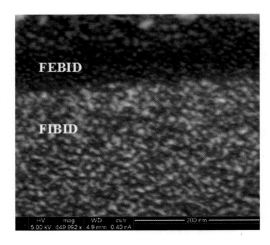

FIGURE 17.3: Cross-section of Pt layers sequentially deposited by FIB and FEB.

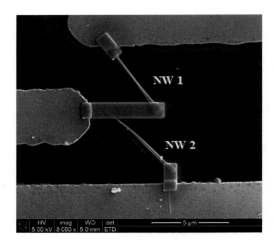

FIGURE 17.4: Nanowires of InAs contacted using Pt FEBID. Courtesy of M. Martins, IFGW, UNICAMP.

In spite of higher resistances, FEBID Pt deposits are frequently used for nanocontacting, as this process is much less damaging than the Pt deposition by energetic Ga ions. In particular, FEBID is used with success for fixation and soldering of carbon nanotubes and various types of NWs [31]. An example is shown in Figure 17.4, where 0.8 μm high and 0.8–1.0 μm wide FEBID Pt contacts to metal (Au/Ge/Ni) electrodes were fabricated to measure resistances of ~100 nm diameter InAs nanowires. In this case, resistances of contacts were estimated to be $R_c \sim 10^4 \Omega$, being much smaller as compared with the measured NW resistances (8×10^6 and $1.6 \times 10^7 \Omega$, for NW 1 and 2, respectively). Note that in the case of highly conductive CNTs ($\sim 10^3 - 10^4 \Omega/\mu$m, depending on quality), lower contact resistances are needed.

Langford et al. [25] reported two ways for considerable improvement of FEBID Pt-C films conductivity: (1) thermal annealing at 500°C for 1 hour; (2) implantation of Ga using FIB in thick FEB-ID produced films (depth of implantation ~30 nm). For the former, the final resistivity dropped by a factor of 10, grain sizes increased to 5–7 nm, and the carbon content reduced by ~15 at.%. For the latter, for the implanted region (30 nm) the estimates showed a 20-fold resistivity reduction for a Ga dose of 5×10^{16} ions/cm^{-2}.

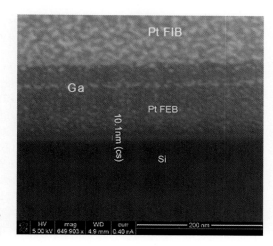

FIGURE 17.5: Cross-section of sequentially deposited FEB and FIBID Pt layers.

It is also possible to deposit a thin, but more conductive FIBID Pt layer over a previously FEBID deposited layer. In this case implantation of Ga forms a thin metal-rich layer at a depth of 70–80 nm, as can be seen in Figure 17.5.

2.3 FIBID OF TUNGSTEN FILMS

Specific interest to this material is in particular due to the possibility of creating super-conductive nanostructures [29; 32]. Typical deposition conditions are: 30 kV Ga$^+$ ions, 10–20 pA, $W(CO)_6$ precursor. The atomic concentrations of C, Ga, and W using EDX were found to change as a function of the ion-beam current from C:Ga:W=65:10:25 at.% to 37:23:40 at.% as current increased from 10 to 10^3 pA [32]. High resolution TEM analysis indicated the possible presence of nanocrystallite clusters with size of ~1 nm, being much smaller that that found for Pt-C films. Higher W content (up to 75%) has been reported in another study [33].

The resistivity as low as 110 $\mu\Omega$ cm was measured on 250 nm wide W-C wires deposited at low ion dose (15 nC/μm^2), that is only 20 times higher than the bulk one, 5.6 $\mu\Omega$ cm [30], and significantly better than that for Pt-C. A contact resistance to aluminium electrodes was estimated as $r_c = 1.64\,\mu\Omega/cm^2$ (for a dose 25 nC/μm^2). Low values of room temperature resistivity, in the range 100–600 $\mu\Omega$ cm, have been reported by de Teresa et al. [29].

The superconducting transition temperature T_c of the W-C FIBID films was shown to change in the range 5.0–6.2 K [32] and 4.6–5.4 K [29]. These results suggest that FIB direct-writing of W composites might be a potential approach to fabricate mask-free superconducting devices as well as to explore the role of reduced dimensionality on superconductivity [32].

2.4 FIBID OF NARROW PT AND W NANOWIRES

The deposition of narrow Pt and W wires was analyzed by Reguer et al. [34]. Ion energy of 30 kV and beam current of 20 pA were used to deposit 35–125 nm (W) and 60–145 nm (Pt) wide wires. Depositions were performed over thin DLC (diamond-like carbon) or silicon nitride membranes (~20 nm thick), transparent for an electron beam and allowing for *ex situ*

TEM observations. The size of the deposited structures was found to exceed the ion beam diameter. After a nucleation time (just 5 sec for W and bigger time for Pt), wire widths were observed to increase with the deposition times, growing rapidly until the saturation. This was explained by analyzing the role of secondary particles (sputterred atoms, ions, and secondary electrons) that can emerge from the lateral flanks of the growing NW and dissociate the adsorbed gas molecules, resulting in lateral expansion of NWs. For the same experimental conditions, the FIBID platinum wires were found to be wider than the tungsten ones, with the difference being within ~20–25 nm during the process. This was attributed to the fact that a Pt-C material has larger content of light elements (C and O) than W-C. Thus, the ion and electron mean free paths are longer in the Pt-C wire than in the W-C one, and effect of the lateral NW expansion during growth is more pronounced for Pt-C. Resistivities of 300 $\mu\Omega$ cm and 600 $\mu\Omega$ cm for as-deposited tungsten and platinum wires, respectively, were measured.

Effect of thermal treatment of beam-deposited metal films that can be beneficial for their resistivities was reported earlier [25]. Note that in the case of nanowires, heating can be realized without external treatment (that is not always possible) just by applying electric current, due to Joule effect [34]. This was done by applying a slow ramp voltage to the W and Pt nanowires. The voltage was varied from 0 to 2.5 V by 0.1 V steps. For small biases, the resistance decreases slightly with increasing voltage. Above a threshold of 2.5 V, the W wire resistivity decreased drastically to 50 $\mu\Omega$ cm (only an order of magnitude higher than that for bulk). The resistance of the treated W-C wire was observed to increase with temperature, contrary to the as-deposited wires, indicating the classical metallic behavior. Another important observation, made by *in situ* SEM, was the formation of single big droplets apparently composed of liquid Ga segregated from a W-rich wire (these two elements are not miscible and do not form any stable compound) as a result of treatment. HRTEM analyses showed formation of crystallites sized from 5 to 20 nm. For Pt wires, formation of many smaller droplets (in contrast to W, Pt can form stable compounds with Ga) and undesirable increase of resistance were observed.

2.5 FIBID AND FEBID FABRICATION OF 3D STRUCTURES

In a work of Li and Warburton [35], "flying" tungsten nanorods (not supported by the substrate) were fabricated using low-current (1 pA) FIBID. The growth rate and shape were found to be dependent upon both the gas precursor nozzle position and the ion-beam focus. Vertical rods with height up to 45 μm were fabricated with smooth and vertical sidewalls. Beam nudging (small lateral displacement) was employed to form 3D tungsten nanofeatures. Under a certain deposition condition with a constant beam-nudging distance, the tilt angle is inversely proportional to the pixel dwell time. The resistivity of the "flying" tungsten wiring fabricated using an ion-beam current of 1 pA is about 550 $\mu\Omega$ cm (comparable with 100–300 $\mu\Omega$ cm obtained for wires fabricated directly on substrates).

3D structures such as bridges were fabricated by FIBID for electrical connection between two Au pads in an optoelectronic device (over Si_3N_4/InGaAs substrate, see Figure 17.6) using 3-step FIB induced Pt deposition with changing ion beam directions. Directions of the ion beam during FIBID are shown in Figure 17.6b. In the first two steps, the beam incidence angles were 52° to normal, and in the last step, 0° to normal. Deposition parameters were: 30 keV, 10 pA, deposition area of 0.25×0.25 μm^2 (in fact, wider deposits especially at the bottom were formed, due to Ga^+ ion scattering inside the growing pillar; see Figure 17.6b), total process duration of 100 sec. The total bridge resistance (total length of ~6 μm) was measured to be 30 Ω.

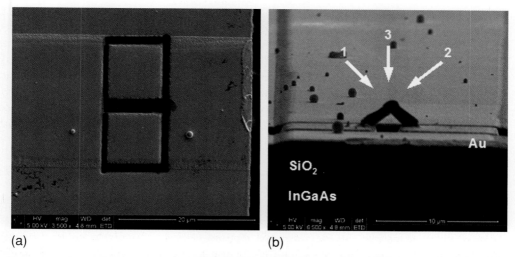

(a)   (b)

FIGURE 17.6: Bridge fabricated using a tilted three-step FIBID process. (a) Substrate after pads fabrication by Ga beam milling. (b) Directions of ion beam Pt deposition and the produced bridge. Courtesy of L. Barea, IFGW, UNICAMP.

In Chapter 22 of this book, Gazzadi and Frabboni [36] demonstrated the deposition of suspended Pt-containing structures using FEBID and controlled lateral shift during deposition. Relatively high resistances (10^{-4} Ωm), characteristic for electron beam induced processes, were obtained.

In order to improve conductivity of FEBID 3D structures like bridges, FEBID process can be followed by FIBID to create a thin layer of higher conductance [25; 31]. An example is shown in Figure 17.7, where two 0.5×0.5 μm wide Pt-C pillars (tilted 50° to normal direction) and a connection between them were fabricated by FEBID (5 kV, 0.4 nA, dwell time 1 μs, total time 9 min.). The process was done in three steps when the sample was tilted consequently at different angles, to form a bridge over the gap in a metal (Ti) electrode (see Figure 17.7(a-c)). After this, a thin 0.2×0.2 μm Pt-C line was deposited by FIBID (30 kV, 30 pA, 0.2μs dwell time, 50 s total time) over the bridge as shown in Figure 17.7(d-f). The bridge resistance (total length of ~5 μm) was measured to decrease about 5 times after the FIBID step, from 9.6 kΩ to 1.9 kΩ. Note that the FEBID pillar cross-section remains constant during deposition, in contrast to FIBID pillars (Figure 17.6) that tend to widen due to ion scattering in the pillar body and lateral deposition (see below the discussion of the "halo effect").

2.6 FIBID FABRICATION OF PARALLEL NANOWIRES, LEAKAGE TESTS

The so called "halo" effect or extended metal film deposition, is known to occur around NWs fabricated using FIBID [34; 37; 38]. This is an important issue in situations when a number of contacts with small separations is formed. The halo effect is much less pronounced for FEBID [38].

The reasons of halo formation are not completely clear at present and may include: ion beam profile with non-Gaussian extended tails of micrometer size (resulting in induced deposition around the beam focus point), redeposition, forward scattering and diffusion of

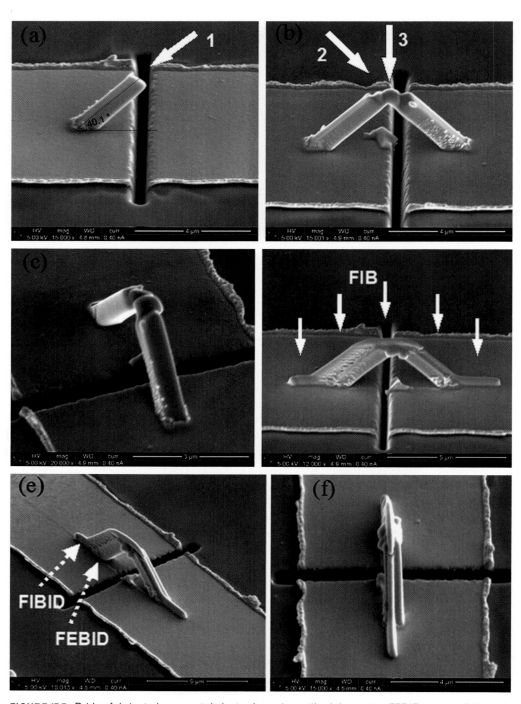

FIGURE 17.7: Bridge fabricated over metal electrodes using a tilted three-step FEBID process followed by one-step FIBID process. (a, b, c): FEBID steps, arrows show the e-beam direction. (d, e, f): after FIBID step, different views of fabricated bridge, arrows show the FIB direction.

Table 17.1 Leakage tests with parallel 20 μm long line resistors with 5 μm overlap, fabricated by 30 kV and 10 pA Ga⁺ FIBID, (CH₃)₃Pt(CpCH₃) precursor. Modified from [4].

Thickness (nm)	Interline distance (nm)	Halo resistance as deposited (MΩ)	Halo resistance after O₂ plasma processing (MΩ)
30	200	0.24	0.29
	400	0.32	>10³
	1000	>10³	>10³
100	200	0.019	0.033
	400	0.022	0.078
	1000	46	>10³

by-products over the substrate surface, which could be enhanced by ions. In a study by Tripathi et al. [19] (Pt deposition with 10–30 pA and 30 keV Ga⁺ FIB), a leak current was observed between Pt pads during their fabrication over an oxidized Si substrate. Lateral dimensions of pads were 10 × 1 μm, and gaps between pads ranged from 1 to 3 μm. A leak was found to appear soon after deposition started. EDS analysis showed the presence of Pt, Ga, and C in the gap region. For Pt pads, leakage current was found to decrease an order of magnitude with gap increasing from 1 to 3 μm (for 1 μm gap, 0.25 μA at 0.2 mV bias). Significantly lower leak currents were detected for W contact pads where gaps varied from 0.5 to 1 μm (no detectable currents for gaps ≥1μm, 0.25 μA for 0.5 μm gap). The strong extended deposition was explained on the basis of molecular dissociation of the adsorbed precursor gas molecules by the forward scattered Ga ions from the edges of growing Pt nanostructures [19]. Calculations were made to show that the deposition profile due to Ga ion scattering from the growing pad edge corresponds well to the experimentally measured profile. The difference between W and Pt was explained by different dissociation cross-sections of the precursor molecules [19]; however, different scattering of ions within W-C and Pt-C compounds (the latter has higher C content) could be also responsible as discussed by Reguer et al. [34].

Morphological and electrical characterization of a "leakage" type device, in which two identical parallel resistors (deposited by 30 kV Ga⁺ FIB) were separated by variable gaps decreasing from 2 μm to 250 nm, was carried out by Notargiacomo et al. [26]. The resistors had measured thickness and width equal to 90 nm and 220 nm, respectively. The halo layer thickness t_h measured by AFM was shown to increase with decreasing gap separation. For halo layer thickness larger than 6 nm, the halo resistivity was found to be close to that of the Pt resistor, possibly showing that the two have similar composition and microscopic morphology [25]. The morphological similarity between the material in the halo region and the main deposit is also evidenced by Figure 17.1(c).

Using 30 kV Ga⁺ FIBID, Vaz et al. [4] fabricated parallel linear Pt-C structures (NWs) with 5 μm overlapping to study current leakage due to the halo effect. Some results can be seen in Table 17.1, indicating fast increase of leakage with gap reduction and NW thickness. It can be seen that for 1 μm separation, the leakage between line resistors (with the thickness up to 100 nm) is relatively low. This agrees well with the data shown in Figure 17.1, where the halo thickness becomes negligibly small for 0.3–0.5 μm distances from the pattern edge. For smaller separations, electrical isolation between parallel nanocontacts is poor, especially for thicker deposits. To sum up, the data presented here and above clearly indicate that FIBID Pt

NWs, with thickness not exceeding 100 nm, present a reasonable trade-off between the contact resistivity and lateral deposition (leakage), when a number of parallel nanocontacts have to be formed.

Finally, in order to remove the impurities causing leakage between FIB-deposited nanocontacts, a low pressure O_2 plasma can be used. The capacitively coupled plasma is known to produce oxygen radicals, as well as a variety of negative and positive ions, and the latter can bombard the exposed surfaces that are charged negatively due to self-bias in a plasma. Thus oxygen plasma processing can remove the carbon deposit and oxidize or sputter thin metal (Pt, Ga) deposits at the halo region. The results of leakage measurements after processing in O_2 plasma (capacitively coupled asymmetric plasma reactor, 150 mTorr, 30 W, 200 V self-bias, 2 min.) are also shown in Table 17.1. The results confirm the efficiency of this method for processing of thin halo deposits, resulting in notably lower leakage currents. Of course, if nanostructures to be tested are sensitive to O_2 processing, this method should be used before their deposition over the contacts, e.g., by dielectrophoresis [4; 15].

3 CONCLUSIONS

In this chapter, various FIB and FEB induced deposition processes employing commercially available precursors (Pt and W) have been reviewed. These processes were successfully employed for fabrication of metallic nanowires, films, and 3 D structures like columns or bridges. An important application of these structures is for nanocontacting purposes. In this case, optimization of process characteristics with respect to electrical properties of deposits (low resistance), minimal width (cross-section) and lateral deposition (halo), which can result in considerable leakage between nanocontacts, is of extreme importance. In terms of contact resistance, better results are achieved with FIBID, however, halo and damaging are inherent for ion-based processes, but can be avoided using FEBID. The minimal width of nanocontacts produced so far by FIBID is usually close to 50–100 nm, depending on the substrate. The ways of reducing lateral deposition have been also discussed. Interestingly, reduced halo effect and smaller NW resistivities are achieved for W depositions, while the most of reports still deal with Pt processes. Post-deposition processing (conventional or dynamic annealing by electric current, and treatment by low-density plasmas) has shown to be beneficial in some cases to improve resistance and reduce halo effects. Combination of techniques (FEBID followed by FIBID) has a number of advantages, especially when it is necessary to reduce ion-induced damaging by creating intermediate protective layers and to reduce the contact resistance. Other ways to reduce ion damaging for a variety of specific applications and materials are considered in various chapters of this book.

REFERENCES

[1] Kelsall, R. W., I. W. Hamley, and M. Geoghegan, eds. *Nanoscale science and technology*. Chichester, UK: J. Wiley & Sons, 2005.

[2] Yan, Yehai, Mary B. Chan-Park, and Qing Zhang. Advances in carbon-nanotube assembly. *Small* 3 (2007): 24–42.

[3] Krupke, R., S. Linden, M. Rapp, and F. Hennrich. Thin films of metallic carbon nanotubes prepared by dielectrophoresis. *Advanced Materials* 18 (2006): 1468–1470.

[4] Vaz, A. R., M. Macchi, J. Leon, S. A. Moshkalev, and J. W. Swart, Platinum thin films deposited on silicon oxide by focused ion beam: Characterization and application. *Journal of Materials Science* 43 (2008): 3429–3434.

[5] Pauzauskie, P. J., A. Radenovic, E. Trepagnier, H. Shroff, P. Yang, and J. Liphardt. Optical trapping and integration of semiconductor nanowire assemblies in water. *Nature Materials* 5 (2006): 97–101.

[6] Ohta, A. T., S. L. Neale, H.-Y. Hsu, J. K. Valley, and M. C. Wu. Parallel assembly of nanowires using lateral-field optoelectronic tweezers. *2008 IEEE/LEOS Internat. Conf. Optical MEMS and Nanophotonics*, OPT MEMS, No. 4607801, pp. 7–8. Freiburg, Germany, 2008.

[7] Ye, H., Z. Gu, T. Yu, and D. H. Gracias, Integrating nanowires with substrates using directed assembly and nanoscale soldering. *IEEE Transactions on Nanotechnology* 5 (2006): 62–66.

[8] Thelander, C., and L. Samuelson. AFM manipulation of carbon nanotubes: Realization of ultra-fine nanoelectrodes. *Nanotechnology* 13 (2002): 108–113.

[9] Cui, Q., F. Gao, S. Mukherjee, and Z. Gu, Joining and interconnect formation of nanowires and nanotubes for nanoelectronics and nanosystems. *Small* 5 (2009): 1246–1257.

[10] Mølhave, K., D. N. Madsen, S. Dohn, and P. Bøggild. Constructing, connecting and soldering nanostructures by environmental electron beam deposition. *Nanotechnology* 15 (2004): 1047–1053.

[11] Dockendorf, C. R. D., M. Steinlin, D. Poulikakos, and T.-Y. Choi, Individual carbon nanotube soldering with gold nanoink deposition. *Appl. Phys. Lett.* 90 (2007): 193116.

[12] Madsen, D. N., K. Mølhave, R. Mateiu, A. M. Rasmussen, M. Brorson, C. J. H. Jacobsen, and P. Bøggild. Soldering of nanotubes onto microelectrodes. *Nano Lett.* 3 (2003): 47–49.

[13] Peng, Y., T. Cullis, and B. Inkson. Bottom-up nanoconstruction by the welding of individual metallic nanoobjects using nanoscale solder. *Nano Lett.* 9 (2009): 91–96.

[14] Stern, E., G. Cheng, J. F. Klemic, E. Broomfield, D. Turner-Evans, C. Li, C. Zhou, and M. A. Reed. Methods for fabricating Ohmic contacts to nanowires and nanotubes. *J. Vac. Sci. Technol.* B 24 (2006): 231–236.

[15] Leon, J., A. R. Vaz, A. Flacker, C. Verissimo, M. B. Moraes, and S. A. Moshkalev. Electrical characterization of multi-walled carbon nanotubes. *J. of Nanosci. and Nanotechnol.* 10 (2010): 6234-6239.

[16] Vaz, A. R., C. Veríssimo, F. P. Rouxinol, R. V. Gelamo, and S. A. Moshkalev. Characterization of nanostructured carbon materials using FIB. Chapter 25 of this book.

[17] Schwamb, T., B. R. Burg, N. C. Schirmer, and D. Poulikakos. On the effect of the electrical contact resistance in nanodevices. *Appl. Phys. Lett.* (2008): 243106.

[18] Dai, H., E. W. Wong, Ch. M. Lieber. Probing electrical transport in nanomaterials: Conductivity of individual carbon nanotubes. *Science* 272 (1996): 523–526.

[19] Tripathi, S. K., N. Shukla, N. S. Rajput, and V. N. Kulkarni. The out of beam sight effects in focused ion beam processing. *Nanotechnology* 20 (2009): 275301.

[20] Matsuda, Y., W.-Q. Deng, and W. A. Goddard, III. Contact resistance properties between nanotubes and various metals from quantum mechanics. *J. Phys. Chem.* C 111 (2007): 11113–11116.

[21] Zhang, Y., N. W. Franklin, R. J. Chen, and H. Dai. Metal coating on suspended carbon nanotubes and its implication to metal-tube interaction. *Chemical Physics Letters* 331 (2000): 35–41.

[22] Tham, D., C.-Y. Nam, and J. E. Fischer. Microstructure and composition of focused-ion-beam-deposited Pt contacts to GaN nanowires. *Adv. Mater.* 18 (2006): 290–294.

[23] Barrya, J. D., M. Ervin, J. Molstad, A. Wickenden, T. Brintlinger, P. Hoffmann, and J. Melngailis, Electron beam induced deposition of low resistivity platinum from $Pt(PF_3)_4$, *J. Vac. Sci. Technol.* B 24 (2006): 3165–3168.

[24] Smith, S., A. J. Watson, S. Bond, A. W. S. Ross, J. T. M. Stevenson, and A. M. Gundlach. Electrical characterization of platinum deposited by focused ion beam. *IEEE Trans. Semicond. Manufact.* 16 (2003): 199–206.

[25] Langford, R. M., T.-X. Wang, and D. Ozkaya, Reducing the resistivity of electron and ion beam assisted deposited Pt. *Microelectronic Engineering* 84 (2007): 784–788.

[26] Notargiacomo, A., L. Di Gaspare, and F. Evangelisti. Ion beam assisted processes for Pt nanoelectrode fabrication onto 1-D nanostructures. *Superlattices and Microstructures* 46 (2009): 149–152.

[27] de Teresa, J. M., R. Córdoba, A. Fernández-Pacheco, O. Montero, P. Strichovanec, and M. R. Ibarra. Origin of the difference in the resistivity of as-grown focused-ion- and focused-electron-beam-induced Pt nanodeposits. *Journal of Nanomaterials* V (2009): 936863.

[28] Fernández-Pacheco, A., J. M. de Teresa, R. Córdoba, and M. R. Ibarra. Metal-insulator transition in Pt-C nanowires grown by focused-ion-beam-induced deposition. *Phys. Rev.* B 79 (2009): 174204.

[29] de Teresa, J. M., A. Fernández-Pacheco, R. Córdoba, and M. R. Ibarra. Electrical transport properties of metallic nanowires and nanoconstrictions created with FIB. Chapter 23 of this book.

[30] Kang, H. H., C. Chandler, and M. Weschler, Gas assisted ion beam etching and deposition. In *Focused Ion Beam Systems*, ed. N. Yao. Cambridge: Cambridge University Press, 2007.

[31] Utke, I., P. Hoffmann, and J. Melngailis. Gas-assisted focused electron beam and ion beam processing and fabrication. *J. Vac. Sci. Technol.* B 26 (2008): 1197–1276.

[32] Li, W., J. C. Fenton, Y. Wang, D. W. McComb, and P. A. Warburton. Tunability of the superconductivity of tungsten films grown by focused-ion-beam direct writing. *J. Appl. Phys.* 104 (2008): 093913.

[33] Prestigiacomo, M., L. Roussel, A. Houe, P. Sudraud, F. Bedu, D. Tonneau, V. Safarov, and H. Dallaporta. Studies of structures elaborated by focused ion beam induced deposition. *Microelectronic Engineering* 76 (2004): 175–181.

[34] Reguer, A., F. Bedu, D. Tonneau, H. Dallaporta. M. Prestigiacomo, A. Houel, and P. Sudraud. Structural and electrical studies of conductive nanowires prepared by focused ion beam induced deposition. *J. Vac. Sci. Technol.* B 26 (2008): 175–180.

[35] Li, W., and P. A. Warburton. Low-current focused-ion-beam induced deposition of three-dimensional tungsten nanoscale conductors. *Nanotechnology* 18 (2007): 485305.

[36] Gazzadi, G., and Stefano Frabboni. Growth and characterization of FEB-deposited suspended nanostructures. Chapter 22 of this book.

[37] Gopal, V., V. R. Radmilovic, C. Daraio, S. Jin, P. Yang, and E. A. Stach. Rapid prototyping of site-specific nanocontacts by electron and ion beam assisted direct-write nanolithography. *Nano Lett.* 4 (2004): 2059.

[38] Park, Y. K., T. Nagai, M. Takai, C. Lehrer, L. Frey, and H. Ryssel. Comparison of beam-induced deposition using ion microprobe. *Nuclear Instruments and Methods in Physics Research* B 148 (1999): 25–31.

FIB ETCHING FOR PHOTONIC DEVICE APPLICATIONS

Martin. J. Cryan, Y-L. Daniel Ho, Pavel S. Ivanov, Peter J. Heard, Judy Rorison, and John G. Rarity

1 INTRODUCTION

The field of photonics is a very broad one that encompasses numerous multibillion-dollar industries such as optical communications, displays, and optical data storage. Photonics is following the well-trod path of its close technology cousin—electronics—of "faster, smaller, cheaper," and a whole host of new technologies are vying to become the next generation of global industries. These include low-cost light emitting diode (LED) lighting, photovoltaics, optical processors, photonic integrated circuits and nanoscale sensors. An important feature of many of these new technologies is the requirement for state-of-the-art processing at the nanoscale. Here photonics has again followed on from electronics and uses many of the material-processing techniques that were developed for electronic integrated circuits. The workhorse technology is electron beam lithography, which allows wafer scale definition of nanoscale features down to a few 10's of nanometers. Photonics has the extra requirement that photons are much more difficult to control than electrons: a metal track will guide electrons easily under most conditions. However, to guide photons is a much more delicate task, as anyone who has bent an optical fiber has witnessed as the light streams from the outer edge of the bend. Thus to control light at the nanoscale requires control of surface roughness down the nanometer scale. A number of other material parameters such as refractive index profile and optical loss are also extremely important, and loss in particular will be discussed later in this chapter.

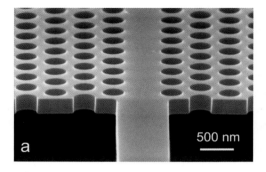

FIGURE 18.1: An air membrane 2D photonic crystal waveguide produced by IBM in 2003. Light is guided down the center of the waveguide, confined laterally by the photonic crystal regions on either side and vertically by total internal reflection. With permission from [1.7]. Copyright (2003), OSA.

A number of large multinational corporations are engaged in photonics research at the nanoscale, termed *nanophotonics,* and this has led to unprecedented improvements in material-processing techniques. For example, there is a major branch of nanophotonics that uses a new type of artificial material termed a *photonic crystal* (PhC), which enables strong control and confinement of photons, which in principle allows them to be guided as well as electrons in a metal wire. Photonic crystals are typically formed from periodic arrays of a few 100 nm diameter holes etched into an optically transparent material such as silicon. They employ the well known phenomenon of Bragg reflection [1.1] which in 1D is used to form many components such as thin film filters and laser distributed Bragg reflector (DBR) mirrors [1.2]. The breakthrough in PhCs came with the seminal papers of Yablonovitch and John, who both realized that with 3D periodic structures Bragg reflection could in principle trap light completely [1.3; 1.4]. The technological challenge here was that the periodicity needed to be roughly half the wavelength of light in the material, so in silicon (n = 3.4) at 1,550 nm this requires 1,550/2*3.4 = 228 nm. It took almost 10 years before photonics fabrication could achieve the processing quality required for these types of structures [1.5]. It was found that 2D periodicity could be fabricated more easily, with the third dimension of confinement using the more conventional approach of total internal reflection [1.6]. Figure 18.1 shows a state-of-the-art E-beam lithography defined and dry etched air membrane 2D photonic crystal waveguide fabricated by IBM in 2003 [1.7].

It can be seen in Figure 18.1 that the level of surface roughness is incredibly low and is in the order of <5 nm, which if one bears in mind the "size" of a silicon atom to be around 0.1 nm is a breathtaking material-processing achievement.

One of the main issues with processing devices such as that shown in Figure 18.1 is that this is done at the wafer scale and is essentially a one-shot process. To alleviate some the risks involved in this, a technique termed *lithographic tuning* [1.8] has been developed whereby some important physical dimensions in the devices are varied across the wafer in the hope that one set of the finally produced devices will have the desired set of physical dimensions.

The dream of nanophotonics researchers would be to be able to make each one of these devices to the exact dimensions they require, perform a set of measurements and possibly change some physical dimension, and repeat the measurements on the same device. Remarkably, this is possible using the techniques of FIB processing, which will be outlined in this chapter.

FIB etching of photonic devices has a long history with some of the first examples of facet polishing and etching in 1986 and 1987 [1.9; 1.10]. This was soon followed by waveguide-laser integration work in 1990 [1.11]. These early examples showed that FIB etching can be used with active devices such as lasers without serious degradation of their characteristics. Detailed studies of damage effects in MQWs have been undertaken [1.12] which showed that large areas of damage can be created by diffusion of non-equilibrium defects during etching, far beyond the beam radius. Control of such damage is a recurring theme in FIB etching and in [1.12] it was shown how operating at low temperatures can reduce this effect, and an alternative approach of operating at elevated temperature [1.13] has also been explored. A further temperature-based approach to damage reduction is post-etch annealing, which has been used in both active [1.14; 1.15] and passive devices [1.16].

This chapter will cover four different aspects of FIB processing in photonics. First, the technological constraints placed on processing by the FIB process will be discussed. Second, a number of examples of nanophotonic devices, in particular photonic crystal devices, that have been processed using FIB will be discussed. Third, another growing area for nanophotonic processing is that of quantum information science, which requires control of photons at the nanometre scale in devices such as the single photon sources. Finally, FIB processing has been used to post-process more conventional photonic devices such as lasers, to enhance or adapt their performance, some of the more notable examples will be discussed.

There is a growing community of researchers employing FIB for photonics, and this has resulted in an international workshop series, the first of which occurred in 2008 [1.17] and will be followed by a second in Cambridge, United Kingdom, in 2010 [1.18]. It is felt that with continued research in the field, the dream of nanophotonic fabrication on a device-by-device basis for advanced device prototyping can be achieved.

2 THE USE OF FOCUSED ION BEAMS FOR THE FABRICATION OF PHOTONIC DEVICES: INSTRUMENT CONSIDERATIONS

Focused ion beams from liquid-metal ion sources can be used to directly fabricate structures on the nanometer scale. Although a serial technique, and therefore not suitable for mass production, individual devices or arrays of devices may be fabricated for the purposes of rapid-turnaround prototyping or proof of principle work. This is of particular importance at the time of writing for photonic devices, for which there is considerable developmental effort.

The most important species of ion used for the purpose is the gallium ion, from the gallium liquid-metal ion source, although this is not the only option available. Indeed, other single-element metals and alloys can be used, for example gold, gold-silicon, gold-silicon-beryllium, and boron-platinum alloys. For the latter examples, a mass separator is required for the selection of the desired species of ion from all those emitted from the source, whereas for the gallium source, by far the most abundant emitted species is the singly-charged Ga^+ ion, and so a mass separator is unnecessary. This factor and the technological simplicity of the source construction have meant that the gallium ion gun is by far the most common. Hence, by a combination of electrostatic optics and deflection, a gallium ion beam may be delivered to an arbitrary point on the sample, typically with an energy of 25–50 keV and beam current of 1 pA to tens of nanoamperes, and a corresponding beam diameter from a few nanometers to a few hundred nanometers.

The beam may be raster scanned over a region of the sample, and secondary electrons collected, with which to form an image of the sample. The beam can then be accurately directed to the parts of the sample to be sputter etched in order to fabricate the desired structures, whether they be facets, gratings, photonic bandgap structures, or waveguides.

2.1 CHALLENGES IN FIB ETCHING OF PHOTONIC STRUCTURES

It may be thought that the issues presented in the fabrication of optical devices are the same as for other materials processing, but the nature of photonic materials and their use present particular difficulties for focused ion beam etching. The structures to be etched usually need to be accurate to approximately 10 nm over an extended region. Furthermore, the structures to be etched are often deep (several microns) with high aspect ratio (10:1 or more), and with vertical sidewalls (for example, photonic bandgap structures). The shape of the beam profile, effects of redeposition of material, the insulating nature of the materials to be processed, stage or beam drift and the surface damage associated with etching and imaging with a focused ion beam will all hinder the fabrication of such structures. Hence special precautions and techniques are often required for their fabrication.

The electrically insulating nature of the materials will mean that the sample will charge positively with respect to ground as etching proceeds. Unlike the situation in the scanning electron microscope, it is not possible to alleviate charging by tilting the sample or by reducing the ion energy, and furthermore, it is unlikely that coating the sample with gold or carbon will be acceptable in the finished devices. The effects of charging are, in the first instance, poor imaging due to the reduced flux of secondary electrons reaching the detector, but more importantly deflection of the incoming ion beam, leading to poor etching accuracy and "drift," or smearing of the patterns to be etched.

The profile or shape of the focused ion beam is often assumed to be Gaussian; however, there is, in addition, a broad background profile that extends many beam diameters from the center, often modeled as a Lorentzian term. For the etching of single shapes, this is rarely a problem, but for large arrays of small features that are close together, the broad tail of the beam distribution may integrate to give an appreciable flux of ions outside the intended etch regions. For photonic devices that are sensitive to surface damage and implantation, this is undesirable. The Gaussian profile part of the beam distribution also limits the profile of the shapes that can be etched with the beam. For example, if a rectangular profile etch were made with such a beam, the sidewall of the pit would not be vertical, but rather, sloping at an angle determined by the diameter of the Gaussian profile. Similarly, photonic bandgap structures would not have vertical sidewalls. In practice, sidewall angles of a few degrees are common in such attempted structures.

A further complication, to add to the growing list, is the redeposition of the sputtered material. Sputtered atoms have an energy of a few eV, and will travel away from the etch site in straight lines until they hit something. While most will escape the sample, some will hit other parts of the sample, where they may stick, so that a layer of redeposited material builds up. The effects of redeposition become increasingly severe during the etching of high aspect-ratio holes (as, for example, required in photonic bandgap structures), for as the hole becomes deeper, the probability of escape of atoms from the bottom of the hole diminishes so that, as time goes on, the depth of the hole does not increase linearly with time. In practice, it becomes difficult to increase the aspect ratio of the hole (ratio of depth to diameter) beyond about 5. For the fabrication of any arbitrarily shaped feature, redeposition effects may determine the scan strategy to be used; that is, the order in which the pixels in the shape are to be scanned. For example, a round hole may be scanned by moving the beam across in a raster-like fashion, or it may be made to move in a spiral, to cover the hole. Furthermore, that speed of the scan may be important.

Most of the current FIB instruments operate with a beam energy in the range 25–50 keV. In operation, the FIB column will give the maximum current density at its maximum operating energy due to chromatic aberration effects within the lens column. Hence, although the energy may be reduced, most work will be done at this maximum energy. When the ion beam strikes the sample, many processes are put into motion; besides the sputtering of neutral particles and the emission of secondary electrons, secondary ions and excited particles are emitted, gallium ions are implanted into the material and any lattice structure will be disrupted so that amorphization or damage will occur. The effects of implantation and damage may be studied using a Monte Carlo simulation program such as SRIM (www.srim.org; Ziegler) in which the trajectories of the implanted ions and displaced sample atoms are calculated from known parameters. The program shows that the range of 30keV gallium ions into diamond at normal incidence for example is approximately 26nm, with damage events occurring to at least twice this depth. The damage and implantation effects are dependent on the beam energy and the angle at which the beam strikes the sample, but for many photonic devices such as resonant cavities and photonic bandgap structures, optical losses that occur due to surface damage may be unacceptably high, rendering the devices inoperable. This effect may even be severe enough to occur when imaging the region with the ion beam.

The final issue with FIB etching of photonic devices is the time required for the etch, coupled with the spatial stability of the system. Any spatial drift in either the source or sample position will cause movement of the ion beam relative to the sample, and may cause smearing of the structures to be etched. This issue will be most severe for the etching of a large array of small features (such as photonic bandgap structures). In order to etch the features with sufficient spatial resolution, a small ion beam is required, and hence a small ion current must be used. The total time for the etch is inversely proportional to the beam current, hence the etch time may be considerable. Any drift in source or sample position over this time will integrate over the etching time, making spatial acuity of the etched structures difficult to achieve.

2.2 STRATEGIES FOR SUCCESS

Many of the above issues may be addressed by a combination of instrument add-on, design, and careful selection of beam parameters.

The electrically insulating nature of the sample may be accommodated by adding a low-energy flood electron gun (charge neutralizer) to the system. Alternatively, on combined FIB/SEM instruments, the electron beam may be used to perform neutralization during sputter etching.

Beam shape/profile issues may be addressed to a certain extent by instrument design. The most recent FIB gun designs have sought to reduce the long-range tails of the beam distribution. On older instruments however, it may still be possible to optimize the beam shape by focusing the beam, not to optimize the apparent resolution in the image, but to minimize the beam tail distribution. The overall diameter of the beam should be taken into account when attempting to prepare vertical faces such as laser facets. By tilting the sample, the diameter of the beam may be offset, so that the face becomes vertical. The degree of tilt required depends on the etch rate of the material, but usually falls in the range of $1°–3°$.

Redeposition issues may be addressed using a combination of etch enhancement gases and scan strategy. For many materials, FIB etch rates may be enhanced by the deliberate introduction of a gas using a capillary needle, close to the etch site. While the gas does not etch the sample on its own, atoms removed from the sample during sputtering are likely to combine with the gas to form a volatile product that is simply pumped away by the instrument vacuum system. This not only speeds up the etching, but prevents, to a large extent, redeposition of material onto other parts of the sample. Examples of etch enhancement gases are iodine or chlorine for the enhancement of metal etching, xenon diflouride for oxides and nitrides, and water for diamond and polymers. The technique appears to offer a double advantage in the enhancement of etch rates and the minimization of redeposition effects, but the enhancement tends to be non-linear with ion current density on the sample, so that the effects of beam tails can be enhanced too. Typically, gases are introduced through a capillary tube of about 1 mm diameter, with the end of the tube positioned within 100 μm or so of the etch site, to maximize the local pressure while minimizing the pump loading and the pressure around the secondary electron detector. By using gas enhancement in this way, holes with aspect ratios of 20:1 or more may be achieved. The scanning parameters for the ion beam are of importance while performing gas-assisted etching, and the dwell time (per pixel) in the scanned shape must be chosen so that the cycle time for the pattern delineation matches the gas arrival rate.

Scan strategies are also of importance in the etching of photonic bandgap structures. Redeposition effects may cause the hole profiles to be asymmetrical if a raster scan pattern is used. However, a spiral scan configuration for each hole will generate symmetrical shapes. For the etching of laser facets, a scan strategy is chosen such that the beam is scanned rapidly in a line, moving toward the facet direction. Hence, material redeposited onto the face of the facet is removed again immediately.

Damage issues represent, perhaps, the most serious challenge associated with FIB etching of photonic structures. Gas assisted etching will help a great deal in reducing the effects during etching because the total ion dose imparted to the sample will be reduced. Surface effects will remain, however, and will be caused by sample imaging as well as etching. More recent combined FIB/SEM instruments have moved some way to solving this problem. By imaging with the electron beam only (and etching with the ion beam), the damage associated with imaging may be completely avoided as long as the electron and ion beams are sufficiently well registered to each other spatially. A further development of recent FIB columns is their ability to operate at very low energies (below 1keV in some cases) while retaining usable spatial resolution. This still does not eliminate damage and implantation completely, but it greatly reduces the depth over which damage occurs.

Finally, etch time and drift issues may be addressed with a combination of factors. Gas-assisted etching will be of benefit in that the etch rate will be enhanced (perhaps by an order of magnitude or more), so that the total etch time, and hence drift, will be reduced accordingly. It is in principle possible to compensate for drift by imaging some reference marks within the field of view, possibly themselves etched with the ion beam (this approach is often used for automated TEM foil preparation), but for single beam FIB instruments, damage issues would be encountered again. The combined FIB/SEM approach again helps with this issue, as it is possible to monitor the progress of the etch with the electron beam as it proceeds, so that drift compensation can be made if necessary. It is for this and other reasons that the modern combined FIB/SEM instrument is likely to be a significant enabling technology for the future of photonics materials research.

3 FIB ETCHING OF PHOTONIC CRYSTAL DEVICES

3.1 INTRODUCTION AND HISTORICAL BACKGROUND

In terms of photonic crystal device fabrication, one of the first examples was shown in macroporous silicon in 2000 [3.1]. Here a 3D Yablonovite photonic crystal structure was created by a combination of plasma or chemical etching with FIB etching. This was one of the first examples where full optical characterization was carried out and reflection data were shown to agree well with the calculated stop bands. Around the same time, two examples of more straightforward 1D PhCs were shown. First, two FIB etched grooves were added near the facet of a QCL and decreased threshold current was observed [3.2]. Similarly in [3.3] a two-section 1D PhC was introduced into a GaN laser with a 13% reduction in threshold current (more details of this will be given in Section 5). In terms of 2D PhCs, one of the first structures to be etched was shown in [3.4], where two sections of periodic arrays of holes were etched on either side of the ridge of a Fabry-Perot laser. This device did not rely on a photonic band gap (PBG) effect, but rather produced a coupled cavity effect which resulted in single mode emission (further details of this device will be given in Section 5). Following from this work, a detailed studied of 2D PhCs was carried out in indium phosphide material [3.5; 3.6], and this will be described in more detail later in this section. In 2004 [3.7] 2D PhCs were etched into a GaAs/AlGaAs heterostructure using fluorine and evidence of amplified spontaneous emission was noted.

One of the main advances in PhC devices has been the move to air-membrane structures where a thin membrane of material is patterned with periodic arrays of holes [3.8]. These tend to give excellent performance since the light is strongly confined in the vertical direction by total internal reflection and laterally by the PhC regions. Interestingly, this type of structure is ideally suited to FIB etching since the holes are no longer blind and material can easily be sputtered out from the bottom of the hole. One of the first examples of such a structure was shown in 2005 in chalcogenide glass [3.9]. Here freestanding films were patterning with up to 46,000 holes in a periodic array with diameters around 360 nm. Optical characterization showed evidence of Fano resonances, a signature of strong control of guided modes in the structure. Later in this section, a recent example of a membrane structure where both the membrane and the PhC are fabricated using FIB will be shown [3.10]. As well as air-membrane structures another popular material system is known as silicon-on-insulator (SOI). Here a thin, 200–300 nm layer of silicon is formed on top of a 1–3 μm thick buried layer of silica which sits on a few 100 μm thick silicon wafer. This is very similar to a membrane in that there is a large refractive index contrast in the vertical direction (3.45 for silicon and 1.45 for silica), but it can be much easier to process this type of structure than an air-membrane. For FIB processing, SOI retains the fact that holes need only be 200–300 nm deep but blind holes are now required, slightly lessening the ease of fabrication. The first example of FIB processing of SOI was shown in 2003 [3.11] where hexagonal arrays of hexagonal holes were etched in the silicon layer. This was followed by an optical characterization of a uniform PhC in 2006 [3.12] which will be discussed later in this section. More recently, a detailed study of annealing in SOI [3.13] has shown how FIB induced losses can be dramatically reduced, meaning that FIB fabricated structures can start to compete directly with E-beam fabricated devices. The flexibility of FIB processing is highlighted in another recent paper, where a vertically oriented membrane [3.14] has been fabricated with a PhC defect containing QWs in order to confine light in TM modes in a QCL device.

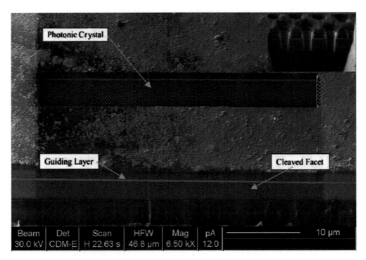

FIGURE 18.2: Photonic crystal fabricated by direct FIB etching in Indium Phosphide. Inset: a FIB cross-section of holes. Reprinted with permission from [3.5]. Copyright (2005), IEEE.

As well as direct FIB processing, a number examples of hybrid FIB processing have been shown where FIB is only used to pattern a mask layer and subsequently conventional wet or dry etching is used to transfer this pattern into the lower layers [3.5, 3.15]. This completely removes the issue of FIB damage, but at the penalty of longer and more complex processing. Later in this section some details of the method described in [3.5] will be discussed.

This section has given a brief historical overview of the development of FIB processing of periodic structures. There are numerous publications that have not been discussed here, but an excellent assessment of state-of-the-art can be found in the proceedings of a recent meeting held in the Netherlands [3.16]. The following sections will describe in more detail some of the examples mentioned in the introduction.

3.2 DIRECT WRITING OF 2D PHOTONIC CRYSTALS IN INDIUM PHOSPHIDE

The fabrication of PhCs in III-V semiconductors has numerous applications, including defect cavity lasers, very small-scale photonic integrated circuits, and quantum information processing. Whilst PhC devices are normally fabricated at the wafer scale by E-beam lithography it can be very useful to work on a device-by-device basis where the structure can be measured, dimensions altered and then remeasured. To this end, FIB etching of indium phosphide has been studied [3.5; 3.6], since this is material of choice for photonic applications in the telecommunications bands around $\lambda = 1550$ nm. Figure 18.2 shows a 38 μm wide array holes, which were etched with an FEI FIB201 machine with a beam current of 150 pA to a depth of 1 μm.

The holes are nominally 225 nm in diameter with a spacing of 375 nm between nearest neighbors (lattice constant, a) and there are 16 rows of holes in the structure. The FIB image shows the optical guiding layer, which has a slightly higher refractive index than the rest of the material, and the edge of the sample has been cleaved to be optically flat to allow coupling of light

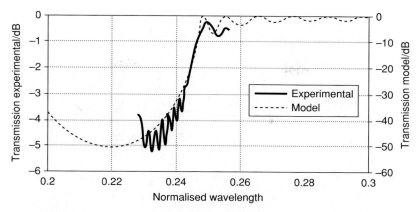

FIGURE 18.3: Measured and modeled transmission results for structure of Figure 18.2. Reprinted with permission from [3.5]. Copyright (2005), IEEE.

into the structure. The inset shows a zoom-in on the hole shape and the poor verticality of direct FIB etched hole in this material can be seen. These holes have been measured to have a sidewall angle of between 6–7 degrees; this leads to high losses and poor performance of photonic crystal devices. This structure has been measured using a fiber coupled test setup and tunable laser. The transmission through the structure (left-hand scale) is shown in Figure 18.3 using what is known as a normalized wavelength (λ_n) scale.

This is commonly used in PhC devices where $\lambda_n = a/\lambda_0$ since a scaling of all dimensions by, say, a factor of two will simply scale the wavelength range by the same factor. Thus $\lambda_n = 0.24$, with a = 375 nm gives $\lambda_0 = 1,562.5$ nm, directly in the telecommunication wavelength range. The figure shows what is known as a band edge where the transmission drops rapidly from strong transmission around 0dB to low transmission of around −4dB in a narrow wavelength range. This is one of the characteristics of PhC devices that is very attractive. The figure also shows the results from a simple electromagnetic model of the structure (right-hand scale) and the band edge position is predicted very well; however, the scale can be seen to be very different from that used for measurement. This is caused by the finite depth of the holes and the poor sidewall verticality. Part of the reason for non-vertical sidewalls is the broad tails of the ion beam. This can be alleviated to some extent by using a masking material. In this case, FIB-deposited platinum was used, and this reduced the sidewall angle by approximately 2 degrees.

3.3 MULTI-STAGE FIB ETCHING OF PHC STRUCTURES

The use of masking layers can dramatically improve the hole shape and reduce damage effects. This approach has been extended to the case where FIB etching is only used for mask definition with no etching directly in the material itself. This section describes a method which employs multiple layers to give excellent final hole profiles and eliminates FIB-induced damage effects [3.5; 3.6]. Here a 500 nm layer of silica is deposited on InP using plasma-enhanced chemical vapor deposition. This is followed by vacuum thermal evaporation of 50 nm of nickel. Figure 18.4(a) shows the long input and output waveguide patterns, which are etched in a two-stage process: first, with a 2,700 pA beam current, followed by a 350 pA etch to smooth

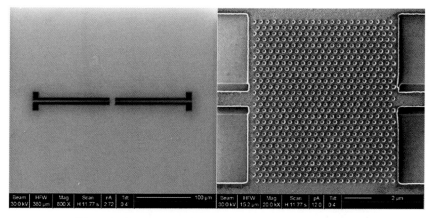

FIGURE 18.4: FIB-etched metal mask layer: (a) input and output waveguide, and (b) central PhC region. reprinted with permission from [3.5]. Copyright (2005), IEEE.

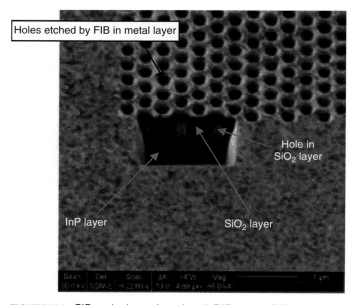

FIGURE 18.5: FIB-etched metal mask with RIE-etched SiO$_2$ layer. Reprinted with permission from [3.5]. Copyright (2005), IEEE.

the pattern sidewalls. Figure 18.4(b) shows an image of the PhC pattern that has been FIB etched into the metal surface.

This is then followed by reactive ion etching of the silica layer as shown in Figure 18.5. Here the very vertical sidewalls of the silica etch can be seen. The nickel layer is then removed and followed by inductively coupled plasma etching of the InP, using the silica layer as a mask. The resulting holes are platinum in-filled using the FIB and cross-sectioned. Figure 18.6 shows that ∼1 μm deep vertical sidewall holes have been obtained. The very deep features are an artifact of the FIB cross-section known as waterfalling.

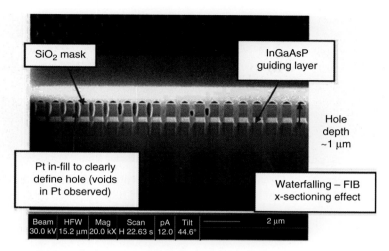

FIGURE 18.6: A photonic crystal fabricated using FIB multi-stage process. Hole diameter = 158 nm; spacing = 480 nm. Reprinted with permission from [3.5]. Copyright (2005), IEEE.

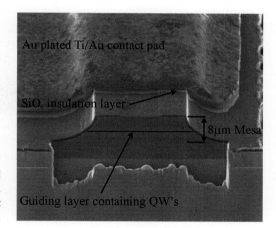

FIGURE 18.7: GaInSb/AlGaInSb strained multiple QW laser facet after FIB polishing. Reprinted with permission from [3.10]. Copyright IOP, 2009, J. Opt. A: Pure Appl. Opt.

As discussed above, air-membrane structures are very attractive for a number of reasons, both in terms of fabrication and optical performance. This section describes a recently fabricated structure aimed at applications in the mid InfraRed band from 3–5 µm. Figure 18.7 shows an FIB cross-section of a mid IR Fabry-Perot laser facet. This work uses the F-P laser as a host for the PhC membrane such that we can obtain a device with a QW at the center of the membrane, which enables both optical and electrical pumping of the device to be explored. The figure shows the position of optical guiding layer within the ridge waveguide of the F-P laser.

3D electromagnetic modeling is used to design a PhC defect cavity with a resonant mode in the region of 3.4 µm. The Q factor of this mode is calculated to be 26,000 for a membrane thickness of 1 µm. We are undertaking a study of the effect of membrane thickness on Q factor

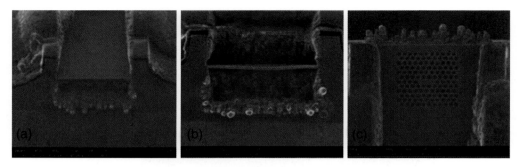

FIGURE 18.8 a-c: Images of the 400 nm thick PhC membrane taken by the FIB. (a) A view of the top side of the membrane after etching away the bulk material; (b) a side view of the membrane; and (c) a plan view of the PhC structure etched through the membrane. Reprinted with permission from [3.10]. Copyright IOP, 2009, J. Opt. A: Pure Appl. Opt.

and thus have initially fabricated a thin membrane of 400 nm thickness. Figure 18.8 shows some of the stages of the fabrication process.

In order to create the membrane and the PhC in one FIB session, a 30 degree angled mount is used. This allows the facet to be placed perpendicular to the ion beam for creating the membrane and then for the device to be rotated such that the membrane is perpendicular to the beam for PhC etching. To begin, high current 2,700 pA box etches are used to remove the bulk of the material from above and below the membrane. Then lower current 70 pA etches are used to gradually polish the surfaces close to the desired thickness, as shown in Figure 18.8(a). A stream file is used to define the PhC pattern and a beam current of 70 pA was used with a dwell time of 800 µs. Figure 18.8(b) shows that after etching there is a downward deflection of the membrane by ∼350 nm. The likely cause of this is tensile strain in the narrow regions between holes induced by the presence of a non-uniform implanted dose of gallium when etching the PhC pattern. Lower beam currents or spiral etching [3.13] could reduce this problem. Figure 18.8(c) shows a top view of the pattern, again highlighting the uniformity and circularity of holes that can be obtained.

3.5 FIB ETCHING OF PHCS IN SILICON-ON-INSULATOR

Here we show how FIB etching can be used to pattern a thin, 205 nm, layer of silicon which is formed on top of a silica layer [3.12]. FIB etching of such thin layers results in much improved hole quality, which should result in much improved PhC performance. This, coupled with the recent advances in annealing of SOI FIB etched devices [3.13] means that FIB etched devices could compete in performance with state-of-the-art E-beam fabricated devices. Figure 18.9 shows an FIB cross-section of the layer structure showing the thin silicon layer and dark silica layer. This insulating layer can result in charging effects, so the top silicon layer needs to be grounded; here we used FIB platinum deposition to form a ground contact.

In this case the whole device, including input and output waveguide and facets, is created using FIB etching. The input and output waveguides can be seen in Figure 18.10(a). They are 150 µm long and 5 µm wide. They were etched with 1nA beam current with iodine gas-assisted etching. The iodine gas reduces the overall etch time from 55 to 10 minutes.

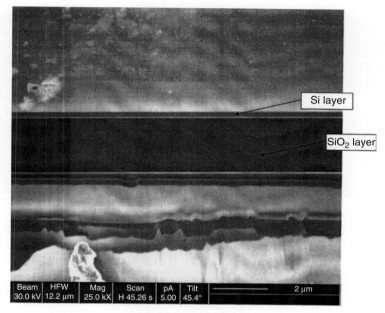

FIGURE 18.9: Focused ion beam etched facet in SOI. Reprinted with permission from [3.12]. Copyright (2006), AIP.

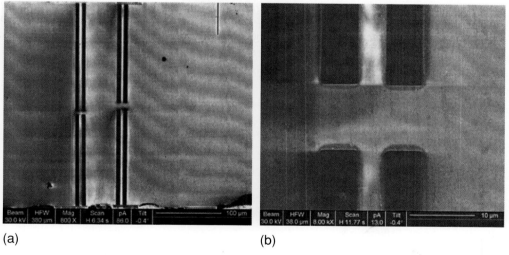

(a)   (b)

FIGURE 18.10: (a) Input and output photonic wire waveguides and (b) zoom-in on region where PhC pattern will be etched. Reprinted with permission from [3.12]. Copyright (2006), AIP.

In order to etch the PhC pattern, the magnification was increased to ×8,000; this gives a viewable area of $38 \times 30.9 \, \mu m$. The device was designed to work near to 1550 nm and the lattice constant, $a = 390$ nm and hole radius, $r = 100$ nm. In order to get the best possible resolution, a very low beam current of 11 pA was used with no added gas. A typical patterning time of 10 minutes is required for a $5 \times 5 \, \mu m$ PhC. Figure 18.11 shows the PhC etched pattern and the

FIGURE 18.11: Photonic crystal with lattice constant 390 nm and $r/a = 0.256$. Reprinted with permission from [3.12]. Copyright (2006), AIP.

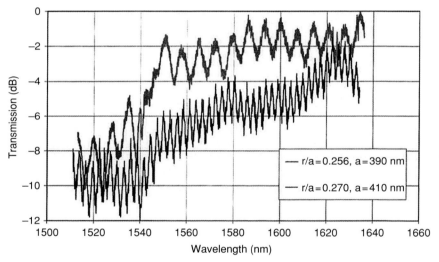

FIGURE 18.12: Measured transmission response of PhCs with different r/a and a. Reprinted with permission from [3.12]. Copyright (2006), AIP.

inset shows a zoom on the holes, highlighting the excellent uniformity and circularity that can be obtained with FIB etching.

The device was then measured using a fiber coupled setup and the transmission response is shown in Figure 18.12 for two different r/a ratios and lattice constants. The band edge feature is seen very clearly in the $a = 390$ nm case, and less so in the $a = 410$ nm case. The band edge shapes are found to agree well with those predicted by 2D and 3D electromagnetic modeling. As observed above in Figure 18.3, the depth of the measured band edge is much less than that predicted by modeling. The hole shapes obtained here are not ideal, but 3D modeling of non-cylindrical holes has show that this should not cause a reduction in band edge depth. Thus it is felt that FIB-induced losses in the holes are causing the degradation in the band edge profile, and it is hoped that annealing will be able to improve the performance of these devices.

3.6 CONCLUSIONS

This section has given a historical perspective on the developments in FIB etching for photonic crystal devices. Dramatic advances have been witnessed over the past few years, up to the point where in certain materials FIB-etched devices could compete with those fabricated by more conventional approaches. The issue of damage remains to be completely addressed, but again there are a numbers of techniques, including gas-assisted etching and post-etch annealing, that can alleviate this problem. It must be noted that different material systems may require very different approaches and while the general principles may be transferred, exact details and recipes must be evaluated in each case. The section has shown a number of examples of state-of-the-art devices, along with optical characterization in some cases. It is felt that FIB etching of photonic crystal devices will continue to develop in coming years and will eventually take its place alongside other fabrication techniques as a rapid nanoscale prototyping and tuning tool that will be invaluable to workers in the field.

4 FIB FABRICATION OF DEVICES FOR QUANTUM INFORMATION PROCESSING APPLICATIONS

All etching approaches are known to cause surface damage, which can lead to enhanced surface recombination and reduced performance in active optical devices such as lasers and LEDs. For this reason, the fabrication of active nanophotonic devices is often limited by such surface effects, and because FIB etching is quite aggressive, its use is often restricted to passive regions of active devices. However, one area where surface recombination problems are not so debilitating is in the study of quantum dot devices. Quantum dots (QD) are nanometer scale islands of a narrower gap semiconductor embedded within a wider gap material. Electron and hole recombination then is confined to within the spatially localized quantum dots, which due to quantum confinement effects lead to narrow excitonic transitions. Etching through a quantum dot layer will damage dots close to the etch surface and increase surface recombination rates, but in the bulk of the material the recombination can remain strongly radiative.

In effect, single quantum dots can behave like two-level "macro-atoms," making them good candidates for single photon sources. Exciting a quantum dot by optically pumping electrons into its upper state leads to the emission of single photons at the excitonic recombination energy with internal efficiency close to unity. The main problem then is to efficiently collect this light from the high refractive index semiconductor material. The out-coupling efficiency can be significantly improved by coupling the emitted light into the resonant cavity mode of a micro-pillar microcavity. Planar microcavities can be grown by molecular beam epitaxy (MBE) of III/V materials. Distributed Bragg reflectors (DBR) are grown as a series of quarter wavelength thick layers of alternating composition and refractive index, while a cavity can be introduced at the center of a stack by increasing the layer thickness. When this planar cavity is subsequently etched into wavelength scale micropillars, the waveguiding in the high refractive index pillar and the resonances of the cavity combine to produce a narrow band single-mode resonance. The light emitted into this single mode can only escape from the top of the pillar, while the sidewall leakage is mostly prevented by total internal reflection, thus making an efficient photon source [4.1–4.3].

Micro-pillar microcavities have been etched from planar cavity structures using photo-lithography and electron beam lithography, followed by a variety of chemically based etching methods [4.4–4.6] to produce pillars in the 0.5–5 μm diameter range. As the diameters of the pillars are reduced, the surface losses from etch damage reduce the Q-factor of the optical resonance, even when using the least aggressive etching methods. Hence the challenge of using the focused ion beam (FIB) direct write process for making optically active micropillars is one of minimizing the optical sidewall damage. Recently [4.7] it was shown that the use of FIB combined with iodine gas-assist reduces the surface damage effects, and the etching of micro-pillar microcavities containing single quantum dots is possible. High-quality factor resonances, comparable to inductively coupled plasma (ICP) etched samples, and single-photon emission were reported.

The steps to etching a micropillar microcavity from an MBE grown planar cavity structure are shown in Figure 18.13(a)-(c) while a pillar etched using conventional ICP etching is shown in Figure 18.13(d). In ICP the different etch rates of the different composition layers leave a ribbed edge, clearly illustrating the cavity structure. This consists of a GaAs λ-cavity surrounded by a lower mirror consisting of a 27-pair alternating GaAs/AlGaAs quarter-wave distributed Bragg reflector (DBR) and an upper DBR mirror of 20 quarter wave pairs. The cavity resonant wavelength is close to 0.96 μm. The center of the cavity contains a layer of InAs self-assembled QDs.

A FEI Strata FIB201 instrument was used with a resolution limit of 8 nm (30 keV, 1 pA) Gallium ion beam. The etching is assisted by the introduction of iodine gas (from solid iodine heated to 32°C) introduced through a needle ~100 μm from the sample surface. The etch rate is enhanced by the formation of volatile iodine compounds at the etching surface, thus strongly suppressing any local redeposition processes. At low beam currents (<70 pA) etching is strongly enhanced, showing up to $24(+/-3)$ μm^3/nC etch rate, which is ~ 27× enhancement above simple sputter etching. Effectively we are removing ~6 atoms/ion without gas and ~170 atoms/ion with gas.

In this work we set the machine to etch square pillars using an automated pattern file routine. The processing starts with a rough etch defining deep circular patterns with inner radius of 5 μm and outer radius of 15 μm at an ion beam current of 1 nA and no gas assist. This gives pillars with a slightly conical shape as shown in Figure 18.13(a). The height of the pillar is about 8 μm. The sample is then tilted to 3.8° and the magnification is increased to 5,000×, which allows smooth, vertical, high-resolution faces to be etched on each side of the square pillar. It also limits the penetration depth of the Ga$^+$ ions to <30 nm into the sidewall, as estimated by Monte Carlo simulation software [4.8]. The pillar required is defined by four rectangular boxes, as shown in Figure 18.13(b). The boxed areas are then etched away using beam currents of 350 pA. Two further etch steps with beam currents of 70 pA and <10 pA are performed with gas assist. The last etch uses the smallest spot size and leads to very smooth sidewalls with surface roughness <10 nm, as shown in the inset of figure Figure 18.13 (c). The etch rate is enhanced by ~27×, and thus the total ion dose is reduced by this factor, leading to reduced surface damage and little evidence of FIB-induced redeposition of gallium.

Square micro-pillar microcavities with dimensions from 0.6 μm to 5 μm containing InAs quantum dots were fabricated by the above technique. The samples were then measured using a low-temperature (4.3 K) micro-photoluminescence (μ-PL) confocal microscope with a 0.55 m monochromator. The spectra contain an enhanced photoluminescence peak due to the cavity resonance, and the relative width $\lambda/\Delta\lambda$ (λ is the central wavelength) gives an estimate of the cavity Q-factor.

To measure the photon statistics, and particularly the anti-bunching, the filtered output light from the exit slit was directed through a 50:50 non-polarizing beamsplitter, and on to two fiber-coupled silicon single photon counting modules (SPCMs), and a time interval

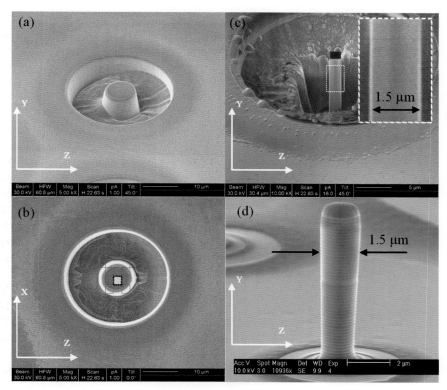

FIGURE 18.13: (a) Focused ion beam FIB image of rough etched pillar with a conical shape. (b) Its top view with the definition of rectangular box patterns. (c) FIB image of 1.5 μm square pillar after performing a final milling by ion beam currents of 11 pA; the inset shows a close-up of pillar to highlight the surface quality. (d) Electron microscope image of the RIE-ICP etched micropillar with circular cross-section and a diameter of 1.5 μm. Reprinted with permission from [4.7; 4.12], Copyright (2007), AIP.

measurement system (HP53310A). This latter arrangement is commonly known as a Hanbury Brown and Twiss interferometer [4.9] and is the standard method for confirming single photon emission. Effectively single photons cannot be detected in both detectors, so we see a suppression of coincident detections when we have a single photon source.

Figure 18.14 shows the spectra obtained for six different pillar sizes. We identify the fundamental cavity mode by arrows. The other peaks are due to higher order modes (large pillars) or individual quantum dots (small pillars). As we shrink the size of the square micropillar microcavity, the modal volume (V_{eff}) is reduced, which increases the coupling of the quantum dot emission to the microcavity mode and strongly enhances spontaneous emission [4.10]. Also, the cavity mode blue-shifts to higher energy or shorter wavelength. However, the field at the cavity wall becomes stronger for small pillars. Even in ideal pillars, we see loss due to mode mismatching between the GaAs spacer-layer cavity fundamental mode and the corresponding fundamental modes of the upper and lower mirror [4.11]. Also, as more of the field resides outside the pillar, the mirror reflectivity is reduced. Both of these fundamental mechanisms lead to a reduction of the Q-factor in thin pillars. However, in "real" etched pillars the losses are augmented by surface-damage-induced absorption and

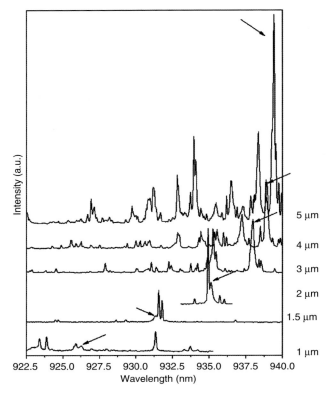

FIGURE 18.14: Low temperature photoluminescence spectra of laterally structured microcavities of different lateral sizes. The dimensions of the square micropillar microcavities are as indicated. The resonances of the HE11 cavity mode are marked by arrows. Other sharp lines correspond to higher order modes or in the smaller dimension pillars to emission from individual dots. Reprinted with permission from [4.7]. Copyright (2007), AIP.

surface-roughness-induced scattering, both of which further reduce the measured Q-factors in thin pillars.

We present our experimentally measured Q-factors in Figure 18.15, comparing GAE-FIB etched pillars with pillars etched without the gas assist and with circular pillars etched using reactive ion etching (inductively coupled plasma, ICP). The measured Q-factors without gas assist are low. With gas assist the measured Q-factors are comparable to the ICP etched pillars from the same sample, and in the case of 1.5 μm we actually see a slightly higher Q-factor. However, this may be in part because we are comparing square and circular pillars. When we look at the image of the pillar sidewalls (Figure 18.13(c)) we see that the surface is smoother than typical RIE etched samples (Figure 18.13(d)), suggesting that we should see less sidewall scattering from small pillars. All the measured Q- factors are lower than theoretical values calculated using the finite difference time domain method (FDTD) [4.10]. We speculate that the main loss mechanism in our FIB-GAE etched pillars is not scattering but residual surface absorption due to damage and adsorbed gallium metal.

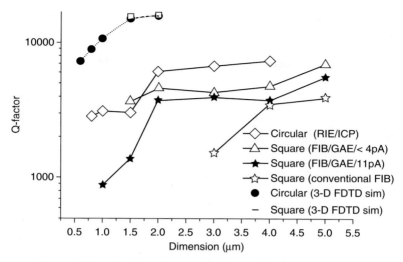

FIGURE 18.15: Pillar-size dependence of Q factors. This graph compares the Q factors from theory [4.10] (filled circles and open squares) with Q factor measurements on RIE-ICP etched circular pillars (open diamonds), FIB etched square pillars (open stars), and gas-assisted FIB etched square pillars (open triangles, filled stars) at varying beam currents, as shown in the inset. Reprinted with permission from [4.7]. Copyright (2007), AIP.

Individual narrow peaks in the spectra can be identified with quantum dot emission and can be seen in the spectra from our smaller micro-pillars when pumping with low power. The narrow quantum dot emission peaks can be tuned in and out of resonance with the cavity by changing the temperature. It was possible to isolate suitable dots in the 1.5 μm and 2 μm pillars. Using these dots as sources of light and pumping with a pulsed laser, it was possible to show from photon statistics measurements that each pulse led to the generation of one single recombination photon. The collection efficiency of these single photons at the end of the pillar was of order 5%.

This work clearly shows that Ga$^+$ focused ion beam etching with iodine gas-assist can be a useful method of fabricating pillar microcavities. Although single pillars may take 20 minutes to make, they can be immediately measured the same afternoon. This is a clear advantage over the multi-step lithography and chemical etching methods [4.12]. Of course the latter has the advantage of producing many hundreds of pillars. The surface smoothness of FIB and GAE-FIB etched samples suggest that Q-factors are limited by surface absorption from damage and adsorbed gallium rather than surface scattering. It may be possible to improve on these results by further surface processing such as annealing.

5 FIB POSTPROCESSING OF PHOTONIC DEVICES

FIB etching has proved a very useful tool in post-processing existing devices to investigate a novel idea such as the incorporation of a grating or PhC structure to improve or alter device performance. At Bristol we have used this practice to introduce even and odd grating structures onto the ridge of edge-emitting GaN-based blue-emitting devices to improve the edge-emitting characteristics (odd-order) or to create coherent surface emission (even-order). In addition, we have introduced PhC structures adjacent to the ridge in an edge-emitting laser and into the top DBRs of a VCSEL. Both of these additions improved the output characteristics of the beam. These devices will be considered below.

5.1 GRATINGS IN GAN STRUCTURES

Gallium nitride (GaN) and its associated alloys (GaInNAs) are relatively novel III-V semiconductor materials that have been investigated for blue, UV, and green emission in LEDs and lasers and for high-power electronic devices. GaN has a large band gap (3.2 eV), with a high binding energy. This material is very slow to etch using FIB but can show excellent aspect ratio for etching of holes. The edge-emitting InGaN/GaN laser is shown in Figure 18.16.

GaN edge-emitting lasers have many problems, which has made their performance difficult to improve. Wurzite GaN is not generally grown on GaN substrates, making cleaving of the facets difficult. This results in the facets having surface roughness, which effectively increases the losses as the light is scattered at the facets. We investigated using FIB etching to smooth the facet and reduce the scattering losses. Facet polishing led to a 4% reduction in the threshold current through reducing the RMS roughness at the surface to about 1–3 nm. The modeled results for facet reflectivity as a function of surface roughness are shown in Figure 18.17 [5.1].

Figure 18.18 shows the laser facet before and after FIB polishing, while Figure 18.19 shows the laser threshold before and after etching. The FIB etching results in a threshold reduction from 640 mA–615 mA, a reduction of 4% as shown in figure 18.19.

In an attempt to reduce mode competition and thereby achieve CW lasing, the incorporation of a GaN/air DBR within the laser ridge was investigated [5.2]. A two-period air/semiconductor/air/semiconductor/air 5th order λ/4 Bragg reflector was introduced into the ridge structure as shown in figure 18.20. The thickness was 509 nm of air and 202 nm of GaN and the depth of the etch was about 1 micron.

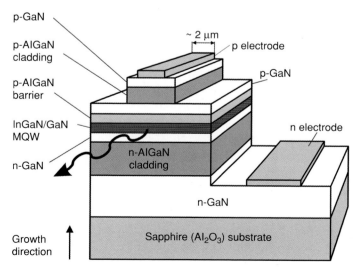

FIGURE 18.16: Schematic of edge emitting GaN laser.

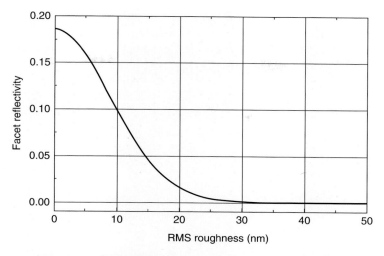

FIGURE 18.17: Facet reflectivity as a function of surface roughness.

The threshold of the laser was reduced from 725 mA to 630 mA, a reduction of 13%. In addition, the mode spectra were altered, resulting in a 1 nm red-shift. The near-field was broadened slightly due to light diffraction and field expansion in the air gaps. The role of the depth of the etch was investigated as a means of controlling the transverse modes. Through modeling it was found that a reduced-depth air-nitride grating could selectively enhance and provide higher reflectivity for one transverse mode, which is most tightly confined within the active area and allows lower reflectivity for high-order modes, as described in Figure 18.21. The fundamental mode of the laser experiences higher confinement within the active area and lower diffraction in air-gaps, resulting in its higher reflection, while a high-order mode faces the higher diffraction, leading to its lower reflection from the facet.

Figures 18.22 and 18.23 show near-field distributions before and after shallow and depth etching of the sample. The transverse near field distribution of the deeply etched sample has side-lobes, suggesting multiple-mode generation of the sample. However, the near field of the sample with the shallow etch does not exhibit any visible side-lobes of high-order modes.

Thus, a reduction in the etch depth resulted in single-peak emission from the laser. This was a very powerful result and was one of the first demonstrations of single mode operation of a GaN based laser diode. This is a crucial requirement of GaN lasers if they are to be used for CD/storage or printing applications.

5.2 PHCS IN EDGE-EMITTING LASERS

It is well-known that etching a 1D grating adjacent to the ridge of a laser can select a certain longitudinal optical mode commensurate with it and degrade others. This is the basis of distributed Bragg reflectors (DBRs). In this study we used the FIB to etch a short section of PhC structure adjacent to the ridge of an InGaAsP/InP FabryPerot laser designed to emit at 1.3 micron wavelength, as shown in Figures 18.24 and 18.25 [5.3].

A 50 micron section of a PhC with a diameter to pitch ratio of 0.33 and a depth comparable with the active region was etched using FIB. Figure 18.26 shows the pre- and post-etch spectra.

(a)

(b)

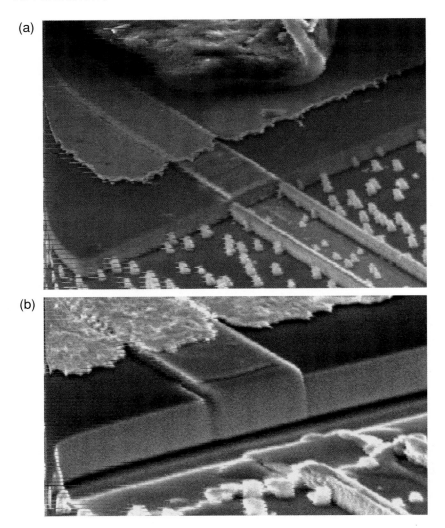

FIGURE 18.18: GaN laser facet (a) before and (b) after FIB polishing.

A SMSR>30 dB was observed after the PhC structure was added. The introduction of a 2D periodic array of holes adjacent to the ridge-waveguide of an Fabry-Perot laser was used to create single-contact, single longitudinal mode laser. The single mode nature of the device was shown to be due to the overlap of the evanescent mode with the etched grating, which acted to separate the main laser cavity into two coupled cavities. Using a 2D PhC rather than a 1D grating allowed the coupling constant between the cavities to be varied much more than is possible using a 1D grating.

Modeling of device reflectivity has shown that only a small modulation of refractive index within the grating is sufficient to allow single-mode operation when the grating is placed at a distance of 50 μm from one facet. Modeling of device gain has shown that a small amount of gallium implantation would account for the blue shift observed in lasing wavelength after etching. Implantation also results in an increase in tensile strain in the quantum wells, which is responsible for the higher power in TM polarized emission after etching. The effect of etching

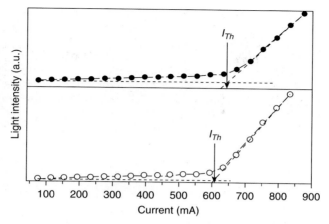

FIGURE 18.19: Laser threshold before (top) and after (bottom) FIB polishing.

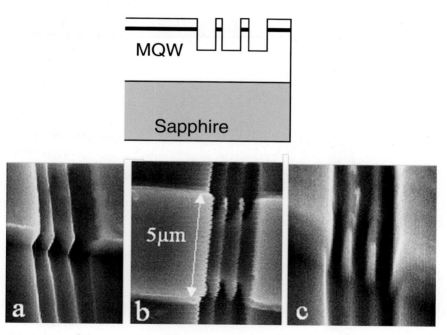

FIGURE 18.20: Schematic of the DBR etch and pictures (a-c) of the completed etch made by FIB etching.

on the guiding properties of the laser are shown, through modeling, to be the result of a reduction in average refractive index around the guiding ridge. Calculated spectra show that it would be reasonable to expect an SMSR >25 dB as a result of introducing the grating. Direct modulation of this device at a data rate of 10 Gb/s has yielded open eyes for an extinction ratio of 6.5 dB. The spectra remain stable under large signal modulation. This simple and repeatable post-processing technique could readily be implemented using reactive ion etching during fabrication, hence providing promising low-cost devices for data communications applications.

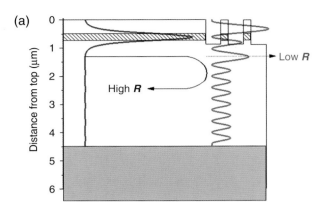
(a)

High *R*

Low *R*

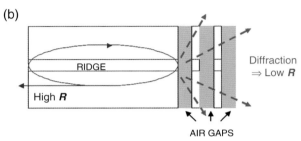
(b)

RIDGE

High *R*

AIR GAPS

Diffraction
⇒ Low *R*

FIGURE 18.21: Schematic (a) side view and (b) top view of the structure showing the use of the DBR gratings as selective filters for different transverse modes. Blue and red lines correspond to fundamental and high-order modes.

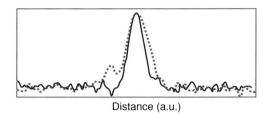

Distance (a.u.)

FIGURE 18.22: Near field intensity distribution without (solid) and with (dashed) DBR Etch-Deep Etch.

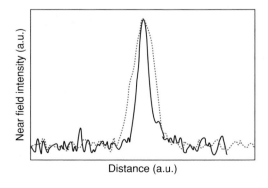

Distance (a.u.)

FIGURE 18.23: Near field intensity distribution with (solid) and without (dashed) DBR Etch-shallow Etch.

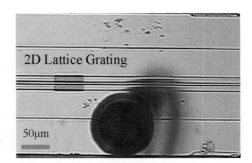

FIGURE 18.24: Top down view of etched device (dark circle is the bond wire).

FIGURE 18.25: SEM picture of holes.

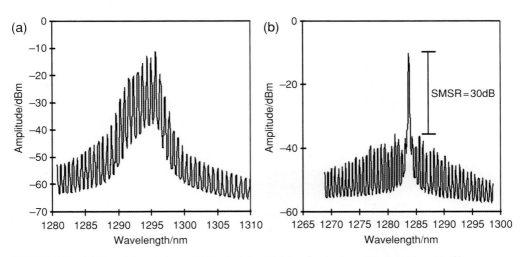

FIGURE 18.26: (a) Pre-etch spectrum (b) Post-etch spectrum for device with single facet 0.1% Anti-Reflection coating.

5.3 PHOTONIC CRYSTALS IN VERTICAL CAVITY SURFACE EMITTING LASERS

Semiconductor vertical-cavity surface emitting lasers (VCSELs) were initially designed to reduce longitudinal mode competition due to the design of a short cavity (a few wavelengths) [5.4]. Because of their design, the longitudinal mode must be reflected many times within the cavity so that lasing can occur, implying the need for high reflectivity mirrors that define the longitudinal cavity. These have been very successful in creating low threshold lasers with no longitudinal mode competition. They can be modulated at high speed, due in part to their small size, and they produce a circular beam ideal for injection into fibers. In addition, it is possible to make large arrays in parallel and to test them on wafer, making them very cheap and useful for parallel interconnects. It was found that the output beam changes its shape as the current is increased due to competition between transverse modes that exist because the VCSEL has a large lateral size. The lateral size of VCSELs has been controlled by proton implantation or by the introduction of an oxide confining layer to control the current flow. Both methods introduce transverse optical waveguiding, allowing a reduction in the mode number by reducing the transverse optical confinement radius; however, this makes the area from which gain is extracted small, reducing the output power. Self-heating caused by current crowding also reduces the output power of the VCSEL [5.5; 5.6].

In 1998, large modal area single-mode photonic crystal (PhC) "holey fibers" were invented [5.7]. These fibers allow single-mode waveguiding inside wide cores due to effective index waveguiding. It was proposed that the introduction of this structure into the VCSEL cavity should allow transverse control of the VCSEL modes and enhancement of the fundamental transverse mode, thereby reducing transverse mode competition. Ideally it would be preferable to grow or fabricate the PhC structure within the cavity where the optical mode has the largest intensity, but it has proved easier to post-fabricate by introducing the PhC holes into the DBR structures where some optical mode intensity exists. Several groups have etched holes into the top DBR using FIB etching, although E-beam etching and dry-etching have also been used [5.8–5.10]. Investigation of the PhC period, hole diameter and hole depth has been carried out by several groups [5.8–5.13]. Generally, it was found experimentally that single transverse mode conditions can be established, particularly as the etch depth is increased through the DBR, but that the output power decreases substantially. We have fabricated PhC waveguides by etching holes using a commercial Focused Ion Beam Ga+ (FIB) FEI200 machine into the top DBR of commercially available oxide-confined VCSELs [5.12, 5.14]. These VCSELs emit light of 850 nm wavelength with a threshold current of 1.2 mA and output power of about 2.6 mW (at an injected current of 10 mA). Iodine gas has been injected into the FIB chamber during the etching process in order to enhance the etch rate and reduce redeposition of the sputtered material. The geometry of the etched areas was defined with stream files imported into the FIB graphical user interface. Figure 18.27 shows an FIB image of the VCSEL with the PhC.

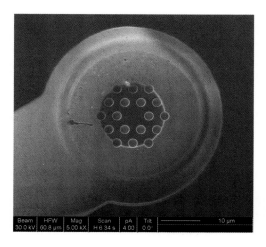

FIGURE 18.27: An FIB image of a PhC-VCSEL. The PhC hole diameter is 2 μm, the lattice constant is 4 μm. The aperture is 17 μm.

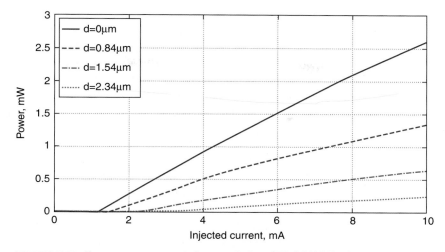

FIGURE 18.28: Power versus current characteristics of PhC-VCSELs for samples with hole depth *d* ranging from 0 to 2.34 μm.

Table 18.1 PhC-VCSEL parameters.

Etching depth d, μm	Threshold current, mA	Series resistance*, Ohm
0	1.2	298.73
0.84	1.6	299.05
1.54	2.4	302.35
2.34	3.3	307.61

*Measurements were taken at $I = 6$ mA.

Output power versus injected current was measured as a function of etch depth, as shown in Figure 18.28. These VCSELs have demonstrated single mode operation, but this resulted in a decrease in output power. Some previous work using FIB etching to introduce uniaxial stress into VCSEL to pin their polarization had found that post-etching annealing restored the output power due to re-adjusting the ion positions after being damaged by Ga ion bombardment. This was confirmed by micro-photoluminescence and micro-raman investigation. We investigated the series resistance of the VCSEL after the PhC etching, as shown in Table 18.1. The increase in resistance with the PhC hole depth is observed. The increase of the threshold current is expected due to the lowered reflection coefficient of the top etched DBR. However, the increase of the resistance results from the FIB etching and may also increase the threshold current. Since the FIB etching was not optimized, some Ga+ ions were implanted into areas around the etched holes and changed the resistivity of the surrounding material. The resistivity may be reduced back to its initial value by annealing the samples as suggested in [5.15; 5.16].

The high-speed nature of FIB etching provides a powerful tool for post-processing of VCSELs and fabricating PhC waveguides for advanced transverse mode control and single-mode operation. The FIB software and hardware allow precise control of the etching depth and an arbitrary etching geometry may be defined either with stream or pattern files. FIB etching requires no physical mask, making the etching process very rapid. The etching time of one VCSEL ranges from 4 minutes to 4 hours and thus is suitable for processing individual devices, not parallel processing of a number of samples as with conventional mask lithography.

Redeposition of the sputtered material can be reduced by applying precursor gas. The FIB etching process must be optimized to reduce damage caused by the FIB, but the damage may be minimized by annealing of the etched sample afterward.

6 CONCLUSIONS

This chapter has given an overview of FIB processing for photonic devices. We have shown a number of examples of devices that have been fabricated within the Photonics research group at the University of Bristol. There are many other research groups around the world working on FIB processing of photonic devices, and we hope that the references we have included here reflect some of their work. We hope this chapter has shown the many exciting possibilities that FIB processing offers in the areas of nanophotonics and more conventional photonics devices.

It is hoped that further advances in FIB systems will help to reduce the effects of ion damage; in particular, dual beam FIB/SEM systems with the possibility of low energy beams appear to be very promising.

In terms of future directions for device work, we are looking at photonic crystals in diamond membranes for single photon sources and nanobeams in III-V semiconductors for optomechanical applications. There are also possibilities of developing FIB-based lithography schemes where the thin damage layer is used as a mask for wet etching techniques.

Overall, it is felt that FIB processing will have an important role to play in the development of future photonic devices. The ability to process on a device-by-device basis, performing measurements after each processing step, is hugely attractive and provided damage can be controlled, will ensure that FIB processing of photonic devices will remain a fruitful area for both fundamental research and device development.

REFERENCES

[1.1]   Yariv, A., and P. Yeh. *Optical waves in crystals.* New York: Wiley, 1984.

[1.2]   Coldren, L. A., and S. W. Corzine. *Diode lasers and photonic integrated circuits.* New York: Wiley, 1995.

[1.3]   Yablonovitch, E. Inhibited spontaneous emission in solid-state physics and electronics. *Phys. Rev. Lett.* 58 (1987): 2059–2062.

[1.4]   John, S. Strong localisation of photons in certain disordered dielectric superlattices. *Phys. Rev. Lett.* 59 (1987): 2955–2958.

[1.5]   Cheng, C. C., and A. Scherer. Fabrication of photonic band-gap crystals. *J. Vac. Sci. Technol.* B 13.6 (1995): 2696–2700.

[1.6]   Krauss, T. F., R. M. De La Rue, and S. Brand. Two-dimensional photonic-bandgap structures operating at near-infrared wavelengths. *Nature* 383 (1996): 699–702.

[1.7]   McNab, J., N. Moll, and Y. A. Vlasov. Ultra-low loss photonic integrated circuit with membrane-type photonic crystal waveguides. *Optics Express* 11.22 (2003): 2927–2939.

[1.8]   Painter, O., et al. Lithographic tuning of a two-dimensional photonic crystal laser array. *IEEE Photonics Technology Letters* 12.9 (2000): 1126–1128.

[1.9]   Puretz, J., et al. Focused-ion-beam micromachined AlGaAs semiconductor-laser mirrors. *Electron. Lett.* 22.13 (1986): 700–702.

[1.10] Takado, N., et al. Chemically enhanced focused ion beam etching of deep grooves and laser-mirror facets in GaAs under Cl2 gas irradiation using a fine nozzle. *Appl. Phys. Lett.* 50.26 (1987): 1891–1893.

[1.11] Remiens, D., et al. GaAs/AIGaAs double heterostructure laser monolithically integrated with passive waveguide using focused ion beam etching. *Electronics Letters* 25.20 (2000): 1400–1402.

[1.12] Schneider, M. Focused ion beam patterning of III–V crystals at low temperature: A method for improving the ion-induced defect localization. *J. Vac. Sci. Technol.* B 18.6 (2000): 3162–3167.

[1.13] Sugimoto, Y., et al. Reduction of induced damage in GaAs processed by Ga^+ focused-ion-beam-assisted Cl_2 etching. *J. Appl. Phys.* 68.5 (1990): 2392–2399.

[1.14] Sargent, L. J., et al. Investigation of polarization-pinning mechanism in deep-line etched vertical-cavity surface-emitting lasers. *Applied Physics Letters* 76.4 (2000): 400–402.

[1.15] Sargent. L. J., et al. Investigation of 2-D-lattice distributed reflector lasers. *IEEE J. Quantum Electron* 38.11 (2002)pp1485-1492

[1.16] Hopman, W. C. L., et al. Focused ion beam milling strategies of photonic crystal structures in silicon. *Proc. ECIO* 2007, 25–27 April 2007, Copenhagen, Denmark, paper FA1.

[1.17] http://fibphoton.ewi.utwente.nl/.

[1.18] http://www.bris.ac.uk/eeng/research/pho/fib4photonics.html.

[3.1]   Chelnokov, A., et al. Near-infrared Yablonovite-like photonic crystals by focused-ion-beam etching of macroporous silicon. *Applied Physics Letters* 77.19 (2000): pp. 2943–2945.

[3.2]   HvozdaraL., et al. Quantum cascade lasers with monolithic air–semiconductor Bragg reflectors. *Applied Physics Letters* 77.9 (2000): pp. 1241–1243.

[3.3]   Marinelli, C., et al. Threshold current reduction in an InGaN MQW laser diode with $\lambda/4$ air/semiconductor Bragg reflectors. *Electron. Lett.* 36 (2000): 1706–1707.

[3.4]   Sargent, L. J., et al. Investigation of 2-D-lattice distributed reflector lasers. *IEEE J. Quantum Electron* 38.11 (2002): pp. 1485–1492.

[3.5]   Cryan, M. J., et al. Focused ion beam based fabrication of nanostructured photonic devices. *IEEE Journal of Selected Topics in Quantum Electronics* 11.6 (2005): 1266–1277.

[3.6]   Hill, Matthew. The study of the fabrication of photonic crystal structures in GaN and InP by focused ion beam etching. PhD thesis, University of Bristol, 2005.

[3.7]   Cabrini, S., et al. Focused ion beam lithography for two dimensional array structures for photonic applications. *Microelectronic Engineering* 78–79 (2005): 11–15.

[3.8]   McNab, S. J., N. Moll, and Y. A. Vlasov. Ultra-low loss photonic integrated circuit with membrane-type photonic crystal waveguides. *Optics Express* 11.22 (2003): 2927–2939.

[3.9]   Freeman, D., S. Madden, and B. Luther-Davies. Fabrication of planar photonic crystals in a chalcogenide glass using a focused ion beam. *Optics Express* 13 (2005): 3079–3086.

[3.10] Pugh, J. R., et al. Design and fabrication of a midinfrared photonic crystal defect cavity in indium antimonide. *J. Opt. A: Pure Appl. Opt.* 11 (2009): 054006.

[3.11] Bostan, C. G., R. M. de Ridder, V. J. Gadgil, L. Kuipers, and A. Driessen. Line-defect waveguides in hexagon-hole type photonic crystal slabs: design and fabrication using focused ion beam technology. *BeneLux IEEE LEOS Meeting*, 2003.

[3.12] Balasubramanian, K., P. J.Heard, and M. J. Cryan. Focused ion beam fabrication of 2D photonic crystals in silicon-on-insulator. *Journal of Vacuum Science and Technology* B 24.6 (2006): 2533–2537.

[3.13] Hopman, W. C. L., et al. Focused ion beam milling strategies of photonic crystal structures in silicon. *Proc. ECIO* 2007, 25–27 April 2007, Copenhagen, Denmark, paper FA1.

[3.14] Lonèar, Marko, et al. Design and fabrication of photonic crystal quantum cascade lasers for optofluidics. *Optics Express* 15.8 (2007): 4499–4514.

[3.15] Kim, Y. K., et al. Focused ion beam nanopatterning for optoelectronic device fabrication. *IEEE Journal of Selected Topics in Quantum Electronics* 11.6 (2005): 1292–1298.

[3.16] http://fibphoton.ewi.utwente.nl/.

[4.1]   Michler, P., A. Kiraz, C. Becher, W. V. Schoenfeld, P. M. Petroff, Lidong Zhang, and E. Hu, A. Imamoglu. A quantum dot single-photon turnstile device. *Science* 290 (2000): 2282.

[4.2]   Gérard, J. M., D. Barrier, J. Y. Marzin, R. Kuszelewicz, L. Manin, E. Costard, V. Thierry-Mieg, and T. Rivera. Quantum boxes as active probes for photonic microstructures: The pillar microcavity case. *Appl. Phys. Lett.* 69 (1996): 449.

[4.3]   Pelton, M., J. Vučković, G. S. Solomon, A. Scherer, and Y. Yamamoto. Three-dimensionally confined modes in micropost microcavities: quality factors and Purcell factors. *IEEE J. of Quantum Electronics* 38 (2002): 170.

[4.4]   Santori, C., D. Fattal, J. Vučković, G. S. Solomon, and Y. Yamamoto. Indistinguishable photons from a single-photon device. Nature 419 (2002): 594.

[4.5]   Pelton, M., C. Santori, J. Vučković, B. Zhang, G. S. Solomon, J. Plant, and Y. Yamamoto. Efficient source of single photons: a single quantum dot in a micropost microcavity. *Phys. Rev. Lett.* 89 (2002): 233602.

[4.6]   Moreau, E., I. Robert, J. M. Gérard, I. Abram, L. Manin, and V. Thierry-Mieg. Single-mode solid-state single photon source based on isolated quantum dots in pillar microcavities. *Appl. Phys. Lett.* 79 (2001): 2865.

[4.7]   Ho, Y.-L. D., R. Gibson, C. Y. Hu, M. J. Cryan, I. J. Craddock, C. J. Railton, J. G. Rarity, P. J. Heard, D. Sanvitto, A. Daraei, M. Hopkinson, J. A. Timpson, A. Tahraoui, P. P. S. Guimaraes, and M. S. Skolnick. Focused ion beam etching for the fabrication of micropillar microcavities made of III-V semiconductor materials. *J. Vac. Sci. Technol.* B 25 (2007): 1197.

[4.8]   Ziegler, J. F. Stopping of energetic light ions in elemental matter. *J. Appl. Phys.* 85 (1999): 1249.

[4.9]   Hanbury Brown, R., and R. Q. Twiss. Correlation between photons in two coherent beams of light. *Nature* 177 (1956): 27–29.

[4.10] Ho, Y.-L. D., T. Cao, P. S. Ivanov, M. J. Cryan, I. J. Craddock, C. J. Railton, and J. G. Rarity. Three-dimensional FDTD simulation of micro-pillar microcavity geometries suitable for efficient single-photon sources. *IEEE J. Quantum Electron.* 43 (2007): 462.

[4.11] Lecamp, G., P. Lalanne, J. P. Hugonin, and J.-M. Gerard. Energy transfer through laterally confined Bragg mirrors and its impact on pillar microcavities. *IEEE J. Quantum Electron.* 41 (2005): 1323.

[4.12] Timpson, J. A., S. Lam, D. Sanvitto, D. M. Whittaker, H. Vinck, A. Daraei, P. S. S. Guimaraes, M. S. Skolnick, A. M. Fox, C. Hu, Y.-L. D. Ho, R. Gibson, J. G. Rarity, A. Tahraoui, M. Hopkinson, P. W. Fry, S. Pellegrini, K. J. Gordon, R. E. Warburton, and G. S. Buller. Single photon sources based upon single quantum dots in semiconductor microcavity pillars. *J. Mod. Opt.* 54 (2007): 453.

[5.1]   Marinelli, C. Techniques for improved-performance InGaN multi-quantum-well laser diodes. PhD thesis, University of Bristol, 2001.

[5.2]   Marinelli, C. Threshold current reduction in InGaN MQW laser diode with $\lambda/4$ air/semiconductor Bragg reflectors. *Electronics Letters* 36.20 (2000): pp. 1706–1707.

[5.3]   Sargent, L. J., A. B. Massara, M. Gioannini, J. C. L. Yong, P. J. Heard and J. M. Rorison. Investigation of 2-D-lattice distributed reflector lasers. *IEEE J. Quantum Electron* 38.11 (2002): pp. 1485–1492.

[5.4]   Iga, K. Surface-cavity surface-emitting laser: its conception and evolution. *Jpn. J. of Appl. Phys.* 47 (2008): 1–10.

[5.5]   Ouchi, T. Thermal analysis of thin-film vertical-cavity surface-emitting lasers using finite element method. *Jpn. J. Appl. Phys.* 41 (2002): 5181–5186.

[5.6]   Piprek, J. Electro-thermal analysis of oxide-confined vertical-cavity lasers. *Phys. Stat. Sol.* (a) 188 (2001): 905–912.

[5.7]   Knight, J. C., T. A. Birks, R. F. Cregan, P. St. J. Russell, and J.-P. de Sandro. Large mode area photonic crystal fibre. *Electronic Letters* 34 (1998): 1347.

[5.8]   Zhou, W. D., J. Sabarinathan, B. Kochman, E. Berg, O. Qasaimeh, S. Pang, and P. Bhattacharya. Electrically injected single-defect photonic bandgap surface-emitting laser at room temperature. *Electronics Lett.* 36 (2000): 1541–1542.

[5.9]   Unold, H., M. Golling, R. Michalzik, D. Supper, and K. Ebeling. Photonic crystal surface-emitting lasers: Tailoring waveguiding for single-mode emission. *Proc. European Conference on Optical Communication* (2001): 520–521.

[5.10] Song, D.-S., S.-H. Kim, H.-G. Park, C.-K. Kim, and Y.-H. Lee. Single-fundamental-mode photonic-crystal vertical-cavity surface-emitting lasers. *Appl. Phys. Lett.* 80 (2002): 3901–3903.

[5.11] Yokouchi, N., A. J. Danner, and K. D. Choquette. Etching depth dependence of the effective refractive index in two-dimensional photonic-crystal-patterned vertical-cavity surface-emitting laser structures. *Appl. Phys. Lett.* 82.9 (2003): 1344–1346.

[5.12] Sargent, L. J. Performance enhancement in vertical-cavity surface-emitting lasers using focused ion beam etching. PhD thesis, University of Bristol, 2001.

[5.13] Optowell Co. Ltd. Available online: www.optowell.com.

[5.14] Ivanov, P. S., Y. Zhu, M. J. Cryan, P. J. Heard, Y.-L. D. Ho, and J. M. Rorison. Static and dynamic properties of vertical-cavity surface-emitting semiconductor lasers with incorporated two-dimensional photonic crystals. in *Proc. SPIE* 6989, Richard M. De La Rue, Ceferino López, Michele Midrio, Pierre Viktorovitch (eds.), 2008.

[5.15] Miyake, H., Y. Yuba, K. Gamo, and S. Namba. Defects induced by focused ion beam implantation in GaAs. *J. Vac. Sci. Technol.* B 6 (1988): 1001–1005.

[5.16] Vallini, F., D. S. L Figueira, P. F. Jarschel, L. A. M. Barea, A. A. G. Von Zuben, A. S. Filho, and N. C. Frateschi. Effects of Ga+ milling on InGaAsP quantum well laser with mirrors etched by focused ion beam. Available online: arxiv.org/pdf/0904.0964.

CHAPTER 19

FIB ETCHING OF InP FOR RAPID PROTOTYPING OF PHOTONIC CRYSTALS

Victor Callegari, Urs Sennhauser, and Heinz Jaeckel

1 INTRODUCTION

The interest of using the direct-bandgap InP/InGaAsP material system for photonic crystal (PhC) structures lies in the optical integration of active devices. Passive PhC components are currently feasible; however, active PhC structures such as optical amplifiers (SOA) and lasers are still under investigation due to the additional complexity encountered when applying metal contacts on a PhC structure and the behavior of charge-carriers (decreased lifetime due to crystal damages) in the PhC structure, including damaged hole surfaces due to the fabrication process. As an additional interesting feature, active PhC devices can make use of the so-called "slow-light" regime, which significantly enhances the light-matter interaction [1] and thus reduces the size of these devices for equal gain in comparison to conventional SOA and lasers.

The main motivation to carry out this research was to locally modify pre-fabricated (dry-etched) PhC structures, but also to investigate the fabrication quality of FIB compared to more standard fabrication processes like electron-beam lithography (EBL) and plasma reactive ion etching (RIE). This question is interesting, as it is clear that FIB is a more flexible technique, since it combines the above-mentioned processing-steps in a single step. Therefore, the question of potential disadvantages of FIB has to be answered.

The need for high aspect ratio (>10), cylindrical holes of about 200 nm diameter in the InP/InGaAsP material system cannot be met by sputtering due to redeposition effects in the holes. Therefore, gas-assisted etching with iodine as the etch gas combined with sputtering of InP/InGaAsP was investigated, and the process was optimized by varying the substrate temperature, the beam dwell time, and the beam overlap to maximize the etch rate.

2 ETCHING OF InP

Chemical etching of InP was investigated since it was found that sputtering results in aspect ratios limited to about 4–6 due to redeposition in the holes [2]. This aspect ratio is insufficient to reach the entire optical mode in the InP/InGaAsP material system. Other serious limitations of sputtering are the distorted hole shape and the low processing rate (the sputter rate is 1.27 $\mu m^3/nC$), which translates into long fabrication times in the order of minutes for a single hole. If entire PhC structures were to be fabricated using FIB, this would result in hour-long fabrication times and thus the influence of drifts would become important. It is known from the literature [3] that with a suitable etching gas, gas-assisted etching (GAE) results in higher processing rates and less redeposition of sputtered material as compared to physical sputtering. This is due to the formation of volatile chemical reaction products on the surface and their larger vapor pressures. Hence, etching of InP/InGaAsP was investigated with the aim to:

- Increase the processing rate in order to reduce fabrication time;
- Increase the volatile reaction products with the aim to reduce redeposition and increase the aspect ratio.

The surface reactions of iodine (I_2) with InP were investigated by X-ray photoelectron spectroscopy (XPS). Iodine was preferred over the more conventional use of chlorine (Cl) because it can be stored in solid form and indium iodides exhibit a higher vapor pressure than indium chlorides [4]. The most important process parameters are:

Beam dwell time;
Beam overlap;
Line and pattern refresh times (corresponding to the time between subsequent irradiations of a spot).

All above-mentioned parameters influence and change the etched shapes as well as the etch rates. If the beam dwell time is long, depletion of surface-adsorbed gas molecules results and GAE reverts into physical sputtering, with its adverse effects of lower material removal rate and redeposition of material removal material. Beam overlap ol, defined as:

$$ol = (d - p)/d, \tag{1}$$

with d the beam diameter and p the step size (with $pmin$ corresponding to 1 pixel, the equivalent step size p depending on the field-of-view and the total number of pixels addressable by the beam scanning electronics), can be chosen as positive or negative. Positive beam overlaps ($>36\%$) distribute the ion fluence homogeneously over the fabrication area. However, due to the high current density of an FIB (some A/cm^2), coupled with the limited flux of reactive gas to maintain an acceptable vacuum in the chamber, beam overlap is usually chosen as negative to avoid too quick gas molecule depletion and a corresponding transition from GAE to sputtering.

If the refresh time is too short (below some ms), molecule surface coverage will be sub-optimal and GAE-to-sputtering transition will occur. If the refresh time is too long (in the order of some tens of ms), and considering Langmuir adsorption kinetics (single-monolayer adsorption), one monolayer of adsorbed etch gas will have formed before the beam irradiates the surface, which results in etching gas not contributing to the etching mechanism and being wasted. Additionally, the fabrication time increases.

2.1 ETCHING AT ROOM TEMPERATURE

Iodine was inserted through a gas needle (500 μm inner diameter) into the vacuum chamber with a base pressure in the low 10^{-6} mbar range. The distance to the wafer surface was $\sim$150 μm. Solid iodine, stored in a crucible outside the vacuum chamber, was heated to 35°C, resulting in an iodine flux J of $\sim$3 $\times$ 10^{17} molecules/cm^{2}s on the substrate.

To determine the etch rate, squares with a side length of 20 μm were etched at ion fluences ranging from 7.9×10^{12} ions/cm^2 to 6.3×10^{14} ions/cm^2. The dwell time was set to 1 μs, the overlap to 0% and the ion current to 300 pA. FIB-GAE etching rate was enhanced 15-fold (to 19 μm^3/nC) compared to physical sputtering, measured by comparing the depth of the holes from FIB-fabricated cross-sections. As expected, the enhancement strongly depends on the scanning conditions (dwell time, overlap, refresh times) and the ion current.

The enhanced etch rate when using iodine is caused by the creation of an In-rich, amorphous zone of rough surface with numerous dangling bonds, which act as adsorption sites. Ion intermixing accelerates diffusion of the etchant into the substrate and lowers the surface binding energies [5], which further increases the sputtering yield. Reaction products are mainly InI$_3$, GaI$_3$ and PI$_3$ at room temperature. From vapor pressure data for the reaction products [6–10], it can be concluded that the reaction-limiting factor is desorption of InI$_x$ ($x = 1 - 3$) and its corresponding dimers. To remove this layer at room temperature, energy must be provided by raising the substrate temperature, or to break bonds by electron-stimulated desorption [11], or, as in the present case, to remove InI$_x$ by sputtering.

Iodine reacts spontaneously with InP only on areas previously irradiated by the Ga$^+$ beam, as shown in Figure 19.1. In Figure 19.1, the wafer area lying beneath the nozzle in the upper left part of the image was protected from ion irradiation that etched the square in the middle of the image. Therefore, iodine could not react with the InP surface.

The native oxide on the InP surface plays the role of an etching mask by inhibiting spontaneous reactions of iodine with InP. The thickness of the native oxide was measured by microellipsometry to be 1.5 nm, agreeing well with previously published results [12; 13]. InI$_x$ remains on the irradiated InP surface and forms islands at room temperature, shown in Figures 19.1 and 19.2. InP was irradiated with increasing ion fluences (Figure 19.2) and it was found that at the lowest fluence of 7.9×10^{12} ions/cm^2, iodine did not react with InP (based on the observation of the appearance of InI$_x$ islands by SEM), whereas for higher fluences, InI$_x$ islands were clearly visible. The threshold fluence to remove the native oxide layer was evaluated to be $\sim$1 $\times$ 10^{13} ions/cm^2. Considering a 300 pA Ga$^+$ beam focused to a spot size of 25 nm, the time required to reach a fluence of 1×10^{13} ions/cm^2 is 26 ns and the equivalent depth sputtered in InP will be 0.02 nm, assuming a cylindrical beam shape.

The sensitivity of the I$_2$-InP reaction to contaminations on the wafer surface was investigated. Irradiating the sample with a fluence exceeding the threshold to remove the native oxide and waiting one minute prior to opening the I$_2$ needle valve did not inhibit the spontaneous reaction, whereas first storing for 24 hours in air inhibited the reaction. Thus, the quality of the oxide layer plays an important role both as a passivating layer and a diffusion barrier for I$_2$. The threshold Ga$^+$ fluence for amorphization of InP was reported

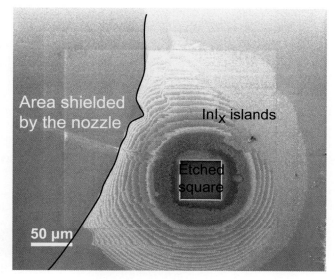

FIGURE 19.1: SEM micrograph of the area surrounding a square with 50 μm side length etched at room temperature showing the spontaneous reaction of I_2 with InP wherever the native oxide was sputtered [2]. The nozzle through which I_2 is inserted into the vacuum chamber prevents sputtering of the native oxide layer. Reprinted with permission from [2]. Copyright (2007), Elsevier.

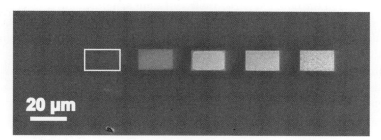

FIGURE 19.2: Etched squares with ion fluences from 7.9×10^{12} ions/cm^2 to 6.3×10^{14} ions/cm^2 from left to right. For the lowest fluence, the native oxide was not completely removed, thus preventing the spontaneous reaction [2]. Reprinted with permission from [2]. Copyright (2007), Elsevier.

to be 1.7×10^{13} ions/cm^2 [14; 15]. Considering the same beam as described above, the time to reach this fluence is 44 ns, corresponding to an equivalent depth of 0.03 nm in InP. Hence, native oxide removal and amorphization are simultaneous phenomena: when the native oxide is removed by sputtering, the underlying InP will be amorphous and In-rich.

Squares of 50 μm side length were etched with I_2 and an ion fluence of 6.3×10^{17} ions/cm^2 and XPS and AFM measurements were performed in them. The beam current was 314 pA, dwell time was 400 ns and overlap 0%. XPS measurements in Figure 19.3 show a P-depletion on the surface and segregation in a near surface layer, analogous to As-segregation for Cl_2

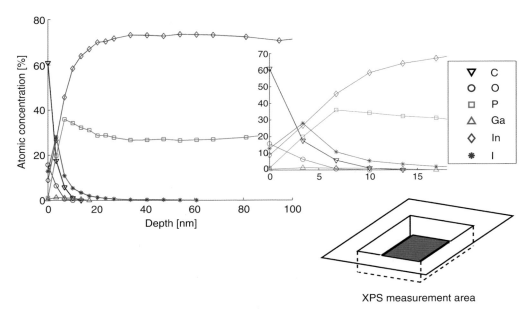

FIGURE 19.3: XPS depth profile of an area etched at room temperature. The maximum Ga-concentration is below 2 at.% as shown in the inset. The bulk composition of InP deviates from 50 at.%/50 at.% due to preferential sputtering of P by the Ar^+ ions used to acquire the depth profile.

etching of GaAs [16]. Concurrently, P evaporates from the surface as PI_3, resulting in an In-rich InP surface. FIB-GAE at room temperature results in detectable Ga-implantation, both at the surface and in the bulk (~2 at.%, Table 19.1), but at a much lower level as compared to sputtering (>20 at.% [2]); Ga is concentrated closer to the surface. The reduced content of Ga and its peak concentration closer to the surface are due to the larger material removal rate, as well as to the spontaneous evaporation of GaI_3.

Accordingly, the depth of the damaged zone is reduced by FIB-GAE [17]. AFM measurements of the etched InP surface were performed. The rms surface roughness was measured to be 2.7 nm.

2.2 ETCHING AT TEMPERATURES ABOVE ROOM TEMPERATURE

Heating the InP surface eventually results in the sublimation of the remaining InI_3. This occurs when a surface temperature is reached at which the vapor pressure of InI_3 equals the pressure in the vacuum chamber. At this temperature, sputtering of InI_3 is no longer required. Using a heating stage, the temperature of the stage surface was varied in steps of 50°C from room temperature to 350°C. Heating to larger values would result in surface depletion of P [18]. Sputter and etch rates were determined by irradiating 20 μm wide squares with an ion current of 313 pA and a dwell time of 100 ns. The ion fluence was 5.3×10^{16} ions/cm². Beam overlaps *ol* of −200%, 0% and 36.6% were tested. It was experimentally confirmed that sputter rates do not vary measurably with temperature (Figure 19.4).

InI_3 starts to desorb above ~80°C [6] and a maximum etching rate of 83.2 μm³/nC occurs at 200°C stage temperature, a 66-fold increase over the sputter rate. This increase is explained by the rapid spontaneous thermal desorption of InI_3 from the surface.

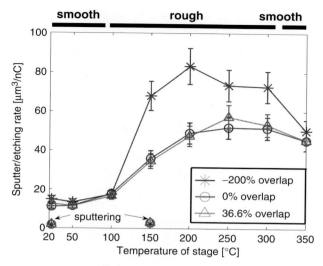

smooth   rough   smooth

FIGURE 19.4: Evolution of the etch-rate as a function of stage temperature and beam overlap. The quality of the surface is indicated on top of the figure. A good compromise between etch rate and surface roughness is found at 150°C stage temperature. Reprinted with permission from [19]. Copyright (2007), American Institute of Physics.

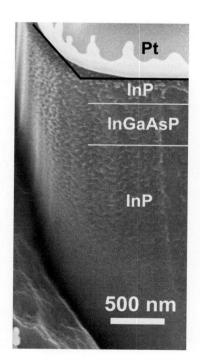

FIGURE 19.5: Etched double cross-section through the InP/InGaAsP waveguide structure: no differential etching can be detected.

At temperatures above 200°C, the etching rate decreases (Figure 19.4). A very low residence time of the I_2 molecule on the hot surface (thermal desorption before a chemical reaction can occur) is assumed to be the cause. Above 330°C, InI_3 decomposes into non-volatile InI and InI_2 and volatile I_2 [6]. $InI_x(x = 1, 2)$ forms a passivating layer, which leads to a decreased etching rate at 350°C because of the lower chemical contribution to the fabrication process. Figure 19.4 also shows that the beam overlap influences the etching rate considerably, especially between −200% and 0%. The beam diameters correspond roughly to the FWHM of a Gaussian distribution. In reality, the beam diameter is larger due to the contribution of the beam tails, which are not taken into account in Equation (1). At 0% calculated beam overlap, in reality $ol > 0$, which explains the small differences in the measured processing rates for $ol = 0\%$ and $ol = 36.6\%$ in Figure 19.4. By increasing the overlap, the ion fluence increases and more Ga^+ impinge per pixel. The gas flux remaining constant, a decreased gas-to-ion ratio results and consequently a faster transition to the sputtering regime occurs. In order to minimize fabrication times while maintaining a good surface quality, a beam overlap of 0%, 100 ns dwell time, and a stage temperature of 150°C were fixed for the process employed to fabricate the test structures. At this temperature, the etch rate of InP is 51-fold larger than the sputter rate. The beam overlap of 0% is an acceptable compromise between the homogeneity of the ion distribution and the processing rate. The stage temperature of 150°C is a compromise between the processing rate and the resulting surface roughness. Stage temperatures from 130°C–170°C in steps of 10°C were found to only slightly change the etching rate and the surface roughness, with larger temperatures exhibiting larger etch rates at the expense of an increased surface roughness.

A critical criterion of the quality of the PhC etch process is its *non*-selectivity between InP and InGaAsP to avoid unwanted undercuts. Therefore, the process was investigated by etching through a waveguide structure and performing cross-sections by FIB. The SEM micrograph in Figure 19.5 does not show a measurable effect pointing to differential etching between InP and InGaAsP. Iodine reacts with all elements of the InP/InGaAsP material system.

The surface roughness was determined by etching squares with a side length of 50 μm at different stage temperatures at a fluence of 2.3×10^{15} ions/cm^2 and characterized using SEM. The surface roughness increased with temperature up to 300°C (Figure 19.4). This was attributed to the spontaneous formation and desorption of InI_3, PI_3, P_2, and GaI_3, supported by ion intermixing. By raising the stage temperature, a more efficient thermal desorption of the reaction products occurred, which resulted in craters and a larger surface roughness due to the larger spontaneous etch rate. Since the chemical etching rate is ~50 times larger than the physical sputtering rate, the surface roughness is no longer tightly controlled by the beam overlap and does not vary considerably with the latter. Above 300°C, the passivating $InI_x(x = 1, 2)$ layer prevents strong chemical etching and the surface becomes smoother again. At 150°C, the surface roughness was measured with AFM to be 12.6 nm rms at the bottom of the 250 nm deep square over an area of 25 μm^2. Although this surface roughness is larger than

Table 19.1 Overview of the fabrication processes.

Process	Temperature [°C]	Peak Ga [at. %]	Sputter rate [μm³/nC]
Sputtering	23	20	1.27
GAE	23	2	19
GAE	150	<1	64.8

Table 19.2 Overview of the surface roughness of all investigated fabrication processes.

Process	Temperature [°C]	Location	rms roughness [nm]
Sputtering	23	Bottom	1.3
GAE	23	Bottom	2.7
GAE	150	Bottom	12.6
GAE	150	Sidewall	5.0

for sputtering and etching at room temperature, it is comparable to values reported in literature for RIE [20]. Table 19.2 summarizes the findings on surface roughness for all investigated fabrication processes.

For PhC holes the sidewall roughness is more important than the roughness at the bottom of the hole. It was experimentally measured by fabricating a lamella by FIB, using the same process as for the PhC holes and measuring the surface of the lamella by AFM. The surface roughness decreased to 5 nm rms, a considerable improvement over the 12.6 nm rms measured for the bottom surface of an etched square. The decrease can be attributed to the shorter range of ions in the sidewalls and hence less efficient ion intermixing and spontaneous etching. As a comparison, for PhC air holes, 5 nm rms roughness is close to the state-of-the-art 3 nm rms obtained by ICP-RIE for InP/InGaAsP [21] and SOI PhC slab waveguides [22].

The XPS measurement performed at the bottom of an etched square in Figure 19.6 shows that a surface layer of 20 nm thickness is P-enriched. This is explained by a favored reaction of I_2 with In. For depths larger than 70 nm, the In and P concentrations are constant but In rich; a result of the preferential removal of P_2 from InP due to the physical sputtering by Ar^+ to acquire the XPS depth profile [23]. Iodine is detected up to a depth of 50 nm within this depth profile. The iodine signal is associated with trapped $InI_x(x = 1 - 3)$ and remaining $InI_x(x = 1 - 2)$ reaction products on the surface, which sublimate only at higher temperatures. Ga atomic concentration is below XPS detection threshold at all depths

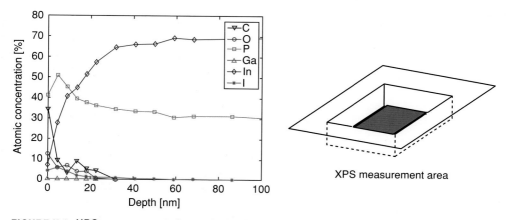

FIGURE 19.6: XPS measurement of an etched InP surface at 150°C. No Ga is detected. The bulk composition of InP deviates from 50 at.%/50 at.% due to preferential sputtering of P by the Ar^+ ions used to acquire the depth profile.

(<1 at.%, Table 19.1). As already discussed, this finding is due to the 50-fold enhanced etching rate and the spontaneous thermal desorption of GaI_3. Since the Ga-implantation (6.4×10^{12} Ga/cm^2, assuming a surface concentration of 0.9 at.% Ga and a monolayer density of 7.0×10^{14} InP/cm^2 on the InP surface) is below the amorphization threshold of InP (1.7×10^{13} Ga/cm^2), it is concluded that the etching process at 150°C stage temperature does not result in the amorphization of InP.

3 FABRICATION OF TRENCH WAVEGUIDES

A critical step to perform the entire rapid-prototyping process (i.e., an efficient fabrication of different PhC designs for optical characterization) using FIB is the fabrication of access waveguides. The waveguides must have a length in the mm-range due to the cleaving process of the 300 μm thick wafers. Cleaving is necessary to perform the optical measurements. Using thinner wafers would allow a decrease of the minimum length of the waveguides; however, the fragility of these wafers would lower the fabrication yields. Hence, the access waveguides are fabricated in several steps, which requires precision alignment of the individual segments constituting the waveguides.

Waveguides with a width and depth of 4 μm and a length of 2 mm were fabricated. Four different fabrication methods were investigated (see Table 19.3). The addition of iodine to the sputtering process, the ion current and the stage temperature constituted the variable process parameters.

All waveguides were fabricated on the same wafer. Irradiation spots were spatially separated by 37 nm. In a first step, trench structures were sputtered with an ion current of 7 nA. Two rectangles (100 μm long and 4 μm wide) were sputtered using a staircase pattern reaching a depth of 4 μm in proximity of the waveguide (see Figure 19.7). After this first step, performed at a large ion current to minimize the fabrication time, the final cleaning in step 2 was performed using a lower ion current. During this step, the ion beam approached the waveguide with single pixel wide lines over a width of 2 μm from each side. Each line was milled to its final depth of 4 μm. The beam dwell time was held constant at 1 μs. The cleaning step was carried out using methods A-D (see Table 19.3). Its purpose is to remove redeposition from the waveguide sidewall that occurs during the initial coarse milling step 1. A sidewall angle of ~2° could be achieved for sputtering and etching and does not result in additional propagation losses [24]. The waveguides were split up in 20 segments, each 100 μm long. To compensate for stage movement errors (with an amplitude of around 1 μm), x-shaped alignment marks were milled and their position in the vertical direction was automatically recorded using image recognition software. After every stage movement, the detected positioning error was corrected

Table 19.3 Overview of all investigated fabrication processes.

Process	Iodine	Ion current [nA]	Stage temperature [°C]
A	No	1	23
B	Yes	1	23
C	Yes	1	150
D	Yes	0.3	150

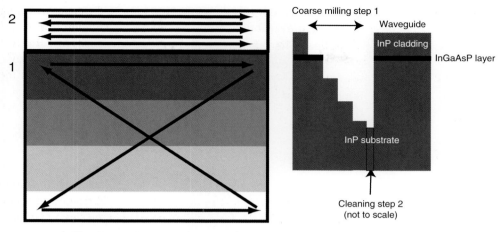

FIGURE 19.7: Top-down (left) and side (right) schematic representation of the ion beam scanning strategy to fabricate the trench waveguides. A staircase pattern is sputtered at 7 nA ion current to quickly remove material. The coarse milling step is followed by a cleaning step at lower ion current (1 nA or 300 pA), which removes redeposition on the waveguide and is carried out either by sputtering or iodine-assisted etching. The milling pattern is mirrored on the other side of the waveguide (not shown).

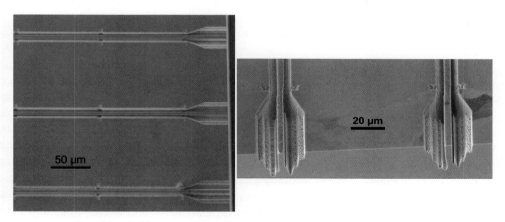

FIGURE 19.8: Top-down (left) overview of 3 trench waveguides with the cleaved facet and the in/out-coupling areas on the right-hand side. Side view (right) of 2 waveguides showing the staircase milling and the cleaning step at the in/out-coupling area.

by electronic beam shift. This resulted in smooth transitions between the individual segments, as shown in Figure 19.8. The stitching errors were determined from SEM micrographs to be comparably small (<100 nm) both at room temperature and at 150°C stage temperature. Prior to the fabrication of the waveguides, the InP/InGaAsP substrate was sputter-coated with a 100 nm thick polycrystalline W layer with a mean grain size of approximately 10 nm, determined from SEM micrographs.

This layer served two purposes: on the one hand, to protect the waveguide from Ga implantation and sputtering of the native oxide (see Section 2.1); on the other hand, the

Table 19.4 Extracted losses of the waveguides, ranging from lowest to highest. The fabrication method is shown in Table 19.3.

Fabrication method (Tab. 19.3)	Extracted loss [dB/cm]
D	49.7
B	61.4
A	75.8
C	82.8

W mask defines the sidewall roughness of the waveguide (approx. 10 nm), as the grain size is larger than the sidewall roughnesses of both sputtering and etching. Hence, in an optical transmission experiment, only optical losses induced by Ga-implantation and the resulting damage to the lattice structure are measured. It was reported that a 10 keV Ga^+ FIB implantation at room temperature results in a $\sim$1.3 μm wide layer, which completely quenches the photoluminescence of AlGaAs/GaAs quantum wells [25]. Considering the width of 4 μm of our waveguides, the extension of the damaged layer in the order of 1 μm from each side, together with the sidewall roughness, will dominate the propagation loss.

The waveguides at 150°C stage temperature were fabricated first to ensure that temperature-enhanced diffusion of defects did not occur for the structures fabricated at room temperature. The I_2-enhanced FIB etched waveguides were fabricated prior to the purely sputtered waveguides to avoid contamination of the sputtered waveguides with I_2. The W mask was wet-etched in 30 vol. % H_2O_2 during 15 minutes at room temperature. Finally, the waveguides were optically characterized using an end-fire setup [26]. TE-polarized light with a wavelength λ varying from 1549.5–1550.5 nm in steps $\Delta\lambda = 0.005$ nm was coupled into the trench waveguides using a lensed fiber. The waveguide losses were measured using the Hakki-Paoli method [27]. This approach allows the separation of the varying in- and out-coupling losses from the losses of the waveguides. Table 19.4 summarizes the results. The waveguides etched at 150°C stage temperature at 300 pA ion current had the lowest propagation losses of 49.7 dB/cm. The etch process at room temperature has the second-lowest losses and has a higher transmission than the sputtered waveguide. Surprisingly, the waveguide etched at 150°C with 1 nA has the highest loss of 82.8 dB/cm.

In line with the amount of implanted Ga (see Figure 19.6), the lateral extension of the damaged layer has decreased for iodine-assisted etching at 150°C and 23°C. This explains the lower losses of fabrication methods D and B. The lower etch rate at 23°C stage temperature is at the origin of the larger propagation loss compared to etching at 150°C. The temperature-enhanced diffusion of defects [25] is counteracted by the larger etch rate.

The large loss for method C is a result of the large beam diameter of around 100 nm. The W-mask was sputtered, and its removal was followed by the rapid etching of the waveguide layers. This finding demonstrates the necessity of a mask, which protects the waveguide from ion irradiation.

The remaining relatively large losses of FIB-fabricated trench waveguides as compared to reactive ion etched waveguides are due to the large surface roughness of the protecting W mask.

By using a mask with smaller grains, propagation losses will approach levels of ICP-RIE etched waveguides (around 3 dB/cm [28]).

4 PHOTONIC CRYSTAL Y-SPLITTER

Photonic crystal Y-splitters have a small footprint (see Figure 19.9) and permit the equal distribution of light from the input branch into the two output branches. The central hole (see Figure 19.10) serves to reduce reflections as the light enters the PhC structure, and the splitters are theoretically lossless. To etch high aspect ratio holes, the gas flux was directed toward the substrate using a nozzle with a 300 μm-diameter hole coaxial to the ion beam and situated ~100 μm above the InP surface (see Figure 19.11). This nozzle geometry enhances the flux of iodine molecules into the holes, hence delaying the transition of the etching process to physical sputtering. Another advantage of the coaxial nozzle is the symmetrical gas flux with respect to the hole axis. The standard tilted nozzle results in distorted holes due to its inhomogeneous gas flux. The main disadvantages of the coaxial nozzle are the blocked

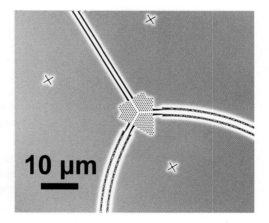

FIGURE 19.9: Overview of a PhC Y-splitter fabricated by electron beam lithography (EBL) and inductively coupled plasma reactive ion etching (ICP-RIE).

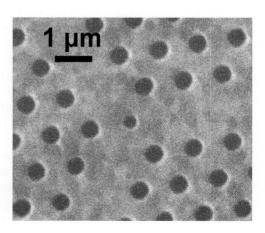

FIGURE 19.10: Close-up view of the splitter with the central hole fabricated by FIB iodine-assisted etching.

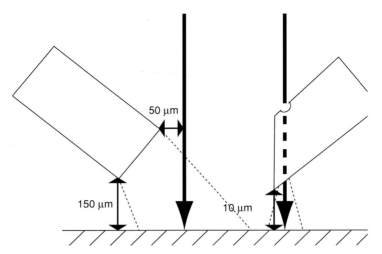

FIGURE 19.11: Schematic view of a conventional (left) gas nozzle and a coaxial (right) gas nozzle.

view from the SEM and the reduced mean free path of the ions as they travel through the iodine vapor.

The central hole was fabricated by FIB in PhC Y-power splitters that had been previously fabricated by electron beam lithography (EBL) and transferred into the semiconductor by inductively coupled plasma reactive ion etching (ICP-RIE) [26]. Details of the ICP-RIE fabrication process can be found in [29; 30]. Fully ICP-RIE fabricated power splitters were compared with power splitters where the central hole was fabricated by iodine-enhanced FIB etching. The filling factor of the splitters was kept constant at $r/a = 0.23$ for the PhC holes surrounding the splitter, r being the hole radius and a the lattice constant of the PhC. For the central hole, $r/a = 0.19$ was used to minimize reflections. To prevent the formation of a funnel, a W layer was sputtered on the pre-structured wafer surface. The thickness of the W layer was 113 nm.

Each hole was fabricated in two steps: first, the hole shape was transferred into the mask by sputtering during 3 s at 10 pA with a dwell time of 52.1 μs and 36.6% overlap. This ensured a uniform ion distribution. Then iodine was added and the hole was exposed to I_2 and Ga^+ during 2 s. The dwell time was 100 ns at 10 pA and 0% overlap. W was subsequently selectively removed by using H_2O_2.

High aspect ratio holes, fabricated using the above-mentioned process, are shown in Figure 19.12. Almost vertical sidewalls are achieved due to the greatly reduced redeposition in the hole. The limiting factor for the aspect ratio was under-etching due to ion scattering and not the removal of etch products from the hole. An aspect ratio of about 11 for a hole diameter of 270 nm was reproducibly achieved.

All devices were optically characterized using the end-fire setup. By using lithographic tuning [31], which consisted of varying the lattice constant a from 375–500 nm in 25 nm steps, the reduced wavelength range $u = \lambda/a = 0.23 - 0.34$ was covered. Figure 19.13 shows the end-fire measurement results for both output branches with FIB-fabricated and EBL/ICP-RIE fabricated central holes, respectively. Except for the tuning step with the lattice constant $a = 375$ nm, the two fabrication methods yield equivalent transmission levels.

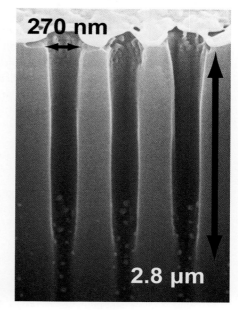

FIGURE 19.12: FIB iodine-assisted etching yields holes with aspect ratio exceeding 11.

FIGURE 19.13: Transmission of FIB- (blue dots and triangles) and EBL/ICP-RIE- (red dots and triangles for upper and lower branch, respectively) Y-splitters. The transmission between the branches and fabrication techniques for a given *a* is equal, showing that the fabrication quality of FIB iodine-assisted etching and EBL followed by ICP-RIE are comparable.

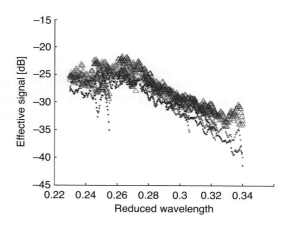

5 CONCLUSIONS

FIB iodine-assisted etching was investigated for variable substrate temperatures and beam overlaps. For etching at higher temperatures, a sacrificial W hard mask was used to limit the reaction to the area etched by the central Gaussian part of the ion beam. At 150°C stage temperature and using an optimized scanning strategy, a 51-fold enhancement over the physical sputtering rate was achieved. Holes with aspect ratios >11 were etched and FIB cross-sections and AFM measurements showed that the hole sidewalls were nearly vertical and had an rms surface roughness amounting to 5 nm. Ga-implantation was below the detection threshold of XPS, i.e., <1 at.%, a strong indication of the absence of an amorphized layer. An automated approach was used to fabricate the mm-long access waveguides required for the cleaving process of the 300 μm-thick InP wafer, and the results showed that a fabrication process based exclusively on FIB is indeed suitable for rapid-prototyping of PhC structures. FIB-modification of PhC-Y-splitters in InP/InGaAsP was investigated to characterize the hole quality. FIB-GAE permits the fabrication of large PhC (hundreds of holes) arrays in less than 1 h and limits the influence of electronic and mechanical drifts. FIB iodine-assisted etching is thus equivalent to EBL/ICP-RIE fabrication of PhC structures in terms of maximum aspect ratio, surface roughness, and shape of the holes.

ACKNOWLEDGMENTS

The authors would like to acknowledge partial funding from the ETH-CDM-INIT project and Philipp M. Nellen and Peter Kaspar for helpful discussions.

REFERENCES

[1] Baba, T. Slow light in photonic crystals. *Nat. Photonics* 2 (2008): 465–473.

[2] Callegari, V., P. M. Nellen, T. H. Yang, R. Hauert, U. Muller, F. Hernandez-Ramirez, and U. Sennhauser. Surface chemistry and optimization of focused ion beam iodine-enhanced etching of indium phosphide. *Appl. Surf. Sci.* 253 (2007): 8969–8973.

[3] Jacobi, K., G. Steinert and W. Ranke. Iodine etching of GaAs(1bar1bar1bar) arsenic surface studied by LEED, AES, and mass-spectroscopy. *Surf. Sci.* 57 (1976): 571–579.

[4] Cho, H., J. Hong, T. Maeda, S. M. Donovan, C. R. Abernathy, S. J. Pearton, and R. J. Shul, Novel plasma chemistries for highly selective dry etching of In_xGaN_{1-x}: Bi_3 and BBr_3. *Mat. Sci. Eng. B-Solid* 59 (1999): 340–344.

[5] Dieleman, J., F. H. M. Sanders, A. W. Kolfschoten, P. C. Zalm, A. E. Devries, and A. Haring. Studies on the mechanism of chemical sputtering of silicon by simultaneous exposure to Cl_2 and low-energy Ar^+ ions. *J. Vac. Sci. Technol. B* 3 (1985): 1384–1392.

[6] Brunetti, B., A. Giustini and V. Piacente. Vapour pressures and sublimation enthalpy of solid indium(iii) iodide. *J. Chem. Thermodyn.* 29 (1997): 239–246.

[7] Kuniya, Y., and T. Chino. Studies on the vapor phase reaction in the system $Ga-I_2$. *Denki Kagaku* 40 (1972): 858–864.

[8] Hillel, R., J. M. Letoffe, and J. Bouix. Thermic stability and thermodynamic properties of phosphorus iodides in condensate and gaseous states. *J. Chim. Phys. Phys.-Chim. Biol.* 73 (1976): 845–848.

[9] Fainer, N. I., Y. M. Rumyantsev, and F. A. Kuznetsov. Kinetics of the reaction of gallium-arsenide with gaseous gallium triiodide. *Inorg. Mater.* 16 (1980): 921–925.

[10] Titov, V. A., T. P. Chusova, and Y. G. Stenin. On thermodynamic characteristics of In-I system compounds. *Z. Anorg. Allg. Chem.* 625 (1999): 1013–1018.

[11] Mokler, S. M., P. R. Watson, L. Ungier, and J. R. Arthur. Electron-stimulated desorption of chlorine from GaAs(100) surfaces. *J. Vac. Sci. Technol. B* 8 (1990): 1109–1112.

[12] Shibata, N., and H. Ikoma. X-ray photoelectron spectroscopic study of oxidation of InP. *Jpn. J. Appl. Phys. 1* 31 (1992): 3976–3980.

[13] Shibata, N., and H. Ikoma. X-ray photoelectron spectroscopic study of oxidation of InP ii -thermal oxides grown at high-temperatures. *Jpn. J. Appl. Phys. 1* 32 (1993): 1197–1200.

[14] Menzel, R., T. Bachmann, and W. Wesch. Physical sputtering of III-V-semiconductors with a focused Ga^+-beam. *Nucl. Instrum. Meth. B* 148 (1999): 450–453.

[15] Menzel, R., K. Gartner, W. Wesch, and H. Hobert. Damage production in semiconductor materials by a focused Ga^+ ion beam. *J. Appl. Phys.* 88 (2000): 5658–5661.

[16] Ameen, M. S., and T. M. Mayer. Surface segregation during reactive etching of GaAs and InP. *J. Appl. Phys.* 59 (1986): 967–969.

[17] Yamaguchi, A., and T. Nishikawa. Low-damage specimen preparation technique for transmission electron-microscopy using iodine gas-assisted focused ion-beam milling. *J. Vac. Sci. Technol. B* 13 (1995): 962–966.

[18] Farrow, R. F. C. Evaporation of InP under Knudsen (equilibrium) and Langmuir (free) evaporation conditions. *J. Phys. D Appl. Phys.* 7 (1974): 2436–2448.

[19] Callegari, V., P. M. Nellen, J. Kaufmann, P. Strasser, and F. Robin. Focused ion beam iodine-enhanced etching of high aspect ratio holes in InP photonic crystals. *J. Vac. Sci. Technol. B* 25 (2007): 2175–2179.

[20] Yu, J. S., and Y. T. Lee. Parametric reactive ion etching of InP using Cl_2 and CH_4 gases: effects of H_2 and Ar addition. *Semicond. Sci. Tech.* 17 (2002): 230–236.

[21] Strasser, P., F. Robin, C. F. Carlstrom, R. Wuest, R. Kappeler, and H. Jackel. Sidewall roughness measurement inside photonic crystal holes by atomic force microscopy. *Nanotechnology* 18 (2007): 405703.

[22] Kuramochi, E., M. Notomi, S. Hughes, A. Shinya, T. Watanabe, and L. Ramunno. Disorder-induced scattering loss of line-defect waveguides in photonic crystal slabs. *Phys. Rev. B* 72 (2005): 161318.

[23] Yu, W., J. L. Sullivan, S. O. Saied, and G. A. C. Jones. Ion bombardment induced compositional changes in GaP and InP surfaces. *Nucl. Instrum. Meth. B* 135 (1998): 250–255.

[24] Wang, J., M. Li, and J. J. He. Analysis of deep etched trench in planar optical waveguide by FDTD method. *IEEE Proc. Opt. Fib. Comm. Conf.* (2007): 218–220.

[25] Schneider, M., J. Gierak, J. Y. Marzin, B. Gayral, and J. M. Gerard. Focused ion beam patterning of III-V crystals at low temperature: A method for improving the ion-induced defect localization. *J. Vac. Sci. Technol. B* 18 (2000): 3162–3167.

[26] Nellen, P. M., V. Callegari, J. Hofmann, E. Platzgummer, and C. Klein. FIB precise prototyping and simulation. *Mater. Res. Soc. Symp. P.* 960 (2007): 0960-N10-03-LL06-03.

[27] Revin, D. G., L. R. Wilson, D. A. Carder, J. W. Cockburn, M. J. Steer, M. Hopkinson, R. Airey, M. Garcia, C. Sirtori, Y. Rouillard, D. Barate, and A. Vicet. Measurements of optical losses in mid-infrared semiconductor lasers using Fabry-Perot transmission oscillations. *J. Appl. Phys.* 95 (2004): 7584–7587.

[28] Docter, B., E. J. Geluk, T. deSmet, F. Karouta, and M. K. Smit. Reflectivity measurement of deeply etched DBR gratings in InP-based double heterostructures. *IEEE Leos. Ann. Mtg.* (2007): 111–114.

[29] Wuest, R., P. Strasser, F. Robin, D. Erni, and H. Jackel. Fabrication of a hard mask for InP based photonic crystals: Increasing the plasma-etch selectivity of poly(methylmethacrylate) versus SiO_2 and SiN_x. *J. Vac. Sci. Technol. B* 23 (2005): 3197–3201.

[30] Strasser, P., R. Wuest, F. Robin, D. Erni, and H. Jackel. Detailed analysis of the influence of an inductively coupled plasma reactive-ion etching process on the hole depth and shape of photonic crystals in InP/InGaAsP. *J. Vac. Sci. Technol. B* 25 (2007): 387–393.

[31] Ferrini, R., D. Leuenberger, M. Mulot, M. Qiu, J. Moosburger, M. Kamp, A. Forchel, S. Anand, and R. Houdre. Optical study of two-dimensional InP-based photonic crystals by internal light source technique. *IEEE J. Quantum Elect.* 38 (2002): 786–799.

CHAPTER 20

APPLICATIONS OF FIB FOR RAPID PROTOTYPING OF PHOTONIC DEVICES, FABRICATION OF NANOSIEVES, NANOWIRES, AND NANOANTENNAS

Vishwas J. Gadgil

1 INTRODUCTION

The development of focused ion beam technology has been driven to a large extent by the semiconductor industry. FIB was primarily used for process control, for mask repairs, and for circuit edits. With rapid developments in nanotechnology, dedicated machines for research purposes were developed and became available in the market. They offer more flexibility and versatility than the machines used in the semiconductor industry.

The traditional approach to fabrication micro- or nanostructures is based on the well-established route of photo lithography. When a structure is designed, a mask or several masks have to be generated. A silicon wafer is coated with photoresist and then exposed to light. The wafer is processed further to create the micro structure, which can be tested for its properties. There are a few limitations to this process. Making a mask is a time-consuming and expensive process. The materials that can be processed are predominantly silicon or glass. The substrates have to be flat. If at the end of the fabrication process the device does not perform as per expectations and changes have to be made, the whole process must be repeated, causing delays and the whole exercise needs additional expense of new masks. With the availability of FIB it became possible in some cases to skip the photolithography process altogether and directly fabricate the prototype device for testing and evaluation. The additional advantage is that the structure can be fabricated on almost any type of material without restrictions on the shape of the substrate that is used. In this chapter we will discuss the rapid prototyping of photonic devices and the fabrication of nanopores, nanoslits and nanosieves.

2 FABRICATION OF PHOTONIC DEVICES

As the semiconductor industry moves to further miniaturization, it is rapidly approaching the physical limits of materials. With 32 nm technology it is now possible to fit up to 1.9 billion transistors on a single chip. Further miniaturization of lithography introduces several problems, such as dielectric breakdown, hot carriers, and short channel effects. All of these factors combine to seriously degrade device reliability. Further developments in technology may temporarily solve these physical problems, but we will ultimately again face them further down the path. However, optical devices offer a completely different alternative. Optical interconnections and optical integrated circuits can provide a way out of these limitations to computational speed and complexity inherent in conventional electronics. Optical computers will use photons traveling on optical fibers or thin films instead of electrons to perform the appropriate functions. In the optical computer of the future, electronic circuits and wires will be replaced by a few optical fibers and films, making the systems more efficient with no interference, more cost effective, lighter, and more compact. Optical components do not need to have insulators as needed between electronic components. Light with multiple frequencies (or different colors) can travel through optical components without interfacing with each other, allowing photonic devices to process multiple streams of data simultaneously. Optical interconnections and optical integrated circuits have several advantages over their electronic counterparts. They are immune to electromagnetic interference, and are free from electrical short circuits. They have low-loss transmission and provide large bandwidth, which means multiplexing capability—the capability to communicate in several channels in parallel without interference. They are capable of propagating signals within the same or adjacent fibers with essentially no interference or cross talk. They are compact, lightweight, and inexpensive to manufacture, and more facile with stored information than magnetic materials.

Focused ion beam (FIB) milling is emerging as an attractive nanostructuring technique and provides an alternative to UV and electron-beam lithographic methods [1]. The main advantage of FIB milling is that it is, to some extent, material independent. Several applications involve materials or material combinations that are hard to etch using more conventional chemical methods. Since FIB can mill almost any material, it is very convenient for fabricating a device from a combination of such materials. In comparison to other technologies for nanofabrication, such as those based on e-beam lithography, FIB has the advantage of enabling fast prototyping, resulting in considerable reduction of the design-fabrication-characterization cycle time [2].

One of the examples of rapid prototyping is the fabrication of Bragg diffraction gratings in Al_2O_3 channel waveguides reported by Ay et al. [2]. These sub-μm period surface relief reflection gratings were fabricated on dielectric channel waveguides. A FEI make Nova 600 dual-beam FIB machine was used for this fabrication. The acceleration voltage was set to 30 kV and the milling current to 93 pA. In order to analyze the effect of milling parameters on the

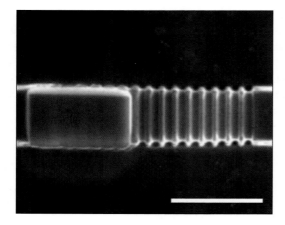

FIGURE 20.1: Grating with platinum deposited. Scale bar – 3 μm.

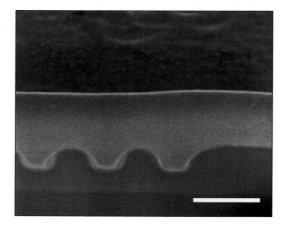

FIGURE 20.2: Cross-section of the grating. Scale bar – 500 nm.

fabricated structures, cross-sections were made with the well-established procedure. First, a layer of Pt was deposited on the grating, as can be seen in Figure 20.1. This is to prevent the profile of the grating from being affected while cross-section milling is carried out. To cut down milling time, the cross-section is made using high current. The cross-section is then cleaned using a lower current of 28 pA and the program for cleaning cross-section which mills line by line. This removes the artifacts that might cover the true profile of the cross-section. The dual beam FIB also has an incorporated SEM, and the cross-section can be directly viewed with the high-resolution SEM without damaging it. The cross section of the grating can be seen in Figure 20.2.

The milling of the grating itself is carried out by generating a stream file. This file contains a pixel-by-pixel address where the ion beam should be directed. At the same time, the file also contains the dwell time at each pixel. Dwell time is the time for which the ion beam will be stationary at a particular spot.

One of the biggest problems encountered in milling of nanostructures with FIB is the problem of redeposition. In the milling process, Ga ions collide with the surface atoms of the substrate. The collision physically ejects the surface atoms. Some of the gallium atoms are embedded in the surface, but most are also ejected along with the substrate atoms.

Depending on the trajectory and energy imparted to the atoms, they either deposit on the inner walls of the specimen chamber, are removed by the vacuum system, or redeposit adjacent to the area of milling. Most redeposition finds a place within the area being milled, and this causes distortion of the desired feature. If a square is milled using FIB, initially the depth will rapidly increase as the material is removed. As the depth of the hole increases, less and less material will escape from the pit, redepositing on the sidewalls. The size of the pit will start decreasing until no further material can the removed and everything being milled redeposits in the pit. The milling process will come to a standstill.

As a simple demonstration of this effect, a square of 1 micron was milled in silicon for different times starting at 1 minute to 10 minutes. The depth of the pits was determined by making cross-sections with the well-established procedure already explained.

The cross-sections after 1 minute of milling and 10 minutes of milling can be seen in Figures 20.3 and 20.4. Cross section after 1 minute of milling in Figure 20.3 does not show a large amount of redeposition. However, the walls of the square are not completely vertical. This indicates some amount of redeposition along the walls. Figure 20.4 shows that the shape of the original pattern is not retained at all with depth. As depth increases, the pattern narrows down

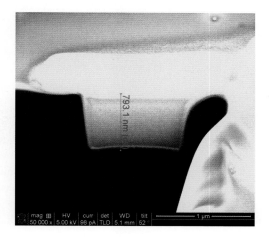

FIGURE 20.3: Cross-section after 1 minute.

FIGURE 20.4: Cross-section after 10 minutes.

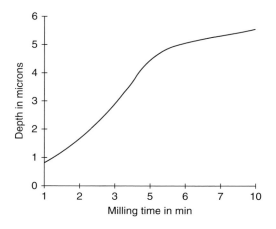

FIGURE 20.5: Depth of milling with milling time.

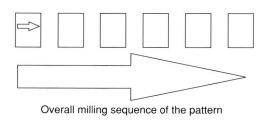

Overall milling sequence of the pattern

FIGURE 20.6: Schematic diagram showing individual scanning direction and milling sequence.

almost to a point. With 10 minutes of milling, there is almost no further removal of material. The depth of milling determined with cross-sections, with time can be seen in Figure 20.5. As it is evident from this figure, the milling almost comes to a standstill. The milling in this case was carried out with a raster scan of the beam in the area to be milled. The effect seen here over an area of 1 micron is also present at each pixel. The same mechanism is working. This means that the dwell time of the beam at any position will affect the redeposition.

The ion beam with a low dwell time will make several passes over the same pixel. In the very first pass, some of the material that has been milled will escape, while some will redeposit. In the second pass, the beam will mill the redeposited material and some virgin material. This process will continue over and over with every pass or loop, as it is called. When the dwell time is higher, the amount of material attempted to be removed is larger at the same time, and the redeposited material is also larger. With the second pass at high dwell time, more redeposition will take place at the beginning of the feature, while the last part of the feature is being milled. This results in a distorted shape in the depth of the feature to be milled.

To evaluate the effect of dwell time (and number of loops) on a pattern, stream files were programmed. The schematic representation of the pattern is shown in Figure 20.6.

The dwell time was programmed at 0.001 ms, 0.01 ms, and 0.1 ms. The dose of gallium ions was kept constant by adjusting the number of loops. The scan direction for every feature was maintained from left to right. For all experiments the substrate material was Si. The effect of dwell time is clearly seen in Figure 20.7. When the dwell time is very low, the area between the two features is also milled, as seen in Figure 20.7 (left). This is because when the dwell time is low (0.001 ms in this case), the number of loops is very high. During each loop the beam does not switch off between the two features but travels in nano seconds to the next feature.

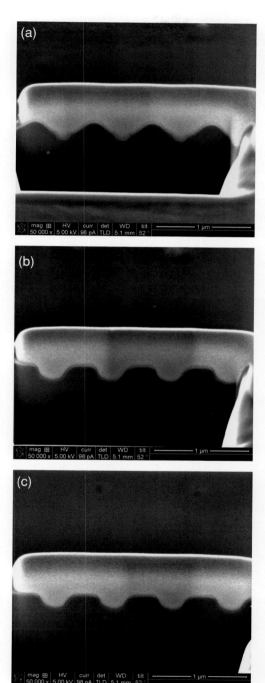

FIGURE 20.7: Cross-sections with dwell time 0.001 ms (a), dwell time 0.01 ms (b), and dwell time 0.1 ms (c).

Therefore during each loop the area between two features is exposed to the beam, which results in the milling of that area. In Figure 20.7 (right) this effect is not visible. This is because the dwell time in this case was 1 ms; thus the number of loops was small. The intermediate area was not milled and the pattern appears to be better defined.

With dwell time of 0.001 ms, the pit is deeper in the middle than on the edges. This is an indication that the dwell time was insufficient to remove the redeposited material that accumulates on the sidewalls. With increasing dwell time, the profile changes, becoming more rectangular. It is clear from this experiment that the dwell time must be adjusted to an optimum to fabricate nanostructures better conformed to the intended design.

The same strategy can be applied to milling of good quality photonic devices. In order to obtain uniform and smooth sidewalls of the grating structures, F. Ay et al. [2] carried out a study for minimization of the redeposition effects in Al_2O_3 milling. The dwell time and number of loops were varied while keeping the total dose constant to achieve similar milling depths. The grating lengths were about 23 μm and waveguides with widths between 2.0 and 3.8 μm were used. The grating period was about 550 nm and the milled depths varied between 150 and 200 nm. The initial thickness of the Al_2O_3 channel waveguides was about 550 nm. The gratings were milled using a stream file that contains milling time, pixel information, and pixel sequence for the desired geometry. Flexibility provided by the stream-file-based patterning allows us to choose the pathway by which the grating is defined on the waveguide. Figure 20.8(a) shows the

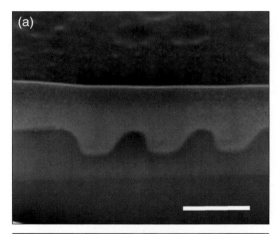

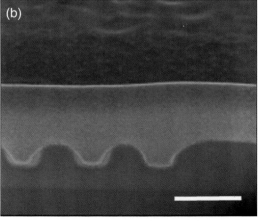

FIGURE 20.8: Cross-section profiles of grating structures obtained with (a) a dwell time of 0.1 ms and 8 loops, (b) a dwell time of 0.001 ms and 800 loops. Grating with platinum deposited. Scale bars – 500 nm.

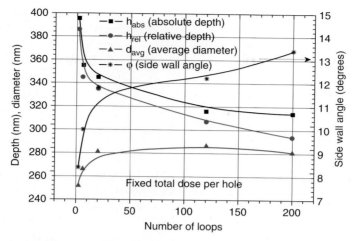

FIGURE 20.9: Milling results for a hole in an array, at a constant dose of 69 pC per hole. The designed hole diameter was 250 nm and the current was fixed at 48 pA.

cross-section profile of a grating structure obtained with a dwell time of 0.1 ms and 8 loops; in Figure 20.8(b) the corresponding profile obtained with a dwell time of 0.001 ms and 800 loops is depicted. The deteriorating effects of redeposition are clearly identified in the first case, dictating that the fabrication process can be optimized by using small dwell times and a higher number of loops, thus smoothing out the effects of redeposition. Previous results show that these findings may depend on both the milled geometry and material used [2–4].

From Figure 20.8 it is clear that with dwell time of 0.1 ms, the walls of the grating are not parallel. The right-hand sidewall is more inclined than the left. With decrease in dwell time, the profiles are much improved. The lower dwell time (higher number of loops) results in removal of the redeposited material.

The effect of the process parameters was further explored by Hopman et al. [5]. Patterns were milled where the dose was kept constant at 69 pC per hole while varying the combination loop number and dwell time. The loop number was varied from 2 to 200 and the dwell time from 6 μs to 1 ms, respectively.

The results of the experiments can be seen in Figure 20.9. It can be seen that the most favorable conditions for milling an array are at low loop numbers and consequently at large dwell times of around 1 ms for the studied dose. With respect to the sidewall angle, it is clear that a high number of loops produce higher side wall angles at a dose of 69 pC. From these experiments one can conclude that not only the dwell time but the dose per pixel that should also be optimized.

With optimizing the parameters it is possible to mill optical devices of good quality, as can be seen in Figure 20.10, where a Bragg grating obtained in Al_2O_3 is shown. Fabrication of two-dimensional photonic crystals can encounter some other problems. Most of the commercially available FIB machines have three choices of the beam scanning. These are raster, serpentine, and spiral. Once the pattern to be milled has been designed, it can be programmed in the machine either by using specially generated stream files or by loading the pattern as a bmp file. Figure 20.11 shows such a pattern. In most of the milling operations, the raster scan is used. In this type of scanning, as the ion beam travels from one assigned position to the other;

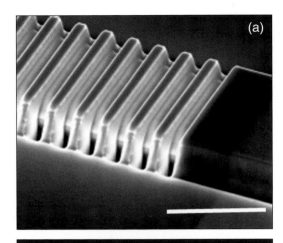

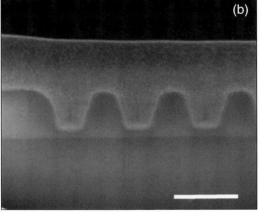

FIGURE 20.10: Bragg diffraction grating fabricated with FIB (a) and its cross–section (b). Scale bars – 2 μm (a) and 500 nm (b). Reprinted with permission from [2], University of Twente.

it is not switched off. This means that the area between the two holes, which we will refer to as the wall, will be unintentionally milled. If the wall thickness is small, the walls get milled every time the beam passes over them. The final result is that the walls are lowered due to milling and the entire pattern sinks into the surface. A better method for milling arrays of holes is to use spiral scanning. A schematic drawing of the raster scan and spiral scan can be seen in Figure 20.12. The spiral scan significantly reduces the amount of milling of the areas between two holes. The spiral scan produces more vertical sidewalls, because locally redeposited material is immediately milled away to a large extent.

Hopman et al. [5] also conducted extensive experiments on scan type and its effect on the hole depth and geometry. The results of the experiments can be seen in Figure 20.13. It can be clearly observed that the sidewall angle is strongly influenced by the type of scan. With a raster scan, one side of the hole is distinctly deeper than the other side. The angle ϕ is larger for the raster scan than the spiral scan. However, there is no considerable difference in the average diameter (D_{avg}). The height measured from the surface of the rest of the unprocessed substrate is denoted as h_{abs}, while the height measured from the top of the wall between two holes is denoted as h_{rel}. There is considerable difference in these parameters between the spiral scan and raster scan. As expected, the raster scan has lower h_{rel}. This is because of milling of the intervening walls of the pattern. The spiral scan shows minimal difference in h_{abs} and h_{rel}. It can

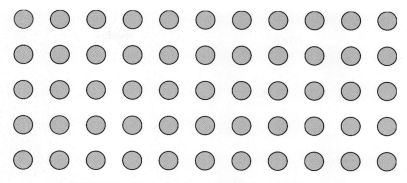

FIGURE 20.11: A two-dimensional array for photonic crystal.

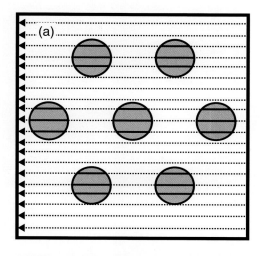

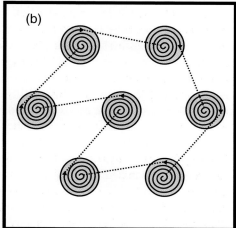

FIGURE 20.12: Schematics diagram showing raster
scan and spiral scan.

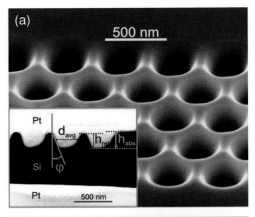

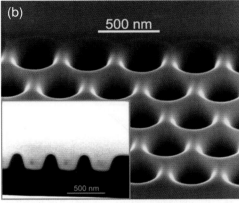

FIGURE 20.13: Array of holes milled with raster scan (a) and spiral scan (b).

be concluded that the spiral scan is a better method for fabricating two-dimensional photonic structures.

Another way of fabricating photonic structures with more vertical walls is described by Bostan et al. [6]. Here the FIB is used to mill the structures in an intermediate metal layer used as a hard mask. This can be achieved by the following steps. First, a thin chromium layer (70 nm) is sputtered on the SOI wafer. Optical lithography with wet chemical etching is used to define the waveguides. With FIB the pattern is milled in the thin chromium layer. This is followed by reactive ion etching (RIE) of the Si and SiO_2 layers using a two-step process. Finally, the chromium mask is removed. In this method, the common problem of tapered sidewalls that occurs when the aspect ratio is close to unity is eliminated. Another problem with FIB milling is the implantation of Ga atoms in the structure. In this process, the "contaminated" layer is removed in the subsequent RIE process. Figure 20.14 shows a structure on which the such a pattern is milled.

3 FABRICATION OF NANOSIEVES

Fluid separation membranes have been known for a very long time. Traditionally, most of these membranes consist of a relatively thick layer (several micrometers) of organic

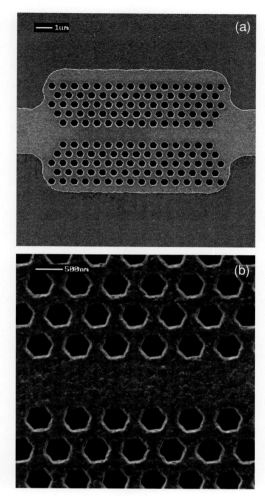

FIGURE 20.14: FIB-etched photonic crystal pattern in chromium layer (a), detailed view of a pattern with hexagonal holes (b).

or inorganic material with nanoporous perforations. Nanomembranes with perforations essentially of one size and cylindrical pore shape will have many potential applications like sterile filtration (bacteria as well as viruses) and size-exclusion-based separations [7–10], and can also serve as templates for nanosensors [11], and masks for stencils [12]. Nanoporous membranes with arrays of cylindrical pores and uniform pore size between several tens down to below 10 nm are referred to as nanosieves.

The nanopores are often milled in a suspended low-stress silicon-nitride (SiN) membrane [7; 11–16]. This is due to the availability of SiN membranes with standard microsystem processing like KOH etching of a silicon support. So far, electron beam lithography followed by reactive ion etching (RIE) [10; 11], or FIB etching [17–20], has been used to create pores in SiN membranes. When a relatively thick membrane is milled by FIB, a higher dose is required to mill the hole, so that the diameter of the pore increases. It is reported that typical pore diameters are 50 nm or greater [17–20]. A simple strategy to directly drill small pores by FIB is to reduce the thickness of the SiN membrane. For fabricating the sieves from a thin layer of SiN, extremely short milling time is required. Typically, a current of 1 pA is used. At 1 pA the

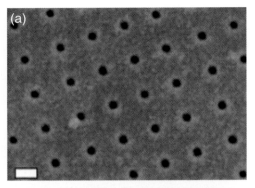

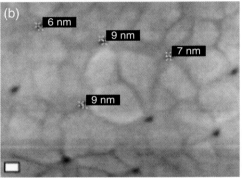

FIGURE 20.15: (a) Nanosieve with pores, about 25 nm in diameter, directly milled by FIB (110 kX, scale bar: 50 nm). (b) Size of the pores further reduced below 10 nm by coating with SiN (150 kX, scale bar: 20 nm).

beam diameter is ca. <10 nm full-width half-maximum (FWHM) corresponding to a current density of ca. 1.2 A/cm^2. The FIB process is controlled by a stream file generated using Matlab program. The stream file used in fabrication had a programmed dwell time of 10 μs. The samples were coated with 2 nm chromium (Cr) for reducing charging during exposure to electron and ion beams. A nanosieve fabricated by FIB was examined by a LEO Gemini 1550 (2 nm lateral resolution) high-resolution scanning electron microscope. As can be seen in Figure 20.15, the pores are circular with an average diameter of around 25 nm, and a pore pitch of around 115 nm. The pores fabricated with FIB are uniform in size (<10% variation) over the whole 50*50 μm^2 patterned area. The membranes were examined from both sides. It was found that there is almost no difference between the top and bottom pore diameters. This indicates that the pore wall has a slope angle of nearly 90° to the lateral direction. The result is consistent with the reports in the literature when small currents are used [21].

The effect of the dose on hole milling has been studied. It has been found that smaller dose does not produce smaller pores but fails to perforate the material. The effects of variation in the thickness of the conducting Cr layer, variations in the alignment of the ion beam, profile of the beam itself, and other parameters as the quality of vacuum are basically unknown. It was therefore not possible to mill smaller pores by just reducing the milling time. However, it is possible to reduce the size of the pores by other means. The FIB-processed nanosieves were treated in oxygen plasma to remove hydrocarbon contaminations from FIB and SEM processing. The thin conducting layer of Cr was removed in a Cr etch solution. The specimens were cleaned in de-ionized water. The FIB-milled pores were then coated with another ultrathin layer of LPCVD SiN. Selecting SiN to coat the sample has the goal of minimizing the possibility of developing stress due to thermal mismatch. Moreover, it exhibits a conformal step coverage, and a high uniformity. The initial pores milled with FIB with a diameter of around 25 nm can be seen in Figure 20.15(a). The SiN deposition unit had a deposition rate of 0.13 nm/s. After coating with 10 nm SiN, the HRSEM image of the coated pores is shown in Figure 20.15(b); all the pores are now below 10 nm in size.

4 FABRICATION OF NANOWIRES

Once the thin SiN membranes could be reliably fabricated, it was possible to think of other applications. The process of making the SiN membrane was identical to the one used for nanosieve membranes fabrication. The details of the process can be found in [20]. Instead of a circular freestanding membrane of SiN, a dumbbell shape membrane was fabricated.

This was done so that the contact pads were integral to the deposition of the nanowire. The contact pads were first milled with FIB using 93 pA current. After that, using a lower current of 9.2 pA, a thin line was milled between the contact pads. This can be seen in Figure 20.16.

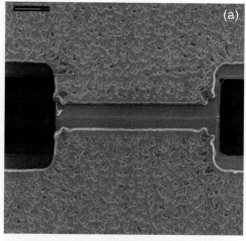

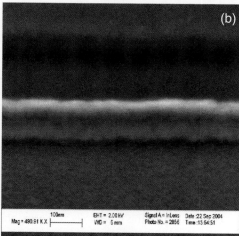

FIGURE 20.16: (a) Nanoslit milled with FIB. (b) gold nanowire.

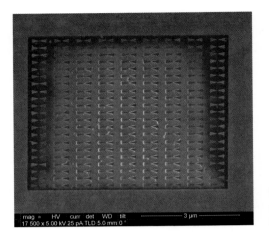

FIGURE 20.17: Array of nanoantennas fabricated with FIB, courtesy of F. B. Segerink of OS group of University of Twente [23].

Using the shadow mask, a gold was deposited with electron beam sputtering. The gold deposited through the nanoslit formed a nanowire which can be seen in Figure 20.16(b).

5 FABRICATION OF NANO ANTENNAS

Antennas are an integral part of wireless communication. Antennas make it possible to transmit and receive electromagnetic radiation. The size and design of these antennas is driven by the type of electromagnetic radiation that it has to function with. Now, it has been shown that the concept of an antenna is equally well applied to small wavelengths in the visible spectrum. However, for an antenna to work with visible light, it has to be millions of times smaller than a conventional antenna. This field has generated considerable interest in the optical properties of metallic nanoantennas. Depending on the metal and on the design of the nanoantennas, their optical response can be tuned. When a nanoantenna is exposed to light of a particular wavelength, it generates a resonant motion of the electrons called localized surface plasmon resonance. There are several possible routes to fabricate these structures. The challenge is to fabricate the designed structures in a quick and efficient way so that the design can be evaluated quickly. FIB offers the possibility of fast prototyping. Apart from milling, it is well-known that FIB can also deposit metals, as explained elsewhere [22].

Recently, with new developments in precursor gas chemistry, it is possible to deposit gold (an appropriate material for nano-antennas) with FIB. However, the accuracy of the pattern is not high and there is contamination of the deposit with Ga atoms. Another method developed by F. B. Segerink from the OS group of the University of Twente [23] is to fabricate the antennas by milling from a thin layer of gold deposited on glass. The problem here is that gold is very sensitive to exposure with Ga ion beam. The layer immediately starts to form lumps and the surface is destroyed within a very short time. This can be avoided by exposing the surface only during milling and to avoid any imaging altogether. Figure 20.17 shows an array of antennas fabricated with the FIB.

A glass substrate was coated with a 20 nm thick gold layer. No intermediate bonding layer was used. Since the gold layer was very thin, an extremely low dose was needed for milling. A current of 1.5 pA was used. The structure was programmed as a stream file. In order to avoid exposure of gold to the beam, a single pass (loop) was used. Dwell time was set at 750 ms and a step size of 2.5 nm was used. Scanning was carried out using a serpentine pattern in the longitudinal direction of the antennas.

6 CONCLUSIONS

Recent advances in patterning with commercially available FIB machines have made it possible to fabricate nanostructures directly by milling with FIB. In many cases, FIB processing eliminates time-consuming and expensive photolithography steps in fabrication. Improved understanding of the milling processes makes it possible to optimize parameters to generate high-accuracy structures. These can serve as proof of concept devices for research and development.

REFERENCES

[1] Cryan, M. J., M. Hill, D. C. Sanz, P. S. Ivanov, P. J. Heard, L. Tian, S. Y. Yu, and J. M. Rorison. Focused ion beam-based fabrication of nanostructured photonic devices. *IEEE J. Select. Topics Quantum Electron.* 11 (2005): 1266–1277.

[2] Ay, F., A. Uranga, J. D. B Bradley, K. Wörhoff, R. M. de Ridder, and M. Pollnau. Focused ion beam nano-structuring of Bragg gratings in Al_2O_3 channel waveguides. *Proceedings of the First International Workshop on FIB for Photonics,* Eindhoven, The Netherlands (2008): 48–50.

[3] Ay, F., J. D. B. Bradley, R. M. de Ridder, K. Wörhoff, and M. Pollnau. Bragg gratings in Al_2O_3 Channel waveguides by focused ion beam milling. *Proceedings of the 13th Annual Symposium of the IEEE LEOS Benelux.* Enschede, The Netherlands (2008): 215–217.

[4] Gadgil, V. J. Strategies for fabricating nanostructures with focused ion beam technology. *Nanotech conference* May 3–7, 2009, Houston, Texas, *Technical Proceedings,* vol 1., pp. 37–40.

[5] Hopman, W. C. L., F. Ay, W. Hu, V. J. Gadgil, L. Kuipers, M. Pollnau, and R. M. de Ridder. Focused ion beam scan routine, dwell time and dose optimizations for submicrometre period planar photonic crystal components and stamps in silicon. *Nanotechnology* 18 (2007): 195305; pp. 11.

[6] Bostan, C. G., R. M. de Ridder, V. J. Gadgil, H. Kelderman, L. Kuipers, and A. Driessen. Design and fabrication of line-defect waveguides in hexagon-type SOI photonic crystal slabs. *Photonics Europe* 2004, SPIE Conference Strasbourg, France, April 2004.

[7] Rijn, C. J. M., and M. Elwenspoek. Micro filtration membrane sieve with silicon micromachining for industrial and biomedical applications. *IEEE Conf MEMS'95* (1995): 83.

[8] Freitas, R. A. Nanomedicine. *Stud. Health. Techno. Information.* 80 (2002): 45.

[9] Desai, A. T., D. J. Hansford, L. Kulinsky, A. H. Nashat, G. Rasi, J. Tu, Y. Wang, M. Zhang, and M. Ferrari. Nanopore technology for biomedical applications. *Biomedical Microdevices* 2.1 (1999): 11.

[10] Mulder, M. *Basic principles of membrane technology.* Dordrecht: Kluwer Academic Publishers, 1998.

[11] Chen, J., M. A. Reed, A. M. Rawlett, and J. M. Tour. Large on-off ratios and negative differential resistance in a molecular electronic device. *Science* 286 (1999): 1550.

[12] Deshmukh, M. M., D. C. Ralph, M. Thomas, and J. Silcox. Nanofabrication using a stencil mask. *Appl. Phys. Lett.* 75 (1999): 1631.

[13] Schmidt, C., M. Mayer, and H. Vogel. A chip-based biosensor for the functional analysis of single ion channels. *Angew. Chem. Int. Ed.* 39 (2000): 3137.

[14] Geoch, J. E. M., M. W. McGeoch, D. J. D. Carter, R. F. Schuman, and G. Guidotti. Biological-to-electronic interface with pores of ATP synthase subunit C in silicon nitride barrier. *Medical. Bio. Eng. Comput.* 38 (2000): 113.

[15] Schenkel, T., V. Radmilovic, E. A. Stach, S. J. Park, and A. Persaud. Formation of a few nanometer wide holes in membranes with a dual beam FIB. *Inter. Conf. on Electron, Ion and Photon Beam Technology and Nanofabrication* (2003).

[16] Li, J., D. Stein, C. McMullan, D. Branton, M. J. Azis, and J. A. Golovchenko. Ion-beam sculpting at nanometre length scales. *Nature* 412 (2001): 166.

[17] Tong, H. D., J. W. Ervin Berenschot, M. J. De Boer, J. G. E. Gardeniers, H. Wensink, H. V. Jansen, W. Nijdam, M. Elwenspoek, F. C. Gielens, and C. J. M. Rijn. Microfabrication of palladium–silver alloy membranes for hydrogen separation. *J. Microelectromech. Syst.* 12 (2003): 622.

[18] Gardeniers, J. G. E., C. C. G. Visser, H. A. C. Tilmans. LPCVD silicon-rich silicon nitride films for applications in micromechanics, studied with statistical experimental design. J. *Vac. Sci. Technol.* A 14 (1996): 2879.

[19] Harriott, L. R. Digital scan model for focused ion beam induced gas etching. *J. Vac. Sci. Technol.* B 11 (1993): 2012.

[20] Lipp, S., L. Frey, G. Franz, E. Demm, S. Petersen, and H. Ryssel.Local material removal by focused ion beam milling and etching. *Nuclear. Instr. Meth. Phys. Res.* B. 106 (1995): 630.

[21] Gadgil, V. J., H. D. Tong, Y. Cesa, and M. L. Bennink. Fabrication of nano structures in thin membranes with focused ion beam technology. *J. Surface and Coatings Technology* 203.17–18 (2009): 2436.

[22] Gadgil, V. J., and F. Morrissey. Applications of focused ion beam in nanotechnology. In *Encyclopedia of nanoscience and nanotechnology*, H. S. Nalwa, ed. Vol. 1, pp. 101–110. Los Angeles: American Science Publishers, 2004.

[23] Segerink, F. B. OS group, University of Twente, private communication.

CHAPTER 21

FOCUSED PARTICLE BEAM INDUCED DEPOSITION OF SILICON DIOXIDE

Heinz D. Wanzenboeck

1 CHEMICAL VAPOR DEPOSITION OF SILICON DIOXIDE

1.1 BULK MATERIAL

Silicon dioxide is a solid, hard material with a high thermal stability. Due to its dielectric properties, silicon dioxide is the preferred insulator in microelectronics. In optics the transparent forms of this material are used for UV-transparent optical devices and for photomasks. SiO_2 is a solid material that may occur in a crystalline or an amorphous form.

1.1.1 Material properties

Silicon (IV) dioxide is an insoluble solid with a density of 2.2 to 2.6 g/cm^3. Silicon dioxide is white in a powder form and transparent or opaque in a lump form. Silicon dioxide has a refractive index of 1.46 (thermal oxide) and amorphous forms (quartz glass). Transparent forms provide a wide transparency range down below 180 nm.

With a dielectric constant of 3.9 an energy gap of 9 eV at 300 K and a breakdown field larger than $1 \times 10^{+6}$ V/cm (CVD oxides), silicon dioxide is an excellent electrical insulator. The material has a melting point of 1713°C, a low thermal conductivity of 0.01 W/cm K (bulk), and a low coefficient of thermal expansion of 0.5 ppm/K. Silicon dioxide has a high chemical stability against all organic and inorganic acids except hydrofluoric acid. The material is slowly etched by inorganic basic media such as KOH or NaOH. More than 10 crystalline forms of silicon oxide are known, most of which are based on tetrahedral SiO_4. The only stable form under normal conditions is α-quartz with a density of 2.648 g/cm^3 and a Si-O bond length of 161 pm.

With incomplete oxidation of silicon, a silicon(II) monoxide also can be synthesized. SiO is an amorphous solid that is stable at room temperature, but experiences gradual disproportionation to silicon and silicon dioxides at temperatures of 400°C–800°C. Silicon monoxide has a melting point above 1702°C. SiO is brown in a powder form or black in a lump form. SiO coatings are used for optical applications such as in reflectors and mirrors, and can reduce reflection in the near infrared.

1.1.2 Fabrication

On silicon wafers the silicon dioxide is grown by thermal oxidation of silicon. "Thermal oxides" have the highest purity and act as superior electric insulators. Alternatively, silicon dioxide can be formed from silicic acid solution ("spin-on-silica") [1]. Silicon dioxide can also be synthesized by chemical vapor deposition (or atomic layer deposition) using a Si-precursor in combination with an O-precursor. For blanket layer deposition the activation energy for the chemical reaction is usually introduced by thermal heating or by plasma. With beam-induced deposition the activation energy is added by inelastic effects of the focused particle beam. Preparation of silicon monoxide coating is performed by evaporation or sputtering from lump material.

1.2 CVD OF SILICON DIOXIDES

The deposition of silicon oxide by a focused particle beam represents only an unusual case of chemical vapor deposition. The energy for activation of the reaction (Table 21.1) is provided by the focused particle beam impinging on a very restricted area of the sample surface with an energy of several keV. In order to crack the molecular bonds of siloxane precursors, energies only in the range below 5 eV are required. This energy is provided by secondary effects.

Table 21.1 Characteristic bond strengths of bond types occurring in precursors for SiO_2 [2].

Bond type	Bond strength [eV]	Bond dissociation enthalpy [kJ/mol]	Bond dissociation wavelength [nm]
Si-O	4.5	434	276
C-O	3.6	347	344
C-H	4.1	396	302
Physisorption (surface-molecule)	<1.7	<164	729
Chemisorption (surface-molecule)	>1.7	>164	729

Chemical reactions in a CVD process can occur either (1) entirely in gas phase (homogeneous reaction) or (2) on the surface of the sample (heterogeneous reaction). For beam assisted CVD the heterogeneous reactions are of major relevance. With two precursors—a silicon-precursor and an oxygen-precursor—a complex multi-molecular heterogeneous reaction occurs in which decomposition products also are involved as adsorbed surface species.

With heterogeneous reactions the following reaction steps have to be considered:

- *Adsorption of precursors.* One (Eley-Rideal-mechanism) or both precursors (Langmuir-Hinshelwood-mechanism) may adsorb from the gas phase on the surface. The surface coverage with each precursor depends on the sticking coefficient of the particular precursor and on the vapor pressure in the process chamber. A simple adsorption model is described by the Langmuir isotherm.
- *Surface diffusion of precursors.* Physisorbed precursors are weakly bound to the surface (0.2–1.7 eV) and may diffuse on the surface. This is relevant in the case of diffusion-limited beam induced deposition.
- *Surface reaction of the precursors and reaction intermediates.* Upon activation by the energy of the impinging particle beam, the adsorbed precursor species react to yield the deposited material and volatile precursor fragments.
- *Desorption of volatile precursor fragments.* After reaction of the precursor, the solid material remains permanently bonded to the surface, while other physisorbed ligands (volatile fragments of the precursor) desorb from the surface. While adsorbed on the surface, reaction products may experience further decomposition by the particle beam. This contributes to the carbon contamination of deposits.
- *Desorption of unreacted precursors.* Adsorption and desorption of precursor molecules is a dynamic equilibrium process. The physisorbed precursors may desorb unchanged when not decomposed.

For the mechanism of focused beam induced deposition of silicon oxide, no reaction pathway has been postulated and no specific data are available. However, findings with thermal- or plasma-assisted CVD can provide insights into reaction pathways. For conventional CVD several reaction mechanisms have been investigated.

Silicon oxide may be deposited from:

- carbon-free precursors: silane (SiH_4), silicon tetrachloride ($SiCl_4$);
- organometallic precursors: alkoxy-silanes (TEOS), cyclic alky-siloxanes (TMCTS).

For a long time, deposition of silicon oxide from carbon-free precursors such as silane was the preferred process. For chemical vapor deposition (CVD) of silicon dioxide (SiO_2) from silane with nitrous oxide, a chain reaction initiated by the decomposition of N_2O and a reaction with SiH_4 leading to SiH_3OH was suggested by Giunta et al. [3]. SiH_3OH presumably acts as the film precursor, which is rapidly oxidized and dehydrogenated on the growth surface. At lower N_2O concentrations silicon-rich oxide films, SiOx, are deposited. Disilane Si_2H_6 also has been used as precursor generating silanone (SiH_2O) as oxide film precursor [3]. With this precursor a strong dependence of the growth rate on the initial disilane concentration was observed. Silane precursors are self-inflammable and require high safety precautions. For focused beam systems these precursors are rarely used.

A large number of CVD processes utilize organometallic silioxane precursors due to their easy handling and the low safety risk involved. Many of these CVD processes are performed at higher temperatures [4]. As an exemplary Si-precursor, tetraethoxysilane (TEOS) will be briefly discussed. TEOS is also frequently used for beam-induced deposition. The low-pressure CVD (LPCVD) of SiO_2 from TEOS is used widely in semiconductor fabrication. At low temperatures an oxidant is required (Equation 1), while at higher temperatures a decomposition of the precursor occurs (Equation 2).

Plasma enhanced chemical vapor deposition using TEOS at approximately 400°C:

$$Si(OC_2H_5)_4 + 12\,O_2 \rightarrow SiO_2 + 10\,H_2O + 8\,CO_2. \tag{1}$$

Decomposition of tetraethoxysilane (TEOS) at 680°C–730°C:

$$Si(OC_2H_5)_4 \rightarrow SiO_2 + H_2O + 2C_2H_4. \tag{2}$$

The surface coverage is a critical parameter. On TiO_2-surfaces tetraethoxysilane (TEOS) was reported to stick with almost a probability of 1 in the temperature range from 130 K to 300 K [5]. At 450 K, TEOS was found to react with a silicon surface, forming an intermediate mixture of adsorbed di- and triethoxysilanes [6], while ethylene, water, and hydrogen were found as decomposition products. The presence of water as oxygen-precursor accelerated the decomposition rate of TEOS. At high temperatures the formation of an intermediate in the gas phase was observed. This activated species adsorbs on Si [7]. With TEOS and O_2 the oxygen radicals and molecular oxygen ions play a role in the deposition process. The O atoms react with TEOS molecules [8]. If the deposition is performed in an RF reactor at high oxygen concentrations, the growth rate is governed by the intermediates generated from TEOS [9].

A well-researched process is the deposition of TEOS with ozone (O_3) at lower temperatures. For CVD from TEOS and O_3, two different mechanisms have been proposed (Figure 21.1). Mechanism 1 is based on the direct oxidation of TEOS by the ozone molecule, resulting in a first-order dependence from the ozone concentration. Mechanism 2 is based on the ozone decomposition and the subsequent reaction of the oxygen atom and ozone with TEOS resulting in a second-order dependence from the ozone concentration [10].

The rate determining process depends on the reaction parameters. At 450°C under atmospheric pressure, the deposition follows a second-order kinetics [11]. The oxidation of TEOS is initiated stepwise by oxygen atoms originating from ozone decomposition. Based on results from *in situ* infrared spectroscopy, a parasitic process with a sequence of five reactions (Figure 21.2) was proposed. Spectroscopy also indicate the formation of silanol deposition intermediates, as well as organic products. The ratio of ethoxysilanol formed versus the amount of ozone consumed is about 1:3. A radical chain oxidative mechanism involving direct reaction of TEOS and ozone is proposed for formation of highly reactive silanol film growth intermediate.

The basic reaction concepts displayed in Figure 21.2 can provide a rough guideline for the approximation of processes during focused particle beam induced deposition of silicon oxide.

Mechanism 1		Mechanism 2	
Ozone decomposition	TEOS decomposition	Ozone decomposition	TEOS decomposition
$O_3 + M \leftrightarrow O_3^* + M$	$TEOS + O_3^* \rightarrow I + R$	$O_3 + M \leftrightarrow O_2 + O + M$	$TEOS + O + O_3 \rightarrow I + R$
$O_3^* \rightarrow O_2 + O$	$I \rightarrow$ Products	$O + O_3 \rightarrow 2O_2$	$I \rightarrow$ Products
$O + O_3 \rightarrow 2O_2$			

FIGURE 21.1: Process steps involved in the TEOS/ozone deposition process. Reprinted with permission from [10]. Copyright (2000), Elsevier.

$$O_3 + M \underset{k_2}{\overset{k_1}{\longleftrightarrow}} O_2 + O + M \qquad \text{(I)}$$

$$TEOS + O \overset{k_4}{\longrightarrow} \alpha TEOS + OH \qquad \text{(III)}$$

$$O + O_3 \overset{k_3}{\longrightarrow} 2O_2 \qquad \text{(II)}$$

$$\alpha TEOS + O_2 \overset{k_5}{\longrightarrow} I + R \qquad \text{(IV)}$$

$$I \overset{k_6}{\longrightarrow} Products \qquad \text{(V)}$$

FIGURE 21.2: A five-step parasitic reaction of the intermediate products for the TEOS/O3 reaction. Reprinted with permission from [11]. Copyright (2000), American Chemical Society.

1.3 SILICON PRECURSORS

The silicon precursors used throughout CVD processing of blanket layers can also be applied to focused particle beam induced deposition. A sufficiently high vapor pressure is one of the primary selection criteria for the precursor. The safety during manipulation is another criterion. The most common silicon precursors are listed in Tables 21.2 and 21.3.

Table 21.2 Silicon precursors frequently used in CVD.

SiH_4	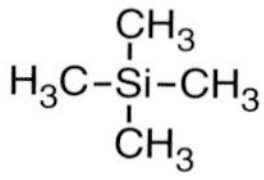	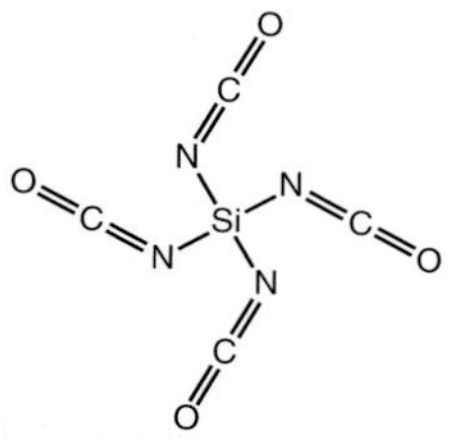	SiH_4
Silane (SiH_4) CAS 7803–62–5	Tetramethylsilane (TMS) CAS 75–76–3	Tetraisocyanatosilane (TICS) CAS 3410–77–3	Silicon tetrachloride $(SiCl_4)$ CAS 10026–04–7
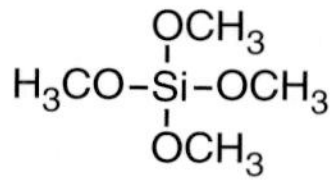	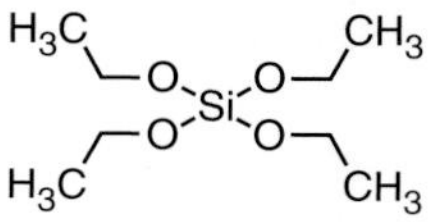	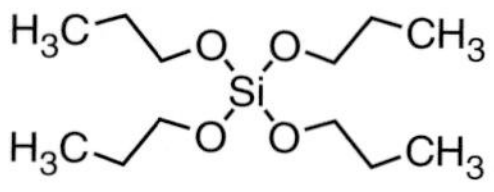	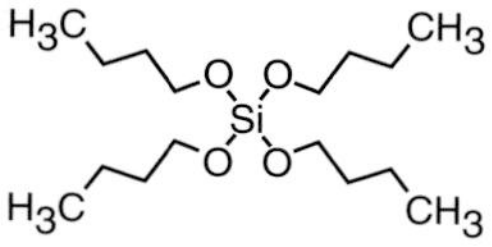
Tetramethoxysilane (TMOS) CAS 681–84–5	Tetraethoxysilane (TEOS) CAS 78–10–4	Tetrapropoxysilan (TPOS) CAS 682–01–9	Tetrabutoxysilane (TBOS) CAS 4766–57–8
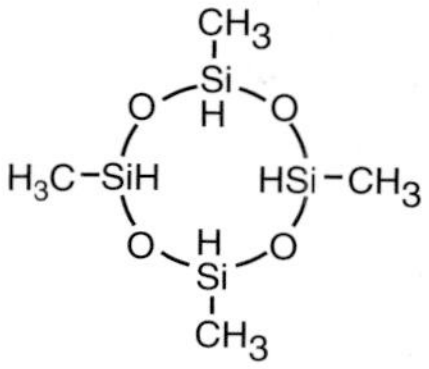	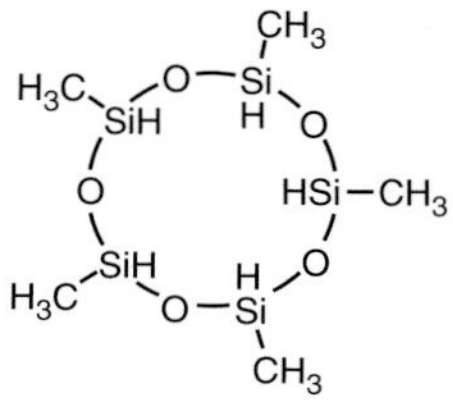	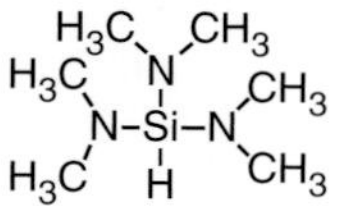	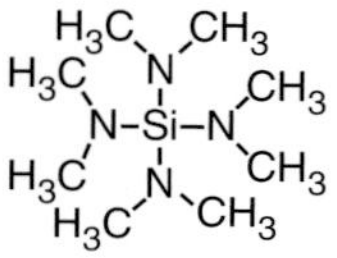
Tetramethylcyclo-tetrasiloxanes (TMCTS) CAS 2370–88–9	Pentamethylcyclo-pentasiloxane (PMCPS) CAS 6166–86–5	Tris(dimethylamido) silane (TriDAS) CAS 15112–89–7	Tetrakis (dimethylamido) silane (TetraDAS) CAS 1624–01–7

Table 21.3 Properties of silicon precursors frequently used in CVD.

Precursor	Physical condition at 20°C	Boiling point (b.p.)	Vapor pressure	Extreme caution
SiH_4	gas	−112°C	–	pyrophorous
TMS	liquid	27°C	231 mbar @20°C	pyrophorous
TICS	liquid	186°C	<3 mbar @25°C	moisture sensitive, toxic
$SiCl_4$	liquid	57,6°C	560 mbar @37°C	moisture sensitive, toxic
TMOS	liquid	121°C	116 mbar @20°C	moisture sensitive, toxic
TEOS	liquid	168°C	<1 mbar @20°C	moisture sensitive
TPOS	liquid	94°C@7 mbar	~0.1 mbar @20°C	moisture sensitive
TBOS	liquid	275°C	~0.06 mbar @20°C	moisture sensitive
TMCTS	liquid	134°C	~14 mbar @20°C	–
PMCPS	liquid	54°C@13 mbar	~1.8 mbar @20°C	–
TriDAS	liquid	145°C	~9 mbar @20°C	–
TetraDAS	liquid	196°C	~1.4 mbar @20°C	–

Table 21.4 Oxygen precursors frequently used in CVD.

oxygen (O_2)	ozone (O_3)	Nitrous oxide (N_2O)	Water (H_2O)	Hydrogen peroxide (H_2O_2)
$O=O$ 120.74 pm	O 1.278Å 116.8°	$N≡N-O^-$ / $^-N=N^+=O$	O 95.84 pm 104.45° H H	H O—O H
CAS 7782–44-7	CAS10028–15-6	CAS10024–97-2	CAS 7732–18-5	CAS 7722–84-1

Table 21.5 Properties of oxygen precursors frequently used in CVD.

Precursor	Physical condition at 20°C	Boiling point	Vapor pressure at room temp.	Extreme caution
O_2	gas	b.p. −112°C	–	
O_3	gas	b.p. −112°C	–	decomposes, not storable
N_2O	gas	b.p. −88°C	5,15 bar @20°C	
H_2O	liquid	b.p. 100°C	23 mbar @20°C	
H_2O_2	liquid	b.p. 150°C	2 mbar @20°C	explosive vapors

1.4 OXYGEN PRECURSORS

Some silicon precursors are already sufficiently coordinated with oxygen, but carbon contamination from the ligand is an issue. Other silicon precursors, such as silane, contain no oxygen at all to allow the formation of silicon oxide. For this reason an additional oxygen species is often introduced to achieve a complete and quantitative conversion to contamination-free silicon oxide. The most popular oxidizing agents are listed in Tables 21.4 and 21.5.

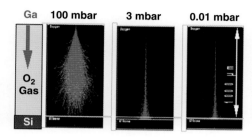

FIGURE 21.3: "TRIM" Simulation of 50 kV Ga$^+$ beam traveling through 1 mm oxygen gas. Up to 100 μm scattering of ions by gas molecules is probable and depends on the pressure [15].

2 FIB INDUCED DEPOSITION OF SILICON OXIDE

2.1 PROCESS DEVELOPMENT

Focused ion beams are intensively used for device modification by local material removal and ion beam induced metal deposition. The first FIB-induced formation of a silicon oxide was reported by Komano et al. in 1989 [12]. They used a 60 kV Si^{2+} focused ion beam (FIB) and a mixture of TMOS and oxygen. In the 1990s Ga$^+$ focused ion beams were increasingly used to deposit silicon oxide, extending the Si-precursors to TMCTS and TEOS.

The secondary electron emission caused by the impinging ions is stated as deposition mechanism [13]. The achievable resolution in deposition was investigated by an adsorption rate model [14]. The lateral resolution depends on the energy and mechanism of molecule dissociation, on the surface diffusion, and on the adsorption-desorption rate of the precursor. The scattering of ions in the precursor gas also affects the resolution (Figure 21.3).

Deposition with the precursor TMOS (flux $1.4 \times 10^{+17}$ molecules/cm^2) was performed by a 30 kV Ga$^+$ beam with beam currents between 6 and 7,000 pA [16]. The density of one adsorbed monolayer of TMOS was estimated to be $2 \times 10^{+14}$ molecules/cm^2 and the cross-section for deposition σ was 2×10^{-13} cm^{-2}. For deposition with a Ga$^+$ beam, the chemical composition of these deposits was determined as 43% Ga, 22% Si, 20% O, and 15% C, utilizing Auger electron spectroscopy.

The deposition process can also be influenced by the scan process. With a digital scan generator the loop time, the pixel spacing, and the dwell time (exposure time per pixel per scan) can be varied. A sufficiently long loop time was found to be essential to ensure full precursor saturation of the scan area between scan cycles (Figure 21.4). With increasing dwell time the sputter process became dominant due to the consumption of adsorbed precursor (Figure 21.5). Dwell times exceeding 5 μs resulted in a sputtered crater. With TMOS a dwell time below 0.2 μs and a loop time above 4 ms was found to be the optimum, yielding a maximum deposition rate of 0.33 μm^3/nC [16]. Similar results were also observed for TMCTS precursor. A longer dwell time on one spot reduced the deposition rate. For the identical ion dose per area irradiated, the smaller pixel spacing resulted in lower deposition rates (Figure 21.6). This is an indicator that with TMCTS at a chamber pressure of 1×10^{-5} mbar for a precursor, limited growth regime prevails.

2.2 MATERIAL PURITY

To achieve a high material purity and a stoichiometric composition of Si:O $= 1:2$, the implantation of ions originating from the focused beam is a critical factor. Currently, most

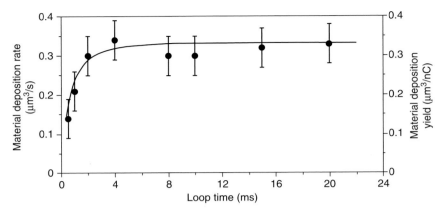

FIGURE 21.4: Silicon oxide deposition rate vs. loop time for ion beam induced deposition with dwell time of 0.2 μs, a 0% beam overlap, and a current density of 13 A/cm². Reprinted with permission from [16]. Copyright (1996), AIP.

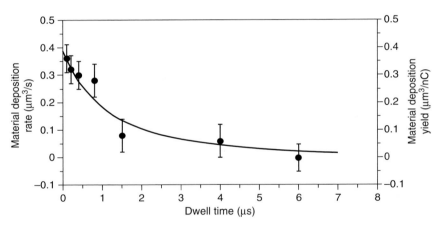

FIGURE 21.5: Silicon oxide deposition rate vs. dwell time for ion beam induced deposition (loop time = 10 μs, current density 13 A/cm²). The dots show experimental values, the solid line represents model calculations. Reprinted with permission from [16]. Copyright (1996), AIP.

commercial FIB systems are based on a liquid metal ion source generating a Ga^+ beam. The implantation depth of the 30 keV Ga^+-ions in SiO_2 exceeds 60 nm [17]. When depositing silicon oxide films, the atomic intermixing also contaminates the deposited material with backscattered substrate material. The SIMS-spectrum in Figure 21.7 shows the 100 nm wide interface zone of As-enriched siliconoxide at the interface to the GaAs substrate. A constant level of Ga originating from the ion beam also was found throughout the entire layer of Ga^+-FIB deposited silicon oxide.

With pure TMCTS as precursor but without additional oxygen, Ogasawara et al. [26] obtained SiO_x with a ratio of Si:O = 1:1.7. For a precursor gas mixture of TMCTS:O_2 = 0.35:1.46, the solid material had an ideal ratio of Si:O = 1:2. Gallium inclusions were below

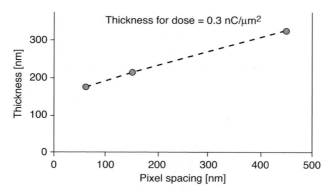

FIGURE 21.6: Correlation of deposited thickness of FIB-deposited silicon oxide with the pixel spacing. Thickness for dose = 0.3 nC/μm2. Modified from [17].

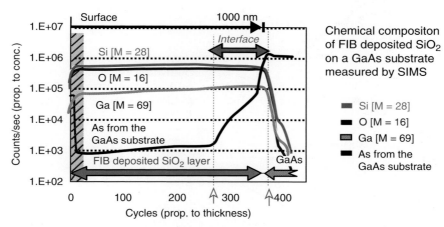

FIGURE 21.7: Depth profile of an FIB-deposited silicon oxide pad on top of a GaAs substrate. The analysis was performed by SIMS. The interface layer displays species of both materials. Modified from [17].

6% and carbon contamination was below the limit of detection of AES. It was found that the carbon content depended on oxygen used as additive gas.

With the same precursor TMCTS, the chemical composition in dependence on the process parameters was investigated [17]. The Si/O-ratio was comparatively stable at varying beam-related parameters, but Ga and C contamination was significantly influenced. For a larger pixel spacing, a lower concentration of Ga contamination was observed. The lowest carbon content of deposited material was observed for a short dwell time and a large pixel spacing (Figure 21.8). With a higher precursor saturation, also higher deposition yields were observed.

The distribution of Si and O in the deposition was related to the mixture ratio of TMCTS and oxygen in the gas phase. Control of the gas composition allows one to tailor the Si:O ratio of the FIB-deposited material (Figures 21.9 and 21.10). Using pure siloxane without additional oxygen, a non-stoichiometric silicon oxide was obtained. The addition of oxygen resulted in the growth of thicker silicon oxide layers with a higher oxygen content. A higher total pressure in the chamber and a higher siloxane partial pressure allowed the production of silicon oxides

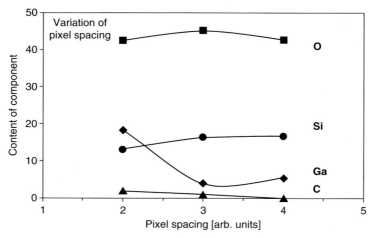

FIGURE 21.8: Chemical composition of FIB-deposited silicon oxide as a function of distance between exposed spots (= deposition parameter "pixel spacing"). Modified from [17].

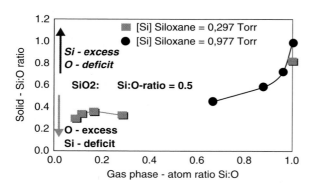

FIGURE 21.9: Effect of mixing ratio of TMCTS:O_2 in the gas phase on the composition of FIB deposited silicon oxide. Modified from [17].

with an over-stoichiometric Si concentration (Figure 21.9). Further addition of oxygen in the gas phase led to a saturation effect and had no significant changes on the Si:O ratio. The minimum of the gallium concentration was observed for a O_2/TMCTS = 15:1 ratio in the gas phase (Figure 21.10). This correlated with the fastest deposition rate where implantation of Ga is lowest.

A high material purity of silicon oxide was only achieved with an Si^{2+} ion beam generated from an Au-Si metal ion source utilizing a mass filter. Silicon dioxide was deposited using an Si^{2+}-ion beam in oxygen atmosphere [18]. The focused 15 kV beam was decelerated down to 100 V before the impact on the sample. For direct deposition the Si is provided by the ion beam itself and no Si-precursor is required. Without oxygen added, almost pure silicon was deposited. In a 10^{-8} to 10^{-7} mbar oxygen atmosphere, silicon oxide was formed. The O/Si ratio increased with higher oxygen background pressure and longer irradiation exposure. Oxygen in the film was chemically bonded with silicon and, according to Raman spectroscopy, it formed silicon-oxygen bonds. Due to the absence of precursor ligands and of Ga^+-ions, a high material purity of the silicon oxide was achieved.

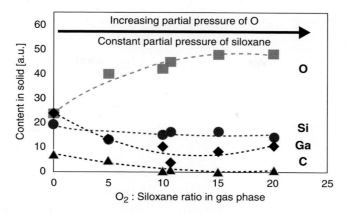

FIGURE 21.10: Effect of mixing ratio of TMCTS:O_2 in the gas phase on the composition of solid silicon oxide deposited by FIB. Modified from [17].

2.3 ELECTRICAL PROPERTIES

Silicon oxide as insulator is desired for FIB-induced repair of integrated circuits. For silicon oxide deposited with an Si^{2+} FIB from TMOS and oxygen, a resistivity of 2.5 MΩ.cm was measured for a 100 nm thick silicon oxide layer, and the breakdown voltage was 40 V. The silicon oxide contained little carbon contamination [12]. For silicon oxide deposited by a Ga^+ FIB using TMOS as precursor, the incorporated gallium impairs the electrical properties. An electrical resistivity of only $5 \times 10^{+3}$ Ωcm was evaluated with a $50 \times 2 \times 2\,\mu m^3$ track. A high Ga concentration of 43% was found in the deposited solid. The low resistivity is presumably caused by the high gallium content [16]. Silicon oxide deposited with a 25-kV gallium focused ion beam from TMCTS and oxygen consisted of mainly silicon and oxygen [19]. Micro-Auger electron spectroscopy indicated a silicon to oxygen ratio of 1:2 and less than 5% gallium and no detectable carbon. The resistivity of 28–79 MΩcm was remarkably high. Electrical characteristics were also measured with a SiO_2/DLC hetero-structure. The resistivity of this SiO_2 was measured to be in the range of 2.03 MΩcm [20] using a silicon oxide air-wire as test vehicle.

The atomic intermixing at the interface to a conductive material compromised the insulating properties in the interface layer of the silicon oxide. A silicon oxide layer from TMCTS and oxygen was deposited on an Al contact pad and covered with a FIB-deposited tungsten top electrode [21]. With this metal-insulator-metal capacitor structure, a satisfactory insulation is obtained for structures exceeding 300 nm thickness (Figure 21.11). Images of cross-sections and SIMS depth profiles show that an interface layer enriched with metal atoms of the electrodes was formed. Impurities in the deposited silicon oxide are mainly caused by implantation of Ga originating from the ion beam and by carbon caused by the incomplete decomposition of organic precursors. Silicon oxide layers exceeding 300 nm thickness showed an increasing resistance of the layer with growing thickness (Figure 21.11).

2.4 OPTICAL PROPERTIES

Silicon dioxide in the form of amorphous quartz glass is used in photolithography masks. The deposition is a desirable process for phase shift mask repair (Figure 21.12). Ga-staining

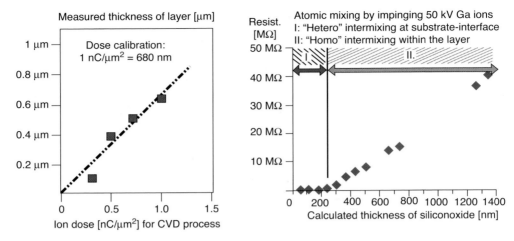

FIGURE 21.11: Capacity and leakage current through a FIB-deposited MIM-capacitor. Modified from [21].

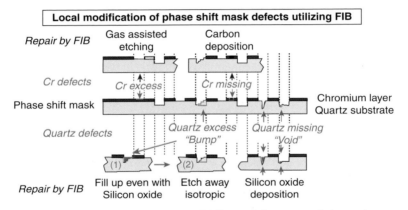

FIGURE 21.12: Schematic illustration of different defect types of alternating aperture phase shift masks and corresponding repair strategies utilizing a focused ion beam (FIB). While the repair of Cr defects by FIB is an established technology, repair of quartz defects is still a challenge.

reduces the transmission of quartz photomasks, but Ga-staining could be reduced by gas-assisted etching with XeF_2 [22; 23; 24]. A vacuum heat treatment at 1400°C for 10 minutes also could reduce the contamination caused by the Ga^+-FIB processing [25].

Silicon oxide deposited from TMCTS and oxygen was characterized for its optical transparency at 248 nm by an Aerial Image Measurement System (AIMS). Narrow pixel spacing with overlapping beam spots produces a uniform material thickness and a smooth surface (Figure 21.13 and 21.14), but the deposition efficiency is lower and the Ga content is higher with the small pixel spacing. In contrast to highly transparent original quartz areas, low transparency in FIB-processed areas also was observed (Figure 21.15). The low transparency occurred both with quartz milled by the Ga^+ ion beam and with FIB-deposited silicon oxide film. This indicates that the reduction of transmission is a problem caused by the implantation of Ga into the substrate.

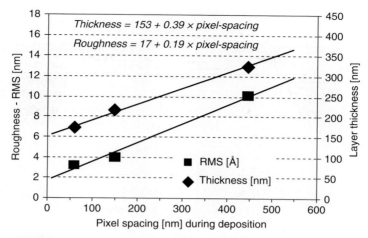

FIGURE 21.13: Correlation between the selected pixel spacing, resulting layer roughness, and film thickness.

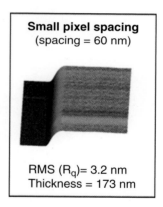

Small pixel spacing
(spacing = 60 nm)

RMS (R_q)= 3.2 nm
Thickness = 173 nm

FIGURE 21.14: AFM-image of silicon oxide deposited with narrow pixel spacing.

Ogasawara et al. reported on the deposition of oxide (SiO_x) film with high deep-UV transmittance [26]. The transmittance of silicon oxide from TMCTS added with a constant pressure of ∼0.5 torr was measured with a microscopic spectrometer. The transmittance depends on both the source gas and ion beam irradiation conditions (see Figures 21.16 and 21.17). A higher transmittance was achieved when the ion energy is lower and the source gas pressure is lower. For a total pressure of 1.5 torr, material deposited by a 5 keV beam showed a transmittance of 88%. Absorption at (5.5 eV) and at (4.7 eV) is attributed to an oxygen deficit in the Si–O–Si network [51]. An absorption peak at 390 nm was suggested to relate to the inclusion of gallium. Thermal annealing in nitrogen and hydrogen at 900°C strongly improved the optical transmission at 250 nm up to 90% for a 300 nm thick silicon oxide film.

2.5 STRUCTURES AND APPLICATIONS

Silicon oxide is used as an insulating structure in microelectronic devices (Figure 21.18). FIB-induced deposition of silicon oxide was first reported by Nakamura et al. [27] for filling a hole. Either an Si^{2+} or a Be^{2+} ion beam was scanned through the hole area, resulting in an extruded

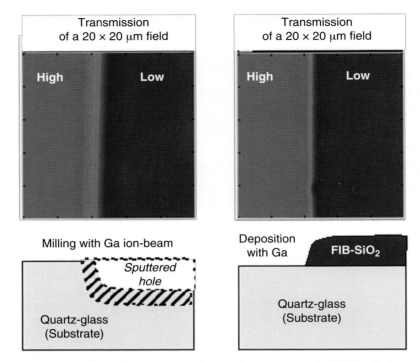

FIGURE 21.15: Optical transparency of pure substrate next to (left) FIB-milled silicon dioxide and (right) FIB-deposited silicon oxide, measured by AIMS at 248 nm. Modified from [17].

FIGURE 21.16: Transmittance spectra of silicon oxide films. For direct comparison the transmittance is recalculated to a normalized film thickness of 0.3 μm. Solid line = after deposition; broken line = after annealing in 400°C O_2; broken–dotted line = annealing in 800°C O_2, dotted line = annealing in H_2/N_2 at 900°C. Reprinted with permission from [26]. Copyright (1996), AIP.

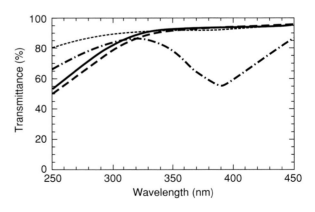

deposit resembling a pent roof. Vacancy formation for silicon oxide was observed and was attributed to the high growth rate of the extruded deposit. A field emission electron source was fabricated by deposition with a particle beam [28]. By chemical reaction of TEOS, an insulator of ring shape was deposited. The Pt gate structure was deposited in the insulator ring and the Pt emitter was deposited into the ring gate opening.

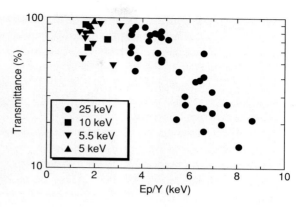

FIGURE 21.17: Correlation between the transmittance at 250 nm wavelength and the normalized ion energy. The transmittance is recalculated as Figure 21.16. The ion energy and the current are 25 keV, 10–75 pA (●), 10 keV, 10–20 pA (■), 5.5 keV, 20–68 pA (▼), and 5 keV, 10 pA (▲). Reprinted with permission from [26]. Copyright (1996), AIP.

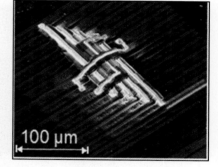

FIGURE 21.18: Typical lines of an interconnect layer with FIB processed multilevel rewiring. For illustrative purpose in this image, the dielectric layers are only partially covering the rewiring metal lines. Feature size of dielectric depositions is in the 1 μm range, adapting to the given architecture, but features down to 200 nm are feasible. Modified from [29].

Furthermore, mechanical nanomanipulators made of three-dimensional (3-D) silicon oxide nanostructures were fabricated by FIB deposition [20]. Nanotools, such as a bio-nanoinjector and an electrostatic nanomanipulator, were also produced from FIB-deposited SiO_2. An electrostatic 3D nanomanipulator with SiO_2/DLC hetero-structure was fabricated by FIB-CVD.

3 FEB INDUCED DEPOSITION OF SILICON OXIDE

3.1 PROCESS DEVELOPMENT

Fundamental investigations on the electron-induced deposition of silicon oxides reach back to the early 1980s, where mainly reaction mechanisms were studied with broad electron beams (Table 21.6). Thompson et al. have demonstrated the deposition via electron impact dissociation from SiH_4 and N_2O with a mixture ratio of $N_2O:SiH_4$=25:1 to 100:1 on heated Si substrates [30]. The current density of 0.04–0.40 A/cm^2 is low in comparison to modern field emitter scanning electron microscopes with current densities exceeding 400 A/cm^2. The dissociation of reactant gases occurred primarily in the region of the electron beam created plasma. Higher beam acceleration voltages resulted in higher electron beam current densities but also lower electron collision cross-sections, which finally led to a small decrease of the deposition rate [31; 32].

Table 21.6 Deposition parameters from several research papers in electron beam induced deposition of silicon oxide.

Lit. Ref.	Beam kV	Current (density)	Precursor Gas	Chamber pressure	Gas flux mol/cm²	Substrate and temp.
Thompson, 1983 [30]	5 kV	40–400 mA/cm²	SiH_4/N_2O	0.1–1.0 torr	n.a.	150°C to 450°C
Kunz, 1987 [33]	<1 kV	22–28 μA	oxygen	10^{-4} torr	n.a.	Si
Matsui, 1988 [36]	10 kV	1 μA	SiH_2Cl_2	7×10^{-7} torr	n.a.	
Lipp, 1996 [16]	10 kV	10 A/cm²	TMOS	n.a.	10^{17}/s	
Perentes, 2004 [46]		10 pA–1 nA	TEOS		n.a.	Si
Perentes, 2007 [37]	10 kV	125 nA	TMOS, TEOS, TMS, TICS	$<10^{-6}$ mbar	n.a.	Au on Si
Wanzenboeck, 2006 [39; 48]	1–10kV	1.4 nA	TMOS, TEOS, TMCTS, PMCPS	$<10^{-5}$ mbar	n.a.	Si

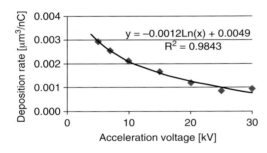

FIGURE 21.19: Deposition rate of silicon oxide correlated to the primary electron energy.

The electron beam induced oxidation of silicon substrates with electron beam energies below 1 keV has been presented by Kunz et al. using a broad electron beam. The thickness of the oxidized silicon was found to correlate to the secondary electron yield (see Figure 21.19) [33]. The maximum thickness growth (for identical doses) was observed at 10 eV, as this was the energy closest to the maximum of the dissociative electron attachment cross-section of oxygen at 6.7 eV [34].

Low acceleration voltage in the 1 kV regime also delivered the highest deposition rates in a scanning electron microscope. At 1 kV, silicon oxide deposition from TEOS resulted in a deposition rate of 0.009 μm³/nC (Figures 21.19 and 21.20) [35]. The interaction cross-section of adsorbed precursor molecules increases with decreasing energy of primary electrons, which indicates an involvement of secondary electrons.

The deposition of silicon with a growth rate of 0.03 nm/min was performed by decomposition of the inorganic precursors dichlorosilane SiH_2Cl_2 [36]. At the extremely low base pressure of 2×10^{-9} torr, carbon contamination by hydrocarbons in the residual gas was not an issue. 500 nm wide lines with nearly atomic level thickness control were demonstrated.

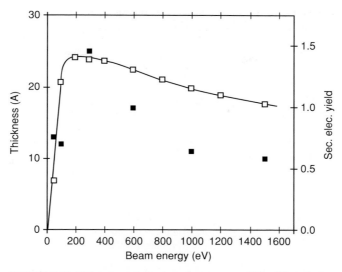

FIGURE 21.20: Thickness of electron-beam grown SiO_x (■) and of the secondary electron yield of the grown suboxide surface (□) plotted versus the electron beam energy. Oxides were grown at a constant electron dose of 0.02 C/m^2. The O_2 pressure was 10^{-4} torr [35], obtained from TEOS and O_2 as precursors. Reprinted with permission from [33]. Copyright (1987), AIP.

Lipp et al. [16] have investigated the electron beam induced deposition with TMOS. The precursor was heated to 40°C. The yield of silicon oxide deposition was dependent on the acceleration voltage and on the beam scanning parameters. For a 10 kV electron beam scanning with 6 ms loop time and 0.2 µs dwell time, the observed deposition yield was 0.016 $\mu m^3/nC$. For a 1 keV beam energy, a deposition rate of 0.15 $\mu m^3/nC$ was predicted for TMOS.

Also with TMOS the material deposition rate increased with decreasing beam energy according to the energy-dependence of the secondary electron yield. With increasing loop time (and 0.2 µs constant dwell time) the deposition yield increased but leveled off above 4 ms loop time (Figure 21.21). This indicates that a precursor gas flux of 10^{17} molecules/cm²s saturates the surface within 4 ms. With increasing dwell time (and a constant loop time of 10 ms and a gas flux of 1.4×10^{17} molecules/cm²s) the silicon oxide deposition yield decreases slightly due to the depletion of gas molecules on the surface (Figure 21.22).

Fischer et al. [35] have used TEOS to deposit silicon oxide. At small pixel spacing, the beam overlap resulted in a lower precursor coverage, resulting in lower deposition rates. Increasing the pixel spacing could triple the deposition rate (Figure 21.23). The deposition rate decreased with increasing dwell time (Figure 21.24). For longer exposure of a single spot, the precursor consumption is faster than precursor replenishment.

The influence of the TEOS gas flow was investigated by varying the pressure in the GIS supply line (Figure 21.25). In the regime of small precursor flows, an increase in the flow resulted in higher deposition rates. Above a threshold value in the large flow regime, the further increase in precursor flow had no further effect on deposition rates.

A comparison between the Si-precursors TEOS, TMOS, TMCTS, and PMCPS was performed. The precursors have a different sticking coefficient and a different number of

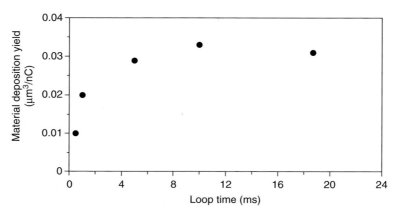

FIGURE 21.21: Material deposition yield vs. loop time for electron beam induced silicon oxide deposition with a dwell time of 0.2 μs, 0% beam overlap, and 10 keV beam energy. Reprinted with permission from [26]. Copyright (1995), AIP.

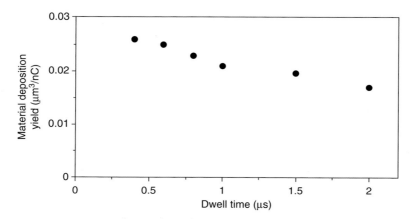

FIGURE 21.22: Material deposition rate vs. dwell time for electron beam induced silicon oxide deposition with a loop time of 10 ms, 0% beam overlap, and 10 keV beam energy. Reprinted with permission from [16]. Copyright (1996), AIP.

FIGURE 21.23: Dependence of the deposition rate of silicon oxide on the pixel spacing [35]. TEOS was deposited with a 5 kV electron beam (4 nA) with a dwell time of 0.5 μs.

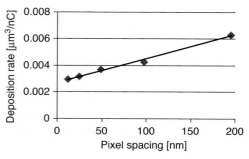

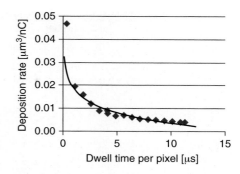

FIGURE 21.24: Dependence of the deposition rate of silicon oxide on the dwell time. TEOS was deposited with a 5 kV electron beam (4 nA) with 184 nm pixel spacing. Modified from [35].

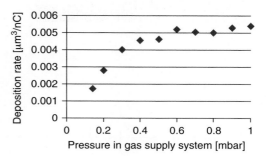

FIGURE 21.25: Dependence of the TEOS precursor flow on the deposition rate. The precursor flow of TEOS was adjusted by the level of gas pressure in the supply lines of the gas injection system. (5 kV electron beam with 4 nA beam current; pixel spacing was 46 nm, dwell time was 0.5 μs; total electron dose 10.7 μC.)

Table 21.7 Comparison of deposition rates with different Si-precursors. As oxygen precursor, O_2 was added to all precursor mixtures used to deposit silicon oxide.

	TEOS	TMOS	TMCTS	PMCPS
Si:O–atom ratio (in precursor mixture)	1:6	1:6	4:6	5:6
thickness of deposited layer [nm]	44	89	208	315
deposition rate [μm³/nC]	0.039	0.078	0.183	0.278
deposition speed [μm³/min]	2.32	4.68	15.38	23.2

Deposition parameters: 1 kV acceleration voltage, 0.375 μs dwell time, 152 nm pixel spacing, 50.000 loops, Si-precursor = 0.5 mbar, O-precursor = 1.5 mbar.

Si-atoms per molecule. The identical flow of molecular oxygen was added in all experiments. With PMCPS the highest deposition rate was achieved (see Table 21.7).

Besides the alkoxysilanes TMOS and TEOS, the alkylsilane TMS and isocyanatosilane TICS also were investigated [37]. The residual water in the SEM chamber was responsible for side reactions leading to carbon etching and oxygen incorporation during the electron-induced deposition. The deposition mechanism is based on a model of competitive coadsorption of (1) Si-precursor, (2) molecular oxygen, and (3) residual water. Si-precursor, H_2O, and O_2 compete for adsorbtion sites. Important parameters of each component are the molecule flow, the sticking coefficient [38], and the chemical reactivity (see Table 21.8).

The residual water is permanently present in the chamber with a stable base pressure of 3×10^{-7} mbar (re-adsorption rate 1 ML in 33 sec). TMS and TICS react rapidly with water

Table 21.8 Data on reactivity of Si-precursors extracted from [37].

Substance	O_2	H_2O	TMOS	TEOS	TMS	TICS
Residual gas pressure	3×10^{-8} mbar	3×10^{-7} mbar				
Flow [molecules. $cm^{-2}.s^{-1}$]	$9.5 \times 10^{+12}$ or <10 sccm	$1.1 \times 10^{+14}$	1.8×10^{18} -8.8×10^{20}	1.9×10^{18} -5.3×10^{19}	1.5×10^{20} -2.8×10^{21}	8.4×10^{17}
Sticking coefficient	low	very high	high	high	low	
Reaction with						
alkoxysilanes	high	low	–	–	–	–
alkylsilanes	low	medium	–	–	–	–
with O_2	–	–	high	high	low	low
water (H_2O)	–	–	medium	medium	high	very high

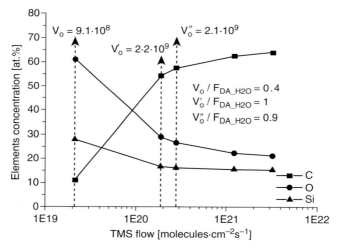

FIGURE 21.26: Chemical composition of silicon oxide as a function of the TMS flow. V_O is the oxygen deposition rate (atoms.s^{-1}) at different precursor flows, and $F_{DA}(H_2O)$ the residual water collision flow in the deposition area (molecules.s^{-1}). Reprinted with permission from [37]. Copyright (2007), AIP.

but have a low reactivity with molecular oxygen. Hence, the oxygen found in the deposit of an O_2-free deposition from TMS originates from residual water vapor. At high TMS flows, the water adsorbed on the surface is displaced by TMS, so that the oxygen content in the deposit decreases. The presence of water also reduced the carbon contamination (Figure 21.26). The addition of O_2 had negligible impact on the deposition from TMS and from TICS.

With TMOS and TEOS, but also TMCTS and PMCPS [39], the addition of O_2 led to a significant increase of deposition rate. Up to a 7-fold increase of silicon oxide deposition rate was observed due to the addition of O_2 (Figure 21.27). An O_2:Si-precursor ratio was

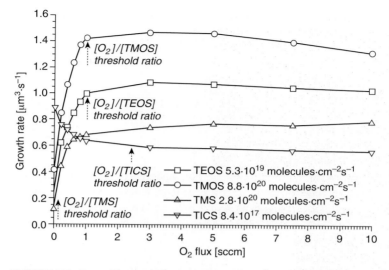

FIGURE 21.27: Deposition growth rates of $40 \times 50\,\mu m^2$ films as a function of the additional O_2 flow at fixed Si-precursor flow injected. The arrows indicate the O_2 flow above which no more C is detected in the FEBID material. Reprinted with permission from [37]. Copyright (2007), AIP.

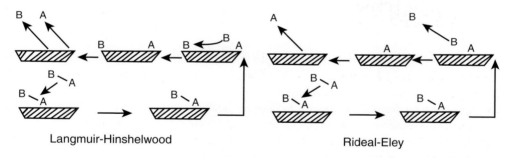

FIGURE 21.28: Surface reaction mechanisms for reaction $A - B \rightarrow A + B$. Modified from [40].

determined as threshold (Figure 21.27), so that at higher O_2 concentrations no further change of deposition rate was experienced. Only with TICS any adsorbed oxygen rather blocks surface sites and leads to a rate decrease [37].

With TEOS and TMS, the dwell time and the replenishment time had no significant effect on the C contamination, but with addition of oxygen the carbon content was significantly reduced. For the precursors TMOS, TEOS, TMS, and TICS, a longer dwell time always resulted in a higher C contamination—even in the presence of oxygen. Additional oxygen reduces carbon in the deposit.

With the pure alkoxysilanes TMOS or TEOS (without additional O_2), the chemical composition of the deposit was independent of the precursor pressure. With alkoxysilanes, water vapor plays no dominant role with deposition reaction, but the addition of molecular oxygen O_2 up to a threshold of 1 sccm led to a significant increase of the deposition rate (see Figure 21.27). Above 1 sccm, a saturation was observed so that no change of deposition rate was observed by further increasing the flow above this threshold level. This indicates a Langmuir-Hinshelwood mechanism (Figure 21.28).

3.2 MATERIAL PURITY

The decomposition of volatile precursors often results in the deposition of materials contaminated by ligands or residual gas species. Frequent impurities are carbon, oxygen, hydrogen and halogenides [41; 42].

To avoid carbon contamination from the ligands, carbon-free precursors were used. Deposition from dichlorosilane as silicon precursor without additional oxygen resulted in a composition of 89 at.% Si, 9 at.% O and 2 at.% Cl. The absence of an intentionally added oxygen precursor resulted in a composition that is far from a stoichiometric silicon oxide [36]. Silicon oxide was also deposited from a carbon-free SiH_4 precursor with N_2O as oxygen precursor [30]. The presence of characteristic SiO_2 peaks and the absence of the Si-O-H peak (3650 cm^{-1}) and the Si-H peak (2270 cm^{-1}) in the infrared spectrum indicate a complete reaction turnover, as no unsaturated bonds were detected. Although no elemental analysis is available, electrical and optical data indicate that a silicon oxide of reasonable purity was fabricated from this carbon-free SiH_4/N_2O precursor mixture. Under stoichiometric silicon oxide SiO_x ($x < 2$) could also be obtained without using any Si-precursor at all, just by electron beam induced oxidation of solid silicon in the presence of oxygen (O_2 pressure 10^{-4} torr) [33]. Carbon contamination from the chamber atmosphere was incorporated as impurity in the oxidized silicon.

Frequently, organometallic precursors are used for SiO_2 deposition. The organic ligands tend to cause contamination, especially with carbon, which influences the refractive index [43] as well as the electrical or insulating properties of the deposit produced [44; 45]. As Si-precursor siloxanes, alkylsilanes and isocyanatosilanes were used. The addition of a suitable oxygen precursor has always proven to decrease carbon contamination of the deposit. The composition is strongly influenced by the selected precursor, by beam parameters (acceleration voltage, beam current density, scan parameters), as well as by precursor-related parameters (gas flow, pressure range, nozzle distance, precursor coverage of the surface) that varied strongly among the experiments reported in the literature. Table 21.9 gives an exemplary overview of material purities reported in the literature; however, a direct comparison is not feasible due to different process parameters used.

For silicon oxide deposited from TMOS the chemical composition of the deposited silicon oxide was determined as 42% silicon, 40% oxygen, and 18% carbon by Auger electron spectroscopy [16]. Another study utilizing TMOS without additional oxidizing agents resulted in a strong presence of carbon in the deposited structure [43]. However, the absence of absorption peaks between 2800 cm^{-1} and 3000 cm^{-1} in the infrared spectrum indicates very low concentrations of hydrocarbon groups (i.e., $-CH_3$) in the deposit so that the 33 at% carbon are assumably present as silicon carbide (SiC) or silicon oxycarbide (SiOC).

With TEOS as precursor, 500 nm high pillars with 320 nm diameter were deposited at a chamber pressure of 2×10^{-5} mbar. EDX analysis revealed an atomic percentage of near 16% Si, 46% O, and 39% C [46]. With TMOS and TEOS, the residual water in the chamber cannot provide a sufficient oxygen supply. Addition of O_2 as oxygen precursor allowed the near elimination of the carbon contamination in the deposit (Figure 21.29). A comprehensive study on TMOS, TEOS, TMCTS, and PMCPS as Si-precursors proved that with an excess of

Table 21.9 Overview of selected silicon oxide depositions performed with a focused electron beam.

Literature reference	Precursor	% Si	% O	%C	Others	Si:O
Matsui, 1988 [36]	$SiCl_2H_2$	89	9	–	–	1:0.1
Thompson, 1993	SiH_4/N_2O	n.a.	n.a.		–	–
Lipp, 1998 [16]	TMOS	42	40	18	–	1:0.95
Perentes, 2006 [43]	TMOS	18	49	33	–	1:2.7
Wanzenboeck, 2006 [39]	$TMOS + O_2$	33	66	1	–	1:2
Perentes, 2004 [46]	TEOS	16	46	39	–	1:2.9
Perentes, 2007 [46]	TEOS	n.a.	n.a.	1	–	
Wanzenboeck, 2006 [39]	TEOS	27	56	17	–	1:2.0
Wanzenboeck, 2006 [39]	$TEOS + O_2$	32	67	1	–	1:2.0
Perentes, 2007 [47]	TICS	14	42	30	14% N	1:3
Perentes, 2007 [47]	$TICS + O_2$	30	69	1	–	1:2.3
Perentes, 2007 [37]	TMS ($2 \times 10^9 s^{-1}$)	15	22	63	–	1:1.5
Perentes, 2007 [37]	TMS ($9 \times 10^8 s^{-1}$)	28	61	11	–	1:2.2
Perentes, 2007 [37]	TMS + O2	n.a.	n.a.	16	–	
Wanzenboeck, 2006 [39]	$TMCTS + O_2$	32	66	1	–	1:2
Wanzenboeck, 2006 [39]	$PMCPS + O_2$	33	66	1	–	1:2

O_2 added to the process, all four of these Si-precursors allowed the deposition of silicon oxide with an almost stoichiometric composition of 33% Si and 67% O [35; 48].

With TMS a composition of 29% Si, 61% O, and only 10% carbon was obtained without additional oxygen (Figures 21.26 and 21.29). Deposition from TICS was performed with a controlled oxygen flux simultaneously injected [47]. A 300x excess of $[O_2]$ over [TICS] was required to obtain pure SiO_2. The chemical composition depends on the $[O_2]/[TICS]$ flux ratio (Figure 21.29). The hydrogen content was below 1 at.%.

With all precursors, the carbon concentration became lower with an increasing replenishment time and a correspondingly decreasing dwell time. The least carbon residues in the deposits were obtained at large replenishment times (above 50 μs) and low dwell times (below 15 μs) (see Figure 21.30) [37].

3.3 ELECTRICAL PROPERTIES

For testing of the electrical characteristics of the silicon oxide, a metal-insulator-metal (MIM) capacitor setup was frequently used. A dielectric constant of $\varepsilon_0 = 3.8$ obtainable with thermal oxides can be considered as reference value.

Silicon oxide deposited at a 250 mTorr from SiH_4/NO_2 precursor had a dielectric constant of $\varepsilon_0 = 3.47$ to 3.61 [30]; the best silicon oxide ($\varepsilon_0 = 3.61$) was obtained at 350°C substrate temperature. An overview is given in Table 21.10 [30]. With electron beam induced deposition, the process is typically performed under ambient conditions (20°C) as heating stages are not standard in an SEM.

Silicon oxide deposited from TMOS (10^{17} molecules/cm²s gas flux) at ambient conditions also displayed an electrical conductivity of 1×10^6 Ωcm [16]. From a capacity measurement

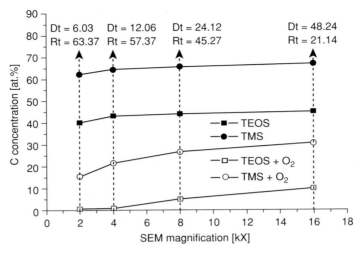

FIGURE 21.29: Carbon contamination after deposition with/without O_2. Precursor flow of: TEOS $= 5.9 \times 10^{19}$ molecules.cm^{-2}.s^{-1}; TMS $= 2.8 \times 10^{20}$ molecules.cm^{-2}.s^{-1} The dwell and replenishment times were varied. Reprinted with permission from [37]. Copyright (2007), AIP.

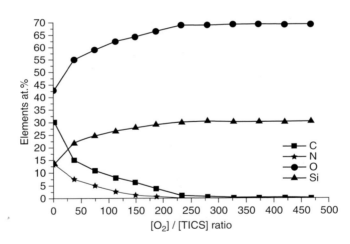

FIGURE 21.30: Chemical composition of FEBID silicon oxide as a function of the $[O_2]/[TICS]$ molecular ratio. Values obtained by EDX investigations, given at ±5 at.%. Reprinted with permission from [37]. Copyright (2007), AIP.

of a metal-insulator-metal-capacitor setup with a $100 \times 100\,\mu$m top electrode and a 170 nm thick dielectric layer of silicon oxide deposited from TMOS and O_2, an average capacitance of 2.41 pF was measured. The calculated $\varepsilon_r = 4.6$ is higher than the reference $\varepsilon_r = 3.9$ of bulk silicon oxide, which is due to the electric field at the edge of small capacitor electrodes [39].

Brezna et al. [49] have used a microstructured MOSCAP device to characterize the dielectric properties of electron beam deposited silicon dioxide with scanning capacitance microscopy. Silicon dioxide was deposited with TMCTS and O_2 on a p-doped Si wafer

Table 21.10 Electrical properties of SiO$_2$ films deposited from SiH$_4$/N$_2$O on heated substrates, with permission from [30], Appl. Phys. Lett., 43, 777. Copyright (1983), AIP.

Substrate temp (°C)	MIS sample location	N_{sub} (cm^{-3}) ×10^{-14}	T_{na} (Å)	C_{FB} (F/cm^2) ×10^{-8}	V_{FB} (V)	N_{fixed} (#/cm^2) ×10^{10}	Pinhole density (#/cm^2)	V_B (MV/cm)	$\epsilon_{0\%}$	$n_{0\%}$
250	1	10.9	1415	1.77	1.11	28	600	1.4	3.58	1.467
250	2	10.10	1458	1.71	1.39	31			3.58	1.465
250	3	8.8	1430	1.62	1.09	25			3.31	1.467
250	4		1420					3.3		
350	1	6.8	1949	1.32	−0.13	8	760	2.1	3.61	1.469
350	2	6.9	1974	1.23	−0.36	5			3.35	1.467
350	3		1978					3.8		
450	1	8.4	1599	1.53	2.55	41	760	4.4	3.47	1.468
450	2	6.6	1646	1.48	2.41	39			3.54	1.466
450	3	7.3	1665	1.49	2.59	41			3.57	1.467
450	4		1665					2.8		

(acceptor concentration $N_A = 9.4 \times 10^{14}$ cm^{-3}) as dielectric layer. With a conductive AFM-cantilever acting as top electrode, locally resolved scanning capacitance spectroscopy could be performed (Figures 21.31 and 21.32). The total oxide charge density of SiO$_2$ was determined. Silicon oxide as deposited had a significant number of trapped charges in the oxide layer. The total oxide charge is all negative, indicating that electrons may be injected and trapped in the oxide. After a short annealing procedure (600°C for 30 seconds), the negative (electron) charges remain trapped inside the oxide and the total oxide charge number density (N_{tot}) changed only barely. Only thermal annealing above 600°C reduced the number of negative oxide charges trapped in the silicon oxide (see Figure 21.33). Only for thin oxide layers and annealing at 800°C could the oxide charges be significantly reduced.

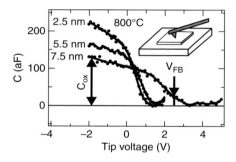

FIGURE 21.31: C(V) spectra recorded on the three SiO$_2$ pads after the second annealing step (60 s at 800°C, forming gas). The inset illustrates how the measurement was performed with an AFM tip as the top electrode.

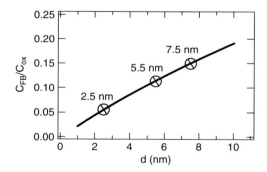

FIGURE 21.32: Normalized flat band capacitance vs. oxide thickness d. One-dimensional MOS theory was applied to calculate the flat band (FB) capacitance values. C$_{ox}$ is the oxide capacitance.

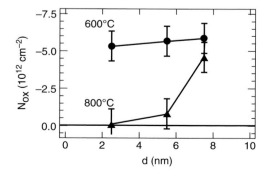

FIGURE 21.33: Total oxide charge number density N_{tot} for the three measured SiO$_2$ pads after the first annealing step for 30 s at 600°C and after the second step for 60 s at 800°C, both in forming gas.

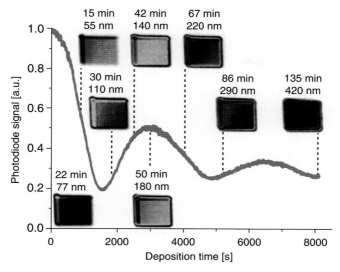

FIGURE 21.34: Reflectometry curve as a function of time for deposition with TMOS. Insets are microscope images of the deposits ($140 \times 120\,\mu m^2$) with corresponding deposition time and thickness. Reprinted with permission from [43]. Copyright (2006), AIP.

3.4 OPTICAL PROPERTIES

The optical properties of deposited silicon oxide are relevant for repair and modification of photo masks and nano-imprint lithography stamps. Transparent properties in UV and deep UV (193 nm) are of particular interest [31]. The incorporation of precursor ligands—especially carbon contaminations—affects adversely the optical transmission. Also, the refractive index n_0 of the deposited material with $n_0 = 1.54$ of bulk quartz as reference value is of relevance.

Silicon oxide deposited on a hot substrate from SiH_4/N_2O had a refractive index of 1.46+/−5%, which is close to the value of stoichiometric SiO_2 [30]. A comparison for different depositions is given in Table 21.9, but SiH_4/N_2O mixture ratios from 30:1 to 100:1 had negligible effects on the optical properties. With wet etching in 5:1 buffered HF, the silicon oxide obtained by electron beam induced deposition displayed much higher etch rates (3.6 to 6 nm/sec.) than thermally grown SiO_2 (1.0–1.5 nm/s).

FEB-induced deposition from TMOS as sole precursor (without additional oxygen) resulted in silicon oxide with up to 33% carbon content. The refractive index at 514 nm wavelength was measured as n $= 1.56$ (± 0.06) $+ i0.14$ (± 0.03) in vacuum and as n $= 1.67 + i0.18$ measured under ambient conditions. Without oxygen added during the deposition, the reflectometry curve of silicon oxide shows a high absorption: 90% of the signal is lost already at the second-order constructive peak (Figure 21.34). The presence of oxygen or residual water during the deposition process can improve the quality of the deposit in these cases, because water vapor reduces the co-deposition of carbon.

Silicon oxide deposited from TMCTS and oxygen also displayed a good transparency. The optical properties were investigated at 248 nm by an aerial image measurement system (AIMS) (see Figure 21.35). A homogeneous but decreased transmission in the deposited area was observed. The peaks and valleys at the side slopes were caused by geometric effects

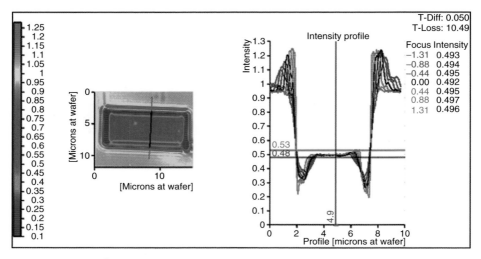

FIGURE 21.35: AIMS measurement at 248 nm showing the transmission of a rectangular-shaped silicon-oxide layer on a quartz carrier. The material was e-beam deposited with a silicon precursor with additional oxygen added. The thickness of the layer was 1045 nm. Modified from [48].

of the sample. Deposition from siloxane with a surplus of oxygen was 1045 nm thick and displayed an intensity loss of 50%. The highest optical transmission at 193 nm was obtained with the organometallic precursors TMS deposited in the presence of water. Adding different amounts of oxygen to these precursors resulted in pure silicon oxide with a high transmission at 193 nm wavelength. With TICS, also a hydrogen contamination in the deposits could be avoided.

3.5 STRUCTURES AND APPLICATIONS

Electron beam induced deposition of silicon oxide was also used to fabricate micro- and nanostructured devices. With the deposition of silicon oxide pillars (320 nm diameters, 500 nm height) from TEOS, a dependence of the structure on the beam current and on the beam diameter was observed [46]. For increasing beam current, the growth rate increases only by a 0.25 power law relation. At a 2 pA electron current, a growth rate of 20 nm/s was observed, while a 1 nA beam current resulted in a growth rate of 8 nm/s (see Figure 21.36). The deposition is in the electron-limited regime.

A two-dimensional photonic band gap (PBG) structure consisting of silicon oxide pillars (Figure 21.37) has been produced by FEBID with TEOS as precursor [46]. To avoid beam drift on the insulating silicon oxide substrate, the surface was covered with a 10 nm thin conductive antimony-doped tin oxide layer. The structure was written in a spiral manner. For better optical properties, the structures were annealed in a furnace at 1000°C for 10 hours. At $\lambda = 632$ nm a complex refractive index of $n = 1.85 \pm i3.9 \times 10^{-4}$ was obtained.

Nanopore-based sensors with locally functionalized nanochannels have been fabricated by direct deposition to silicon oxide [50]. Silicon oxide deposition from TEOS and water was used to reduce the diameter of previously drilled holes (Figure 21.38). From a slope of 0.5 nm per scan, the variation of the dose allowed the adjustment of the diameter of the nanopores.

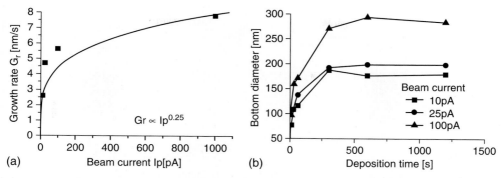

FIGURE 21.36: Dynamic and geometrical properties of FEBID tips with TEOS on silicon substrate. (a) Growth rate as function of beam current and (b) bottom diameter as function of deposition time and beam current. Reprinted with permission from [46]. Copyright (2004), Elsevier.

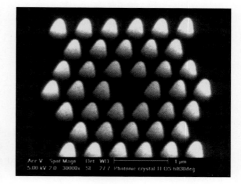

FIGURE 21.37: SEM image of the PBG structure (30° tilted view angle). The hexagonal lattice and micro-cavity are well observable. Reprinted with permission from [46]. Copyright (2004), Elsevier.

By retracting the gas nozzle away from the sample, a lower surface coverage rate was obtained so that a slower, more controlled deposition was achieved. These nanopore structures may find application in high-speed DNA-sequencing.

4 COMPARISON BETWEEN FIB- AND FEB-INDUCED DEPOSITION

The principle of silicon oxide deposition with a focused ion beam (FIB) and a focused electron beam (FEB) are widely identical, as a locally confined chemical vapor deposition is initiated by secondary effects of the impinging particle beam. Differences between ions and electrons in their yield of inelastic effects such as secondary electron emission, secondary ion emission, backscattering, and beam heating are the reason for the different deposition rates. In general, FIB-induced processes provide a higher yield in silicon oxide deposition, but sufficient surface coverage with the precursor is a precondition as mechanical sputtering and hence material removal by ion bombardment is a permanent side reaction with FIB processes. The yield of electron beam induced deposition of silicon oxide films depends on the acceleration voltage and on the beam scanning parameters. For a 10 keV electron beam

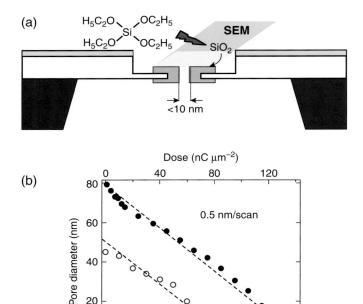

FIGURE 21.38: Nanopore fabrication by FEBIP showing (a) schematic illustration of the site-selective silicon oxide deposition of a FIB-milled nanopore; (b) diameter of nanopores as a function of the dose for two individual experiments. Reprinted with permission from [50]. Copyright (2006), American Chemical Society.

scanning with 6 ms loop time and 0.2 μs dwell time and a current density of 10 A/cm, the observed deposition yield was 0.016 μm^3/nC. With the same system using a focused ion beam a 20-times higher deposition rate of 0.33 μm^3/nC was obtained [16].

The advantages of Si oxide FIB deposition are a) higher deposition rates than compared to FEB deposition and b) if a Si^{2+} ion beam would be used, the deposition of silicon oxide would proceed without the contamination with additional elements from the focused ion beam (as would be the case for a gallium FIB).

The advantages of FEB-induced deposition comprise a) a well-defined interface between substrate and deposited silicon oxide (no atomic intermixing as caused by ion impingement) and b) no implantation of extraneous material originating from the particle beam.

As precursors, carbon-free precursors (such as SiH_4 or $SiCl_2H_2$) or organometallic precursors such as alkylsilanes (TMS, TICS) or alkoxysilanes (TMOS, TEOS, TMCTS, PMCPS) are frequently used for silicon oxide deposition. The precursors are in general both applicable for FEB- and FIB-induced deposition. For selection of the precursor, the different sticking coefficients of the precursors (but also of their ligands) should be considered. Especially with small structures, the actual deposition rate is limited not by the yield of the deposition process but by the gas replenishment. Hence, a high precursor flow and polar precursor molecules (such as alkoxysilanes) with a high sticking coefficient on polar surfaces are advantageous.

Hydrocarbons and organic ligands of the precursor are sources of carbon, so that contamination of the deposited material by carbon is a widespread nuisance. With almost all precursors, the addition of a suitable oxygen precursor has proven to be advantageous for reduction or even elimination of carbon impurities in the deposited material. Water vapor present in the residual gas has to be considered as an oxygen species that contributes to the deposition process [37].

Optimization of process parameters has allowed the deposition of material with a stoichiometric Si:O=1:2 ratio both with FIB [17] and with FEB [37; 39; 47]. However, with FIB-induced deposition using a Ga^+ ion beam, a 10 to 45 at.% Ga-contamination of the deposited silicon oxide cannot be avoided. Thermal post-deposition treatment could slightly relieve this problem [26]. With FEB the additional implantation of Ga can be avoided [16; 19; 51]. Ion-induced deposition with TMOS precursor yielded 43% gallium and 15% carbon contamination, leaving only 22% silicon and 20% oxygen content of the deposited material [16].

The gallium contamination with FIB-deposited silicon oxide also influences the electrical and the optical parameters of the deposited silicon oxide. Silicon oxide deposited by FEB from TMOS at ambient conditions displayed an electrical conductivity of 1×10^6 Ωcm [16]. For comparison, deposition with FIB resulted in a 200-times higher conductivity of 5×10^3 Ωcm with FIB-deposited material. For optical devices the "Ga-staining" also reduces the optical transmission. Thermal annealing can improve the transmission. However, the high demands for deep-UV masks cannot be met currently by FIB-induced deposition.

With focused electron beams, the emergence of electron microscopes with high beam currents above 10nA and the future prospect for multibeam tools offer an opportunity for the advance of FEB-induced deposition of silicon oxide. With focused ion beams, a new generation of He-ion-microscopes and a renaissance of Si^{2+} ion beam systems can offer potential for future developments.

REFERENCES

[1] Kuntmana, Ayten, Rifat Yenidünyaa, Ahmet Kagözb, and Hakan Kuntmanc. A new study on spin-on-silica for multilevel interconnect applications. *Microelectronics Journal* 30.2 (1999): 127–131.

[2] *CRC Handbook of chemistry and physics.* Boca Raton, FL: CRC Press, 2001.

[3] Giunta, Carmen J., Jonathan D. Chapple-Sokol, and Roy G. Gordon. Kinetic modeling of the chemical vapor deposition of silicon dioxide from silane or disilane and nitrous oxide. *J. Electrochem. Soc.* 137.10 (1990): 3237–3253.

[4] Doering, Robert, and Yoshio Nishi. *Handbook of semiconductor manufacturing technology.* CRC Press, 2007.

[5] Gamble, L., M. B. Hugenschmidt, C. T. Campbell, T. A. Jurgens, J. W. Rogers, Jr. Adsorption and reactions of tetraethoxysilane (TEOS) on clean and water-dosed titanium dioxide (110). *Journal of the American Chemical Society* 115.25 (1993): 12096–12105.

[6] Cromwell, J. E., L. L. Tedder, H. C. Cho, F. M. Cascarano, M. A. Logan. *J. Electron. Spectrosc. Relat. Phenom.* 54/55 (1990): 1097.

[7] Desu, Seshu B., Decomposition chemistry of tetraethoxysilane. *Journal of the American Ceramic Society* 72.9 (1989): 1615.

[8] Janča, J., A. Tálský, and V. Zvoníček. Kinetics of O_2 + TEOS gas-phase chemical reactions in a remote RF plasma reactor with electron spin resonance. *Plasma Chemistry and Plasma Processing* 16.2 (1996): 187–194.

[9] Kim, Min Tae. Deposition kinetics of silicon dioxide from tetraethylorthosilicate by PECVD. *Thin Solid Films* 360 (2000): 60–68.

[10] Coltrin, M. E., P. Ho, H. K. Moffat, and R. J. Buss. Chemical kinetics in chemical vapor deposition: Growth of silicon dioxide from tetraethoxysilane (TEOS). *Thin Solid Films* 365.2 (2000): 251–263.

[11] Flores, Lucio D., and John E. Crowell. Boundary layer chemistry probed by in situ infrared spectroscopy during SiO2, deposition at atmospheric pressure from tetraethylorthosilicate and ozone, *J. Phys. Chem.* B 109 (2005): 16544–16553.

[12] Komano, Haruki, Youji Ogawa, and Tadahiro Takigawa. Silicon oxide film formation by focused ion beam (FIB)-assisted deposition. *Japanese Journal of Applied Physics*, Part 1 28.11: 2372–2375.

[13] Ohyaa, Kaoru, and Tohru Ishitani. Monte Carlo study of secondary electron emission from SiO2 induced by focused gallium ion beams. *Applied Surface Science* 237 (2004): 606–610.

[14] Utke, I., V. Friedli, M. Purrucker, and J. Michler. Resolution in focused electron- and ion-beam induced processing. *Journal of Vacuum Science and Technology* B 25.6 (2007): 2219–2223.

[15] Wanzenboeck, Heinz D., Stefan Harasek, Helmut Langfischer, Emmerich Bertagnolli, Herbert Hutter, and Herbert Störi. Local deposition on semiconductor surfaces with a focused beam. *Presentation at the 12. Meeting of Solid State Analytics*, Vienna, 22–24 September 2003.

[16] Lipp, S., L. Frey, C. Lehrer, B. Frank, E. Demm, S. Pauthner, and H. Ryssel. Tetramethoxysilane as a precursor for focused ion beam and electron beam assisted insulator (SiOx) deposition. *Journal of Vacuum Science and Technology* B 14.6 (1996): 3920–3923.

[17] Wanzenboeck, H. D., M. Verbeek, W. Maurer, and E. Bertagnolli. FIB based local deposition of dielectrics for phaseshift mask modification. *SPIE Proceedings* 4186, 20th Annual BACUS Symposium on Photomask Technology Proceedings (2001). Monterey, CA, USA.

[18] Yanagisawa, J., H. Nakayama, O. Matsuda, K. Murase, and K. Gamo. Direct deposition of silicon and silicon-oxide films using low-energy Si focused ion beams. *Nuclear Instruments and Methods in Physics Research*, Section B 127/128 (1997): 893–896.

[19] Komano, H., H. Nakamura, M. Kariya, and M. Ogasawara. Silicon oxide deposition using a gallium focused ion beam. *Proceedings of SPIE* (The International Society for Optical Engineering) 2723 (1996): 46–53.

[20] Kometani, R., T. Hoshino, K.-I. Nakamatsu, K. Kondo, K. Kanda, Y. Haruyama, T. Kaito, J.-I. Fujita, T. Ichihashi, Y. Ochiai, and S. Matsui. Fabrication of nano-manipulator with SiO2/DLC hetero-structure by focused-ion-beam chemical-vapor-deposition. *Digest of Papers – Microprocesses and Nanotechnology* 2004 (2004): 296–297.

[21] Wanzenboeck, H., S. Harasek, B. Eichinger, A. Gruen, M. Karner, M. Hetzl, and E. Bertagnolli. Direct-write deposition: A maskless prototyping technique. *Proceedings of Mikroelektronik* 2003 (Vienna).

[22] Nakamura, Hiroko, Haruki Komano, and Munehiro Ogasawara. Focused ion beam assisted etching of quartz in XeF_2 without transmittance reduction for phase shifting mask repair. *Japanese Journal of Applied Physics* 31.12 B (1992): 4465–4467.

[23] Matsumoto, Masami, Tsukasa Abe, Toshifumi Yokoyama, Hiroyuki Miyashita, Naoya Hayashi, and Hisatake Sano. Development of focused-ion-beam repair for quartz defects on alternating phase-shift masks. *Proceedings of SPIE – The International Society for Optical Engineering* 3096 (1997): 308–321.

[24] Fu, Yongqi and N. K. A. Bryan, Investigation of physical properties of quartz after focused ion beam bombardment. *Applied Physics* B80 (2005): 581–585.

[25] Youn, Sung Won, Masaharu Takahashi, Hiroshi Goto, and Ryutaro Maeda. A study on focused ion beam milling of glassy carbon molds for the thermal imprinting of quartz and borosilicate glasses. *J. Micromech. Microeng.* 16 (2006): 2576–2584.

[26] Ogasawara, M., M. Kariya, H. Nakamura, H. Komano, S. Inoue, K. Sugihara, N. Hayasaka, K. Horioka, T. Takigawa, H. Okano, I. Mori, Y. Yamazaki, M. Miyoshi, T. Watanabe, and K. Okumura. Beam induced deposition of an ultraviolet transparent silicon oxide film by focused gallium ion beam. *Applied Physics Letters* 732 (1995): 116724 (3p.)

[27] Nakamura, Hiroko, Haruki Komano, Kenji Norimatu, and Yoshio Gomei. Silicon oxide deposition into a hole using a focused ion beam. *Japanese Journal of Applied Physics*, Part 1 30.11B (1991): 3238–3241.

[28] Murakami, K., W. Jarupoonphol, K. Sakata, and M. Takai. Fabrication of nano electron source using beam-assisted process. *Japanese Journal of Applied Physics*, Part 1 42.6B (2003): 4037–4040.

[29] Wanzenboeck, H. D., S. Harasek, H. Langfischer, A. Lugstein, E. Bertagnolli, J. Brenner, C. Tomastik, and H. Störi. Rapid prototyping by local deposition of silicon oxide and tungsten nanostructures for interconnect rewiring. Presented at 47th International Symposium of the American Vacuum Society, Boston, Oct. 2000, p. 114.

[30] Thompson, L. R., J. J. Rocca, K. Emery, P. K. Boyer, and G. J. Collins, Electron beam assisted chemical vapor deposition of SiO2. *Appl. Phys. Lett.* 43.8 (1983): 777.

[31] Gressier, J. P. M., M. Krishnan, G. de Rosny, and J. Perrin. Production mechanism and reactivity of the SiH radical in a silane plasma. *Chemical Physics* 84.2 (1984): 281–293.

[32] Perrin, J., J. P. M. Schmitt, G. de Rosny, B. Drevillon, J. Huc, and A. Lloret. Dissociation cross-sections of silane and disilane by electron impact. *Chemical Physics* 73.3 (1982): 383–394.

[33] Kunz, R. R., and T. M. Mayer. Surface reaction enhancement via low energy electron bombardment and secondary electron emission. *Journal of Vacuum Science and Technology* B 5.1 (1987): 427–429.

[34] Rapp, D., and D. D. Braglia. Total Cross Sections for Ionization and Attachment in Gases by Electron Impact. II. Negative Ion Formation. *Journal of Chemical Physics* 43 (1965): 1480.

[35] Fischer, M., H. D. Wanzenboeck, J. Gottsbachner, S. Müller, W. Brezna, M. Schramboeck, and E. Bertagnolli. Direct-write deposition with a focused electron beam. *Microelectronic Engineering* 83.4–9 (2006): 784–787.

[36] Matsui, S., and M. Mito. Si deposition by electron beam induced surface reaction. *Applied Physics Letters* 53.16 (1988): 1492–1494.

[37] Perentes, A., and P. Hoffmann. Oxygen assisted focused electron beam induced deposition of Si-containing materials: Growth dynamics. *Journal of Vacuum Science and Technology* B 25.6 (2007): 2233–2238.

[38] de Boer, J. H. *The dynamic character of adsorption*, 2nd ed. Oxford: Clarendon, 1968.

[39] Wanzenboeck, H. D. Internal report of Vienna University of Technology, unpublished.

[40] Masel, Richard I. *Principles of adsorption and reaction on solid surfaces*, 1st ed. New York: Wiley Interscience, 1996.

[41] Utke, I., P. Hoffmann, B. Dwir, K. Leifer, E. Kapon, and P. Doppelt. Electron beam induced deposition of metallic tips and wires for microelectronics applications. *J. Vac. Sci. Technol.* B 18 (2000): 3168.

[42] Cicoira, F., K. Leifer, P. Hoffmann, I. Utke, B. Dwir, D. Laub, P. A. Buffat, E. Kapon, and P. Doppelt. Electron beam induced deposition of rhodium from the precursor $[RhCl(PF_3)_2]_2$: Morphology, structure and chemical composition. *J. Cryst. Growth* 265 (2004): 619.

[43] Perentes, A., T. Bret, I. Utke, P. Hoffmann, and M. Vaupel. Real-time reflectometry-controlled focused-electron-beam-induced deposition of transparent materials. *Journal of Vacuum Science and Technology* B 24 (2006): 587–592.

[44] Nakamura, K., T. Nozaki, T. Shiokawa, K. Toyoda, and S. Namba. Formation of submicron isolation region in GaAs by Ga focused ion beam implantation. *J. Vac. Sci. Technol.* B 5 (1987): 203.

[45] Edinger, K., J. Melngailis, and J. Orloff. A Study of Precursor Gases for Ion Beam Insulator Deposition. *J. Vac. Sci. Technol.* B 16 (1998): 3311.

[46] Perentes, A., A., Bachmann, M. Leutenegger, I. Utke, C. Sandu, and P. Hoffmann. Focused electron beam induced deposition of a periodic transparent nano-optic pattern. *Microelectronic Engineering* 73–74 (2004): 412–416.

[47] Perentes, A., P. Hoffmann, and F. Munnik. Focused electron beam induced deposition of DUV transparent SiO2. *Proceedings of SPIE - International Society for Optical Engineering* 6533 (2007): art. no. 65331Q.

[48] Wanzenboeck, H. D., M. Fischer, R. Svagera, J. Wernisch, and E. Bertagnolli. Custom design of optical-grade thin films of silicon oxide by direct-write electron-beam-induced deposition. *Journal of Vacuum Science and Technology* B 24.6 (2006): 2755–2760.

[49] Brezna, W., M. Fischer, H. D. Wanzenboeck, E. Bertagnolli, and J. Smoliner. Electron-beam deposited SiO2 investigated by scanning capacitance microscopy. *Applied Physics Letters* 88.12 (2006): art. no. 122116.

[50] Danelon, C., C. Santschi, J. Brugger, and H. Vogel. Fabrication and functionalization of nanochannels by electron-beam-induced silicon oxide deposition. *Langmuir* 22.25 (2006): 10711–10715.

[51] Duraud, J. P., F. Jollet, Y. Langevin, and E. Dooryhee, Surface modifications of crystalline SiO2 and Al_2O_3 induced by energetic heavy ions. *Nucl.Instrum.Methods* B 32 (1988): 248.

GROWTH AND CHARACTERIZATION OF FEB-DEPOSITED SUSPENDED NANOSTRUCTURES

Gian Carlo Gazzadi and Stefano Frabboni

1 INTRODUCTION

While the standard beam-induced deposition approach is to grow deposits on a substrate, deposition of self-supporting suspended structures can be obtained by slowly moving the ion or electron beam laterally from an elevated edge, in order to grow an overhanging deposit behind the beam path. An example of nanofabrication by lateral focused electron beam induced deposition (FEBID) is given in Figure 22.1, where a suspended nanowire (SNW) has been deposited between two vertical pillars [1]. The structure is obtained (Figure 22.1(a)) by moving the electron beam from the top of the left pillar toward the right pillar at a scan speed of 28 nm/s, with discrete steps of 5 nm, under Pt-organometallic gas flow. As shown by the lateral dimensions in Figure 22.1(b) and Figure 22.1(c), suspended deposition allows to obtain higher lateral resolution compared to on-substrate deposition, because of the lack of secondary emissions from the substrate, which enlarge the deposit footprint well above the beam spot size and are responsible for the so-called "halo effect."

A major capability offered by lateral deposition is three-dimensional (3D) nanofabrication. Matsui and colleagues demonstrated the potentiality of this technique with several examples of both functional (nanocoils, electrical circuit elements, nanogrippers) and artwork shapes (nanoglass) produced with focused ion beam induced deposition (FIBID) of carbon precursor [2].

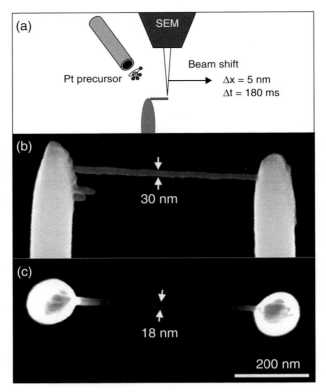

FIGURE 22.1: Pt SNW grown between pillars by lateral-FEBID.
(a) deposition procedure; (b) side-view SEM image;
(c) top-view SEM image. Reprinted with permission from
Reference 1.

By means of 3D FEBID, functional nanotools and nanodevices have been fabricated, including nanohooks for the manipulation of DNA fibers made with carbon SNWs [3], carbon-tip nanotweezers [4], and electron field-emission devices based on organometallics deposits [5]. Several works have been devoted to the study of the growth mechanism. Lateral resolution of 5 nm-width SNWs grown with a scanning electron microscope (SEM) from carbon contamination was compared to that of vertical pillars deposited on substrate, and it was concluded that the higher resolution in the former was due to the reduced secondary electrons generation within the structure [6]. The geometry of SNWs grown from W precursor with a transmission electron microscope (TEM) was studied as a function of the lateral scan speed, and upward-sloped, parallel, and downward-sloped SNWs were obtained as the speed increased [7]. Lateral FEBID on bulk substrates has also been investigated. The inclination of pillars deposited from Au and Mo precursors [8] and the periodicity of arch-like structures grown from Cu precursor [9] were studied as a function of the lateral scan speed. The geometry of suspended structures grown from Au precursor with an environmental SEM has also been investigated [10].

In the following paragraphs we will focus on lateral FEBID of suspended nanostructures (SNSs) from Pt-organometallic precursor, performed with an SEM. This precursor is commonly employed in dual-beam nanofabrication workstations for the deposition of conductive material.

We will cover the growth mode as a function of electron beam parameters, and the structural and compositional characterization of the deposited material by TEM analysis, along with a C removal treatment by high-current TEM irradiation. Finally, the electrical characterization of SNWs will be presented, involving dramatic structural modifications (Pt aggregation, C-matrix graphitization) when high current densities are reached.

2 LATERAL FEBID OF SUSPENDED NANOSTRUCTURES FROM Pt PRECURSOR

The growth of SNSs from Pt-organometallic precursor has been studied as a function of the electron beam energy (5, 10, and 15 keV) and of the electronic charge deposited per unit length (CDL, 1–9 pC/nm range), a parameter linked to the beam scan speed [11].

Depositions are performed with a Schottky-field emission SEM in a Dual Beam workstation (FEI Strata DB235M), combining a focused ion beam (FIB) and an SEM on a single platform. The apparatus is equipped with a gas injector system (GIS) containing trimethyl-methyl-cyclopentadienyl-platinum, $(CH_3)_3CH_3C_5H_4Pt$, precursor. Chamber pressure during depositions is $P_{Pt} = 5 \times 10^{-6}$ mbar, and GIS is kept in retracted position, about 4 cm distant from the sample, in order to reduce the gas flux at the sample for a better control on the growth rate, and also to avoid anisotropic growth effects from the flow direction. SNSs deposition is started by focusing the beam on the top of a Pt pillar oriented perpendicular to the electron beam (differently from Figure 22.1(a), where the pillar is oriented parallel to the beam). This geometry is chosen because it gives a minimum substrate volume (the 30 nm-diameter tip) to the impinging beam. The beam is shifted laterally over a range of 600 nm, with discrete steps (Δx) of 5 nm, and beam dwell time (Δt) is selected in order to obtain the desired CDL value. This quantity is the beam current divided by the scan speed, $CDL = I/(\Delta x/\Delta t)$ [pC/nm], and is adopted because, contrary to the simple scan speed, it allows a direct comparison between depositions with different beam energies, that have different beam currents. In fact, since we operated the SEM at a nominally constant spot size (~4 nm Gaussian width), beam current decreased with energy and was 41, 32, and 24 pA at 15, 10, and 5 keV, respectively. The spanned CDL range is between 1 and 9 pC/nm, and corresponds to lateral scan speeds between 27 and 3 nm/s, respectively.

In Figure 22.2, Pt SNSs grown at 15 keV, in a CDL range between 1.5 and 7 pC/nm, are shown under side view in Figure 22.2(a), i.e., along the pillar axis, and under top view in Figure 22.2(b), i.e., along the electron beam incidence during deposition. Let us consider Figure 22.2(a). For CDLs up to 2.2 pC/nm, corresponding to faster beam shifts, the deposit grows as horizontal SNWs, with length that progressively increases and approaches the full beam shift range as CDL increases. Average SNWs thickness is also increasing with CDL from 20 to 40 nm. The SNW lateral profile is thinner at the end, indicating that the amount of deposited material progressively reduces as the shift proceeds. As beam dwell time increases, and CDL is above 2.2 pC/nm, SNSs show a progressive inclination and thickening, reaching a slope of 55°, measured with respect to the horizontal axis, and a thickness of 65 nm for 7 pC/nm. Length, because of slope, exceeds the beam shift range. This shape evolution indicates that two growth regimes are present: a horizontal growth regime, below 2 pC/nm, and a sloping growth regime above. Top view images, in Figure 22.2(b), do not reveal the presence of this shape evolution: they just show an increasing width of the structures, from 15 to 30 nm, over

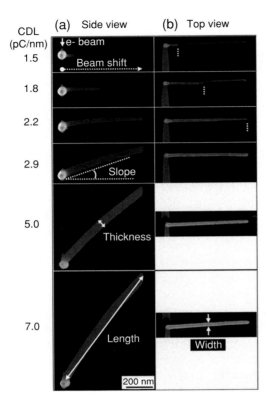

FIGURE 22.2: SEM images under side (a) and top (b) view of Pt SNSs deposited with electron beam energy of 15 keV and increasing CDL values (decreasing beam shift speed). Definitions of the geometrical parameters are indicated. Reprinted with permission from [11], J. Phys. Chem. B 1009, 16544. Copyright (2005), ACS.

the CDL range. Length approaches to the full beam shift range between 1.5 and 2.2 pC/nm (dashed lines mark the end of SNWs from the deposit below, on the substrate), and then keeps constant—as a horizontal projection—for higher CDLs.

Structures grown at lower beam energy, 5 and 10 keV, show the same qualitative trend and shape evolution as a function of CDL, with geometrical parameters all expanding as the energy decreases.

These parameters are summarized in Figure 22.3 as a function of CDL and energy. The occurrence of a transition between horizontal and sloping growth is particularly evident in the length data (Figure 22.3(a)). There are two distinct linear trends below and above the value of beam shift range (600 nm), which, for the 15 keV data, occurs at CDL ~2 pC/nm. Below this value, in the horizontal growth mode, length increases at a much higher rate than above, where sloping growth sets on. The horizontal/sloping transition is shifted to lower CDL as the energy decreases: around 1.5 pC/nm for 10 keV and 1.0 pC/nm for 5 keV. In Table 22.1, the conditions for horizontal and sloping growth are reported. Lengths in the sloping growth region increase with linear rates that are higher the lower the energy.

Width and thickness data are displayed in Figure 22.3(b). Widths start with similar values at low CDLs, then they follow distinct trends, slowly increasing with CDL and having higher values the lower the energy. Thickness data display a strong increase at the beginning, followed by a saturation to a steady value that, contrary to other geometrical parameters, is larger at higher energy. Slope data, in Figure 22.3(c), show absolute values that are inversely proportional to the energy, with a marked increase at low CDL that progressively weakens as CDL grows. We point out that these curves, when plotted as a function of beam shift speed (nm/s),

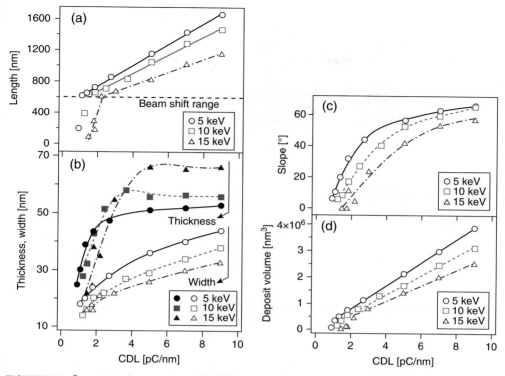

FIGURE 22.3: Geometrical parameters of Pt SNSs deposited at 5, 10 and 15 keV, displayed as a function of CDL: length (a), width and thickness (b), slope (c), and deposited volume (d). Lines in (a) and (d) are linear fits, curves in (b) and (c) are guidelines for the eyes. Reprinted with permission from Reference 11.

Table 22.1 Electron beam parameters (energy and current) and beam scan conditions (CDL and corresponding beam shift speed) for horizontal and sloping growth of the SNSs.

Beam Energy, Current	Horizontal growth		Sloping growth	
	CDL (pC/nm)	Speed (nm/s)	CDL (pC/nm)	Speed (nm/s)
$E = 5$ keV, $I = 24$ pA	≤ 0.9	≥ 26.5	1.1–9.0	22–2.5
$E = 10$ keV, $I = 32$ pA	≤ 1.2	≥ 26.5	1.5–9.0	21–3.5
$E = 15$ keV, $I = 41$ pA	1.4–2.0	30–20	2.7–9.0	15–4.5

are similar to those reported in the literature for similar suspended structures [10]. Finally, deposited volume (Figure 22.3(d)), calculated as the product of length, width and thickness, has a linear trend with CDL and increases as the energy is lowered, indicating that deposition efficiency is inversely proportional to the primary electron energy. Average deposition yields, calculated from data on the high-CDL side, are 710, 570, and 470 nm^3/pC for 5, 10, and 15 keV, respectively.

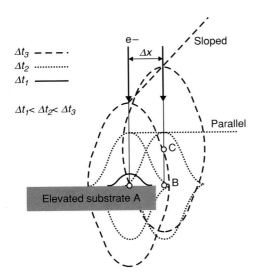

FIGURE 22.4: Model of suspended deposition by lateral FEBID. A Gaussian electron beam impinges on a thin edge with increasing dwell times (Δt) and then is shifted to the right by a fixed amplitude (Δx). No suspended depostion (Δt_1), parallel (Δt_2), and upward-sloped (Δt_3) depositions are obtained as dwell time is increased.

The growth mechanism at the basis of suspended deposition is sketched in Figure 22.4. We follow a similar approach to the one proposed by Liu et al. [7], but in this case we consider a fixed beam shift amplitude and variable dwell time. A Gaussian electron beam impinges at point A on a thin edge with increasing dwell times ($\Delta t_1 < \Delta t_2 < \Delta t_3$), i.e., increasing CDLs, and then is moved to the right by a constant amplitude Δx. As the dwell time increases, volume and width of the deposit increase, and shape changes from Gaussian-like (Δt_1, Δt_2) to pillar-like (Δt_3). Deposit below the edge is due to primary and secondary electron emission from the bottom face of the thin substrate. When the beam is shifted laterally by Δx it strikes on nothing for Δt_1, on the edge of the deposit (point B) for Δt_2 and well inside the deposit profile (point C) for Δt_3, giving rise to no suspended deposition, parallel deposition, or upward-sloped deposition, respectively. The growth process is ruled by the interplay of several parameters such as beam dwell time, lateral shift amplitude, deposition rate, current, energy, and shape of the beam.

Analyzing the data of Figure 22.3, we observe that overall dimensions (except thickness) and the deposited volume increase as electron energy is decreased. This can be explained by the enhanced contribution of low-energy secondary electrons to the deposition process. These electrons, whose intensity is enhanced as the primary energy is reduced, are very efficient in dissociating the precursor molecules since their energy is peaked around the typical molecular binding energies (few tens of eV). The same qualitative trend of deposition yield vs. beam energy has been reported for on-substrate growth [12]. The thickness trend is opposite because it follows the penetration range of primary electrons that increases with energy. An increase of the geometrical parameters is also observed as CDL increases. At very low CDL (i.e., for faster beam shifts), two effects are present: SNW length is below the beam shift range, and lateral profile becomes thinner at the end. These observations suggest that surface diffusion of precursor molecules over the suspended structure may represent a limiting factor for growth, as the length of the structure increases [13]. The condition for growing horizontal SNWs, of particular interest for the fabrication of suspended wirings, occurs for CDL values between 1–2 pC/nm. The transition from horizontal to sloping growth mode and the thickness increase as CDL increases are well explained by the model in Figure 22.4, where longer beam dwell times produce wider and taller deposits.

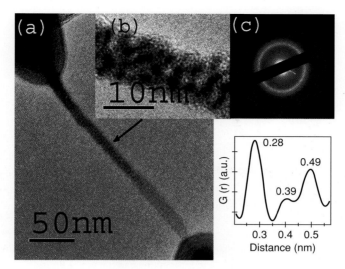

FIGURE 22.5: (a) BF TEM image of as-deposited Pt SNW.
(b) Structural details from higher magnification BF TEM image.
(c) SAED pattern typical of amorphous/nanocrystalline material.
(d) RDF extracted from the SAED pattern with atomic distances
labeling the peaks. Reprinted with permission from Reference 14.
Copyright (2006), American Institute of Physics.

3 CHARACTERIZATION OF Pt SUSPENDED NANOWIRES BY TEM ANALYSIS

Structural and compositional characterization of Pt SNWs have been performed by high-resolution (HR) TEM analysis, using selected area electron diffraction (SAED), electron energy loss spectroscopy (EELS), and energy dispersive X-ray spectroscopy (EDS) [14].

TEM analysis is performed with a JEOL 2010 microscope (point resolution 0.19 nm, LaB_6 emitter) equipped with energy-filtering system for EELS, and an EDS system. All the analyses are performed at 200 keV beam energy. EELS spectra are acquired in diffraction mode, with spectrometer dispersion of 0.3 eV/pixel, acceptance angle of 10 mrad, and beam convergence angle of 6 mrad. In order to enable the TEM measurements, sample is prepared as follows. Sub-micron wide slits are opened by FIB milling on an ultrathin Si substrate; then, pillars are deposited on the opposite sides of the slit and the SNW is deposited between them, thus exposing its full length to TEM observation through the slit. Depositions are performed with beam energy of 15 keV, using the conditions for horizontal growth described in Section 2.

In Figure 22.5(a) a low magnification bright field (BF) TEM image of the wire is shown: the wire is about 250 nm long and 20 nm wide. A magnified image of the central part of the wire and the relative SAED pattern are shown in Figure 22.5(b) and Figure 22.5(c), respectively. BF images show the morphology and structure of the nanowires: they are formed by a low electro-optical density amorphous matrix (bright contrast) and nano-areas (~2nm wide) with high electro-optical density (dark contrast). The SAED pattern displays broad diffraction rings typical of amorphous/nanocrystalline materials. Similar morphology and structure have been already observed in thicker films prepared with substrate-mediated FEBID [15; 16]. In order to obtain quantitative structural data, we performed analysis of the SAED pattern of Figure 22.5(c)

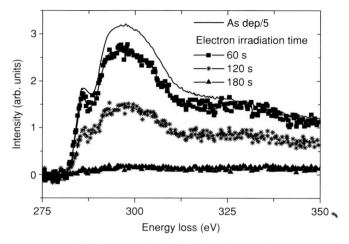

FIGURE 22.6: EELS spectra of the C K edge taken on the as-deposited Pt SNW and after increasing times of electron irradiation. Vanishing of the carbon signal after 180 s irradiation is evident. Reprinted with permission from Reference 14. Copyright (2006), American Institute of Physics.

in terms of radial density function (RDF), G(r), which is reported in Figure 22.5(d). The RDF can be derived by Fourier sine transform of diffracted intensities and describes the fluctuation of the local density with respect to its average value. Therefore, RDF peaks reveal the most probable atomic distances. We measure a first neighbor distance of (0.28±0.01) nm, a second neighbor distance of (0.39±0.01) nm, and a third neighbor distance of (0.49±0.01) nm. Since all these distances reproduce well the first, second, and third neighbor distances of Pt fcc cubic unit cell (a = 0.3919 nm), we can deduce a nanocrystalline arrangement of the Pt atoms inside the grains dispersed into the amorphous carbon (a-C) matrix. EDS analysis provides an estimation of the mean composition to be 70 at.% for C and 30 at.% for Pt, which is in agreement with results for on-substrate deposition [15]. SAED patterns provide little information on the amorphous C material due to the big difference in total elastic scattering cross-section between C and Pt. On the contrary, EELS is the elective tool for this kind of elemental/structural analysis.

In Figure 22.6, the C K-edge of an as-deposited wire (solid line) is reported. A sharp pre-edge peak (285 eV), followed by a broader one (290–302 eV), is clearly visible in the spectrum. The pre-edge peak is ascribed to the $1s \rightarrow \pi^*$ transitions typical of sp^2-hybridized C, whereas the broad peak is ascribed to $1s \rightarrow \sigma^*$ transitions [17]. The edge shape looks very similar to those observed in evaporated amorphous carbon films characterized by large sp^2/sp^3 ratio (>80%). A sequence of spectra taken at the center of the wire after 60 seconds electron irradiation steps, using a current density of $50A/cm^2$ and a spot size of about 50 nm, is also reported in Figure 22.6. A decrease of the C-K signal without appreciable edge modification is clearly detected after the first two irradiation steps. After 180 seconds irradiation, the C K-edge has vanished. Energy filtered images (taken with the zero loss peak) showing the structural modification of the wire during irradiation are reported in Figure 22.7.

In particular, Figure 22.7(a) has been recorded after 60 seconds irradiation of the central part of the wire (arrowed): the removal of a-C on the lateral edge and the growth of Pt grains are evident. The image recorded after five 180 seconds irradiations, performed on neighboring areas, in order cover the whole wire length, is reported in Figure 22.7(b).

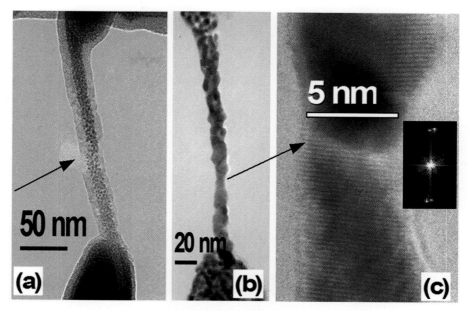

FIGURE 22.7: (a) TEM image after 60 s irradiation at the wire center (arrowed). (b) TEM image after a sequence of five 180 s irradiations on neighboring areas, in order to remove a-C over the whole wire length. (c) High resolution image and corresponding Fourier transform of a grain junction showing the (111) Pt lattice fringes. Reprinted with permission from Reference 14. Copyright (2006), American Institute of Physics.

The complete removal of the a-C matrix and the formation of a continuous polycrystalline Pt wire are clearly observed. In Figure 22.7(c), an HRTEM image shows lattice fringes spaced by (0.231±0.006) nm, a mean value from different grains that agrees quite well with {111} Pt crystallographic planes spacing (0.226 nm). The experimental evidence of a-C removal and growth of the Pt grains can be explained invoking different mechanisms based on the energy transfer between the high-energy electrons and the wire: heating, knock-on effects on the organic matrix [18], and formation of C-O compounds with residual oxygen gas or water vapor.

4 ELECTRICAL CHARACTERIZATION OF Pt SUSPENDED NANOWIRES

A critical issue of the FEBID technique is the low purity of metal deposits, especially those obtained from organometallic precursors. The deposited material generally consists of metal nanograins embedded into a carbonaceous matrix, due to incomplete dissociation of the molecules, and this determines an electron-hopping conduction mechanism that dramatically reduces the conductivity and is responsible for non-metallic behavior at low temperature [19]. Electrical characterization of Pt-organometallic deposits has been carried out mainly on micron- and submicron-size wires deposited on substrates, often in combination with thermal treatments for C removal [20; 21; 22]. These studies have shown a linear current (I)-voltage (V) behavior at room temperature (RT) for narrow bias ranges (hundreds of mV), with resistivity (ρ)

ranging between 10^{-1} and 10^{-3} Ωm, to be compared with the 10^{-7} Ωm of pure Pt. Thermal annealing in vacuum or C-reactive atmosphere (oxygen, forming gas) have shown to improve ρ between 10^{-3} and 10^{-5} Ωm due to C removal and Pt grains aggregation. Promising results have been obtained by depositing organometallic Au in a high water vapor pressure (0.4 Torr) with an environmental SEM. Polycristalline Au wires with resisitvity just half the one of pure Au were produced [10].

In this section we present an electrical characterization of Pt SNWs, exploring first the low voltage range (± 0.1 V) and the effect of water vapor addition during depositon [23]. Afterward, we investigate the electrical behavior over an extended bias range (up to 3–5 Volts) and show that a dramatic structural evolution occurs in the SNWs when high current densities are involved [24].

4.1 LOW VOLTAGE RANGE

To enable electrical measurement, pairs of contact electrodes are patterned on Au-coated silicon nitride membrane by iodine-assisted FIB patterning [25]. Electrodes are separated by a submicron-wide slit opened through the membrane by FIB, and the SNW is grown across the slit on short Pt pillars. This sample design, besides TEM analysis, allows cleaner electrical measurements as it eliminates spurious deposition (halo effect) into the electrodes' gap. SNW growth from Pt-precursor is performed with the same conditions described previously. Water vapor is added in some depositions, at a pressure of 5×10^{-5} mbar. I–V measurements are performed *in situ* at RT, with a two-probe configuration, using W probes attached to nanomanipulators.

A typical SNW deposited from Pt precursor with the addition of water vapor is shown in Figure 22.8(a). It has average dimensions of 590 nm in length, 29 nm in thickness, and 18 nm in width. Deposited material looks homogeneous because the limited SEM resolution does not resolve the nanogranular structure. I–V measurements, performed by sweeping the voltage over a $(0, 0.1, -0.1, 0$ V) loop, are shown in Figure 22.8(d). The first I–V cycle shows Ohmic behavior with a resistance (R) of 190 kΩ, corresponding to a resistivity of 1.8×10^{-4} Ωm. The SEM image taken after measurement (Figure 22.8(b)) reveals changes in the SNW. An homogeneous grainy structure becomes evident, with bright Pt grains of average size around 5 nm, embedded into the dark C-matrix. The second I–V cycle shows again a linear characteristic with decreased resistance (90 kΩ, $\rho = 8 \times 10^{-5}$ Ωm). From a structural point of view, as shown in Figure 22.8(c), Pt grains undergo further coalescence, becoming as large as 15–20 nm and separated by even larger gaps. The C matrix maintains its profile, and it is clear that it represents the supporting frame of the SNW. We notice that the larger Pt grains are concentrated at the center of the wire, while at both ends they are smaller and uniformly distributed. From the inset of Figure 22.2c, showing a top-view image, a marked bending of the SNW at the centre is also evident. These observations strongly call for a Joule heating effect, where the maximum temperature is expected at the center of the wire [26]. Electrical behavior of SNW deposited without water vapor is strictly similar, with higher resistivity values of 4.0×10^{-4} and 2.5×10^{-4} Ωm measured after first and second I–V cycle, respectively. However, structural modifications are much less evident: Pt grain coalescence is absent after the first I–V cycle and slightly visible after the second one. All the measured resistivities are much lower than those typically reported for on-substrate depositions, and become comparable to the values reached after carbon removal treatments. Water vapor addition is found to slightly improve conductivity and to favor the structural evolution of the SNWs under I–V cycling,

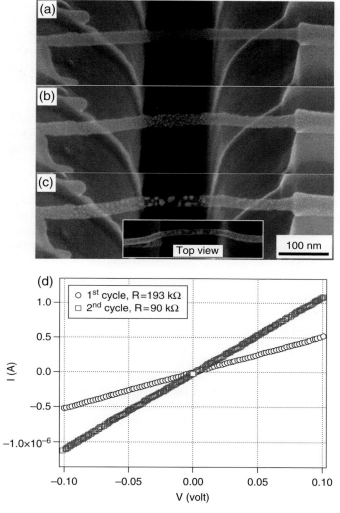

FIGURE 22.8: Tilted-view SEM images of a SNW deposited from Pt precursor with water vapour assistance: (a) as-deposited, (b) after the first and (c) second I–V cycle, shown in (d). I–V cycles are measured over (0, 0.1, −0.1, 0 V) voltage loop. Inset in (c) shows a top-view image of the wire. Reprinted with permission from [23]. Copyright © 2008 Elsevier B.V.

though not in the direction of obtaining a carbon-free metal wire. To this end, much higher vapor pressures should be used to compare with results on Au [10]. A major question arises if the improving conductivity for subsequent cycles is considered against the structural evolution. How can the conductivity improve when the metal grains are evolving toward more isolated units instead of a more continuous structure? There must be a primary role of the C matrix in the electrical conduction. This is also suggested by works on amorphous carbon nanowires with embedded metal nanoparticles, reporting a C graphitization process (graphitic carbon (g-C) is conductive whereas a-C is insulating) when the wires are subject to high current density [27] or temperature [28].

4.2 EXTENDED VOLTAGE RANGE: STRUCTURAL EVOLUTION AND GRAPHITIZATION PROCESS

To investigate the structural evolution, and to clarify the role of the C matrix in the electrical conduction, we performed live SEM imaging during the electrical test, extending the bias range to few volts. In addition, we employed TEM analysis (*ex situ*) to observe the structural details after the most significant transition stages.

In Figure 22.9, the structural evolution of a pillar-supported SNW under electrical measurement is shown through a sequence of video frames taken from real-time SEM imaging. I–V curve (Figure 22.9(a)) is measured in the (0, 4 V) range, with a sweep rate of 0.03 V/s at steps of 0.03 V. Figure 22.9(b) shows the SNW at the beginning of measurement: it is 450 nm long and has a 20 nm average diameter. The deposited material presents a bright homogeneous look, and in the inset the nanogranular structure is shown by the TEM image. I–V curve has a linear trend up to 1 V, then it bends up following the typical shape of hopping-transport systems, but no changes in the wire structure from the as-deposited condition are apparent. At (12.9 μA, 2.1 V) a step of 4.7 μA is recorded in current, and, as shown in the corresponding video snapshot (Figure 22.9(c)), a structural transition occurs in the SNW. Platinum aggregates into large nanograins (10–15 nm), mostly located in the central region, where a bending of the wire can also be noticed. The coalescence mechanism is sudden, and completes within a single I–V reading (1 s). Above 2.1 V, current keeps increasing with a polynomial dependence, and structure of the SNW remains stable over a wide voltage range (until 29.0 μA, 2.7 V), as shown in Figure 22.9(d). Here another transition is observed: the aggregated grains are progressively depleted by Pt migration toward the wire ends. The video snapshot in Figure 22.9(e) is taken a few seconds after onset of depletion process: the larger grains have shrunk considerably but are still visible, whereas the smaller ones have disappeared. We notice that the absolute position of the grains is not changing during the process, indicating that material depletion occurs from the periphery of the grain. The final stage of depletion process is shown in Figure 22.9(f): the central region is completely empty of platinum, and new big grains have formed at both ends, as the bright spots indicate. From the live video it can be appreciated that Pt migration is mainly symmetrical, from the center toward the ends, but isolated grains moving in the same direction of the electron current are also present. Further increase in voltage leads to SNW breakdown at (51.1 μA, 3.2 V), with formation of a few nm gap in the middle, as shown in Figure 22.9(g). We point out that it is a breakdown and not a burn-out process, as the sharp gap cut and the stiffness of the broken wire indicate.

To investigate the structural details after the main transition stages, we performed *ex situ* TEM analysis. In Figure 22.10, an SNW after coalescence of the Pt grains is shown. TEM imaging was performed at 100 keV to prevent structural modifications by primary knock-on effects on the C atoms. In Figure 22.10(a), big Pt grains are aligned in a row along the wire axis and occupy almost the entire cross-section; size distribution is fairly homogeneous around 13 nm. As magnification is increased, in Figure 22.10(b), the crystallographic details become evident. The carbon matrix between the grains shows clear evidence of graphitization, as revealed by the interference fringes matching the 0.349 nm spacing of (0002) graphite planes. The Pt grain on the left displays a homogeneous fringing with 0.224 nm periodicity, corresponding to Pt (111) crystal orientation.

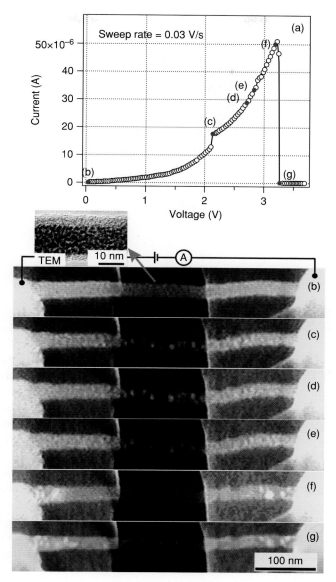

FIGURE 22.9: (a) Current-voltage characteristic of a Pt SNW; (b)–(g) video frames from real-time SEM imaging at those points labeled in the I–V curve. (b) As-deposited Pt SNW (inset shows a TEM image of the SNW); (c) sudden Pt coalescence; (d) persistence of Pt coalescence; (e) onset of Pt depletion; (f) Pt-depleted C-SNW; (g) C-SNW breakdown with formation of a nanosize gap in the middle. Reprinted with permission from Reference 24. Copyright (2009), American Institute of Physics.

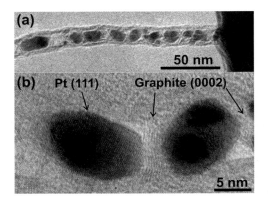

FIGURE 22.10: TEM images of a Pt SNW after coalescence of the Pt grains. (a) Overall view showing large Pt grains with average size of 13 nm, (c) close-up view revealing monocrystallinity of a Pt grain (Pt (111) fringes) and (0002) graphite planes in between the grains. Reprinted with permission from Reference 24. Copyright (2009), American Institute of Physics.

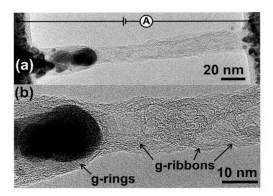

FIGURE 22.11: TEM images of a Pt SNW after Pt depletion stage. (a) Overall view of the Pt-depleted SNW with Pt accumulation on the left (anode) side; (b) close-up view of the left side with (0002) graphite rings surrounding the big Pt grain and wavy graphite ribbons running along the wire. Reprinted with permission from Reference 24. Copyright (2009), American Institute of Physics.

In Figure 22.11, we show the TEM images of an SNW after Pt-depletion stage, a condition achieved after an I–V measurement up to 5 V, where a current density of 5×10^7 A/cm^2 was reached. In this case, TEM imaging was performed at 200 keV. In Figure 22.11(a), a low-magnification image of the SNW is presented. The wire is fully depleted of platinum, all migrated toward the positive electrode on the left side, where big dark grains are present. The migration direction, as it will be discussed later, is in agreement with electromigration effect. In Figure 22.11(b), a magnification of the left side is presented, clearly showing that the C-matrix is fully graphitized. The arrangement of the graphitic structures is quite disordered, but wavy graphite ribbons running along the wire can be recognized, and graphite rings surrounding the big Pt grain are also evident.

At the basis of the structural evolution shown in Figure 22.9, there are physical effects related to the electrical current (Joule heating, electro- and thermo-migration), and geometrical factors represented by the suspended geometry and the nanoscale one-dimensional nature of the conductor. The first transition stage, the coalescence of the Pt grains, is a thermal-induced effect as a consequence of the Joule heating, as indicated by grains' distribution in size and position along the SNW. Grains' coalescence is more pronounced at the center of the wire with respect to the sides, and this directly recalls the temperature distribution along a Joule-heated wire: a bell-shaped curve with the maximum centered in the middle of the wire [26]. The presence of g-C between the aggregated grains (Figure 22.10(b)) is an indication of the high temperature reached. In fact, the formation of graphite rings around Pt grains has been observed in evaporated Pt-C films after annealing above 1000°C [29]. In our case, such a high temperature can be attained because of the suspended geometry, dramatically reducing the thermal dissipation in the wire. Several SNWs have been tested over the same bias range, and Pt-grains coalescence was observed in all cases. Transition onset was mostly located after the initial I–V linear region, above 1.5 V and 15 μA; however, as shown in Section 4.1, there were also cases in which coalescence started at tenths of volts and few μA, in the linear portion of the I–V curve. This variation of the transition onset could be due to a variable thermal and electrical conductance at the contact between the Pt-C deposit and the Au electrode, for different samples.

The second transition stage is the depletion of aggregated grains by Pt migration. The process takes place either symmetrically, with Pt migrating from center to sides, as shown in Figure 22.9, or by motion of entire grains in one direction. The first process can be explained by thermomigration [30], i.e., mass diffusion in a temperature gradient, and, according to the aforementioned temperature distribution, we observe Pt transport from hot (center) to cold (ends) sites. The second diffusion mechanism can be interpreted as electromigration [30], the motion of atoms in the same direction of electron current under the "electron wind" force. The full-graphitization of the C-matrix, evident after Pt depletion, is similar to results reported for a-C SNWs with embedded Fe nanoparticles playing the role of catalyzers of the graphitization process, under high-temperature thermomigration [28] or electromigration [27].

5 CONCLUSIONS

Suspended nanostructures (SNSs) grown by lateral FEBID of Pt-organometallic precursor have been characterized in the growth mode, structure, composition, and electrical behavior.

Deposition as a function of beam energy (5, 10, 15 keV) and linear electron dose (1–9 pC/nm) shows that SNSs grow as horizontal nanowires for doses up to 2 pC/nm and with increasing slope (up to 60°) as the dose increases. Linear SNSs dimensions and deposited volume increase as the beam energy is reduced.

TEM analysis of suspended nanowires (SNWs) reveals that the deposit consists of crystalline Pt nanograins (2 nm) embedded into a carbonaceous amorphous matrix. It is shown that high current density TEM irradiation induces C removal and aggregation of the Pt grains into a continuous polycrystalline wire.

Electrical measurements at low voltage (−0.1, 0.1 V), on as-deposited SNWs, show a linear I–V behavior, with resistivity around 1×10^{-4} Ohm m. Measurements over extended voltage range (up to 3–5 V) show non-linear I–V behavior typical of hopping transport systems.

As low voltage cycles are repeated, Pt aggregation into big isolated grains (10–15 nm) is observed with a decrease in resistance. Investigation by live SEM imaging during extended measurements reveals a two-stage transition: Pt aggregation is followed by Pt depletion under thermo- and electro-migration effects, when current density approaches 10^7 A/cm². Graphitization of the C matrix is observed between the aggregated grains, and, after Pt depletion, the SNW is fully graphitized. The transition from a-C to g-C, shown by TEM analysis, accounts for the improved SNW conductivity and the high current density sustained.

This structural evolution, not reported before, to our knowledge, for Pt deposits on-substrate, is a result of the high thermal insulation and high temperatures attainable in suspended structures. Interesting perspectives are opened, such as the catalytic growth of C nanotubes with the site-specificity of FEBID, and the possibility to study isolated metal nanoparticles, for example of magnetic material, by using magnetic organometallic precursors.

REFERENCES

[1] Frabboni, S., G. C. Gazzadi, and A. Spessot. TEM study of annealed Pt nanostructures grown by electron beam-induced deposition. *Physica E* 37 (2007): 265.
[2] Matsui, S. Focused-ion-beam deposition for 3-D nanostructure fabrication. *Nucl. Instrum. Meth. Phys. Res. B* 257 (2007): 758.
[3] Ooi, T., K. Matsumoto, M. Nakao, M. Otsubo, S. Shirakata, S. Tanaka, and Y. Hatamura. 3d nano wire-frame for handling and observing a single DNA fiber. *MEMS 2000 Proceedings*, pp. 580–583.
[4] Bøggild, P., T. M. Hansen, K.Mølhave, A. Hyldgård, M. O. Jensen, J. Richter, L. Montelius, and F. Grey. Customizable nanotweezers for manipulation of free-standing nanostructures. *IEEE-NANO 2001 Proceedings*, pp. 87–92.
[5] Floreani, F., H. W. Koops, and W. Elsasser. Operation of high power field emitters fabricated with electron beam deposition and concept of a miniaturised free electron laser. *Microelectron. Eng.* 57–58 (2001): 1009.
[6] Fujita, J., M. Ishida, T. Ichihashi, Y. Ochiai, T. Kaito, and S. Matsui. Carbon nanopillar laterally grown with electron beam-induced chemical vapor deposition. *J. Vac. Sci. Technol. B* 21 (2003): 2990.
[7] Liu, Z. Q., K. Mitsuishi, and K. Furuya. Features of self-supporting tungsten nanowire deposited with high-energy electrons. *J. Appl. Phys.* 96 (2004): 619.
[8] Koops, H. W. P., J. Kretz, M. Rudolph, and M. Weber. Constructive three-dimensional lithography with electron-beam induced deposition for quantum effect devices. *J. Vac. Sci. Technol. B* 11 (1993): 2386.
[9] Bret, T., I. Utke, C. Gaillard, and P. Hoffmann. Periodic structure formation by focused electron-beam-induced deposition. *J. Vac. Sci. Technol. B* 22 (2004): 2504.
[10] Mølhave, K., D. Nørgaard Madsen, S. Dohn, and P. Bøggild. Constructing, connecting and soldering nanostructures by environmental electron beam deposition. *Nanotechnology* 15 (2004): 1047.
[11] Gazzadi, G. C., S. Frabboni, and C. Menozzi. Suspended nanostructures grown by electron beam-induced deposition of Pt and TEOS precursors. *Nanotechnology* 18 (2007): 445709.
[12] Lipp, S., L. Frey, C. Lehrer, E. Demm, S. Pauthner, and H. Ryssel. A comparison of focused ion beam and Electron beam induced deposition processes. *Microelectron. Reliab.* 36 (1996): 1779.
[13] Liu, Z. Q., K. Mitsuishi, and K. Furuya. Three-dimensional nanofabrication by electron-beam-induced deposition using 200-keV electrons in scanning transmission electron microscope. *Appl. Phys. A* 80 (2005): 1437.
[14] Frabboni, S., G. C. Gazzadi, L. Felisari, and A. Spessot. Fabrication by electron beam induced deposition and transmission electron microscopic characterization of sub-10-nm freestanding Pt nanowires. *Appl. Phys. Lett.* 88 (2006): 213116.
[15] Koops, H. W. P., A. Kaya, and M. Weber. Fabrication and characterization of platinum nanocrystalline material grown by electron-beam induced deposition. *J. Vac. Sci. Technol. B* 13 (1995): 2400.
[16] Gazzadi, G. C., and S. Frabboni. Fabrication of 5 nm gap pillar electrodes by electron-beam Pt deposition. *J. Vac. Sci. Technol. B* 23 (2005): L1.
[17] Egerton, R. F. *Electron energy loss spectroscopy*, 2nd ed. New York: Plenum, 1996.
[18] Williams, D. B., and C. Barry Carter. *Transmission electron microscopy: A textbook for materials science.* Vol. I. New York: Kluwer Academic/Plenum Publishers, 1996.
[19] Koops, H. W. P., C. Schossler, A. Kaya, and M. Weber. Conductive dots, wires, and supertips for field electron emitters produced by electron-beam induced deposition on samples having increased temperature. *J. Vac. Sci. Technol. B* 14 (1996): 4105.
[20] Rotkina, L., S. Oh, J. N. Eckstein, and S. V. Rotkin. Logarithmic behavior of the conductivity of electron-beam deposited granular Pt/C nanowires. *Phys. Rev. B* 72 (2005): 233407.

[21] Botman, A., J. J. L. Mulders, R. Weemaes, and S. Mentink. Purification of platinum and gold structures after electron-beam-induced deposition. *Nanotechnology* 17 (2006): 3779.

[22] Langford, R. M., T.-X. Wang, and D. Ozkaya. Reducing the resistivity of electron and ion beam assisted deposited Pt. *Microelectron. Eng.* 84 (2007): 784.

[23] Gazzadi, G. C., S. Frabboni, C. Menozzi, and L. Incerti. Electrical characterization of suspended Pt nanowires grown by EBID with water vapour assistance. *Microelectron. Eng.* 85 (2008): 1166.

[24] Gazzadi, G. C., and S. Frabboni. Structural evolution and graphitization of metallorganic-Pt suspended nanowires under high-current-density electrical test. *Appl. Phys. Lett.* 94 (2009): 173112.

[25] Gazzadi, G. C., E. Angeli, P. Facci, and S. Frabboni. Electrical characterization and Auger depth profiling of nanogap electrodes fabricated by I_2-assisted focused ion beam. *Appl. Phys. Lett.* 89 (2006): 173112.

[26] Durkan, C., M. A. Schneider, and M. E. Welland. Analysis of failure mechanisms in electrically stressed Au nanowires. *J. Appl. Phys.* 86 (1999): 1280.

[27] Jin, C. H., J. Y. Wang, Q. Chen, and L.-M. Peng. In situ fabrication and graphitization of amorphous carbon nanowires and their electrical properties. *J. Phys. Chem. B* 110 (2006): 5423.

[28] Ichihashi, T., J. Fujita, M. Ishida, and Y. Ochiai. In situ observation of carbon-nanopillar tubulization caused by liquidlike iron particles. *Phys. Rev. Lett.* 92 (2004): 215702.

[29] Lamber, R., and N. I. Jaeger. Electron microscopy study of the interaction of Ni, Pd and Pt with carbon. *Surf. Sci.* 289 (1993): 247.

[30] Huntington, H. B. Effect of driving forces on atom motion. *Thin Solid Films* 25 (1975): 265.

C H A P T E R 2 3

ELECTRICAL TRANSPORT PROPERTIES OF METALLIC NANOWIRES AND NANOCONSTRICTIONS CREATED WITH FIB

Jose María De Teresa, Amalio Fernández-Pacheco, Rosa Córdoba, and Manuel Ricardo Ibarra

1 INTRODUCTION

The use of focused ion beams in the nanometric range opens the route for fascinating applications in nanotechnology, as well as for basic studies in condensed-matter physics. In this chapter, we will discuss recent examples developed in our group regarding the use of a focused Ga beam for the creation of metallic nanowires (NWs) and nanoconstrictions (NCs). Our interest lies in the study of electrical transport in confined geometries, such as those provided in NWs and NCs. Roughly speaking, NWs are defined as one-dimensional objects with lateral dimension of the order of 100 nm or smaller, whereas NCs imply the confinement of the electrical current to lateral dimensions of a few nanometers, even a single atom (atomic-sized constriction). In general, NWs and NCs can be useful in topics such as nanoelectronics, spintronics, information storage, sensors, superconducting circuits, etc. [1–5]. Due to this broad range of fields of interest, new approaches to the creation of metallic NWs and NCs are appealing.

When using Ga ions accelerated up to high voltage (typically of the order of 30 kV), the kinetic energy is high enough to sputter atoms from the sample of interest, eventually shaping the material with the desired form. This approach can be considered a *subtractive method*. With this method, it has been possible to create functional NWs and NCs down to about 50 nm width [6], even though it is necessary to consider that high ion doses can deteriorate the intrinsic sample properties [7]. In the present contribution, we will describe our approach to push the limits of this *subtractive method* and create atomic-sized constrictions, with lateral size of the order on one single atom. At that size scale, quantum features dominate the electrical transport behavior of the sample and new phenomena emerge.

On the other hand, the ion beam can be used to dissociate precursor molecules adsorbed on the surface of the sample of interest, producing the local growth of a new material. This approach is an *additive method*, and is usually called FIBID (focused ion beam induced deposition). With this method it has been possible to grow a wide variety of functional NWs [8]. The results obtained in our group using two different types of precursor molecules, one of them based on Pt, and another one on W, will be described. It will be shown that the grown Pt-based NWs undergo an insulator-metal transition as a function of the sample thickness, whereas the W-based NWs show superconductivity below 5.4 K, opening applications across several research fields. NWs can be also grown with the "sister" technique named FEBID (focused electron beam induced deposition), and comparison between both techniques will be given.

For the experiments reported hereafter, a commercial "dual beam" instrument (Nova 200 NanoLab from FEI) has been used. It integrates a 30 kV field-emission electron column and a Ga-based 30 kV ion column placed forming 52° with coincidence point at about 5 mm away from the electron-column pole piece and about 20 mm away from the ion-column pole piece. The processes (either subtractive or additive) carried out inside our equipment sometimes require control of the sample resistance. For that, we use two different methods. In the first one, we use electrical microprobes (from Kleindiek) connected via a feed-through to our external current-source/voltmeter instruments. The movement of such microprobes can be controlled with nanometric resolution, which allows placing the probes on the targeted points of interest. The lead resistance of the electrical circuit containing these microprobes is only about 13 Ω. In Figure 23.1(a), a picture of the microprobe station with the four available probes on top of the sample is shown. In Figure 23.1(b), a SEM image of the tips of the four microprobes approaching the circuit of interest in the sample is given. In the second approach for *in situ* resistance monitoring of our sample, we use a home-designed sample holder (see Figure 23.1(c)) with wire connections between the chip carrier of our sample and a connector whose wiring is taken out of the chamber via the same feed-through mentioned before. The electrical contact between the chip carrier and our sample is realized via micro-contacts, as shown in the SEM image given in Figure 23.1(d). For *in situ* electrical measurements, we typically perform four-probe measurements by means of a 6220 DC current source-2182A nanovoltmeter combined Keithley system, which warrants no influence of any contact resistance.

2 CREATION OF METALLIC NANOCONSTRICTIONS BY FIB

In a macroscopic metallic conductor, the Ohm's law ($G = \sigma A/L$) connects its conductance, G, with the sample dimensions, L = length, A = area, implying that the sample's dimensions are much greater than the carrier mean free path, l. This is called the *diffusive* conduction regime.

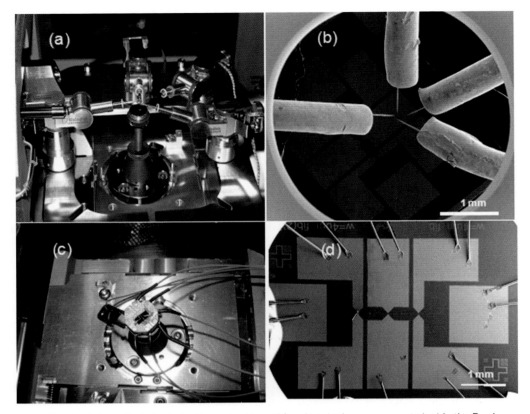

FIGURE 23.1: Pictures that illustrate the methods used for electrical measurements inside the Dual Beam chamber: (a) electrical microprobes stage; (b) SEM image of the microprobes inside the chamber, with a sample just below; (c) home-made chip carrier holding a sample for *in situ* electrical measurements; (d) SEM image of a sample with microcontacts connected to the chip carrier.

If the carriers are forced to flow through a constriction with dimensions smaller than l, they can flow without scattering, also named *ballistic* conduction regime. A semi-classical theoretical treatment by Sharvin [9] concluded that in the ballistic conduction regime the conductance is governed by the formula:

$$G = \frac{2e^2}{h} \left(\frac{\pi a}{\lambda_F} \right)^2 , \qquad (1)$$

where a is the radius of the constriction and λ_F is the carrier Fermi wavelength. In the case of metallic materials, λ_F is of the order of 1 Å, which means that, when approaching atomic-sized constrictions where $a \sim$ a few Å, changes in the radius of the constriction of just one atom imply significant changes in G. Thus, a step-wise behavior in G, caused by changes in the constriction radius of the order of one atom, is a typical signature of the formation of atomic-sized constrictions [10]. Moreover, G in metallic atomic-sized constrictions will show values of the order of $G_0 = 2e^2/h = (1/12.9) \text{ k}\Omega^{-1}$ as deduced from formula (1). G_0 is called the conductance quantum for the reason explained hereafter.

This semiclassical approach is, however, not valid when the contacts are indeed atomic-sized, with only a few atoms contributing to the conduction of electrons. In this case, the problem is more realistically treated within the formalism developed by Landauer [11], as we enter into the *full quantum limit*. In the quantum regime the conductance is well described by the formula:

$$G = \frac{2e^2}{h} \sum_{i=1}^{N} \tau_i,$$ (2)

where only a few channels (N) contribute to the transmitted current and where τ_i, ranging between 0 and 1, is the transmission coefficient of the channel i (i = 1...N). A beautiful example is that of gold, where only one channel (monovalent metal) contributes to the current when the contact is determined by a single atom and conductance histograms show a clear peak at $G \approx G_0$ [10]. The physics of metallic atomic-sized constrictions is very rich and could bring about applications in powerful sensors and platforms for single-molecule research, which at the moment is an active field of investigation.

So far, only a few methods have been successfully used for the creation of such atomic-sized constrictions with metallic conduction. These methods normally work at helium liquid temperature, and in ultra-high vacuum conditions, which limits the applications [10]. We refer to the *mechanical break junction* method, where a metallic film is broken in a controllable way by bending a flexible substrate on which the material has previously been deposited; the method with a *scanning tunneling microscope* (STM), where the STM tip is indented onto a surface of the same material forming a neck that is stretched to rupture; the *electrical break junction* method, where atoms are electromigrated by the application of large current densities on a previously patterned neck of a metallic thin film; the *electrochemical junction* method, where in an electrochemical bath, a previously patterned gap is filled by electrodeposition; and the method with a *high-resolution transmission electron microscope* (HRTEM), where a focused electron beam produces holes and metallic bridges.

Hereafter, we show a novel technique to establish NCs and atomic-sized contacts at room temperature in metallic materials, based on a dual beam equipment. In order to illustrate the procedure, the creation of controlled atomic-sized constrictions based on chromium/aluminium electrodes as well as on iron electrodes are described. We expect that this technique will be applicable to many types of materials. Starting from a micrometric electrode, previously patterned by optical lithography, we create a progressive decrease of conductance by means of focused Ga-ion milling. The *in situ* monitoring of the resistance using electrical microprobes during the milling process allows the observation of a step-wise structure in the electrical resistance, with steps in the range of $1/G_0$, just prior to an abrupt jump, which occurs at the crossover to the tunneling regime. The simultaneous SEM imaging during the process enables us to establish the correlation between the transport measurements and the microstructure of the etched material. We have also performed current-versus-voltage curves (I-V) at determined stages of the milling process, which allows one to clearly distinguish between the metallic (ohmic) and tunneling (non-ohmic) regimes. First, we chose a Cr metallic layer to demonstrate the feasibility of this novel technique for the fabrication of constrictions due to its good adhesion to the substrate. Afterward, we applied the same technique to Fe metallic layers due to their potential applications in spin electronics. The procedure is expected to be applicable to many different metallic materials, being of general interest, and the results shown hereafter have been published in References 12 and 13.

We first draw attention to the results obtained on the Cr-based electrodes, which were covered by Al to produce a rather anisotropic milling, favoring the formation of narrow channels. The Ga-ion milling process was done in two consecutive steps. First, starting from 15 Ω, with the ion-column set at 5 kV and 0.12 nA, aiming to shorten the time of the experiment, the milling was stopped when the resistance reached 500 Ω, still far from the metallic quantum limit. In the second step, the ion-column was set at 5 kV and 2 pA, for about 12 minutes. During this period we monitored the variation of the resistance and simultaneously 10 kV-SEM images were collected. This allowed us to monitor the correlation between resistance and microstructure. In other experiments, the milling was stopped at determined resistance values, at which current versus voltage curves were measured. The minimum bias current was selected in order to avoid the deterioration of the device due to heating effects.

In Figure 23.2, we show a typical resistance-versus-milling time result at a starting value of 500 Ω. The resistance increases continuously for about 10 minutes, up to a point, in which a jump of three orders of magnitude is seen. Just before this abrupt change, the resistance shows a discrete number of steps. This step-wise structure is a clear indication of the formation of point contacts in the metal electrode before the complete breaking of the formed NWs. This is similar to what it is seen in experiments done with other techniques [10], as explained earlier. From tens of experiments, we can conclude the reproducible existence of steps in the conductance in the verge of the metal-tunneling crossover and certain dispersion in the specific values of the conductance plateaus, which we always observe below 4 G_0. In the inset of Figure 23.2, an SEM image of the initial electrode with the electrical microprobes is shown. In the bottom panel, SEM images of the patterned microelectrode at three different moments of the milling process (initial, intermediate, final) are displayed together with the corresponding measured resistance. As can be noticed from the current-versus-voltage results shown in the inset, the dependence is linear when the constriction is not broken and metallic conductance takes place, whereas it becomes non-linear when the contact is broken and tunneling conductance occurs.

The second type of NCs performed on Fe-based electrodes, will be presented hereafter. Such a material has been chosen due to its magnetic character and its potential application in spin electronics. In fact, one exciting research line today is that of atomic-sized constrictions based on magnetic materials [14–17], where sizable effects such as ballistic anisotropic magnetoresistance or tunneling anisotropic magnetoresistance have been found. In the particular case of Fe, experiments by Viret et al. with the mechanical break junction technique performed at 4.2 K indicate that in the vicinity of the conductance quantum the magnetoresistive effects can approach giant values [16]. Aiming at the creation of stable Fe-based atomic-sized constrictions at room temperature, we applied the described Ga-milling technique to Fe electrodes (in this case, we did not cover with Al, as the milling on Fe was inhomogeneous enough to form narrow channels). One typical experiment is shown in Figure 23.3, already represented in the Y scale as G/G_0. During the first 2 minutes, the Ga milling process is performed using 5 kV and 70 pA, the resistance approaching 2 kΩ at the end. Then, we continue the Ga milling process using 5 kV and 0.15 pA. The process is slower and allows the detection of step-wise changes in the resistance before the jump to the tunneling regime, similar to the observed behavior in the case of the Cr-based electrodes. Clear steps are observed that are also interpreted as evidence of the formation of atomic-sized metallic contacts just before entering the tunneling regime. This opens a fascinating route to create stable NCs at room temperature, which could have functional use.

We have found that this type of atomic-sized constrictions become oxidized under ambient conditions, the resistance increasing several orders of magnitude. This is why we

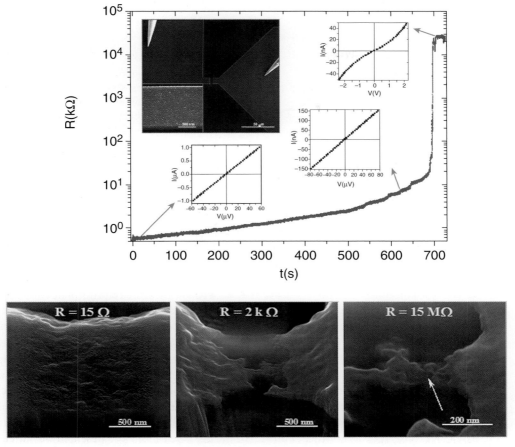

FIGURE 23.2: Top panel: Resistance versus time in a typical ion milling process of the Al/Cr electrode, where discrete jumps are observed. The insets show I-V curves measured at fixed resistance values. A linear dependence, as expected for metallic conduction is observed except for the last stages of milling, where the tunneling regime has been reached. The inset shows an artificially colored SEM image of the microelectrode after milling together with the electrical microprobes. The inset of the inset shows the electrode before the milling process.

Bottom panel: Three SEM images tilted at 52° at three different moments of the milling process with the respective values of resistance. The substrate is 300 nm of SiO_2 on Si. The content of this figure has been partially reprinted from [12], with permission. Copyright (2008), IOP.

decided to cover the constrictions with a semi-insulating material once they had been created. With this aim, a ~10 nm-thick layer of FEBID Pt-C was deposited by electrons on top of the NC, using $(CH_3)_3Pt(CpCH_3)$ as precursor. The high resistance of this material, because of the high amount of carbon, guarantees a resistance in parallel to the constriction of the order of tens of MΩ [18], which is confirmed by the negligible change of R while the deposit is made. Once the constriction had been fabricated, the sample was put in contact with the atmosphere during some minutes, and transferred to a closed cycle refrigerator that can reach a minimum temperature of 12 K, combined with an electromagnet. The measurements of the resistance at room temperature showed an increase of the resistance by a factor of 10 (R≈100 kΩ), evidencing the departure from metallic conduction in the NC. This value is significantly above

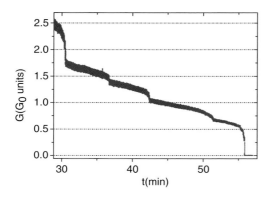

FIGURE 23.3: Conductance (in $G_0=2e^2/h$ units) versus milling time in the process where a typical constriction is formed from an Fe microelectrode. In this stage of milling, steps of the order of some fraction of G_0 are seen, corresponding to discrete thinning of the contact area of the order of one atom. The last step is followed by a sharp decrease of conductance, corresponding to the crossover to the tunneling regime. Reprinted with permission from [12]. Copyright (2008), IOP.

the resistance corresponding to the quantum of conductance, 12.9 kΩ. The non-linearity of the current-versus-voltage curves indicates that the constriction is no longer in the metallic regime, but in the tunneling one. The high reactivity to oxidation of the nanostructure, due to its large surface-to-volume ratio, seems to be the most reasonable explanation of this effect. The resistance increased up to 120 kΩ at T = 24 K. We measured the magnetoresistance for low temperatures with the magnetic field (H) parallel to the current path.

In Figure 23.4(a) the evolution with H and T is shown. MR ratios of the order of 3.2% are obtained. This value is a factor 30 times higher than in Fe non-etched samples, used as reference, where we observed a MR = −0.11% in this configuration (note the different sign). As it is schematically explained by the red arrows in the graph, the evolution of R is understood by a change from a parallel (P) to an anti-parallel (AP) configuration of the ferromagnetic electrodes, separated by an insulator (in the present context, by electrode we mean the magnetic material in the vicinity of one side of the NC). More in detail, starting from saturation, a first continuous increase in R is observed, of the order of 0.8%. This seems to be caused by a progressive rotation of the magnetization (M) in one of the electrodes. At around H = 700 Oe, an abrupt jump of the resistance occurs (MR = 2.5%), since the magnetization of one electrode switches its direction, and aligns almost AP to the M of the other electrode (intermediate state: IS). The MR becomes maximum at H = 1 kOe, when the magnetization in both electrodes is AP. At H≈2 kOe, the M in the hard electrode also rotates, resulting again in a low-R state (P). As T increases, the IS, previously explained, disappears. IS is probably caused by the pinning of the M by some defects present in the soft electrode. Thus, the increase of the thermal energy favors the depinning of M. When the temperature was increased above 35 K, the NC became degraded, likely due to the flow of the electrical current. Before this, it was possible to measure the change of MR with the bias voltage, as shown in Figure 23.4(b).

The diminishing of the resistance with the voltage is a typical feature of tunnel junctions, attributed to several factors such as the increase of the conductance with bias, excitation of magnons, or energy dependence of spin polarization due to band structure effects [19]. As the magnetostrictive (change of length with the magnetization) state of parallel and antiparallel electrodes is the same, it seems that magnetostriction is not the cause of the observed TMR effect. On top of this tunnel magnetoresistance effect, we also observed a giant anisotropic magnetoresistance effect, which is discussed in Reference 13. Summarizing, our first attempts to measure the magnetoresistance of these NCs out of the chamber are promising, but more experiments are necessary to further the research in atomic-sized constrictions using this approach.

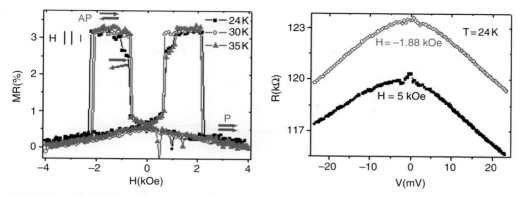

FIGURE 23.4: The left-hand figure shows for one of the created Fe constrictions the magnetoresistance ratio as a function of the magnetic field at low temperatures. The MR is positive, with a value of 3.2% at 24 K, and can be understood by the decoupling of the magnetic electrodes separated by a tunneling barrier. The right-hand figure shows the resistance as a function of bias voltage for the same constriction in the two magnetic configurations (parallel and antiparallel). Reprinted from Reference 13(a).

3 GROWTH OF METALLIC NANOWIRES WITH FIBID

In other chapters of this book, the physical mechanisms behind FIBID have been discussed. In the following, we will apply this technique to the growth of two types of metallic NWs, based on Pt and W, respectively, and a brief comparison with NWs grown with FEBID will be given. Stress will be put on the functionality of these NWs. In Figure 23.5 we show the SEM images of the protocol followed for the growth of these NWs when we need to monitor the NW resistance while the deposit is being performed. In Figure 23.5(a), an SEM image shows the experimental configuration for the deposition, where the four microprobes are contacted to the pads, and the GIS needle is inserted near the substrate. By applying a constant current between the two extremes of the NW, and measuring the voltage drop in two of the intermediate pads, we are able to measure the resistance of the NW that is being grown, once it is below 1 GΩ. The probe current (I_{probe}) was changed during the measurement, in order to optimize the signal-to-noise ratio, and trying to minimize heating effects of the device while it was created. Typical values for I_{probe} range from 5 nA at the beginning of the monitoring (R~1 GΩ) up to 10 μA at the end (R~1 kΩ). In Figure 23.5(b) the transversal deposit is an example of a studied NW. Before its deposition, additional "perpendicular-to-the-NW" deposits were performed (horizontal deposits) for four wire measurements, avoiding non-linear effects associated with the resistance of the contacts. These extra-connections were done using several growth parameters, finding no differences in the resistance measurements of the NWs. This rules out any influence on the results of the halo associated with the deposition, which could be the source of spurious effects [20]. Thermally oxidized silicon (~200 nm of SiO_2) was used as substrate, and aluminium pads were previously patterned by optical lithography.

3.1 Pt-BASED NANOWIRES

The main reason to study these Pt nanowires in detail is the great variety of published results on their electrical properties, which could cast doubts on the degree of control and

FIGURE 23.5: (a) SEM image of the experimental configuration for four-wire electrical measurements during the growth of a NW by FIBID. The four microprobes are contacted to Al micrometric pads, while the deposit is carried out. For deposition, the GIS needle is inserted near the substrate. (b) SEM image of one deposited Pt-C NW (transversal line). The additional two horizontal lines are done before the vertical NW, to perform four wire measurements (see text for details). Reprinted with permission from [21]. Copyright (2009), APS.

reproducibility achieved in the performed growth. However, it should be noted that in some groups the NWs are deposited with cyclopentadienyl-trimethyl platinum, $(CH_3)_3PtCp$, whereas in other groups methylcyclopentadienyl-trimethyl platinum, $(CH_3)_3Pt(CpCH_3)$ is used [21]. Deposits using the first precursor (one carbon less present in the molecule) have the lowest resistivity reported (only about 7 times higher than that of bulk Pt) and a positive thermal dependence. Also, the width and thickness of the NWs could play a role in the observed differences, as well as other growth parameters. Our aim is to study in depth the role played by the NW thickness while keeping all the other parameters constant, which has led to the finding of a metal-insulator transition as a function of thickness, explaining in part the diversity in the published results.

In our case we use the following deposition parameters: $(CH_3)_3Pt(CpCH_3)$ precursor gas, $(T_G) = 35°C$ precursor temperature, $(V_{FIB}) = 30\,kV$ beam voltage, $(I_{FIB}) = 10\,pA$ beam current, substrate temperature = room temperature, dwell time = 200 ns, chamber base pressure = 10^{-6} mbar, process pressure = 4×10^{-6} mbar, beam overlap = 0%, refresh time = 0 s, distance between GIS needle and substrate $(L_D) = 1.5$ mm. Under these conditions, a dose of 3×10^{16} ion/cm^2 min roughly irradiates the sample. We must emphasize that the values for T_G (about 10°C below the usual operation temperature) and L_D ($\sim$1.35 mm higher than usually) were chosen to decrease the deposition rate, allowing detailed measurements of the evolution of the resistance as the deposit was realized. This is required to get fine control, especially during the initial stages, where a very abrupt change in the resistance occurs in a very short period of time. We obtained approximately the same final resistivity when depositing NWs using normal conditions [18], so these changes do not seem to affect substantially the properties of the NWs. The studied NWs have length (L) = 15 μm, width (w) = 500 nm, thickness (t) = variable with process time.

Figure 23.6 shows how the resistivity (ρ) of a NW evolves as a function of time in a typical experiment. The highest value for ρ corresponds to $\sim$1 GΩ in resistance, and to $\sim$1 kΩ for the lowest value. We have correlated the process time with the deposited thickness by doing cross-section inspections of NWs created at different times, so that the results can be expressed in terms of resistivity. We can see that under these conditions the resistivity of the NWs starts from a value above $10^8\ \mu\Omega$ cm. An abrupt decrease is produced as the thickness increases, lowering its resistivity in four orders of magnitude when the thickness reaches 50 nm. From that moment on, ρ decreases slightly, saturating to a constant value of $\sim$700 $\mu\Omega$ cm (around 65 times higher than the value for bulk Pt) for t$>$150 nm. From these experiments, we conclude that the resistivity at room temperature is highly dependent on the NW thickness. This result explains the large diversity of values for ρ of FIBID-Pt NWs reported in the literature, as discussed in Reference 21. The resistivity as a function of thickness crosses the value reported by Mott for the maximum metallic resistivity for a non-crystalline material, $\rho_{max} \approx 3000\ \mu\Omega$ cm [22] at $\approx$ 50 nm (see dashed line in Figure 23.6(a)). Taking into account this criterion, we would roughly expect a non-metallic conduction for samples with $\rho > \rho_{max}$, and a metallic conduction for $\rho < \rho_{max}$.

To study the nature of the conductivity in deposits having different thickness, we have fabricated several NWs, by means of stopping the growth process when a determined value of the resistivity is reached. Hereafter, these NWs will be generically referred to as: M_0 ($\sim$$10^7\ \mu\Omega$ cm), M_1 ($\sim$$10^5\ \mu\Omega$ cm), M_2 ($\sim$ $10^4\ \mu\Omega$ cm), M_3 ($\sim$2 $\times 10^3\ \mu\Omega$ cm) and M_4 ($\sim$700 $\mu\Omega$ cm) (see squares in Figure 23.6(a)). The resistance in all cases is stable under the high vacuum atmosphere, once the deposit has stopped. Current versus voltage (I-V) measurements have been performed *in situ*, at room temperature. In Figure 23.6(b) we show examples of the differential conductance (G = dI/dV, obtained numerically) as a function of voltage. For M_4 (Figure 23.6(b_5)) and M_3 (Figure 23.6(b_4)), I-V is linear (G constant), so a metallic character is inferred. However, for M_2 (Figure 23.6(b_3)), G(V) is slightly parabolic. The same non-linear effect is even much stronger for M_1 (Figure 23.6(b_2)). A different behavior in G(V) occurs for the most resistive sample, M_0 (Figure 23.6(b_1)), where a peak in the differential conductance is found at low bias, decreasing from $\sim$10 nΩ^{-1} to a constant value of 4 nΩ^{-1} for voltages higher than 0.5 V. G(V) measurements thus confirm Mott's criterion for the maximum resistivity for metallic conduction in non-crystalline materials [22]. The transport mechanisms of the different NWs, M_0 to M_4, have been discussed in detail in Reference 21.

We can understand these strong differences for the electrical transport properties at room temperature considering a gradient in composition for the NWs as a function of thickness as discussed hereafter. It is well reported through TEM measurements that the microstructure of the Pt nanodeposits consist of nanometric Pt crystallites in an amorphous carbonaceous matrix [18]. We performed chemical analysis by energy-dispersive x-ray spectroscopy (EDS) on NWs with different thicknesses. The composition was found to be homogeneous over the entire deposited surface. The results are shown in Figure 23.7. For t = 25 nm and 50 nm, the deposit is highly rich in carbon: $\sim$68%, with Pt $\sim$18% and Ga $\sim$14% (atomic percentages). A residual fraction of oxygen (less than 1%) is also present. For higher thickness, the metal content increases gradually, until approximately saturated values: C $\sim$40–45%, Pt $\sim$30–35%, Ga $\sim$25%. We can understand the gradient in composition with thickness taking into account an important decrease in the secondary electrons yield in SiO_2 that occurs when the SiO_2 is irradiated with ions in the keV range [23] because of the large energy gap present in SiO_2. The SE are considered to be the main cause of the dissociation of the precursor gas molecules, as discussed in other chapters of the book. Thus, at the beginning of the growth process, a smaller amount of SE is emitted in comparison with the subsequent stages, when the effective substrate

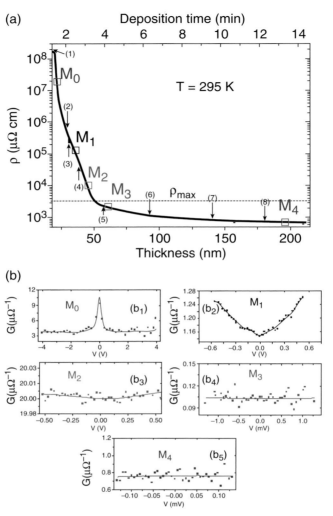

FIGURE 23.6: Resistivity as a function of deposition time and thickness for a typical growth of a Pt NW by FIBID. The resistivity varies by more than four orders of magnitude as a consequence of a change in composition. A thickness of 50 nm marks the transition between a non-metallic conduction (t<50 nm) and metallic one (t>50 nm). The resistivity saturates to a constant value for t>150 nm, $\rho \approx 700\,\mu\Omega$cm (about 65 times higher than bulk Pt). The maximum resistivity for metallic conduction in non-crystalline materials (ρmax), as calculated by Mott, is marked. This line separates the insulator NWs from the metallic ones. The numbers in parentheses correspond to different probe currents during the measurements. (1): 5 nA, (2): 10 nA, (3): 30 nA, (4): 100 nA, (5): 1 μA, (6): 3 μA, (7): 5 μA, (8): 10 μA (b): Differential conductivity versus voltage for samples in different regimes of conduction, marked in the ρ(t) curve (figure 23.6(a)). G(V) has been obtained by a numerical differentiation of current versus voltage curves. In M_0, an exponential decrease in the differential conductance is present. In M_1, a positive tendency for the G(V) curve is observed, this trend being hard to see for M_2. In M_3 and M_4, G(V) is constant (see text for details). The lines are guides to the eye. Reprinted with permission from [21]. Copyright (2009), APS.

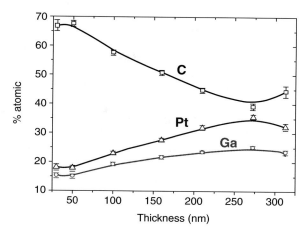

FIGURE 23.7: Atomic percentage compositions of Pt NWs by FIBID as a function of their thickness, determined by EDS. We observe a clear difference in composition for the NWs smaller than 50 nm in comparison with the others, where a higher metallic content is present. Oxygen is always below 1%. Reprinted with permission from [21]. Copyright (2009), APS.

becomes the initial C-Pt-Ga deposit. This thesis is in agreement with the penetration range of Ga at 30 kV in a SiO_2 and Pt-C substrate (of the order of 30–50 nm). Other effects cannot be discarded but are more unlikely [21]. The EDS results were supported by the XPS measurements shown in Reference 21, which indicated two main results: oxidation of the surface of the Pt nanodeposit when exposed to ambient conditions; and a shift in the binding energy of Pt 4f electrons as a function of the deposit thickness that was interpreted as due to the increase in Pt cluster size as the deposit thickness increases.

Summarizing, our work explains the discrepancy existing for previous Pt-C nanostructures created by FIB, where in some cases a metallic character was reported, in contrast to others, where insulating conduction was observed. It also gives a counterpoint to the traditional work when growing metal nanostructures by FIB, since we demonstrate the possibility of depositing materials with different conduction characteristics, making it, in principle, feasible to fabricate insulator or metal nanowires with the same technique and precursor, just by controlling and changing the growth parameters.

3.2 W-BASED NANOWIRES

The control in the growth and patterning of superconducting materials in the nanometer scale is opening interesting research fields in condensed-matter physics. Basic studies regarding one-dimensional superconductivity [24], vortex confinement [25], vortex pinning [26], etc., have been tackled recently. The fabrication of nanoSQUIDS is one of the most appealing applications of such nanosuperconductors [27]. Normally, complex nanolithography techniques such as electron-beam lithography, involving several process steps, have been used to create such superconducting nanostructures. It has been recently found that FIBID is able to produce superconducting nanodeposits with relatively high $T_C \sim 5.4$ K using $W(CO)_6$ as the precursor gas [28]. These nanodeposits closely follow the BCS theory with a well-defined Abrikosov vortex lattice [29]. As previously found with XPS measurements [29], the nanodeposit composition is mainly formed by W ($\approx 43\%$), C ($\approx 40\%$), Ga ($\approx 10\%$) and O($\approx 7\%$). TEM studies indicate that the nature of the deposit is amorphous [30], which explains the high T_C in comparison with crystalline W.

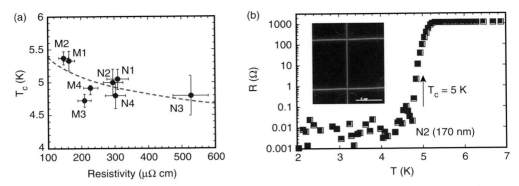

FIGURE 23.8: (a) T_C versus resistivity of all the W FIBID microwires (M1, M2, M3, M4) and nanowires (N1, N2, N3, N4) studied. T_C ranges between 4.7 and 5.4 K, with a slight tendency to decrease with the resistivity. Error bars come from the transition temperature width (y-axis) and in the determination of the wire section (x-axis). (b) Resistance (in log scale) versus temperature of N2 (width = 170 nm), indicating $T_C = 5.0\pm0.2$ K. The inset is a SEM image of this nanowire (vertical line with white contrast) and the two horizontal wires for the measurement of voltage during the resistance measurements, where the current flows through the nanowire. Reprinted with permission from [31]. Copyright (2009), SPIE.

We have studied two types of superconducting NWs using $W(CO)_6$ as the precursor gas. The first type consists of microwires with width around 2 μm and thickness around 1 μm. The second type consists of NWs with width in the range of 100–250 nm and thickness in the range of 100–150 nm. The aim of these experiments is to characterize "bulk-type" deposits in the micrometer range in order to investigate if narrow wires in the nanometer range still keep the same performance as the bulk ones. A full report on these wires has been published in Reference 31. Here, we summarize the main results. As can be noticed in Figure 23.8(a), where microwires are denoted starting with the letter "M" and nanowires with the letter "N," all the investigated wires display a large T_C, in the range 4.6–5.4 K, despite the relatively large dispersion in the value of room-temperature resistivity, in the range 100–600 μΩ cm. As an example, the results obtained in one of the NWs, with width ~170 nm, are shown in Figure 23.8(b). A sharp superconducting transition is observed at $T_C \sim 5$ K, indicating good sample homogeneity. The critical current found in these NWs is in the range of 10^5 A/cm^2 at 2 K, which opens the route to use these NWs for several applications in nanotechnology based on superconducting materials, as well as for basic studies of low-dimensional superconductors. In particular, these FIBID W nanodeposits have been used as a playground for basic studies of the mechanisms for the depinning and melting of the vortex lattice in 2D superconductors [32]. Applications are under way in the fabrication of micro/nano-SQUIDs at targeted places for high-resolution magnetometry at the nanoscale, in the fabrication of resistance-free nanocontacts for dedicated transport experiments, and in nanocontacts for the study of interesting phenomena like Andreev reflection and measurements of the spin polarization of ferromagnets.

3.3 COMPARISON WITH FEBID NANOWIRES

So far, we have outlined the wide range of application of NWs grown with the FIBID technique and illustrated with the examples of Pt and W-based NWs. At this point, it is tempting to compare the FIBID technique with the sister technique named FEBID, where electrons are used instead of ions for triggering the precursor molecule dissociation in order to grow NWs. One obvious difference of FEBID NWs with respect to FIBID is the absence of Ga implantation. This can be an advantage or a disadvantage, depending on the functionality desired in the NW and will be illustrated hereafter with two examples, those of FEBID NWs grown using Pt-based and Co-based precursors.

3.3.1 FEBID of Pt nanowires

In Figure 23.9, we compare the *in situ* resistance of Pt samples with similar width (1 μm) and thickness grown by FEBID and FIBID using $(CH_3)_3Pt(CpCH_3)$ as precursor gas. In the case of the shown FEBID deposits, we have used the conditions: beam current = 0.54 nA, beam energy = 10 kV, dwell time = 1 μs. After total deposition time of 21 minutes, the final resistance is found to be 23.5 MΩ. In the case of the FIBID deposits shown, we have used the conditions: beam current = 10 pA, beam energy = 30 kV, dwell time = 200 ns. After total deposition time of 12 minutes, the final resistance is found to be 1 kΩ. Even though the final thickness of all the deposits is roughly the same ($\sim$160 nm), the resistance is four orders of magnitude lower for Pt by FIBID. For these nanodeposits of thickness 160 nm, the resistivity value in the case of the Pt by FEBID is about 10^7 $\mu\Omega$ cm, whereas it is about 800 $\mu\Omega$ cm in the case of Pt by FIBID, in any case much higher than for bulk Pt, 10.8 $\mu\Omega$ cm.

The differences in resistivity of the FEBID and FIBID samples are explained by the difference in the nanodeposit composition as shown by the EDS measurements [18]. As mentioned earlier, the microstructure of these deposits consists of Pt inclusions in a carbonaceous matrix. In the case of the FEBID deposit, EDS indicates Pt content lower than 20% and, consequently, the intrinsic semiconducting behavior of the carbonaceous matrix seems to dominate. This is the reason for the high resistivity of such deposits, which is at room temperature *six orders of magnitude* larger than that of bulk Pt. This demonstrates that the conduction does not take place through the Pt inclusions, which act only as doping centers of the carbonaceous matrix creating localized states in the band gap or in band tails. However, in the case of Pt by FIBID, the observed behavior resembles that of a highly doped semiconductor [22], which is not unexpected owing to the large amount of Pt inclusions and Ga impurities, altogether representing about 30–40% of the deposit content. The increase of disorder in this case, caused by the ion implantation, can produce tail states, decreasing the forbidden band gap. Thus, the measured room-temperature resistivity in the Pt by FIBID is only a factor of 100 higher than in bulk Pt and its temperature dependence is only slightly semiconducting, which indicates the presence of a large amount of available states in the gap and the proximity to metallic conduction.

After having established the high resistivity of the FEBID Pt NWs, one can wonder about the functionality of such deposits. Indeed, such high-resistance deposits can be useful in applications needing nanomaterials able to protect other layers from oxidation or contamination but at the same time requiring semi-insulating electrical behavior (see, for instance, Section 2 of this chapter). Thus, the in-principle bad news about the low conductivity of FEBID Pt deposits in applications such as for electrical contacts can be turned into good news for certain applications requiring low conductivity. Moreover, the bias voltage dependence of the conductivity of FEBID Pt deposits is peculiar, with a negative differential conductivity behavior at low voltage, which could have applications in high-frequency nanoelectronic devices [21].

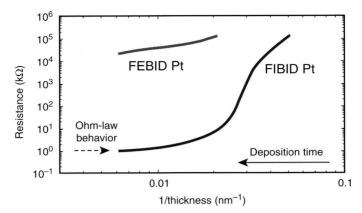

FIGURE 23.9: Representation of the resistance as a function of the inverse of thickness for one selected FEBID Pt NW and one selected FIBID Pt NW, which illustrates that the linear relationship expected if the Ohm-law applies only occurs for the FIBID Pt deposits above ≈ 100 nm.

3.3.2 FEBID of Co nanowires

An unexpected favorable behavior has been found in Co-based NWs grown with the FEBID technique [33–35]. Previous results of Co nanodeposits based on the $Co_2(CO)_8$ gas precursor indicated that relatively high Co content in the deposit was possible (<80%) using high beam currents ($\sim\mu A$) but with limited quality of their magnetic properties compared to pure Co [20; 36]. By using a field-emission electron emitter and high currents, it was possible to grow NWs with Co purity around 95% and excellent magnetic properties. This topic will be developed in detail in Chapter 16 of this book. Here, we just show the temperature dependence of resistivity in one of the grown NWs. The growth conditions are: room temperature, chamber pressure around 2.5×10^{-6} mbar, beam energy of 10 kV, and beam current of 2 nA. Typical dimensions for the deposits are length = 9 μm × width = 500 nm × thickness = 200 nm, with deposition time of the order of a few minutes when using high currents. As can be noticed in Figure 23.10, the temperature dependence of resistivity is the typical one of metallic systems, and the low-temperature resistivity, $\rho(0) \approx 27\,\mu\Omega$ cm, can be explained with the bulk parameters of Co and a short mean free path of 4 nm, which coincides with the grain size measured by TEM, suggesting that the scattering is mainly produced at the grain boundaries between crystalline grains. The reason for the high purity of these Co-based NWs is thought to lie on the relatively low temperature needed to dissociate the $Co_2(CO)_8$ precursor molecule, as found recently [37].

4 CONCLUSIONS AND OUTLOOK

In conclusion, we expect to have illustrated with a few examples the wide range of applications of focused ion/electron beam techniques for the creation of metallic nanowires and nanoconstrictions and the investigation of interesting electrical transport properties. Moreover, the commercial equipments where focused ion/electron beam sources are installed (as occurs in our case) allow the use of *in situ* characterization techniques, which favor the study of

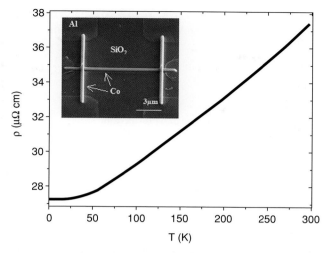

FIGURE 23.10: Resistivity of a FEBID Co NW grown at 10 kV beam energy and high current, 2 nA, as a function of temperature. Fully metallic behavior is observed. The inset is an SEM image of one of the created devices. The current flows through the horizontal deposit, whereas transversal deposits are performed for four-probe measurements. Deposits are done on top of a Si//SiO$_2$ substrate in which Al pads were previously patterned by conventional optical lithography techniques. Reprinted with permission from [33]. Copyright (2009), IOP.

correlations between the growth parameters and the physical properties. In particular, we have used *in situ* transport characterization while the nanowires and nanoconstrictions are being created in order to finish the process in the desired value of resistance. This approach has been crucial to study constrictions with atomic size and to obtain Pt nanowires in diverse values of resistance. Besides, by measuring the device resistance first inside the chamber and second after exposure to ambient conditions, it is possible to know the impact of the oxidation or contamination on the device resistance.

We can envisage fascinating applications of focused ion and electron beams in many applications of nanotechnology, and we think that the research with relatively new tools like "dual beam" systems is in its infancy and that in the coming years this topic will flourish. In this chapter we have emphasized the possibilities for creating very narrow constrictions, including atomic-sized constrictions, using the milling properties of focused ions. In addition, we have shown that nanowires with high functionality (metallicity or non-metallicity, superconductivity, ferromagnetism) can be grown using gas precursors and focused electron/ion beams to dissociate them, with an appropriate selection of the growth parameters.

ACKNOWLEDGMENTS

We acknowledge our collaboration with J. Sesé, S. Vieira, H. Suderow, and I. Guillamón in the topic of FIBID W nanodeposits. We acknowledge our collaboration with J. V. Obona for the creation of nanoconstrictions. This work was supported by the Spanish Ministry of

Science (through projects MAT2008–03636-E, MAT2005–05565-C02, and MAT2008–06567-C02, including FEDER funding), and the Aragon Regional Government.

REFERENCES

[1] Ferry, D. K. Materials science: Nanowires in nanoelectronics. *Science* 319 (2008): 579.

[2] Parkin, S. S. P., M. Hayashi, and L. Thomas. Magnetic domain-wall racetrack memory. *Science* 320 (2008): 190.

[3] Cowburn, R., and D. Petit. Spintronics: Turbulence ahead. *Nature Materials* 4 (2005): 722.

[4] Lu, W., and C. Lieber. Nanoelectronics from the bottom up. *Nature Materials* 6 (2007): 841.

[5] Doudin, B., and M. Viret. Ballistic magnetoresistance? *J. Phys.: Cond. Matter.* 20 (2008): 083201.

[6] (a) Faulkner, C. C., D. A. Alwood, and R.P. Cowburn. Tuning of biased domain wall depinning fields at Permalloy nanoconstrictions. *J. Appl. Phys.* 103 (2008): 073914; (b) Khizroev, S., et al. Focused-ion-beam-fabricated nanoscale magnetoresistive ballistic sensors. *Appl. Phys. Lett.* 86 (2005): 042502; (c) Céspedes, O., et al. Magnetoresistance and electrical hysteresis in stable half-metallic $La_{0.7}Sr_{0.3}MnO_3$ and Fe_3O_4 nanoconstrictions. *Appl. Phys. Lett.* 87 (2005): 083102.

[7] Khizroev, S., and D. Litvinov. Focused-ion-beam-based rapid prototyping of nanoscale magnetic devices. *Nanotechnology* 15 (2004): R7.

[8] Utke, I., P. Hoffmann, and J. Melngailis. Gas-assisted focused electron beam and ion beam processing and fabrication. *J. Vac. Sci. Technol B* 26.4 (2008): 1197 and references therein.

[9] Sharvin, Yu V. A possible method for studying Fermi surfaces. *Sov. Phys.-JETP* 21 (1965): 655.

[10] Agraït, N., A. L. Yeyati, and J. M. Ruitenbeek. Quantum properties of atomic-sized conductors. *Phys. Reports* 377 (2003): 81.

[11] Landauer, R. Spatial variation of currents and fields due to localized scatterers in metallic conduction. *IBM J. Res. Dev.* 1 (1957): 223.

[12] Fernández-Pacheco, A., J. M. De Teresa, R. Córdoba, and M. R. Ibarra. Exploring the conduction in atomic-sized metallic constrictions created by controlled ion milling. *Nanotechnology* 19 (2008): 415302.

[13] (a) Fernández-Pacheco, A., J. M. De Teresa, R. Córdoba, and M. R. Ibarra. Tunneling and anisotropic-tunneling magnetoresistance in iron nanoconstrictions fabricated by focused-ion-beam. *Mater. Res. Soc. Symp. Proc.*, 1181 (2009): 1181-DD02-03; (b) Obona, J. V., J. M. De Teresa, R. Córdoba, A. Fernández-Pacheco, and M. R. Ibarra. Creation of stable nanoconstrictions in metallic thin films via progressive narrowing by focused-ion-beam technique and in-situ control of resistance. *Microelectronic Engineering* 86 (2009): 639.

[14] Velev, J., R. F. Sabirianov, S. S. Jaswal, and E. Y. Tsymbal. Ballistic anisotropic magnetoresistance. *Phys. Rev. Lett.* 94 (2005): 127203.

[15] Bollotin, K. I., et al. From ballistic transport to tunnelling in electromigrated ferromagnetic breakjunctions. *Nanoletters* 6 (2006): 123.

[16] Viret, M., M. Gaboureac, F. Ott, C. Fermon, C. Barreteau, G. Autes, and R. Guirado-Lopez. Giant anisotropic magneto-resistance in ferromagnetic atomic contacts. *Eur. Phys. J. B* 51 (2006): 1.

[17] Sokolov, A., C. Zhang, E.Y. Tsymbal, J. Redepenning, and B. Doudin. Quantized magnetoresistance in atomic-size contacts. *Nature Nanotechnology* 2 (2007): 171.

[18] De Teresa, J. M., R. Córdoba, A. Fernández-Pacheco, O. Montero, P. Strichovanec, and M. R. Ibarra. Origin of the difference in the resistivity of as-grown focused-ion- and electron-beam-induced Pt nanodeposits. *Journal of Nanomaterials* (2009): 936863.

[19] Moodera, J. S., and G. Mathon. Spin polarized tunneling in ferromagnetic junctions. *J. Magn. Mag. Mat* 200 (1999): 248–273.

[20] Boero, G., I. Utke, T. Bret, N. Quack, M. Todorova, S. Mouaziz, P. Kejik, J. Brugger, R. S. Popovic, and P. Hoffmann. Submicrometer Hall devices fabricated by focused electron-beam-induced deposition. *Appl. Phys. Lett.* 86 (2005): 042503.

[21] Fernández-Pacheco, A., J. M. De Teresa, R. Córdoba, and M. R. Ibarra. Conduction regimes of Pt-C nanowires grown by focused-ion–beam induced deposition: Metal-insulator transition. *Phys. Rev. B* 79 (2009): 174204.

[22] Mott, N. F., and E. A. Davis, *Electronic processes in non-crystalline materials.* New York: Oxford University Press, 1971.

[23] Ohya, K., and T. Ishitani. Monte Carlo study of secondary electron emission from SiO_2 induced by focused gallium ion beams. *Appl. Surf. Sci.* 237 (2004): 606.

[24] Yu, K., Arutyunov, D. S. Golubev, and A. D. Zaikin. Superconductivity in one dimension. *Physics Reports* 464 (2008): 1.

[25] Chibotauro, L. F., A. Ceulemans, V. Bruyndoncx, and V. V. Moschalkov. Symmetry-induced formation of antivortices in mesoscopic superconductors. *Nature* 408 (2000): 408; Grigorieva, I. V., et al. Long-range nonlocal flow of vortices in narrow superconducting channels. *Phys. Rev. Lett.* 92 (2004): 237001.

[26] Velez, M., et al. Superconducting vortex pinning with artificial magnetic nanostructures. *J. Magn. Magn. Mater.* 320 (2008): 2547.

[27] (a) Granata, C., E. Exposito, A. Vettoliere, L. Petti, and M. Russo. An integrated superconductive magnetic nanosensor for high-sensitivity nanoscale applications. *Nanotechnology* 19 (2008): 275501; (b) Wu, C. H., et al. Fabrication and characterization of high-Tc YBa2Cu3O7-x nanoSQUIDs made by focused ion beam milling. *Nanotechnology* 19 (2008): 315304.

[28] Sadki, E. S., S. Ooi, and K. Hirata. Focused-ion-beam-induced deposition of superconducting thin films. *Appl. Phys. Lett.* 85 (2004): 6206.

[29] (a) Guillamón, I., et al. Nanoscale superconducting properties of amorphous W-based deposits grown with focused ion beam. *New Journal of Physics* 10 (2008): 093005; (b) Guillamón, I., et al. Superconducting density of states at the border of an amorphous thin film grown with focused ion beam. *Journal of Physics: Conference Series* 150 (2009): 052064.

[30] Luxmoore, I. J., et al. Low temperature electrical characterization of tungsten nano-wires fabricated by electron and ion beam induced chemical vapour deposition *Thin Solid Films* 515 (2007): 6791.

[31] De Teresa, J. M., et al. Transport properties of superconducting amorphous W-based nanowires fabricated by focused-ion-beam-induced-deposition for applications in nanotechnology. *Mater. Res. Soc. Symp. Proc.* 1180 (2009): 1180–CC04–09.

[32] Guillamón, I., et al. Direct observation of melting in a 2-D superconducting vortex lattice. *Nature Physics* 5 (2009): 651.

[33] Fernández-Pacheco, A., J. M. De Teresa, R. Córdoba, and M. R. Ibarra. High-quality magnetic and transport properties of cobalt nanowires grown by focused-electron-beam-induced deposition. *J. Phys. D: Appl. Phys.* 42 (2009): 055005.

[34] Fernández-Pacheco, A., et al. Domain wall conduit behavior in cobalt nanowires grown by focused-electron-beam induced deposition. *Appl. Phys. Lett.* 94 (2009): 192509.

[35] Fernández-Pacheco, A., et al. Magnetization reversal in individual cobalt micro- and nano-wires grown by focused-electron-beam-induced-deposition. *Nanotechnology* 20 (2009): 475704.

[36] Utke, I., J. Michler, Ph. Gasser, C. Santschi, D. Laub, M. Cantoni, A. Buffat, C. Jiao, and P. Hoffmann. Cross-section investigations of compositions and Sub-structures of tips obtained by focused electron beam deposition. *Adv. Eng. Mater.* 7 (2005): 323.

[37] Córdoba, R., J. Sesé, J. M. De Teresa, and M. R. Ibarra. In-situ purification of cobalt nanostructures grown by focused-electron-beam-induced deposition at low beam current. *Microelectronic Engineering* 87 (2010): 1550.

C H A P T E R   2 4

STRUCTURE-PROPERTY RELATIONSHIP OF ELECTRONIC TRANSPORT IN FEBID STRUCTURES

Michael Huth

1 INTRODUCTION

At large, structures prepared by FEBID fall into the class of disordered electronic materials in which disorder exists in varying degree, depending on the process parameters and the used precursor, ranging from a few impurities in an otherwise well-ordered polycrystalline host, to the strongly disordered limit of amorphous materials. In between these extremes the material can have the microstructure of a nanocomposite and is formed of reasonably well ordered nanocrystals embedded into a carbon-rich dielectric matrix. For those FEBID structures which fall into the weak disorder limit, electronic transport can be described by the scattering of Bloch waves by impurities. The theoretical framework for calculating the transport coefficients is the Boltzmann equation for the quasiparticles. At sufficiently low temperatures the electrical conductivity can be written as:

$$\sigma(T) = \sigma_0 - A\sigma_0^2 T^n \tag{1}$$

where σ_0 is the residual conductivity due to impurity scattering, A is a material-dependent positive constant, and $n \geq 2$ is a positive integer. If electron-electron scattering is dominating,

n equals 2. With increasing temperature, additional scattering channels become relevant, such as electron-phonon scattering. For nanocomposite and, naturally so, for amorphous materials it is not possible to use this conceptual framework. Disorder must be included in the theoretical analysis right from the beginning. This implies that two additional aspects must be taken into account. The first aspect is Anderson localization, which is related to the (spatial) structure of the wave function for a single electron in the presence of a random potential. The second aspect deals with interactions between electrons in the presence of this same random potential. Electron propagation for highly disordered materials is diffusive, which leads to substantial modifications to the view derived from Landau's Fermi liquid theory.

From most of the electronic transport experiments on FEBID structures that can be found in the literature it becomes readily apparent that such Anderson localization effects, which are well understood in the weak-disorder limit for uncorrelated electrons, are by far dominated by much stronger effects due to the interplay of disorder and interactions. For the latter, there is no complete theoretical framework available. On this account only a brief introduction of basic aspects of Anderson localization will be given here. The focus will be on some new developments in the field of disordered electronic materials, and in particular, on granular electronic systems to which most of the FEBID structures can be assigned. The compilation of recent theoretical results, as presented in this chapter, is not specific to materials prepared by FEBID but covers certain aspects of granular electronic metals in general. For an in-depth study of electronic transport in granular electronic systems, for which as yet no textbooks are available, the reader is referred to the recent review by Beloborodov and collaborators [14] and references therein.

1.1 BASICS OF ANDERSON LOCALIZATION

In 1958 Anderson pointed out that the electron's wave function in a sufficiently strong random potential may become localized [5]. This means that the envelope of the wave function decays exponentially from some localization point $\vec{r}_0$

$$\psi(\vec{r}) \sim exp(-|\vec{r} - \vec{r}_0|/\xi) \tag{2}$$

with the localization length ξ. The propensity for localization depends strongly on dimensionality and it can be rigorously shown that in one dimension all states are localized, even if the random potential is arbitrarily weak. An important point is the structure of the wave function for intermediate disorder in two and three dimensions. Coming from the itinerant picture of extended states one can argue that the states in the band tails, for which the dispersion is flat and the group velocity becomes small, have the strongest tendency to become localized in the (deepest) wells of the random potential. Mid-band states stand the best chance to remain extended in the presence of disorder. Consequently, there exists a critical energy at which a change of the wave function character from localized to extended takes place. This energy is called mobility edge [35]. If the chemical potential μ (Fermi energy) is in the range of localized states, the conductivity will vanish at zero temperature. The conductivity will remain finite, if μ lies in the range of extended states. The question arises whether the transition from zero to finite conductivity at zero temperature is continuous as μ crosses the mobility edge, e.g., by varying the degree of disorder or changing μ. Mott [37] stated, by referring to an idea from Ioffe and Regel [29], that the transition should be discontinuous in three dimensions. His argument assumes that the result from Boltzmann theory for weak scattering

$$\sigma = \frac{ne^2\ell}{mv_F} \qquad (3)$$

still holds at the transition. ℓ is the elastic mean free path. This would then imply a jump of the conductivity from zero to

$$\sigma^{(min)} \sim \frac{ne^2}{\hbar k_F^2} = \frac{n^{1/3}e^2}{(3\pi^2)^{2/3}\hbar} \qquad (4)$$

by setting $\ell \sim 1/k_F$. This argument was later extended to two dimensions, resulting in a length scale independent minimal conductivity $\sigma^{(min)} \sim 0.1e^2/\hbar$. In contrast, the scaling theory of localization states a continuous transition in three dimensions. In two dimensions it predicts that all states are localized for arbitrarily weak disorder. An important issue, however, must be kept in mind. Scaling concepts were developed within the model of non-interacting electrons. They are a direct consequence of the proper inclusion of quantum interference effects in disordered systems. On the other hand, in the zero-disorder limit it is known that Coulomb correlations can drive a metal-insulator transition as well [34]. This has been at the focus of intense research over many decades [28]. So, the relevant question to ask is: What is nature of the metal-insulator transition in disordered systems in the presence of Coulomb correlations? This question is not answered yet.

1.2 PHYSICAL QUANTITIES CHARACTERIZING GRANULAR MATERIALS

Granular metals constitute one-, two-, or three-dimensional arrays of (mesoscopically different) metallic particles—or grains—which are subject to an intergranular electron coupling due to a finite tunneling probability. The arrangement of the particles, with a typical size range from a few nm to 100 nm, can be regular or irregular. For FEBID structures the tunnel coupling is provided by the carbon-rich matrix. The matrix may also contain individual metal atoms or few-atom clusters. It represents itself a disordered electronic system and can give rise to additional conductance channels due to activated transport between localized states in the matrix. This has to be taken into account for FEBID structures with a very small volume fraction of metallic particles. In most instances it can be neglected.

Depending on the zero-temperature limit of the conductivity one discriminates *metallic* samples, showing $\sigma(T = 0) > 0$, and *insulating* samples, for which $\sigma(T = 0) = 0$. The effects of disorder in the grain positions and in the strength of tunnel coupling are less important for *metallic* samples which are characterized by strong intergranular coupling. For low tunnel coupling, i.e., for *insulating* samples, the effects of irregularities become crucial and have a direct influence on the temperature dependence of the conductivity. As a consequence of the formation processes in FEBID the obtained samples are highly disordered.

Electric transport within the metallic grains can be considered to be diffusive. In general, the grains will have internal defects or defects located at their surface. Trapped charges in the matrix will change the local potential of individual grains. Even if the elastic mean free path inside the grains exceeds the grain diameter, multiple scattering at the grain surface leads to chaotic motion of the electrons, which is equivalent to assuming diffusive transport inside the grains due to intragranular disorder [17]. Nevertheless, the mean spacing

δ between the one-electron levels inside the grains is still a well-defined quantity. It is given by

$$\delta = \frac{1}{N_F V} \qquad (5)$$

where $V \sim r^3$ is the grain volume and N_F denotes the density of states at the chemical potential. For grains with a diameter of a few nm, as is typically the case for FEBID structures, δ is of the order of 1 K for metallic grains with density of states of the order of 1 eV^{-1}nm^{-3}. Accordingly, quantum size effects due to the discrete energy levels are only relevant at very low temperatures. For semiconducting grains δ amounts to several eV even for grain sizes of several 10 nm due to the low density of states. In the following we will focus on granular materials with metallic grains.

The key parameter governing most of the electronic properties of granular metals is the average tunnel conductance G between neighboring grains. This is most conveniently expressed as a dimensionless quantity $g = G/(e^2/\hbar)$ by normalization to the quantum conductance. Broadly speaking, *metallic* behavior will be observed, if $g \geq 1$, while samples with $g \leq 1$ show *insulating* behavior. The normalized conductivity within a grain is denoted as g_0, not to be confused with the quantum conductance $e^2/\hbar$, and the notion *granular metal* implies that $g_0 \gg g$.

Another important parameter is the single-grain Coulomb charging energy E_C

$$E_C = \frac{e^2}{2C} \qquad (6)$$

where $C \propto r$ is the capacitance of the grain with radius r. E_C is equal to the change in electrostatic energy of the grain when one electron is added or removed. For *insulating* samples, charge transport is suppressed at low temperatures due to this charging energy. In this respect, the *insulating* state is closely related to the well-known Coulomb blockade effect of a single grain connected via tunneling to a metallic reservoir. As can be seen from Equations 5 and 6, the average level spacing can become larger than the charging energy for small grains $r < r^*$, where r^* represents the grain radius which separates the regimes for which either the condition $E_C > \delta$ or $E_C < \delta$ holds. For FEBID structures prepared from the precursor $W(CO)_6$ and assuming bcc-W metallic particles with the corresponding density of states of bulk bcc-W one obtains $r^* \ll 1$ nm for any reasonable value of the static dielectric constant of the insulating matrix. In general, the assumption $E_C \gg \delta$ is well justified for metallic grains.

2 ELECTRONIC TRANSPORT IN GRANULAR ELECTRONIC MATERIALS

2.1 TRANSPORT REGIMES OF GRANULAR ELECTRONIC METALS

In the following, a short review will be given on the electronic transport properties of granular systems consisting of normal metallic grains. Examples for granular metals with superconducting grains prepared by FEBID are not known to the author. Recently, there has been some interesting work on superconducting structures prepared by focused (Ga) ion beam induced deposition (FIBID) employing the $W(CO)_6$ precursor [25]. Presently, it is not known whether a specific crystallographic phase of tungsten and its carbides of the grains, possibly including gallium, or a specific enhancement effect in the pair interaction is responsible for the observed critical temperature of about 5 K. It has been speculated that an increase of the electronic screening length and the associated increase in the electron-phonon coupling constant may be responsible for the relatively large critical temperature [31]. The increase of the electronic screening length would be a consequence of the proximity to a metal-insulator transition for samples prepared by FIBID. However, this issue is currently not settled and granular systems with superconducting grains will not be discussed any further in this contribution. For a recent review on superconductivity in granular systems we refer to [14].

The transport regimes of granular metals are classified according to the intergrain coupling strength g. In the strong-coupling limit, $g \gg 1$, a granular array has *metallic* properties. In the opposite regime, $g \ll 1$, the array is *insulating*. The insulating state is a consequence of the strong Coulomb correlations associated with the single-electron tunneling. Moreover, quantum interference effects may also play an important role in this regime, as has been briefly mentioned in the previous section. In the *metallic* regime and at sufficiently high temperatures neither Coulomb correlations nor quantum interference effects are relevant. Within a model of identical metallic grains on the sites of a cubic lattice with lattice constant a in one, two, or three dimensions d the global sample conductivity σ_0 is given by the Drude formula [14]

$$\sigma_0 = \frac{2e^2}{\hbar} g a^{2-d} \tag{7}$$

The factor 2 stems from the spin degeneracy. This equation is valid for grains with internal disorder, which causes a break of momentum conservation as the electrons tunnel from grain to grain. The intergrain conductance g is best considered as a phenomenological, effective parameter which controls the behavior of the system. For FEBID structures g cannot be derived from first principles for several reasons, such as the unknown grain size and coupling strength distribution and the ill-defined electronic properties of the matrix.

As temperature is reduced, Coulomb correlation and interference effects become important also in the *metallic* regime. As a consequence, the simple Drude relation does not hold anymore and the properties of granular metals may differ considerably from those observed in homogeneously disordered metals.

2.1.1 Insulating regime

The *insulating* regime of granular systems with metallic grains is the regime to which most of the existing work on FEBID structures can be assigned to. Figure 24.1 shows the temperature-dependent conductivity of two FEBID structures prepared from the $W(CO)_6$ precursor with a metal content of about 20 at.%. This same qualitative behavior is also observed for FEBID structures prepared with the platinum precursor $(CH_3)_3PtCpCH_3$.

The conductivity does not follow a simple Arrhenius law

$$\sigma(T) \sim \exp\left[-\left(\Delta/k_B T\right)\right] \tag{8}$$

as could be expected if a hard energy gap Δ in the excitation spectrum was present at the chemical potential. Such an activated behavior is very rarely observed in FEBID structures. Much

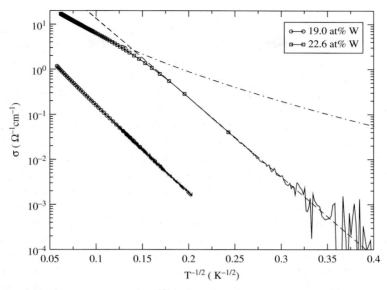

FIGURE 24.1: Plot of ln σ vs. $1/\sqrt{T}$ for two FEBID samples with the following composition (W:C:O): #1: (19.0:67.1:13.8) and #2: (22.6:56.0:21.4), respectively. Beam parameters during deposition have been (E:I:pitch:dwell time): #1: (10.0 keV:2.5 nA:20 nm:100 μs) and #2: (5.0 keV:3.3 nA:20 nm:100 μs), respectively. The dashed lines indicate fits according to Equation 9. Fit parameters (σ_0', T_0): #1: (15.5 Ω^{-1}cm^{-1}, 2090 K), #2: (770 Ω^{-1}cm^{-1}, 1638 K) for $T < 40$ K. For $T > 80$ K the fit for sample #2 follows Equation 10 with parameters (714 Ω^{-1}cm^{-1}, 50254 K).

more frequently the conductivity follows a stretched exponential temperature dependence of the form

$$\sigma(T) = \sigma_0' \exp\left[-\left(T_0/T\right)^{1/2}\right] \qquad (9)$$

This functional dependence was derived by Efros and Shklovskii for doped semiconductors [18]. Until very recently it remained a puzzle why this same functional form should be obeyed by disordered granular metals in the *insulating* regime. An early attempt to explain this behavior based on capacitance disorder due to the grain-size dispersion [6] was discarded for theoretical reasons. Capacitance disorder cannot fully lift the Coulomb blockade of an individual grain, so that a finite gap in the density of states at the chemical potential must remain which would necessarily lead to an Arrhenius behavior at low temperature [39, 53]. Experimentally, the same stretched exponential was also observed in periodic arrays of quantum dots [52] and periodic granular arrays of gold particles with very small size dispersion [48]. In these systems, capacitance disorder was very weak. From this arises the assumption that another type of disorder, unrelated to the grain-size dispersion, is necessary for lifting the Coulomb blockade of a single grain and can lead to a finite density of states at the chemical potential. In recent theoretical work it was suggested that electrostatic disorder, most probably caused by charged defects in the insulating matrix (or substrate), is responsible for lifting the Coulomb blockade [53]. Carrier traps in the insulating matrix at energies below the chemical potential are charged at sufficiently low temperature. They induce a potential of the order $e^2/\kappa r$ on the closest granule at a distance r. κ denotes the (static) dielectric constant of the matrix. For two-dimensional granular arrays one can also assume that random potentials are induced by charged defects in the substrate.

In order to fully account for the observed stretched exponential there must furthermore be a finite probability for tunneling to spatially remote states close to the chemical potential. This is analogous to Mott's argument in deriving his *variable range hopping* (VRH) law for disordered electronic systems in the non-correlated case [36]

$$\sigma(T) = \sigma_0' \exp\left[-\left(T_0/T\right)^{1/4}\right] \quad \text{for} \quad d = 3 \tag{10}$$

Hopping transport over distances exceeding the average distance between adjacent granules in sufficiently dense granular arrays can in principle be realized as tunneling via virtual electron levels in a sequence of grains, which is also called elastic and inelastic *co-tunneling*. This transport mechanism was first considered by Averin and Nazarov as a means to circumvent the Coulomb blockade effect in single quantum dots [4]. In elastic co-tunneling the charge transfer of a single electron via an intermediate virtual state in an adjacent grain to another state in a grain at a larger distance is at fixed energy. In inelastic co-tunneling the energies of the initial state and final state differ, so that the electron leaves behind in the granule electron-hole excitations as it tunnels out of the virtual intermediate state. The intermediate states are, as the qualification *virtual* indicates, not classically accessible. The inelastic co-tunneling channel dominates at temperatures above $T_1 \sim \sqrt{E_C\delta}/k_B$. The co-tunneling mechanism was generalized to the case of multiple co-tunneling through several grains and it was shown that the tunneling probability falls off exponentially with the distance or the number of grains involved [13, 20, 48]. This is equivalent to the exponentially decaying probability of tunneling between states near the chemical potential in the theory of Mott, Efros, and Shklovskii for doped semiconductors [18, 46] and eventually leads to the expression given in Equation 9. T_0 is a characteristic temperature which depends on the microscopic characteristics of the granular material in the *insulating* regime. Explicit expressions for T_0 in the elastic and inelastic co-tunneling regime can be found in Section II.G of a recent theoretical review by Beloborodov et al. [14].

The hopping length for transport via virtual electron tunneling decays as the temperature increases. At some temperature $\tilde{T}$ it will be reduced to the average grain size, so that only tunneling to adjacent grains is possible. The variable range hopping scenario of Mott, Efros, and Shklovskii no longer applies. It was suggested that the conductivity should follow an Arrhenius behavior for $T \geq \tilde{T}$ [14]. This does not appear to be the case for sample #2 shown in Figure 24.1 which rather shows a crossover to Mott's variable range hopping with exponent $1/4$.

2.1.2 Metallic regime

As the metal content in FEBID structures increases, the situation gets more complicated. In principle one can enter the regime of percolation at a critical metal volume fraction $y = y_c$ in the deposits. Such a percolating path of directly touching metallic grains can short-circuit the remaining part of sample which has activated transport behavior. Experimentally, this can be studied by a critical exponent analysis of the dependence of the conductivity vs. metal volume fraction $\sigma(y)$ at the smallest accessible temperature. From energy dispersive X-ray spectroscopy (EDX), most often used for element analysis of FEBID structures, the atomic abundance of the metal component x is obtained. In order to deduce the metal volume fraction y, the density and molar mass of the metal, ρ_M and M_M, and the insulating matrix, ρ_I and M_I, have to be known

$$y = \left(1 + \frac{1-x}{x}\frac{\rho_M M_I}{\rho_I M_M}\right)^{-1} \tag{11}$$

For the insulating matrix the density is very often not well-known. Depending on the degree of sp^2 to sp^3 hybridization of the carbon atoms the matrix can be diamand- or graphite-like. Furthermore, other contributions, such as oxygen in many instances, need to be taken into account. As a rough starting point for the matrix density, the range from 1.8 g/cm³ (graphite-like, amorphous) to 3.5 g/cm³ (diamond-like) using an averaged molar mass derived from the matrix composition can be used. In order to estimate whether percolation can play a role for a given volume fraction a suitable microstructural model has to be applied. For metallic grains in an amorphous matrix, a model of large metallic spheres in a matrix consisting of very small insulating spheres can be used [7]. This model predicts a critical volume fraction of $y_c = 0.64$.

As an example, Figure 24.2 shows the temperature-dependent conductivity of three samples prepared with the $W(CO)_6$ precursor with metal content above 30 at.%. The metal volume fraction can be estimated to lie between 37 vol.% and 57 vol.%, depending on the assumed matrix density. This is in either case below the percolation threshold y_c within the model assumption made above. Furthermore, the behavior of $\sigma(T)$ shows a qualitative change from activated to non-activated behavior as the metal content increases. This is in fact indicative of a insulator-metal transition as a function of metal content, as is shown in Figure 24.3 for a series of samples over a wide range of metal contents. Apparently, the curvature of $\sigma(T)$ changes sign as the metal-insulator transition is crossed. From these results it can be concluded

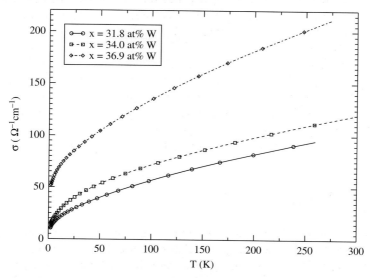

FIGURE 24.2: Temperature-dependent conductivity of samples in the *metallic* regime prepared from the precursor $W(CO)_6$. Sample compositions are (W:C:O): (31.8:44.4:23.8), (34.0:44.3:21.7), (36.9:35.6:27.5), respectively. Beam parameters during deposition have been (E:I:pitch:dwell time): (5.0 keV:3.7 nA:20 nm:100 μs), (5.0 keV:6.6 nA:20 nm:100 μs), (5.0 keV:3.9 nA:20 nm:100 μs), respectively.

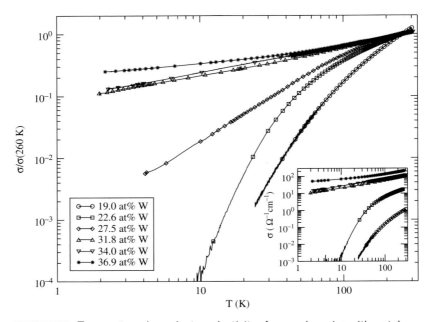

FIGURE 24.3: Temperature-dependent conductivity of a sample series with metal content *x* as indicated. The conductivity is shown normalized to its corresponding value at 260 K in log-log representation due to the large difference in conductivity between the low-metal content and large metal-content samples. Inset: Absolute values of the conductivity for the samples. The data for the sample with 27.5 at.% metal content is not shown since the sample cross section could not be determined sufficiently accurately.

that percolation of directly touching metallic particles is most likely not the reason for the clear change in the transport mechanism implied by the $\sigma(T)$ behavior of samples with larger metal content. It is very likely that the microstructure that forms as the metal content is increased is one in which (inelastic) tunneling prevails to large metal concentrations because the metallic nanocrystals do not touch directly. Rather, the tunneling probability grows with metal content because of grain size growth and a reduction of the intergrain spacing. The crossover to a different transport regime, which we discuss next, is not simply percolative but is tunneling, albeit in a more complex form which may need to take into account higher order effects in tunneling and also correlations. Microstructural aspects which have to be kept in mind are that the nanocrystals may often have a core-shell structure, with insulating shell, which hinders a direct percolative path to be formed. It would then be appropriate to assume the existence of a tunnel-percolation path, a concept that has recently received support in conductance atomic force microscopy measurements on granular Ni in a SiO_2 matrix [47].

We now turn to a brief discussion of the present theoretical understanding of the *metallic* transport regime of granular metals. In this regime the time τ_0 spent by an electron inside a grain is very much reduced due to the strong intergrain conductance g. From this arises a broadening Γ of the discrete energy levels inside the grains which is related to the tunnel conductivity

$$\Gamma = g\delta \tag{12}$$

In the limit $g \gg 1$, the level broadening very much exceeds the level spacing. The electron motion on length scales much larger than the average grain diameter is always diffusive and can be described by an effective diffusion constant

$$D_{eff} = \Gamma r^2 / \hbar \qquad (13)$$

with r as an average intergrain distance. Using Einstein's relation one obtains for the conductivity

$$\sigma_0 = 2e^2 N_F D_{eff} \qquad (14)$$

which is equivalent to Equation 7 for the ordered granular lattice, but is also applicable to disordered granular systems with proper definition of D_{eff}. From homogeneously disordered metals, two major causes for changes to the classical conductivity σ_0 have been identified, namely electron-electron interaction [2, 30] and quantum interference effects [3]. This can be used as a guiding principle for searching for corresponding corrections in granular metals.

We first discuss the consequences of electron-electron interaction. It turns out that the level broadening Γ defines an energy scale that separates two temperature regions in $\sigma(T)$. For $T > \Gamma/k_B$, Beloborodov et al. [9] find a logarithmic correction $\delta\sigma_1(T)$ for ordered granular arrays in one, two, and three dimensions

$$\delta\sigma_1(T)/\sigma_0 = -\frac{1}{2\pi dg} \ln\left(\frac{gE_C}{\max(k_B T, \Gamma)}\right) \qquad (15)$$

For $T < \Gamma/k_B$ this corrections saturates and $\sigma(T)$ is governed by a second temperature-dependent correction $\delta\sigma_2(T)$

$$\delta\sigma_2(T)/\sigma_0 = \begin{cases} \frac{\alpha}{12\pi^2 g} \sqrt{k_B T/\Gamma} & : \quad d = 3 \\ -\frac{1}{4\pi^2 g} \ln(\Gamma/k_B T) & : \quad d = 2 \\ -\frac{\beta}{4\pi g} \sqrt{\Gamma/k_B T} & : \quad d = 1 \end{cases} \qquad (16)$$

with $\alpha \approx 1.83$ and $\beta \approx 3.13$ being numerical constants. The full temperature-dependent conductivity is now given as

$$\sigma(T) = \sigma_0 + \delta\sigma_1(T) + \delta\sigma_2(T) \qquad (17)$$

The $\delta\sigma_2$ correction is quite analogous to the result obtained for homogeneously disordered metals [1]. It reflects the universal character of the large-scale behavior of disordered systems. The first correction $\delta\sigma_1$ is unique to granular metals. It results from the weakened screening in granular metals, which enhances correlation effects, as compared to homogeneously disordered metals. In particular, in three-dimensional granular metals the correlation effects can become the main driving force of a metal-insulator transition.

Presently, unambiguous experimental verification of this predicted behavior appears not to be available. The logarithmic correction was speculated to be observed in conductivity measurements on FEBID structures employing a platinum precursor [43]. The results shown in Figure 24.2 can indeed be fit quite well using a modified power law of the form [26]

$$\sigma(T) = \sigma_0' + bT^\gamma \tag{18}$$

with $\gamma \approx 1/2$. However, the $\delta\sigma_2(T)$ correction for $d = 3$ should be small, as is evident from the prefactor in Equation 16 for $g > 1$. What is observed experimentally is a change of conductivity by a factor of 4 to 10 in the temperature range from about 2 K to 300 K. It remains to be shown whether proper inclusion of disorder in the theoretical model will shed more light on this unsolved issue.

In the last part of this section, some remarks concerning quantum interference effects or weak localization are in order. In the *metallic* regime these can be accounted for by a systematic perturbation expansion in the small parameter $1/g$. In leading order, electron-electron interaction and weak-localization corrections are independent. The weak-localization correction is a consequence of quantum interference of self-intersecting electron trajectories. It is relevant for one- and two-dimensional disordered systems. Since electron motion is not coherent over arbitrarily large distances, a dephasing length L_Φ and corresponding dephasing time τ_Φ with $L_\Phi = \sqrt{D_{eff}\tau_\Phi}$ is introduced. Quantum interference corrections are relevant, if L_Φ exceeds the average intergrain distance. For granular thin films ($d = 2$) and wires ($d = 1$) it is found [12, 8]

$$\delta\sigma_{wl}/\sigma_0 = -\frac{1}{4\pi^2 g}\ln(\tau_\Phi\Gamma/\hbar) \quad (d = 2) \quad \text{with} \quad \tau_\Phi = \hbar g/k_B T$$

$$\delta\sigma_{wl}/\sigma_0 = -\frac{1}{2\pi g}\sqrt{\tau_\Phi\Gamma/\hbar} \quad (d = 1) \quad \text{with} \quad \tau_\Phi = \left(\hbar^3 g/k_B^2 T^2\delta\right)^{1/3} \tag{19}$$

The weak-localization corrections can be readily suppressed by applying a (weak) magnetic field. This can be used to decide whether quantum interference effects are relevant for a given sample. Presently, the author is not aware of experimental work on weak-localization effects in FEBID structures.

2.2 METAL-INSULATOR TRANSITION IN GRANULAR ARRAYS

Figure 24.3 shows that FEBID structures with varying metal content can exhibit a metal-insulator transition. The nature of this transition, in particular, whether it is of first or higher order at $T = 0$, is not known. This is only in part due to the inherent disorder in these systems. Also for the periodic models, which assume equal-size grains and disorder free intergrain conductance, the order of the metal-insulator transition cannot yet be predicted. Nevertheless, it is instructive to briefly review what is presently known for periodic granular metals.

For periodic granular arrays in the limit $g \ll 1$, the conductivity can be derived via a gauge transformation technique and it was shown that the electron excitation spectrum exhibits a hard energy gap, so that for the conductivity follows [19]

$$\sigma(T) = 2\sigma_0 \exp\left[-\left(E_C/k_B T\right)\right] \tag{20}$$

The prefactor 2 stems from the fact that the conductivity is governed by the presence of electrons and holes. If finite intergrain tunneling is taken into account, the activation energy or *Mott gap*

Δ_M is reduced from the pure charging Energy E_C, because of virtual tunneling processes of the electrons to neighboring grains. As a result, from perturbation theory Beloborodov et al. obtain [13]

$$\Delta_M = E_C - \frac{2zg}{\pi} E_{eh} \ln 2 \qquad (21)$$

where z is the coordination number in the granular array and E_{eh} is the energy necessary to create an electron-hole pair by moving one electron from a grain to its neighbor grain. Approximately this is equivalent to $2E_C$. From Equation 21 it is evident that for larger coupling strength $gz \sim 1$ the Mott gap is suppressed. However, in this case the perturbation expansion does not hold. From renormalization group analysis, one can obtain an exponentially suppressed Mott gap for $gz \sim 1$ [14]

$$\Delta_M \sim E_C \exp\left[-\pi gz\right] \qquad (22)$$

If the escape rate from a grain $(\Gamma/\hbar)^{-1}$ is larger than the rate set by the Mott gap $(\Delta_M/\hbar)^{-1}$, the equation does not hold anymore. By equating $\Delta_M \sim \Gamma$ and solving for g, one can obtain a *critical conductance* g_c, which separates the *insulating* from the *metallic* regime.

The theoretical analysis for either the *metallic* or the *insulating* side of the metal-insulator transition at $g \sim g_c$ for periodic arrays cannot be used in the vicinity of the transition itself, so the critical behavior of the Mott gap cannot be found. The order of the Mott transition in three-dimensional granular arrays is also not known. A weakly first-order transition cannot be excluded. One can speculate whether indications of a first order phase transition, i.e., one with a finite jump of the conductivity at $T = 0$, obtained for the Hubbard model for $d \geq 3$ can also be expected for granular arrays, because the physics is similar. However, there are differences. In granular arrays the electron motion is diffusive, both within the grains and on distances larger than the intergrain spacing (for $g > 1$). In the Hubbard model, electron movement is through a periodic lattice, which implies that the states can be labeled by their quasimomenta. Moreover, the intragrain energy level parameter δ has no correspondence in the Hubbard model. For two- and one-dimensional granular metals there is no Mott transition, as the conductivity corrections (Equation 16) diverge for $T \to 0$. In this case $g \sim g_c$ marks a crossover between regimes of activated transport for $g < g_c$ to weakly insulating behavior for $g > g_c$.

2.3 SHORT SUMMARY OF TRANSPORT REGIMES

For ease of reference Table 24.1 gives an overview of the different temperature-dependent conductivity regimes which have been deduced from theoretical analysis of granular electronic metals, both ordered and disordered, as presented in the previous sections. In either case a short statement is given whether the predicted behavior has been observed in samples prepared by FEBID. The region very close to the metal-insulator transition is out of the range of validity of current theories.

2.4 CURRENT-VOLTAGE CHARACTERISTICS

For the temperature-dependent conductivity measurements presented so far, care must be applied to take the data at small and fixed bias voltage, ideally in the linear-response regime.

Table 24.1 Reference scheme for electrical conductivity regimes of granular metals and their experimental observation in FEBID structures. The question marks indicate an undecided issue.

	Ordered	Experimental	Disordered	Experimental
Insulating	Equation 20	[44]	Eq. 9	e.g., [26]
Metallic	Eqs. 15, 16	[44]?	Eq. 15?	Eq. 15 in [43], Eq. 16 in [26]?

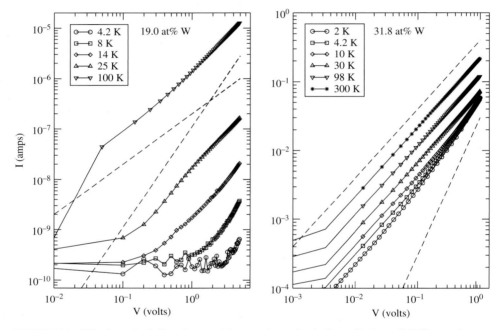

FIGURE 24.4: $I(V)$-characteristics at several temperatures (as indicated) of two FEBID samples prepared using the $W(CO)_6$ precursor. The dashed lines are for reference purposes only. They indicate linear and quadratic $I(V)$-dependencies. Left: Sample in the *insulating* regime. At low temperature a voltage threshold V_T is apparent. The $I(V)$-characteristic shows a steeper rise the lower the temperature. A simple linear or quadratic $I(V)$-dependence is not obeyed. Right: Sample in the *metallic* regime. At small bias and for all temperatures the sample shows a linear $I(V)$-characteristic. For larger bias and at low temperature a crossover behavior to a quadratic $I(V)$-characteristics is detectable. See text for details.

This is possible for *metallic* FEBID structures, as can be seen in Figure 24.4 (right). For samples with small metal content, one observes a voltage threshold V_T, so that current flow is observable only for $V > V_T$ at low temperatures, as is shown in Figure 24.4 (left).

We focus first on the sample in the *insulating* regime. For the data shown the maximum applied electric field amounts to 2,500 V/cm, which is rather small so that field-induced charge carrier release effects of the Poole-Frenkel type are not relevant. The current increase above the threshold voltage, which is rounded due to the distribution of Coulomb blockades in the disordered grain array, follows a power law of the form

$$I \propto (V/V_T - 1)^\zeta \tag{23}$$

The exponent ζ is temperature-dependent with values larger than 2 at low temperatures and saturating to about 1.4 at bias voltages above about 0.2 V and at higher temperatures. Such a behavior is in qualitative agreement with a model of capacitively coupled grains, allowing for tunnel currents and in the presence of electrostatic disorder, as was introduced by Middleton and Wingreen [38]. In this approach the charge and current distribution within the granular array was studied in one and two dimensions seeking for the bias voltage-dependent solution with minimal electrostatic energy. For large array sizes a critical exponent of 5/3 was predicted in two dimensions. Although this analysis is not directly applicable to the present situation, because of its dimensional limitation and its negligence of higher order tunneling effects, it is still relevant. It predicts the formation of individual current paths at small bias above threshold. These current paths show branching as the bias voltage is increased, which is associated with switching behavior in the $I(V)$-characteristic, as was experimentally observed for FEBID structures in the *insulating* regime [24] and also in recent experiments on artificial nano-dot arrays prepared by FEBID [44, 41]. A purely quantum mechanical approach that neglects the current-branching effects but takes multiple co-tunneling into account predicts in the elastic channel [48]

$$I \propto V \tag{24}$$

and in the inelastic channel

$$I \propto g^{N-1} V \left[\frac{(eV)^2 + (k_B T)^2}{E_C^2} \right]^{N-2} \quad (N \geq 2) \tag{25}$$

where N is the number of grains involved in the co-tunneling event. For $N = 1$ co-tunneling is replaced by simple activated transport. Co-tunneling is only relevant at low temperatures and would, in lowest order, predict a linear ($N = 2$) or cubic ($N = 3$) $I(V)$-characteristic. As evident from the prefactor g^{N-1}, co-tunneling is quickly suppressed for weak intergrain coupling. One could thus speculate that it might be negligible in the present case for samples in the *insulating* regime. However, this would be in contradiction to the current theoretical understanding of the reasons for the Efros-Shklovskii–type temperature-dependent conductivity, as detailed above. More work in this direction appears to be necessary.

We now turn to the sample in the *metallic* regime. As is apparent from Figure 24.4 (right), the $I(V)$-characteristic is linear for all temperatures at small bias. At low temperatures and bias voltages above 0.2 V, a crossover to a quadratic dependence is observed. The latter might be due to local heating effects. In the present case current densities above 10 kA/cm² at 1 V bias were reached at 2 K. The linear behavior at small bias and at low temperatures might be due to multiple inelastic co-tunneling involving an average of $N = 2$ neighboring grains. However, one has to consider that the co-tunneling scenario neglects the electrostatic optimization developed by Middleton and Wingreen in the classical regime [38]. In more general terms, if samples in the *metallic* regime are indeed similar to homogeneously disordered metals [14], then a linear $I(V)$-characteristic would be expected in any case.

2.5 HIGHER-ORDER TRANSPORT COEFFICIENTS

Investigations on higher-order transport coefficients of FEBID structures are scarce. For magnetic structures Hall effect measurements have been performed with a view on deducing the magnetization from the anomalous Hall effect contribution [21]. However, these measurements have been performed for FEBID structures prepared with the $Co_2(CO)_8$ precursor and their metal content was, in one instance, close to 100% [21]. In this case the particularities of granular materials are not exhibited. Other transport coefficient, such as the thermopower or heat conductivity, are difficult to measure. The measurements require that one establish an appreciable temperature gradient over the sample, which is hard to realize for sample, areas in the few 10 μm^2 range or less, as is typical for FEBID.

In collaboration with the Institute for Microtechnologies, a chip design and experimental procedure has been devised which allows for measuring the thermopower of FEBID structures [51]. This design is also the basis for planned measurements of the heat conductivity. In Figure 24.5 preliminary results for the thermopower of a FEBID structure prepared using the $W(CO)_6$ precursor is shown. For comparison, the temperature-dependent conductivity is also depicted which follows Mott's VRH law, according to Equation 10, over the temperature range for which thermopower data could be obtained. At low temperatures correlation effects become important and the Efros-Shklovskii–type behavior, according to Equation 9, is observed. The temperature-dependent Seebeck coefficient is negative and follows an approximate $\sqrt{T}$-behavior in the temperature range for which data are available.

We now turn to a brief discussion of the present theoretical understanding of the Seebeck coefficient in granular metals. Until very recently, no analytical results obtained from a

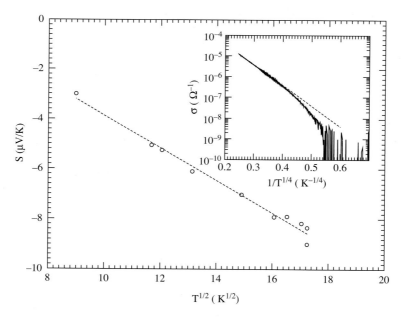

FIGURE 24.5: Temperature-dependent Seebeck coefficient of FEBID structure prepared using the $W(CO)_6$ precursor. Metal content is approximately 22 at.%. The dashed line indicates a $\sqrt{T}$ temperature dependence. See text for details. The inset shows the temperature-dependent conductivity of the same FEBID sample. A crossover from Mott to Efros-Shklovskii–type VRH is observed.

microscopic model for coupled grains were available. Based on the grain lattice model, developed by Beloborodov et al. [14], Glatz and Beloborodov have considered the thermopower in the *metallic* regime [23]. They derived the following expression for the thermopower $S(T)$

$$S(T) = S^{(0)} \left(1 - \frac{\pi - 2}{4\pi g d} \ln\left(\frac{g E_C}{k_B T} \right) \right) \tag{26}$$

where $S^{(0)} = -\pi^2/6e \cdot k_B^2 T/E_F$ is the thermopower in the absence of Coulomb interactions. Phonon contributions have been neglected which become relevant only at higher temperature. To the author's knowledge, thermopower measurements on FEBID structures in the *metallic* regime have not been done so far. For the insulating regime, in which Coulomb correlations are particularly important, no microscopic theory exists that takes Coulomb correlations and disorder into account on an equal basis. Extensions of the so-called percolation model of the VRH problem in the Mott regime (see, e.g., [33] and references therein) predict that in the presence of Coulomb interactions the Seebeck coefficient should become temperature-independent. For FEBID structures this has not yet been verified. In the Mott VRH regime the percolation model yields (see, e.g., [15])

$$S(T) = \frac{5}{42} \frac{k_B^2 \sqrt{T_0 T}}{e} \left[\frac{d \ln N(E)}{dE} \right]_{E=E_F} \tag{27}$$

where $N(E)$ denotes the energy-dependent density of states. The preliminary data shown in Figure 24.5 appears to support the predicted $\sqrt{T}$ temperature dependence, although further experiments will have to corroborate this conclusion. In particular, it will be highly interesting to extend these measurements to lower temperatures and to FEBID samples covering a wide range of metal content.

This section is concluded with a brief remark on the thermal conductivity of granular metals. In granular metals the vibronic contribution to heat conduction is strongly suppressed as compared to fully crystalline materials. As a consequence, the electronic part dominates and can yield valuable insight into the electronic transport mechanisms which are at work in granular materials. In particular, experiments on thermal transport are important since they allow for a direct observation of deviations from ordinary Fermi liquid theory. Recent theoretical work has focused on studying the thermal conductivity in the *metallic* regime in ordered granular lattices [11]. Deviations from the Wiedemann-Franz law are found, if Coulomb correlation effects are taken into account

$$\delta\kappa = \kappa - L_0 T \sigma = \begin{cases} \gamma T/a & : \quad d = 3 \\ \frac{T}{3} \ln\left(\frac{g E_C}{k_B T} \right) & : \quad d = 2 \end{cases} \tag{28}$$

where L_0 is the Lorentz number and $\gamma \approx 0.355$ is a numerical constant.

Experimentally it is a challenge to measure the electronic heat conductivity in a mesoscopic system at low energy. This would be necessary to verify the predicted deviations from Fermi liquid behavior. Work along these lines for FEBID structures is in progress in the author's group.

3 MICROSTRUCTURAL ASPECTS OF FEBID STRUCTURES DEDUCED FROM *IN SITU* CONDUCTIVITY MEASUREMENTS

The FEBID process involves different types of electrons (PE, BSE, SE) whose abundance depends in part on the substrate material. This substrate changes as the deposit grows. For thin deposits it is mainly the substrate proper. For deposits of a few 100 nm thickness it is the previously deposited material. For intermediate thicknesses it is both. Since the substrate strongly determines the microstructure of the growing deposit, a thickness-dependent microstructure in the deposit might result. In addition, the growing material is constantly bombarded with electrons as the deposition process goes on. As long as volatile residual atomic or molecular species are embedded in the FEBID structure, this electron bombardment can cause further changes. *In situ* electrical conductivity measurements on FEBID structures are valuable in this regard. They provide information about transient effects in the growth process and the response of the material to post-irradiation. With a view to applications of FEBID structure in electronic transport, they also help to elucidate aging effects as the materials are exposed to air. This has been appreciated recently [10, 50, 40, 49, 22]. Some precautions must be taken when doing this type of measurements. The data should be taken at fixed and low bias voltage of typically 1 mV to 10 mV. Under constant-current conditions the highly resistive state of the deposits in the early growth stage will result in bias voltages that can be in the several volts range. As a consequence, preferential conductance paths can be formed [24], so that the results of measurements on the growing deposits are hard to interpret.

In the left part of Figure 24.6 exemplary conductivity measurements during the deposition of FEBID structures with final metal content between 8 at.% and 39 at.% using the $W(CO)_6$ precursor are shown. The deposition starts when the precursor enters the vacuum chamber at $t = 10$ s. The conductivity then changes at a rate $\sigma' = \sigma/t$. It grows monotonically with time. The sample with the largest metal content shows a rapid increase of σ' in the first seconds of deposition following a power law of the form $\sigma' \propto t^\alpha$ with $\alpha = 3$. After that, σ' drastically decreases, tending to become constant, as is to be expected, if the thickness of the deposit grows linearly with time and the conductivity itself shows no transient effects. The exponent α is smaller for deposits with lower metal content, amounting to 1.86 and 1.43 for the samples with 8.1 at.% and 14.7 at.% metal content, respectively. For the sample with the lowest metal content, this power-law behavior is observed until the end of the deposition process. The growth process terminates with the simultaneous shut-off of the electron beam and the precursor gas supply. Correspondingly, the conductivity starts to decrease. The relaxation of the conductivity can be attributed to the migration of excess electrons injected by the electron beam toward the electrodes. This relaxation follows a logarithmic time dependence.

In general, a linear increase of the time-dependent conductivity can be explained by supposing a linear time-dependent growth of the thickness. However, the measurements performed during growth show an increase of the conductivity versus time faster than linear. Therefore, an additional mechanism which contributes to the enhancement of the conductivity has to be considered. Possible additional mechanisms for the increase of the conductivity include an enhancement of the electron carrier density and of the metal particles' size during deposition, both due to the electron beam irradiation. In [40] a shell-growth model of the metallic nanocrystals was developed which is based on the Lifshitz-Slyozov coarsening law [32]. In conjunction with an assumed variable range hopping conduction mechanism according to

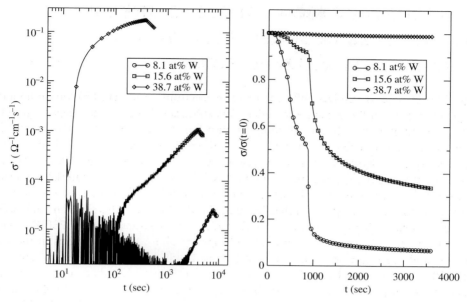

FIGURE 24.6: Left: Electrical conductivity measurements during the growth of FEBID structures using the W(CO)$_6$ precursor. Metal content as indicated. The deposition was started at $t = 10$ s. The electrical conductivity of each sample decreases once the deposition process is terminated, as indicated by the kinks in the data. Right: Electrical conductivity versus time as the SEM is vented. $t = 0$ corresponds to about one hour after completion of the deposition process. At this time any further relaxation effects in the conductivity proved negligible. During the venting procedure first nitrogen enters the chamber. After approximately 6 minutes the door of the microscope slightly opens; 8 minutes later the door is fully open. The conductivity decreases with time at a rate which depends on the material composition. See text for details.

Equation 10, this model can account for the observed time-dependence of σ'. The observed logarithmic time dependence of post-growth relaxation finds an explanation in the frame of hierarchically constrained dynamics which is imposed by the Coulomb interaction between the tunneling electrons, as detailed in [40].

Conductivity measurements versus time as the FEBID structures are exposed to air during the SEM venting process appear in Figure 24.6 on the right. The conductivity degradation rate is smaller as one considers deposits prepared at larger beam current (power) and, thus, having a larger density of the matrix, as detailed in [40]. In order to better describe the degradation process, the experimental data for times $t \geq 900$ s, when the SEM chamber door is fully opened, were fit with a sum of two exponential functions

$$\sigma(t) = a_1 \exp(-t/\tau_1) + a_2 \exp(-t/\tau_2) \tag{29}$$

In [40] the first relaxation time is associated with irreversible changes in the insulating matrix influencing such tunnel processes which involve matrix states. The second relaxation time is thought to be reflecting irreversible changes of the tunnel probability g between the grains. In any case, the pronounced aging effects, in particular for samples with small metal content, have to be taken into account in interpreting transport data.

4 OUTLOOK

Charge transport in granular metals has been at the focus of intense research for many years. Recently, with the advent of new and advanced theoretical approaches to the interplay of correlation and disorder, interest in these materials has been heightened again. FEBID can play an important role in this regard. It allows for the precise fabrication of granular metals with lateral resolution on the nanometer scale. Consequently, structures can be prepared covering the range from one to three dimensions. At the same time, it is possible to change the correlation strength by many orders of magnitude by means of the average grain size and intergrain coupling. For selected precursors this can be extended toward the metal-insulator transition in granular metals, which is considered to be one of the most difficult theoretical problems in solid state physics. Very recent work on electron irradiation effects on deposits made with the precursor $(CH_3)_3Pt(CpCH_3)$ represents a particularly useful way to address the issue of the metal-insulator transition [42]. Via prolonged electron irradiation and associated changes in the inter-granular coupling strength it proved possible to tune the Pt-C-deposits from insulating to metallic. This has also been shown to have a direct positive impact for using FEBID structures as sensor elements in strain sensing applications [27, 45]. Recent developments toward magnetic FEBID structures covering the range from the granular *metallic* regime to the coherent metallic regime, with metal contents approaching 100%, will allow us to extend these studies toward systems with additional spin degrees of freedom. Moreover, ferromagnetic FEBID structures offer a particularly elegant pathway to micromagnetic studies, as discussed elsewhere in this book, as well as to investigations of superconductor-ferromagnet hybride systems [16].

ACKNOWLEDGMENTS

Financial support by the *NanoNetzwerkHessen*, the Beilstein-Institut Frankfurt am Main through the research collaboration NanoBiC and by the *Bundesministerium für Bildung und Forschung* under grant 0312031C is gratefully acknowledged.

REFERENCES

[1] B. L. Altshuler and A. G. Aronov. Zero bias anomaly in tunnel resistance and electron-electron interaction. *Solid State Communications*, 30(3): 115–117, 1979.

[2] B. Altshuler and A. Aronov. *Electron-electron interaction in disordered conductors*. North Holland, Amsterdam, 1985.

[3] E. Abrahams, P. W. Anderson, D. C. Licciardello, and T. V. Ramakrishnan. Scaling theory of localization: Absence of quantum diffusion in two dimensions. *Phys. Rev. Lett.*, 42(10): 673–676, Mar 1979.

[4] D. V. Averin and Yu V. Nazarov. Virtual electron diffusion during quantum tunneling of the electric charge. *Phys. Rev. Lett.*, 65(19): 2446–2449, Nov 1990.

[5] P. W. Anderson. Absence of diffusion in certain random lattices. *Phys. Rev.*, 109(5): 1492–1505, Mar 1958.

[6] B. Abeles, P. Sheng, M. D. Coutts, and Y. Arie. Structural and electrical properties of granular metal films. *Advances in Physics*, 24: 407, 1975.

[7] I. Balberg and N. Binenbaum. Scher and zallen criterion: Applicability to composite systems. *Phys. Rev. B*, 35(16): 8749–8752, Jun 1987.

[8] C. Biagini, T. Caneva, V. Tognetti, and A. A. Varlamov. Weak localization effects in granular metals. *Phys. Rev. B*, 72(4): 041102, Jul 2005.

[9] I. S. Beloborodov, K. B. Efetov, A. V. Lopatin, and V. M. Vinokur. Transport properties of granular metals at low temperatures. *Phys. Rev. Lett.*, 91(24): 246801, Dec 2003.

[10] A. Botman, M. Hesselberth, and J. J. L. Mulders. Investigation of morphological changes in platinum-containing nanostructures created by electron-beam-induced deposition. *Journal of Vacuum Science and Technology B*, 26: 2464–2467, 2008.

[11] I. S. Beloborodov, A. V. Lopatin, F. W. J. Hekking, R. Fazio, and V. M. Vinokur. Thermal transport in granular metals. *EPL (Europhysics Letters)*, 69(3): 435–441, 2005.

[12] I. S. Beloborodov, A. V. Lopatin, G. Schwiete, and V. M. Vinokur. Tunneling density of states of granular metals. *Phys. Rev. B*, 70(7): 073404, Aug 2004.

[13] I. S. Beloborodov, A. V. Lopatin, and V. M. Vinokur. Coulomb effects and hopping transport in granular metals. *Phys. Rev. B*, 72(12): 125121, Sep 2005.

[14] I. S. Beloborodov, A. V. Lopatin, V. M. Vinokur, and K. B. Efetov. Granular electronic systems. *Reviews of Modern Physics*, 79(2): 469, 2007.

[15] S. Boutiche. Unified thermopower in the variable range hopping regime. *arXiv [cond-mat.mtrl-sci]*, 0902.0200v1: 1–10, 2009.

[16] O. V. Dobrovolskiy, M. Huth, V. A. Shklovskij. Anisotropic magnetoresistive response in thin Nb films decorated by an array of co stripes. *Supercond. Sci. Techn.* 23, 125014, 2010.

[17] K. Efetov. *Supersymmetry in disorder and chaos*. Cambridge University Press, 1999.

[18] A. L. Efros and B. I. Shklovskii. Coulomb gap and low temperature conductivity of disordered systems. *Journal of Physics C: Solid State Physics*, 8(4): L49–L51, 1975.

[19] K. B. Efetov and A. Tschersich. Coulomb effects in granular materials at not very low temperatures. *Phys. Rev. B*, 67(17): 174205, May 2003.

[20] M. V. Feigelman and A. S. Ioselevich. Variable-range cotunneling and conductivity of a granular metal. *JETP Letters*, 81: 277, 2005.

[21] A. Fernandez-Pacheco, J. M. De Teresa, R. Cordoba, and M. R. Ibarra. Magnetotransport properties of high-quality cobalt nanowires grown by focused-electron-beam-induced deposition. *Journal of Physics D: Applied Physics*, 42(5): 055005 (6pp), 2009.

[22] A. Fernandez-Pacheco, J. M. De Teresa, R. Cordoba, and M. R. Ibarra. Metal-insulator transition in pt-c nanowires grown by focused-ion-beam-induced deposition. *Physical Review B (Condensed Matter and Materials Physics)*, 79(17): 174204, 2009.

[23] A. Glatz and I. S. Beloborodov. Thermoelectric properties of granular metals. *Physical Review B (Condensed Matter and Materials Physics)*, 79(4): 041404, 2009.

[24] C. H. Grimm, D. Klingenberger, and M. Huth. Tunnel percolation and current path switching in a granular metal. *Journal of Physics: Conference Series*, 150(2): 022019 (3pp), 2009.

[25] I. Guillamon, H. Suderow, S. Vieira, A. Fernandez-Pacheco, J. Sese, R. Cordoba, J. M. De Teresa, and M. R. Ibarra. Nanoscale superconducting properties of amorphous w-based deposits grown with a focused-ion-beam. *New Journal of Physics*, 10(9): 093005 (10pp), 2008.

[26] M. Huth, D. Klingenberger, Ch. Grimm, F. Porrati, and R. Sachser. Conductance regimes of w-based granular metals prepared by electron beam induced deposition. *New Journal of Physics*, 11(3): 033032 (17pp), 2009.

[27] M. Huth. Granular metals – from electronic correlations to strain-sensing applications. *J. Appl. Phys.* 107, 113709: 2010.

[28] M. Imada, A. Fujimori, and Y. Tokura. Metal-insulator transitions. *Rev. Mod. Phys.*, 70(4): 1039–1263, Oct 1998.

[29] A. F. Ioffe and A. R. Regel. *Progress in Semiconductors*, 4: 237, 1960.

[30] P. A. Lee and T. V. Ramakrishnan. Disordered electronic systems. *Rev. Mod. Phys.*, 57(2): 287–337, Apr 1985.

[31] W. Li, J. C. Fenton, P. A. Warburton. Focused-Ion-Beam Direct-Writing of Ultra-Thin Superconducting Tungsten Composite Films. *IEEE Transactions on Applied Superconductivity*, 19(3): 2819–2822, 2009.

[32] I. M. Lifshitz and V. V. Slyozov. The kinetics of precipitation from supersaturated solid solutions. *Journal of Physics and Chemistry of Solids*, 19(1–2): 35–50, 1961.

[33] N. V. Lien and D. D. Toi. Coulomb correlation effects in variable-range hopping thermopower. *Physics Letters A*, 261(1–2): 108–113, 1999.

[34] N. F. Mott. The basis of the electron theory of metals, with special reference to the transition metals. *Proceedings of the Physical Society. Section A*, 62(7): 416–422, 1949.

[35] N. F. Mott. Electrons in disordered structures. *Advances in Physics*, 16: 49, 1967.

[36] N. F. Mott. Conduction in non-crystalline materials – III. Localized states in a pseudogap and near extremities of conduction and valence bands. *Philosophical Magazine*, 19: 835, 1969.

[37] N. F. Mott. *Electronics and structural properties of amorphous semiconductors*. Academic, London, 1973.

[38] A. A. Middleton and N. S. Wingreen. Collective transport in arrays of small metallic dots. *Phys. Rev. Lett.*, 71(19): 3198–3201, Nov 1993.

[39] M. Pollak and C. J. Adkins. Conduction in granular metals. *Philosophical Magazine B*, 65: 855, 1992.

[40] F. Porrati, R. Sachser, and M. Huth. The transient electrical conductivity of W-based electron-beam-induced deposits during growth, irradiation and exposure to air. *Nanotechnology*, 20(19): 195301 (10pp), 2009.

[41] F. Porrati, R. Sachser, M. Strauss, I. Andrusenko, T. Gorelik, U. Kolb, L. Bayarjargal, B. Winkler, and M. Huth. Artificial granularity in two-dimensional arrays of nanodots fabricated by focused electron beam induced deposition. *Nanotechnology* 21: 375302, 2010.

[42] F. Porrati, R. Sachser, C. H. Schwalb, A. Frangakis, and M. Huth. Tuning the electrical conductivity of Pt-containing granular metals by postgrowth electron irradiation. *J. Appl. Phys.* (2011) 109: 063715.

[43] L. Rotkina, S. Oh, J. N. Eckstein, and S. V. Rotkin. Logarithmic behavior of the conductivity of electron-beam deposited granular pt/c nanowires. *Phys. Rev. B*, 72(23): 233407, Dec 2005.

[44] R. Sachser, F. Porrati, M. Huth. Hard energy gap and current-path switching in ordered two-dimensional nanodot arrays prepared by focused electron-beam-induced deposition. *Phys. Rev. B*, 80: 195416, 2009.

[45] Ch. H. Schwalb, Ch. Grimm, M. Baranowski, R. Sachser, F. Porrati, H. Reith, P. Das, J. Müller, F. Völklein, A. Kaya, and M. Huth. *A tunable strain sensor using nanogranular metals Sensors* 10, 9847: 2010.

[46] B. I. Shklovskii. *Electronic properties of doped semiconductors*. Springer, New York, 1988.

[47] D. Toker, D. Azulay, N. Shimoni, I. Balberg, and O. Millo. Tunneling and percolation in metal-insulator composite materials. *Phys. Rev. B*, 68(4): 041403, Jul 2003.

[48] T. B. Tran, I. S. Beloborodov, X. M. Lin, T. P. Bigioni, V. M. Vinokur, and H. M. Jaeger. Multiple cotunneling in large quantum dot arrays. *Phys. Rev. Lett.*, 95(7): 076806, Aug 2005.

[49] J. M. De Teresa, R. Cordoba, A. Fernandez-Pacheco, O. Montero, P. Strichovanec, and M. R. Ibarra. Origin of the difference in the resistivity of as-grown focused-ion- and focused-electron-beam-induced pt nanodeposits. *Journal of Nanomaterials*, 2009: 936863, 2009.

[50] I. Utke, P. Hoffmann, and J. Melngailis. Gas-assisted focused electron beam and ion beam processing and fabrication. *Journal of Vacuum Science & Technology B: Microelectronics and Nanometer Structures*, 26(4): 1197–1276, 2008.

[51] F. Völklein, H. Reith, M. C. Schmitt, M. Huth, M. Rauber, and R. Neumann. Microchips for the investigation of thermal and electrical properties of individual nanowires. *J. Electr. Mat.* 39: 1950, 2010.

[52] A. I. Yakimov, A. V. Dvurechenskii, A. I. Nikiforov, and A. A. Bloshkin. Phononless hopping conduction in two-dimensional layers of quantum dots. *JETP Letters*, 77: 376, 2003.

[53] J. Zhang and B. I. Shklovskii. Density of states and conductivity of a granular metal or an array of quantum dots. *Phys. Rev. B*, 70(11): 115317, Sep 2004.

CHARACTERIZATION AND MODIFICATION OF NANOSTRUCTURED CARBON MATERIALS USING FIB AND FEB

Alfredo R. Vaz, Carla Veríssimo, Rogério V. Gelamo, Francisco P. Rouxinol, and Stanislav A. Moshkalev

1 INTRODUCTION

Carbon nanotubes (CNTs) have been the subject of intense studies since their discovery in 1991, and this interest is due to the unique combination of exceptional electrical, mechanical, thermal, and other properties, and numerous potential applications. Single-wall carbon nanotubes (SWCNTs) can be metallic or semiconducting, while multi-walled nanotubes (MWCNTs) are basically metallic. Semiconducting SWCNTs can be used in nanotube-based field effect transistors, while metallic SWCNTs and MWCNTs can be employed for electrical and thermal interconnections, in sensors and other micro- and nanodevices [1]. However, the integration of nanotubes in microcircuits is a very challenging task that requires development of reliable and compatible technologies for controlled synthesis, characterization, accurate positioning, and contacting of nanotubes and their arrays in new devices. Mechanisms of electrical and thermal conductivity in individual nanotubes, in their bundles and films, the role of defects, formation of contacts with metals, and many other important issues have to be investigated thoroughly.

In this chapter, some examples will be given to demonstrate the wide use of FIB (and to some extent, FEB) in investigations involving nanotubes and more recently, graphene (or few-layer graphite, FLG), another nanocarbon material with great potential for applications [2]. The areas of studies include: characterization, irradiation for modification of properties, templating of substrates for controlled growth, and fabrication of devices.

In part, increasing use of particle beams for processing of nanocarbons is due to very specific character of energetic particles interaction with thin graphitic (or graphenic) layers, as shown in a recent review by Krasheninnikov and Banhart [3]. Irradiating solids with energetic particles is usually thought to induce disorder, normally considered undesirable. But it can also have beneficial effects, tailoring the structure and properties of nanosystems. In general, the irradiation-induced generation of defects in nanosystems is different from bulk materials due to the limited size of the system. In many cases, irradiation can lead to self-organization or self-assembly in nanostructures. Carbon nanoforms show very peculiar behavior under irradiation (at low doses), due to the ability of graphenic networks to reorganize their structures. Ions can traverse thin graphitic layers without inelastic interaction so that little energy is deposited, contrary to bulk systems where all the energy is eventually absorbed. The probability for energy loss of an impinging ion decreases for thinner layers and for higher energies.

Similarly to bulk materials, irradiation-induced defects in carbon nanostructures can be divided into point defects, such as interstitial-vacancy pairs, and defects of higher dimensions like dislocations in graphene sheets or mechanically strained nanotubes [3]. Carbon sp^2 nanostructures develop an extended reconstruction of the atomic network near the vacancy by saturating two dangling bonds and forming a pentagon. Di-vacancies and multi-vacancies also reconstruct, so that curved graphitic structures such as carbon nanotubes can be referred to as self-healing materials under irradiation. Along with vacancies and interstitials, defects of other types like pentagon/heptagon Stone-Wales defects may be formed in nanotubes. More complex defects like inter-shell covalent bonds in MWCNTs and defect-mediated covalent bonds between adjacent SWCNTs in the bundles may also exist. Finally, it is important that much of the irradiation-induced damage can be annealed when the irradiation is carried out at temperatures higher than 300°C through dangling bond saturation and an increased mobility of point defects [3], and *in situ* annihilation of irradiation-induced defects has been observed in carbon nanotubes heated to lower temperatures (200°C) under the electron irradiation [4].

Recently, Gierak et al. have demonstrated that extremely high resolution can be achieved using ions beams for processing of graphene layers [5]. In this study, 5 nm diameter 30 keV Ga+ ion beam for direct patterning of suspended graphene layers, while preserving the interesting physical properties of this material. The damaging by ions is dramatically reduced in extremely thin layers (that can be made ~1 nm or thinner in the case of SWCNTs or graphene, compared with ~22 nm projected range for incoming ions) due to much smaller contribution of both ion implantation and scattering effects. This makes thin graphene layers very attractive for studies of new physical phenomena in low-dimensional materials that can be patterned *in situ* by focused beams. Other applications of ultrathin graphite layers modified by FIB are considered below.

2 ELECTRICAL CHARACTERIZATION OF CARBON NANOTUBES AND OTHER NANOCARBONS USING FIB

As SWCNTs are essentially one-dimensional structures, and in the absence of defects they are characterized by a ballistic transport of electrons (without scattering) at moderate current densities for nanotube lengths up to a few micrometers [6]. This is in striking contrast to metal (copper) wires where the mean free pass (MFP), determined by the mean grain size, is in the range of a few tens of nanometers. That is the reason that nanotubes are considered as an alternative for copper in interconnects in microelectronics [7].

The resistance of a SWCNT (or a shell in a MWCNT) has three components [8; 9]: (1) a contact resistance associated with one-dimensional systems, given by a quantum resistance $G_0^{-1} = h/2e^2 = 12.9$ kΩ corresponding to one conducting state (a factor of two is added due to two possible spin states); (2) an intrinsic resistance due to scattering, which is length dependent; and (3) an additional contact resistance associated with imperfect contacts between a metal electrode and CNTs. In practice, the contact resistance, depending on the contact quality and the metal used, may be considerably higher that the intrinsic CNT resistance. As shown in experiments with gas sensors using CNTs and Pt nanowires [10], contact resistances R_c can be much higher (one-two orders of magnitude) than intrinsic resistances of these nanostructures. This can affect seriously the performance of the whole device, especially when Joule heating by current, passing in the functional element, is significant. In that work [10], gas sensors based on electrothermal effect (change of resistance due to cooling by gas) were studied, and R_c was shown to depend strongly on the local contact temperature. As heating depends on the resistance (power dissipated $\sim I^2R$), parasitic heating of contacts appeared to be very strong. It was shown that superposition of two opposite effects: negative temperature coefficient of resistance (TCR) for the contact area (reduction of R_c upon heating) and positive TCR for the functional element (increase of NW resistance by heating) could lead to erroneous results in gas sensing.

It is very challenging, however, to evaluate precisely the contribution of contact resistances due to evident experimental difficulties. Comparison between different experiments is also not straightforward because of wide variation of conditions and particular geometries used for studies (for example, side- and end- contacts, presence of surfactants and other contaminants, metal grain sizes, and degree of metal annealing after its deposition over nanotubes, etc.).

Metals that form carbides (like Ti) are believed to be an optimal contact electrode material, as it is expected that carbides ensure better electrical coupling with nanotubes [8]. Theoretical studies have indicated that Ti contacts have lower resistances followed by Pd, Pt, Cu, and Au [9]. This basically confirms the experimentally observed trends. For example, Zhang et al. [11] studied depositions of various transition metals on carbon nanotubes to gain an understanding of metal-tube interactions. Depending on the metal type, continuous metal (Ti, Ni, Pd) nanowires or isolated particles (Au, Al, Fe) were formed on nanotube substrates. Ti and Ni were found to interact strongly with the sidewall of nanotubes, likely due to partial covalent bonding between the metals and carbon atoms. Unfortunately, at present none of these two metals is available commercially for FIBID. Au and Al were found to interact weakly with SWCNTs through van der Waals forces.

Measuring the contact resistance is a very challenging task, and this usually requires a great number of experiments to give statistically averaged results. An interesting approach to measure the contact properties between an individual multi-wall nanotube and thin metal layer by using FIB has been recently developed by Lan et al. [12]. For this, sequential cuts by a ~ 10 nm diameter focused ion beam in the area of contact (sequentially reducing the contact length) were utilized. Electrical measurements were performed inside the FIB. Then, from the measured dependence of two-terminal resistance on the contact length, both specific nanotube resistance and contact resistance can be evaluated. For PE-CVD grown MWCNTs with diameters in the range of 50–60 nm and thin Ag metal film deposited by evaporation, following parameters were obtained: (1) nanotube resistances ~4.5 kΩ/μm, (2) specific contact resistances r_c were shown to depend strongly on the thickness of the Ag film, being of 38 kΩ μm and 1.6 kΩ μm (i.e., 6.4 kΩ for 2 contacts of 0.5 μm length each) for Ag layers of 23 and 63 nm, respectively. Further, from the relation $r_c = \rho_c/(\pi d/2)$, where ρ_c is the specific contact resistivity for a nanotube of diameter d, ρ_c values were determined: 35 and 1.3 $\mu\Omega$ cm^2, for 23 and 63 nm thick layers of Ag, respectively. Using these data, it was possible to conclude that the contribution of contact resistances (inversely proportional to its length) to a total two terminal resistance becomes insignificant for contact lengths exceeding 1 μm. Much higher values for thinner metal films are due to non-complete coverage of the nanotubes. These results establish a guide for optimization of the thickness and length of metal contacts. For layers thicker than the nanotube radius, low-resistance contacts were created.

Comparison of various methods to make electrical contact to nanowires (100 nm diameter, Au) and carbon nanotubes (50 nm diameter) was done by Langford et al. [13]. A number of techniques were compared in terms of the success yield, contact resistance, complexity of the fabrication steps, and potential for creating novel device structures and architectures. The methods compared were: (1) *in situ* micromanipulation of wires onto pre-patterned electrodes, (2) ion and electron beam assisted deposition, (3) electron beam lithography, and (4) drop casting of wires from solution onto pre-patterned electrodes. The ion and electron beam assisted deposition methods required the fewest process steps and were relatively faster. The room temperature contact resistances to the Au wires and nanotubes for the different methods and the calculated Au wire and MWCNT resistances were found to vary considerably. In particular, the contact resistances to MWCNTs were found to vary by up to a factor of 10^3 in different methods; the lowest resistance was reported for FIB ID Pt films (200 nm thick). In contrast, the highest contact resistance was obtained to Au NWs using FIB. This study, especially the great variation of results obtained using different methods, highlights the importance of development of reliable techniques for manipulation and electrical characterization of nanotube samples.

In another study, Brintlinger et al. [14] compared the quality of gold contacts to SWCNT produced by two methods: (1) electron beam lithography (EBL) and lift-off, (2) FEBID of gold film using AuClPF$_3$ precursor (25 kV, 100 pA beam current). The high quality of FEBID gold contacts (150 nm wide, 20 nm thick) was demonstrated, yielding comparable or even lower contact resistance compared with EBL (10 and 54 kΩ, respectively).

In addition to the methods compared above [13; 14], the technique of di-electrophoresis (DEP), using AC electric fields that align and attract polarizable objects to a pair of electrodes, proved to be very efficient for controlled deposition of carbon nanotubes and other nanoobjects from solutions [15–18]. By controlling the process conditions (concentration of the solution, time of deposition, AC field frequency), both nanotube films and individual tubes can be deposited, as can be seen in Figure 25.1(a) and (b). The method proved to be very selective: at low frequencies (500 Hz), impurities present in a solution (amorphous carbon, nanoparticles) are attracted to electrodes, while for frequencies higher than 0.1 MHz, mainly nanotubes are deposited under the same process conditions (compare Figure 25.1(b) and (c)). FIBID process was used to prepare four-terminals measurements of MWCNT electrical properties (Figure 25.1(d)).

Besides supported nanotubes as shown in Figure 25.1, FIB milling was used in DEP experiments to prepare suspended carbon nanostructures as shown in Figure 25.2. Before deposition, 5 μm deep and 0.5–2 μm wide cuts by FIB were made to produce separate metal (Au, Ti, W) electrode structures. In this case, it was possible to deposit not only suspended nanotubes (Figure 25.2(a)) but also individual thin (10–20 nm) layers of graphite or graphite

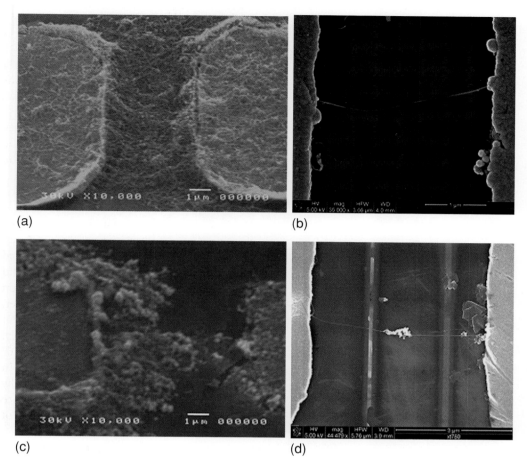

(a)   (b)

(c)   (d)

FIGURE 25.1: Results of deposition by di-electrophoresis from aqueous solutions with sodium dodecyl sulfate (SDS) surfactant. (a) MWCNT film, (b) individual MWCNT, (c) amorphous carbon, (d) MWCNT deposited over intermediate Pt metal electrodes prefabricated by FIB, for four-terminals measurements. AC field frequency: 10 MHz (a,b,d) and 500 Hz (c). Material of main electrodes fabricated by lithography: Pd.

oxide (Figure 25.2(b, c)). The former was prepared from HOPG (highly ordered pyrolytic graphite) by sonication and the latter by graphite powder oxidation in H_2SO_4 acid solutions using the Hummers method [19]. It is interesting that annealing in vacuum (650°C, 1 hour) resulted in partial reduction of the graphite oxide layers, as can be seen from I-V curves shown in Figure 25.2(d) before and after annealing, with the measured two-terminal resistance falling more than 10^4 times. It should be noted that deposition of thin graphite and graphite oxide layers was possible in a suspended configuration (above gaps) using the FIB prepared 3D electrode structures, while the conventional (supported on SiO_2 substrates) configuration presented relatively low yield for such processes.

The suspended configuration was employed to study heat transfer between a substrate and nanotubes self-heated due to Joule effect by current passing through nanotubes [20] and thin graphite layers [21]. In particular, it was shown that self-heating of MWCNTs due to Joule effect is much stronger for suspended nanotubes compared to those supported on SiO_2

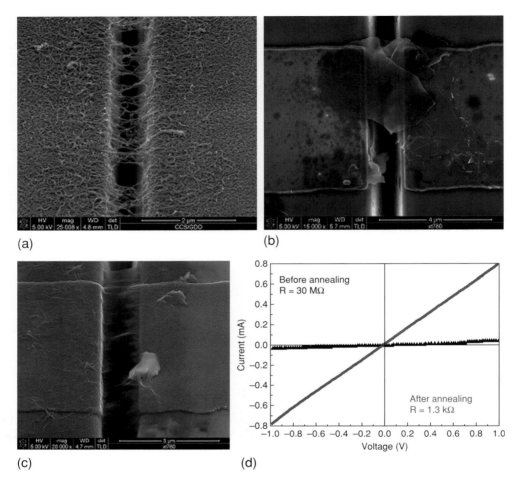

FIGURE 25.2: (a) Suspended carbon nanotubes, (b) thin graphite layer, (c) graphite oxide layers deposited by AC DEP (0.1 MHz) from dimethylformamide (DMF) solutions. (d) I-V curves for graphite oxide layer before and after thermal annealing in vacuum. Gap between metal (Ti) electrodes is made by FIB milling.

substrates [20]. Under low current conditions (a few μA per nanotube, at 0.5 V bias between electrodes), nanotubes were self-heated up to 30°C–50°C and 300°C–400°C for supported and suspended configurations, respectively [20]. This self-heating was shown to enhance strongly sensitivity to gases like O_2 for gas sensors based on carbon nanotubes decorated with metal (Ti) nanoparticles, as chemically active nanoparticles sitting on nanotubes were also heated to elevated temperatures.

Using supported configuration, Vaz et al. [16] carried out electrical characterization of multi-walled carbon nanotubes grown by thermal CVD. For reliable evaluation of the CNT electrical parameters, four-terminal measurements were performed using Pt nanocontacts formed by both focused ion and electron beams. Two different setups were used: (1) individual nanotubes deposited by AC dielectrophoresis over pre-fabricated system of four electrodes configuration (two main Pd electrodes fabricated by lift-off and two intermediate narrow FIBID Pt electrodes, as shown in Figure 25.1(d)); and (2) after DEP deposition over two

Pd electrodes, two narrow Pt contacts were fabricated over CNTs using less damaging FEBID process. Intrinsic resistances for a number of multi-walled nanotubes using both configurations were measured, and values of 90 ± 20 kΩ/μm were obtained for nanotubes with diameters in the range of 25–35 nm. More detailed discussion of electrical properties of multi-walled nanotubes, including the effect of intershell conductance, can be found elsewhere [16–18].

In another study, Rykaczewski et al. [22] also used AC di-electrophoresis to prepare samples with MWCNTs for analysis of contribution of various components to the total resistance of interconnects between metal electrodes and nanotubes. Nanotubes, 50 to 150 nm in diameter, were deposited over 200 nm thick chrome electrodes. After deposition, contacts with metal electrodes were improved using FEBID deposition of carbon joints (carbon precursor–residual hydrocarbons) over the nanotubes in the area of contact, with the beam parameters: 20–30 kV, 5–140 pA, 5 min. The influence of the FEBID-made carbon joints on the electrical resistance of the MWCNT-to-metal interconnects was evaluated, during and after the deposition of joints with varying shapes and volumes. It was found that the electron beam contributes to the measured current through an interconnect, in a two-terminal configuration. The secondary electrons generated in the course of inelastic collisions of primary electrons as they scatter in the material during the deposition process alter the *in situ* measured current significantly. The FEBID-deposited material and the SiO_2 layer were found to act as capacitors, "discharging" through the nanotubes during and after the electron beam irradiation. The analysis of different components contributing to the total joint resistance showed that the bulk resistance of the carbon joint (R_{aC}) is ultimately the most dominant factor, and the bulk carbon resistivity ρ_{aC} can be reduced via annealing. As fabricated, the total resistance was the sum of (1) the metal electrode–carbon joint interfacial resistance, (2) carbon joint resistance, and (3) the MWCNT–carbon joint interfacial resistance. After deposition of a sufficiently large FEBID amorphous carbon joint, the total resistance was limited solely by the R_{aC} value. The quality of contact was improved by partial graphitization in the carbon joints induced by Joule heating, raising the bias between electrodes up to 2.5 V. This was found to lower the resistivity of the deposited amorphous carbon by two orders of magnitude, from $\sim$2 to $\sim$0.05 MΩ. The final resistivity was estimated to be 5×10^3 Ω cm, still much higher than that of bulk graphite. In this case, the resistance of the MWCNT–metal interconnect was found to be limited by the resistance of the outer shell of the MWCNT. Further reduction in the total resistance of the MWNCT interconnect can only be achieved through establishing an electrical contact to the inner shells of the nanotube. Interestingly, in this study the electron beam was used for both formation of the metal/MWCNT contacts and testing the electrical parameters of these contacts.

In the study performed by Vaz et al. [16], the effect of Ga^+ ion beam irradiation on the nanotube resistance was investigated. For ion irradiation, repeated scans in an ion-imaging mode (30 kV, 30 pA) were applied for nanotubes deposited over narrow Pt contacts fabricated by FIBID (similar to Figure 25.1(d)). The single imaging dose was estimated to be $\sim$5 $\times$ 10^{15} ions/cm^2. After each irradiation cycle, four-terminals measurements were performed. The results show that a single scan in an ion imaging mode has a relatively small effect on the nanotube resistance ($\sim$10% rise), however strong changes were observed for accumulated irradiation doses (Figure 25.3).

Experiments with suspended multi-walled nanotubes have shown that defects produced by Ga^+ beam irradiation cause a visible tensile strain in a nanotube (straightening) already at a dose of 5×10^{15} ions/cm^2 (see Figure 25.4). For higher doses, the tube gets more and more narrow (Figure 25.4(c)), before eventual cutting. For the sequence shown in Figure 25.4, the nanotube diameter decreased from $\sim$50–60 nm to $\sim$30 nm (Figures 25.4(a)

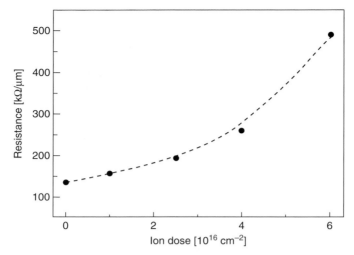

FIGURE 25.3: Effect of 30 keV Ga$^+$ ion irradiation on resistance of MWCNT.

and (c), respectively). The decrease in the nanotube diameter is consistent with increasing resistance, as shown in Figure 25.3.

One more application of FIB related to carbon nanotubes is demonstrated in Figure 25.5. In the study of mechanisms, responsible for nucleation of MWCNT in a process of catalytic chemical vapor deposition using CH$_4$ as a carbon source and Ni as catalyst [23; 24], the shape of nanoparticles was shown to be crucial. Here, fine FIB cross-sectioning was used to evaluate the shape of Ni nanoparticles formed over oxidized Si surface by thermal annealing, before CH$_4$ admission into the chamber (see Figure 25.5(a)). Cross-sectioning was performed using protective Pt-C layer previously deposited by FEBID at 0.4 nA current and 5 kV beam energy (Figure 25.5(b)). In a conventional cross-sectioning process through the protective layer (Figure 25.5(c)), the formation of so called curtaining below Ni particles (traces of Ga atoms forward scattered by the particle material) is evident. This phenomenon results in blurred images in the area of the particle-surface interface. To avoid this, samples were also cross-sectioned in the opposite direction (from the side of a silicon substrate, near its edge, with samples slightly inclined with respect to the normal direction), and then rotated 180° to obtain a normal image (see Figure 25.5(d)). No curtaining was formed in this case. Almost spherical shape of particles can be seen, with a large contact angle due to weak wetting of SiO$_2$ surface by Ni at the process temperature (800°C–900°C).

3 MODIFICATION OF CARBON NANOTUBES AND GRAPHENE USING FIB

Another group of studies is focused on site-selective modification and functionalization of nanotubes and graphene using focused ion beams [25; 26]. New hybrid structures based on nanocarbons allow access to new electronic, electrical, and magnetic properties and provide additional degrees of freedom that enable the assembly and interconnection of nanostructures

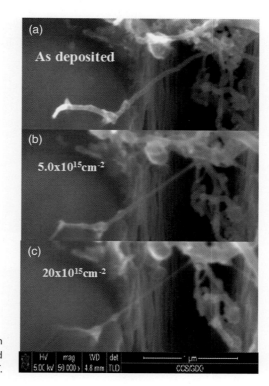

FIGURE 25.4: Effect of 30 kV, 30 pA Ga ion beam irradiation on structural properties of suspended MWCNT.

into architectures for nanodevice fabrication [25]. For example, decoration of carbon nanotubes by metal or metal oxide nanoparticles (NP) provides selectivity in gas sensing using CNT/NP nanostructures [20]. However, lack of control over the functionalization location in the CNTs remains a major challenge. Several strategies have been devised to functionalize CNTs in specific chemically active locations to harness them for applications such as CNT-filled composites, field emitters, sensors, and catalyst supports; the most known is a wet-chemical oxidation. The type and location of defects or adsorption sites created using these methods, however, are difficult to control. As an alternative, FIB irradiation and subsequent mild chemical treatments to functionalize multi-walled CNTs at preselected locations have been recently reported [25]. The CNT arrays rastered by FIB irradiation were air-exposed and treated with PDADMAC, a cationic polyelectrolyte known to enable electrostatic immobilization of Au nanoparticles on carboxylated CNTs. The Au nanoparticles attached only to the irradiated segments of the CNT, indicating selective carboxylation. No observable attachment was detected in the unirradiated regions. The nanoparticles were not dislodged from the irradiated segments, despite repeated washing and rinsing, indicating strong electrostatic anchoring. Typically, nanoparticle anchoring was observed when CNTs are irradiated with doses in the range $10^{15}-10^{17}$ cm^{-2} of Ga$^+$ ions at 10–30 keV, whereas dosing CNTs with $\leq 10^{13}$ cm^{-2} did not result in any attachment.

Ion beam nanopatterning in graphite was explored by Melinon et al. [26]. The main objective of this study was to create well-ordered 2D arrays of magnetic CoPt nanoparticles for a next generation of ultra-high density recording media. As a template, highly oriented pyrolithic graphite (HOPG) surface was used, and FIB was employed to functionalize the HOPG surface with a periodic array of identical controlled extended defects. Such functionalized

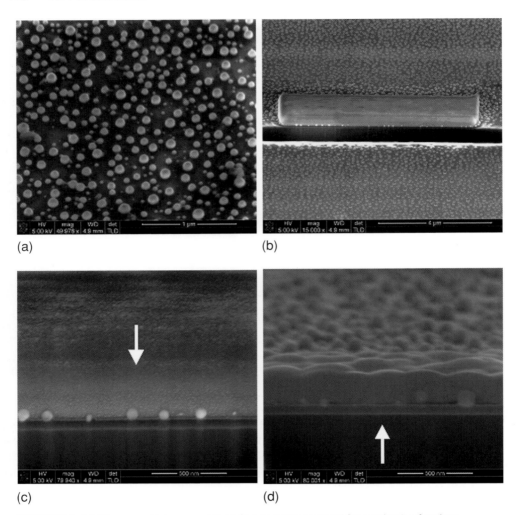

(a)   (b)

(c)   (d)

FIGURE 25.5: (a) Ni nanoparticles on oxidized Si substrate prepared for synthesis of carbon nanotubes using catalytical chemical vapor deposition process, tilt 52°. (b) Sample partly covered with Pt protective layer and cross-sectioned using FIB. (c) and (d) Cross-sections by FIB milling, arrows show ion beam directions.

graphite surfaces can be used as templates to trap nanoparticles on surfaces with controlled geometry and periodicity. The morphology and the electronic structure of a single focused ion-beam-induced artificial extended defect were probed by several methods, including micro-Raman spectroscopy, atomic force and scanning tunneling microscopies, and Monte Carlo and/or semi-analytical simulations.

In the experiment, square arrays of defects with lattice parameters of 300 and 1000 nm were patterned onto the samples with the ion dose varying from 10 ions/dot (2×10^{13} ions cm^{-2}) up to 10^6 ions/dot (2×10^{18} ions cm^{-2}). Cobalt-platinum magnetic clusters were synthesized in a standard laser vaporization source.

With confocal micro-Raman spectroscopy (laser spot diam. <1 μm, photon energy 2.41 eV), isolated extended defects could be studied for the 1000 nm lattice array [26]. It was shown that this method is very sensitive to defects created in HOPG, detectable even at lower doses (<100 ions/dot or 2×10^{14} ions/cm^{-2}) when AFM/STM observations did not reveal any deviation for flat surfaces. The so-called D band (1350 cm^{-1}) associated with defects appeared at doses as low as of 10 ions/dot, as well as other smaller features as shoulders for G peak at 1620 cm^{-1}. A careful analysis of the D region splitting and the G peak broadening gives information about the electronic structure of the extended defect, as well as in-plane disorder in the basal plane. Clear correlation between micro-Raman spectra and nanocluster pinning efficiency was found. For the small ion fluences, the multi-featured Raman structure is consistent with a simple softening of the selection rules near the defect, and the extended defect can be assimilated to a weak distortion of the graphite, that was found to be not fully efficient for cluster pinning. At moderate fluences, Raman features are consistent with a large selection rule softening, and an increase of the in-plane disorder inside the defect resulted in irreversible pinning of clusters by the defects (dangling bonds). For higher fluences, the in-plane disorder is large (dangling bonds and/or Ga$^+$ induced chemical effect). Then, clusters are efficiently trapped by the defects, and well-ordered cluster arrays can also be obtained.

In another report by the same group led by J. Gierak [4], fabrication of various test structures in suspended graphene layers such as nanowires and nanorings has been successfully demonstrated. Thin graphene (or few-layer graphite) layers were prepared by mechanical exfoliation of HOPG over prefabricated parallel 1.5 μm wide and 200 nm high gold contacts patterned on a SiO$_2$ surface. The thickness of graphene flakes was estimated to be 1–3 nm through evaluation of optical contrast using a high resolution optical microscope correlated with AFM measurements. Due to the extremely small thickness of flakes (projected range of 30 keV Ga ions in graphite is 22 nm), a very high processing resolution and sharp edges (no scattering and no redeposition) can be achieved. To avoid any problems with possible charging (although graphite is known by its high in-plane conductivity) and stress in the film, multi-scan patterning strategies with low step dose have been applied. After fabrication, two-terminal conductivity measurements were performed for a 1 nm thick, 40 nm wide and 1 μm long suspended graphene nanowire in 230 to 4 K temperature range. The first results demonstrated significant difference in the conductance behavior as compared with supported graphene structures; further studies are necessary. It should be noted that certain difficulties were found in obtaining low-resistance ohmic contacts between Au electrodes and graphene flakes (the discussion about the best metals for contacting graphitic carbon can be found above), so that mild ion irradiation was required to provide intermixing between gold and graphene, reducing contact resistances to values ~0.1 kΩ.

4 CONCLUSIONS

Focused ion and electron beams are increasingly used for characterization, contacting, processing of carbon nanostructured materials and fabrication devices based on them. Energetic Ga$^+$ ions inevitably produce defects in solids; however, damaging by the ions is dramatically reduced in extremely thin layers. Thin graphitic layers (that can be made ~1 nm or thinner in the case of SWCNTs or graphene) interact very weakly with ions, and contribution of both ion implantation and scattering effects is much smaller compared with bulk solids. This makes thin graphene layers very attractive for studies of new physical phenomena in low-dimensional materials that can be patterned *in situ* by focused beams.

REFERENCES

[1] Meyyappan, M., ed. *Carbon nanotubes: science and applications.*, Boca Raton, FL: CRC Press, 2005.

[2] Geim, A. K., and K. S. Novoselov. The rise of graphene. *Nature Materials* 6 (2007): 183–191.

[3] Krasheninnikov, A. V., and F. Banhart. Engineering of nanostructured carbon materials with electron or ion beams. *Nature Mater.* 6 (2007): 723–733.

[4] Lucot, D., J. Gierak, A. Ouerghi, G. Faini, and D. Mailly. Deposition and FIB direct patterning of nanowires and nanorings into suspended sheets of graphene. *Microelectron. Eng.* 86 (2009): 882–884.

[5] Hulman, M., V. Skákalová, A. V. Krasheninnikov, and S. Roth. Effects of ion beam heating on Raman spectra of single-walled carbon nanotubes. *Appl. Phys. Lett.* 94 (2009): 071907.

[6] Graham, A. P., G. S. Duesberg, R. Seidel, M. Liebau, E. Unger, F. Kreupl, and W. Honlein. Towards the integration of carbon nanotubes in microelectronics. *Diam. Relat. Mater.* 13 (2004): 1296–1300.

[7] Naeemi, A., and J. D. Meindl. Carbon nanotube interconnects. *Proc. of the 2007 Internat. Symp. Physical Design*, ISPD'07, pp. 77–84, March 2007, Austin, Texas, 2007.

[8] Tan, C.W., and J. Miao. Transmission line characteristics of a CNT-based vertical interconnect scheme. *Proc. IEEE 57th Electronic Components & Technology Conference*, 1936–1941. Sparks, Nevada, 2007.

[9] Matsuda, Y., W.-Q. Deng, and W. A. Goddard III. Contact resistance properties between nanotubes and various metals from quantum mechanics. *J. Phys. Chem* C 111 (2007): 11113–11116.

[10] Schwamb, Timo, Brian R. Burg, Niklas C. Schirmer, and Dimos Poulikakos. On the effect of the electrical contact resistance in nanodevices. *Appl. Phys. Lett.* 92 (2008): 243106.

[11] Zhang, Y., N. W. Franklin, R. J. Chen, and H. Dai. Metal coating on suspended carbon nanotubes and its implication to metal-tube interaction. *Chem. Phys. Lett.* 331 (2000): 35–41.

[12] Lan, C., D. N. Zakharov, and R. G. Reifenberger. Determining the optimal contact length for a metal/multiwalled carbon nanotube interconnect. *Appl. Phys. Lett.* 92 (2008): 213112–3p.

[13] Brintlinger, T., M. S. Fuhrera, J. Melngailis, I. Utke, T. Bret, A. Perentes, P. Hoffmann, M. Abourida, and P. Doppelt. Electrodes for carbon nanotube devices by focused electron beam induced deposition of gold. *J. Vac. Sci. Technol.* B 23 (2005): 3174.

[14] Langford, R. M., T.-X. Wang, M. Thornton, A. Heidelberg, J. G. Sheridan, W. Blau, and R. Leahy. Comparison of different methods to contact to nanowires. *J. Vac. Sci. Technol.* B 24 (2006): 2306–2311.

[15] Krupke, R., F. Hennrich, H. B. Weber, D. Beckmann, O. Hampe, S. Malik, M. M. Kappes, and H. V. Lohneysen. Contacting single bundles of carbon nanotubes with alternating electric fields. *Appl. Phys. A* 76 (2003): 397–400.

[16] Vaz, A. R., M. Macchi, J. Leon, S. A. Moshkalev, and J.W. Swart. Platinum thin films deposited on silicon oxide by focused ion beam: characterization and application. *J. Mater. Sci.* 43 (2008): 3429.

[17] Moshkalev, S.A., J. Leon, C. Verissimo, A. R. Vaz, A. Flacker, M. B. de Moraes, and J. W. Swart. Controlled deposition and electrical characterization of multi-wall carbon nanotubes. *J. Nano Res.* 3 (2008): 25–32.

[18] Leon, J., C. Verissimo, A. R. Vaz, A. Flacker, M. B. de Moraes, and S. A. Moshkalev. Electrical characterization of multi-walled carbon nanotubes. *J. Nanosci. Nanotechnol.* 10 (2010): 6234-6239.

[19] Jannuzzi, S. A. V., S. Dubin, M. J. Allen, J. Wassei, S. Moshkalev, and R. B. Kaner. Two different chemical routes to obtain graphene or few-layer graphite. *Smart Nanocomposites* 1 (2010): 1–12.

[20] Gelamo, R. V., F. P. Rouxinol, C. Verissimo, A. R. Vaz, M. A. B. de Moraes, and S. A. Moshkalev. Low-temperature gas and pressure sensor based on multi-wall carbon nanotubes decorated with Ti nanoparticles. *Chemical Physics Letters* 482 (2009): 302–306.

[21] Rouxinol, F. P., R. V. Gelamo, R. G. Amici, A. R. Vaz, and S. A. Moshkalev, Low contact resistivity and strain in suspended multilayer graphene. *Applied Physics Letters.* 97 (2010):.253104.

[22] Rykaczewski, K., M. R. Henry, S.-K. Kim, A. G. Fedorov, D. Kulkarni, S. Singamaneni, and V. V. Tsukruk. The effect of the geometry and material properties of a carbon joint produced by electron beam induced deposition on the electrical resistance of a multiwalled carbon nanotube-to-metal contact interface. *Nanotechnology* 21 (2010): 035202.

[23] Moshkalev, S. A., C. Verissimo. Nucleation and growth of carbon nanotubes in catalytic chemical vapor deposition. *Journal of Applied Physics*, 102 (2007): 044303.

[24] Verissimo, C., M. R. Aguiar, and S. A. Moshkalev, Formation of catalyst nanoparticles and nucleation of carbon nanotubes in chemical vapor deposition. *Journal of Nanoscience and Nanotechnology* 9 (2009): 4459–4466.

[25] Raghuveer, M. S., A. Kumar, M. J. Frederick, G. P. Louie, P. G. Ganesan, and G. Ramanath. Site-selective functionalization of carbon nanotubes. *Adv. Mater.* 18 (2006): 547–552.

[26] Mélinon, P., A. Hannour, L. Bardotti, B. Prével, J. Gierak, E. Bourhis, G. Faini, and B. Canut. Ion beam nanopatterning in graphite: Characterization of single extended defects. *Nanotechnology* 19 (2008): 235305.

ELECTRON BEAM CONTROLLED PATTERNING OF MOLECULAR LAYERS: FUNCTIONAL SURFACES AND NANOMEMBRANES

Armin Gölzhäuser

1 INTRODUCTION

Lithography is the preferred patterning process for the fabrication of microelectronic circuitry and micromechanical devices as well as for biosensors and biochips [1; 2]. A lithographic exposure tool utilizes radiation: photons, ions, or electrons to locally modify a material, typically a polymeric resist on a semiconductor substrate. The modified resist may be used as a nanostructure of its own, but is mostly applied as a mask for a subsequent pattern transfer. The resolution of a lithographic process is determined by the size of the exposure beam as well as by the dimension of the interaction area of the beam and the exposed material. These do not necessarily coincide, as one might combine small or large exposure spots with small or large resist molecules. To optimize lithographic nanofabrication, one thus needs to match an exposure tool with a resist material.

Electron beam lithography tools have very small spot sizes, and their beams can be focused in areas with diameters below ~2 nm. However, the typical resolution in e-beam lithography (~30 nm) is far from this limit, as the resolution of the created pattern is mainly determined by secondary electron processes in the polymeric resists. Forward scattering and proximity effects widen the beam, which leads to additional exposures, see the chapter by Melngailis and co-workers in this volume. To minimize these effects and to achieve a higher resolution, novel materials for e-beam lithography should thus not only show a specific sensitivity to electrons but should also be thin and composed of small molecular units.

Materials that are highly suitable for ultrahigh-resolution electron beam patterning are self-assembled monolayers (SAMs). SAMs are homogeneous, highly ordered films of organic molecules covalently anchored to a surface with a typical thickness of 1–2 nm and an intermolecular spacing of 1–0.5 nm [3; 4; 5]. SAMs can be tailored to adsorb on various materials, among them noble metals, semiconductors, and oxides. Since the adsorption of a SAM changes the surface properties, SAMs are used as model systems to study wetting, corrosion, lubrication, and adhesion. In biological applications, SAMs are used as linker molecules to couple cells and proteins to surfaces [6; 7; 8]. SAMs are routinely patterned by micro contact printing (μCP) [9], a versatile method in which a patterned elastomeric stamp prints SAM structures on a surface with a spatial resolution below 50 nm [10]. By wetting the tip of an atomic force microscope with molecules, self-assembled monolayer patterns with lateral dimensions down to $\sim$15 nm have been created [11].

SAMs also have been patterned by ionizing radiation [12; 13; 14], among them neutral atoms [15], ions [16], photons [17], and electrons [18]. Amino terminated aliphatic SAMs have been patterned with $\sim$80 nm resolution by electron-induced cleavage of the amino group and were used as templates for the directed adsorption of colloids and small particles on the areas not exposed to the beam [19]. Using focused electron beams, patterns with lateral dimensions down to $\sim$6 nm could be written into aliphatic SAMs [18].

SAMs thus combine a potential for highest resolution e-beam patterning and a variety of diverse surface functionalization applications, from corrosion protection to bio-interfaces. In this short contribution, I shall first review some basics on SAMS and their specific inter-actions with electrons. Then I shall show examples of how SAMs patterns are used as functional surfaces in nanofabrication, bio-surfaces. Finally, the fabrication of novel freestanding 1 nm thin carbon nanomembranes from electron irradiated SAMs and their applications are described.

2 SELF-ASSEMBLED MONOLAYERS

Self-assembled monolayers (SAMs) formed of amphiphilic molecules spontaneously form highly ordered films on solid surfaces. A molecule in a SAM consists of three building blocks: a head group that binds strongly to a substrate, a tail group that constitutes the outer surface of the film, and a spacer that connects head and tail (cf. Figure 26.1). The spacer essentially controls the intermolecular packing and the degree of order in the film, whereas the tail group determines the macroscopic surface properties, i.e., whether the SAM is hydrophilic or hydrophobic. The adsorption of SAMs allows the chemical functionalization (NH_2, NO_2, CH_3, OH, COOH, ...) of surfaces by using molecules with appropriate tail groups.

The first SAMs were alkyl silanes on Al_2O_3 and glass surfaces and have been investigated by Sagiv in 1978 [20]. In the early 1980s, Nuzzo and Allara [21] prepared the first alkanthiol-SAMs on gold surfaces, which ignited enormous research efforts, as the adsorption of thiols on noble metal surfaces is a simple way to produce highly ordered molecular films. After immersion of a gold surface into a thiol solution, a well-organized SAM spontaneously forms. Whitesides et al. [22] systematically studied applications of thiol SAMs, such as patterning [23] and protein resistance. Currently there are a large number of groups working in this field, and during the last 20 or so years, the SAM literature has grown exponentially.

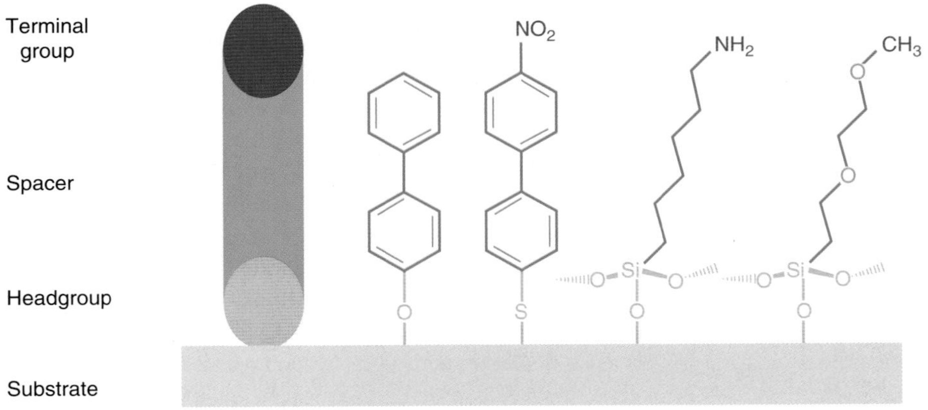

Substrate : Au, Ag, Si, SiO$_2$, Fe, Ti, ITO, Cr ...

Head group : –SH, –OH, –SiCl$_3$ ...

Spacer : Alkyl, Aryl, Ethylenglycol ...

Terminal group : –H, –COOH, –NO$_2$, –NH$_2$, –OH ...

FIGURE 26.1: Self-assembled monolayers (SAMs) formed from amphiphilic molecules that consist of the functional units: head group, tail group, spacer group.

Introductions to SAMs are found in several review articles [4; 5] and the monograph by Ulman [3]. Aromatic SAMs of biphenyl molecules were first described by Rubinstein [24]. Ulman [25; 26] investigated the structure of functionalized biphenyl SAMs with infrared spectroscopy and X-ray diffraction. Biphenyl-SAMs with different alkyl spacers were studied by STM in air and identified several adsorbate phases, whose structures depended on the length of the spacers [27].

2.1 INTERACTIONS OF SELF-ASSEMBLED MONOLAYERS WITH ELECTRONS

The response of a SAM to electron irradiation depends on the structure of its molecular constituents. By choosing the right molecules, the effect of the e-beam exposure can be tailored to modify the surface. The electron irradiation of a SAM affects the entire ensemble of self-assembled molecules and all groups (head, tail, and spacer) within each molecule. In a series of spectroscopic investigation, different aliphatic and aromatic SAMs were irradiated with electrons or extreme ultraviolet (EUV) light and the occurring changes were monitored by surface analytical techniques, including X-ray photoelectron spectroscopy (XPS), ultraviolet photoelectron spectroscopy (UPS), near edge X-ray absorption spectroscopy (NEXAFS), temperature programmed desorption (TPD), and infrared spectroscopy (IR) [28; 29; 30]. It was found that the spacer and terminal groups are predominantly affected by the irradiation. The spacer group basically controls the overall structure that results from the exposure, and the tail group can be modified in a way to clearly change the surface properties. In short, three types of electron induced modifications, as shown in Figure 26.1, could be identified.

In aliphatic SAMs, like hexadecanethiol (HDT), the electrons induce a cleavage of C–H bonds that leads to orientational and conformational disorder of the chains, the desorption of

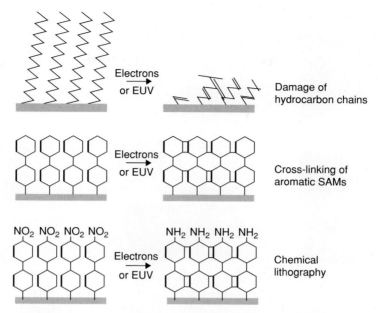

FIGURE 26.2: Classification of interaction of electrons or extreme UV photons with SAMs. Aliphatic SAMs are destroyed, while aromatic SAMs are cross-linked upon irradiation. Nitroterminated aromatic SAMs can be chemically functionalized, i.e., nitro groups are reduced to amino groups by the electron exposure via "chemical lithography".

material, and the formation of C=C double bonds in the fragments that remain on the surface. In a wet-chemical process, these effects lower the etching resistance of the irradiated regions and aliphatic SAMs are thus used as *positive tone resists*.

In aromatic biphenylthiol (BPT), Geyer et al. found that upon irradiation with electrons, C-H cleavage occurs, which is then followed by cross-linking between neighboring phenyl units [28]. During this process, the cross-linked molecules maintain their preferred orientation, and only little material desorbs. Hence the electrons generate a molecularly well-ordered *and* cross-linked monolayer. It was shown that this cross-linked aromatic layer has an increased wet-etching resistance against KCN/KOH and could be used as the first *negative tone electron resist* based on a monolayer.

In a 4-nitrobiphenylthiol (NBT) SAM, the hydrogen atoms liberated from the aromatic cores by electron irradiation locally reduce the nitro to amino groups which can be further chemically modified by electrophilic agents. Hence, electron irradiation converts the terminal functionality, while the underlying aromatic layer is dehydrogenated and cross-linked. By employing a local electron exposure, one thus can define regions of cross-linked 4-aminobiphenyl-thiol (cABT) in a NBT matrix in a *single* processing step. This NBT/cABT pattern can then be used as a template for a *site selective molecular coupling*. This process has been named *Chemical Lithography*, as the lithographic exposure directly affects the surface chemistry of the SAM.

These three types of reaction clearly show the potential of SAMs in the modification of surfaces. Very different and distinct irradiation-induced modifications of monolayers can by generated in by an appropriate choice of SAM.

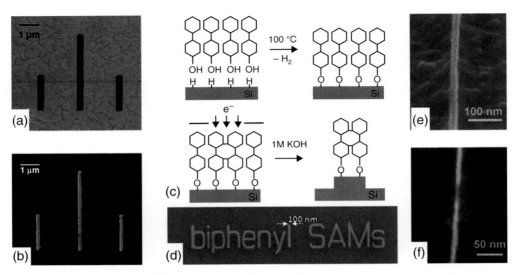

FIGURE 26.3: Application of SAMs as resists: (a) 100 nm wide lines were written in aliphatic hexadecanethiol positive resist on gold. The pattern transfer was performed by a wet-etching. (b) 100 nm wide lines were written in aromatic biphenylthiol negative resist on gold. The pattern transfer was performed by a wet-etching. (c) Scheme of the binding of hydroxybiphenyl (HBP) negative resist on silicon. (d) 100 nm wide letters displaying the name biphenyl SAM in silicon fabricated by wet-etching of HBP. (e) 30 nm and (f) 10 nm lines in silicon fabricated by wet-etching of HBP, demonstrating the limits of the resolution with SAMs.

3 APPLICATION OF SAMs AS RESISTS IN THE NANOFABRICATION OF GOLD AND SILICON

To test the performance of aromatic and aliphatic SAMs as resist materials in nanofabrication, line patterns were written in hexadecanthiol (HDT) and biphenylthiol (BPT) using electron beam lithography (2.5 keV; area dose: 25 mC/cm^2), and transferred into the gold substrate by a wet-etching process [31]. The scanning electron micrographs in Figure 26.3(a, b) shows gold patterns that were written with a line width of 100 nm. One recognizes at first sight that for the HDT-coated sample the lines are etched away, whereas for the BPT-coated sample all gold except the 100 nm wide cross-linked regions is removed by the etching. Hence, HDT acts as a positive resist, whereas BPT acts as a negative tone resist. Using AFM, it was found that the lines have the height of the initial gold film. Thus cross-linked BPT completely protects the exposed regions. For area doses below 15 mC/cm^2 we observe smaller heights, indicative of incomplete cross-linking of BPT.

Hydroxybiphenyl SAMs were employed to pattern silicon [32]. To form a SAM, a reaction of alcohol head groups with hydrogenated silicon surfaces is used. These were produced cleaning boron-doped silicon wafers in a H_2SO_4/H_2O_2 mixture (3:1) and subsequent etching with 48% hydrofluoric acid. The surface is immersed in a degassed and dry 0.05 molar toluene solution of 4-hydroxy-1,1'-biphenyl (HBP) and kept there at 100°C for 16 hours in a dry nitrogen atmosphere. HBP self-assembles into well-ordered monolayers that can be cross-linked under e-beam irradiation. The exposed regions show an increased etching stability against 1 M KOH. Figure 26.3(c) shows a schematic of this process.

Wide line patterns, such as the words "biphenyl SAMs" were written with 100 nm line width to demonstrate the suitability of SAMs for silicon patterning. To characterize the resolution that can be achieved with a particular resist material, single lines as well as patterns of lines and spaces were written, as proximity effects show up more clearly when the exposed regions are densely packed. Figure 26.3(d, e) shows the SEM image of single lines. The smallest individual silicon lines that have been created by wet chemical pattern transfer with HBP have lateral dimensions of $\sim$10 nm. Line gratings with 40 nm wide lines (not shown) have also been transferred into silicon from an e-beam exposed HBP monolayer (area doses between 20 and 80 mC/cm^2) [32]. The pattern is clearly developed and shows good transfer. By optimization of the radiation dose and the etching parameters, it can be assumed that the quality of the pattern transfer can still be improved.

4 FUNCTIONAL SURFACE PATTERNS BY E-BEAM PATTERNING OF SAMs

In SAMs of 4'-nitro-1,1'-biphenyl-4-thiol (NBT) on gold, the nitro groups are converted to amino groups by electron irradiation, while the underlying aromatic layer is dehydrogenated and cross-linked. The irradiation thus results in a direct modification of the surface chemistry, creating reactive sites of cross-linked aminobiphenylthiol (cABT) within a nitro terminated matrix that can act as a template for the coupling of other molecules. Figures 26.4(a–d) show the utilization of this effect for an electron-induced "chemical nanolithography" in which different molecules couple to distinct surface locations. Sequential electron beam exposures, followed by coupling of electrophilic molecules to the amino-terminated regions, can be used for the fabrication of surface patterns of different molecular components. Since the reactive amine sites are newly created with each electron exposure, the repeatability of the procedure for the immobilization of further molecules is obvious.

In a pioneering effort, fluorinated carboxylic acid anhydrides with different lengths, i.e., trifluoro acetic acid anhydride (TFAA) and perfluoro butyric acid anhydride (PFBA), were coupled to the amino-terminated lines by immersion into the respective solutions. X-ray photoelectron spectroscopy confirmed the coupling of the molecules by the detection of the appropriate amounts of fluorine. Non-contact atomic force microscopy (AFM) was used to investigate line patterns of locally irradiated SAMs before and after coupling. Figure 26.4(e) shows images of 100 wide lines together with averaged line profiles. The "average heights" determined from these line profiles are $\sim$10 Å for the irradiated 100 nm line. These heights do not only contain topographical information, but also reflect the chemical composition of the terminal groups and the change of the elastic film properties due to cross-linking of the biphenyl. These averaged heights are characteristic for the irradiated region of the SAM and are compared before and after the molecular coupling reactions. Figure 26.4(f) shows 100 nm wide lines after immersion in TFAA and PFBA solutions, respectively. Their heights increased to $\sim$16 Å and $\sim$23 Å after coupling of TFAA and PFBA, respectively. A direct relation of these numbers to the size of the immobilized molecules is difficult, since the surface terminal group (CF_3) as well as the elastic properties of the film are altered simultaneously. However, the systematic differences between the irradiated and the chemically modified lines indicate successful molecular coupling and allow the distinction of different molecules [33].

Amino patterns were written into NBT SAMs and several other molecules were bound to them. Along this path dyes [34], proteins [7], polymers [35; 36], as well as metal clusters [37]

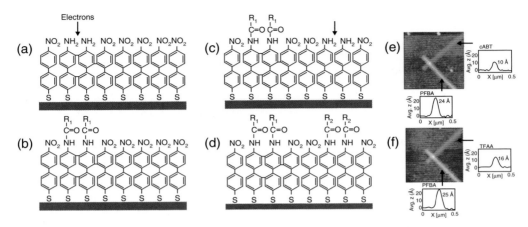

FIGURE 26.4: Chemical Lithography: (a) by locally exposing an aromatic Nitrobiphenylthiol SAM regions of cross-linked Aminobiphenylthiol (cABT) are generated. (b) Molecules (R₁) can couple to the amino terminated region. (c, d) In a second exposure, other cABT regions are generated to which other molecules (R₂) can bind. (e, f) AFM micrographs demonstrating these reactions by a height change of the cABT lines to cABT+TFAA lines.

were also bound to chemically patternend SAMs. The amino termination of the electron irradiated SAM was also used to act as nucleation site for the electroless deposition of palladium [38].

Chemical lithography with electron beams was utilized to fabricate large-area highly parallel protein chip templates with features from the micro- to the nanorange. Here a combination of electron-beam or EUV lithography, molecular self-assembly, and the use of biochemical tweezers (N-nitrilotriacetic Acid, NTA) for highly specific affinity capturing of His-tagged proteins was used. The fabrication scheme in Figure 26.5 presents the elemental steps of the protein chip assembly [8]. First, a NBT SAM on a gold surface is formed (i). Then, e-beam (or EUV) lithography is used to locally reduce the terminal nitro groups into amino groups and to cross-link the underlying aromatic substrate (ii). Multivalent chelators (tris-NTA) are grafted to the prepatterned amino-terminated areas. The carboxylic functionalities of the NTA-groups are protected. After their deprotection, the areas of the pristine NBPT SAM are exchanged in solution with protein-repellent 16-mercaptohexadecanoic acid (triethylene glycol) ester, to prevent unspecific adsorption of proteins (iii). In the regions with the grafted multivalent chelators, the loading with metal ions results in the specific, high affine, oriented and reversible immobilization of His-tagged proteins or protein complexes (iv).

The functionality of the protein chips has been demonstrated by specific, homogeneous, oriented, and reversible immobilization of His-tagged proteasome proteins [8]. The specific affinity of the protein arrays is achieved by multivalent interactions in NTA/His-tag pairs. The fabricated chips were used in multiple loading and regeneration cycles with different His-tagged proteins. This work has demonstrated that diverse and specific functional surfaces can be made by a combinations of a choice of appropriate SAMs, chemical lithography, and molecular exchange. Both electron and EUV lithography are suitable to fabricate SAM nanostructures with lateral dimensions down to 10 nm and chemical lithography with SAMs. The concept can thus be developed for the functional immobilization of structured proteins arrays on solid substrates down to single molecule resolution.

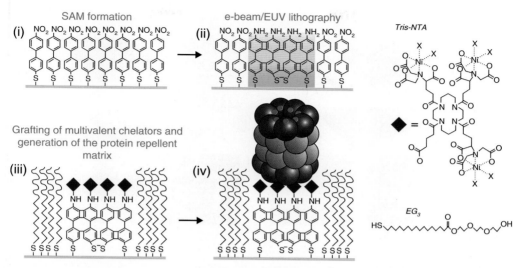

FIGURE 26.5: Fabrication of a protein biochip by chemical lithography: After nitrobiphenylthiol SAM formation (i) and a localized e-beam (or EUV) exposure (ii) multivalent chelators (tris-NTA) are coupled to the generated amino groups (iii). The non cross-linked nitrobiphenylthiols are exchanged with protein resistent oligo-ethyleneglycol (EG3) SAMS. Tris-NTA provides localized binding sites for the immobilization of His-tagged proteins (iv).

5 NANOMEMBRANES FROM E-BEAM CROSS-LINKED SAMs

Cross-linked aromatic layers are mechanically and thermally very stable, so that it is possible to release of the cross-linked film from its substrate. A schematic is shown in Figure 26.6. Due to their high mechanical stability, cross-linked SAMs can form macroscopically extended ultrathin carbon nanomembranes (CNMs) with the thickness of a single molecule. CNMs even freely suspend across holes with diameters of up to 300 μm (cf. Figure 26.6(d)). When such nanomembranes are supported by a solid substrate, they can be made with much larger sizes, CNMs with macroscopic dimensions, such as 3×3 cm^2, have recently been fabricated (cf. Figure 26.6(c)).

Comparable only to graphene, CNMs are only 1–2 nm thick. There are many similarities as well as many differences between CNMs and graphene. Both two-dimensional materials consist mainly of carbon and both have a thicknesses of ~1 nm. However, whereas graphene is perfectly crystalline and harder, CNMs are more polymer-like and softer. Due to the oriented polar nature of the underlying monolayer, preparation of CNMs with defined surface functional groups, i.e., with two chemically different sides is straightforward, whereas a chemical functionalization of graphene is rather difficult. Freely suspended carbon nanomembranes are easily obtained by local dissolution of the substrate under cross-linked Biphenylsilane SAMs that have been prepared on 30 nm thin silicon nitride (SiN$_x$) window structures that suspend over openings (frames) in silicon structure.

A way to transfer nanomembranes onto arbitrary surfaces uses a polymeric transfer medium, such as photoresist. After a SAM is cross-linked on one substrate, a layer of a photoresist is deposited onto the cross-linked aromatic SAM that acts as a transfer medium

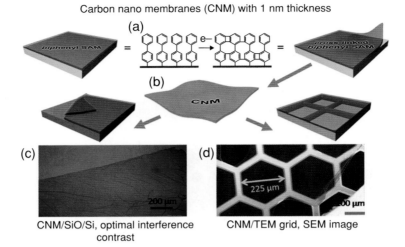

Carbon nano membranes (CNM) with 1 nm thickness

(a)

(b)

(c) CNM/SiO/Si, optimal interference contrast

(d) CNM/TEM grid, SEM image

225 μm

200 μm

00 μm

FIGURE 26.6: Schematic for the fabrication of freestanding carbon nanomembranes (CNMs): An electron cross-linked biphenyl monolayer (a) is detached from the surface forming a CNM (b) that can be transferred to arbitrary surfaces. (c) Optical micrograph of a CNM on a 300 nm SiO layer on a Si wafer. The CNM is visible due to Raleigh interference contrast. (d) Scanning electron micrograph of a CNM transferred onto a TEM grid. Note that openings of ∼200 nm diameter are completely covered by freestanding CNM.

to mechanically stabilize the CNM. Then the substrate is dissolved and the CNM adheres to the photoresist. The CNM/photoresist sandwich structure is placed onto another surface. One needs to assure that the CNM maintains its shape during this procedure, i.e., that it does not fold or rupture in an uncontrolled manner. Finally, the photoresist is dissolved so that a CNM of a cross-linked biphenyl monolayer has been transferred onto the new substrate. This preparation of freestanding CNM membranes allows much flexibility. CNMs of arbitrary size and shapes can be placed on many different surfaces, including those of flexible and holey structures. CNMs have been successfully used as ultrathin sample supports in transmission electron microscopy [39]. This widens the possibilities for optical examination, characterization, and application as a solid support for other molecules.

Recently, it was found that even atomically thin carbon films [40; 41] become visible when they are located on an oxidized (SiO_2) surface of a silicon wafer. CNMs were placed onto an oxidized silicon wafer and imaged with a light microscope [42]. Figure 26.6(c) shows an edge of a transferred CNM with a total size in the range of a few mm^2. We observe a color contrast that results from interference between light reflected from the SiO_2 layer and the CNM. This phase shift depends on the thickness of the SiO_2 layer and of the CNM, and we find that the purple color of the 300 nm thick SiO_2 layer shifts toward blue. The thickness of the transferred CNM has been measured by atomic force microscopy (AFM) to be ∼1.5 nm. So far, we have transferred CNMs with sizes up to 2 × 2 cm^2 and much larger areas may also be feasible.

To quantify the mechanical properties of CNMs, their Young's modulus was measured by bulge tests [43]. Freely suspended CNMs were mounted onto a sealed cell. A well-defined pressure was applied, and the resulting membrane deflection was measured by AFM as schematically displayed in Figure 26.7(A).

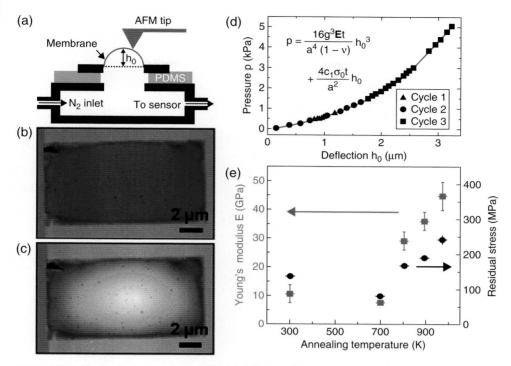

FIGURE 26.7: Mechanical properties of CNM. (a) Schematic representation of the bulging test: A pressure cell is mounted into an AFM that measured the membrane deflection. (b) AFM image of a CNM without and (c) with an applied pressure of 1.8 kPa. (d) Young's modulus determination for one CNM. The data are fitted to the equation, which yields the modulus E. (e) Change of Young's modulus with temperature after annealing of CNM.

Figure 26.7(B) shows an AFM image of a CNM that is pushed down ~75 nm by the tip but remains intact, even after several AFM scans. By applying a pressure of ~1.8 kPa to the sealed cell under the membrane, an upward deformation (bulging) occurs (Figure 26.7(C)). This deformation is quantified by recording the AFM tip-height at the membrane centre as function of the applied pressure, cf. Figure 26.7D. All measured data points lie on one curve, i.e., the deformation is elastic without any permanent change [42]. A model for the elastic deformation of thin membranes under tension contains two contribution: (1) membrane stretching leads to a cubic dependence of the pressure to the height; (2) membranes tension at zero pressure due to residual stress results in a linear pressure-height-dependence. The experimental data fits very well to this model (cf. Figure 26.7(D)). Curve fitting yields a Young's modulus ~10 GPa.

The electrical conductivity of CNMs has also been measured [42]. It was found that the sheet resistance of the CNMs depends on its thermal treatment. When a CNM is made, it is almost a perfect insulator. After annealing in UHV (pyrolysis), the sheet resistance drops and the CNM becomes conducting. The reason for this unusual behavior is a structural transformation of the carbon atoms in the CNM. Whereas at lower temperatures, they are best characterized as a cross-linked polymer, after annealing, they transform into a graphitic phase, as is clearly documented by Raman spectroscopy [42]. The sheet resistance value obtained by annealing CNMs is more than a factor of 10 lower than that of chemically reduced graphene [44]

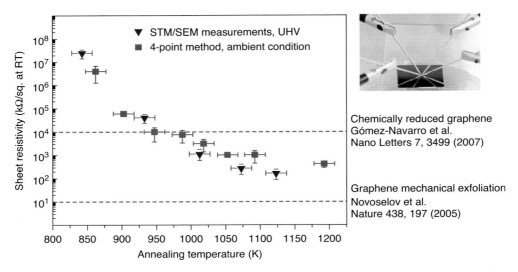

FIGURE 26.8: Plot of the CNM sheet resistivity as a function of annealing temperature. Upon annealing at temperatures above 800 K the CNM becomes conductive due to a phase transformation from cross-linked aromatic molecules into nanocrystalline graphene.

and only another factor of 10 times higher than of single crystalline graphene sheets [45]. Hence, upon annealing CNMs transform into nanocrystalline graphene and thus are two-dimensional materials with a tunable electrical resistance (cf. Figure 26.8).

Functionalized SAMs can be made by chemical lithography, i.e., the electron or UV induced conversion of terminal groups of SAMs [46]. Chemical lithography has been employed to reversibly immobilize functional proteins onto patterned aromatic SAMs [8]. Functionalized SAMs can also be cross-linked and converted into CNMs. The experience with chemical lithography of SAMs can thus be transferred to build functionalized CNMs. Methods to produce functionalized CNMs are currently tested and a chemical modification of both sides with dyes as well as polymers has recently been demonstrated [46–48]. These "Janus-type" membranes can be utilized for a variety of different applications: functionalized TEM supports, (bio)sensors, and ultrathin dielectrics.

6 CONCLUSION

Electron beam controlled modification of self-assembled monolayers is a powerful method to modify surfaces as well as to fabricate novel two-dimensional entities with a manifold of technological applications.

ACKNOWLEDGMENTS

Fruitful cooperation and valuable discussions with A. Beyer, M. Buck, W. Eck, K. Edinger, W. Geyer, M. Grunze, R. Jordan, L. Kankate, A. Küller, H. Muzik, C. T. Nottbohm, M. Schnietz, V. Stadler, R. Tampé, A. Tinazli, A. Turchanin, B. Völkel, T. Weiman, Ch. Wöll, X. Zhang,

M. Zharnikov and Z. Zheng are acknowledged. The reviewed work was financially supported by the Volkswagenstiftung, the Deutsche Forschungsgemeinschaft, the Bundesministerium für Bildung und Forschung, and the European Science Foundation through the COST Action Electron Controlled Chemical Lithography.

REFERENCES

[1] Madou, M. J. *Fundamentals of microfabrication.* Boca Raton, FL: CRC Press, 2002.

[2] Timp, G. *Nanotechnology.* New York: Springer, 1999.

[3] Ulman, A. *An introduction to ultrathin organic films: From Langmuir-Blodgett to self-assembly.* San Diego: Academic Press, 1991.

[4] Schreiber, F. Structure and growth of self-assembled monolayers. *Progress in Surface Science* 65 (2000): 151.

[5] Love, J. C., et al. Self-assembled monolayers of thiolates on metals as a form of nanotechnology. *Chem. Rev.* 105 (2005): 1103.

[6] Mrksich, M., and G. M. Whitesides. Using self-assembled monolayers to understand the interactions of man-made surfaces with proteins and cells. *Annual Review of Biophysics and Biomolecular Structure* 25 (1996): 55.

[7] Biebricher, A., et al. Controlled three-dimensional immobilization of biomolecules on chemically patterned surfaces, *J. Biotechnol.* 112 (2004): 97.

[8] Turchanin, A., A. Tinazli, M. Desawy, H. Großmann, M. Schnietz, H. Solak, R. Tampé, and A. Gölzhäuser. Molecular self-assembly chemical lithography and biological tweezers: A path fort he fabricatioof functional nanometer-scale protein arrays, *Adv. Mater.* 20 (2008): 471.

[9] Xia, Y. N., and G. M. Whitesides, Soft lithography. *Annual Review of Materials Science* 28 (1998): 153; Y. N. Xia and G. M. Whitesides. *Angew. Chemie Int. Ed. in English* 37 (1998): 551.

[10] Li, H. W., et al. Nano contact printing: A route to sub-50-nm-scale chemical and biological patterning. *Langmuir* 19.6 (2003): 1963.

[11] Piner , R. D., et al. "Dip-pen" nanolithography, *Science* 283 (1999): 661.

[12] Calvert, J. M. Lithographic patterning of self-assembled films. *J. Vac. Sci. Technol.* B 11 (1993): 2155.

[13] Seshadri, K., K. Froyd, A. N. Parikh, D. L. Allara, M. J. Lercel, and H. G. Craighead. Electron-beam-induced damage in self-assembled monolayers. *Journal of Physical Chemistry* 100 (1996): 15900.

[14] Olsen, C., and P. A. Rowntree. Bond-selective dissociation of alkanethiol based self-assembled monolayers absorbed on gold substrates, using low-energy electron beams. *Journal of Chemical Physics* 108 (1998): 3750.

[15] Berggren, K. K., A. Bard, J. L. Wilbur, J. D. Gillaspy, A. G. Helg, J. J. Mcclelland, S. L. Rolston, W. D. Phillips, M. Prentiss, and G. M. Whiteside. Microlithography by using neutral metastable atoms and self-assembled monolayers. Science 269 (1995): 1255.

[16] Ada, E. T., L. Hanley, S. Etchin, J. Melngailis, W. J. Dressick, M. S. Chen, and J. M. Calvert. Ion beam modification and patterning of organosilane self-assembled monolayers. Journal of Vacuum Science and Technology B 13 (1995): 2189.

[17] Dulcey, C. S., J. H. Georger, M. S. Chen, S. W. Mcelvany, O. F. Ce, V. I. Benezra, and J. M. Calvert. Photochemistry and patterning of self-assembled monolayer films containing aromatic hydrocarbon functional groups. *Langmuir* 12 (1996): 1638.

[18] Lercel, M. J., H. G. Craighead, A. N. Parikh, K. Seshadri, and D. L. Allara. Sub-10 nm lithography with self-assembled monolayers. *Applied Physics Letters* 68 (1996): 1504.

[19] Harnett, C. K., K. M. Satyalakshmi, and H. G. Craighead. Low-energyelectron-beam patterning of amine-functionalized self-assembled monolayers. *Applied Physics Letters* 76 (2000): 2466.

[20] Polymeropoulos, E. E., and J. Sagiv. Electrical-conduction through adsorbed monolayers. *Journal of Chemical Physics* 69.5 (1978): 1836.

[21] Nuzzo, R. G., and D. L. Allara. Adsorption of bifunctional organic disulfides on gold surfaces. *J. Am. Chem. Soc.* 105.13 (1983): 4481.

[22] Laibinis, P. E., G. M. Whitesides, D. L. Allara, Y. T. Tao, A. N. Parikh, and R. G. Nuzzo. Comparison of the structures and wetting properties of self-assembled monolayers of norma-alkanethiols on the coinage metal-surfaces, Cu, Ag Au. *Journal of the American Chemical Society* 113.19 (1991): 7152.

[23] Xia, Y. N., and G. M. Whitesides. Soft lithography. *Annual Review of Materials Science* 28 (1998): 153.

[24] Sabatani, E., J. Cohenboulakia, M. Bruening, and I. Rubinstein. Thioaromatic monolayers on gold–A new falily of self-assembling monolayers. *Langmuir* 9.11 (1993): 2974.

[25] Leung, T. Y. B., P. Schwartz, G. Scoles, F. Schreiber, and A. Ulman. Structure and growth of 4-menthyl-4'-mercaptobiphenyl monolayers on Au(111): A surface diffraction study. *Surface Science* 458.1–3 (2000): 34.

[26] Kang, J. F., A. Ulman, S. Liao, R. Jordan, G. H. Yang, and G. Y. Liu. Self-assembled rigid monolayers of 4'-substituted-4-mercaptobiphenyls on gold and silver surfaces. *Langmuir* 17.1 (2001): 95.

[27] Cyganik, P., M. Buck, T. Strunskus, A. Shaporenko, Jdet Wilton-Ely, M. Zharnikov, and C. Wöll. Competition as a design concept: Polymorphism in self-assembled monolayers of biphenyl-based thiols, *Journal of the American Chemical Society* 128.42 (2006): 13868; P. Cyganik, M. Buck, J. Wilton-Ely, and C. Wöll. Stress in self-assembled monolayers omega-biphenyl alkane thiols on Au(111). *Journal of Physical Chemistry* B 109.21 (2005): 10902; P. Cyganik and M. Buck. Polymorphism in biphenyl-based self-assembled monolayers of thiols. *Journal of the American Chemical Society* 126.19 (2004): 5690; P. Cyganik, M. Buck, W. Azzam, and C. Wöll. Self-assembled monolayers of omeg-biphenylalkanethiols on Au(111): Influence of spacer chain on molecular packing. *J. Phys. Chem.* B 108.16 (2004): 4989; W. Azzam, P. Cyganik, G. Witte, M. Buck, and C. Wöll. Pronounced odd-even changes in the molecular arrangement and packing density of biphenyl-based thiol SAMs: A combined STM and LEED study. *Langmuir* 19.20 (2003): 8262; W. Azzam, C. Fuxen, A. Birkner, H. T. Rong, M. Buck, and C. Wöll. Coxestence of different structural phases in thioaromatic monolayers on Au(111). *Langmuir* 19.12 (2003): 4958.

[28] Geyer, W., V. Stadler, W. Eck, M. Zharnikov, A. Gölzhäuser, M. Grunze. Electron-induced crosslinking of aromatic self-assembled monolayers: Negative resists for nanolithopraphy. *Appl. Phys. Lett.* 75 (1999): 2401.

[29] Zharnikov, M., W. Geyer, A. Gölzhäuser, S. Frey, and M. Grunze. Modification of alkanethiolate monolayers on Au-substrates by low energy electronic irradiation: Alkyl chains and the S/A interface. *Physical Chemistry Chemical Physics* 1 (1999): 3163; K. Heister, S. Frey, A. Gölzhäuser, A. Ulman, and M. Zharnikov. An enhanced sensitivity of alkaaaanethiolate self-assembled monolayers to electron irradiation through the incorporation of a sulfide antity into the alkyl chains. *Journal of Physical Chemistry* B 103 (1999): 11098.

[30] Eck, W., V. Stadler, W. Geyer, M. Zharnikov, A. Gölzhäuser, and M. Grunze. Generation of surface amino groups on aromatic self-assembled monolayers by low energy electron beams–A first step towards chemical lithography. *Advanced Materials* 12 (2000): 805.

[31] Gölzhäuser, A., W. Geyer, V. Stadler, W. Eck, M. Grunze, K. Edinger, T. Weimann, and P. Hinze. Nanoscale patterning of self-assembled monolayers by e-beam lithography. *Journal of Vacuum Science & Technology* B 18 (2000): 3414; T. Weimann, W. Geyer, P. Hinze, V. Stadler, W. Eck, and A. Gölzhäuser. Nanoscale patterning of self-assembled monolayers by e-beam lithopraphy, Microelectronic Engineering 57–8, 903 (2001).

[32] Küller, A., W. Eck, V. Stadler, W. Geyer, and A. Gölzhäuser. Nanostructuring of silicon by electron beam lithography of self-assembled hydroxybiphenyl monolayers. *Appl. Phys. Lett.* 82 (2003): 3776; A. Küller, M. El-Desawy, V. Stadler, W. Geyer, W. Eck, and A. Gölzhäuser. Electron beam lithography with self-assembled hydroxybiphenyl monolayers on silicon surfaces. *J. Vac. Sci. Technol.* B 22.3 (2004): 1114.

[33] Gölzhäuser, A., W. Eck, W. Geyer, V. Stadler, Th. Weimann, P. Hinze, and M. Grunze. Chemical nanolithography with electron beams. *Adv. Mater.* 13 (2001): 806.

[34] Geyer, W., V. Stadler, W. Eck, A. Gölzhäuser, M. Grunze, M. Sauer, Th. Weimann, and P. Hinze. Electron induced chemical nanolithography with self-assembled monolayers. *J. Vac. Sci. Technol.* B 19 (2001): 2732.

[35] Schmelmer, U., R. Jordan, W. Geyer, W. Eck, A. Gölzhäuser, M. Grunze, and A. Ulman. Surface-initiated polymerization on self-assembled monolayers. Amplification of patterns on the micrometer and nanometer scale. *Angew. Chem. Int. Ed.* 42 (2003): 559.

[36] Schmelmer, U., A. Paul, A. Küller, M. Steenackers, A. Ulman, M. Grunze, A. Gölzhäuser, and R. Jordan. Nanostructured polymer brushes. *Small* 3.3 (2007): 459.

[37] Mendes, P. M., S. Jacke, K. Chrichtley, J. Plaza, K. Nikitin, R. E. Palmer, J. A. Preece, S. D. Evans, and D. Fitzmaurice. Gold nanoparticle patterning of silicon wafers using chemical e-beam lithography. *Langmuir* 20 (2004): 3766.

[38] Nottbohm, C. T., A. Turchanin, A. Gölzhäuser. Metallization of organic monolayers: Electroless deposition of Cu onto cross-linked aromatic self-assembled monolayers. *Z. Phys. Chem*, 222 (2008): 917–926.

[39] Nottbohm, C. A., A. Beyer, A. S. Sologubenko, I. Ennen, A. Hütten, H. Rösner, W. Eck, J. Mayer, and A. Gölzhäuser. Novel carbon nanosheets as support for ultrahigh-resolution structural analysis of nanoparticles. *Ultramicroscopy* 108 (2008): 885.

[40] Novoselov, K. S., et al. Electric field effect in atomically thin carbon films. *Science* 306 (2004): 666–669.

[41] Casiraghi, C., A. Hartschuh, E. Lidorikis, H. Qian, H. Harutyunyan, T. Gokus, K. S. Novoselov, and A. C. Ferrari. Rayleigh imaging of graphene and graphene layers. *Nano Lett.* 7 (2007): 2711.

[42] Turchanin, A., A. Beyer, C. T. Nottbohm, X. Zhang, R. Stosch, A. Sologubenko, J. Mayer, P. Hinze, T. Weimann, and A. Gölzhäuser. One nanometer thin carbon nanosheets with tunable conductivity. *Adv. Mater.* 21 (2009): 1233.

[43] Vlassak, J. J., and W. D. Nix. A new bulge test technique for the determination of young modules and poisson ratio of thin-films. *Journal of Materials Research* 7 (1992): 3242.

[44] Gómez-Navarro, A. S., et al. Electronic transport properties of individual chemically reduced graphene oxide sheets. *Nano. Lett.* 7 (2007): 3499.

[45] Novoselov, K. S., et al. Two-dimensional gas of massless Dirac fermions in graphene. *Nature* 438 (2005): 197.

[46] Zheng, Z., et al. Janus nanomembranes: A generic platform for chemistry in two dimensions, *Angewandte Chemie* 49 (2010): 8493.

[47] Amin, I., et al. Polymer Carpets, *Small* 6 (2010) 1623.

[48] Amin, I., et al. Patterned Polymer Carpets, *Small* 7 (2011): 683.

CHAPTER 27

NANOFABRICATION USING ELECTRON BEAM LITHOGRAPHY PROCESSES

Mariana Pojar, Simone Camargo Trippe, and Antonio Carlos Seabra

1 INTRODUCTION

Nanoscience and nanotechnology have become keywords for scientific development in many areas of research. In the race to build devices with resolution better than 100 nm, the electron beam lithography (EBL) has been one of the most widely used techniques. Among its biggest advantages are the resolution, the registration accuracy, and the large depth of focus. Structures with dimensions smaller than 100 nm using conventional resists can be routinely obtained, and structures even smaller than 20 nm are routine when special resists and field emission gun (FEG) sources are used.

In this chapter, electron beam lithography processes which aim to obtain high resolution patterning will be considered, with emphasis on methodologies that enable to achieve resolution below 20 nm. The choice of the resist, the exposure, and development processes and methodology for pattern transfer to the sample influence the final resolution becoming imperative to optimize all of them. For best discussion of each step, this chapter is divided into sections. In Section 2, the main properties of high-resolution resists whose nominal resolutions are less than 20 nm are presented. In Section 3, a description of the exposure process emphasizes the intrinsic factors that must be considered for achieving nanopatterning. During the exposure process, the electron beam suffers backscattering, both in the resist and the sample, which generates undesired shapes due to the so-called proximity effects. In Section 4, some methodologies are discussed that are used for the proximity effects correction. The development processes and the methodologies for pattern transfer to the sample are described in Sections 5 and 6, respectively. Depending on the goal to be achieved, bilayer or trilayer resist systems can be utilized, as shown in Section 7. Finally, in Section 8, fabrication of a SQUID sensor application obtained with the electron beam lithography technique is shown.

2 MAIN HIGH-RESOLUTION ELECTRON BEAM RESISTS

Resists are generally polymer organic materials with properties tailored for specific applications. They can be classified as positive or negative, depending on the response to exposure. Four main properties summarize the characteristics of a given resist: resolution, contrast, sensitivity, and resistance to etching. For e-beam nanolithography, the final resolution is determined by the electron scattering and by the resist molecular size and weight and its chemical structure. In order to minimize the forward scattering of the primary beam, the resist layer must be very thin, in the order of 50 nm.

The polymethylmethacrylate (PMMA) was one of the first resists developed for EBL and remains the most commonly positive resist used, both in academic research and in industrial production. PMMA has a moderate glass transition temperature (114°C) and it is usually purchased in two high molecular weight forms (495 K or 950 K). It offers high resolution, good contrast ($\sim$4), and an acceptable sensitivity (200 μC/cm^2 at 20 keV) leading to intermediate exposure times. Patterns with features sized less than 20 nm can be routinely achieved in PMMA. Its ultimate resolution has been demonstrated to be less than 10 nm [1; 2]. The major limitations of PMMA are its relatively poor dry etch resistance (which makes the replication of high-density patterns with nanometer resolution more difficult) and moderate thermal stability. In this chapter, PMMA will be used as a basis of comparison for new high-resolution resists that have been developed and are increasingly used for nanofabrication.

In recent years, ZEP-520 has been developed as an alternative positive e-beam resist. ZEP-520 consists of a virtual 1:1 copolymer of α-chloromethacrylate and α-methylstyrene, whose molecular weight is about 55,000. It is regarded as a classical chain scission-type resist. ZEP-520 has sensitivity 10 times higher than PMMA (around 30 μC/cm^2 at 20 keV) but similar resolution, as can be seen in Table 27.1(a–b). Contrasts are about the same for both resists (>4) [3; 4]. Its main advantage is the much better dry etch resistance when compared to PMMA. The main applications are concentrated in the construction of quantum nanowires [5], production of optical components [6], and manufacturing of recording media [7; 8].

E-beam processes are simplified when negative resists are used. Recently, hydrogen silsesquioxane (HSQ) and calixarene derivatives have attracted much attention as negative-tone electron beam resists.

Calixarenes are a group of macrocyclic molecules in which phenolic units are linked via methylene bridges in *ortho*-position to the phenolic hydroxyl groups [9; 10]. In other words, they are cyclic phenol resins, but their chemical and physical properties are very different from those of linear phenol resins (novolaks). Calixarenes are available by an easy one-pot condensation of p-tert-butylphenol and formaldehyde under alkaline conditions [11; 12]. This reaction leads to a calix[n]arene consisting of n phenolic units (n = 4 − 20) in a cyclic arrangement. When compared to most organic compounds, calixarenes have a high chemical stability. The high resolution of calixarene results from their small molecular size of about 1 nm,

Table 27.1 Summary of main characteristics of some high-resolution e-beam resists.

	Resists	Tone	Composition	Mw	Dose	Resolution	Thickness	Contrast
a-)	PMMA	Positive	Polymethyl-methacrilate	495 K–950 K	0,1 C/cm^2 at 25 kV	10 nm	30 nm	>4
b-)	ZEP-520	Positive		55000	30 μC/cm^2 at 20 kV	15 nm	120 nm	>4
c-)	Calixarene	Negative	MC6AOAc	972	7 mC/cm^2 at 50 kV	7 nm	30 nm	2.5
			CMC6AOMe	–	700 μC/cm^2 at 50 kV	–	30 nm	–
d-)	HSQ	Negative	Hydrogen Silsesquioxane	11000	5519 C/cm^2 at 50 KV	6 nm	20 nm	–
					300 μC/cm^2 at 70 kV	7 nm	100 nm	–

allowing structures smaller than 10 nm to be produced. The small size of the molecule also allows one to obtain smoother pattern sidewalls. Unfortunately, calixarene resists have low sensitivity with critical doses ranging between 0.7 and 7 mC/cm^2 (see Table 27.1(c)). Compared to PMMA, calixarene resists require long writing times. Keep in mind that exposure times routinely take about one hour, which means that a resist 10 times less sensitive may take 10 hours of exposure (and machine) for the same job. Among all existing derivatives of calixarenes, the most widely used has been the hexaacetate p-methylcalixarene (which will be referred to as MC6AOAc) first reported by Fujita et al. [13–15]. The MC6AOAc leads to structures with sub-10 nm dimensions, unlike the other derivatives that provide sub-30 nm resolution. The main applications include fabrication of electronic devices [16–18], wave guides [19; 20] and media for high-density recording [21].

Recently, HSQ became a serious candidate as negative high-resolution e-beam resist. Its chemical composition does not contain hydrocarbons or organic groups such as vinyl groups, which provide high sensitivity and low contrast. Spectroscopy analyses have revealed that this material is only composed of $H–SiO_{3/2}$ units. It is believed that these units form a three dimensional configuration, with molecular weight of 11,000. During the incidence of electron beam a break mainly in the SiH bonds (weaker than the SiO bond) occurs [22; 23]. Its main features are small line edge roughness and relatively small molecular size. Patterning HSQ with 70 kV electron beam lithography, Namatsu [24] demonstrated lines with a width of 7 nm and an aspect ratio of 10. Apart from its high-resolution capability, HSQ has a better dry etch resistance than PMMA, which makes it a desirable resist for processing at nanometer dimensions (see Table 27.1) [25; 26]. HSQ is also an excellent resist for testing e-beam machine resolution limits because HSQ lines on silicon can be imaged directly in a SEM without the need of gold evaporation for conduction or "lift-off" techniques. The main applications are concentrated in the construction of electronic devices [27–29] and for the study of plasmon resonance [30].

3 EXPOSURE PROCESS

The choice of beam energy is critical for production of high-resolution structures. High beam energies (~100 keV) have as an advantage less electron scattering in the resist, leading to reduction of the proximity effects, but at the cost of more damages in the sample due to heating caused by the beam. For low energies (~2 keV), very thin resists should be used, correction of the proximity effect is not required, and damage by heating is minimal. An alternative method is to work in an intermediate range of e-beam energy (near 30 keV) where there is a tolerable heating of the sample and the unique disadvantage is the necessity to correct for the proximity effect.

For nanolithography the use of low current values (~10 pA) is unanimous. The main advantages are weak Coulomb interactions and high depth of focus. The disadvantages are the increased time of exposure and difficulty of focusing (and correction of astigmatism) due to the increased noise/signal ratio. It is well-known that the current value is directly correlated with the beam diameter, which in turn influences the resolution of the process. Because of its importance, several methods and models have been developed in order to specify (with the lowest possible uncertainty) the size of the electron beam [31–34]. Most of the methods are based on the combination of patterns and structures specially designed in order to develop the metrology of the beam. In practice the choice of the current value should take into account the optimization of the whole exposure process. A larger diameter is used in the exposure of large structures (large fields of exposure) because of the shorter time necessary (in a process where high resolution is not required). A smaller diameter of the beam is typically used for exposure of smaller structures (small fields of exposure) resulting in both high resolution and exposure time.

The working distance (between the sample surface and the upper part of the objective lens) must be adjusted in order to obtain the best trade-off between the resolution and depth of focus. Usually 5 mm is utilized as a standard working distance.

After performing the initial steps of firmly attaching the sample and focus/astigmatism correction, the exposure process itself begins. As the sample surface is covered with resist (which is sensitive to the electron beam) focus/astigmatism correction should be performed on a scratch or special structure on the sample surface. The scanning exposure of the structures can be done in two ways: the line mode and meander mode. In the line mode, the beam travels along the structure and exposure is always in the same direction with a period of time to change from one line to another (flyback time). In the meander mode, the beam fills in the structure in alternating directions. The electron beam is kept on during the exposure of the desired structure, being off when transiting between the structures. The e-beam scanning process along the field under exposure consists of constant displacements (step grid) both in horizontal and vertical directions.

For any software that controls the exposure process there are three main menus that define the characteristics of the process: the field, the time of exposure, and the alignment.

In the menu of field of exposure basically two parameters are defined: the area of exposure and the step grid. The step grid provides the formation of a virtual grid that is correlated to the constant displacements of the e-beam during the scan. The sizes utilized for the step-grid both in x and y directions should have independent values from each other, but it is more common to the use the same value for both. The choice of the step grid value deserves special attention because it is directly correlated to the beam diameter. For a homogenous exposure an appropriate value for the step grid is half of the electron beam diameter. For a FEG (field emission gun) that has a beam diameter around 2 nm, the maximum step grid must be 1 nm, and for a conventional SEM (scanning electron microscope) with beam diameter of 30 nm the maximum step grid will be 15 nm. In a lithography system composed by a 16-bit digital-to-analog (DAC) board, these ideal conditions can be only achieved utilizing maximum fields of exposure of 65 μm and 800 μm, respectively. In normal applications (especially in the case of FEG) it is commonly required to work with fields of exposure greater than 65 μm, for which the step grid size should be set to values larger than the beam diameter. For systems with precision x-y stages (such as interferometer stages) a common solution is to create large fields by precisely abutting subfields (by mechanically moving the precision stage). Alternative methods to avoid this problem are either to defocus the electron beam (resolution loss) or to increase the current used for exposure.

In the menu of time, the values of the dwell time (interval of time that the beam stops at every point of exposure—generally in order of microseconds), the dose, and the beam current are defined [35]. The expressions below show the calculation of the dose to the exposure of a structure of type area, line or point:

$$AreaDose = \frac{I_{beam} T_{dwell}}{S^2} (\mu As/cm^2) \tag{3.1}$$

$$LineDose = \frac{I_{beam} T_{dwell}}{S} (\mu As/cm) \tag{3.2}$$

$$DotDose = I_{beam} T_{dwell} (\mu AS) \tag{3.3}$$

The alignment menu is necessary when the exposure should be precisely positioned over structures already defined in the sample. To this effect it is necessary to use an orthogonal correction of the structure in the software. The correction is done by adjusting the angles x and y and the scale factor in x and y-axes beforehand, with help of a well-known set of structures (e.g., a 1 μm checkerboard).

At the end of the definition of exposure parameters, the magnification should be adjusted. The magnification is related to the size of the field to be exposed. For a structure that has $100 \times 100 \, \mu m^2$ field, the magnification must be set approximately to 500×, depending on working distance.

For some applications, as already mentioned, the desired structure has large areas of exposure, making it necessary to break it into small parts, the subfields. This fragmentation creates an error of stitching, which occurs when adjacent subfields are not precisely matched, generating small discontinuities at the border of the fragmented structure. There are different ways to correct this effect. The Moiré method [36] can be a simple option as well as using specific algorithms [37].

4 PROXIMITY EFFECT CORRECTION

The proximity effect is caused by elastic scattering (small angle or direct) of electrons in the resist and backscattering in the substrate during the sample exposure. These two processes form secondary electrons with low energy that may expose the resist outside of the expected region, which can span to a few tens of microns [38] for accelerating voltages around 50 kV. The change of the trajectory of the electron beam due to scattering events is a function of initial kinetic energy of electrons and of the composition of the sample [39].

To correct the proximity effect, it is necessary to use software algorithms that calculate the changes in dose distribution in the exposed region so that the final structures will have the desired shape and size [40–44]. In exposure terms, using the proximity correction information, the exposure software will locally (step by step) vary the dwell time to locally change the dose. Several algorithms for the proximity effect correction can be found in the literature [45–49].

For example, there are models where the proximity effect is calculated by a linear combination of two or more Gaussian functions [38; 50; 51]:

$$f(r) = \frac{1}{\pi(1+\eta)}\left(\frac{1}{\alpha^2}\exp\left[-\left(\frac{r}{\alpha}\right)^2\right] + \frac{\eta}{\beta^2}\exp\left[-\left(\frac{r}{\beta}\right)^2\right]\right) \tag{4.1}$$

where r is the diameter of the beam, α the dispersion of the beam due to direct scattering of electrons, β describes the dispersion of the beam due to the substrate backscattering of electrons, and η describes the relative contribution of two effects.

For exposure patterns of the order of nanometers [52], a correction of the effect is more rigorous so that a third term has to be added to Equation 4.1:

$$f(r) = \frac{1}{\pi(1+\eta+\nu)}\left[\frac{1}{\alpha^2}\exp\left[-\left(\frac{r}{\alpha}\right)^2\right] + \frac{\eta}{\beta^2}\exp\left[-\left(\frac{r}{\beta}\right)^2\right] + \frac{\nu}{2\gamma^2}\exp\left[-\left(\frac{r}{\gamma}\right)\right]\right]$$
$$\tag{4.2}$$

where γ is the decay constant of the exponential function introduced for better fitting of the profile at the 10–100 nm vicinity of the beam and ν is the efficiency factor. In Equations (4.1) and (4.2) the parameters α, β, η, γ and ν are determined experimentally through tests with different designs of structures [53–58].

Normally the value of α is below 0.1 μm and this is a value usually difficult to measure accurately because it is related to the incident beam diameter, which varies from machine to machine and time to time. Due to the same reason, the measurement of η is also difficult. These parameters can be determined beforehand or during proximity calculation, depending on the algorithm.

A popular proximity software is PROXECCO [47], which adjusts the dose for each fraction of the design of the structure. This adjustment takes into account the composition of photoresist and substrate, the thickness of the resist, and the substrate. But for structures smaller than 75 nm [46], results are not totally consistent unless an alternative 3 or 4 Gaussian model is used.

PROX-in [59] achieves better proximity correction (higher resolution) by better adjustment of the dose of exposure (point to point) or through the analysis and interpretation of the distortions in the geometry of exposure. This understanding aims to determine the best shapes to reconstruct the structures considering the proximity effects. The PROX-in also uses a linear combination of 2 or 3 exponential or Gaussian functions to correct the proximity effect. This software has advantages of easy, fast, and efficient correction of the effects of proximity and has been used successfully for production of masks and structures of about 90 nm and is also used to write structures with resolution below 50 nm [60].

Determination of α, β, and η parameters before proximity correction by software should be carried for exactly the same substrate, resist (type and thickness), and energy as will be used during exposure.

One procedure to determine these parameters is using specific software such as [38]. In PROXY software the α parameter is determined by making the exposure of lines with different widths ranging from α to 5α up in the horizontal axis and varying the value of the dose of exposure on the vertical [50]. Each of these lines is separated by a small distance of approximately 0.1 μm and the dose increases 0.95 step by step. For a given value of α, the PROXY software calculates the value of the appropriate dose for the exposure of the vertical lines. The value of α is accepted if it is small compared to the value calculated by PROXY. The uncertainty on the value of α is around 10%. Following similar procedures, PROXY can estimate values for β and η. The value of β is large compared to the value of α. In the case of a silicon substrate, β is 3.9 μm at an energy of 30 keV and accuracy of 10% and $\eta \sim 0.7$.

Other methods do not need special software for determination of the proximity parameters. One such method often used to correct proximity is the exposure of points [61; 62] (dots) of different sizes, based on the diameter of electron beam in SEM—in this case, the resolution of the measured point is limited by the beam diameter. These points are spaced so that there is interaction between them during the test exposure; in this situation the dose varies, depending on the size of the point and time of exposure. By optical or SEM inspection, the value of alpha can be determined.

The value of the parameter β is obtained from the slope of the logarithmic plot of the dose depending on radius of the points exposed. The value of η, although it varies significantly for different substrates and different resists, is still valid for this method, but it must be remembered that the nearest two or more Gaussian may not be the best option for the correction of this parameter using this method.

Another very popular method uses pattern structures in the form of rings (doughnuts) [63] with internal and external radii R1 and R2, respectively. To correct the parameters of a proximity matrix is to be exposed to different values of R1 and different doses of exposure varying in direction x and y of the matrix of exposure. The value of R2 is kept constant and it is greater than the backscattering of the incident beam. After the exposure of this matrix, one has to identify the dose which completely exposed the internal radius R1 (for positive resists). The adjustment is done by bringing the rays by internal and external polynomial function [63] so that you can play the internal region of the radii of doughnuts. This method is simple and easy to use and proves to be quite appropriate to correct the parameters of proximity for nanolithography [64]. Table 27.2 gives a summary of parameters for proximity effect correction for different types of resists.

5 DEVELOPMENT PROCESSES

Under e-beam irradiation, in positive resists a scission of its carbon chains (chain-scission) occurs resulting in a reduction in the molecular weight. This reduction causes an increase in the solubility of specific solvents that are usually organic. The exposed regions are removed and non-exposed regions remain on the sample. An opposite process occurs with negative resists. They are based on cross-linking of polymer chains (chemical process that combines the polymer chains). The consequence of this cross-linking is a decrease in solubility of irradiated areas [65].

The development process consists of the immersion of the sample in a solution for a predefined time. A solvent for the resist and another non-solvent part composes this solution. The ratio between these two components determines the solubility that can be optimized. In practice, after the sample being immersed in the solvent it can be immediately immersed in a

Table 27.2 Summary of proximity parameters for different experimental conditions.

Substrate	Resist	Beam Diam. (nm)	Accel. Voltage (kV)	Spot size	Beam Current (pA)	Approximation	α (μm)	β (μm)	η (μm)	γ (μm)	ν	Refs
		Equipment				Approximation	Parameters					Refs
Si/SiO2	PMMA		25	10 nm		2- Gaussian	0.2–0,5	1.45–1.6	0.7			53
Si/SiO2	ZEP-520A (30 nm)		25	10 nm		2- Gaussian	0.505	2.99	1.054			53
spin valve/ Si/SiO2	ZEP-520A (30 nm)		25	10 nm		2- Gaussian	1.021	2.96	0.855			53
Si	HSQ (40 nm)	10 nm	50			Gaussian		8				54
Si	HSQ (40 nm)	10 nm	100			Gaussian		26				54
Si	PMMA (10 nm)				62.5	2- Gaussian	0.2	1.9	0.45			55
Si	PMMA (10 nm)				250	2- Gaussian	0.25	1.9	0.45			55
Si	PMMA (10 nm)				1000	2- Gaussian	0.4	1.9	0.450			55
Si	PMMA (10 nm)				4000	2- Gaussian	0.5	2	0.35			55
Si	PMMA (25 nm)				62.5	2- Gaussian	0.25	2	0.5			55
Si	PMMA (25 nm)				250	2- Gaussian	0.3	2	0.55			55
Si	PMMA (25 nm)				1000	2- Gaussian	0.45	2	0.7			55
Si	PMMA (25 nm)				4000	2- Gaussian	0.5	2	0.5			55
Si	PMMA (45 nm)				62.5	2- Gaussian	0.35	2	0.75			55
Si	PMMA (45 nm)				250	2- Gaussian	0.3	2	0.85			55
Si	PMMA (45 nm)				1000	2- Gaussian	0.35	2	0.9			55
Si	PMMA (45 nm)				4000	2- Gaussian	0.4	2	0.8			55
Si	HSQ (300 nm)		50			3- Gaussian	0.03	0.29	0.12	10.99	0.67	56
Si.	ZEP-520A (30 nm)		30			2- Gaussian	0.0226	3.6	0.57			57
InP	PMMA(220 nm)		30			Gaussian+ exponential	1.59		0.72	0.76	0.58	58

solution that finishes the action of the solvent (stopper) or it can be washed in deionized water or IPA (rinse). Finally, the sample is dried with N_2.

In Table 27.3 the solutions used for development of the same high-resolution resists cited in Section 2 are shown. Some studies give examples of significant improvement in both resolution and quality of structures obtained by optimization of the solvent used. For example, the HSQ has contrast ~5 for the TMAH (tetramethylammonium hidroxide) solvent, while for KOH (potassium hydroxide) it is as high as 10 [24; 28]. Henschel characterized the development process of the HSQ varying the concentration of TMAH solvent, as also shown in Table 27.3 [66]. Using NaOH (1:5) as a solvent, Grigorescu was able to fabricate dense structures with 5 nm resolution spaced by 20 nm from each other [67]. For positive resists similar studies were also performed. Namatsu demonstrated that the use of hexyl acetate as a solvent for ZEP520 results in an increase in resolution and reduction of size variation for desired structures [68] when compared with xylene (standard solvent) as shown in Table 27.3. For PMMA, studies indicate a significant improvement in contrast when using only IPA (isopropyl alcohol) when compared with the standard solution of MIBK:IPA (1:3). This feature deserves attention because generally IPA is used as a rinse in the process of resist development [69]. This means that exposed regions of PMMA are supposed to be almost insoluble for IPA. However, recent studies have shown that pure IPA has considerable potential as a solvent for PMMA [70]. Table 27.3(a) also presents the data, showing the increase in contrast through the use of IPA instead of MIBK:IPA. Another research conducted by Yasin [71] also demonstrated that IPA diluted in small proportions of deionized water behaved as a solvent for PMMA. Individually, IPA and water are non-solvents for PMMA at room temperature, but the mixture of them has better synergy, and a composition of 3:7 water:IPA is a optimum solvent due to the combination of high sensitivity and good contrast, see Table 27.3(a). In practice, in most laboratories, a single standard solvent is used for each resist. Therefore, the development temperature becomes a variable that can be varied in a greater range and has become widely used to obtain structures with high resolution. Yang et al. [72] studied the influence of temperature for the HSQ and ZEP 520. In the case of ZEP520 (xylene as solvent), a significant increase in contrast occurred, from 2.1 to 8.8, while varying the temperature from 40°C to 10°C. However, the cost of this increase in contrast is the decrease in sensitivity of 3 mC/cm^2 to 12 mC/cm^2. In turn, HSQ demonstrated an opposite behavior compared to ZEP 520 with TMAH as a solvent. For this negative resist, the sensitivity decreases (but the contrast increases) with increasing temperature. Chen [73] showed for HSQ an 8-fold increase of contrast when the development temperature with TMAH varied from 18°C to 50°C. For PMMA, Hu [69] showed that the combination of higher doses with a reduction of the development temperature of PMMA generated an increase in the sensitivity (the contrast did not change significantly), allowing fabrication of structures with sub-10 nm resolution.

Development is a limiting factor to obtain structures with sub-100 nm resolution and should be carefully studied. During e-beam irradiation, the transition region is formed between exposed and non-exposed areas. The choice of suitable solvent should be made, aiming at a complete removal of the exposed region while having minimal impact on non-irradiated areas. The purpose is to prevent the lateral swelling of the resist structures that may cause distortions in patterns. In particular, closely spaced lines may swell, causing collapse of adjacent structures. Negative resists are generally more susceptible to swelling at the development process compared to positive resists. This swelling is due to the fact that during the development process, the cross-linking absorbs a significant amount of solvent and consequently increases the size of structures. The volume of solvent absorbed and thus swelling is proportional to the time of development.

Table 27.3 Summary of development processes for some high resolution e-beam resists.

	Resists	Thickness	Developer	Stopper	Rinse	Contrast	Sensitivity ($\mu C/cm^2$)
a-)	PMMA	70 nm	MIBK:IPA (1:3) por 30 s	–	IPA	4.3	100
		70 nm	IPA por 30s	–	–	6.1	250
		1 μm	MIBK:IPA (1:3) por 30 s	–	–	5.4	520
		1 μm	Water:IPA (3:7) por 30 s	–	–	6.4	380
b-)	ZEP-520	100 nm	Xylene	–	1:3 MIBK::IPA	4	–
		100 nm	hexyl acetate	–	IPA	5.3	–
c-)	HSQ	50 nm	2.38% TMAH	TMAH: H2O=1:9	Deionized water	5	–
		200 nm	2.5% TMAH			3	123
		200 nm	25% TMAH			7	261
		50 nm	20% KOH			10	–
d-)	Calixarene		Xylene	–	IPA		
			Methylisobutyl-ketone (MiBK)	–	IPA		

In order to circumvent this problem, an alternative method of ultrasonically assisted development (UAS) has been proposed [74]. It is well-known that intermolecular forces present in both exposed and non-exposed areas limit the conventional development process. The alternative UAS proposes to overcome these intermolecular forces. This process can be done in a beaker suspended in a bath with controlled temperature similarly to a conventional development except for the ultrasonic agitation applied. The UAS process requires smaller development times, resulting in lower swelling. However, the ultrasonic agitation tends to affect resist adhesion. Using this technique, Chen demonstrated a resolution of 5–7 nm for PMMA [74], compared with 10 nm obtained conventionally. Using this method for calixarene (MC6AOAC) [75; 76], Yasin obtained resist lines of 6 nm with 5 seconds of development. For ZEP520, the development with ultrasonic agitation showed an increase of 4% in sensitivity (but no change in contrast) when compared to the traditional method [77]. In turn, HSQ has demonstrated no significant increase in contrast although found it useful in the removal of a residual layer of resist [78].

6 TRANSFERRING PATTERNINGS TO THE SAMPLE

To transfer the structures created by e-beam lithography to the sample, two main techniques are employed: etching or lift-off.

6.1 ETCHING

The etching main characteristics are: etch directionality, uniformity, selectivity, etch rate, undercut and substrate damage. In practice the process can be done by wet [79–81] or dry etching [82–84]. Here, only the reactive ion etching (RIE) technique is considered, being one of the mostly used processes of pattern transfer with sub-100 nm resolution after e-beam nanolithography.

In RIE, the combination of physical and chemical effects (due to energetic ions and reactive radicals, respectively) allows one to obtain a good anisotropy and a high selectivity, thus being well suited for nanofabrication purposes.

Considering the quality of resists as protection masks, it is important to choose high-resolution resists highly resistant to plasma etching, especially because extremely fine resists are used to produce structures with sub-100 nm resolution. In Table 27.4 etch rates in different reactive plasmas are given for the same high-resolution resists cited in Section 2.

ZEP-520 provides a plasma etching resistance 6 times higher than PMMA for CF_4 while calixarenes were found almost 4 times more resistant than PMMA under the same conditions [13; 15; 3]. The same behavior was verified for SF_6 and C_2F_6. The etch rates for Si and SiO_2, materials widely used as substrates in the production of nanodevices, are also shown for comparison.

Some procedures, physical and chemical, have been developed to increase the plasma resistance of resists without compromising their patterning potential [85; 86]. Researches developed in order to achieve this goal are highlighted in gray in Table 27.4. The most common is a high temperature treatment or baking, aiming to increase the material hardness. For example, HSQ baking at 400°C for 60 minutes [87] increased the selectivity from 3 to 4 (Table 27.4(c)). Among chemical processes, most common is silylation. This process incorporates silicon in the resist, although other elements such as In or Ti can be also used. The incorporation can be realized by implantation, or from gas or liquid phases. Ruderish et al. [9] demonstrated that etching resistance of calixarene can be strongly improved by silylation of resists, with the etch rate changing from 2.5 nm/s to 0.3 nm/s (Table 27.4(d)). Another procedure performed with HSQ was incorporation of oxygen atoms by O_2 RIE (Table 27.4(c)) [86].

Using an unconventional approach, Yang [88] increased the HSQ resistance to plasma through an e-beam re-exposure of structures after they had passed the traditional process of exposure/development. The re-exposure was done using a very high dose (50 mC/cm^2). This procedure elevated the etch rate in comparison to a traditional process. In turn, Hu [89] has incorporated small percentages of silica particles (with 1–2 nm diameters) in ZEP-520. The addition of nanoparticles contributed to decrease secondary scattering of electrons. As silica has an atomic number greater than the polymer, it has consequently a higher electronic stopping power. The main feature of this process was a considerable decrease in etch rate of ZEP as can be seen in Table 27.4(b).

6.2 LIFT-OFF

Lift-off is a method of pattern transfer to the sample whose main application is fabrication of metal contacts. This technique is also crucial when the lithographic process requires the pattern transfer into some metals thin films hardly removable by etching. The structure is formed by electron beam in a positive resist (which may be composed of a single layer,

Table 27.4 Etch rates for several high resolution resist for different plasmas.

	Resists	Tone	Gas	Flux (sccm)	Pressure (mTorr)	Power/Voltage	Etching Rate or Selectivity	Physical or Chemical Processes
a-)	PMMA	Positive	CF_4		37.5	50 W/200 V	38 nm/min	–
			C_2F_6		15	100 W	30 nm/min	–
			SF_6		15	100 W	130 nm/min	–
b-)	ZEP-520	Positive	CF_4		37.5	50 W/200 V	5.4 nm/min	–
			C_2F_6		15	100 W	20 nm/min	–
			SF_6		15	100 W	75 nm/min	–
			O_2	30	60	0.25 W/cm^2	197.6 nm/min	Incorporation of 7 wt% n-SiO2 in ZEP 520
			O_2	30	60	0.25 W/cm^2	10 nm/min	
c-)	HSQ	Negative	$SiCl_4$/Ar	4/20	20	100 W	Selectivity 3	–
			$SiCl_4$/Ar	4/20	20	100 W	Selectivity 4	Temperature Treatment
			$SiCl_4$/Ar	4/20	20	100 W	Selectivity 7	RIE O_2
			SF_6/He	12.5/10	7.5	40 W	0.48 nm/s	–
			CF_4	15	15	100 W	7:1	Re-exposure Process
			CF_4	15	15	100 W	4:1	
d-)	Calixarene MC6AOAc Tetraally-calix[4]arene	Negative	CF_4		37.5	50 W/200V	10 nm/min	–
			O_2	10	75	0.25 W/cm^2	2.5 nm/s	–
			O_2	10	75	0.25 W/cm^2	0.3 nm/s	Silylation Process
e-)	Si		CF_4		37.5	50 W/200 V	9 nm/min	–
			SF_6/He	12.5/10	7.5	40 W	2.7 nm/s	–
f-)	SiO_2		C_2F_6		15	100 W	23 nm/min	–

bi-layers, tri-layers, or multilayers) placed on a substrate. Then, a deposition of a thin film of the desired material is made over the entire substrate. The lift-off development is done by immersing the sample in heated solvents. Among the most used lift-off solvents are the SU8 Remover, NMP and acetone. The deposited material must have good adherence to the substrate.

The procedure for a bilayer system, frequently used for lift-off processes with sub-100nm resolution, is considered in more detail as can be seen in Figure 27.1(a). The top layer is composed of a high-resolution electron beam resist, which defines the final pattern resolution. The bottom layer is composed by a resist that generally is not sensitive to electron beam exposure. The lithography process follows a standard procedure of exposure/development for the top layer. Then, for the bottom layer, an appropriate solvent is used that etches the resist, leading to an undercut formation. This overhang prevents sidewall thin metal film deposition and makes it easier for the lift-off solvent to access the bottom layer during the lift-off development. For nanolithography, the production of high-resolution objects requires a precise control of the undercut length. If it is too large, it may impede the fabrication of high-resolution dense structures.

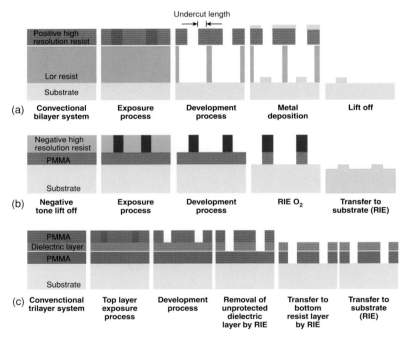

FIGURE 27.1: Mostly used bilayer and trilayer resists systems. (a) The top layer is composed by a high-resolution e-beam resist, which defines the final pattern resolution. The bottom layer is composed by a resist that generally is not sensitive to electron beam exposure. (b) A bilayer system is composed by a negative high resolution and PMMA creating negative tone lift-off. (c) A trilayer system where a dielectric material is used as the intermediate layer, while the top and bottom layers are composed by PMMA resist.

7 BILAYER AND TRILAYER SYSTEMS

The most widely used bilayers systems are composed by PMMA/PMMA-MMA [90; 91] and ZEP520/PPMA [92; 93] resists. However, for applications with dimensions smaller than 100 nm they present some restrictions. This is caused by the undercut profile that becomes not well defined when the resist thickness is decreased for achieving the sub-100 nm resolution. This limitation has been overcome with special lift-off resists (LOR). Using a PMMA/LOR system, Chen produced 20–30 nm constrictions for spintronic devices [94; 95]. To further optimize the lift-off process, the use of ZEP520 resist instead of PMMA as top layer was investigated, and using ZEP520/LOR, Tu et al. [96] produced Cr lines with resolution less than 70 nm.

Another bilayer system is composed by HSQ/PMMA creating negative tone lift-off [97–99]. The main objective for this system is to avoid long periods of exposure needed with positive tone lift-off. It combines the high resolution of HSQ and good lift-off quality of PMMA, see Figure 27.1(b). Initially, the structures are defined on a top HSQ resist layer by electron beam lithography and are subsequently transferred to the bottom layer of PMMA by oxygen RIE.

In trilayer systems, dielectric materials are used, usually as the intermediate layer, while the top and bottom layers are composed by PMMA resist (Figure 27.1(c)). In the top layer, patterns are defined by e-beam exposure. After development of the exposed area of PMMA, the unprotected part of the dielectric material is removed by reactive ion etching. The final transfer to the bottom resist layer is also performed by RIE [100]. In some cases, trilayer systems do not include an intermediate dielectric layer making use the positive resists system like ZEP520/PMMA-MMA/PMMA [101].

8 APPLICATIONS: MICROSQUID SENSORS

Nanodevices oriented for applications in semiconductor lead the ranking of high-resolution structures obtained by electron beam lithography. Efforts made for the construction of double gate MOSFET [5; 29; 32] or multiple-gate transistors [102], NEMS [103; 104], sensors [105] can be pointed as main industrial applications. However, the e-beam lithography has its application

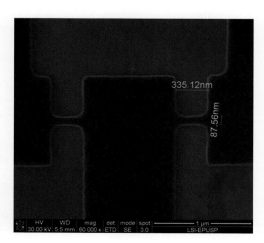

FIGURE 27.2: MicroSQUID device with 80 nm Dayen junction produced by electron beam lithography utilizing SNR 200 1:4 resist with thickness of 110 nm.

in areas far beyond the conventional electronic devices. Researches have demonstrated the potential of this technique applied in areas such as biology, biomedical applications, optics, and magnetism.

The main application discussed in this chapter is the construction of microSQUID sensors. MicroSQUIDs are highly sensitive magnetic sensors that are suitable for the study of nanometer magnetic objects [106]. The SQUID device usually consists of a superconducting ring interrupted by one or two Josephson (or Dayen) junctions. In Figure 27.2 a microSQUID device is shown with 80 nm Dayen junction produced by electron beam lithography in LSI/PSI/EPUSP. In the process we used the SNR200 1:4 negative resist with a thickness of 110 nm.

REFERENCES

[1] Craighead, H. G., Howard R. E., et al. 10-nm linewidth electron beam lithography on GaAs. *Appl. Phys. Lett.* 42.1 (January 1983): 38–40.

[2] Wei, C., and Ahmed, H. Fabrication of high aspect ratio silicon pillars of <10 nm diameter. *Appl. Phys. Lett.* 63.8 (August 1993): 1116–1118.

[3] Nishida, T., Notomi, M., et al. Quantum wire fabrication by e-beam lithography using high resolution and high sensitivity e-beam resist ZEP520. *Jpn. J. Appl. Phys.* 31 (1992): 4508–4514.

[4] Yang, H., Fan, L., et al. Low-energy electron-beam lithography of ZEP-520 positive resist. *Proceedings of the 1st IEEE International Conference on Nano/Micro Engineered and Molecular Systems*, 2006, China.

[5] Dreeskornfild, L., et al. Fabrication of ultra-thin-film SOI transistors using the recessed channel concept. *Microelectronic Engineering* 78–79 (2005): 224–228.

[6] Yasuoka, Y., Suzuki, K., and Abe Y. Fabrication of dielectric lens supported and antenna coupled and warm carrier devices. *Microelectronic Engineering* 73–74 (2004): 496–501.

[7] Hosoka, S., Sano, H., Itoh, K., and Sone, H. Possibility to form an ultrahigh packed fine pit and dot arrays for future storage using EB writing. *Microelectronic Engineering* 83 (2006): 792–795.

[8] Sbiaa, R., Tan, E. L., et al. Sub-50-nm track pitch mold using electron beam lithography for discrete track recording media. *J. Vac. Sci. Technol.* B 26.5 (2008): 1666–1669.

[9] Ruderisch, A., Sailer, H., Schuring, V., and Kern, D. P. Studies on sensitivity and etching resistance of calix[4]arene derivatives as negative tone electron beam resists. *Microelectronic Engineering* 67–68 (2003): 292–299.

[10] Sailer, H., Ruderisch, A., Kern, D. P., and Schuring, V. Evaluation of calyx[4]arene derivatives as high resolution tone electron beam resists. *J. Vac. Sci. Technol.* B 20.6 (2002): 2958–2961.

[11] Gutsche, C. D., Levine, J. A., and Sujeeth, P. K. *Functionalized calixarenes: the claisen rearrangement route. J. Org. Chem.* 50 (1985): 5802.

[12] Gutsche, C. D., and Lin, L. G. The synthesis of functionalized calixarenes.*Tetrahedron* 42 (1986): 1633–1640.

[13] Fujitaj, Ohnishi, Y., Ochiai, Y., and Matsui, S. Ultrahigh resolution of calixarene negative resist in electron beam lithography. *Appl. Phys. Lett.* 68.9 (1996): 1297–1299.

[14] Fujitaj, Ohnishi Y., et al. Calixarene electron beam resist for nano-lithography. *Jpn. J. Appl. Phys* 36 (1997): 7769–7772.

[15] Buhlmann, S., Muralt, P., et al. Electron beam lithography with negative Calixarene resists on dense materials: Taking advantage of proximity effects to increase pattern density. *J. Vac. Sci. Technol.* B 23.5 (2005): 1895–1900.

[16] Kokubo, A., Hatorri, T., et al. A 25-nm-pitch GaInAs_InP buried structure using calixarene resist. *Jpn. J. Appl. Phys.* 37 (1998): 827–829.

[17] Weber, W., Ilicali, G., et al. Electron beam lithography for nanometer-scale planar double-gate transistors. *Microelectronic Engineering* 78–79 (2005): 206–211.

[18] Kretz, J., Dreekorfeld, L., et al. Realization and characterization of nano-scale FinFET devices. *Microelectronic Engineering* 73–74 (2004): 803–808.

[19] Ollinger, C., Fuhse, C., Jarre, A., and Salditt, T. Two-dimensional X-ray waveguides on a grating. *Physyca* B 357 (2005): 53–56.

[20] Knop, M., Richter, M., et al. Preparation of electron waveguide devices on GaAs_AlGaAs using negative-tone resist calixarene. *Semicond. Sci. Technol.* (2005): 814–818.

[21] Hosaka, S., Mohamed, Z., et al. Nano-dot and -pit arrays with a pitch of 25 nm verseus 25 nm fabricated by EB drawing, RIE and nano-imprinting for 1 Tb/in2 storage. *J. Vac. Sci. Technol.* B 85 (2008): 774–777.

[22] Grigorescu, A. E., Krogt, M. C., Hagen, C. W., and Kruit, P. 10 nm lines and spaces written in HSQ using electron beam lithography. *Microelectronic Engineering* 84 (2007): 822–824.

[23] Namatsu, H., Yamaguchi, T., et al. Nano-patterning of a hydrogen silsesquioxane resist with reduced linewidth fluctuations. *Microelectronic Engineering* 41–42 (1998): 331–334.

[24] Namatsu, H., Takahashi, Y., et al. Three-dimensional siloxane resist for the formation of nanopatterns with minimum linewidth fluctuations. *J. Vac. Sci. Technol.* B 16.1 (1998): 69–76.

[25] Word, M. J., Adesida, I., and Berger, P. R. Nanometer-period gratings in hydrogen silsesquioxane fabricated by electron beam lithography. *J. Vac. Sci Technol.* B 21 (2003): 12–15.

[26] Falco, C. M. M., Weterings, J. P., et al. Hydrogen silsesquioxaneNovolak bilayer resist for high aspect ratio nanoscale ebeam lithography. *J. Vac. Sci Technol.* B 18.6 (2000): 3419–3423.

[27] Zegaoui, M., Choueib, N., et al. A novel bondpad report process for III–V semiconductor devices using full HSQ properties. *Microelectronic Engineering* 86 (2009): 68–71.

[28] Lauvernier, D., Vilcot, J. P., François, M., and Decoster, D. Optimisation of HSQ e-beam lithography forthe patterning of FinFET transistors. *Microelectronic Engineering* 75 (2004): 177–182.

[29] Trellenkamp, St., Moers, J. et al., *Patterning of 25-nm-wide silicon webs with an aspect ratio of 13*, Microelectronic Engineering 67–68 (2003): 376–380.

[30] Shi, X., Cleary, A., et al. Multiple plasmon resonances at terahertz frequencies from arrays of arsenic doped silicon dots. *Microelectronic Engineering* 86 (2009): 1111–1113.

[31] Babin, S., M. Gaevski, D. Joy, M. Machin, and A. Martynov. Automatic measurement of electron-beam diameterand astigmatism: BEAMETR. *Physics Procedia* 1 (2008): 113–118.

[32] Yamazaki, K., and H. Namatsu. Electron-Beam Diameter Measurement Using a Knife Edge with a Visor for Scattering Electrons. *Jpn. J. Appl. Phys.* 42 (2003): L 491–L 493.

[33] Babin, S., S. Cabrini, S. Dhuey, B. Harteneck, M. Machin, A. Martynov, and C. Peroz. Fabrication of 20 nm patterns for automatic measurement of electron beam size using BEAMETR technique. *Microelectronic Engineering* 86 (2009): 524–528.

[34] Sidorkin, A., A. van Run, A. van Langen-Suurling, A. Grigorescu, and E. van der Drift. Towards 2–10 nm electron-beam lithography: A quantitative approach. *Microelectronic Engineering* 85 (2008): 805–809.

[35] Tan, J., and Z. Jin. Depth of focus estimation based on exposure dose distribution. *Optics Communications* 281 (2008): 2233–2237.

[36] Murphy, T. E., M. K. Mondol, and H. I. Smith. Characterization of field stitching in electron-beam lithography using moiré metrology. *J. Vac. Sci. Technol.* B 18.6 (2000): 3287–3291.

[37] Koek, B. H., T. Chisholm, J. Romijn, and A. J. van Run. Sub 20 nm stitching and overlay for nano lithography applications. *Jpn. J. Appl. Phys.* 33 (1994): 6971–6975.

[38] Ogino, K., H. Hoshino, Y. Machida, M. Osawa, H. Arimoto, T. Maruyama, and E. Kawamura. Three-dimensional proximity effect correction for multilayer structures in electron beam lithography. *Jap. Journal of Appl. Phys.* 43.6B (2004): 3762–3766.

[39] Aparshina, L. I., S. V. Dubonos, S. V.Maksimov, A. A. Svintsov and S. I. Zaitsev. Energy dependence of proximity parameters investigated by fitting before measurement tests. *J.Vac. Sci. Technol.* B 15.6 (1997): 2298–2302.

[40] Rizvi, Syed, ed. *Handbook of photomask manufacturing techonology.* Taylor & Francis Group. Boca Raton, FL: CRC Press, 2005. Chapter 4.

[41] Rizvi, Syed, ed. *Handbook of photomask manufacturing techonology.* Taylor & Francis Group. Boca Raton, FL: CRC Press, 2005. Chapter 2.

[42] Hannink, R. H. J., and A. J. Hill. *Nanostructure control of materials.* Woodhead Publishing Limited Cambridge. Australia: CSIRO Manufacturing and Infrastructure Technology, 2006. Chapter 12.

[43] Osawa, M., K. Ogino, H. Hoshino, Y. Machida, F. Koba, H. Yamashida, and H. Arimoto. Evalution of performace of proximity effect correction in electron projection lithography. *Japanese Journal of Applied Physics* 44.7B (2005): 5535–5539.

[44] Babin, S. V., and A. A. Svintsov. Effect of resist development process on the determination of proximity function in electron lithography. *Microelectronic Engineering* 17 (1992): 417–420.

[45] Wind, S. J., P. D. Gerber, and H. Rothuizen. Accuracy and efficiency in electron beam proximity effect correction. *Journal of Vacuum Science & Technology* B16.6 (1998): 3262–3268.

[46] Manakli, S., K. Docherty, L. Pain, J. Todeschini, M. Jurdit, B. Icard, M. Chomat, and B. Minghetti. New electron beam proximity effect correction approach for 45 and 32 nm nodes. *Jap. Journal of Appl. Phys.* 45.8A (2006): 6462–6467.

[47] Eisenmamm, H., T. Waas, and H. Hartmann. PROXECCO–Proximity effect correction by convolution. *Journal of Vacuum Science & Technology* B 11 No. 6 (1993) p. 2741–2745.

[48] Aristov, V. V., A. I. Erko, B. N. Gaifullin, A. A. Svintsov, S. I. zaitsev, R. R. Jede, and H. F. Raith. Proxy–A new approach for proximity correction in electron beam lithography. *Microelectronic Engineering* 17 (1992): 413–416.

[49] Hofmann, U., et al. Hierarchical proximity correction using CAPROX. *Proceedings of SPIE* 2621 (1995): 558–567.

[50] Dubonos, S. V., B. N. Gaifullin, H. F. Raith, A. A. Svintsov, S. I. Zaitsev. Evaluation, verification and error determination of proximity parameters α, β and η in electron beam lithography. *Microelectronic Engineering* 21 (1993): 293–296.

[51] Chang, A. L. Nanostructure technology. *IBM J. Res. Develop.* 32.04 (1988): 462–493.

[52] Robin, F., S. Costea, G. Stark, R. Wüest, P. Strasser, H. Jäckel, A. Rampe, M. Levermann, and G. Piaszenski. Accurate proximity-effect correction of nanoscale structures with NanoPECS.Raith – Application Note.

[53] Yi, H., J. Chang. Proximity-effect correction in electron-beam lithography on metal multi-layers. *J. Mater. Sci.* 42 (2007): 5159–5164.

[54] Anderson, E. H., D. L. Olynick, W. Chao, B. Harteneck, and E. Veklerov. Influence of sub-100 nm scattering on high-energy electron beam lithography. *Journal of Vacuum Science & Technology* B 19.6 (2001): 2504–2507.

[55] Stevens, L., R. Jonckheere, E. Froyen, S. Decoutere, and D. Lanneer. Determination of the proximity parameters in electron beam lithography doughnuts structures. *Microelectronic Engineering* 5 (1986): 141–150.

[56] Keil, K., K.-H. Choi, C. Hohle, J. Kretz, L. Szikszai, and J.-W. Bartha. Detailed characterization of hydrogen silsesquioxane for e-beam applications in a dynamic random access memory pilot line environment. *Journal of Vacuum Science & Technology* B 27.1 (2009): 47–51.

[57] Seo, E., B. K. Choi, and O. Kim. Determination of proximity effect parameters and the shape bias parameter in electron beam lithography. *Microelectronic Engineering* 53 (2000): 305–308.

[58] Wuest, R., P. Strasser, M. Jungo, F. Robin, D. Erni, and H. Jackel. An efficient proximity-effect correction method for electron-beam patterning of photonic-crystal devices. *Microelectronic Engineering* 67–68 (2003): 182–188.

[59] Rishton, S. A., and D. P. Kern. Point exposure distribution measurement for proximity correction in electron beam lithography on a sub-100 nm scale. *Journal of Vacuum Science & Technology* B 5.1 (1987): 135–141.

[60] Hudek, P., and Beyer, D. Exposure optimization in high- resolution e-beam lithopraphy. *Microelectronic Engineering* 83 (2006): 780–783.

[61] Rishton, S. A., and D. P. Kern. Point exposure distribution measurement for proximity correction in electron beam lithography on a sub-100 nm scale. *Journal of Vacuum Science & Technology* B 5.1 (1987): 135–141.

[62] Anderson, E. H., D. L. Olynick, W. Chao, B. Harteneck, and E. Veklerov. Influence of sub-100 nm scattering on high-energy electron beam lithography. *Journal of Vacuum Science & Technology* B 19.6 (2001): 2504–2507.

[63] Stevens, L., R. Jonckheere, E. Froyen, S. Decoutere, and D. Lanneer. Determination of the proximity parameters in electron beam lithography doughnuts structures. *Microelectronic Engineering* 5 (1986): 141–150.

[64] Yi, H., and J. Chang. Proximity-effect correction in electron-beam lithography on metal multi-layers. *J. Mater. Sci.* 42 (2007): 5159–5164.

[65] Medeiros, D. et al. Recent progress in electron-beam resists for advanced mask-making. *IBM J. Res. & Dev*, 45 (5): 2001: 639–650.

[66] Henschel, W., Georgiev, Y. M., Kurz. Study of a high contrast process for hydrogen silsesquioxane as a negative tone electron beam resist. *J. Vac. Sci. Technol.* B 21.5 (2003): 2018–2025.

[67] Grigorescu, A., Krogt, M. C., Cees, W., and Hagen, Kruit P. Influence of the development process on ultimate resolution electron beam lithography, using ultrathin hydrogen silsesquiaoxane resist layers. *J. Vac. Sci. Technol.* B 25.6 (2007): 1998–2003.

[68] Namatsu, H., Nagase, M., Kurihara, K., Iwadate, K., Furuta, T., and Murase, K. Fabrication of sub-10 nm silicon lines with minimum fluctuation. *Electronic Materials Letters* 3.1 (2007): 1–5.

[69] Hu, W., Sarveswaram, K., Lieberman, M., and Bernstein, G. Sub-10 nm electrons beam lithography using cold development of poly(methylmethacrylate). *J. Vac. Sci. Technol.* B 22.4 (2004): 1711–1716.

[70] Wi, J. S., Lee, H. S., and Kim, K. B. Enhanced development properties of IPA (isopropyl alcohol) on the PMMA electron beam resists. *Electronic Materials Letters* 3.1 (2007): 1–5.

[71] Yasin, S., Hasko, D. G., and Ahmed, H. Comparison of MIBK/IPA and water/IPA as PMMA developers for electron beam nanolithography. *Microelectronic Engineering* 61–62 (2002): 745–753.

[72] Yang, H., Jin, A., Luo, Q., Gu, C., and Cui, Z. Comparative study of e-beam resist processes at different development temperature. *Microelectronic Engineering* 84 (2007): 1109–1112.

[73] Chen, Y., Yahng, H., and Cui, Z. Effects of developing on the contrast and sensitivity of hydrogen silsesquioxane. *Microelectronic Engineering* 83 (2006): 1119–1123.

[74] Chen, W., and Ahmed, H. Fabrication of 5–7 nm wide etched lines in silicon using 100 keV electron beam lithography and polymethylmethacrylate resist. *Appl. Phys. Lett.* 62.13 (1993): 1499–1501.

[75] Yasin, S., Hasko, D., and Carcenac, F. Nanolithography using ultrasonically assisted development of calixarene negative electron beam resist. *J. Vac. Sci. Technol.* B 19 (1) jan/feb 2001.

[76] Yasin, S., Hasko, D., and Ahmed, H. Comparison of sensitivity and exposure latitude for polymethylmethacrylate, UVIII, and calixarene using conventional dip and ultrasonically assisted development. *J. Vac. Sci. Technol.* B 17.6 (1999): 3390–3393.

[77] Lee, K. L., Bucchignano, J., Gelorme, J., Viswanathan, R. Ultrasonic and dip resist development processes for 50 nm device fabrication. *J. Vac. Sci. Technol.* B 15.6 (1997): 2621–2626.

[78] Regonga, S., Aryal, M., and Wenchuang, H. Stability of HSQ nanolines defined by e-beam lithography fr Si nanowire field effect transistors. *J. Vac. Sci. Technol.* B 26.6 (2008): 2247–2251.

[79] Tiron, R., Mollard, L., Louveau, O., and Lajoinie, E. Ultrahigh resolution pattern using electron beam lithography HF wet etching. *J. Vac. Sci. Technol.* B 25.4 (2007): 1147–1151.

[80] Zaborowski, M., Grabiec, P., and Rangelow, I. W. Chromium nano-width ribbons by standard lithography and wet etching. *Microelectronic Engineering* 73–74 (2004): 588–593.

[81] Xingquan, L., Mingfang W., Bo Zhang, et al. PL Study of QDs manufactured by visible light lithography and wet etching. *SPIE* 3175 (1998): 412–414.

[82] Wolf, E. D., Adesida, I., and Chinn, J. D. Dry etching for submicron structure. *J. Vac. Sci. Technol.* A 2.2 (1984): 464–469.

[83] Gokan, H., Esho, S., and Ohnishi, Y. Dry etch resistance of organic materials. *J. Electrochem. Soc.: Solid-State Science and Technology* 130 (1983): 143–146.

[84] Asakawa, K., and Sugimoto, Y. Dry etching and nanofabrication technology perspective for ovel devices. *SPIE Proceedings Series* 4532 (2001): 300–313.

[85] Vourdas, N., Boudouvis, A. G., and Gogolides, E. Increased plasma etch resistance of thin polymeric and photoresist films. *Microelectronic Engineering* 78–79 (2005): 474–478.

[86] Zelentsov, S. V., Kolesov et al. Enhancing the dry-etching durability of photoresist masks: A review of the main approaches. *Russian Microelectronics* 36.1 (2007): 40–48.

[87] Lauvernier, D., Garidel, S., et al. Realization of sub-micron patterns on GaAs using a HSQ etching mask. *Microelectronic Engineering* 77 (2005): 210–216.

[88] Yang, J. K. W., et al. Enhancing etch resistance of hydrogen silsesquioxane via post-develop electron curing. *J. Vac. Sci. Technol.* B 24.6 (2006): 3157–3161.

[89] Hu, Y., Wu, H., Gonsalves, K., and Merhari, L. Nanocomposite resits for electron beam nanolithography. *Microelectronic Engineering* 56 (2001): 289–294.

[90] Park, Y. D., Caballero, J. A., Cabbibo, A., Childress, J. R., Hudspeth, H. D., Schultz, T. J., and Sharifi, F. Fabrication of nanometer-sized magnetic structure using e-beam patterned deposition masks. *J. Appl. Phys.* 81 (1997): 4717.

[91] Ishii, T., Nozawa, H., and Tamamura, T. Nanocomposite resist system. *Appl. Phys. Let.* 70 (1997): 1110.

[92] Lihua, A., Zheng, Y., Li, K., Luo, P., and Wu, Y. Nanometer metal line fabrication using a ZEP520/50 k PMMA bilayer resist by e-beam lithography. *J. Vac. Sci Technol.* 23.4 (2005): 1603–1606.

[93] Maximov, I., Sarwe, E. L., Beck, M., Deppert, K., Graczyk, M. Magnusson, M. H., and Montelius, L., Fabrication of Si-based nanoimprint stamps with sub-20 nm features. *Microelectronic Engineering* 61–62 (2002): 449–454.

[94] Chen, Y., Peng, K., Cui, Z. A lift-off process for high resolution patterns using PMMA/LOR resist stack. *Microelectronic Engineering* 73–74 (2004): 278–281.

[95] Chen, Y., Lu, Z., Wang, X., Cui, Z., Pan, G., et al. Fabrication of ferromagnetic nanoconsrictions by electron beam lithography using LOR/PMMA bilayer technique. Microelectronic Engineering 84 (2007): 1499–1502.

[96] Tu, D., Liu, M., Shang, L., Xie, C., and Zhu, X. A ZEP520-LOR bilayer lift-off process by e-beam lithography for nanometer pattern transfer. *Proceedings of the 7th IEEE- International Conference on Nanotechnology* 2007.

[97] Yang, H., Jin, A., Luo, Q., Li, J., Gu, C., and Cui, Z. Electron beam lithography of HSQ/PMMA bilayer resists for negative tone lift-off. *Microelectronic Engineering* 85 (2008): 814–817.

[98] Arnal, T., Soulimane, R., Aassime, A., Bibes, M., Lecoeur, Ph., et al. Magnetic nanowires patterned in the $La_{2/3}Sr_{1/3}MnO_3$ half-metal. *Microelectronic Engineering* 78–79 (2005): 201–205.

[99] Magdenko, L., Gaucher, F., Aassime, A., Vanwolleghem, M., Lecoeur, P., Dagens, B. Sputtered metal lift-off for grating fabrication on InP based optical devices. *Microelectronic Engineering* 86 (2009): 2251–2254.

[100] Dubos, P., Charlat, P., Crozes, Th., Paniez, P., and Pannetier,B. Thermostable trilayer resist for niobium lift-off. *J. Vac. Sci. Technol.* 18.1 (2000): 122–126.

[101] Kim, S. C., Li, B. O., Lee, H. S., Shin, D. H., Kim, S. K., Park, H. C., and Rhee, J. K. Sub-100 nm T-gate fabrication using a positive resist ZEP520/P(MMA-MMA) trilayer by double exposure at 50kV e-beam lithography. *Materials Science in Semiconductor Processing* 7 (2004): 7–11.

[102] Cornu, F. F., Penaud, J., Dubois, E., François, M., and Muller, M. Optimisation of HSQ e-beam lithography for patterning of FinFET transistors. *Microelectronic Engineering* 83 (2006): 776–779.

[103] Castellanos A., Leffew J., and Moreno, W. FIB and E-beam cross-linked poly(N-isopropylacrylamide) patterning for BioMEMS/NEMS. *Proceedings of the 7th International Caribbean Conference on Devices* (2008): 28–30.

[104] Gemma Rius, Llobet J., Arcamone, J., Borrisé, X., and Murano, F. P. Electron and ion beam lithography for the fabrication of nanomechanical devices integrated on CMOS circuits. *Microelectronic Engineering* 86 (2009): 1046–1049.

[105] Candeloro, P., Carpentiero, A., Cabrini, S., Fabrizio, E. D., Comini, E., et al. SnO2 wires for gas sensors. *Microelectronic Engineering* 78–79 (2005): 178–184.

[106] Ootuka, Y., Yamagishi, T., Suzuki, K., Miyagawa, Y., Maekawa, T., and Kanda, A. A micro SQUID using ultra-small tunnel junctions. *Journal of Physics: Conference Series* 150 (2009): 12033.

PART III

PROSPECTIVES

CHAPTER F-1

FOCUSED BEAM PROCESSING—NEW BEAM TECHNOLOGIES—NEW CHALLENGES IN PROCESS DEVELOPMENT AND NANOFABRICATION

John Melngailis, Stanislav A. Moshkalev, and Ivo Utke

1 NOVEL FIB SOURCES AND FIB/FEB MULTIBEAM SYSTEMS

Almost all focused ion beam (FIB) tools use a gallium liquid metal ion source. The focused electron beam (FEB) tools can be thought of as scanning electron microscopes modified to include gas feed for beam-induced material addition or beam-induced etching. Applications that have been explored but have not been adopted in practice include *in situ* processing and direct implantation of semiconductor devices.

In the past few years three developments have occurred in focused ion or electron beams:

- gas field ion source FIB systems, (He ion microscope) [1];
- plasma source-based FIB system [2];
- multibeam systems of both electrons and ions [3].

These systems, particularly with further development, will likely enable many new micro/nano applications or alter existing ones. Envisioning the applications often motivates and directs system development. Schematics of the three systems are shown in Figure F.1(a-c).

An ion source development that has not yet been demonstrated in systems, but which may open new nanofabrication possibilities is the magneto-optical-trap-based ion source [5; 6]. It promises ultrasmall energy spread and the possibility of extracting single atoms one at a time in a controlled fashion [7].

1.1 GAS FIELD ION SOURCE

The gas field ion source development has a long history, even longer than the liquid metal ion source. Even though helium ion sources with relatively long lifetime (160 hrs.), ultra-high brightness ($\sim 5 \times 10^9$ A/cm^2 sr, 1,000$\times$ higher than Ga), and low energy spread ($\sim$1 eV vs. 4 eV for Ga) were demonstrated more than 15 years ago [8], it was not until fairly recently that the ALIS Corp. (now part of Zeiss) announced the helium ion microscope [1], see figure F1a. The main application of the helium ion microscope is, as the name implies, microscopy. The beam diameter is in the sub nanometer range. Since the secondary electrons are emitted closer to the point of helium ion impact than they are from the point of electron beam impact in an SEM, the helium ion beam will have higher resolution, i.e., it will be limited more by beam diameter than by the beam surface interaction volume. In addition, it has better material contrast and larger depth of focus. However, the beam current is low, in the pico-ampere range, so that imaging is slow. When imaging with ions, sputtering has to be considered. But even for the heavier gallium ions, an image can be formed at the cost of less than a monolayer. The sputter yield of helium is at least 10 times lower than that of heavier ions. However, implantation during extended exposure can cause material swelling, thus distorting dimensions. A natural extension of this tool is, of course, direct patterning. Resist has already been exposed yielding gratings with a half pitch of 10 nm [9].

Direct fabrication, such as beam-induced deposition and beam-induced etching with a precursor gas, may also yield features that are smaller than can be achieved with gallium ions or with electrons. The mechanism by which the precursor gas is activated is likely to involve more secondary electron processes than is the case with the heavier gallium ions. For heavier ions the dominant process is collision cascades, just as it is in sputtering [10].

1.2 PLASMA SOURCES FOR FIB

The first demonstrations of focused ion beam fabrication in 1973 were done at Hughes Research Laboratories by focusing the beam from an ion implanter [11]. This is in effect like using a plasma source, see figure F1b. The brightness was four orders of magnitude lower than of the gallium source that was developed six years later. After that there was little incentive to develop a plasma ion source for focused ion beams. Nevertheless, the potential of sample contamination with gallium during processing was a nagging concern and the plasma source for FIB applications was examined [12]. Still, the brightness of the source was more than three orders of magnitude less than that of gallium. However, recent advances have pushed the brightness of xenon ion plasma sources within two orders of magnitude of gallium [2]. As a result, the beam spot size for a beam current of 10 pA is only about a factor of three larger than that of a gallium ion system. Moreover, at currents higher than about 50 nA, the xenon ion source has a smaller beam diameter than a gallium ion source because of the reduced spherical aberration.

Many of the sample sectioning applications in failure analysis require sizable volumes of material removal with modest resolution. So for beam diameter larger than 300 nm, the xenon

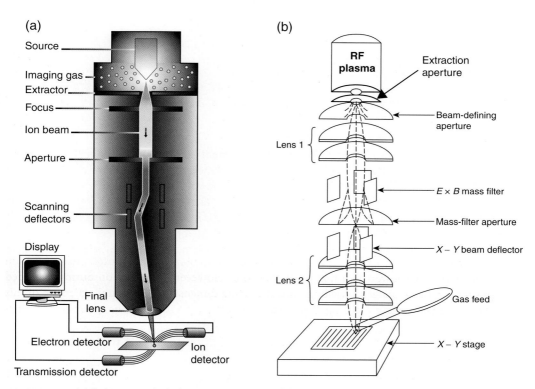

(a)
Source
Imaging gas
Extractor
Focus
Ion beam
Aperture
Scanning deflectors
Display
Final lens
Electron detector
Ion detector
Transmission detector

(b)
RF plasma
Extraction aperture
Beam-defining aperture
Lens 1
$E \times B$ mass filter
Mass-filter aperture
$X - Y$ beam deflector
Lens 2
Gas feed
$X - Y$ stage

FIGURE F.1: (a) Schematic of a helium ion microscope. The ion source is a (111) oriented single crystal tungsten needle with three atoms at the apex. Helium atoms from the gas are predominantly ionized at these three tungsten atoms, and one of the three beams of ions is arranged to be steered down the column. The ions are accelerated to about 40 kV and focused on the sample. The beam can be scanned back and forth on the sample. If gas-assisted beam processing is to be used, a gas injection system needs to be added; see Chapter 3 by Friedli et al. in this book. Reprinted with permission from [4]. Copyright 2006, Cambridge University Press.

(b) Schematic of a focused ion beam system with an RF plasma source mounted in place of the usual liquid metal ion source. The ion column is shown with mass separation, which would be used if the plasma source emits more than one species of ion. If the plasma source is used only with noble gases, the mass separation would not be needed. However, if compound gases are used to yield ions necessary for semiconductor implantation, the mass separation would be needed.

The gas feed can be used optionally for local gas-assisted ion beam deposition or etching.

(c) Schematic of the ion paths in the multibeam system. The ions that pass through the condenser are incident on the aperture plate systems (APS) at normal incidence and well collimated. The 200X demagnification necessitates two beam crossovers. The horizontal dimension is greatly exaggerated. All beams are much closer to the vertical axis.

Reprinted with permission from [3]. Copyright 2008, American Vacuum Society.

(c)

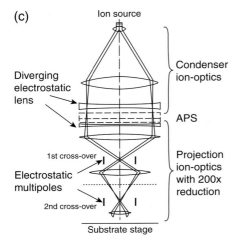

FIGURE F.1: (Continued)

ion plasma source FIB system would deliver more current than gallium liquid metal ion source based systems, and in addition the sputter yield of the xenon ions is higher than that of gallium: a milling speed of about 500 um³/s was achieved for normal incidence on Si with a 30 keV, 128 nA and 500 nm focused Xe-ion beam [2] being a factor twenty to hundred higher than Ga-FIB milling. For final high-resolution trimming or for imaging, the lower beam current that the plasma source delivers compared to gallium FIB may not be a serious disadvantage since these are not time-consuming operations. Moreover, one has a choice of ions, for example xenon for milling, helium for imaging.

The choice of ion species may have wider implications. One of the original envisioned applications of focused ion beams was direct implantation of semiconductor devices without mask or resist to define geometries. Alloy liquid metal ion sources were developed, such as Pd/As/B for implantation into silicon and Au/Si/Be for implantation into gallium arsenide. Many interesting and potentially useful devices were demonstrated, but the very low throughput (more that later) and a lack of a stable, easy to use ion source were serious impediments. The plasma ion source can be designed to yield the desired dopants for semiconductor devices [13]. Of course, such a source would have to be followed by a mass filter because the gases used are likely to be chemical compounds.

1.3 MAGNETO-OPTICAL-TRAP-BASED ION SOURCE

In this source a laser with wavelength appropriately matched to the electronic energy levels of neutral atoms is used to produce ultra-cold atoms confined in the magneto optical trap. Another laser is used to ionize atoms, which are then extracted by an applied field. In this way, ions with an energy spread below 100 meV can be extracted yielding brightness as high as 2.25 Acm^{-2} sr^{-1} eV^{-1} [5]. So far only chromium [5] and rubidium [6] atoms have been reported, but in principle many species could be available depending on the matching of the laser wavelengths to the atomic levels. Another important feature is the ability to extract single ions in a controlled fashion [7]. This opens the possibility of, for example, implanting single ions in materials at precise locations which may have application in quantum optics or quantum computing. In addition, as the dimensions of MOS transistors shrink, the number of dopant atoms in the channel region may become so low that statistical fluctuations become important. Thus the implantation of single ions in a controlled fashion may be of interest at least in studying such phenomena.

1.4 ION OR ELECTRON MULTIBEAM SYSTEMS

The high cost of photolithography masks in integrated circuit manufacturing has motivated development of maskless lithography. Of course, electron beam lithography is already used as a maskless process for the direct writing of prototype devices and for the making of photomasks. Unfortunately, the throughput is much too low for manufacturing. Various multibeam schemes have been considered. The most advanced at the moment are the MAPPER, which uses an array of miniature electron columns [14], and the IMS multibeam, which uses the projection of the image of a programmable aperture plate [3; 15]. The MAPPER is being developed for resist exposure as a replacement for optical photography. The IMS multibeam systems are also being developed for resist exposure but in addition can be used for direct resistless fabrication with both ions and electrons and are therefore of more interest for this discussion. Moreover, of the three new beam technologies, the multibeam represents the most radical departure from current technology and merits more attention.

1.4.1 How does the IMS multibeam tool work?

As shown in Figure F.1(c), the tool (also known as the "charged particle nanopatterning" (CHARPAN) tool) consists of a plasma ion source (or electron source), an optical column, and a sample stage. The aperture plate is back illuminated with a collimated beam of ions (or electrons) forming an array of beamlets that can individually be turned on and off by applying (or not) an electric field normal to the direction of travel. The deflected beamlets fail to pass through an aperture further down in the column. The image of this aperture plate is demagnified 200 times using electrostatic lenses and projected on the sample surface [16]. The aperture plate built so far has 43,000 apertures, each producing a beamlet 2.5 μm × 2.5 μm, which will form a spot on the sample 12.5 nm × 12.5 nm. The beamlets from the aperture plate are in an array measuring a 5.76 mm × 6.72 mm and are spaced by 30 μm. Thus there is a lot of dead space between them. So how can one irradiate the entire sample surface? And how can one program 43,000 apertures to turn on and off as needed?

The apertures are arranged in an offset array, as shown in Figure F.2. The space between the apertures is occupied by shift register circuits. These circuits synchronously toggle the programming information from aperture to aperture in a given horizontal row. Thus each point on the sample is redundantly exposed by as many apertures as there are in one horizontal row. The programmed image information is read into one side, say the left side, of the aperture plate [17]; as the image moves across the aperture plate, the sample stage moves into synchronism. Thus the writing mechanism can be thought of as a dot matrix nano printer.

Two fundamental limitations have to be considered. Because of the large demagnification, the optical column requires two beam crossovers where all beamlets intersect in close proximity. Thus coulomb repulsion between particles that are at random distances from one another will produce a stochastic blur, which limits the maximum possible current. In addition, the brightness of the plasma source that illuminates the aperture plate also limits the total

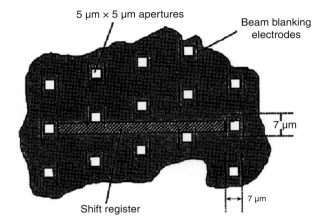

5 µm × 5 µm apertures

Beam blanking electrodes

7 µm

7 µm

Shift register

FIGURE F.2: Schematic of the offset layout of the beam defining apertures. The space between the apertures is occupied by shift register circuits, which apply voltage to the beam blanking electrodes at program times. The image in effect moves horizontally across the aperture plate as the sample moves in synchronism on the stage below the column. Reprinted with permission from [17]. Copyright 1997, American Vacuum Society.

current because the beam incident on the aperture plate has to be collimated within a fairly small angle.

1.4.2 Performance and projected performance of multibeam tool

So far the efforts at IMS in Vienna, Austria, appear to have been directed toward demonstrating the smallest beam diameters and resist exposure at ever finer dimensions. Sub 20 nm lines in gratings have been exposed in HSQ resist using 10 keV H_3^+ ions at a total current of 0.35 nA [16]. This is ion source current limited and not stochastic blur limited since simulations [18] show that at 1 nA, stochastic blur is the same as at 0.35 nA. Other simulations show that at 50 kV, H^+ ions 90 nA of beam current are possible before stochastic blur degrades the performance. The stochastic blur, δ_{stoch}, is largely due to the ion-ion interaction at the beam crossover points. To extrapolate these results to other currents, energies, and ion masses, the following formula can be used [19]

$$\delta_{stoch} = const \cdot (I^{1/2} \cdot f \cdot m^{1/4})/(E^{5/4} \cdot \alpha_0 \cdot D^{1/2}),$$

where I is the ion beam current, f is the focal length of the imaging lens, m is the ion mass, E is the ion energy, α_0 is the convergence half angle of the cluster of beamlets at the crossover, and D is the diameter of the beam waist at the crossover. Thus if we were to extrapolate from the 50 kV hydrogen ion to a 50 kV argon ion and keep the blur the same, we would need to reduce the current from 90 to 14.2 nA. On the other hand, one can envision applications of this multibeam writing scheme that do not require the ultimate resolution. Thus if we allow twice as much blur we can have four times as much current. Also if we double the ion energy, we can increase the current by a factor of 5.7 and keep the blur constant.

The more serious performance limitation therefore appears to be the ion source current. Of the current incident on the aperture plate system, only 0.7% passes through assuming all apertures are open. In the present IMS system, which achieves sub 20 nm dimensions on the sample, the total beam current passing through the apertures is 1 nA. The simulations suggest that much more current could be used. The ion source current incident on the aperture plate system can be increased by increasing the diameter of the extraction apertures of the plasma source. For example, doubling the extraction apertures diameter would yield four times as much current. But since the beam crossover is in effect an image of the ion source, D in the above formula would also double. Thus the contribution to the stochastic blur would only be a factor of $2^{1/2}$ rather than 4. The likely limitation on how large the source can be made is the chromatic aberration. Just as in an FIB system, the contribution to the beam diameter due to chromatic aberration is given by $d = C \cdot \alpha \cdot \Delta E/E$, where C is the chromatic aberration coefficient, α is the convergence angle in each beamlet and will be proportional to the diameter of the extraction aperture of the source (this is not the same as the αo above), ΔE is the energy spread of the ions from the source, and E is the ion energy.

So for the sake of discussing applications, let us assume that we can have a current of 10 nA for any ion species with 30 to 50 nm resolution. That means we give up a factor of about three in resolution to get the source current up by a factor of 10. The total current, and potential fabrication throughput, is at least a factor of 100 higher than for a single FIB.

What about electrons? Similar considerations apply to electrons. When both electrons and ions were considered for static mask projection lithography, the higher current that could be obtained with electrons before stochastic blur effects became dominant was offset by the fact that the charge density needed to expose resist was one or two orders of magnitude higher than for ions. The same considerations will likely hold in the application of electrons versus ions in maskless and resistless direct fabrication. For example, the deposition yield, namely the number of atoms deposited from the dissociation of an absorbed precursor gas per incident particle, is typically 2 to 10 for ions and 100 times smaller for electrons.

The electron multibeam projection has so far been tested in a limited setting with 2,500 beamlets of 12.5 nm beam size on the wafer. Extrapolating this to a full aperture mask yields a total current of 0.9 µA. At this current and 16 nm half-pitch lithography requiring an exposure dose of 120 micro-Coulombs per square centimeter, the exposure time for a 12 inch wafer would be one day [20]. The stochastic blur limit is projected to be 5 µA of beam current. This can be reduced by further development of the aperture plate system to yield many 100 thousands of beams and further by adopting a multi-axis column approach. Thus, the potential throughput of multi-axis projection electron multi-beam systems is as high as five 12 inch wafers per hour for the 16 nm hp technology node with extendibility beyond [20].

1.5 APPLICATIONS OF ELECTRON AND ION MULTIBEAMS

1.5.1 Material addition or removal

Both electrons and ions have been used to induce surface reactions in absorbed precursor gases [21]. For example, if the precursor gas is an organometallic such as tungsten carbonyl or dimethyl-gold-trifluoro-acetylacetonate, the incident electron beam or ion beam will cause these compounds to disassociate leaving a metal deposit on the surface, usually heavily contaminated with carbon if special measures were not taken (see the section below). This kind of deposition has been used in photomask repair, local integrated circuit rewiring, AFM tip fabrication, and the making of contacts to carbon nanotubes, nanofibers, or nanowires. Conversely, if an appropriate reactive gas is introduced, such as xenon difluoride, iodine, or water vapor, the substrate is etched. Focused ion beams can, of course, also remove material by sputtering without the use of reactive gas, but this is usually a factor of 10 or so slower than gas-assisted process.

The multibeam provides 100 times more current, so what new applications might be enabled? Note that the multibeam will not be useful in producing a single feature, for example growing an AFM tip, since only one beamlet at that time could be used, i.e., the beamlets would have to address a single point in sequence. Consider the following example to get an idea of how the multibeam could be applied using material addition or removal. In an early version of RFID tags, each tag was marked with a separate 120 digit binary identity number. This was done by either cutting or not cutting each of the sequence of 120 connections per 1 mm 2 chip using a programmed laser beam. They considered focused ion beam milling as a cleaner alternative, but the time to mill these cuts for 30,000 chips per wafer turned out to be astronomically large. Suppose we use a multibeam. And since the resolution is not at all demanding, we give up another factor of three in beam diameter and achieve total current of 100 nA. Suppose we have to remove 0.1 μm^3 of material per cut or about 5×10^9 atoms. If we use a gas-assisted process, which is typically a factor of 10 faster than milling, we can reasonably remove 25 atoms/ion. Thus the 3.6 million sites per wafer could be addressed in 20 minutes—much closer to being practical. This is a special example of a step where batch processing cannot be used because all chips have to be different not only within one wafer but from wafer to wafer.

One can also envisage fabrication steps where a relatively small amount of material has to be added or removed in selected spots over a wafer, which can be done with a multibeam, eliminating the need to make a mask, spin resist, expose, develop, evaporate or etch, and remove resist.

1.5.2 Ion implantation

Ion implantation into semiconductor devices to make special structures was one of the early applications considered for focused ion beams. In fact, more than 30 focused ion beam systems operating at 100 to 200 keV with mass separation and alloy liquid metal ion sources were built worldwide. Numerous unique devices that generally could not be built otherwise were demonstrated. None was put into practice because the fabrication times were unacceptably long and in some cases the alloy ion sources were difficult to operate. The multibeam systems combined with a plasma ion source can reduce the writing time by 2 to 3 orders of magnitude and provide a stable ion current. So it may be appropriate to consider new applications of direct maskless resistless implantations, which allow one to vary the ion dose from device to device as well as create gradients of dose within a given device. A few examples of devices built in the past are:

- 4 bit 1 GHz flash analog to digital converter included 16 transistors per device with precisely adjusted threshold voltages [22].
- CCD with a longitudinally graded storage gate implants. Charge transfer efficiency and clocking speed improved by more than an order of magnitude [23].
- MOFET with doping gradients in the channel reduced hot electron effects and increased early voltage [24].
- GaAs MESFETs with grated channel doping resulted in higher power ratings and improved trans-conductance and pinch-off voltage [25].
- Tunable Gunn diodes in GaAs with doping gradients in the direction of current flow could be tuned from 25 to 41 GHz [26].
- Junction field effect transistor (JFET) where all fabrication steps are carried out with FIB [27].

With the new multibeam technology, it is possible that some of these devices or similar ones that one can conceive of would be practical. To build a multibeam system for a specific application for a particular ion at a particular energy would require a large investment. To justify such investment, one would first have to build and test the device. For this, an ordinary single beam FIB system operating with a plasma source, with mass separation, and variable ion energy would be useful.

1.5.3 *In situ* processing

In situ processing was an actively researched topic some 20 years ago. The dream was to have a single vacuum chamber where wafers are introduced at one end and all fabrication steps are carried out in a single vacuum chamber or series of chambers and finished devices would emerge at the other end [28; 29]. Another potentially related area of endeavor is the combining of focused electron or ion beams with molecular beam epitaxy in a single or joined vacuum chamber for the fabrication of compound semiconductor device structures [30; 31; 32]. The early *in situ* processing activity spawned the so-called cluster tools where several processing steps consisting of material addition or removal are carried out in connected vacuum chambers with no human handling in between. However, no full fabrication systems were developed. The missing step was fast patterning. Direct patterning with lasers, electrons, and ions was discussed but always proved to be too slow or otherwise inadequate. Wet resist processing seemed to be unavoidable and incompatible with the vacuum.

Now that multibeam tools promise to increase the fabrication speed for both electrons and ions by several orders of magnitude, some of these concepts should be revisited. Clearly, fabrication of complete integrated circuit chips is still far out of reach. However, specialized structures such as MEMS, sensors, or optical devices may require smaller area patterning and may be candidates for in-us situ processing.

1.6 WHAT LIES AHEAD?

Whenever a radically new fabrication technology becomes available, many new applications, often unforeseen, emerge. This was the case with conventional FIB. We now have three new and interrelated technologies, and one more on the horizon. The gas field ion source-based systems will allow us to explore fabrication at ultra-small dimensions, which may potentially be exploited using the multibeam systems. The plasma source-based focused ion beam systems provide faster material removal and an opportunity to explore fabrication with a variety of ions prior to committing to multibeam fabrication. The magneto optical ion trap promises unprecedented source performance, including the possibility of controlled single ion delivery. The multibeam electron and ion systems provide an opportunity for direct maskless, resistless patterned fabrication at three or more orders of magnitude higher speeds than hitherto available. One of the issues in patterning by beam-induced deposition has always been the high concentration of carbon in the deposit. In addition, when a focused gallium ion beam is used, gallium contamination is unavoidable. With the potential of other ion species becoming available, the source of contamination will be removed. First results of focused helium-ion induced deposition of Pt-C nanopillars were reported [106]. Thus the search for a higher purity material by beam-induced processes is likely to be more rewarding.

2 THE PURITY ISSUE IN GAS-ASSISTED FEB AND FIB DEPOSITION

Gas-assisted FEB and FIB deposition relies on *non-thermal* dissociation of adsorbates by electrons or ions. In contrast to thermal dissociation, the dissociation via electronic excitation can result in several non-volatile reaction products, some of them containing unwanted elements leading to "non-pure" deposit material. The unwanted elements originate from the adsorbate molecule itself and/or from the residual background pressures of hydrocarbons and water of the microscope vacuum chamber.

For gas-assisted FIB deposition, the process of ion implantation adds to the purity issue. Depending on the ion species used for deposition, it can have advantageous or detrimental effects. When using gallium ions, about 10–20at.% gallium remains as an implant in the deposited material.

The evident approach to tackle carbon co-deposition is using carbon-free molecules in combination with ultra-high vacuum (UHV) systems. Unfortunately, there are not many carbon-free molecules suitable for gas-assisted FEB and FIB deposition; for lists and discussion see the recent reviews of Botman et al. [33] and Utke et al. [21]. In addition, UHV systems are expensive, time-consuming to operate, and are not compatible with many materials such as polymers.

Therefore, high-vacuum (HV) solutions are also being pursued, and most device proto-typing with functional molecules was indeed tested within such microscopes. The most promising approach against co-deposition of hydrocarbons is co-injection of reactive gases with organometallic molecules. The reactive gas is meant to etch away both the carbon atoms from the molecule and hydrocarbons from the HV chamber background, leaving the pure metal behind. Also chamber cleaning with plasma and substrate baking prior to gas-assisted FEB and FIB processing reduces considerably the hydrocarbon sources in HV microscopes.

We will present the most prominent results in the following.

2.1 CONVENTIONAL HIGH VACUUM MICROSCOPES

Most of the scanning electron or ion microscopes are equipped with chambers featuring high-vacuum conditions, i.e., a background pressure of about 10^{-5}–10^{-7} mbar. At 10^{-6} mbar pressure the impingement rates of residual water vapor and of hydrocarbons are both roughly one monolayer per second. The material obtained by FEB deposition from several hydrocarbon adsorbates under HV conditions was characterized in detail [34] and had a generic chemical composition of $C_{0.75}H_{0.17}O_{0.08}$. The challenge is to find ways of controlling and avoiding the co-deposition of hydrocarbons present in the system.

A straightforward approach to achieve high metal contents in high-vacuum microscopes relies on the concept of reducing the sources of hydrocarbon contamination, namely "dirty" chamber and pipe walls and substrates. This can be achieved via special cleaning protocols in conjunction with an oil-free pump system. Especially effective is baking the sample and a plasma-cleaning of the chamber to remove hydrocarbons prior to each experiment [35] or a combined ultraviolet light irradiation with ozone (O_3) treatment of the chamber [107]. However, the residual water pressure still remains in the HV system, and water molecules will impinge on the substrate.

A very recent approach consists in driving the deposition process in an exposure regime where hydrocarbons from the chamber background become depleted while there are still enough injected functional adsorbates to be dissociated. Controlled metal contents were obtained ranging from 30 at.% of up to 70 at.% for $Co_2(CO)_8$ assisted FEB deposition by simply varying the exposure dwell time [36].

2.1.1 Organometallic molecules

FEB induced deposition using organometallic compounds under high-vacuum conditions results roughly in 10–30 at.% metal content dispersed in a carbonaceous matrix. The carbon matrix itself can contain further elements, like hydrogen, oxygen, or fluorine, when these elements were present in the molecule species injected. Additionally, carbon, oxygen, and hydrogen in the matrix can also originate from the chamber background gases impinging at roughly one monolayer per second. How can the metal content be increased under such conditions?

In situ and *ex situ* substrate heating can considerably increase the metal content above 60 at.%. For an excellent overview, we refer to the topical review of Botman et al. [33].

Local electron beam heating during FEB deposition and local thermal decomposition to pure metals was observed mostly for molecules with low thermal dissociation temperature ($<100°C$), like $Co_2(CO)_8$ [37; 38; 39], (hfa)CuVTMS [40], $Fe(CO)_5$ [41] in conjunction with substrate structures having low thermal dissipation, e.g. membranes. An extensive discussion concerning heat effects in FEBID can be found in the review of Utke et al. [21].

In contrast to FEB deposition, Ga^+-FIB induced deposition from organometallic precursors under high-vacuum conditions results in considerably larger metal contents; roughly >60 at.% metal content dispersed in a carbonaceous matrix. This fact is attributed to the simultaneous physical sputtering of the carbonaceous matrix by the impinging ions but could also be a result of local ion beam heating.

2.1.2 Carbon-free molecules

Carbon-free molecules were introduced to reduce the co-deposition of carbon from the molecule itself. The remaining carbon content can then be attributed to the residual hydrocarbon pressure in the microscope or to sample surface contamination.

Interestingly, the results obtained so far for carbon-free molecules classify into four categories when irradiated with electrons; see also Table F.1.

a) There is one molecule that results in pure metal deposits without taking special care in cleaning protocols of substrates or microscope chambers: $AuClPF_3$ assisted FEB deposition results in pure Au deposition (>95 at.%) [42; 43]. Presently, due to the very limited stability of the compound, an industrial use does not appear easily feasible.

b) There are numerous carbon-free molecules that result in metal contents depending on background pressure; see Table F.1 for carbon and oxygen contents in the deposits. All these molecules are expected to give high-purity deposits in UHV conditions.

c) There are phosphine-based molecules that deposit a considerable amount of phosphorous matrix. These comprise $Ni(PF_3)_4$ [44], $Pt(PF_3)_4$ [45], $Rh_2Cl_2(PF_3)_4$ [46; 47], and $Ir_2Cl_2(PF_3)_4$ [46; 47].

d) There are molecules decomposing into compounds. Pure GaN FEBID was obtained by using the perdeuterated gallium azide molecule D_2GaN_3 [48]. It decomposes

Table F.1 Deposit composition obtained by electron- or ion-induced dissociation of carbon-free molecules in HV- systems. Elements detected in the deposit (and not being part of the molecule) originate from the residual hydrocarbon and water pressure in the microscope chamber. Hydrogen can be present in all deposits but is generally not detected by standard compositional analysis employing energy dispersive X-ray spectroscopy (EDX) or Auger electron spectroscopy. Fourier transform infrared spectroscopy (FTIR) can detect hydrogen bonds.

Molecule/Irradiation	Composition	Remarks	Ref.
$Ni(PF_3)_4$ 10–25 keV electrons	Ni : O : C : P : F 36 : 22 : 14 : 17 : 11	P ≈ 10^{-6} mbar	[44]
$Pt(PF_3)_4$ 10 keV electrons	Pt : P : F 83 : 13 : 4	P ≈ 10^{-6} mbar C-content not specified.	[50]
$AuClPF_3$ 25 keV electrons	Au >95at.%	P ≈ 10^{-6} mbar	[43; 51]
$Rh_2Cl_2(PF_3)_4$ 25 keV electrons	Rh : P : Cl : O : N 60 : 18 : 7 : 8 : 7	P ≈ 10^{-6} mbar	[52; 53]
$Ir_2Cl_2(PF_3)_4$ 25 keV electrons	Ir : Cl : P 0.33 : 0.33 : 0.33	P ≈ 10^{-6} mbar C-content not specified.	[47]
WCl_6 15 keV electrons	W : Cl : C : O 58 : 16 : 8 : 18	P ≈ 10^{-6} mbar (not specified: oil diffusion pump with LN_2 trap).	[54]
$Ti(NO_3)_4$ 25 keV electrons	Ti : O : N 35 : 57 : 8	P ≈ 10^{-6} mbar No carbon detected (EDX) No hydrogen detected (FTIR).	[49]
Ge_2H_6 200 keV electrons	Ge : C 82 : 18	P ≈ $1.3 \cdot 10^{-7}$ mbar, Si_3N_4 membrane substrates	[55]
D_2GaN_3 200 keV electrons	GaN	P ≈ $2.7 \cdot 10^{-7}$ mbar, Si membrane with native oxide.	[48]
$SnCl_2$ 200 eV electrons	Metallic Sn	P ≈ $2.7 \cdot 10^{-7}$ mbar, quartz substrate, T > −90°C.	[56]
$SiCl_4$ 2 keV electrons 50 keV H_2^+	Si : Cl 52 : 48 58 : 42	P ≈ $2.7 \cdot 10^{-7}$ mbar, silicon substrate, T = 120 K.	[57]

thermally at 150°C into crystalline GaN, and beam heating might have contributed to this result. FEB induced deposition with titanium nitrate $Ti(NO_3)_4$ as precursor [49] resulted in a *carbon-free* TiO_x deposit containing some nitrogen, however.

2.1.3 Co-injection of reactive gases and organometallics

The carbon content of deposits can be decreased by co-injection of molecules that react to volatile compounds with carbon upon electron or ion irradiation. Consequently, the balance of deposition and etching will determine the final metal-to-carbon ratio in the deposit. Co-injected molecules comprise H_2O, O_2, H_2, and XeF_2. H_2O and O_2 electron-triggered dissociation products result in a reasonable carbon etch rate. However, the electron triggered

Table F.2 FEB and FIB induced deposition of metals and silicon oxides with additional reactive gases in HV systems. Gallium contents originate from implantation during Ga-FIB deposition.

Molecule/Irradiation	Composition w/o co-injection	Co-injected molecule	Composition with co-injection	Ref.
$(CH_3)_2$-Au-$(C_5H_7O_2)$ 10 keV electrons	Au-C nano-composite	N_2 (1.3 mbar) H_2O (>0.5 mbar)	Little effect Au core in C	[61]
$(CH_3)_2$-Au-$(C_5HF_6O_2)$ 10–30 keV electrons	Au <12at.%	80%Ar/20%O_2 H_2O Ar	50 at.% Au 20 at.% Au 15 at.% Au	[62; 63]
(CH_3)-Pt-Cp$(CH_3)_3$ 30–40 keV Ga-ions	Pt : C : Ga : O 46 : 24 : 28 : 2	H_2	Little effect	[64]
(CH_3)-Pt-Cp$(CH_3)_3$ 15 keV electrons	Pt : C : O 15 : 70 : 10	H_2O	Pt : C : O 50 : 15 : 35	[65]
(CH_3)-Pt-Cp$(CH_3)_3$ 15 keV Ga-ions	Pt : C : O : Ga 10 : 20 : 10 : 20	H_2O	Pt : C : O : Ga 55 : 15 : 15 : 15	[65]
Cu-$(C_5HF_6O_2)_2$ 5 keV electrons	Cu : C : O : F 11 : 62 : 21 : 6	H_2/Ar plasma	Cu : C : O : F 41 : 45 : 13 : <1	[59]
TEOS Si-(O-$C_2H_5)_4$ 10 keV electrons	$SiO_{2.5}C_{3.2}$	O_2/TEOS = 15	SiO_2	[66; 67]
TMOS Si-(O-$CH_3)_4$ 10 keV electrons	$SiO_{2.5}C_{3.2}$	O_2/TMOS = 0.6	SiO_2	[66; 67]
TMOS Si-(O-$CH_3)_4$ 60 keV Si^{2+} ions	SiO_x x = 0.5–2	O_2/TMOS = 1:2 Substrate: RT–200°C	SiO_x x = 1–2	[68]
Tetramethyl-silane Si-$(CH_3)_4$ 5, 10, 25 keV electrons	$SiO_{1.8}C_{3.7}$	O_2/TMS = 1.7	SiO_2	[66; 67]
Tetra isocyonato-silane Si(NCO)$_4$ 10 keV electrons	SiC_2NO_3	O_2/TIS = 470	SiO_2	[66; 67]

dissociation products of these adsorbates also readily oxidize metals, e.g., platinum or tungsten [33]. H_2 has a very small electron dissociation cross-section and consequently a low carbon etch rate. Furthermore, it was found that H_2 co-injection tripled the H_2O background level in the chamber, probably due to competitive adsorption. In turn the oxidation of the metal increased [44]. Instead, highly reactive atomic hydrogen (H-radical) was directly co-injected [58; 59]. However, not only is the co-deposited carbon reduced, but also all adsorbates within the area of H-radical impingement will be suffering the same fate. It will form a large thin metal film around the local FEB (FIB) deposit if the co-injection cannot be processed in a toggled manner. Electron (ion) beam dissociated XeF_2 also etches carbon reasonably well and could be used as a replacement for H_2O and O_2 if oxidation of the metal is to be avoided [60]. It goes without saying that the metal should not be able to form a fluorine compound. A compilation of results is given in Table F.2.

2.2 ULTRA-HIGH VACUUM MICROSCOPES

The use of ultra-high vacuum electron (ion) microscopes permits one to work at pressures of about 10^{-8}–10^{-10} mbar where the impingement rates of residual water and hydrocarbon vapours are two to four orders of magnitude less than at standard high vacuum systems. For the pressure 10^{-10} mbar the impingement rate is less than 0.3 monolayers per hour. Any co-deposition of carbon from the residual chamber vacuum is thus greatly reduced or fully avoided and purer deposit material is expected. In contrast to the high-vacuum experiments discussed in the previous section, often carefully cleaned *crystal* surfaces of metals or semiconductors are used as substrates in UHV systems, or at least special preparation procedures of the substrate are employed. Clean substrate surfaces can act catalytically and help dissociate the adsorbates. A specific effect for some molecules is auto-catalysis, where the freshly deposited clean metal atoms constitute themselves a very efficient catalyst for the new adsorbate molecules, even to such an extent that dissociation to pure metal proceeds without electron (ion) irradiation [69; 70; 71]. Evidently, the material purity of the early deposition stages (and hence UHV conditions and surface preparation) will determine how dominant auto-catalysis can be become for the deposition process and the final purity.

2.2.1 Carbon-containing molecules

Carbon-containing molecules were investigated recently under UHV conditions. One might argue that the carbon contained in the adsorbate will be co-deposited and that a metal-carbon deposit will be obtained. Interestingly, 15 keV electron irradiation of iron carbonyl adsorbates on clean (100) Si at room temperature resulted in pure iron deposits (>95 at.% Fe). Using trimethyl-methylcyclopentadienyl-platinum as precursor on clean (111) Si did not result in pure metal deposits [72]. Other molecules were not yet investigated.

2.2.2 Carbon-free molecules

According to Table F.3, it appears that predominantly metal-halide molecules were investigated in UHV systems for electron-induced deposition studies. The most prominent halide used for FEBID seems to be tungsten-hexafluoride WF_6. Surprisingly high amounts of oxygen and carbon were co-deposited in some UHV experiments with WF_6 [54] which might be attributed to (unwanted) water and hydrocarbon contamination sources in the gas injection system. The same study revealed that with SiO_2 as substrate etching occurred above 50°C, whereas below this temperature deposition of 60 at.% pure W was obtained [54]. It was argued that there is a competition between electron-induced WF_6 adsorbate dissociation into non-volatile tungsten and the follow-up reaction of fluorine adsorbate products with the substrate to volatile SiF_4 (and its thermally induced desorption). However, the freshly deposited tungsten should have passivated the substrate surface from being etched by fluorine adsorbates and thus it seems likely that the background gases played a more decisive role in this specific experiment. The importance of hydrocarbon decontamination of the substrates prior to UHV-FEBID of WF_6 has been demonstrated by an O_2 plasma cleaning process carried out at 300°C substrate temperature [73; 74; 75].

FEB-induced pure silicon (Si) deposition from disilane (Si_2H_6) was reported in a UHV system. In contrast, dichlorosilane ($SiCl_2H_2$) FEB deposition revealed about 2% Cl and 9% O contamination in the silicon deposit [76].

Table F.3 FEB and FIB induced deposition in UHV systems with various molecules. The residual chamber pressure P is indicated.

Molecule/Irradiation	Composition	Remarks	Ref.
$Fe(CO)_5$ 15 keV electrons	Fe (better 95 at.%)	$P \approx 2 \cdot 10^{-10}$ mbar, (100) Si at room temperature.	[72]
WF_6 10 keV electrons	W : C : O : F $57:10:31:2$ @$-110°C$ $66:19:14:1$ @$+25°C$	$P \approx 1.3 \cdot 10^{-8}$ mbar or better, SiO_2 substrate, Deposition below 50°C, Etching of SiO_2 above 50°C.	[54]
WF_6 10 keV electrons	W : F : O $85:7.5:7.5$	$P \approx 2.7 \cdot 10^{-9}$ mbar, substrate (100) Si.	[77; 78]
WF_6 120 keV electrons	β – W phase	$P \approx 4 \cdot 10^{-8}$ mbar, native oxide SiO_2.	[77; 78]
WF_6 0.5 keV Ar^+	W : F : O $93:9:2$ no carbon detected (XPS)	$P < 1.3 \cdot 10^{-7}$ mbar, substrate: Si_3N_4, Si. 2 keV Ar^+ substrate cleaning	[79]
SiH_2Cl_2 10 keV electrons	Si : O : Cl $89:9:2$	$P \approx 2.7 \cdot 10^{-9}$ mbar, (100) Ge substrate	[76]
Si_2H_6 5 keV electrons	Si	$P \approx 5.3 \cdot 10^{-10}$ mbar, Si and Ge substrates.	[80]
CrO_2Cl_2 3 kV electrons	Cr : O : Cl $1 : 2.2 : 1.1$	$P \approx 2.7 \cdot 10^{-9}$ mbar, substrate: Au or Pd film on oxidized Si	[81]

Table F.4 FEB and FIB induced deposition of metals and metal oxides with additional reactive gases in UHV systems.

Molecule Irradiation	Composition w/o co-injection	Co-injected molecule	Composition with co-injection	Ref.
$Fe(CO)_5$ 30 keV electrons	$Fe_1O_{0.55}C_{0.45}$	H_2O	crystalline Fe_3O_4	[83]
(hfa)Cu-TMVS 0.5 keV Ar^+	67 at.% Cu	Atomic H	Better 99 at.% Cu	[58]
Si_2H_6 0.6–17 keV electrons	Si	O_2	SiO_2	[82]

2.2.3 Co-injection of reactive gases

Pure metals and oxides were obtained by co-injecting reactive gases and functional molecules under UHV conditions (see Table F.4). Adding oxygen to disilane resulted in FEB deposition of good-quality SiO_2 [82].

2.3 POSTDEPOSITION TREATMENTS

In most cases, thermal annealing or plasma treatments are carried out to decrease the carbon content of the deposit. For an excellent review of results we refer the reader to [33].

2.4 WHAT LIES AHEAD?

Summarizing, the purity issue is still a hot topic in gas-assisted FEB and FIB processing. Here we reviewed the deposition results and routes to better purity. The role of residual pressure and residual gas composition in the microscope chamber is increasingly studied to understand variations in deposit composition. UHV systems with the lowest hydrocarbon and water background permit well-defined "single adsorbate" electron or ion induced deposition studies. In contrast, and now becoming increasingly evident, HV experiments constitute a "three-adsorbate" system originating from the injected molecule, the residual water molecule, and the residual hydrocarbon molecule. All the three adsorbates compete in substrate surface sites and in being dissociated by the electron (ion) beam. Moreover, their dissociation fragments can cross-react with each other to form non-volatile compounds or other volatile compounds. In order to get the desired pure material the following routes are being investigated: using reactive gases, to introduce plasma cleaning prior to processing; or to drive the beam exposure in an adsorbate-limited regime for one or two of the three adsorbate species; or to use the "right" adsorbate molecule (in the sense that all dissociation fragments and related cross-reactions originating from the three-adsorbate system work in favour for a pure metal or pure oxide deposit). First impressive purity results were achieved and summarized above. Alternatively, many more results are to be expected when exploiting the "three-adsorbate" system as a handle to *control* the metal-to-carbon (or more generally, metal-to-matrix) ratio in order to deposit nanocomposites with outstanding physical properties [108].

3 NANOFABRICATION USING FIB: TOWARD ULTIMATE RESOLUTION

Many types of devices with nanometer features can be envisioned on thin membranes, including biological sensors, bolometers, or pressure sensors. Moreover, in a thin membrane ion beam-solid interaction occurs in a limited volume, potentially enabling fabrication at smaller dimensions. This discussion will be mostly limited to gallium ion beams since these are the ones most readily available. However, with lighter ions such as helium, discussed above, fabrication on a thin membrane may be even more attractive.

In fact, with regard to lateral resolution, FIB has to compete hard with FEB, which currently leads the lateral resolution record with a dot deposit on a thin SiN membrane using $W(CO)_6$ as a precursor gas. Its size was below *one* nanometer [84] and sub-10-nm lateral resolution now becomes routine in FEB. FEB nano-etching (without gas-assistance) of 4 nm sized holes in membranes of different materials using ≥ 100 keV electron beams was demonstrated [85]. However, we consider FEB and FIB processing as complimentary nanofabrication methods due to their different dominant interaction mechanisms with matter (see [21] for a review of both methods), and it seems that FIB catches up in practical resolutions with its counterpart FEB. For more details on FEB resolution, we refer the reader to Introduction 1 by Utke and Koops in this book.

3.1 ULTRA-THIN MEMBRANES AS TEMPLATES TO ACHIEVE ULTIMATE FIB PROCESS RESOLUTION

In FIB-based nanofabrication, secondary processes resulting from collision cascades generated by an incident ion on a solid (generation of backscattered, sputtered, recoil atoms and ions, secondary electrons), play a critical role in the achievable patterning resolution. These processes effectively result in lateral spreading of the beam energy over areas much wider than the incident beam. In different ways, this affects the resolution obtained in milling, ion-induced etching, and deposition processes. Fabricated structures thus usually have lateral dimensions from tens to hundreds of nanometers, compared to a sub-10 nm diameter of the ion beam.

The loss of beam processing dimensionality occurs for substrates thicker than the projected ion range in the solid (R). For substrates thinner than R, most of effects that deteriorate the FIB pattering resolution can be avoided [86]. The ion range values depend on the material, and are bigger for targets with smaller atomic masses. For 30 keV Ga ions, the typical values are 40 nm and 20 nm for Si and W targets, respectively [87]. The range of helium ions is about seven times larger than the range of gallium in the same materials.

Examples that demonstrate the effect of the membrane thickness on the interaction volume (penetration depth and lateral expansion) are shown in Figures F.3 and F.4, where TRIM simulations of gallium and helium ion beam interactions with thin carbon and platinum layers (light and heavy materials, respectively) are presented. A zero beam diameter was chosen to show clearly the lateral dimensions of the ion-solid interaction volume. Primary ion trajectories are shown in red, and collisions of secondary atoms (collision cascade) are shown in green.

For Ga^+ ions (Figure F.3) on a 50 nm thick Pt target, much of the ion energy is deposited near the surface (short ion range due to strong scattering), resulting in wide features ($\sim$30 nm wide) that can be produced by milling or in FIBID processes. In contrast, deeper ion penetration and a laterally more narrow interaction area is characteristic of a 50 nm thick carbon target, where predominantly small angle scattering occurs for high energy incident ions. For thinner (10 nm) membranes, the difference between two materials is more visible. On the other hand, for 1 nm thick layers, which are much thinner than the ion range in both materials, the difference in the interaction volume is not significant anymore.

In the case of He^+ ion beam (Figure F.4), the reduced ion scattering, especially in carbon, results in much deeper penetration of the beam and dramatic lateral reduction of the interaction volume near the surface. In this case, very narrow holes (a few nm in diameter) can be milled in 10 nm thick C membranes. For 50 nm thick platinum, stronger backscattering of light He ions results in an increased interaction volume. However, for Pt layers as thin as 10 nm, a certain similarlity with 10 nm carbon layers (reduced lateral beam expansion, little energy deposition at the subsurface area), can be seen again. Similarly to the case of Ga^+ ions, for 1 nm thick membranes the difference between C and Pt targets tends to be even smaller.

Finally, the trends observed in such simulations should be confirmed experimentally (see Section 3.3), as a number of effects (e.g., generation of secondary electrons or excited surface atoms – both are important for gas assisted processing) is not considered by TRIM.

The data available in the literature usually refer to amorphous materials. For crystalline targets, the energy losses and ranges for energetic ions depend strongly on the incident angle with respect to the crystal axis and plane. In a channeling condition, when an ion travels along the crystal axis or plane, it may experience very low energy losses and penetrate deeply into the target causing less lattice disorder. Viewed as an undesired effect for ion implantation, this may be interesting in the present case for processes at low ion dose. Of course, as the ion dose increases, the irradiated material becomes amorphous.

In addition to beam-solid interaction, other effects, such as incident ion or sputtered particle scattering or redeposition, may limit the minimum dimensions that can be achieved. For example, in the process of nanowires fabrication using FIB-induced metal deposition,

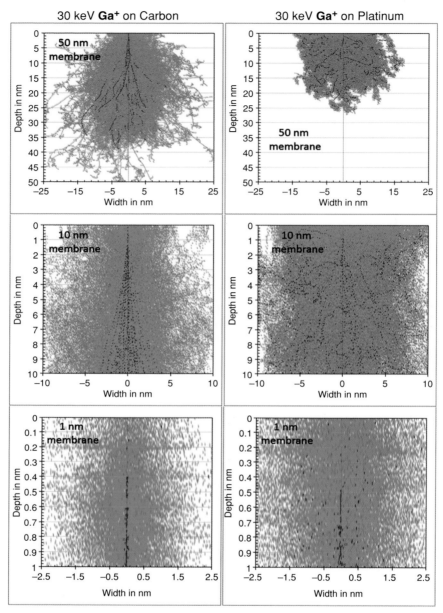

30 keV **Ga⁺** on Carbon　　　30 keV **Ga⁺** on Platinum

FIGURE F.3: Beam broadening: TRIM simulated trajectories of 30 keV Ga⁺ primary ions (zero diameter beam) impinging on carbon (density 2.25 g cm⁻³) and platinum membranes having thicknesses of 50, 10, and 1 nm. Primary ion collisions are shown in red and collision cascades in green. Note the decreasing lateral width of the interaction volume with decreasing thickness of the membrane. Furthermore, with decreasing mass and atomic number of the membrane material, the stopping power decreases and thus the ion penetration range increases; however, the lateral interaction volume becomes smaller in the vicinity of the surface. For comparison: the corresponding beam broadening for a zero diameter 30 keV *electron* beam is much smaller: 0.1 nm and 1 nm for a carbon membrane thickness of 10 and 50 nm, respectively [21].

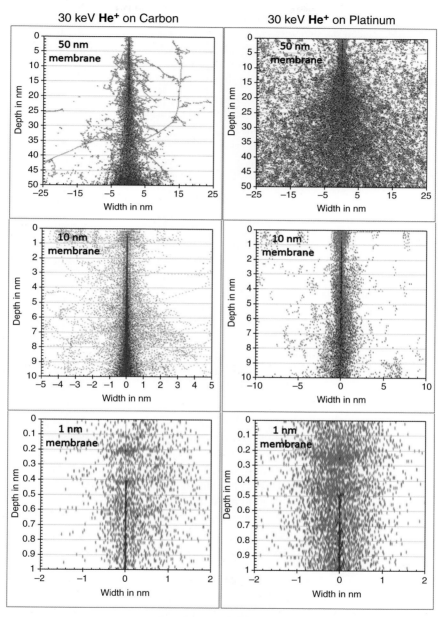

FIGURE F.4: Beam broadening: TRIM simulated trajectories of 30 keV He$^+$ primary ions (zero diameter beam) impinging on carbon (density 2.25 gcm^{-3}) and platinum membranes having varying thickness (50, 10, and 1 nm). Primary ion collisions are shown in red and collision cascades in green. Note the smaller lateral width of the interaction volume as compared with Ga$^+$ ions of the same energy; see Figure F.3.

scattering of incident ions in the flanks of growing nanowire is one of the reasons for a so-called halo effect when a thin conductive film is deposited around the nanowire. The halo area width can reach 300–500 nm in the case of Pt FIB deposition (smaller halo effect is observed in the case of W deposition) [88; 89]. This may result in a considerable leakage between closely placed parallel wires fabricated, e.g., for nanocontacting purposes. To reduce the undesired lateral deposition, Pt nanocontacts with thickness between 50 and 100 nm [90] should be used.

In milling, redeposition of sputtered material onto the walls of milled holes or trenches makes it difficult to obtain vertical walls and does not allow producing narrow holes. These issues can be counter acted by using gas assisted FIB induced etching as is discussed in chapter 19 of Callegari and coworkers in this book.

3.2 MATERIALS FOR SUSPENDED MEMBRANES AND SAMPLE PREPARATION

Based on the above considerations, the list of materials for experiments designed to achieve an ultimate resolution in FIB processing should obviously include C, Si, and their compounds. In particular, thin suspended silicon oxide, nitride, and carbide (SiO_2, SiN_x and SiC) layers or membranes can be readily fabricated using conventional microfabrication methods, such as CVD or sputtering over Si wafers followed by selective wet or dry etching from the wafer back side to prepare suspended membranes. In fact, silicon nitrate member of thickness down to 15 nm can be purchased commercially.

SOI (silicon on insulator) technology [91] has been introduced in the last decade in microelectronics to fabricate thin (tens of nanometers) layers of crystalline Si over buried oxide layers. This technology suggests a method for producing thin *single crystal* membranes for FIB processing.

Thin amorphous carbon layers (20 nm) are also routinely obtained by sputtering. This can be done using Cu TEM grids as a support, and other layers can be further deposited over it by sputtering [92].

Ultra-thin graphite layers (also known as few layer graphene) are obtained usually from highly ordered pyrolytic graphite (HOPG) by exfoliation. It should be noted that the exfoliation method for graphene preparation is time consuming and has a very low yield. An alternative method for controlled deposition of thin graphite layers over metal electrodes from liquid solutions is based on AC di-electrophoresis [93; 88]. This method has been successfully employed first for deposition of carbon nanotubes and, more recently, thin graphite (graphene) or graphite oxide layers [94].

Thin graphite or graphene [95; 96] is especially interesting as a template because of its crystallinity (smaller scattering possible due to channeling) and high in-plane conductivity; the latter can help to eliminate a possible effect of layer charging during the FIB processing.

3.3 EXAMPLES OF FIB PATTERNING IN ULTRATHIN MEMBRANES

A number of experimental studies using thin and ultrathin substrates for FIB processing have been reported, most of them by the group of J. Gierak [97; 98; 99; 100; 101; 102].

The results obtained so far clearly demonstrate the potential of this approach to improve the FIB-patterning resolution.

Penuelas et al. [101] reported fabrication of uniformly distributed magnetic CoPt nanoparticles with controlled size over TEM grids covered by a 5 nm thick amorphous carbon membranes. First, a thin film of a CoPt alloy was deposited in ultra-high vacuum over the carbon layer, and formation of $\sim$1.3 nm diameter particles with surface density of 10^{12} cm^{-2} was confirmed by TEM observation. This was followed by Ga$^+$ irradiation at 30 keV with doses varying from 10^{12} to 10^{16} cm^{-2}. The projected ion range in the substrate was estimated to be between 8 and 20 nm, depending on the composition, thus being larger than the total layer thickness. As a result, no preferential ablation or implantation of Ga in the material was detected. Gradual increase of the mean CoPt particles diameter from 1.3 to 1.6 nm with irradiation was observed, and their density reduced correspondingly. The effect of ion irradiation, which induces metal mobility on the surface, is very close to that of thermal annealing, but it is important that with FIB this can be done *locally*. Besides, structural defects can be also induced by irradiation to tailor their properties. For doses as high as 5×10^{15} cm^{-2} ablation of material was evident, followed by the membrane destruction at doses $\sim 10^{16}$ cm^{-2}.

Direct FIB patterning of nanowires and nanorings in ultrathin (1–3 nm thick) suspended graphene layers, preserving the interesting physical properties of this material, was successfully demonstrated by Lucot et al. [100]. Electrical transport in the fabricated graphene structures at low temperatures was analyzed.

Ion-beam nanopatterning in graphite was performed by Mélinon et al. [97]. It was shown that extended nanodefects created by FIB in the HOPG surface can serve as traps for particles deposited on the surface. Sample imaging using confocal Raman microscopy was shown to be extremely sensitive to defects created in the surface graphite layers. Ion doses as small as 10 ions/dot could be detected in Raman spectra due to the appearance of the D line, absent for clean HOPG samples (no changes detected using AFM and STM). Periodic arrays of 2 nm diameter CoPt nanoclusters were created by their deposition over the FIB-functionalized HOPG surface. Without FIB patterning, distribution of CoPt clusters was not regular.

FIB patterning of HOPG surfaces followed by O_2 plasma etching was employed by Bottcher et al. [103] to create periodic arrays of hexagonal holes, with periods from 300 to 900 nm, in areas up to several mm^2. These arrays can be useful as catalyst supports or in nanopohotonics/plasmonics.

Shiedt et al. [102] used 20–50 nm thick silicon carbide and nitride membranes to fabricate arrays of single nanopores with 50 nm diameter, to be used in biomolecules manipulation and analysis. Smaller diameter pores ($\sim$3 nm), even smaller than the beam diameter (due to autofocusing effects), can be produced in these materials by carefully ajusting the process parameters [98]. Redeposition of material on walls is also reduced for thin milled layers so that pores with sharp edges can be milled.

Deposition of narrow Pt and W wires by FIBID over thin diamond-like carbon (DLC) or silicon nitride membranes ($\sim$20 nm thick) was analyzed by Reguer et al. [104]. Very narrow wires, down to 35 for W and 60 nm for Pt, were deposited using low ion current (20 pA). The size of the deposited structures was found to exceed the ion beam diameter. This was explained by analyzing the role of secondary particles (sputter ions and secondary electrons) that can emerge from the lateral flanks of the growing NW and activate the adsorbed gas molecules in the deposition process, resulting in lateral expansion of NWs. For the same experimental conditions, the platinum wires were found to be wider than the tungsten ones.

3.4 WHAT LIES AHEAD?

Although presently FEB is leading the competition in spatial process resolution (the record is sub-1-nm and controlled deposition at molecular-level is anticipated [105]), the examples presented above show the high potential of nanopatterning using FIB and thin suspended membranes. Further exploration of the ultimate patterning potential achievable with both focused ion beams and focused electron beams certainly will bring new results interesting for both technological and fundamental aspects of the nanoscience.

4 SUMMARY

In this chapter we have discussed three areas of new opportunities and challenges:

1) New beam species (ions other than Ga) and new beam delivery methods (electron or ion multi-beams) offer three orders of magnitude increase in writing speed with electrons and a *variety* of ions. This may open new applications of direct resistless beam processing—many of which were considered in the past but were rejected because the processes were too slow. *In situ* processing, i.e., many micro- and nanofabrication steps carried out in one chamber, may also now be an attractive possibility.

2) The lack of purity of material deposited from a precursor gas by either ion or electron beam induced processes has been a long-standing impediment. Doing deposition in ultrahigh vacuum chambers removes contaminants from background gas and yields very high purity deposits. In some cases, introducing appropriate reactive gases with the precursor helps to remove contaminants in high-vacuum chambers. The role of the three-adsorbate system in high-vacuum conditions (originating from the injected molecule, water, and hydrocarbon background) is increasingly understood and used as a handle to tune the deposit composition. The availability of such tuned nano-composite material as well as higher purity deposits will also open new areas of application.

3) The minimum dimensions of structures achievable by direct electron or ion beam processing is limited in most cases by the beam-solid interaction volume rather than by the beam diameter itself. Thus fabrication on ultrathin membranes, where the beam-solid interaction volume is limited, provides an opportunity to achieve smaller dimensions structures than hitherto possible.

REFERENCES

[1] Ward, B. W., J. A. Notte, et al. Helium ion microscope: A new tool for nanoscale microscopy and metrology. *Journal of Vacuum Science & Technology* B 24.6 (2006): 2871–2874.

[2] Smith, N. S., P. P. Tesch, et al. A high brightness source for nano-probe secondary ion mass spectrometry. *Applied Surface Science* 255.4 (2008): 1606–1609; Smith, N. S., D. E. Kinion, et al. A High Brightness Plasma Source for Focused Ion Beam Applications. *Microscopy and Microanalysis* 13.2 (2007): 180–181.

[3] Platzgummer, E. Maskless Lithography and Nanopatterning with Electron and Ion Multi-Beam Projection. *Proc. of SPIE* 7637 (2010): 763703; M. J. Wieland, G. de Boer et al. MAPPER: High throughput maskless lithography. *Proc. of SPIE* 7637 (2010): 76370F.

[4] Morgan, J. C., J. Notte, et al. An introduction to the helium ion microscope. *Microscopy Today* 14.4 (2006): 24–31.

[5] Hanssen, J. L., S. B. Hill, et al. Magneto-optical-trap-based, high brightness ion source for use as a nanoscale probe. *Nano Letters* 8.9 (2008): 2844–2850.

[6] Reijnders, M. P., P. A. van Kruisbergen, et al. Low-energy-spread ion bunches from a trapped atomic gas. *Physical Review Letters* 102.3 (2009): 034802.

[7] Hill, S. B., and J. J. McClelland. Atoms on demand: Fast, deterministic production of single Cr atoms. *Applied Physics Letters* 82.18 (2003): 3128–3130.

[8] Borret, R., K. Jousten, et al. Long-time current stability of a gas-field ion-source with a supertip. *Journal of Physics D-Applied Physics* 21.12 (1988): 1835–1837.

[9] Winston, D., B. M. Cord, et al. Scanning-helium-ion-beam lithography with hydrogen silsesquioxane resist. *Journal of Vacuum Science and Technology B: Microelectronics and Nanometer Structures* 27.6 (2009): 2702–2706.

[10] Chen, P., H. W. M. Salemink, et al. Roles of secondary electrons and sputtered atoms in ion-beam-induced deposition. *Journal of Vacuum Science & Technology* B 27.6 (2009): 2718–2721.

[11] Seliger, R. L., and W. P. Fleming. Focused ion-beams in microfabrication. *Journal of Vacuum Science & Technology* 10.6 (1973): 1127–1127.

[12] Scipioni, L., D. Stewart, et al. Performance of multicusp plasma ion source for focused ion beam applications. *Journal of Vacuum Science & Technology* B 18.6 (2000): 3194–3197.

[13] Smith, N. Oregon Physics. Personal communication (2009).

[14] Kruit, P. The role of MEMS in maskless lithography. *Microelectronic Engineering* 84.5–8 (2007): 1027–1032.

[15] Ebm, C., E. Platzgummer, et al. Ion multibeam nanopatterning for photonic applications: Experiments and simulations, including study of precursor gas induced etching and deposition. *Journal of Vacuum Science & Technology* B 27.6 (2009): 2668–2673.

[16] Platzgummer, E., and H. Loeschner. Charged particle nanopatterning. *Journal of Vacuum Science & Technology* B 27.6 (2009): 2707–2710.

[17] Berry, I. L., A. A. Mondelli, et al. Programmable aperture plate for maskless high-throughput nanolithography. *Journal of Vacuum Science & Technology* B 15.6 (1997): 2382–2386.

[18] Platzgummer, E., H. Loeschner, et al. Projection Mask-Less Patterning (PMLP) for the fabrication of leading-edge complex masks and nano-imprint templates. *Proceedings of SPIE – The International Society for Optical Engineering SPIE* 6730 (2007): 6730033.

[19] Weidenhausen, A., R. Spehr, et al. Stochastic ray deflections in focused charged-particle beams. *Optik* 69.3 (1985): 126–134.

[20] Loeschner, H., C. Klein, E. Platzgummer. Projection charged particle nanolithography and nanopatterning. *Jap. J. Appl. Phys.* 49.6 Part 2 (2010): 06GE011-06GE015.

[21] Utke, I., P. Hoffmann, et al. Gas-assisted focused electron beam and ion beam processing and fabrication. *Journal of Vacuum Science & Technology* B 26.4 (2008): 1197–1276.

[22] Walden, R. H., and IEEE. Analog-to-digital converters and associated IC technologies. *IEEE CSIS Symposium* (2008): 152–153.

[23] Lattes, A. L., S. C. Munroe, et al. Improved drift in 2-phase, long-channel, shallow buried-channel CCDs with longitudinally nonuniform storage-gate implants. *IEEE Transactions on Electron Devices* 39.7 (1992): 1772–1774.

[24] Shen, C. C., J. Murguia, et al. Use of focused-ion-beam and modeling to optimize submicron MOSFET characteristics. *IEEE Transactions on Electron Devices* 45.2 (1998): 453–459.

[25] Evason, A. F., J. R. A. Cleaver, et al. Fabrication and performance of GaAs-MESFETS with graded channel doping using focused ion-beam implantation. *IEEE Electron Device Letters* 9.6 (1988): 281–283.

[26] Lezec, H. J., K. Ismail, et al. A tunable-frequency Gunn diode fabricated by focused ion-beam implantation. *IEEE Electron Device Letters* 9.9 (1988): 476–478.

[27] De Marco, A. J., and J. Melngailis. Maskless fabrication of JFETs via focused ion beams. *Solid-State Electronics* 48.10–11 (2004): 1833–1836.

[28] Ehrlich, D. J., and V. Tran Nguyen. *Emerging technologies for in situ processing*. Netherlands: Springer, 1988.

[29] Ishikawa, T. In situ electron-beam processing for GaAs/AlGaAs nanostructure fabrications. *Japanese Journal of Applied Physics Part 1-Regular Papers Short Notes & Review Papers* 35.11 (1996): 5583–5596.

[30] Miyauchi, E., and H. Hashimoto. Application of focused ion-beam technology to maskless ion-implantation in a molecular-beam epitaxy grown GaAs or AlGaAs epitaxial layer for 3-dimensional pattern doping crystal-growth. *Journal of Vacuum Science & Technology A-Vacuum Surfaces and Films* 4.3 (1986): 933–938.

[31] Wang, Y. L., H. Temkin, et al. Buried-heterostructure lasers fabricated by in situ processing techniques. *Applied Physics Letters* 57.18 (1990): 1864–1866.

[32] Sazio, P. J. A., G. A. C. Jones, et al. Fabrication of in situ Ohmic contacts patterned in three dimensions using a focused ion beam during molecular beam epitaxial growth. *Journal of Vacuum Science and Technology B: Microelectronics and Nanometer Structures* 15.6 (1997): 2337–2341.

[33] Botman, A., J. J. L. Mulders, et al. Creating pure nanostructures from electron-beam-induced deposition using purification techniques: a technology perspective. *Nanotechnology* 20.37 (2009): 372001.

[34] Bret, T., S. Mauron, et al. Characterization of focused electron beam induced carbon deposits from organic precursors. *Microelectronic Engineering* 78–79 (2005): 300–306.

[35] Botman, A., M. Hesselberth, et al. Improving the conductivity of platinum-containing nano-structures created by electron-beam-induced deposition. *Microelectronic Engineering* 85.5–6 (2008): 1139–1142.

[36] Bernau, L., M. S. Gabureac, et al. Tunable nanosynthesis of composite materials by electron-impact reaction. *Angew. Chem. – Intl Ed.* 49.47 (2010): 8880–8884.

[37] Utke, I., T. Bret, et al. Thermal effects during focused electron beam induced deposition of nanocomposite magnetic-cobalt-containing tips. *Microelectronic Engineering* 73–74 (2004): 553–558.

[38] Utke, I., J. Michler, et al. Cross section investigations of compositions and sub-structures of tips obtained by focused electron beam induced deposition. *Advanced Engineering* Materials 7.5 (2005): 323–331.

[39] Fernandez-Pacheco, A., J. M. De Teresa, et al. Magnetotransport properties of high-quality cobalt nanowires grown by focused-electron-beam-induced deposition. *Journal of Physics D-Applied Physics* 42.5 (2009): 055005.

[40] Utke, I., A. Luisier, et al. Focused-electron-beam-induced deposition of freestanding three-dimensional nanostructures of pure coalesced copper crystals. *Applied Physics Letters* 81.17 (2002): 3245–3247.

[41] Hochleitner, G., H. D. Wanzenboeck, et al. Electron beam induced deposition of iron nanostructures. *Journal of Vacuum Science & Technology* B 26.3 (2008): 939–944.

[42] Hoffmann, P., I. Utke, et al. Focused electron beam induced deposition of gold and rhodium. *Materials Research Society Symposium – Proceedings* 624 (2000): 171–177.

[43] Utke, I., P. Hoffmann, et al. Focused electron beam induced deposition of gold. *Journal of Vacuum Science & Technology* B 18.6 (2000): 3168–3171.

[44] Perentes, A., G. Sinicco, et al. Focused electron beam induced deposition of nickel. *Journal of Vacuum Science & Technology* B 25.6 (2007): 2228–2232.

[45] Wang, S., Y. M. Sun, et al. Electron-beam induced initial growth of platinum films using $Pt(PF_3)_4$. *Journal of Vacuum Science & Technology B (Microelectronics and Nanometer Structures)* 22.4 (2004): 1803–1806.

[46] Cicoira, F., K. Leifer, et al. Electron beam induced deposition of rhodium from the precursor [RhCl(PF3)(2)](2): morphology, structure and chemical composition. *Journal of Crystal Growth* 265.3–4 (2004): 619–626.

[47] Bret, T., I. Utke, et al. Electron range effects in focused electron beam induced deposition of 3D nanostructures. *Microelectronic Engineering* 83.4–9 (2006): 1482–1486.

[48] Crozier, P. A., J. Tolle, et al. Synthesis of uniform GaN quantum dot arrays via electron nanolithography of D_2GaN_3. *Applied Physics Letters* 84.18 (2004): 3441–3443.

[49] Perentes, A., T. Bret, et al. Real-time reflectometry-controlled focused-electron-beam-induced deposition of transparent materials. *Journal of Vacuum Science & Technology* B 24.2 (2006): 587–591.

[50] Barry, J. D., M. Ervin, et al. Electron beam induced deposition of low resistivity platinum from Pt(PF3)(4). *Journal of Vacuum Science & Technology* B 24.6 (2006): 3165–3168.

[51] Brintlinger, T., M. S. Fuhrer, et al. Electrodes for carbon nanotube devices by focused electron beam induced deposition of gold. *Journal of Vacuum Science & Technology* B 23.6 (2005): 3174–3177.

[52] Bret, T., I. Utke, et al. In situ control of the focused-electron-beam-induced deposition process. *Applied Physics Letters* 83.19 (2003): 4005–4007.

[53] Cicoira, F., P. Hoffmann, et al. Auger electron spectroscopy analysis of high metal content microstructures grown by electron beam induced deposition. *Applied Surface Science* 242.1–2 (2005): 107–113.

[54] Matsui, S. and K. Mori. New selective deposition technology by electron-beam induced surface-reaction. *Journal of Vacuum Science & Technology* B 4.1 (1986): 299–304.

[55] Ketharanathan, S., R. Sharma, et al. Electron beam induced deposition of pure, nanoscale Ge. *Journal of Vacuum Science & Technology* B 24.2 (2006): 678–681.

[56] Christy, R. W. Conducting thin films formed by electron bombardment of substrate. *Journal of Applied Physics* 33.5 (1962): 1884–1888.

[57] Funsten, H. O., J. W. Boring, et al. Low-temperature beam-induced deposition of thin tin films. *Journal of Applied Physics* 71.3 (1992): 1475–1484.

[58] Chiang, T. P., H. H. Sawin, et al. Ion-induced chemical vapor deposition of high purity Cu films at room temperature using a microwave discharge H atom beam source. *Journal of Vacuum Science & Technology A – Vacuum Surfaces and Films* 15.5 (1997): 2677–2686.

[59] Miyazoe, H., Utke, I., Kikuchi, H., Kiriu, S., Friedli, V., Michler, J., and Terashima, K. Improving the metallic content of focused electron beam-induced deposits by a scanning electron microscope integrated hydrogen-argon microplasma generator. *J. Vac. Sci. & Technol.* B 28.4 (2010): 744–750.

[60] Miyazoe, H., I. Utke, et al. Controlled focused electron beam-induced etching for the fabrication of sub-beam-size nanoholes. *Appl. Phys. Lett.* 92.4 (2008): 043124.

[61] Molhave, K., D. N. Madsen, et al. Solid gold nanostructures fabricated by electron beam deposition. *Nano Letters* 3.11 (2003): 1499–1503.

[62] Folch, A., J. Tejada, et al. Electron-beam deposition of gold nanostructures in a reactive environment. *Applied Physics Letters* 66.16 (1995): 2080–2082.

[63] Folch, A., J. Servat, et al. High-vacuum versus "environmental" electron beam deposition. *Journal of Vacuum Science & Technology* B 14.4 (1996): 2609–2614.

[64] Tao, T., J. S. Ro, et al. Focused Ion-Beam Induced Deposition of Platinum." *Journal of Vacuum Science & Technology* B 8.6 (1990): 1826–1829.

[65] Langford, R. M., D. Ozkaya, et al. Effects of water vapour on electron and ion beam deposited platinum. *Microscopy and Microanalysis* 10(Suppl. 2) (2004): 1122–1123.

[66] Perentes, A., and P. Hoffman. Oxygen assisted focused electron beam induced deposition of Si-containing materials: Growth dynamics. *Journal of Vacuum Science & Technology* B 25.6 (2007): 2233–2238.

[67] Perentes, A., and P. Hoffmann. Focused electron beam induced deposition of Si-based materials from SiOxCy to stoichiometric SiO2: Chemical compositions, chemical-etch rates, and deep ultraviolet optical transmissions. *Chemical Vapor Deposition* 13.4 (2007): 176–184.

[68] Komano, H., Y. Ogawa, et al. Silicon-oxide film formation by focused ion-beam (FIB)-assisted deposition. *Japanese Journal of Applied Physics Part 1-Regular Papers Short Notes & Review Papers* 28.11 (1989): 2372–2375.

[69] Kunz, R. R., and T. M. Mayer. Catalytic growth-rate enhancement of electron-beam deposited iron films. *Applied Physics Letters* 50.15 (1987): 962–964.

[70] Kunz, R. R., and T. M. Mayer. Electron-beam induced surface nucleation and low-temperature decomposition of metal-carbonyls. *Journal of Vacuum Science & Technology* B 6.5 (1988): 1557–1564.

[71] Walz, M.-M., Schirmer, M., et al. Electrons as "Invisible Ink": Fabrication of nanostructures by local electron beam induced activation of SiOx. *Angew. Chem. – Intl. Ed-* 49–27 (2010): 4669–4673.

[72] Lukasczyk, T., M. Schirmer, et al. Electron-beam-induced deposition in ultrahigh vacuum: Lithographic fabrication of clean iron nanostructures. *Small* 4.6 (2008): 841–846.

[73] Hiroshima, H., and M. Komuro. High growth rate for slow scanning in electron-beam-induced deposition. *Japanese Journal of Applied Physics, Part 1 (Regular Papers, Short Notes & Review Papers)* 36.12B (1997): 7686–7690.

[74] Hiroshima, H., and M. Komuro. Fabrication of conductive wires by electron-beam-induced deposition. *Nanotechnology* 9.2 (1998): 108–112.

[75] Hiroshima, H., N. Suzuki, et al. Conditions for fabrication of highly conductive wires by electron-beam-induced deposition. *Japanese Journal of Applied Physics, Part 1 (Regular Papers, Short Notes & Review Papers)* 38.12B (1999): 7135–7139.

[76] Matsui, S., and M. Mito. Si Deposition by electron-beam induced surface-reaction. *Applied Physics Letters* 53.16 (1988): 1492–1494.

[77] Ichihashi, T., and S. Matsui. Insitu observation on electron-beam induced chemical vapor-deposition by transmission electron-microscopy. *Journal of Vacuum Science & Technology* B 6.6 (1988): 1869–1872.

[78] Matsui, S., T. Ichihashi, et al. Electron beam induced selective etching and deposition technology. *Journal of Vacuum Science & Technology B (Microelectronics Processing and Phenomena)* 7.5 (1989): 1182–1190.

[79] Xu, Z., T. Kosugi, et al. An X-ray photoelectron-spectroscopy study on ion-beam induced deposition of tungsten using WF6. *Journal of Vacuum Science & Technology* B 7.6 (1989): 1959–1962.

[80] Hirose, F., and H. Sakamoto. Low-temperature Si selective epitaxial growth using electron-beam-induced reaction. *Japanese Journal of Applied Physics Part 1-Regular Papers Short Notes & Review Papers* 34.11 (1995): 5904–5907.

[81] Wang, S., Y. M. Sun, et al. Electron induced deposition and in situ etching of CrOxCly films. *Applied Surface Science* 249.1–4 (2005): 110–114.

[82] Nakano, K., T. Horie, et al. Low-temperature growth of SiO2 films by electron-induced ultrahigh vacuum chemical vapor deposition. *Japanese Journal of Applied Physics Part 1-Regular Papers Short Notes & Review Papers* 35.12B (1996): 6570–6573.

[83] Shimojo, M., M. Takeguchi, et al. Formation of crystalline iron oxide nanostructures by electron beam-induced deposition at room temperature. *Nanotechnology* 17.15 (2006): 3637–3640.

[84] van Dorp, W. F., C. W. Hagen, et al. Growth behavior near the ultimate resolution of nanometer-scale focused electron beam-induced deposition. *Nanotechnology* 19.22 (2008): 225305.

[85] Humphreys, C. J., T. J. Bullough, et al. Electron-beam nano-etching in oxides, fluorides, metals and semiconductors. *Scanning Microscopy* (Suppl. 4): 185–192.

[86] Gierak, J., A. Madouri, et al. Sub-5 nm FIB direct patterning of nanodevices. *Microelectronic Engineering* 84.5–8 (2007): 779–783.

[87] Ohya, K., and T. Ishitani. Imaging using electron and ion beams. In *Focused ion beam systems: Basics and applications.* Editor: N. Yao, Cambridge University Press, 2007.

[88] Vaz, A. R., M. M. da Silva, et al. Platinum thin films deposited on silicon oxide by focused ion beam: characterization and application. *Journal of Materials Science* 43.10 (2008): 3429–3434.

[89] Tripathi, S. K., N. Shukla, et al. The out of beam sight effects in focused ion beam processing. *Nanotechnology* 20.27 (2009) : 275301.

[90] Vaz, A. R., and S. A. Moshkalev. Chapter 17 in this book.

[91] See for example, www.soitec.com.

[92] Verissimo, C., A. L. Gobbi, et al. Synthesis of carbon nanotubes directly over TEM grids aiming the study of nucleation and growth mechanisms. *Applied Surface Science* 254.13 (2008): 3890–3895.

[93] Krupke, R., S. Linden, et al. Thin films of metallic carbon nanotubes prepared by dielectrophoresis. *Advanced Materials* 18.11 (2006): 1468–.

[94] Rouxinol, F. P., R. V. Gelamo, R. G. Amici, A. R. Vaz.,and S. A. Moshkalev. Low contact resistivity and strain in suspended multilayer graphene. *Applied Physics Letters.* 97 (2010): 253104.

[95] Geim, A. K., and K. S. Novoselov. The rise of graphene. *Nature Materials* 6.3 (2007): 183–191.

[96] Krasheninnikov, A. V., and F. Banhart. Engineering of nanostructured carbon materials with electron or ion beams. *Nature Materials* 6.10 (2007): 723–733.

[97] Melinon, P., A. Hannour, et al. Ion beam nanopatterning in graphite: Characterization of single extended defects. *Nanotechnology* 19,23 (2008): 235305.

[98] Gierak, J. Focused ion beam technology and ultimate applications. *Semiconductor Science and Technology* 24.4 (2009): 043001.

[99] Gierak, J., E. Bourhis, et al. Exploration of the ultimate patterning potential achievable with focused ion beams. *Ultramicroscopy* 109.5 (2009): 457–462.

[100] Lucot, D., J. Gierak, et al. Deposition and FIB direct patterning of nanowires and nanorings into suspended sheets of graphene. *Microelectronic Engineering* 86.4–6 (2009): 882–884.

[101] Penuelas, J., A. Ouerghi, et al. Local tuning of CoPt nanoparticle size and density with a focused ion beam nanowriter. *Nanotechnology* 20.42 (2009): 425304.

[102] Schiedt, B., L. Auvray, et al. Direct FIB fabrication and integration of 'single nanopore devices' for the manipulation of macromolecules. *Materials and Strategies for Lab-on-a-Chip – Biological Analysis, Cell-Material Interfaces and Fluidic Assembly of Nanostructures* 1191 (2009): 93–100.

[103] Bottcher, A., M. Heil, et al. Nanostructuring the graphite basal plane by focused ion beam patterning and oxygen etching. *Nanotechnology* 17.23 (2006): 5889–5894.

[104] Reguer, A., F. Bedu, et al. Structural and electrical studies of conductive nanowires prepared by focused ion beam induced deposition. *Journal of Vacuum Science and Technology B: Microelectronics and Nanometer Structures* 26.1 (2008): 175–180.

[105] van Dorp, W. F., C. W. Hagen, et al. In situ monitoring and control of material growth for high resolution electron beam induced deposition. *Journal of Vacuum Science & Technology* B 25.6 (2007): 2210–2214.

[106] Chen, P., Van Veldhoven, E., Sanford, C.A., Salemink, H.W.M., Maas, D.J., Smith, D.A., Rack, P.D., Alkemade, P.F.A. "Nanopillar growth by focused helium ion-beam-induced deposition." *Nanotechnology* 21.45 (2010): 455302.

[107] Wanzenboeck, H.D., Roediger, P., Hochleitner, G., Bertagnolli, E., Buehler, W. "Novel method for cleaning a vacuum chamber from hydrocarbon contamination." *J. Vac. Sci. & Technol.* A 28.6 (2010): 1413–1420.

[108] Utke, I. and Gölzhäuser, A. (2010), "Small, minimally invasive, direct: Electrons induce local reactions of adsorbed functional molecules on the nanoscale", *Angewandte Chemie – International Edition* 49 (49), pp. 9328–9330.

INDEX